Telecommunication System Engineering

Telecommunication System Engineering
Fourth Edition

Roger L. Freeman

WILEY-
INTERSCIENCE

A JOHN WILEY & SONS, INC., PUBLICATION

Library of Congress Cataloging-in-Publication Data:

Freeman, Roger L.
    Telecommunication system engineering/Roger Freeman—4th ed.
        p.cm.
    Includes bibliographical references and index.
    ISBN 0-471-45133-9 (cloth)
    1. Telecommunication systems–Design and construction. 2. Telephone systems–Design and construction. 1. Title.

    TK5103.F68 2004
    621.382–dc22

                                                                2003063763

Printed in the United States of America.

10 9 8 7 6 5 4 3 2 1

To my daughters—Rosalind and Cristina

CONTENTS

PREFACE

I have shaped this fourth edition to follow closely in the footsteps of the popular third edition, to be the definitive work on the broad field of telecommunications. The text has been prepared with three overall objectives in mind:

1. To guide the reader through the telecommunication network in a logical, step-by-step fashion
2. To serve as a tutorial on one hand, yet a reference base on the other
3. To impart the practical aspects of designing or troubleshooting a telecommunication system, whether it transports and distributes data or voice

Specifically, it is a treatise on telecommunication system engineering, and it is meant for people who need a technical grounding on the subject. I have tried to reduce both the amount and complexity of the prerequisite mathematics to that of the first-year university level. I would also expect the reader to be able to handle the physics of basic electricity, have some rudimentary knowledge of modulation, and understand such terms as bandwidth, bit error rate (or ratio), and line of sight. These terms are defined in *The IEEE Standard Dictionary of Electrical and Electronic Terms*, 6th ed.

The intended audience for this book includes craft persons in the field of telecommunications, electrical engineers, computer scientists, telecommunication managers, and fourth-year university undergraduates that are either electrical engineering or telecommunication majors. However, I hasten to add that this is not a theoretical university text. I have written the work to impart knowledge of practice in telecommunications. Nevertheless, the inclusion of some theory was unavoidable.

Many changes have taken place in electrical communication technology since the publication of the third edition of this book. More and more the industry reflects a drastic increase in the use of data communications and the digital network, whereas conventional telephony is taking a backseat. The internet has become ubiquitous much like the telephone in the 1920s. These vital changes are reflected in the text.

Yes, we include data communications as a functional part of telecommunications. This fourth edition shows that it is an overwhelming function. There was, and still is, a philosophical difference between voice telephony and data communications in two arenas: *signaling* and *error performance.*

As in the three previous editions, some fairly heavy emphasis is placed on *signaling.* Few other textbooks devote time to this important specialty. Let us say that *signaling* provides an exchange of information specifically concerned with the establishment and control of communication circuits whether connection-oriented or connectionless. We devote two entire chapters to the subject and partial treatment in several other chapters.

In public switched telephony, signaling sets up a circuit in a connection-oriented fashion that is held in that condition until the call is terminated. In the data communications world, a message is broken down into packets, or frames, each of which has a header. The header directs the frame, or packet, to its destination. This is called connectionless service.

Chapter 12 introduces a new topic: *voice over IP,* also called *voice over packet.* This aptly joins the data and voice worlds. We believe that in the next 5 years there will be no hard line distinction between voice and data communications.

The reader will grasp the primary importance of error performance in data communication where we should expect better than 1 bit error in over 1000 million bits. There is clear explanation of why we can let error performance degrade on circuits carrying voice telephony to 1 bit error in 100 bits.

Another aspect of telecommunication systems is so-called *wireless,* still known by some as radio systems, including broadband/ultrabroadband, line-of-sight microwave, satellite, cellular/PCS, and WLAN (wireless local area network). The book describes the evolution of cellular radio from a mobile telephone service to a mobile multimedia digital multiaccess capability.

The text highlights the bulk delivery of information in the form of binary 1s and 0s. An entire chapter is dedicated to the principal bottleneck in such bulk delivery, namely the last mile (or first mile) in the network. The delivery of these signals may be by wire pair, radio (wireless), coaxial cable (as in CATV), or optical fiber. The text compares the advantages and disadvantages of each approach.

The book gives equal billing to ISDN (integrated services digital networks) and B-ISDN (broadband-ISDN). Neither has turned out to be a panacea, but for different reasons. Chapter 20 deals with the all-optical network, which requires some push of the imagination. Chapter 21 covers *network management* or how we may keep the network operating at peak efficiency. We come at this subject from two directions. First, we examine network management from the perspective of the public switched telecommunications network (PSTN). The second direction is network management of an all-data network emphasizing the *simple network management protocol* (SNMP).

Telecommunication standards are richly referenced. The largest standard-setting organization is the *International Telecommunication Union* (ITU), Geneva, Switzerland. This organization underwent a major reorganization that officially took place January 1, 1993. Prior to that date its two principal subsidiary organizations were the CCITT (International Consultive Committee on Telephone and Telegraph) and the CCIR (International Consultive Committee on Radio). After that date, CCITT became the Telecommunication Standardization Sector of the ITU, which I have abbreviated ITU-T. CCIR after that date became the ITU Radiocommunication Bureau, which I have abbreviated ITU-R. Any reference to a standard developed prior to January 1, 1993, carries the older denomination (i.e., CCITT, CCIR), whereas an ITU standard carrying a date after January 1, 1993, uses the new denomination (i.e., ITU-T or ITU-R). Where appropriate other standardization bodies are quoted such as EIA/TIA, ISO, IEEE 802 committees, ETSI, and ANSI.

Another aspect of telecommunications I have covered in all four editions of the book is the fact that there are two worlds out there: North America and Europe. European thinking and standards extend south of the Rio Grande. For example, synchronous optical network (SONET) is North American. Its European counterpart is SDH (synchronous digital hierarchy). We show that even "standard" test tones are different. In the United States and Canada the basic digital network format is T1 (also called DS1), whereas in Europe and much of the rest of the world the format is based on E1. Here the difference is considerable. It is such discussion, we believe, that makes this book unique.

ACKNOWLEDGMENTS

This text was not prepared in a vacuum. Many friends and associates provided help and support. I have been in this business for a very long time, with my present publisher for over 30 years and with the current editor for more than 20 years. There is a hard-core cadre of technical professionals here in the valley, mostly veterans of Motorola. For example, Dr. Ernie Woodward gave me the responsibility to develop the interface control document for frame relay to be used on the Celestri program. Ernie is now with Intel. I relied on Dr. Ken Peterson, one of the early achievers in low Earth orbit (LEO) satellite work. Ken is now with Rockwell-Collins in Cedar Rapids. Jill and Dave Wheeler are a wife and husband engineering team that has taught me some of the rudiments of data security.

I am also deeply indebted to my ex-ITT and GTE friends: John Lawlor and Dr. Ronald Brown, now both independent consultants like me. They bring to the fore a broad depth of experience, between them over 100 years in the telecommunication business. For antennas and propagation I lean on Marshall Cross, a founder of Megawave, which is based in Boylston, MA. Bill Ostaski tries to keep me honest on internet issues, and my son, Bob Freeman, regarding VoIP (Chapter 12). Joe Golden of Hopkinton, Massachusetts, sales engineer at Entrisphere, and professor at Northeastern University brought to light many of the QoS (quality of service) concerns with VoIP.

I also thank my friends and associates at USAF Rome Laboratories (NY) such as Peter Leong and Frank Z (Zawislan), now retired. Frank and I go back to 1962 with the AN/TRC-97 analog mini-tropo. Later it was the AN/TRC-170 digital tropo. Frank reads my material and comments.

Of course, I owe a large debt of gratitude to my wife, Paquita, for her patience and forbearing over the long and arduous days during the shaping of this text.

ROGER L. FREEMAN
Scottsdale, Arizona
January 2004

1

BASIC TELEPHONY

1 DEFINITION AND CONCEPT

Telecommunication deals with the service of providing electrical communication at a distance. The service is supported by an industry that depends on a large body of increasingly specialized scientists, engineers, and craftspeople. The service may be private or open to public correspondence (i.e., access). Examples of the latter are government-owned telephone companies, often called *administrations* or private corporations, that sell their services publicly.

1.1 Telecommunication Networks

The public switched telecommunication network (PSTN) is immense. It consists of hundreds of smaller networks interconnected. There are "fixed" and "mobile" counterparts. They may or may not have common ownership. In certain areas of the world the wired and *wireless* portions of the network compete. One may also serve as a backup for the other upon failure. It is estimated that by 2005 there will be as many wireless telephones as wired telephones, about 5×10^9 handsets worldwide of each variety.

These networks, whether mobile or fixed, have traditionally been based on speech operations. Meanwhile, another network type has lately gained great importance in the scheme of things. This is the *enterprise network*. Such a network supports the business enterprise. It can just as well support the government "enterprise" as a private business. Its most common configuration is a *local area network* (LAN) and is optimized for data communications, The enterprise network also has a long-distance counterpart, called a *WAN* or wide area network. The U.S. Department of Defense developed a special breed of WAN where the original concept was for resource sharing among U.S. and allied universities. Since its inception around 1987, it has taken on a very large life of its

Telecommunication System Engineering, by Roger L. Freeman
ISBN 0-471-45133-9   Copyright © 2004 Roger L. Freeman

own, having been opened to the public worldwide. It is the *internet*. Its appeal is universal, serving its original intent as a resource-sharing medium extending way beyond the boundaries of universities and now including a universal messaging service called *email* (electronic mail).

Some may argue that telecommunications with all its possible facets is the world's largest business. We do not take sides on this issue. What we do wish to do is to impart to the reader a technical knowledge and appreciation of telecommunication networks from a system viewpoint. By *system* we mean how one discipline can interact with another to reach a certain end objective. If we do it right, that interaction will be synergistic and will work for us; if not, it may work against us in reaching our goal.

Therefore, a primary concern of this book is to describe the development of the PSTN and enterprise network and discuss why they are built the way they are and how they are evolving. The basic underpinning of the industry was telephone service. That has now changed. The greater portion of the traffic carried today is data traffic, and all traffic is in a digital format of one form or another. We include wireless/cellular and "broadband" as adjuncts of the PSTN.

Telecommunication engineering has traditionally been broken down into two basic segments: transmission and switching. This division was most apparent in conventional telephony. Transmission deals with the delivery of a quality electrical signal from point X to point Y. Let us say that switching connects X to Y, rather than to Z. When the first edition of this book was published, transmission and switching were two very distinct disciplines. Today, that distinction has disappeared, particularly in the enterprise network. As we proceed through the development of this text, we must deal with both disciplines and show in later chapters how the dividing line separating them has completely disappeared.

2   THE SIMPLE TELEPHONE CONNECTION

The common telephone as we know it today is a device connected to the outside world by a pair of wires. It consists of a handset and its cradle with a signaling device, consisting of either a dial or push buttons. The handset is made up of two electroacoustic transducers, the earpiece or receiver and the mouthpiece or transmitter. There is also a sidetone circuit that allows some of the transmitted energy to be fed back to the receiver.

The transmitter or mouthpiece converts acoustic energy into electric energy by means of a carbon granule transmitter. The transmitter requires a direct-current (dc) potential, usually on the order of 3–5 V, across its electrodes. We call this the *talk battery*, and in modern telephone systems it is supplied over the line (central battery) from the switching center and has been standardized at −48 V dc. Current from the battery flows through the carbon granules or grains when the telephone is lifted from its cradle or goes "off hook."* When sound impinges

---

* The opposite action of "off hook" is "on hook"—that is, placing the telephone back in its cradle, thereby terminating a connection.

on the diaphragm of the transmitter, variations of air pressure are transferred to the carbon, and the resistance of the electrical path through the carbon changes in proportion to the pressure. A pulsating direct current results.

The typical receiver consists of a diaphragm of magnetic material, often soft iron alloy, placed in a steady magnetic field supplied by a permanent magnet, and a varying magnetic field caused by voice currents flowing through the voice coils. Such voice currents are alternating (ac) in nature and originate at the far-end telephone transmitter. These currents cause the magnetic field of the receiver to alternately increase and decrease, making the diaphragm move and respond to the variations. Thus an acoustic pressure wave is set up, more or less exactly reproducing the original sound wave from the distant telephone transmitter. The telephone receiver, as a converter of electrical energy to acoustic energy, has a comparatively low efficiency, on the order of 2–3%.

Sidetone is the sound of the talker's voice heard in his (or her) own receiver. Sidetone level must be controlled. When the level is high, the natural human reaction is for the talker to lower his or her voice. Thus by regulating sidetone, talker levels can be regulated. If too much sidetone is fed back to the receiver, the output level of the transmitter is reduced as a result of the talker lowering his or her voice, thereby reducing the level (voice volume) at the distant receiver and deteriorating performance.

To develop our discussion, let us connect two telephone handsets by a pair of wires, and at middistance between the handsets a battery is connected to provide that all-important talk battery. Such a connection is shown diagrammatically in Figure 1.1. Distance D is the overall separation of the two handsets and is the sum of distances d_1 and d_2; d_1 and d_2 are the distances from each handset to the central battery supply. The exercise is to extend the distance D to determine limiting factors given a fixed battery voltage, say, 48 V dc. We find that there are two limiting factors to the extension of the wire pair between the handsets. These are the IR drop, limiting the voltage across the handset transmitter, and the attenuation. For 19-gauge wire, the limiting distance is about 30 km, depending on the efficiency of the handsets. If the limiting characteristic is attenuation and we desire to extend the pair farther, amplifiers could be used in the line. If the battery voltage is limiting, then the battery voltage could be increased. With the telephone system depicted in Figure 1.1, only two people can communicate. As soon as we add a third person, some difficulties begin to arise. The simplest approach would be to provide each person with two handsets. Thus party A

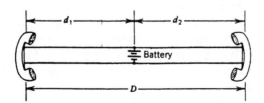

Figure 1.1. A simple telephone connection.

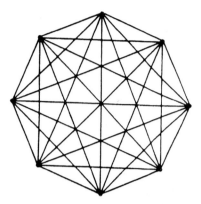

Figure 1.2. An 8-point mesh connection.

would have one set to talk to B, another to talk to C, and so forth. Or the sets could be hooked up in parallel. Now suppose A wants to talk to C and doesn't wish to bother B. Then A must have some method of selectively alerting C. As stations are added to the system, the alerting problem becomes quite complex. Of course, the proper name for this selection and alerting is *signaling*. If we allow that the pair of wires through which current flows is a loop, we are dealing with loops. Let us also call the holder of a telephone station a *subscriber*. The loops connecting them are subscriber loops.

Let us now look at an eight-subscriber system, each subscriber connected directly to every other subscriber. This is shown in Figure 1.2. When we connect each and every station with every other one in the system, this is called a *mesh* connection, or sometimes full mesh. Without the use of amplifiers and with 19-gauge copper wire size, the limiting distance is 30 km. Thus any connecting segment of the octagon may be no greater than 30 km. The only way we can justify a mesh connection of subscribers economically is when each and every subscriber wishes to communicate with every other subscriber in the network for virtually the entire day (full period). As we know, however, most telephone subscribers do not use their telephones on a full-time basis. The telephone is used at what appear to be random intervals throughout the day. Furthermore, the ordinary subscriber or telephone user will normally talk to only one other subscriber at a time. He/she will not need to talk to all other subscribers simultaneously.

If more subscribers are added and the network is extended beyond about 30 km, it is obvious that transmission costs will spiral, because that is what we are dealing with exclusively here—transmission. We are connecting each and every subscriber together with wire transmission means, requiring many amplifiers and talk batteries. Thus it would seem wiser to share these facilities in some way and cut down on the transmission costs. We now discuss this when switch and switching enter the picture. Let us define a *switch* as a device that connects inlets to outlets. The inlet may be a calling subscriber line, and the outlet may be the line of a called subscriber. The techniques of switching and the switch

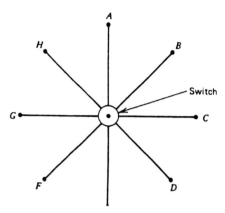

Figure 1.3. Subscribers connected in a star arrangement.

as a concept are widely discussed later in this text. Switching devices and how they work are covered in Chapters 3 and 9. Consider Figure 1.3, which shows our subscribers connected in a *star* network with a switch at the center. All the switch really does in this case is to reduce the transmission cost outlay. Actually, this switch reduces the number of links between subscribers, which really is a form of concentration. Later in our discussion it becomes evident that switching is used to concentrate traffic, thus reducing the cost of transmission facilities.

3 SOURCES AND SINKS*

Traffic is a term that quantifies usage. A subscriber *uses* the telephone when he/she wishes to talk to somebody. We can make the same statement for a telex (teleprinter service) subscriber or a data-service subscriber. But let us stay with the telephone.

A network is a means of connecting subscribers. We have seen two simple network configurations, the mesh and star connections, in Figures 1.2 and 1.3. When talking about networks, we often talk of sources and sinks. A call is initiated at a traffic source and received at a traffic sink. Nodal points or nodes in a network are the switches.

4 TELEPHONE NETWORKS: INTRODUCTORY TERMINOLOGY

From our discussion we can say that a telephone network can be regarded as a systematic development of interconnecting transmission media arranged so that one telephone user can talk to any other within that network. The evolving layout of the network is primarily a function of economics. For example, subscribers share common transmission facilities; switches permit this sharing by concentration.

---

* The traffic engineer may wish to use the terminology "origins and destinations."

Consider a very simplified example. Two towns are separated by, say, 20 miles, and each town has 100 telephone subscribers. Logically, most of the telephone activity (the traffic) will be among the subscribers of the first town and among those of the second town. There will be some traffic, but considerably less, from one town to the other. In this example let each town have its own switch. With the fairly low traffic volume from one town to the other, perhaps only six lines would be required to interconnect the switch of the first town to that of the second. If no more than six people want to talk simultaneously between the two towns, a number as low as six can be selected. Economics has mandated that we install the minimum number of connecting telephone lines from the first town to the second to serve the calling needs between the two towns. The telephone lines connecting one telephone switch or exchange with another are called *trunks* in North America and *junctions* in Europe. The telephone lines connecting a subscriber to the switch or exchange that serves the subscriber are called *lines, subscriber lines,* or *loops.* Concentration is a line-to-trunk ratio. In the simple case above, it was 100 lines to six trunks (or junctions), or about a 16:1 ratio.

A telephone subscriber looking into the network is served by a *local exchange.* This means that the subscriber's telephone line is connected to the network via the local exchange or central office, in North American parlance. A local exchange has a serving area, which is the geographical area in which the exchange is located; all subscribers in that area are served by that exchange.

The term *local area,* as opposed to *toll area,* is that geographical area containing a number of local exchanges and inside which any subscriber can call any other subscriber without incurring tolls (extra charges for a call). Toll calls and long-distance calls are synonymous. For instance, a local call in North America, where telephones have detailed billing, shows up on the bill as a time-metered call or is covered by a flat monthly rate. Toll calls in North America appear as separate detailed entries on the telephone bill. This is not so in most European countries and in those countries following European practice. In these countries there is no detailed billing on direct-distance-dialed (subscriber-trunk-dialed) calls. All such subscriber-dialed calls, even international ones, are just metered, and the subscriber pays for the meter steps used per billing period, which is often one or two months. In European practice a long-distance call, a toll call if you will, is one involving the dialing of additional digits (e.g., more than six or seven digits).

Let us call a network a *grouping of interworking telephone exchanges.* As the discussion proceeds, the differences between local networks and national networks are shown. Two other types of network are also discussed. These are specialized versions of a local network and are the rural network (rural area) and metropolitan network (metropolitan area). (Also consult Refs. 9 and 16–18.)

5 ESSENTIALS OF TRAFFIC ENGINEERING

5.1 Introduction and Terminology

As we have already mentioned, telephone exchanges are connected by trunks or junctions. The number of trunks connecting exchange X with exchange Y is the number of voice pairs or their equivalent used in the connection. One of the most important steps in telecommunication engineering practice is to determine the number of trunks required on a route or connection between exchanges. We could say we are *dimensioning* the route. To dimension a route correctly, we must have some idea of its usage—that is, how many people will wish to talk at once over the route. The usage of a transmission route or a switch brings us into the realm of traffic engineering, and the usage may be defined by two parameters: (1) *calling rate*, or the number of times a route or traffic path is used per unit period, or, more properly defined, "the call intensity per traffic path during the busy hour";* and (2) *holding time*, or "the duration of occupancy of a traffic path by a call,"* or sometimes, "the average duration of occupancy of one or more paths by calls."* A *traffic path* is "a channel, time slot, frequency band, line, trunk, switch, or circuit over which individual communications pass in sequence."* *Carried traffic* is the volume of traffic actually carried by a switch, and *offered traffic* is the volume of traffic offered to a switch.

To dimension a traffic path or size a telephone exchange, we must know the traffic intensity representative of the normal busy season. There are weekly and daily variations in traffic within the busy season. Traffic is very random in nature. However, there is a certain consistency we can look for. For one thing, there usually is more traffic on Mondays and Fridays and a lower volume on Wednesdays. A certain consistency can also be found in the normal workday hourly variation. Across the typical day the variation is such that a 1-h period shows greater usage than any other. From the hour with least traffic to the hour of greatest traffic, the variation can exceed 100 : 1. Figure 1.4 shows a typical hour-by-hour traffic variation for a serving switch in the United States.[†] It can be seen that the busiest period, the *busy hour* (BH), is between 10 A.M. and 11 A.M. From one workday to the next, originating BH calls can vary as much as 25%. To these fairly "regular" variations, there are also unpredictable peaks caused by stock market or money market activity, weather, natural disaster, international events, sporting events, and so on. Normal system growth must also be taken into account. Nevertheless, suitable forecasts of BH traffic can be made. However, before proceeding, consider the five most common definitions of BH:

Busy Hour Definitions (CCITT Rec. E.600)

1. *Busy Hour.* The busy hour refers to the traffic volume or number of call attempts, and is that continuous 1-h period lying wholly in the time interval concerned for which this quantity (i.e., traffic volume or call attempts) is greatest.

2. *Peak Busy Hour.* The busy hour each day; it usually is not the same over a number of days.

---

* Reference Data for Radio Engineers [1], pages 31–38.

[†] The busy hour will vary from country to country because of cultural differences.

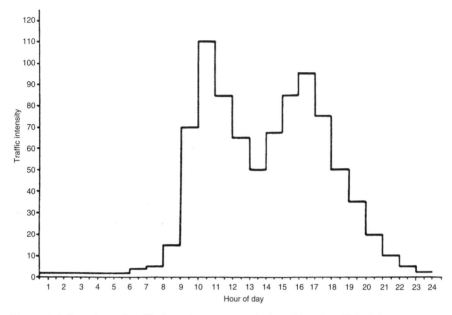

Figure 1.4. Bar chart of traffic intensity over a typical working day (United States, mixed business and residential).

3. *Time Consistent Busy Hour.* The 1-h period starting at the same time each day for which the average traffic volume or call-attempt count of the exchange or resource group concerned is greatest over the days under consideration.

From *Engineering and Operations in the Bell System,* 2nd ed. [23]

4. The engineering period (where the grade of service criteria is applied) is defined as the busy season busy hour (BSBH), which is the busiest clock hour of the busiest weeks of the year.

5. The average busy season busy hour (ABSBH) is used for trunk groups and always has a grade of service criterion applied. For example, for the ABSBH load, a call requiring a circuit in a trunk group should encounter "all trunks busy" (ATB) no more than 1% of the time.

Reference 23 goes on to state that peak loads are of more concern than average loads when engineering switching equipment and engineering periods other than the ABSBH are defined. Examples of these are the highest BSBH and the average of the ten highest BSBHs. Sometimes the engineering period is the weekly peak hour (which may not even be the BSBH).

When dimensioning telephone exchanges and transmission routes, we shall be working with BH traffic levels and care must be used in the definition of busy hour.

5.2 Measurement of Telephone Traffic

If we define *telephone traffic* as the aggregate of telephone calls over a group of circuits or trunks with regard to the duration of calls as well as their number [2], we can say that traffic flow (*A*) is expressed as

$$A = C \times T$$

where *C* designates the number of calls originated during a period of 1 h and *T* is the average holding time, usually given in hours. *A* is a dimensionless unit because we are multiplying calls/hour by hour/call.

Suppose that the average holding time is 2.5 min and the calling rate in the BH for a particular day is 237. The traffic flow (*A*) would then be 237×2.5, or 592.5 call-minutes (Cm) or 592.5/60, or about 9.87 call-hours (Ch).

Ramses Mina [2] states that a distinction should be made between the terms "traffic density" and "traffic intensity." The former represents the number of simultaneous calls at a given moment, while the latter represents the average traffic density during a 1-h period. The quantity of traffic used in the calculation for dimensioning of switches is the traffic intensity.

The preferred unit of traffic intensity is the erlang, named after the Danish mathematician A. K. Erlang [5]. The erlang is a dimensionless unit. One erlang represents a circuit occupied for 1 h. Considering a group of circuits, traffic intensity in erlangs is the number of call-seconds per second or the number of call-hours per hour. If we knew that a group of 10 circuits had a call intensity of 5 erlangs, we would expect half of the circuits to be busy at the time of measurement.

In the United States the term "unit call" (UC), or its synonymous term, "hundred call-second," abbreviated CCS,* generally is used. These terms express the sum of the number of busy circuits, provided that the busy trunks were observed once every 100 s (36 observations in 1 h) [2].

There are other traffic units. For instance: call-hour (Ch)—1 Ch is the quantity represented by one or more calls having an aggregate duration of 1 h; call-second (Cs)—1 Cs is the quantity represented by one or more calls having an aggregate duration of 1 s; traffic unit (TU), a unit of traffic intensity. One TU is the average intensity in one or more traffic paths carrying an aggregate traffic of 1 Ch in 1 h (the busy hour unless otherwise specified). 1 TU = 1 E (erlang) (numerically). The *equated busy hour call* (EBHC) is a European unit of traffic intensity. 1 EBHC is the average intensity in one or more traffic paths occupied in the BH by one 2-min call or an aggregate duration of 2 min. Thus we can relate our terms as follows:

$$1 \text{ erlang} = 30 \text{ EBHC} = 36 \text{ CCS} = 60 \text{ Cm}$$

assuming a 1-h time-unit interval.

---

* The first letter C in CCS stands for the Roman numeral 100.

Traffic measurements used for long-term network planning are usually based on the traffic in the busy hour (BH), which is usually determined based on observations and studies.

The traditional traffic measurements on trunks during a measurement interval are:

- Peg count*—calls offered
- Usage—traffic (CCS or erlangs) carried
- Overflow—call encountering all trunks busy

From these measurements, the blocking probability and mean traffic load carried by the trunk group can be calculated.

Extensive traffic measurements are made on switching systems because of their numerous traffic sensitive components. Usual measurements for a component such as a service circuit include calls carried, peg count, and usage. The typical holding time for a common-control element in a switch is considerably shorter than that for a trunk, and short sampling intervals (e.g., 10 s) or continuous monitoring are used to measure usage.

Traffic measurements for short-term network management purposes are usually concerned with detecting network congestion. Calls offered, peg count, and overflow count can be used to calculate attempts per circuit per hour (ACH) and connections per circuit per hour (CCH), with these measurements being calculated over very short time periods (e.g., 10-min intervals).

Under normal circumstances, ACH and CCH are approximately equal. Examples of abnormal conditions are:

- ACH high, CCH normal—heavy demand, excessive blockage, normal holding times for connected calls indicating that most calls switched are completed, heavy traffic but low congestion [25].
- ACH high, CCH high—heavy traffic, short trunk holding times indicate uncompleted call attempts being switched, congestion [25]. (Consult Ref. 24. Also see Refs. 7, 10, and 19.)

5.3 Blockage, Lost Calls, and Grade of Service

Assume that an isolated telephone exchange serves 5000 subscribers and that no more than 10% of the subscribers wish service simultaneously. Therefore, the exchange is dimensioned with sufficient equipment to complete 500 simultaneous connections. Each connection would be, of course, between any two of the 5000 subscribers. Now let subscriber 501 attempt to originate a call. He/she cannot

---

* A term taken from telephony in the older days where manual switching was prevalent. A peg board was installed by the telephone operator to keep count of offered calls. The present definition is taken from Ref. 23. "A count of all calls offered to a trunk group, usually measured for one hour. As applied to units of switching systems with common control, *peg count*, or *carried peg count*, means the number of calls actually handled."

because all the connecting equipment is busy, even though the line he/she wishes to reach may be idle. This call from subscriber 501 is termed a *lost call* or *blocked call*. He/she has met blockage. The probability of meeting blockage is an important parameter in traffic engineering of telecommunication systems. If congestion conditions are to be met in a telephone system, we can expect that those conditions will usually be met during the BH. A switch is engineered (dimensioned) to handle the BH load. But how well? We could, indeed, far overdimension the switch such that it could handle any sort of traffic peaks. However, that is uneconomical. So with a well-designed switch, during the busiest of BHs we may expect some moments of congestion such that additional call attempts will meet blockage. *Grade of service* expresses the probability of meeting blockage during the BH and is expressed by the letter p. A typical grade of service is $p = 0.01$. This means that an average of one call in 100 will be blocked or "lost" during the BH. Grade of service, a term in the Erlang formula, is more accurately defined as the *probability of blockage*. It is important to remember that lost calls (blocked calls) refer to calls that fail at *first* trial. We discuss attempts (at dialing) later, that is, the way blocked calls are handled.

We exemplify grade of service by the following problem. If we know that there are 354 seizures (lines connected for service) and 6 blocked calls (lost calls) during the BH, what is the grade of service?

$$\text{Grade of service} = \frac{\text{Number of lost calls}}{\text{Total number of offered calls}}$$
$$= \frac{6}{354 + 6} = \frac{6}{360} \tag{1.1}$$

or

$$p = 0.017$$

The average grade of service for a network may be obtained by adding the grade of service contributed by each constituent switch, switching network, or trunk group. The *Reference Data for Radio Engineers* [1, Section 31] states that the grade of service provided by a particular group of trunks or circuits of specified size and carrying a specified traffic intensity is the probability that a call offered to the group will find available trunks already occupied on first attempt. That probability depends on a number of factors, the most important of which are (1) the distribution in time and duration of offered traffic (e.g., random or periodic arrival and constant or exponentially distributed holding time), (2) the number of traffic sources [limited or high (infinite)], (3) the availability of trunks in a group to traffic sources (full or restricted availability), and (4) the manner in which lost calls are "handled."

Several new concepts are suggested in these four factors. These must be explained before continuing.

5.4 Availability

Switches were previously discussed as devices with lines and trunks, but better terms for describing a switch are "inlets" and "outlets." When a switch has full availability, each inlet has access to any outlet. When not all the free outlets in a switching system can be reached by inlets, the switching system is referred to as one with "limited availability." Examples of switches with limited and full availability are shown in Figures 1.5A and 1.5B.

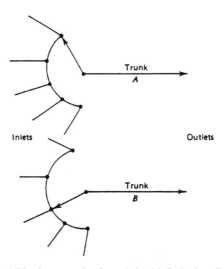

Figure 1.5A. An example of a switch with limited availability.

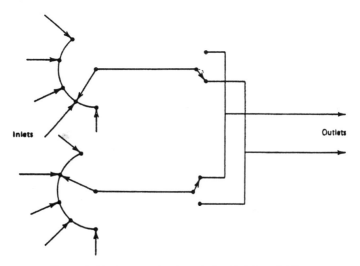

Figure 1.5B. An example of a switch with full availability.

Of course, full availability switching is more desirable than limited availability but is more expensive for larger switches. Thus full availability switching is generally found only in small switching configurations and in many new digital switches (see Chapter 9). *Grading* is one method of improving the traffic-handling capacities of switching configurations with limited availability. Grading is a scheme for interconnecting switching subgroups to make the switching load more uniform.

5.5   "Handling" of Lost Calls

In conventional telephone traffic theory, three methods are considered for the handling or dispensing of lost calls: (1) lost calls held (LCH), (2) lost calls cleared (LCC), and (3) lost calls delayed (LCD). The LCH concept assumes that the telephone user will immediately reattempt the call on receipt of a congestion signal and will continue to redial. The user hopes to seize connection equipment or a trunk as soon as switching equipment becomes available for the call to be handled. It is the assumption in the LCH concept that lost calls are held or waiting at the user's telephone. This concept further assumes that such lost calls extend the average holding time theoretically, and in this case the average holding time is zero, and all the time is waiting time. The principal traffic formula used in North America is based on the LCH concept.

The LCC concept, which is used primarily in Europe or those countries accepting European practice, assumes that the user will hang up and wait some time interval before reattempting if the user hears the congestion signal on the first attempt. Such calls, it is assumed, disappear from the system. A reattempt (after the delay) is considered as initiating a new call. The Erlang formula is based on this criterion.

The LCD concept assumes that the user is automatically put in queue (a waiting line or pool). For example, this is done when the operator is dialed. It is also done on most modern computer-controlled switching systems, generally referred to under the blanket term *stored program control* (SPC). The LCD category may be broken down into three subcategories, depending on how the queue or pools of waiting calls is handled. The waiting calls may be handled last in first out (LIFO), first in first out (FIFO), or at random.

5.6   Infinite and Finite Sources

We can assume that traffic sources are infinite or finite. For the case of infinite traffic sources, the probability of call arrival is constant and does not depend on the state of occupancy of the system. It also implies an infinite number of call arrivals, each with an infinitely small holding time. An example of finite sources is when the number of sources offering traffic to a group of trunks or circuits is comparatively small in comparison to the number of circuits. We can also say that with a finite number of sources, the arrival rate is proportional to the number of sources that are not already engaged in sending a call.

5.7 Probability-Distribution Curves

Telephone-call originations in any particular area are random in nature. We find that originating calls or call arrivals at an exchange closely fit a family of probability-distribution curves following a Poisson distribution. The Poisson distribution is fundamental to traffic theory.

Most of the common probability-distribution curves are two-parameter curves. That is, they may be described by two parameters, mean and variance. The mean is a point on the probability-distribution curve where an equal number of events occur to the right of the point and to the left of the point.

Mean is synonymous with *average*. We define mean as the *x*-coordinate of the center of the area under the probability-density curve for the population. The lowercase Greek letter mu (μ) is the traditional indication of the mean; $\overline{x}$ is also used.

The second parameter used to describe a distribution curve is the dispersion, which tells us how the values or population are dispersed about the center or mean of the curve. There are several measures of dispersion. One is the familiar *standard deviation*, where the standard deviation s of a sample of n observations $x_1, x_2, \ldots, x_n$ is

$$s = \sqrt{\frac{1}{n-1} \sum_{i=1}^{n} (x_i - \overline{x})^2} \tag{1.2}$$

The *variance V* of the sample values is the square of s. The parameters for dispersion s and s^2, the standard deviation and variance, respectively, are usually denoted σ and σ^2 and give us an idea of the squatness of a distribution curve. Mean and standard deviation of a normal distribution curve are shown in Figure 1.6, where we can see that σ^2 is another measure of dispersion, the variance, or essentially the average of the squares of the distances from mean aside from the factor $n/(n-1)$.

We have introduced two distribution functions describing the probability of distribution, often called the *distribution* of *x* or just $f(x)$. Both functions are used in traffic engineering. But before proceeding, the variance-to-mean ratio

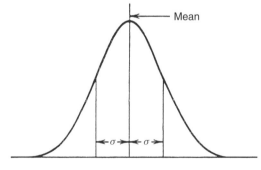

Figure 1.6. A normal distribution curve showing the mean and the standard deviation, σ.

(VMR) is introduced. Sometimes VMR(α) is called the *coefficient of overdispersion*. The formula for VMR is

$$\alpha = \frac{\sigma^2}{\mu} \tag{1.3}$$

5.8 Smooth, Rough, and Random Traffic

Traffic probability distributions can be divided into three distinct categories: (1) smooth, (2) rough, and (3) random. Each may be defined by α, the VMR. For smooth traffic, α is less than 1. For rough traffic, α is greater than 1. When α is equal to 1, the traffic distribution is called *random*. The Poisson distribution function is an example of random traffic where VMR $= 1$. Rough traffic tends to be peakier than random or smooth traffic. For a given grade of service, more circuits are required for rough traffic because of the greater spread of the distribution curve (greater dispersion).

Smooth traffic behaves like random traffic that has been filtered. The filter is the local exchange. The local exchange looking out at its subscribers sees call arrivals as random traffic, assuming that the exchange has not been overdimensioned. The smooth traffic is the traffic on the local exchange outlets. The filtering or limiting of the peakiness is done by call blockage during the BH. Of course, the blocked traffic may actually overflow to alternative routes. Smooth traffic is characterized by a positive binomial distribution function, perhaps better known to traffic people as the *Bernoulli distribution*. An example of the Bernoulli distribution is as follows [6]. If we assume that subscribers make calls independently of each other and that each has a probability p of being engaged in conversation, then if n subscribers are examined, the probability that x of them will be engaged is

$$B(x) = C_x^n p^x (1 - p)^{n-x}, \qquad 0 < x < n$$

$$\text{Its mean} = np \tag{1.4}$$

$$\text{Its variance} = np(1 - p)$$

where the symbol C_x^n means the number of ways that x entities can be taken n at a time. Smooth traffic is assumed in dealing with small groups of subscribers; the number 200 is often used as the breakpoint [6]. That is, groups of subscribers are considered small when the subscribers number is less than 200. And as mentioned, smooth traffic is also used with carried traffic. In this case the rough or random traffic would be the offered traffic.

Let's consider the binomial distribution for rough traffic. This is characterized by a negative index. Therefore, if the distribution parameters are k and q, where k is a positive number representing a hypothetical number of traffic sources and q represents the occupancy per source and may vary between 0 and 1, then

$$R'(x, k, q) = \binom{x + k - 1}{k - 1} q^x (1 - q)^k \tag{1.5}$$

where R' is the probability of finding x calls in progress for the parameters k and q [2]. Rough traffic is used in dimensioning toll trunks with alternative routing. The symbol B (Bernoulli) is used by traffic engineers for smooth traffic and R for rough traffic. Although P may designate probability, in traffic engineering it designates Poissonian, and hence we have "P" tables such as those in Ref. 20, Table 1-1.

The Bernoulli formula is

$$B'(x, s, h) = C_s^x h^x (1 - h)^{s-x} \tag{1.6}$$

where C_s^x indicates the number of combinations of s things taken x at a time, h is the probability of finding the first line of an exchange busy, $1 - h$ is the probability of finding the first line idle, and s is the number of subscribers. The probability of finding two lines busy is h^2, the probability of finding s lines busy is h^s, and so on. We are interested in finding the probability of x of the s subscribers with busy lines.

The Poisson probability function can be derived from the binomial distribution, assuming that the number of subscribers s is very large and the calling rate per line h is low* such that the product $sh = m$ remains constant and letting s increase to infinity in the limit

$$P(x) = \frac{m^x}{x!} e^{-m} \tag{1.7}$$

where

$$x = 0, 1, 2, \ldots$$

For most of our future discussion, we consider call-holding times to have a negative exponential distribution in the form

$$P = e^{-t/h} \tag{1.8}$$

where t/h is the average holding time and in this case P is the probability of a call lasting longer than t, some arbitrary time interval.

Figure 1.7 compares smooth, random, and rough traffic probability distributions.

6   ERLANG AND POISSON TRAFFIC FORMULAS

When dimensioning a route, we want to find the number of circuits that serve the route. There are several formulas at our disposal to determine that number of circuits based on the BH traffic load. In Section 5.3 four factors were discussed that will help us to determine which traffic formula to use given a particular set of circumstances. These factors primarily dealt with (1) call arrivals and holding-time distribution, (2) number of traffic sources, (3) availability, and (4) handling of lost calls.

---

* For example, less than 50 millierlangs (mE).

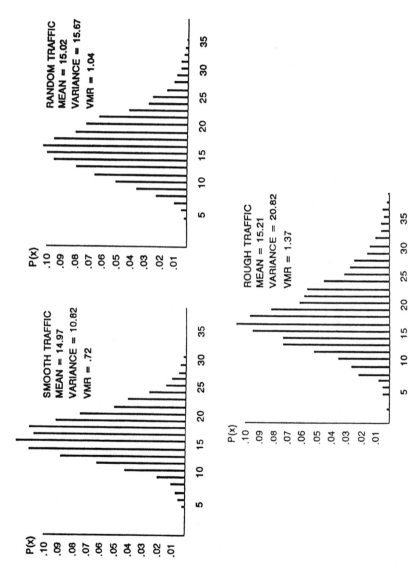

Figure 1.7. Traffic probability distributions: smooth, random, and rough traffic. Courtesy of John Lawlor Associates, Sharon, Massachusetts [25].

The Erlang B loss formula has been widely used today outside of the United States. Loss here means the probability of blockage at the switch due to congestion or to "all trunks busy" (ATB). This is expressed as *grade of service* (E_B) or the probability of finding x channels busy. The other two factors in the Erlang B formula are the mean of the *offered* traffic and the number of trunks of servicing channels available. Thus

$$E_B = \frac{A^n/n!}{1 + A + A^2/2! + \cdots + A^n/n!} \tag{1.9}$$

where n is the number of trunks or servicing channels, A is the mean of the offered traffic, and E_B is the grade of service using the Erlang B formula. This formula assumes the following:

- Traffic originates from an infinite number of sources.
- Lost calls are cleared assuming a zero holding time.
- The number of trunks or servicing channels is limited.
- Full availability exists.

At this point in our discussion of traffic we suggest that the reader learn to differentiate between time congestion and call congestion when dealing with grade of service. *Time congestion*, of course, refers to the decimal fraction of an hour during which all trunks are busy simultaneously. *Call congestion*, on the other hand, refers to the number of calls that fail at first attempt, which we term *lost calls*. Keep in mind that the Erlang B formula deals with offered traffic, which differs from carried traffic by the number of lost calls.

Table 1-2 in Ref. 20 is based on the Erlang B formula and gives trunk dimensioning information for several specific grades of service, from 0.001 to 0.10 and from 1 to 200 trunks. The traffic intensity units are in CCS and erlangs for 1 to 200 trunks. Keep in mind that 1 erlang = 36 CCS (based on a 1-h time interval). As an example of how we might employ an Erlang B table, suppose we wished a grade of service of 0.001 and the route carried 16.68 erlangs of traffic, we would then see that 30 trunks would be required. When sizing a route for trunks or when dimensioning an exchange, we often come up with a fractional numbering of servicing channels or trunks. In this case we would opt for the next highest integer because we cannot install a fraction of a trunk. For example, if calculations show that a trunk route should have 31.4 trunks, it would be designed for 32 trunks.

The Erlang B formula is based on lost calls cleared. It is generally accepted as a standard outside the United States (see CCITT Rec. Q.87). In the United States the Poisson formula (2) is favored. The formula is often called the *Molina formula*. It is based on the LCH concept. Table 1-1 of Ref. 20 provides trunking sizes for various grades of service from 0.001 to 0.10. The units of traffic intensity are the CCS and the erlang.

Table 1-1 of Ref. 20 is based on full availability. Also, we should remember that the Poisson formula also assumes that traffic originates from a large (infinite) number of independent subscribers or sources (i.e., random traffic input) with a limited number of trunks or servicing channels and LCH.

It is not as straightforward as it may seem when comparing grades of service between Poisson and Erlang B formulas (or tables). The grade of service $p = 0.01$ for the Erlang B formula is equivalent to a grade of service of 0.005 when applying the Poisson (Molina) formula. Given these grades of service, assuming LCC with the Erlang B formula permits up to several tenths of erlangs less of traffic when dimensioning up to 22 trunks, where the two approaches equate (e.g., where each formula allows 12.6 erlangs over the 22 trunks). Above 22 trunks the Erlang B formula permits the trunks to carry somewhat more traffic, and at 100 trunks it permits 2.7 erlangs more than for the Poisson formula under the LCH assumption.

6.1 Alternative Traffic Formula Conventions

Some readers may be more comfortable using traffic formulas with a different convention and notation. The Erlang B and Poisson formulas were derived from Ref. 2. The formulas and notation used in this subsection have been taken from Ref. 20. The following is the notation used in the formulas given below:

A = The expected traffic density, expressed in busy hour erlangs.
P = The probability that calls will be lost (or delayed) because of insufficient channels.
n = The number of channels in the group of channels.
s = The number of sources in the group of sources.
p = The probability that a single source will be busy at an instant of observation. This is equal to A/s.
x = A variable representing a number of busy sources or busy channels.
e = The Naperian logarithmic base, which is the constant 2.71828+.
$\binom{m}{n}$ = The combination of m things taken n at a time.
$\sum\limits_{X=m}^{n}$ = The summation of all values obtained when each integer or whole number value, from m to n inclusive, is substituted for the value x in the expression following the symbol.
∞ = The conventional symbol for infinity.

The Poisson formula has the following assumptions: (1) infinite sources, (2) equal traffic density per source, and (3) lost calls held (LCH). The formula is

$$P = e^{-A} \sum_{x=n}^{\infty} \frac{A^x}{x!} \qquad (1.10)$$

The Erlang B formula assumes (1) infinite sources, (2) equal traffic density per source, and (3) lost calls cleared (LCC). The formula is

$$P = \frac{\dfrac{A^n}{n!}}{\displaystyle\sum_{x=0}^{n} \dfrac{A^x}{x!}} \tag{1.11}$$

The Erlang C formula, commonly used with digital switching where one would expect to find queues, assumes (1) infinite sources, (2) lost calls delayed (LCD), (3) exponential holding times, and (4) calls served in order of arrival. Refer to Table 1-3 in Section 1-5 of Ref. 20. The formula is

$$P = \frac{\dfrac{A^n}{n!} \cdot \dfrac{n}{n-A}}{\displaystyle\sum_{x=0}^{n-1} \dfrac{A^x}{x!} + \dfrac{A^n}{n!}\dfrac{n}{n-A}} \tag{1.12}$$

The binomial formula assumes (1) finite sources, (2) equal traffic density per source, and (3) lost calls held (LCH). The formula is

$$P = \left(\frac{s-A}{s}\right)^{s-1} \sum_{x=n}^{s-1} \binom{s-1}{x}\left(\frac{A}{s-A}\right)^x \tag{1.13}$$

(Also consult Ref. 19.)

6.2 Computer Programs for Traffic Calculations

6.2.1 Erlang B Computer Program. Relatively simple programs can be written in BASIC computer language to solve the Erlang B equation shown below:

$$P = \frac{A^N/N!}{1 + A + A^2/2! + A^3/3! + \cdots + A^N/N!}$$

Following is a sample program. It is written using GWBASIC, Version 3.11.

```
10  PRINT "ERLANG B, TRAFFIC ON EACH TRUNK."
20  INPUT "OFFERED TRAFFIC (ERLANG) = ", A
30  LET B = 0: LET H = 0
40  INPUT "MINIMUM GRADE OF SERVICE (DEFAULT = 1) = ", H
50  IF H = 0 THEN LET H = 1
60  INPUT "OBJECTIVE GRADE OF SERVICE (DEFAULT = 0.001) = ", B
70  IF B = 0 THEN LET B = .001
80  LET N = 0: LET T = 1: LET T1 = 1: LET 0 = A: LET Z = 15
90  PRINT "TRUNKS OFFERED CARRIED CUMULATIVE P = G/S"
100 PRINT
```

```
"-----------------------------------------------------------------------------------"
110  IF T1 > 9.999999E + 20 THEN 230
120  LET N = N + 1: LET T = T * A/N: LET T1 = T1 + T: LET P = T/T1
130  LET L = A * P: LET S = A - L: LET C = 0 - L
140  IF P > H THEN LET Z = N + 14: GOTO 110
150  PRINT USING "###          ";N,
160  PRINT USING "###.####    ";0,
170  PRINT USING "    .####    ";C,
180  PRINT USING "###.####    ";S,
190  PRINT USING ".####";P
200  LET 0 = L: IF P < B THEN 240
210  IF N < Z THEN 110
220  STOP: LET Z = Z + 15: PRINT "OFFERED TRAFFIC (ERLANG) IS ", A: GOTO 90
230  LET T = T * 1E - 37: LET T1 = T1 * 1E - 37: GOTO 120
240  END
```

For a given traffic load, this program shows the load offered to each trunk, the load carried by each trunk, the cumulative load carried by the trunk group, and the probability of blocking for a group with "N" trunks. This assumes that the trunks are used in sequential order. The results are presented on a trunk-by-trunk basis until the probability of blocking reaches some minimum value [Pr(Blocking) = 0.001 in this version]. Both the minimum and maximum probability values can be adjusted by entering input values when starting the program.

The variable "Z" is used to stop the calculation routine every "Z" lines so that the data will fill one screen at a time. Calculation will continue when the instruction "CONT" is entered (or by pressing the "F5" key in some computers).

In order to avoid overflow in the computer registers when very large numbers are encountered, the least significant digits are dropped for large traffic loads (see Instruction 110 and Instruction 230). The variables affected by this adjustment (T and T1) are used only to determine the probability ratio, and the error introduced is negligible. As the traffic load A becomes very large, the error increases.

Following is a sample program calculation for an offered traffic load of 10 erlangs:

```
ERLANG B, TRAFFIC ON EACH TRUNK.
OFFERED TRAFFIC (ERLANG) = 10
MINIMUM GRADE OF SERVICE (DEFAULT = 1) =
OBJECTIVE GRADE OF SERVICE (DEFAULT = 0.001) =
```

TRUNKS	OFFERED	CARRIED	CUMULATIVE	P = G/S
1	10.0000	0.9091	0.9091	0.9091
2	9.0909	0.8942	1.8033	0.8197
3	8.1967	0.8761	2.6794	0.7321
4	7.3206	0.8540	3.5334	0.6467
5	6.4666	0.8271	4.3605	0.5640
6	5.6395	0.7944	5.1549	0.4845
7	4.8451	0.7547	5.9096	0.4090
8	4.0904	0.7072	6.6168	0.3383
9	3.3832	0.6511	7.2679	0.2732
10	2.7321	0.5863	7.8542	0.2146
11	2.1458	0.5135	8.3677	0.1632

12	1.6323	0.4349	8.8026	0.1197
13	1.1974	0.3540	9.1566	0.0843
14	0.8434	0.2752	9.4318	0.0568
15	0.5682	0.2032	9.6350	0.0365
16	0.3650	0.1420	9.7770	0.0223
17	0.2230	0.0935	9.8705	0.0129
18	0.1295	0.0581	9.9286	0.0071
19	0.0714	0.0340	9.9625	0.0037
20	0.0375	0.0188	9.9813	0.0019
21	0.0187	0.0098	9.9911	0.0009

6.2.2   Poisson Computer Program.

Relatively simple programs can be written in BASIC computer language to solve the Poisson equation shown below:

$$P = \frac{A^N/N!}{1 + A + A^2/2! + A^3/3! + \cdots} = \frac{A^N/N!}{e^A} = \frac{A^N}{N!}e^{-A}$$

Following is a sample program which can be used for traffic loads up to 86 erlangs. (Loads greater than 86 erlangs may cause register overflow in some computers.) It is written using GWBASIC, Version 3.11.

```
10  PRINT "POISSON (86 ERLANG MAXIMUM)"
20  PRINT "P = PROBABILITY OF N TRUNKS BUSY"
30  PRINT "P1 = PROBABILITY OF BLOCKING"
40  INPUT "OFFERED TRAFFIC IN ERLANGS = ", A: LET E = EXP(A)
50  LET N = 0: LET T = 1: LET T1 = 1: LET T2 = 0: LET Z = 15
60  PRINT "TRUNKS PR (N TRUNKS BUSY) PR (BLOCKING)"
70  PRINT "---------------------------------------"
80  LET P1 = 1 - T2: LET T2 = T1/E: LET P = T/E
90  IF P<.00001 THEN LET Z = N + 14: GOTO 170
100 PRINT USING "###    ";N,
110 PRINT USING "       #.####      ";P,P1
120 IF N<Z THEN 180
130 STOP: LET Z = Z + 15
140 PRINT "OFFERED TRAFFIC IN ERLANGS IS ";A
150 PRINT "TRUNKS PR (N TRUNKS BUSY) PR (BLOCKING)"
160 PRINT "---------------------------------------"
170 IF N<Z THEN 120: LET Z = Z + 15
180 LET N = N + 1: LET T = T * A/N: LET T1 = T1 + T
190 IF P1>.0001 THEN 80
200 END
```

This program calculates both (a) the probability that exactly "N" trunks are busy and (b) the probability of blocking (the probability that "N" or more service requests are received). The results are presented on a trunk-by-trunk basis until the probability of blocking reaches some minimum value [P1 = Pr(Blocking) = 0.0001 in this version]. Both the minimum and maximum probabilities can be adjusted by changing the values of P (Instruction 90) and P1 (Instruction 190).

The variable "Z" is used to stop the calculation routine every "Z" lines so that the data will fill one screen at a time. Calculation will continue when the instruction "CONT" is entered (or by pressing the "F5" key in some computers).

Following is a sample program calculation for an offered traffic load of 10 erlangs:

```
POISSON (86 ERLANG MAXIMUM)
P = PROBABILITY OF N TRUNKS BUSY
P1 = PROBABILITY OF BLOCKING
OFFERED TRAFFIC IN ERLANGS = 10
TRUNKS    PR (N TRUNKS BUSY)    PR (BLOCKING)
```

TRUNKS	PR (N TRUNKS BUSY)	PR (BLOCKING)
0	0.0000	1.0000
1	0.0005	1.0000
2	0.0023	0.9995
3	0.0076	0.9972
4	0.0189	0.9897
5	0.0378	0.9707
6	0.0631	0.9329
7	0.0901	0.8699
8	0.1126	0.7798
9	0.1251	0.6672
10	0.1251	0.5421
11	0.1137	0.4170
12	0.0948	0.3032
13	0.0729	0.2084
14	0.0521	0.1355
15	0.0347	0.0835
16	0.0217	0.0487
17	0.0128	0.0270
18	0.0071	0.0143
19	0.0037	0.0072
20	0.0019	0.0035
21	0.0009	0.0016
22	0.0004	0.0007
23	0.0002	0.0003
24	0.0001	0.0001
25	0.0000	0.0000

6.2.3 Erlang C Computer Program.

Relatively simple programs can be written in BASIC computer language to solve the Erlang C equation shown below:

$$P = \frac{(A^N/N!)(N/(N-A))}{1 + A + A^2/2! + A^3/3! + \cdots + (A^N/N!)(N/(N-A))}$$

Following is a sample program. It is written using GWBASIC, Version 3.11.

```
10 PRINT "ERLANG C CALCULATES DELAY"
20 INPUT "OFFERED TRAFFIC (ERLANGS) = ",A
30 PRINT "TRUNKS    P    D1    D2    Q1    Q2";
```

```
40 PRINT "  P8   P4   P2   P1   PP"
50 PRINT
"------------------------------------------------------------------------"
60 LET N = 1: LET T = 1: LET T1 = 1
70 IF T1 > 1E + 21 THEN 200
80 IF N<=A THEN 190
90 IF N = A + 1 THEN LET Z = A + 15
100 LET T2 = T * (A/N) * (N/(N - A)): LET P = T2/(T1 + T2)
110 LET D2 = 1/(N - A): LET D1 = P * D2: LET Q2 = A * D2: LET Q1 = P * Q2
120 LET PO = P/EXP(2/D2)
130 PRINT USING "###      ";N,
140 PRINT USING ".####   ";P,
150 PRINT USING "###.##  ";D1, D2, Q1, Q2,
160 PRINT USING ".##     ";PO
170 IF N>Z THEN 210
180 IF P<.02 THEN 250
190 LET T = T * A/N: LET T1 = T1 + T: LET N = N + 1: GOTO 70
200 LET T = T * 1E - 37: LET T1 = T1 * 1E - 37: GOTO 80
210 STOP: LET Z = Z + 15
220 PRINT "OFFERED TRAFFIC (ERLANGS) = ",A
230 PRINT "TRUNKS  P    D1   D2    Q1   Q2    PP"
240 PRINT "---------------------------------": GOTO 180
250 END
```

For a given traffic load, this program shows the probability of delay for a group with "N" trunks, with call delays expressed in units of average holding time:

P	probability that a call will be delayed
D1	average delay on all calls, including those not delayed
D2	average delay on calls that are delayed
Q1	average calls in queue
Q2	average calls in queue when all servers are busy
PP	probability of delay exceeding two holding time intervals

The program calculates delay and queue characteristics by using the following relationships derived from the Erlang C formula:

The average delay on all calls, D1, including those not delayed, is given by

$$D1 = P(> 0)(h/(N - A)), \qquad \text{where h is the average holding time}$$

The average delay, D2, on calls delayed is given by

$$D2 = h/(N - A), \qquad \text{where h is the average holding time}$$

The average calls in queue, Q2, when all servers are busy is given by

$$Q2 = Ah/(N - A) = AD2$$

The average calls in queue, Q1, is given by

$$Q1 = P(> 0)(Ah/(N - A)) = P(> 0)Q2$$

The probability of delay exceeding some multiple of the average holding time, PP (PP $> 2 * $ HT is used in the program), is given by

$$PP = P(> 0)/e^{(XHT/D2)}, \quad \text{where X is the number of HT intervals}$$

This assumes that the trunks are used in sequential order. The results are presented on a trunk-by-trunk basis beginning with "A + 1" trunks to ensure that there are sufficient trunks available to carry the busy hour traffic after some delay has occurred. This process continues until the probability of delay is small enough that delays are relatively short [Pr(Delay) = .02 in this version]. The minimum delay probability can be adjusted as desired.

The variable "Z" is used to stop the calculation routine every "Z" lines so that the data will fill one screen at a time. Calculation will continue when the instruction "CONT" is entered (or by pressing the "F5" key in some computers).

In order to avoid overflow in the computer registers when very large numbers are encountered, the least significant digits are dropped for large traffic loads (see Instruction 70 and Instruction 200). The variables affected by this adjustment (T and T1) are used only to determine the probability ratio, and the error introduced is negligible. As the traffic load A becomes very large, the error increases.

Following is a sample program calculation for an offered traffic load of 10 erlangs:

```
ERLANG C CALCULATES DELAY
OFFERED TRAFFIC (ERLANGS) = 10
```

TRUNKS	P	D1	D2	Q1	Q2	PP
11	0.6821	0.68	1.00	6.82	10.00	0.09
12	0.4494	0.22	0.50	2.25	5.00	0.01
13	0.2853	0.10	0.33	0.95	3.33	0.00
14	0.1741	0.04	0.25	0.44	2.50	0.00
15	0.1020	0.02	0.20	0.20	2.00	0.00
16	0.0573	0.01	0.17	0.10	1.67	0.00
17	0.0309	0.00	0.14	0.04	1.43	0.00
18	0.0159	0.00	0.13	0.02	1.25	0.00

Source: The information in Section 6.2 was graciously provided by John Lawlor Associates, Sharon, MA.

7 WAITING SYSTEMS (QUEUEING)

A short discussion follows regarding traffic in queueing systems. Queueing or waiting systems, when dealing with traffic, are based on the third assumption, namely, lost calls delayed (LCD). Of course, a queue in this case is a pool of callers waiting to be served by a switch. The term *serving time* is the time a call takes to be served from the moment of arrival in the queue to the moment of being served by the switch. For traffic calculations in most telecommunication queueing systems, the mathematics is based on the assumption that call arrivals are random and Poissonian. The traffic engineer is given the parameters of offered traffic, the size of the queue, and a specified grade of service and will determine the number of serving circuits or trunks required.

The method by which a waiting call is selected to be served from the pool of waiting calls is called *queue discipline*. The most common discipline is the first-come, first-served discipline, where the call waiting longest in the queue is served first. This can turn out to be costly because of the equipment required to keep order in the queue. Another type is random selection, where the time a call has waited is disregarded and those waiting are selected in random order. There is also the last-come, first-served discipline and bulk service discipline, where batches of waiting calls are admitted, and there are also priority service disciplines, which can be preemptive and nonpreemptive. In queueing systems the grade of service may be defined as the probability of delay. This is expressed as $P(t)$, the probability that a call is not being immediately served and has to wait a period of time greater than t. The average delay on all calls is another parameter that can be used to express grade of service, and the length of queue is another.

The probability of delay, the most common index of grade of service for waiting systems when dealing with full availability and a Poissonian call arrival process, is calculated by using the Erlang C formula, which assumes an infinitely long queue length. Syski [3] provides a good guide to Erlang C and other, more general waiting systems. (Also consult Refs. 13–15.)

7.1 Server-Pool Traffic

Server pools are groups of traffic resources, such as signaling registers and operator positions, that are used on a shared basis. Service requests that cannot be satisfied immediately are placed in a queue and served on a first-in, first-out (FIFO) basis. Server-pool traffic is directly related to offered traffic, server-holding time, and call-attempt factor and is inversely related to call-holding time. This is expressed in equation 1.14.

$$A_S = \frac{A_T \cdot T_S \cdot C}{T_C} \qquad (1.14)$$

where A_S = server-pool traffic in erlangs
  A_T = total traffic served in erlangs
  T_S = mean server-holding time in hours
  T_C = mean call-holding time in hours
  C = call-attempt factor (dimensionless)

Total traffic served refers to the total offered traffic that requires the services of the specific server pool for some portion of the call. For example, a dual-tone multifrequency (DTMF) receiver pool is dimensioned to serve only the DTMF tone-dialing portion of total switch traffic generated by DTMF signaling sources.

Table 1.1 gives representative server-holding times for typical signaling registers as a function of the number of digits received or sent.

The mean server-holding time is the arithmetic average of all server-holding times for the specific server pool. Equation 1.15 can be used to calculate mean server-holding time for calls with different holding-time characteristics.

$$T_S = a \cdot T_1 + b \cdot T_2 + \cdots + k \cdot T_n \qquad (1.15)$$

where
$$T_S = \text{mean server-holding time in hours}$$
$$T_1, T_2, \ldots, T_n = \text{individual server-holding times in hours}$$
$$a, b, \ldots, k = \text{fractions of total traffic served } (a + b + \cdots + k = 1)$$

Consider the following example. Determine the mean DTMF receiver-holding time for a switch where subscribers dial local calls using a 7-digit number and long-distance calls using an 11-digit number. Assume that 70% of the calls are local calls, the remainder are long-distance calls, and the typical signaling register holding times found in Table 1.1 are applicable.

$$T_s = (0.7)(8.1 \text{ s}) + (0.3)(12.0 \text{ s}) = 9.27 \text{ s}$$

Call-attempt factors are dimensionless numbers that adjust offered traffic intensity to compensate for call attempts that do not result in completed calls. Therefore, call-attempt factors are inversely proportional to the fraction of completed calls as defined in equation 1.16.

$$C = 1/k \qquad (1.16)$$

where $C = \text{call-attempt factor (dimensionless)}$
$k = \text{fraction of calls completed (decimal fraction)}$

Here is another example. Table 1.2 gives representative subscriber call-attempt dispositions based on empirical data taken from the large North American PSTN

TABLE 1.1   Typical Signaling Register Holding Times in Seconds

Signaling Register	Number of Digits Received or Sent				
	1	4	7	10	11
Local dial-pulse (DP) receiver	3.7	8.3	12.8	17.6	19.1
Local DTMF receiver	2.3	5.2	8.1	11.0	12.0
Incoming MF receiver	1.0	1.4	1.8	2.2	2.3
Outgoing MF sender	1.5	1.9	2.3	2.8	3.0

TABLE 1.2 Typical Call-Attempt Dispositions

Call-Attempt Disposition	Percentage
Call was completed	70.7
Called subscriber did not answer	12.7
Called subscriber line was busy	10.1
Call abandoned without system response	2.6
Equipment blockage or failure	1.9
Customer dialing error	1.6
Called directory number changed or disconnected	0.4

data base. Determine the call-attempt factors for these data, where 70.7% of the calls were completed ($k = 0.707$).

$$C = \frac{1}{k} = \frac{1}{0.707} = 1.414$$

Source: Section 7.1 is based on Section 1.2.2 of *Traffic System Design Handbook* by James R. Boucher, IEEE Press, 1992 [26].

8 DIMENSIONING AND EFFICIENCY

By definition, if we were to dimension a route or estimate the required number of servicing channels, where the number of trunks (or servicing channels) just equaled the erlang load, we would attain 100% efficiency. All trunks would be busy with calls all the time or at least for the entire BH. This would not even allow several moments for a trunk to be idle while the switch decided the next call to service. In practice, if we engineered our trunks, trunk routes, or switches this way, there would be many unhappy subscribers.

On the other hand, we do, indeed, want to size our routes (and switches) to have a high efficiency and still keep our customers relatively happy. The goal of our previous exercises in traffic engineering was just that. The grade of service is one measure of subscriber satisfaction. As an example, let us assume that between cities X and Y there are 100 trunks on the interconnecting telephone route. The tariffs, from which the telephone company derives revenue, are a function of the erlangs of carried traffic. Suppose we allow a dollar per erlang-hour. The very upper limit of service on the route is 100 erlangs. If the route carried 100 erlangs of traffic per day, the maximum return on investment would be $2400 a day for that trunk route and the portion of the switches and local plant involved with these calls. As we well know, many of the telephone company's subscribers would be unhappy because they would have to wait excessively to get calls through from X to Y. How, then, do we optimize a trunk route (or serving circuits) and keep the customers as happy as possible?

In our previous discussions, an excellent grade of service was 0.001. We relate grade of service to subscriber satisfaction. Turning to Ref. 20, Table 1-2, such a grade of service with 100 circuits would support 75.24 erlangs during the BH. With 75.24 erlangs loading, the route would earn \$75.24 during that one-hour period and something far less that \$2400 per day. If the grade of service was reduced to 0.01, 100 trunks would bring in \$84.06 for the busy hour. Note the improvement in revenue at the cost of reducing grade of service. Another approach to save money is to hold the erlang load constant and decrease the number of trunks and switch facilities accordingly as the grade of service is reduced. For instance, 70 erlangs of traffic at $p = 0.001$ requires 96 trunks and at $p = 0.01$, only 86 trunks.

8.1 Alternative Routing

One method of improving efficiency is to use alternative routing (called *alternate routing* in North America). Suppose that we have three serving areas, X, Y, and Z, served by three switches, X, Y, and Z as illustrated in Figure 1.8.

Let the grade of service be 0.005 (1 in 200 in Table 1-2, Ref. 20). We found that it would require 67 trunks to carry 50 erlangs of traffic during the BH to meet that grade of service between X and Y. Suppose that we reduced the number of trunks between X and Y, still keeping the BH traffic intensity at 50 erlangs. We would thereby increase the efficiency on the $X-Y$ route at the cost of reducing the grade of service. With a modification of the switch at X, we could route the traffic bound for Y that met congestion on the $X-Y$ route via Z. Then Z would route this traffic on the $Z-Y$ link. Essentially, this is alternative routing in its simplest form. Congestion probably would only occur during very short peaky periods in the BH, and chances are that these peaks would not occur simultaneously with peaks in traffic intensity on the $Z-Y$ route. Furthermore, the added load on the $X-Z-Y$ route would be very small. Some idea of traffic peakiness that would overflow onto the secondary route $(X + Z + Y)$ is shown in Figure 1.9.

One of the most accepted methods of dimensioning switches and trunks using alternative routing is the equivalent random group (ERG) method developed by Wilkinson [11]. The Wilkinson method uses the mean M and the variance V.

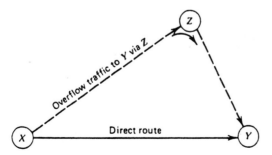

Figure 1.8. Simplified diagram of the alternative routing concept (solid line represents direct route, dashed line represents alternative route carrying the overflow from X to Y).

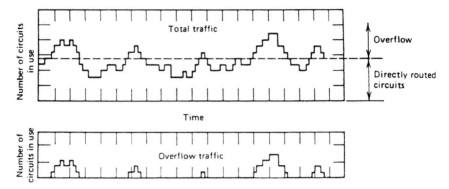

Figure 1.9. Traffic peakiness, the peaks representing overflow onto alternative routes.

Here the *overflow traffic* is the "lost" traffic in the Erlang B calculations, which were discussed earlier. Let M be the mean value of that overflow and A be the random traffic offered to a group of n circuits (trunks). Then

$$V = M \left(1 - M + \frac{A}{1 + n + M - A} \right) \tag{1.17}$$

When the overflow traffic from several sources is combined and offered to a single second (or third, fourth, etc.) choice of a group of circuits, both the mean and the variance of the combined traffic are the arithmetical sums of the means and variances of the contributors.

The basic problem in alternative routing is to optimize circuit group efficiency (e.g., to dimension a route with an optimum number of trunks). Thus we are to find what circuit quantities result in minimum cost for a given grade of service, or to find the optimum number of circuits (trunks) to assign to a direct route allowing the remainder to overflow on alternative choices. There are two approaches to the optimization. The first method is to solve the problem by successive approximations, and this lends itself well to the application of the computer [12]. Then there are the manual approaches, two of which are suggested in CCITT Rec. E.525 [25]. Alternate (alternative) routing is further discussed in Chapter 6.

8.2   Efficiency versus Circuit Group Size

In the present context a *circuit group* refers to a group of circuits performing a specific function. For instance, all the trunks (circuits) routed from X to Y in Figure 1.8 make up a circuit group, irrespective of size. This group should not be confused with the "group" used in transmission-engineering carrier multiplex systems.

If we assume full loading, it can be stated that efficiency improves with circuit group size. From Table 1-2 of Ref. 20, given $p = 0.001$, 5 erlangs of traffic

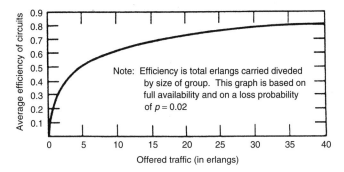

Figure 1.10. Group efficiency increases with size.

requires a group with 11 trunks, more than a $2:1$ ratio of trunks to erlangs, and 20 erlangs requires 30 trunks, a $3:2$ ratio. Note how the efficiency has improved. One hundred twenty trunks will carry 100 erlangs, or 6 trunks for every 5 erlangs for a group of this size. Figure 1.10 shows how efficiency improves with group size.

9   BASES OF NETWORK CONFIGURATIONS

In this section we discuss basic network configurations that may apply anywhere in the telecommunication community. Networks more applicable to the local area are covered in Chapter 2, and those for the long-distance plant are discussed in Chapter 6.

9.1   Introductory Concepts

A network in telecommunications may be defined as a method of connecting exchanges so that any one subscriber in the network can communicate with any other subscriber. For this introductory discussion, let us assume that subscribers access the network by a nearby local exchange. Thus the problem is essentially how to connect exchanges efficiently. There are three basic methods of connection in conventional telephony: (1) mesh, (2) star, and (3) double and higher-order star (see Section 2 of this chapter). The mesh connection is one in which each and every exchange is connected by trunks (or junctions) to each and every other exchange as shown in Figure 1.11A. A star connection utilizes an intervening exchange, called a *tandem exchange*, such that each and every exchange is interconnected via a *single* tandem exchange. An example of a star connection is shown in Figure 1.11B. A double-star configuration is one where sets of pure star subnetworks are connected via higher-order tandem exchanges, as shown in Figure 1.11C. This trend can be carried still further, as we see later on, when hierarchical networks are discussed.

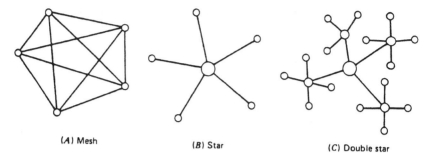

(*A*) Mesh   (*B*) Star   (*C*) Double star

Figure 1.11. Examples of star, double-star, and mesh configurations.

As a general rule we can say that mesh connections are used when there are comparatively high traffic levels between exchanges, such as in metropolitan networks. On the other hand, a star network may be applied when traffic levels are comparatively low.

Another factor that leads to star and multiple-star network configurations is network complexity in the trunking outlets (and inlets) of a switch in a full mesh. For instance, an area with 20 exchanges would require 380 traffic groups (or links), and an area with 100 exchanges would require 9900 traffic groups. This assumes what are called *one-way groups*. A one-way group is best defined considering the connection between two exchanges, *A* and *B*. Traffic originating at *A* bound for *B* is carried in one group and the traffic originating at *B* bound for *A* is carried in another group, as shown in the following diagram:

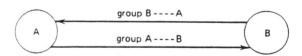

One-way and both-way groups are further discussed in Section 11.

Thus, in practice, most networks are compromises between mesh and star configurations. For instance, outlying suburban exchanges may be connected to a nearby major exchange in the central metropolitan area. This exchange may serve nearby subscribers and be connected in mesh to other large exchanges in the city proper. Another example is the city's long-distance exchange, which is a tandem exchange looking into the national long-distance network, whereas the major exchanges in the city are connected to it in mesh. An example of a real-life compromise among mesh, star, and multiple-star configurations is shown in Figure 1.12.

9.2 Higher-Order Star Network

Figure 1.13 illustrates a higher-order star network. It is simply several star networks of Figure 1.11B stacked on top of each other. Another high-order star

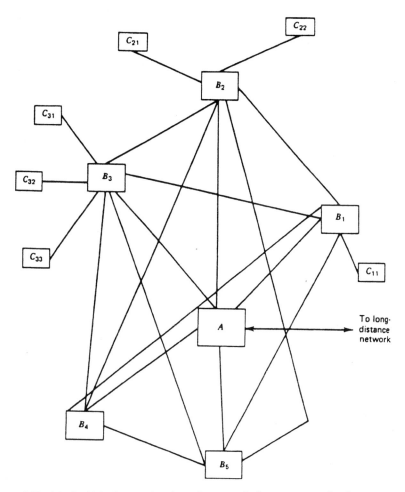

Figure 1.12. A typical telephone network serving a small city as an example of a compromise between mesh and star configuration. *A* is the highest level in this simple hierarchy. *A* might house the "point of presence" (POP) in the U.S. network. *B* is a local exchange. *C* may be a satellite exchange or a concentrator. Consult Ref. 24.

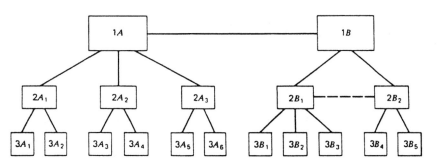

Figure 1.13. Higher-order star network.

network is shown in Figure 1.11C. In these types of networks the higher layers are given more importance than the lower layers. The incisive reader will say that we describe a hierarchical network. The reader is correct, but we wish to reserve our detailed discussion of hierarchical networks for Chapter 6, Sections 6 and 7.

We illustrate the order of importance of the several levels in a high-order star network in Figure 1.13. There are three levels or ranks of exchanges in the figure. The smallest blocks in the diagram are the lowest-ranked exchanges, which have been marked with a "3" to indicate the third level or rank. Note that there are restrictions or rules of traffic flow. As the figure is drawn, traffic from $3A_1$ to $3A_2$ would have to flow through exchange $2A_1$. Likewise, traffic from exchange $2A_2$ to $2A_3$ would have to flow through exchange 1A. Carrying the concept one step further, traffic from any A exchange to any B exchange would necessarily have to be routed through exchange 1A.

The next consideration is the high-usage (HU) route. For instance, if we found that there were high traffic intensities (e.g., >20 erlangs) between $2B_1$ and $2B_2$, trunks and switch gear might well be saved by establishing a HU route between the two (shown by a dashed line in Figure 1.13). Thus we might call the high-usage route a *highly traveled shortcut*. Of course, HU routes could be established between any pair of exchanges in the network if traffic intensities and distances involved proved this strategy economical. When HU routes are established, traffic between the exchanges involved will first be offered to the HU route, and overflow would take place through a last choice route or, as shown in Figure 1.13, up to the next level and down. If routing is through the highest level of higher-order star network, we call this route the final route, a hierarchical network term. (See Chapter 6, Sections 6 and 7.)

10   VARIATIONS IN TRAFFIC FLOW

In networks covering large geographic expanses and even in cases of certain local networks, there may be a variation in time of day of the BH or in the direction of traffic flow. In the United States, business traffic peaks during several hours before and after the noon lunch period on weekdays, and social calls peak in early evening. Traffic flow tends to be from suburban living areas to urban centers in the morning, and the reverse in the evening.

In national networks covering several time zones where the differences in local time may be appreciable, long-distance traffic tends to be concentrated in a few hours common to BH peaks at both ends. In such cases it is possible to direct traffic so that peaks of traffic in one area fall into valleys of traffic in another (noncoincident busy hour). The network design can be made more economical if configured to take advantage of these phenomena, particularly in the design and configuration of direct routes versus overflow.

11   ONE-WAY AND BOTH-WAY (TWO-WAY) CIRCUITS

We defined one-way circuits in Section 9.1. Here traffic from A to B is assigned to one group of circuits, and traffic from B to A is assigned to another separate group. In both-way (or two-way) operation a circuit group may be engineered to carry traffic in both directions. The individual circuits in the group may be used in either direction, depending on which exchange seizes the circuit first.

In engineering networks it is most economical to have a combination of one-way and both-way circuits on longer routes. Signaling and control arrangements on both-way circuits are substantially more expensive. However, when dimensioning a system for a given traffic intensity, fewer circuits are needed in both-way operation, with notable savings on low-intensity routes (i.e., below about 10 erlangs in each direction). For long circuits, both-way operation has obvious advantages when dealing with a noncoincident BH. During overload conditions, both-way operation is also advantageous because the direction of traffic flow in these conditions is usually unequal.

The major detriment to two-way operation, besides its increased signaling cost, is the possibility of double seizure. This occurs when both ends seize a circuit at the same time. There is a period of time when double seizure can occur in a two-way circuit; this extends from the moment the circuit is seized to send a call and the moment when it becomes blocked at the other end. Signaling arrangements can help to circumvent this problem. Likewise, switching arrangements can be made such that double seizure can occur only on the last free circuit of a group. This can be done by arranging in turn the sequence of scanning circuits so that the sequence on one end of a two-way circuit is reversed from that of the other end. Of course, great care must be taken on circuits having long propagation times, such as satellite and long undersea cable circuits. By extending the time between initial seizure and blockage at the other end, these circuits are the most susceptible, just because a blocking signal takes that much longer to reach the other end.

12   QUALITY OF SERVICE

Quality of service appears at the outset to be an intangible concept. However, it is very tangible for a telephone subscriber unhappy with his or her service. The concept of service quality must be mentioned early in any all-encompassing text on telecommunications systems. System engineers should never once lose sight of the concept, no matter what segment of the system they may be responsible for. Quality of service also means *how happy* the telephone company (or other common carrier) is keeping the customer. For instance, we might find that about half the time a customer dials, the call goes awry or the caller cannot get a dial tone or cannot hear what is being said by the party at the other end. All these have an impact on quality of service. So we begin to find that quality of service is an important factor in many areas of the telecommunications business and means different things to different people. In the old days of telegraphy, a rough measure of how well the system was working was the number of service messages received at a switching center. In modern telephony we now talk about service observing (see Chapter 3, Section 17).

The transmission engineer calls quality of service "customer satisfaction," which is commonly measured by how well the customer can hear the calling party. It is called *loudness rating* (see Chapter 2, Section 2.2.1) and is measured in decibels. In our discussion of traffic, lost calls or blockage certainly constitute another measure of service quality. If this is measured in a decimal quantity, one target figure for grade of service would be $p = 0.01$. Other items listed under service quality are:

- Delay before receiving dial tone ("dial-tone delay").
- Postdial(ing) delay (time from completion of dialing a number to first ring of telephone called).
- Availability of service tones (busy tone, telephone out of order, ATB, etc.).
- Correctness of billing.
- Reasonable cost to customer of service.
- Responsiveness to servicing requests.
- Responsiveness and courtesy of operators.
- Time to installation of new telephone, and, by some, the additional services offered by the telephone company [22].

One way or another, each item, depending on service quality goal, will have an impact on the design of the system.

Furthermore, each item on the list can be quantified, usually statistically, such as loudness rating, or in time, such as time taken to install a telephone. In some countries it can be measured in years. Regarding service quality, good reading can be found in ITU-T Recs. E.430, E.432, I.350, and X.140.

REVIEW QUESTIONS

1. Give the standard telephone battery voltage with respect to ground.

2. What is *on hook* and *off hook*? When a subscriber subset (the telephone) goes "off hook," what occurs at the serving switch? List two items.

3. Suppose that the sidetone level of a telephone is increased. What is the natural reaction of the subscriber?

4. A subscriber pair, with a fixed battery voltage, is extended. As we extend the loop further, two limiting performance factors come into play. Name them.

5. Define a mesh connection. Draw a star arrangement.

6. In the context of the argument presented in the chapter, what is the principal purpose of a local switch?

7. What are the two basic parameters that define "traffic"?

8. Distinguish offered traffic from carried traffic.

9. Give one valid definition of the *busy hour*.

10. On a particular traffic relation the calling rate is 461 and the average call duration is 1.5 min during the BH. What is the traffic intensity in CCS, in erlangs?

11. Define *grade of service*.

12. A particular exchange has been dimensioned to handle 1000 calls during the busy hour. On a certain day during the BH 1100 calls are offered. What is the resulting grade of service?

13. Distinguish a full availability switch from a limited availability switch.

14. In traffic theory there are three ways lost calls are handled. What are they?

15. Call arrivals at a large switch can be characterized by what type of mathematical distribution? Such arrivals are_____in nature.

16. Based on the Erlang B formula and given a BH requirement for a grade of service of 0.005 and a BH traffic intensity of 25 erlangs on a certain traffic relation, how many trunks are required?

17. Carry out the same exercise as in question 16 but use the Poisson tables to determine the number of trunks required.

18. Give at least two queueing disciplines.

19. As the grade of service is improved, what is the effect on trunk efficiency?

20. What is the basic purpose of alternative routing? What does it improve?

21. How does circuit group size (number of trunks) affect efficiency for a fixed grade of service?

22. Give the three basic methods of connecting exchanges. (These are the three basic network types.)

23. At what erlang value on a certain traffic relation does it pay to use tandem routing? Is this a maximum or a minimum value?

24. Differentiate between one-way and both-way circuits.

25. What is the drawback of one-way circuits? of both-way circuits?

26. Hierarchical networks are used universally in national and international telephone networks. Differentiate between high-usage (HU) connectivities and final route.

27. Distinguish a tandem exchange from a transit exchange.

28. Name at least five items that can be listed under *quality of service* (QoS).

REFERENCES

1. International Telephone and Telegraph Corporation, *Reference Data for Radio Engineers*, 5th ed., Howard W. Sams, Indianapolis, 1968.
2. R. R. Mina, "The Theory and Reality of Teletraffic Engineering," *Telephony*, a series of articles (April 1971).
3. R. Syski, *Introduction to Congestion Theory in Telephone Systems*, Oliver and Boyd, Edinburgh, 1960.
4. G. Dietrich et al., *Teletraffic Engineering Manual*, Standard Electric Lorenz, Stuttgart, Germany, 1971.
5. E. Brockmeyer et al., "The Life and Works of A. K. Erlang," *Acta Polytechnica Scandinavia*, The Danish Academy of Technical Sciences, Copenhagen, 1960.
6. *A Course in Telephone Traffic Engineering*, Australian Post Office, Planning Branch, 1967.
7. Arne Jensen, *Moe's Principle*, The Copenhagen Telephone Company, Copenhagen, Denmark, 1950.
8. *Networks*, Laboratorios ITT de Standard Eléctrica SA, Madrid, 1973 (limited circulation).
9. *Local Telephone Networks*, The International Telecommunications Union, Geneva, 1968.
10. *Electrical Communication System Engineering Traffic*, U.S. Department of the Army, TM-11-486-2, August 1956.
11. R. I. Wilkinson, "Theories for Toll Traffic Engineering in the USA," *BSTJ*, **35** (March 1956).
12. *Optimization of Telephone Trunking Networks with Alternate Routing*, ITT Laboratories of Standard Eléctrica (Spain), Madrid, 1974 (limited circulation).
13. J. Riordan, *Stochastic Service Systems*, John Wiley & Sons, New York, 1962.
14. L. Kleinrock, *Queuing Systems*, Vols. 1 and 2, John Wiley & Sons, New York, 1975.
15. T. L. Saaty, *Elements of Queueing Theory with Applications*, McGraw-Hill, New York, 1961.
16. J. E. Flood, *Telecommunication Networks*, IEE Telecommunications Series 1, Peter Peregrinus, London, 1975.
17. "Telcordia Notes on the Networks," Telecordia Special Report SR-2275, Issue 4, Telecordia, Piscataway, NJ, October 2000.
18. *National Telephone Networks for the Automatic Service*, International Telecommunication Union—CCITT, Geneva, 1964.
19. D. Bear, *Principles of Telecommunication Traffic Engineering*, IEE Telecommunications Series 2, Peter Peregrinus, London, 1976.
20. R. L. Freeman, *Reference Manual for Telecommunication Engineering*, 3rd ed., John Wiley & Sons, New York, 2002.
21. *Traffic Routing*, CCITT Rec. E.170, ITU Geneva, October 1992.

22. "Telecommunications Quality," *IEEE Communications Magazine* (entire issue), IEEE, New York, 1988.

23. *Engineering and Operations in the Bell System*, 2nd ed., AT&T Bell Laboratories, Murray Hill, NJ, 1983.

24. *BOC Notes on the LEC Networks—1994*, Issue 2, SR-TSV-002275, Bellcore, Piscataway, NJ, April 1994.

25. Private Communications—John Lawlor/John Lawlor Associates, Sharon, MA, December 1994.

26. J. R. Boucher, *Traffic System Design Handbook*, IEEE Press, New York, 1992.

27. *Designing Networks to Control Grade of Service*, CCITT Rec. E.525, ITU, Geneva, 1992.

2

LOCAL NETWORKS

1 INTRODUCTION

The importance of local network design, whether standing on its own merit or part of an overall national network, cannot be overstressed. In comparison to the long-distance sector, the local sector is not the big income producer per capita invested, but there would be no national network without it. Telephone companies or administrations invest, on the average, more than 50% in their local areas. In the larger, more developed countries the investment in local plant may reach 70% of total plant investment.

The local area, as distinguished from the long-distance or national network, was discussed in Section 4 of Chapter 1. In this chapter we are more precise in defining the local area itself. Let us concede that the local area includes the subscriber plant, local exchanges, and the trunk plant interconnecting these exchanges as well as those trunks connecting a local area to the next level of network hierarchy, or the point of presence (POP)* (USA) or primary center (CCITT).

To further emphasize the importance of the local area, consider Table 2.1, which was taken from Ref. 1 (CCITT). Figure 2.1 is a simplified diagram of a local network with five serving exchanges and illustrates the makeup of a typical small local area.

The design of such a network (Figure 2.1) involves a number of limiting factors, the most important of which is economic. Investment and its return are not treated in this text. However, our goal is to build the most economical network assuming an established quality of service. Considering both quality of service and economy, certain restraints will have to be placed on the design. For example, we will want to know:

---

* POP is a point of interface with interexchange carriers (e.g., AT&T, MCI, etc.).

Telecommunication System Engineering, by Roger L. Freeman
ISBN 0-471-45133-9 Copyright © 2004 Roger L. Freeman

TABLE 2.1 Average Percentage of Investments in Public Telephone Equipment

Item	Average for 16 Countries (%)
Subscriber plant	13
Outside plant for local networks	27
Exchanges	27
Long-distance trunks	23
Buildings and land	10

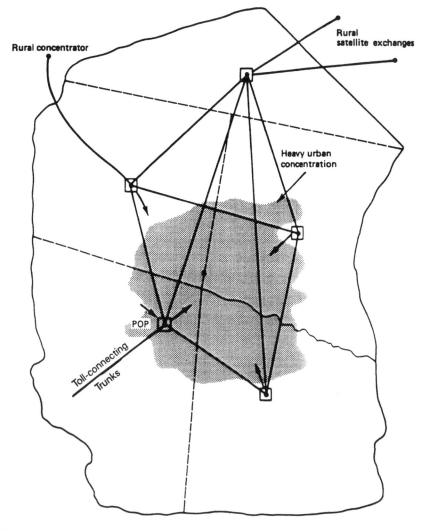

Figure 2.1. A sample local area (arrows represent trunk pull; dashed lines delineate serving areas).

- Geographic extension of the local area of interest.
- Number of inhabitants and existing telephone density.
- Calling habits.
- Percentage of business telephones.
- Location of existing telephone exchanges and extension of their serving areas.
- Trunking scheme.
- Present signaling and transmission characteristics.

Each of these criteria or limiting factors is treated separately, and interexchange signaling and switching per se are dealt with later in separate chapters. Let us also assume that each exchange in the sample will be capable of serving up to 10,000 subscribers. Also assume that all telephones in the area have seven-digit numbers, the last four of which are the subscriber number of the respective serving area of each exchange. The reasoning behind these assumptions becomes apparent in later chapters.

A further assumption is that all subscribers are connected to their respective serving exchanges by wire pairs, resulting in some limiting subscriber loop length. This leads to the first constraining factor dealing with transmission and signaling characteristics. In general terms, the subscriber should be able to hear the distant calling party reasonably well (transmission) and to "signal" that party's serving switch. These items are treated at length in the following section.

2 SUBSCRIBER LOOP DESIGN

2.1 General

The pair of wires connecting the subscriber to the local serving switch has been defined as the *subscriber loop*. It is a dc loop in that it is a wire pair supplying a metallic path for the following:

- Talk battery for the telephone transmitter (Chapter 1, Section 2).
- An ac ringing voltage for the bell on the telephone instrument supplied from a special ringing source voltage.
- Current to flow through the loop when the telephone instrument is taken out of its cradle ("off hook"), telling the serving switch that it requires "access," thus causing a line seizure at that switch.
- The telephone dial that, when operated, makes and breaks the direct current on the closed loop, which indicates to the switching equipment the number of the distant telephone with which communication is desired; alternatively, a touch-tone pad with digit buttons. Unique pairs of audio tones, representing digits 1–0, are transmitted to the serving exchange switching equipment.

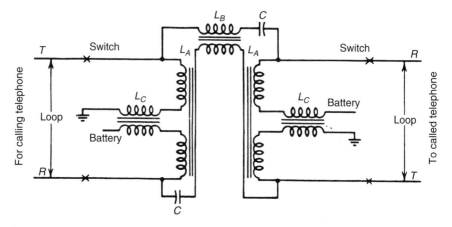

Figure 2.2. Battery feed circuit. Note battery and ground are fed through inductors L_B and L_A through switch to loops. Based on Refs. 22 and 23.

The typical subscriber loop is powered by means of a battery feed circuit at the switch. Such a circuit is shown in Figure 2.2. Telephone battery source voltage has been fairly well standardized at −48 V dc.

2.2 Quality of a Telephone Speech Connection

2.2.1 Loudness Rating. *Loudness rating* is the principal parameter for measuring the quality of a speech connection. CCITT/ITU-T Recs. P.11, P.76, P.78, and P.79 provide information on subjective and objective methods for the determination of loudness ratings (LRs). The currently recommended values of loudness loss in terms of loudness ratings are given in ITU-T Recs. G.111 and G.121 [25, 26]. In simple terms, loudness rating is a measure of speech level (volume). The term replaces *reference equivalent*, used in previous editions and older texts.

2.2.1.1 Customer Opinion. Customer opinion, as a function of loudness loss, can vary with the test group* and particular test design. Table 2.2 gives opinion results for various values of overall loudness rating (OLR) in decibels (dB). These are based on representative laboratory conversation test results for telephone connections in which other characteristics such as circuit noise have little contribution to impairment. [2, 3]

2.2.1.2 Recommended Values of Loudness Rating. Table 2.3 provides further information on selected values of loudness rating which have been recommended by the ITU-T Organization. [3]

---

*Test group: A group of people selected randomly to judge the subjective quality of a telephone connection. [7]

TABLE 2.2 Overall Loudness Rating, Opinion Results

Overall Loudness Rating (dB)	Representative Opinion Results[a]	
	Percent "Good plus Excellent"	Percent "Poor plus Bad"
5 to 15	>90	<1
20	80	4
25	65	10
30	45	20

[a] Based on opinion relationship derived from the transmission quality index (see Annex A).

Source: Table 1/P.11, page 2, CCITT Rec. P.11, [24].

TABLE 2.3 LR Values (dB) Cited in ITU-T Recs. G.111 and G.121

	SLR[a]	CLR[a]	RLR[a]	OLR[a]
Traffic weighted mean values				
Long term	7–9[b]	0–0.5[e]	1–3[b,f]	8–12[e–g]
Short term	7–15[b]	0–0.5[e]	1–6[b,f]	8–21[e–g]
Maximum values for an average-sized country	16.5[c]		13[c]	
Minimum value	−1.5[d]			

[a] As in Figure 2.3.
[b] Clause 1/G.121, ITU-T Rec. G.121.
[c] Subclause 2.1/G.121, ITU-T Rec. G.121.
[d] Clause 3/G.121, ITU-T Rec. G.121.
[e] When the International chain is digital, CLR = 0. If the international chain consists of one analog circuit, CLR = 0.5 and then OLR is increased by 0.5 dB. (If the attenuation distortion with frequency of this circuit is pronounced, the CLR may increase by another 0.2 dB. See A.4.2/G.111, ITU-T Rec. G.111.)
[f] See also the remarks made in 3.2/G.111, ITU-T Rec. G.111.
[g] Subclause 3.2/G. 111, ITU-T Rec. G.111.

Source: Table 2b/P.11, page 4, ITU-T Rec. P.11, March 1993 [24].

2.2.1.3 Determination of Loudness Rating. The designation of loudness ratings (LRs) in an international connection are given in Figure 2.3. Telephone sensitivity must be measured. The measurement can be made using guidelines in ITU-T Recs. P.66, P.76, P.78, and P.79. Telephone sensitivity includes both microphone and earpiece sensitivity. Overall loudness rating (OLR) is then calculated using the following formula:

$$OLR = SLR + CLR + RLR \qquad (2.1)$$

OLR is defined as the loudness loss between the speaking subscriber's mouth and the listening subscriber's ear via a connection.

The send loudness rating (SLR) is defined as the loudness loss between the speaking subscriber's mouth and an electric interface in the network. (The loudness loss here is defined as the weighted decibel average of driving sound pressure to measured sound pressure.)

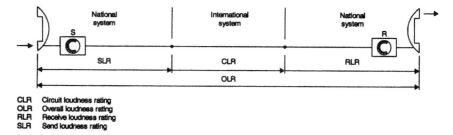

CLR Circuit loudness rating
OLR Overall loudness rating
RLR Receive loudness rating
SLR Send loudness rating

Figure 2.3. Designation of LRs in an international connection. From ITU-T Rec. 1/P.11, page 4, March 1993 [24].

The receive loudness rating (RLR) is the loudness loss between an electric interface in the network and the listening subscriber's ear. (The loudness loss in this case is defined as the weighted decibel average electromotive force to measured sound pressure.)

The circuit loudness rating (CLR) is the loudness loss between two electrical interfaces in a connection or circuit, each interface terminated by its nominal impedance, which may complex. (The loudness loss here is approximately equivalent to the weighted decibel average of the composite electric loss.)

Source: These definitions have been taken from ITU-T Rec. G.111, March 1993 [25].

2.3 Subscriber Loop Design Techniques

2.3.1 Introduction. Consider the following drawing of a simplified subscriber loop:

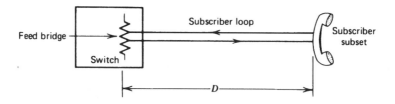

The distance D, the loop length, is a critical parameter. The greater the value of D, the greater the attenuation that the loop suffers, and signal level drops as a result. Likewise, there is a limit to D due to dc resistance, so signaling the local switch can be affected. When we lift the telephone off hook, there

Note: The terms *on hook* and *off hook* derive from old-fashioned telephones. This type of telephone usually had a wooden instrument box hanging on a wall. Extending from the box through a hole on the side was a lever held normally in the "up-position." There was a metal ring on the end of the lever. The telephone instrument had a wire pair connected to it, several feet (~1 meter) long, earpiece on the top and a mouthpiece on the bottom and a hook on the end. When not in use, the must be enough current flow in the loop to actuate the local switch where the loop terminates.

Of course it will follow that the greater the wire diameter of the loop pair, the less resistance there is per unit length; also, the less attenuation there is per unit length. On a particular subscriber loop we must set an attenuation limit and a minimum current flow. The current flow is usually stated as a resistance in ohms. We expect a common battery voltage of −48 V. This is what a high-impedance voltmeter will read anywhere in the loop when no current is drawn, such as a telephone instrument off hook.

When designing a subscriber loop, we would be vitally interested on what its maximum length would be. There are two variables that must be established: (1) *The maximum loop resistance.* This value is a function of the circuit in the switch where the loop terminates. One current value that comes to mind is 2400 Ω. (2) *The maximum loss or attenuation on the loop.* This will be taken from the national transmission plan. In Europe, 6 dB is commonly used for this value. This is 6 dB at the reference frequency of 800 Hz. In North America the reference frequency is 1000 Hz. The loss value may be as high as 9 dB.

Loss values will be the SLR or the RLR from equation 2.1. Then the maximum length of a subscriber loop will be governed by a resistance limit and a loss limit. Which is the more important of the two parameters? The resistance limit wins every time. If we cannot signal over the loop (i.e., cause a line seizure at the serving switch), the loop will not work.

Remember when budgeting parameter values for the resistance limit on a subscriber loop, we must budget something for the telephone subset itself. Use 300 Ω for this value. Again, the maximum loop resistance is set by the local serving switch design. Prior to the days of digital switches in the United States, the value was 1300 Ω. Allow 300 Ω for the telephone subset leaving only 1000 Ω for the loop itself. Some earlier digital switches advanced this value to 1800 Ω; some Northern Telecom switches provide 2400 ohms.

2.3.2 Calculating the Resistance Limit.
To calculate the dc loop resistance for copper conductors, the following formula is applicable:

$$R_{dc} = \frac{0.1095}{d^2} \tag{2.2}$$

where R_{dc} is the loop resistance in ohms per mile (statute) and d is the diameter of the conductor (in inches).

telephone instrument hook engaged the ring on the lever, pulling the lever down with its own weight. When someone wished to use the telephone, he/she unhooked the instrument from its ring, and the spring-loaded lever moved upwards. This caused a contact closure on the loop and current would flow in the loop. This became known as the "off-hook" condition. When the user was finished, he/she would replace the instrument, engaging the ring on the lever with the hook on the instrument, pulling the lever down. The contact now would open and current would stop flowing in the loop. This was called the "on-hook" condition. [7, 8]

If we wish a 10-mile loop and allow 100 Ω per mile of loop (for the stated 1000-Ω limit), what diameter of copper wire would be needed?

$$100 = \frac{0.1095}{d^2}$$

$$d^2 = \frac{0.1095}{100}$$

$$d = 0.033 \text{ in. or } 0.76 \text{ mm (round off to } 0.80 \text{ mm)}$$

Using Table 2.4, we can compute maximum loop lengths for 1000-Ω signaling resistance. Use a 26-gauge loop. We then have

$$\frac{1000}{83.5} = 11.97 \text{ kilofeet or } 11,970 \text{ ft}$$

Let's use the 2400-Ω switch as another example. Subtract 300 Ω for the telephone subset, leaving us with a net of 2100 Ω. We will use a 26-gauge wire pair on the loop, then from Table 2.4 we have

$$\frac{2100}{83.5} = 25.149 \text{ kft or } 25,149 \text{ ft}$$

Thus we are dealing here with what some call the *signaling limit* on a subscriber loop, and not the loss (attenuation) limit, or what some call the *transmission limit*. This is described in Section 2.3.3. Another term we introduce is *resistance design*. This is a method of designing subscriber loops where resistance is the only limiting parameter. If we cover for sufficient resistance, the loop loss will take care of itself.

2.3.3   Calculating the Loss Limit. Attenuation or loop loss is the basis of transmission design of subscriber loops. The attenuation of a wire pair varies with frequency, resistance, inductance, capacitance, and leakage conductance. Also, resistance of the line will depend on temperature. For open-wire lines, attenuation may vary by $\pm 12\%$ between winter and summer conditions. For

TABLE 2.4   Loss and Resistance per 1000 ft of Subscriber Cable[a] [8]

Cable Gauge AWG	Cable Diameter (mm)	Loss per 1000 ft (dB) @1000 Hz	Loss per Kilometer (dB) @ 1000 Hz	Loop Resistance (Ω/1000 ft)	Loop Resistance (Ω/km)
28	0.32	0.666	2.03	142	433
26	0.405	0.51	1.61	83.5	270
24	0.511	0.41	1.27	51.9	170
22	0.644	0.32	1.01	32.4	107
19	0.91	0.21	0.71	16.1	53

buried cable, which we are more concerned with in this context, variations due to temperature are much less.

Table 2.4 also gives losses of some common subscriber cable per 1000 ft. If we are limited to 6 dB (loss) on a subscriber loop, then by simple division we can derive the maximum loop length permissible for transmission design considerations for the wire gauges shown.

$$28 \quad \frac{6}{0.666} = 9.0 \text{ kft}$$

$$26 \quad \frac{6}{0.51} = 11.7 \text{ kft}$$

$$24 \quad \frac{6}{0.41} = 14.6 \text{ kft}$$

$$22 \quad \frac{6}{0.32} = 19.0 \text{ kft}$$

$$19 \quad \frac{6}{0.21} = 28.5 \text{ kft}$$

2.3.4 Loading. In many situations it is desirable to extend subscriber loop lengths beyond the limits described in Sections 2.3.2 and 2.3.3. Common methods to attain longer loops without exceeding loss limits are to increase conductor diameter, use amplifiers and/or range extenders,* and use inductive loading.

Inductive loading tends to reduce transmission loss on subscriber loops and on other types of voice pairs at the expense of good attenuation-frequency response beyond 3000 Hz. Loading a particular voice-pair loop consists of inserting inductances in series (loading coils) into the loop at fixed intervals. Adding load coils tends to decrease the velocity of propagation and increase the impedance. Loaded cables are coded according to the spacing of the load coils. The standard code for load coils regarding spacing is shown in Table 2.5.

TABLE 2.5 Code for Load Coil Spacing [2, 8]

Code Letter	Spacing (ft)	Spacing (m)
A	700	213.5
B	3000	915.0
C	929	283.3
D	4500	1372.5
E	5575	1700.4
F	2787	850.0
H	6000	1830.0
X	680	207.4
Y	2130	649.6

* A *range extender* is a device that increases battery voltage on a loop. This extends its signaling range. It may also contain an amplifier, thereby extending transmission loss limits as well. [2]

Loaded cables are typically designated 19-H-44, 24-B-88, and so forth. The first number indicates the wire gauge (AWG); the letter is taken from Table 2.5 and is indicative of the spacing, and the third set of digits is the inductance of the coil in millihenries (mH). For instance, 19-H-66 was a cable commonly used in Europe for long-distance operation. Thus the cable has 19-gauge voice pairs loaded at 1830-m intervals with coils of 66-mH inductance. The most commonly used coils have spacings of B, D, or H.

Table 2.6 shows the attenuation per unit length of several popular voice-pair cables used in the subscriber plant—for example, 19-H-88 cable (last entry in table), the attenuation per kilometer 0.26 dB (0.42 dB/statute mile of loop). Thus, for our 6-dB loop loss limit, we have 6.0/0.26, limiting the loop length to 23 km (14.3 statute miles).

Load coils add resistance to the loop. Budget 5.5 Ω per coil, as well as 8.5 Ω for each miniload coil. The minimum loop current is 20 mA.

2.3.5   Summary of Limiting Conditions: Transmission and Signaling.

We should be aware that the physical size of an exchange serving area is limited by factors of economy involving signaling and transmission. Signaling limitations are a function of the type of exchange and the diameter of the subscriber pairs and their conductivity, whereas transmission is influenced by pair characteristics. Both limiting factors can be extended, but that extension costs money, particularly when there may be many thousands of pairs involved. The decision boils down to the following:

1. If the pairs to be extended are few, they should be extended.
2. If the pairs to be extended are many, it probably is worthwhile to set up a new exchange area or a satellite exchange or to use an outside plant module in the area.

These economies are linked to the cost of copper. The current tendency is to reduce the wire gauge wherever possible or even resort to the use of aluminum as the pair conductor.

TABLE 2.6   Transmission Loss (Attenuation) Values with and without Load Coils for the More Popular Wire Gauges Found in the Subscriber Plant [2, 8]

Wire Gauge (AWG)	Diameter (mm)	Mutual Capacitance (nF/km)	Type of Loading	Loss (dB/km)
26	0.405	40	None	1.61
26	0.405	40	H-66	1.25
24	0.511	40	None	1.27
24	0.511	40	H-66	0.79
22	0.644	40	None	1.01
22	0.644	40	H-88	0.44
19	0.91	50	None	0.79
19	0.91	50	H-66	0.29

2.3.6 Subscriber Loop Impedances. For a conventional two-wire* switch, the characteristic impedance is 900 Ω. This is called a *compromise impedance*, and it is the impedance looking into the line circuit.

Since the late 1970s with the advent of nearly universal digital transmission, there has been considerable development to improve impedance matching the characterization of a subscriber loop looking into the switch. A family of subscriber loop impedances is shown in Table 2.7.

Most equipment attached to a two-wire loop is considered to have a 600-Ω impedance, typically 26 gauge (see item 2.7d, Table 2.7). Also, termination impedances at each end of a four-wire circuit are considered to be 600-Ω resistive. Both the 600-Ω and 900-Ω values are conventions and compromises. Another value is 735 Ω (item 2.7e). Some test instruments use this value, which is still another compromise, between 600 and 900 Ω.

Consider Figure 2.4. It shows how characteristic impedance varies with frequency. The example in the figure is a 26-gauge nonloaded wire pair. It is also important to remember discussing wire pair impedances that the reference frequency in North America is 1000 Hz, whereas in Europe and in countries under European hegemony it is 800 Hz. The figure shows that at about 1000 Hz the characteristic impedance is indeed 900 Ω.

TABLE 2.7   Standard Termination Impedances

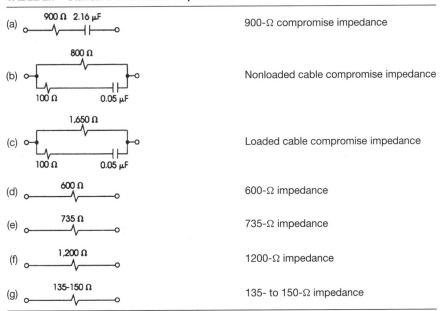

(a) 900 Ω 2.16 μF	900-Ω compromise impedance
(b) 800 Ω / 100 Ω 0.05 μF	Nonloaded cable compromise impedance
(c) 1,650 Ω / 100 Ω 0.05 μF	Loaded cable compromise impedance
(d) 600 Ω	600-Ω impedance
(e) 735 Ω	735-Ω impedance
(f) 1,200 Ω	1200-Ω impedance
(g) 135-150 Ω	135- to 150-Ω impedance

Source: Subscriber Loop Signaling and Transmission Handbook Analog, IEEE Press, Figure 5-6, page 100 [23].

---

* Two-wire transmission and four-wire transmission are described in Chapter 5.

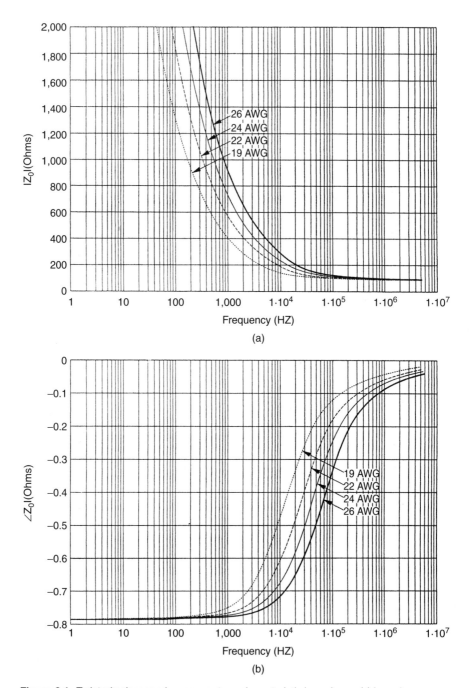

Figure 2.4. Twisted pair secondary parameters, characteristic impedance: (a) Impedance magnitude; (b) Impedance angle. From Figure 7–15, page 339, Ref. 29.

Conventional analog trunk plant often uses heavier-gauge cable, say 22 AWG (American Wire Gauge), where the characteristic impedance (Z_0) approaches 600 Ω at 1000 Hz (Table 2.7, item d). Loaded subscriber loops have a Z_0 approaching 1200 Ω (Table 2.7, item f). Table 2.7, item g, shows impedance values of 135–150 Ω. These lower values were selected because the impedance of the loop may go as low as 100 Ω at the higher frequencies on higher-speed data circuits. (See also Refs. 13, 14, 15, and 16)

3 CURRENT LOOP DESIGN TECHNIQUES USED IN NORTH AMERICA*

Before 1980, U.S. Bell operating companies designed subscriber loops in accordance with one of three loop design plans: (1) Resistance design (RD) 96%, (2) Unigauge design (UG) 1%, and (3) Long route design (LRD), 3%. (Percentages show approximate application usage of a particular plan—that is, percentage of total outside plant.) The objective, of course, is to design subscriber loops on a global basis rather than on an individual basis, which is extremely costly. If outside plant engineers fully comply with these design rules, then no loop will exceed switch signaling limitations (resistance) and there will be an adequate distribution of transmission losses.

A 1980 survey [20] studied loss design in a population of loops having an average length greater than the mean length of the general population. It was found that approximately 4% of the measured losses were greater than 8.5 dB, and 2% exceeded 10 dB. With the design rules in effect at the time, losses exceeding 8.5 dB were expected. Under the new, revised resistance design (RRD) rules, no properly designed loop should display a loss greater than 8.5 dB at 1000-Hz reference frequency.

3.1 Previous Design Rules

3.1.1 Resistance Design (RD)

Resistance: 0–1300 Ω (includes only resistance of cable and load coils).

Load coils: H-88 beyond 18 kft [not including bridged tap (BT)].

End sections (ES) and bridged tap (BT): Nonloaded BT = 6 kft (max); loaded ES + BT = 15 kft (max), 3 kft (min), 12 kft recommended.

Transmission limitations: None.

Cable gauges: Any combination of 19–26 gauge.

3.1.2 Long Route Design (LRD) (Rural Areas)

Loop resistance: 1301–3000 Ω (include here only the resistance of cable and load coils).

---

* Based on Ref. 20, Section 7.

Load coils: Full H-88.

End section (ES) and bridged tap (BT): ES + BT = 12 kft.

Transmission limitations: For loop resistances >1600 Ω, loop gain required.

Cable gauges: Any combination of 19–26 gauge.

3.2 Current Loop Design Rules

3.2.1 Revised Resistance Design (RRD)

Loop resistance: 0–18 kft, 1300 Ω (max) (includes resistance of cable and load coils only).

18–24 kft, 1500 Ω (max) (includes resistance of cable and load coils only).

>24 kft, use digital loop carrier.

[*Note*: Over 24 kft, consider using pair gain—meaning a concentrator (with gain) or a multiplexer (possibly with gain, and a digital multiplexer might use a regenerator every 3 kft).]

Load coils: Full H-88 for loops over 18 kft.

End sections (ES) and bridged tap (BT): Nonloaded total cable + BT = 18 kft, maximum BT = 6 kft. Loaded ES + BT = 3–12 kft.

Transmission limitations: None.

Cable gauges: Two gauge combinations preferred (22, 24, 26 gauge).

3.2.2 Concentration Range Extension with Gain (CREG)

Loop resistance: 0–2800 Ω (includes only resistance of cable and load coils).

Load coils: Full H-88 > 15 kft.

End sections (ES) and bridged tap (BT): Nonloaded cable + BT = 15 kft, maximum BT = 6 kft. Loaded ES + BT = 3–12 kft.

Transmission limitations: Gain range extension required for loop resistance >1500 Ω.

Cable gauges: Two gauge combinations preferred (22, 24, 26 gauge).

3.2.3 Modified Long Route Design (MLRD)

Loop resistance: 1501–2800 Ω.

Load coils: Full H-88.

End sections (ES) and bridged tap (BT) (max): ES + BT = 3–12 kft.

Transmission limitations: Range extension with gain required for loop resistances greater than 1500 Ω.

Cable gauges: Two gauge combinations preferred (22, 24, 26 gauge). (Based on Refs. 20 and 28.)

4   SIZE OF AN EXCHANGE AREA BASED ON NUMBER OF SUBSCRIBERS SERVED

The size of an exchange area (also called a serving area) obviously will depend largely on subscriber (or potential subscriber) density and distribution. Subscriber traffic is another factor to be considered. If statistics on subscriber traffic intensity are not available, use Table 2.8 based on ITU-T data.

Exchange sizes are often in units of 10,000 lines. Although the number of subscribers initially connected should be considerably smaller than when an exchange is installed, 10,000 is the number of subscribers that may be connected when an exchange reaches "exhaust," where it is filled and no more subscribers can be connected.

Ten thousand is not a magic number, but it is a convenient one. It lends itself to crossbar (switch) unit size and is a mean unit for subscriber densities in suburban areas and mid-sized towns in fairly well developed countries. More important, though, is its significance in telephone numbering (the assignment of telephone numbers). Consider a seven-digit number. Now break that down into a three-number group and a four-number group. The first three digits—that is, the first three dialed—identify the local exchange. The last four identify the individual subscriber and is called the subscriber number. Note the breakdown in the following sample:

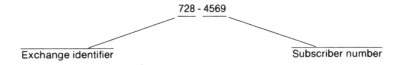

For the subscriber number there are 10,000 number combination possibilities, from 0000 to 9999. Of course there are up to 1000 possibilities for the exchange identifier. Numbering is discussed at length in Chapter 3 as a consideration under switching.

The foregoing discussion does not preclude exchanges larger than 10,000 lines. But we still deal in units of 10,000 lines, at least in conventional telephony. The term "wire center" is often used to denote a single location housing

TABLE 2.8   Average Occupation Time During the Busy Hour per Subscriber Line

Subscriber Type	BH Traffic Intensity (erlangs)
Residence	0.01–0.04
Business	0.03–0.06
PABX	0.1–0.6
Coin box	0.07

one or more 10,000 line exchanges. Some wire centers house up to 100,000 lines with a specific local serving area. Wire centers with an ultimate capacity of up to 140,000 lines can be economically justified under certain circumstances. Subscriber density, of course, is the key. Nevertheless, many exchanges will have extended loops requiring some sort of special conditioning, such as larger-gauge wire pairs, loading, range extenders, amplifiers, and the application of carrier techniques (Chapter 5). Leaving aside rural areas, 5–25% of an exchange's subscriber loops may well require such conditioning or may be called "long loops." (Also see Refs 2, 3, 9, 10 and 12.)

5   SHAPE OF A SERVING AREA

The shape of a serving area has considerable effect on optimum exchange size. If a serving area has sharply angular contours, the exchange size may have to be reduced to avoid excessively long loops (e.g., revert to the use of more exchanges in a given local geographical area of coverage).

There is an optimum trade-off between exchange size, and we mean here the economies of large exchanges (centralization) and the high cost of long subscriber loops. An equation that can assist in determining the trade-off is as follows, which is based on uniform subscriber density, a circular serving area A of radius r such that

$$C = \frac{A}{\pi r^3}(a + bd\pi r^3) + Ad(f + gL) \tag{2.3}$$

where C is the total cost of exchanges, which decreases when r increases and to which is added the cost of subscriber loops, which increases with r, and L is the average loop length, which may be related as $L = (2r/3)$, the straight-line distance. To determine the minimum cost of C with respect to r, the equation is differentiated and the result is set equal to zero. Thus

$$\frac{2Aa}{\pi r^3} = \frac{2Adg}{3} \tag{2.4}$$

and this may be fairly well approximated by

$$r = \left(\frac{a}{dg}\right)^{1/3} \tag{2.5}$$

The cost of exchange equipment is $a + bn$, where n is the number of lines, and the cost of subscriber loops is equal to $(f + gL)$ where, as we stated, L is the average loop length given a uniform density of subscribers, d; and a, b, f, and g are constants.

Since r varies as the cube root, its value does not change greatly for wide ranges of values of d. One flaw is that loops are seldom straight-line distances, and this can be compensated for by increasing g in the ratio (average loop length). This theory is simplified by making exchange areas into circles.

If an entire local area is to be covered, fully circular exchange serving areas are impractical. Either the circles will overlap or uncovered spaces will result, neither of which is desirable. There are then two possibilities: square serving areas or hexagonal serving areas. Of the two, a hexagon more nearly approaches a circle. The size of the hexagon can vary with density with a goal of 10,000 lines per exchange as the ultimate capacity. Again, a serving area could have a wire center of 100,000 lines or more, particularly in heavily populated metropolitan areas.

Besides the hexagon, full coverage of local areas may only be accomplished using serving areas of equal triangles or squares. This assumes, of course, that the local area was *ideally* divided into identical geometric figures and would apply only under the hypothetical situation of nearly equal telephone density throughout. A typical hexagon subdivision is shown in the diagram that follows.

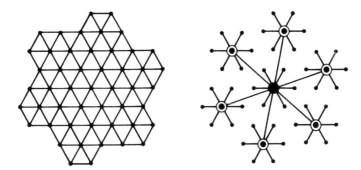

The routing problem then arises. How should the serving areas with their respective local exchanges be interconnected? From our previous discussion we know that two extremes are offered, mesh and star. We are also probably aware that as the number of exchanges involved increases, full mesh becomes very complicated and is not cost-effective. Certainly it is not as cost-effective as a simple star network of two or three levels permitting high-usage (HU) connections between selected nodes. For instance, given the hexagon formation in the preceding diagram, a full-mesh or two-level star network can be derived, as shown in the diagram that follows.

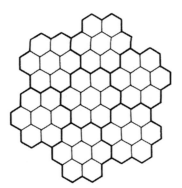

From the routing pattern in the preceding diagram, it should be noted that fan-outs of 6 and 8 allow symmetry, whereas fan-outs of 5 and 7 lead to inequalities. A fan-out of 4 is usually too small for economic routing.

It will be appreciated that some of the foregoing assumptions are rarely found in any real telecommunications environment. Uniform telephone density was one real assumption, and the implication of uniform traffic flow was another. Serving areas are not uniform geometric figures, exchanges seldom may be placed at serving area centers, and routing will end up as a mix of star and mesh. Small local areas serving 5–15 exchanges or even more may well be fully mesh connected. Some cities are connected in full mesh with over 50 exchanges. But as telephone growth continues, tandem routing will become the economic alternative [5, 11, 12].

6   EXCHANGE LOCATION

A fairly simple, straightforward method for determination of the theoretical optimum exchange location is described in Ref. 1, Chapter 6. Basically, the method determines the center of subscriber density in much the same way the center of gravity would be calculated. In fact, other publications call it the center-of-gravity method.

Using a map to scale, a defined area is divided into small squares of 100–500 m on a side. One guide for determination of side length would be to use a standard length of the side of a standard city block in the serving area of interest. The next step is to write in the total number of subscribers in each of the blocks. This total is the sum of three figures: (1) existing subscribers, (2) waiting list, and (3) forecast of subscribers for 15 or 20 years into the future. It follows that the squares used for this calculation should coincide with the squares used in the local forecast. The third step is to trace two lines over the subscriber area. One is a horizontal line that has approximately the same number of total subscribers above the line as below. The second is a vertical line where the number of subscribers to the left of the line is the same as that to the right. The point of intersection of these two lines is the theoretical optimum center or exchange location. A sample of this method is shown in Figure 2.5, where S_1 is the sum of subscribers across a single line and C_s is the cumulative sum.

Now that the ideal location is known, where will the real optimum location be? This will depend considerably on secondary parameters such as availability of buildings and land; existing and potential cable or feeder runs; the so-called trunk pull; and layout of streets, roads, and highways. "Trunk pull" refers to the tendency to place a new exchange near the one or several other exchanges with which it will be interconnected by trunks (junctions). Of course, this situation occurs on the fringes of urban areas where a new exchange location will tend to be placed nearer to the more populated area, thereby tending to shorten trunk routes. This is illustrated in Figure 2.1 and discussed in more detail below.

The preceding discussion assumed a bounded exchange area; in other words, the exchange area boundaries were known. Assume now that exchange locations

EXCHANGE LOCATION

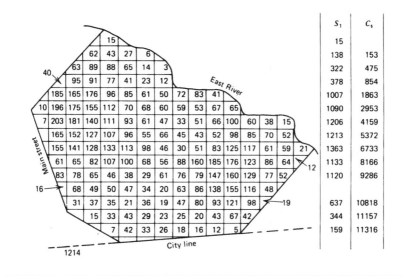

S_1	C_s
15	
138	153
322	475
378	854
1007	1863
1090	2953
1206	4159
1213	5372
1363	6733
1133	8166
1120	9286
637	10818
344	11157
159	11316

S_2	17	1104	1264	1211	1060	885	615	489	556	722	927	1072	770	399	242	33
C_s		1121	3195	3546	4686	5491	6106	6595	7151	7873	8800	9872	10642	11041	11283	11316

Figure 2.5. Sample wire centering exercise using "center-of-gravity" method [9].

are known and that the boundaries are to be determined. What follows is also valid for redistributing an entire area and cutting it up into serving areas. A great deal of this chapter has dealt with subscriber loop length limits. Thus an outer boundary will be the signaling limits of loops as described previously. The optimum cost–benefit trade-off is found when all or nearly all loops in a serving area remain nonconditioned and of small diameter, say, 26 gauge. We note that with H-66 loading the outer boundary will be just under 5 km in this case. It also would be desirable to have a hexagonal area if possible. In practice, however, natural boundaries may well be the most likely real boundaries of a serving area. "Main street," "East River," and "City line" in Figure 2.5 illustrate this point. In fact, these boundaries may set the limits such that they may be considerably greater or less than the maximum signaling (supervisory) limits suggested earlier. Of course, there are two types of serving areas where the argument does not hold. These are rural areas and densely populated urban areas. For the rural areas we can imagine very large serving areas and, for urban areas of dense population, considerably smaller serving areas than those set out with maximum supervisory signaling limits. (See also Ref. 12.)

To determine boundaries of serving areas when dealing with an exchange that is already installed and a new exchange, we could use the so-called *ratio technique*. Again, we employ signaling (supervisory) limits as the basis. As we have been made aware, these limits have been basically determined by the type

TABLE 2.9   Resistance Limits for Several Types of Exchange

Exchange Type	Resistance Limit (Ω)
5-ESS (United States)	1520
Metaconta (local — Europe)	2000
Pentaconta 2000 (Europe)	1250
DMS-100	1900 (up to 4000)
Other digital local exchange	2000–4000 [21]

of exchange and the copper wire gauge utilized for subscriber loops. Table 2.9 gives resistance limits for several of the more common types of local serving exchanges found in practice.

The ratio technique may be described using the following example. Given an existing exchange A and a new exchange B that will be installed on a cable route from A, let's assume that A is a DMS-100 with 1900 Ω resistance and B is a 5ESS extended to 2000 Ω resistance capability. The maximum distance along the cable route from A to B can be calculated by equating the distance to the sum of the resistances of (A-300 Ω) and (B-300 Ω) (or 1700 + 1600 Ω or 3300 Ω). What we are doing here is subtracting 300 Ω from each value for the telephone subset terminating each cable end.

Use 26-gauge cable in this case. From Table 2.4, no loading, the resistance can be equated to 270 Ω/km. Use the sum of the net resistance limits or 3300 Ω and divide by 270. Thus the maximum permissible distance from exchange A to B is 12.2 km (3300/270 = 12.2 km). Let the distance from exchange A be D_A.

$$D_A = 1700 \times 12.2/(1700 + 1600)$$
$$\approx 6.28 \text{ km}$$

The total distance from exchange B will be the difference or 12.2 − 6.28 = 5.92 km. The total distance from B to the boundary line will be 5.92 km; the total distance from A to the boundary line will be 6.28 km. These distances might be apportioned as shown in Figure 2.6.

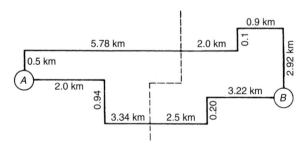

Figure 2.6. Determining serving area boundaries with the ratio method [17].

Continue the exercise and examine the exchange serving area around exchange A for subscriber density. Assume that the area is a square with A at its center. Allowing for *non-crow-fly* feeder routes, we can assume that the square has 10 km on a side (a rough assumption) as shown in the following diagram:

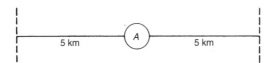

or 100 km^2. If we limited the largest exchange to a maximum of 100,000 subscribers and the smallest to 10,000 subscribers at the end of the forecast period (say 15 years), then subscriber density would be 10,000/100 to 100,000/100 or 100 subscribers/km^2 to 1000 subscribers/km^2.

We would limit the maximum size of an exchange to 100,000 subscribers. At one time it was physical size, enormous. Today, with digital switching, the entry mainframe is much larger than the switch. We think the more important point is survivability. Break the exchange up into smaller exchanges. In case of fire or other disaster, it is a gigantic project to replace a 100,000 line exchange. It is much easier to replace a 20,000-line exchange. Don't put all your eggs in one basket, so to speak. So we'd accept the added expense and install five 20,000-line exchanges rather than one 100,000-line exchange.

Another consideration in exchange location is what is called "trunk pull" discussed briefly above. This is a secondary factor in exchange placement and refers to the tendency, in certain circumstances, to shift a proposed exchange location to shorten trunks (junctions). Trunk pull (see Fig. 2.1) becomes a significant factor only where population fringe areas around urban centers and trunks extend toward the city center and the exchanges in question may well be connected in full mesh in their respective local area. Some possible saving may be accrued by shifting the exchange more toward the center of population, thereby shortening trunks at the expense of lengthening subscriber loops in the direction of more sparse population, even with the implication of conditioning on well over 10% of the loops. The best way to determine if a shift is worthwhile is to carry out an economic study comparing costs in PWAC (present worth of annual charges) [18, 19], comparing the proposed exchange as located by the center-of-gravity method to the cost of the shifted exchange. Basic factors in the comparison are in the cost of subscriber plant, trunk plant, plant construction, and cost of land (or buildings). The various methods of handling the sparsely populated sections of the fringe serving area in question should also be considered. For instance, we might find that in those areas, much like rural areas, there may be a tendency for people to bunch up in little villages or other small centers of population. This situation is particularly true in Europe. In this case the system designer may find considerable savings in telephone-plant costs by resorting to the use of satellite exchanges or concentrators. Trunk connections between satellite or concentrator and main exchange may be made by using

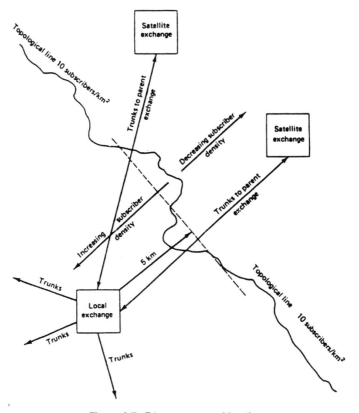

Figure 2.7. Fringe-area considerations.

digital multiplex techniques described in Chapter 8, such as E1 or T1 carrier. Concentrators, outside plant modules, and satellite exchanges are discussed in Chapter 3.

A typical fringe-area situation is shown in Figure 2.7. The "bunching" and the other possibility, "thinning out," are shown in the figure. "Thinning out" is just the population density per unit area decreases as we proceed from an urban center to the countryside. A topological line of population density of 10 inhabitants per square kilometer (26 inhabitants per square mile) is a fair guideline for separation of the rural part of the fringe area from the urban–suburban part. Of course, in the latter part we would have to resort to widespread conditioning, to the use of concentrator–satellites, or to apply subscriber carrier or subscriber pulse-code modulation (PCM) (Chapter 8). (Also see Refs 6–8, 9, 12, 14.)

7 DESIGN OF LOCAL AREA ANALOG TRUNKS (JUNCTIONS)

Exchanges in a common local area are interconnected by trunks. In countries where the telecommunication infrastructure is more developed, these trunks are liable to be digital based on E1 or T1 techniques which we will discuss in Chapter 8. Where the infrastructure of telecommunications is less developed, these trunks and their connected exchanges are analog. In either case there will be a relatively small number of these trunks compared to the number of subscriber lines terminating in the associated local exchanges. Because of their lower number, often there is a greater investment made for these local trunk facilities.

Our interest here is the design of these local analog trunks. One approach used by some telephone companies is to allot 1/3 of the total end-to-end loudness loss to each subscriber loop and 1/3 to the total trunk network. Keep in mind that there are two subscriber loops in a connection, one for the calling subscriber and one for the called subscriber. Figure 2.8 illustrates this concept. For instance, if the transmission plan called for an 18-dB OLR, then 1/3 of 18 or 6 dB would be assigned to the trunk plant. Of this we might assign 3 dB to the four-wire portion of the long distance (toll) network and 3 dB to the trunk plant. Of course our example is highly simplified. For toll-connecting trunks (e.g., those trunks connecting the local network to the toll network), if a good return loss* cannot be maintained on all or nearly all connection, losses on two-wire toll connecting trunks may have to be increased to control echo and/or singing. Sometimes the range of loss may have to be extended to 5 or 6 dB. It is just these circuits into which the four-wire toll network looks directly. Two-wire and four-wire circuits, along with echo and singing, are discussed in Chapter 5. Thus it can be seen that the approach to the design of VF trunks varies considerably from that used for subscriber loop design. Although we must assure that signaling limits are not exceeded, the transmission limits are usually exceeded before the signaling limits. The tendency is to use larger diameter cable on long routes. If inductive loading is to be employed, the first coil is installed at distance $D/2$, where D is the normal separation between load points. Take the case of H-loading, for example. The normal distance between load points is 1830 m (from Table 2.5), but the first load coil from the exchange is placed at $D/2$, or 915 m from the exchange. Then, for circuits that bypass the exchange, a full loaded section exists. This concept is illustrated in Figure 2.9. However, it is more commonly done at the switching center (exchange) because of accessibility of cable pairs at the main frame associated with the center (exchange). When this trunk (junction)

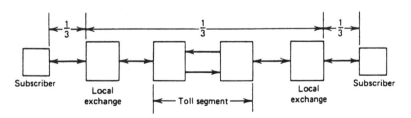

Figure 2.8. One conceptual thumb rule for network loss assignment.

* Return loss is discussed in Chapter 5, Section 5.

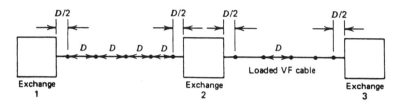

Figure 2.9. Inductive loading of VF trunks (junctions).

plant is upgraded to digital, the load coils are replaced by regenerative repeaters (Chapter 8).

8 VOICE-FREQUENCY REPEATERS

A voice frequency (VF) repeater is an audio amplifier used on a subscriber loop to extend it or on a VF trunk to meet certain loss or gain requirements. There are two types of VF repeaters used in the industry: a four-wire repeater or a two-wire negative impedance repeater. Their application as we describe it is for two-wire operation. For the four-wire repeater a hybrid is used on the input and another on the output to convert two-wire operation to four-wire operation and back again. An example of this type of repeater is shown in Figure 2.10. Historically, these types of repeater were used on long distance 19-gauge telephone circuits. They had a gain in the range of 20–25 dB. Today, trunks using this type of repeater have the gain adjusted to compensate for circuit loss minus a 4-dB loss to provide the necessary singing margin (see Chapter 5). In practice, a repeater is installed at each end of a trunk circuit so simplify maintenance and power feeding.

The negative impedance repeater is a true two-wire repeater. It can provide a gain as high as 12 dB, but 7 or 8 dB is more common in actual practice. It requires and LBO (line build-out) at each port for matching electrical line lengths. The repeater action is based on regenerative feedback of two amplifiers. The major advantage of negative impedance repeaters is that they are transparent to dc signaling. On the other hand, VF repeaters require a composite arrangement to pass dc signaling. This consists of a transformer bypass.

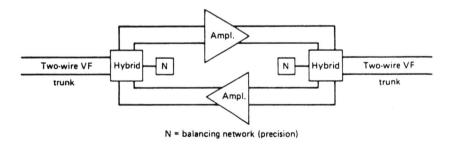

N = balancing network (precision)

Figure 2.10. Simplified functional block diagram of a 4-wire VF repeater.

9   TANDEM ROUTING

The local-area trunking scheme evolutionally has been the mesh connection of exchanges, and in many areas of the world it remains full mesh. We said initially that mesh connection is desirable and viable for heavy traffic flows. As traffic flows reduce, going from one situation to another, the use of tandem routing in the local area becomes an interesting, economical alternative.

Furthermore, it can be shown that a local trunk network can be optimized, under certain circumstances, with a mix of tandem, high-usage (overflow to alternative routes), and direct connection (mesh). We often refer to these three possibilities as THD. The system designer wishes to determine, on a particular trunk circuit, if it should be "tandem," "high-usage," or "direct" (T, H, or D). This determination is based on incremental cost of the trunk (junction), making the total network costs as low as possible. Such incremental cost can be stated:

$$B = c + (bl) \tag{2.6}$$

where c is the cost of switching equipment per circuit, b is the incremental costs for trunks per mile or kilometer, and l is the length of the trunk (or junction) circuit.

To carry out a THD decision for a particular trunk route, the input data required are the *offered* traffic between the local exchanges in question and the grade of service. We can now say that the THD decision is to a greater extent determined by the offered traffic A between the exchanges and the cost ratio ε, where

$$\varepsilon = \frac{B}{B_1 + B_3} \tag{2.7}$$

and where B is the cost for the direct route, with B_1 from exchange 1 to the proposed tandem and B_2 from the proposed tandem exchange to exchange 2, for incremental costs between direct and tandem routing. Of course, before starting such an exercise, the provisional tandem points must be known. Figure 2.11, which was taken from Ref. 4, may be helpful as a decision guide.

To approximate the number of high-usage (HU) trunk circuits required, the following formula may be used:

$$F(n, A) = AE(n, A) - E(n + 1, A) = \varepsilon\psi \tag{2.8}$$

where E is the grade of service, n is the number of high-usage circuits, A is the offered traffic between exchanges, $F(n, A)$ is the "improvement" function [i.e., increase in traffic carried on high-usage trunk group on increasing the number of these trunk circuits from n to $(n + 1)$], ε is the cost ratio (as previously), and ψ is the efficiency of incremental trunks [marginal utilization (of magnitude 0.6–0.8)]. However, a still better approximation will result if the following formula is used:

$$F(n, A) = \varepsilon[1 - 0.3(1 - \varepsilon^2)] \tag{2.9}$$

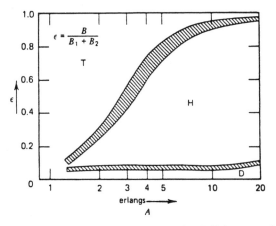

Figure 2.11. THD diagram. Courtesy of the International Telecommunications Union — CCITT [4].

The exercise below emphasizes several practical points on the application of tandem routing in the local area. Assume that there are 120,000 subscribers in a certain local area. If we allow a BH calling rate of 0.5 erlangs per subscriber (see Table 2.8) and suppose there is full-mesh connection, we can assemble Table 2.10. On the basis of this table we assembled, it would not be reasonable to expect any appreciable local area trunk economy through tandem working with less than 30–40 exchanges in the area. Of course, Table 2.10 supposes unity community of interest (see Section 2.11) with the more distant exchanges so that a few attractive tandem routings might exist in the 20–30 exchange range. Areas smaller than those in our sample will naturally reduce the number of exchanges at which tandem working is viable, and a larger area and higher calling rate will increase this number. The results of any study to determine feasibility of employing the tandem exchange technique is very sensitive to the number of exchanges in the local area and considerably less sensitive to the size of the area and the calling rate. Hence we can see that the economy of tandem routing is least in areas of dense subscriber population, where exchanges can be paced

TABLE 2.10 Traffic Table: Full-Mesh Connections (120,000 Subscribers)

Number of Exchanges	Average Size of Exchange	Originated Traffic per Exchange (erlangs)	Average Traffic to Each Distant Exchange (erlangs)
10	12,000	600	60
12	10,000	500	42
15	8,000	400	27
20	6,000	300	15
30	4,000	200	6.7
40	3,000	150	3.8

without exceeding subscriber loop length limits and where the trunks will be short. Relatively sparsely populated areas are more favorable and are likely to show relatively low community-of-interest factors between distant points [1, 4, 5, and 13].

10 DIMENSIONING OF TRUNKS

A primary effort for the system engineer in the design of the local trunk network is the dimensioning of the trunks of that network. Here we simply wish to establish the economic optimum number of trunk circuits between exchanges *X* and *Y*. If we are given the traffic (in erlangs) between the two exchanges and the grade of service, we can assign the number of trunk circuits between *X* and *Y*. As discussed in Chapter 1, it is assumed that all traffic values are BH values. Once given this input information, the erlang value (the number of trunk circuits) can be derived from Table 1-2, Ref. 27, with no overflow assumed.

It can be appreciated by the reader that the trunk traffic intensities used in the design of trunk routes should allow for growth—that is, the increase of traffic from the present with the passage of time. This increase is attributable to several factors: (1) the increase in the number of telephones in the area that generate traffic, (2) the probable increase in telephone usage, and (3) the possibility of a change in character of the area in question, such as from rural to suburban or from residential to commercial. Thus the designer must use properly forecast future traffic values. These values in practice are for the forecast period 5–8 years in the future. The art of arriving at these figures is called *forecasting*. Present traffic values should always be available as a base or point of departure.

Suppose we use a sample local area with five exchanges: *A, B, C, D*, and *E*. For an 8-year forecast period we could possibly come up with the traffic matrix like that shown in Table 2.11. Thus from the traffic matrix in the direction of exchange *C* to exchange *B* there is a traffic intensity of 11 erlangs. Applying the 11 erlangs

TABLE 2.11 Sample Traffic Matrix[a]

	Traffic (erlangs)								
From/to	*A*	*B*	*C*	*D*	*E*	Toll	Total Orig.	Lines Working	Traffic per Line
A	21	20	65	2.5	2.5	1.5	112.5	9,200	0.122
B	22	80	13	6.0	5.0	4.0	130	26,000	0.005
C	5	11	7	2.5	2.5	1.0	29	7,500	0.004
D	2	7	1	0.3	0.2	0.5	11	3,000	0.004
E	2	5	2	0.2	0.3	0.5	10	2,800	0.004
Toll	2	3	1	0.5	0.5	—	7	—	—
Total	53	126	89	12	11	7.5	298.5	48,500	0.006
Total-line	0.006	0.005	0.012	0.004	0.004	0.00015	—	—	—

[a] For 8-year forecast period.
Example: Traffic *A* to *A* is traffic originated at *A* and terminating at *A*.

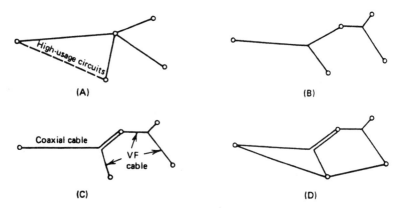

(A)

(B)

Coaxial cable

VF
cable

(C)

(D)

Figure 2.12. Routing diagrams and practical routing (VF stands for voice frequency): (A) link diagram; (B) route diagram; (C) practical routing; (D) routing for service continuity [5].

to Table 1-2 in Ref. 27, we see that 23 circuits would be required. If the traffic intensity were 13 erlangs, 26 trunks would be required. These circuit figures suppose a grade of service of $p = 0.001$. For a grade of service of $p = 0.01$, 19 and 22 circuits would be required, respectively.

The preceding discussion of routing implied an on-paper routing that would probably vary in practice where we were dealing with actual cable lays and facility drawings to something quite different. In simplified terms, these differences are shown in Figure 2.12.

11   COMMUNITY OF INTEREST

We have referred to the community-of-interest concept in passing. This is a method used as an aid to estimate calling rate and traffic distribution for a new exchange and its connecting trunks. The community-of-interest factor K is defined as follows:

$$\text{Traffic } A - B = K \times \frac{\text{Traffic originating at } A \times \text{Traffic originating at } B}{\text{Total originating traffic in area}}$$

(2.10)

If all subscribers are equally likely to call all others, the proportion results:

$$\frac{\text{Traffic originating at } B}{\text{Traffic originating in area}} = \frac{\text{Traffic terminating at } B}{\text{Traffic terminating in area}}$$

(2.11)

This is the expected proportion of all A traffic that is directed to B. In this case $K = 1$, corresponding to equal community of interest between all subscribers. The condition of $K > 1$ or $K < 1$ indicates a greater or lesser interest than average between exchanges A and B. The factor K is affected by the type of area, whether residential or business, as well as the distances between exchanges.

For instance, in metropolitan areas K may range from 4 for calls originating and terminating on the same exchange to 0.25 for calls on opposite sides of the city or metropolitan area.

The K factor is a useful reference for the installation of a new exchange where the serving area of the new exchange will be cut out from serving areas of other exchanges. The community-of-interest factors may then be taken from traffic data from the old exchanges, and the values averaged and then applied to the new exchange. The same principle may be followed for exchange extensions in a multiexchange area [5].

REVIEW QUESTIONS

1. Local networks can be defined in a number of ways. Give the definition of a local network that is provided in the text.

2. Considering both quality of service and economy, give at least five constraints or planning factors that go into the design of a local telephone network.

3. What are the two limiting performance factors in the design of a subscriber loop that constrain length?

4. Loudness rating is a measure of transmission quality. What singular parameter does it measure?

5. What are the three elements (measured in decibels) that, when summed, constitute the value of OLR?

6. What important parameter contained in a national transmission plan is vital in calculating OLR?

7. If we assume that a modern local exchange is designed for a maximum of 2000 Ω of loop resistance, not including the resistance of the end instrument (i.e., the telephone subset), what is the maximum loop length of 26-gauge copper wire that can be used? Assume 83 Ω per 1000 ft of loop.

8. The National Transmission Plan allows a 7-dB maximum subscriber loop loss. Assume a 0.51-dB loss per 1000 ft of loop for a 26-gauge wire pair. What is the maximum subscriber loop length considering only loss?

9. What is the maximum subscriber loop length permissible using inputs from question 7 if we increase the exchange maximum resistance to 2800 ohms?

10. The exchange involved in question 8 must serve some rural subscribers who are over the maximum loop length. Name at least four expedients that can be used to serve these subscribers. Differentiate whether these expedients have an impact on resistance limit or transmission limit.

11. What is a reasonable range of values for subscriber loop impedance?

12. A subscriber connected to a local exchange commonly (not always) is identified by a seven-digit number made up of a three-digit prefix followed by a four-digit number. What does the three-digit prefix identify? What is the four-digit sequence called?

13. How many different individual subscribers can be served by a four-digit number assuming no blocked numbers?

14. Using only subscriber density, describe a method of idealized exchange placement.

15. Both resistance design and revised resistance design are, as the names imply, based solely on loop resistance. Explain why loss seems to be of no concern as a limiting factor.

16. Describe LBO and how it is used on loaded wire-pair trunks.

17. If D is the separation of trunk load coils, why is the first load coil outward from an exchange at $D/2$?

18. Name two types of VF repeaters.

19. What is the philosophy behind tandem routing? Why use it in the first place?

20. In network design, what is the THD decision? What are the three input parameters necessary for this decision?

21. Define community of interest.

22. Describe a traffic matrix and what it is used for.

REFERENCES

1. *Local Telephone Networks*, International Telecommunications Union, Geneva, 1968.

2. International Telephone and Telegraph Corporation, *Reference Data for Radio Engineers*, 6th ed., Howard W. Sams, Indianapolis, 1976.

3. *Transmission Planning Aspects of Speech Service in Digital Public Land Mobile Networks*, ITU-T Rec. G. 173, Helsinki, March 1993.

4. *National Networks for the Automatic Service*, International Telecommunications Union, Geneva, 1968.

5. *Networks*, Telecommunication Planning Documents, ITT Laboratories (Spain), Madrid, 1973.

6. *CCITT Red Books, Malaga-Torremolinos*, 1984, in particular Vols. III and V.

7. *Outside Plant*, Telecommunication Planning Documents, ITT Laboratories (Spain), Madrid, 1973.

8. R. L. Freeman, *Telecommunication Transmission Handbook*, 4th ed., John Wiley & Sons, New York, 1998.

9. Y. Rapp, *Algunos Puntos de Vista Económicos para el Planeamiento a Largo Plazo de la Red Telefónica*, L. M. Ericsson, Stockholm, 1964.

10. *Placement of Exchanges in Urban Areas—Computer Program*, ITT Laboratories (Spain), Madrid, 1974.

11. J. C. Emerson, *Local Area Planning*, Telecommunications Planning Symposium [ITT Laboratories (Spain)], Boksburg, South Africa, 1972.

12. L. Alvarez Mazo and P. H. Williams, *Influence of Different Factors on the Optimum Size of Local Exchanges* [ITT Laboratories (Spain)], Boksburg, South Africa, 1972.

13. J. C. Emerson, *Factors Affecting the Use of Tandem Exchanges in the Local Area* [ITT Laboratories (Spain)], Boksburg, South Africa, 1972.

14. IEEE ComSoc, *The International Symposium on Subscriber Loops and Services*, Atlanta, GA, 1977.

15. IEEE ComSoc, *Second International Symposium of Subscriber Loops and Services*, London, 1976.

16. J. E. Flood, *Telecommunications Networks*, IEE Telecommunications Series 1, Peter Peregrinus, London, 1975.

17. Y. Rapp, *Planning of Exchange Locations and Boundaries in Multi-Exchange Networks*, Vol. 18, Ericsson Technology, 1962, p. 94.

18. *Telecommunications Planning*, ITT Laboratories (Spain), Madrid, 1974.

19. O. Smidt, *Engineering Economics*, Telephony Publishing Co., Chicago, 1970.

20. *BOC Notes on the LEC Networks—1994*, Issuc 2, SR-TSV-002275, Bellcore, Piscataway, NJ, April 1994.

21. *Electronic Switching: Digital Central Offices of the World*, Amos Joel, ed., IEEE Press, New York, 1982.

22. *Transmission Systems for Communications*, 5th ed., Bell Telephone Laboratories, Holmdel, NJ, 1982.

23. W. D. Reeve, *Subscriber Loop Signaling and Transmission Handbook, Analog*, IEEE Press, New York, 1992, p.44.

24. *Effect of Transmission Impairments*, ITU-T Rec. P.11, ITU Helsinki, March 1993.

25. *Loudness Ratings (LRs) in an International Connection*, ITU-T Rec. G.111, ITU Helsinki, March 1993.

26. *Loudness Ratings (LRs) of National Systems*, ITU-T Rec. G.121, ITU Helsinki, March 1993.

27. R. L. Freeman, *Reference Manual for Telecommunication Engineering*, 3rd ed., John Wiley & Sons, New York, 2002.

28. "Telcordia Notes on the Networks," Telcordia Special Report SR-2275, Issue 4, Telcordia, Piscataway NJ, October 2000.

29. Whitham D. Reeve, *Subscriber Loop Signaling and Transmission Handbook—Digital*, IEEE Press, New York, 1995.

3

SWITCHING IN AN ANALOG ENVIRONMENT

1 INTRODUCTION

1.1 Background and Approach

Automatic switching goes back to 1889 and A. B. Strowger, an undertaker in Kansas City, Missouri. The lore tells us that Strowger depended on the newly invented telephone to be informed of a death and a possible business opportunity. Switching at the time was done manually, usually by women. In this case, the young lady attendant had a boyfriend, a competing undertaker. So the story goes, she listened in on calls destined for Strowger and immediately informed her boyfriend. He usually got to the scene before Strowger and took the business away from him.

Strowger gave the problem a lot of thought on how to automate switching, removing the young lady from the scene. He came up with a novel approach, which we call the *step-by-step switch*, which the British call the *Strowger switch*.

Curiously, automatic switching, the step-by-step switch, was not adopted by the U.S. Bell System until 1919, some 30 years after its invention. North American independent telephone companies started to implement the switch by the early 1890s [22, 26].

Today, switching in the public switched telecommunications network (PSTN) is almost entirely digital. Some pockets of analog switching and networks remain, however, in some developing countries. We introduce the subject of switching with a brief description of analog switching concepts. We do this for two reasons:

1. Most of the concepts covered in the analog description hold for digital switching as well.

Telecommunication System Engineering, by Roger L. Freeman
ISBN 0-471-45133-9   Copyright © 2004 Roger L. Freeman

2. The chapter development provides an excellent historical perspective of the evolution of the telecommunications network.

Such factors as functional description, desirable features, and operational requirements, as well as those basic switching concepts, are introduced in the following sections. The discussion will be further carried on in Chapter 9, which covers digital switching.

1.2   Switching in the Telephone Network

A network of telephones consists of pathways connecting switching nodes so that each telephone in the network can connect with any other telephone for which the network provides service. Today there are hundreds of millions of telephones in the world, and nearly each and every one can communicate with any other one. Chapter 1 discussed the two basic technologies in the engineering of a telephone network, namely, transmission and switching. Transmission allows any two subscribers in the network to be heard satisfactorily. Switching permits the network to be built economically by concentration of transmission facilities. These facilities are the pathways (trunks) connecting the switching nodes.

Switching establishes a path between two specified terminals, which we call *subscribers* in *telephony*. The term *subscriber* implies a public telephone network. There is, however, no reason why these same system criteria cannot be used on private or quasipublic networks. Likewise, there is no reason why that network cannot be used to carry information other than speech telephony. In fact, in later chapters these "other" applications are discussed, as well as modifications in design and features specific for special needs.

A switch sets up a communication path on demand and takes it down when the path is no longer needed. It performs logical operations to establish the path and automatically charges the subscriber for usage. A commercial switching system satisfies, in broad terms, the following user requirements:

1. Each user has need for the capability of communicating with any other user.
2. The speed of connection is not critical, but the connection time should be relatively small compared to holding time or conversation time.
3. The grade of service, or the probability of completion of a call, is also not critical but should be high. Minimum acceptable percentage of completed calls during the BH may average as low as 95%, although the general grade of service goal for the system should be 99%* (equivalent to $p = 0.01$).
4. The user expects and assumes conversation privacy but usually does not specifically request it, nor, except in special cases, can it be guaranteed.
5. The primary mode of communication for most users will be voice (or the voice channel).

---

* See CCITT Rec. Q.95; $p = 0.01$ per link on an international connection.

6. The system must be available to the user at any time the user may wish to use it [2–4].

2 NUMBERING, ONE BASIS OF SWITCHING

A telephone subscriber looking into a telecommunication network sees a repeatedly branching tree of links. At each branch point there are multiple choices. Assume that a calling subscriber wishes to contact one particular distant subscriber. To reach that distant subscriber, a connection is built up utilizing one choice at each branch point. Of course, some choices lead to the desired end point, and others lead away from it. Alternative paths are also presented. A call is directed through this maze, which we call a *telephone network*, by a telephone number. It is this number that activates the switch or switches at the "maze" branch point(s).

Actually, a telephone number performs two important functions: (1) It routes the call, and (2) it activates the necessary equipment for proper call charging. Each telephone subscriber is assigned a distinct number, which is cross-referenced in the telephone directory with the subscriber's name and address; in the local serving exchange (switch), this number is associated with a distinct subscriber line.

If a subscriber wishes to make a telephone call, she lifts her receiver "off hook" (i.e., takes the handset out of its cradle) and awaits a dial tone that indicates readiness of her serving switch to receive instructions. These "instructions" are the number that the subscriber dials (or the buttons that she punches) giving the switch certain information necessary to (a) route the call to the distant subscriber with whom she wishes to communicate and (b) set up the call-charging equipment.

A subscriber number is the number to be dialed or called to reach a subscriber in the same local (serving) area. Remember that our definition of a local "serving area" is the area served by a single switch (exchange). The thinking that follows ties that switch capacity in total lines to the number of digits in the telephone number.

If we had a switch with a capacity of 100 lines,

it could serve up to 100 subscribers and we could assign telephone numbers 00 through 99.

If we had a switch with a capacity of 1000 lines,

it could serve up to 1000 subscribers and we could assign telephone numbers 000 through 999.

If we had a switch with a capacity of 10,000 lines,

it could serve up to 10,000 subscribers and we could assign telephone numbers 0000 through 9999.

Thus the critical points occur where the number of subscribers reaches numbers such as 100, 1000, and 10,000.

In most present switching systems there is a top limit to the number of subscribers that can be served by one switching unit. Increase beyond this number is either impossible or uneconomical. A given switch unit is usually most economical when operating with the number of subscribers near the maximum of its design. However, it is necessary for practical purposes to hold some spare capacity in reserve. As we proceed in the discussion of switching, we consider exchanges with seven-digit subscriber numbers, such as

The subscriber is identified by the last four digits, permitting up to 10,000 subscribers, 0000 through 9999, allowing for no blocked numbers, such as

$$746 - 0000$$

The three-digit calling area has a capacity of 999 local exchanges, again allowing for no blocked numbers, such as

000
911 (emergency number—USA)

3 CONCENTRATION AND EXPANSION

One key to switching and network design is concentration. A *local* switching exchange concentrates traffic. This concept is often depicted as shown in the following diagram:

Let us dwell on the term *concentration* a bit more. Concentration reduces the number of switching paths or links *within* the exchange and the number of trunks connecting the local exchange to other exchanges. A switch also performs the function of *expansion* to provide all subscribers served by the exchange with access to incoming trunks and local switching paths.

Consider trunks in a long-distance network. A small number of trunks is inefficient not only in terms of loading (see Chapter 1, Section 8.2) but also in terms of economy. Cost is amortized on a per circuit basis; 100 trunks in a link or traffic relation are much more economical than 10 on the same link.* Tandem exchanges concentrate trunks in the local area for traffic relations (links) from sources of low traffic intensity, particularly below 20 erlangs, improving trunk efficiency.

4 BASIC SWITCHING FUNCTIONS

In a local exchange, means are provided to connect each subscriber line to any other in the same exchange. In addition, any incoming trunk can connect to any subscriber line and any subscriber to any outgoing trunk. The switching functions are remotely controlled by the calling subscriber, whether he is a local or long-distance subscriber. These remote instructions are transmitted to the exchange by "off hook," "on hook," and dial information. There are eight basic functions of a conventional switch or exchange:

1. Interconnection
2. Control
3. Alerting
4. Attending
5. Information receiving
6. Information transmitting
7. Busy testing
8. Supervisory

Consider a typical manual switching center (Figure 3.1) where the eight basic functions are carried on for each call. The important interconnecting function is illustrated by the jacks appearing in front of the operator, subscriber-line jacks, and jacks for incoming and outgoing trunks. The interconnection is made by double-ended connecting cords, connecting subscriber to subscriber or subscriber to trunk. The cords available are always less than half the number of jacks appearing on the board, because one interconnecting cord occupies two jacks (by definition). Concentration takes place at this point on a manual exchange. Distribution is also carried out because any cord may be used to complete a connection to any of the terminating jacks. The operator is alerted by a lamp when there is an incoming call requiring connection. This is the attending–alerting function. The operator then assumes the control function, determining an idle connecting cord and plugging into the incoming jacks. She/he then determines call destination, continuing her control function by plugging the cord into the terminating jack of the called subscriber or proper trunk to terminate her portion of control of the incoming call. Of course, before plugging into the terminating

---

* Assuming an efficient traffic loading in each case.

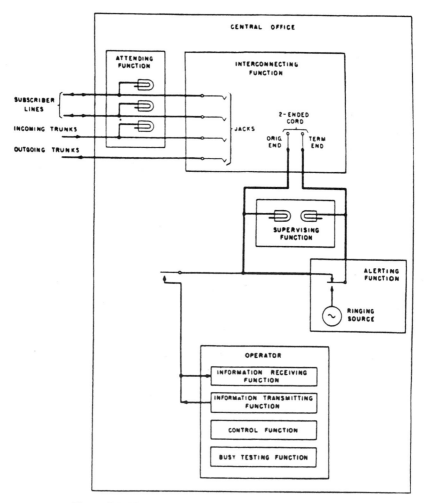

Figure 3.1. Manual exchange illustrating switching functions.

jack, she/he carries out a busy-test function to determine that the called line–trunk is not busy. To alert the called subscriber that there is a call, she/he uses the manual ring-down by connecting the called line to a ringing current source, as shown in Figure 3.1. Other signaling means are usually used for trunk signaling if the incoming call is destined for another exchange. On such a call the operator performs the information function orally or by dialing the call information to the next exchange.

The supervision function is performed by lamps to show when a call is completed and the cord taken down. The operator performs numerous control functions to set up a call, such as selecting a cord, plugging it into the originating jack of the calling line, connecting her headset to determine calling information, selecting (and busy testing) the called subscriber jack, and then plugging the other end of the cord into the proper terminating jack and alerting the called subscriber by ring-down. Concentration is the ratio of the field of incoming jacks to cord positions. Expansion is the number of cord positions to outgoing (terminating) jacks. The terminating jacks and originating jacks can be interchangeable. The called subscriber at another moment in time may become a calling subscriber. On the other hand, incoming and outgoing trunks may be separated. In this case they would be one-way circuits. If not separated, they would be both-way circuits, accepting both incoming and outgoing traffic [2–4, 6].

5 INTRODUCTORY SWITCHING CONCEPTS

All telephone switches have, as a minimum, three functional elements: concentration, distribution, and expansion. Concentration (and expansion) was briefly introduced in Chapter 1, Section 2, to explain the basic rationale of switching. Viewing a switch another way, we can say that it has originating line appearances and terminating line appearances. These are shown in the simplified conceptual drawing in Figure 3.2. Figure 3.2 shows the three different call possibilities of a typical local exchange (switch):

1. A call originated by a subscriber who is served by the exchange and bound for a subscriber who is served by the same exchange (route $A-B-C-D-E$).
2. A call originated by a subscriber who is served by the exchange and bound for a subscriber who is served by another exchange (route $A-B-F$).
3. A call originated by a subscriber who is served by another exchange and bound for a subscriber served by the exchange in question (route $G-D-E$).

Call concentration takes place in B and call expansion at D. Figure 3.3 is simply a redrawing of Figure 3.2 to show the concept of distribution. The distribution stage in switching serves to connect by switching the concentration stage to the expansion stage.

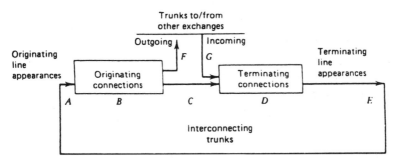

Figure 3.2. Originating and terminating line appearances.

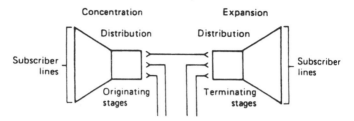

Figure 3.3. The concept of distribution.

The symbols used in switching diagrams are as those in the following diagram, where concentration is shown on the left and expansion is shown on the right.

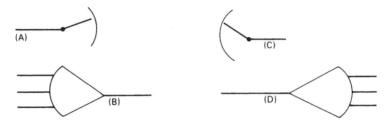

The number of inputs to a concentration stage is determined by the number of subscribers connected to the exchange. Likewise, the number of outputs of the expansion stage is equal to the number of connected subscribers whom the exchange serves. The outputs of the concentration stage are less than the inputs.

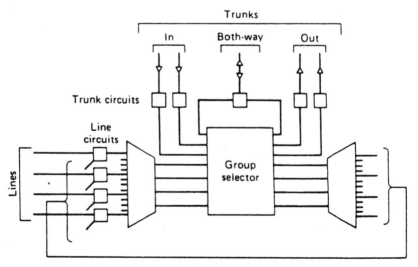

Figure 3.4. The group selector concept where both lines and trunks can be switched by the same matrix.

These outputs are called *trunks* and are formed in groups; thus we refer to *trunk groups*. The sizing or dimensioning of the number of trunks per group is a major task of the systems engineer. The number is determined by the erlangs of traffic originated by the subscribers and the calling rate (see Chapter 1, Sections 6 and 8).

A group selector is used in distribution switching to switch one trunk (between concentration and expansion stages) to another and is often found not only in switches that switch subscriber lines but also where trunk switching is necessary. Such a requirement may be applicable in small cities where a switch may carry out a dual function, both subscriber switching and trunk switching from concentrators or from other local switches. A group selector alone is a tandem switch. Figure 3.4 illustrates the group selector principle [1].

6 ELECTROMECHANICAL SWITCHING

Electromechanical switching was pervasive in the world until about 1983 when the installation of crossbar systems peaked. The step-by-step switch was introduced previously in this chapter. There was a rotary switch which was popular in Europe. Then the crossbar switch appeared. Sometimes people in the field classified it as a coordinate switch or a matrix switch. The matrix concept was carried over to digital switch for the *space switching* function. In the case of the crossbar switch, the crossbar moved and electrical contacts were made mechanically. In other words, a speech-path connection proceeding through a switch is made by crosspoints. In later switches, the crosspoints were reed relays. In this manner a matrix is formed as illustrated by Figure 3.5. With the electronic switch (digital switch), the crosspoints were solid state and there was no mechanical motion whatsoever.

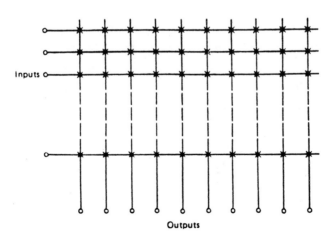

Figure 3.5. Typical diagram of a cross-point matrix.

7   MULTIPLES AND LINKS

A multiple multiplies. It is a method of obtaining several outputs from one input. Thus access is extended (see Figure 3.6A). Or the reverse may be true, where multiple inputs gain access to one output (see Figure 3.6B). Links provide

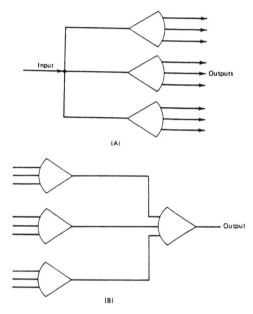

Figure 3.6. Examples of "multiple": (A) single inputs to multiple outputs; (B) multiple inputs to a single output.

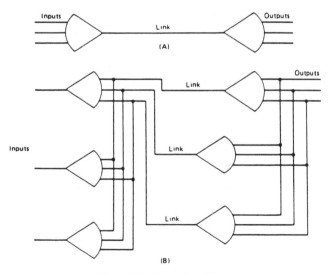

Figure 3.7. Examples of links.

connection for a multiple of switch inputs from one stage to a multiple of switch outputs in another stage (see Figure 3.7) [5].

8 DEFINITIONS: DEGENERATION, AVAILABILITY, AND GRADING

8.1 Degeneration

Degeneration can be expressed by the following ratio:

$$\text{Degeneration on a link} = \frac{\text{Variance of offered traffic}}{\text{Mean of offered traffic}} \tag{3.1}$$

Degeneration is a measure of the extent to which the traffic on a given link varies from pure random traffic. For pure random traffic, degeneration (the preceding ratio) equals 1. For overflow traffic, the variance is equal to or, in the majority of cases, greater than the mean. The more degenerate the traffic, the heavier the demand during peak periods and the greater the number of transmission facilities required.

8.2 Availability

Availability was discussed in Chapter 1, Section 5.4. We now consider availability to introduce *grading*. At a switching array, availability describes the number of outlets that a free inlet is able to reach and test for a busy condition.

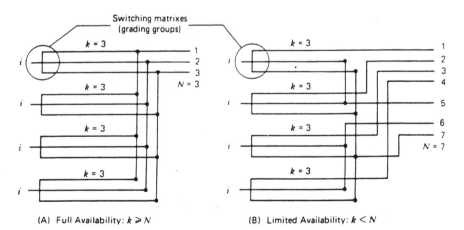

(A) Full Availability: $k \geqslant N$ (B) Limited Availability: $k < N$

Figure 3.8. Full and limited availability. i = number of inlets per switching matrix (grading group), m = number of switching matrices (grading groups) (4, in A and B), N = number of outlets (three in A; seven in B), k = availability [number of outlets per switching matrix (grading group)]. (*Note:* This is a simplified illustration. In a typical switching array, k would equal 10 or even 20.) [5].

1. *Full Availability.* Every free inlet is at all times able to test every outlet.
2. *Limited Availability.* The absence of full availability. The availability at a switching array can be assigned a value, namely, the number of outlets available to each inlet. (See "grading" in Section 8.3 below.)

8.3 Grading

At a switching array with limited availability the inlets are arranged into groups, called *grading groups*. All the inlets in a grading group always have access to the same outlets. Grading is a method of assigning outlets to grading groups in such a way that they assist each other in handling the traffic (see Figure 3.8) [11].

9   THE CROSSBAR SWITCH

Crossbar switching dates back to 1938 and reach a peak of installed lines in 1983. Its life was extended by using stored-program control (SPC) rather than hard-wire control in the older crossbar configuration. Stored-program control means nothing more than a computer-controlled switch.

As we mentioned above, the crossbar is actually a matrix switch used to establish a speech path. Several crossbar grids in series may be involved to establish this path from inlet to outlet. An electrical contact is made by actuating a horizontal and a vertical relay. A typical crossbar matrix concept is shown in Figure 3.9. To make contact at point B_4 on the matrix, horizontal relay B and vertical relay 4 must close to establish connection. Such a closing is momentary but sufficient to cause "latching." Two forms of latching were used in conventional crossbar practice: mechanical and electrical. The latch keeps the speech-path connection until an "on-hook" condition results, freeing the horizontal and vertical relays to establish other connections, whereas connection B_4 in Figure 3.9 has been "busied out."

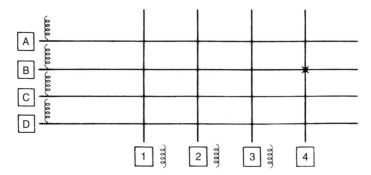

Figure 3.9. The crossbar concept.

10 SYSTEM CONTROL

10.1 Introduction

The basic function of a switch's control system is to establish an appropriate speech path through the switch matrix. To carry out his function, the control system must know the calling and called ports on the matrix and be able to find a free path between them. One artifice used by switch designers was to convert the incoming called telephone number to an internal series of digits that were easier to handle than telephone numbers. Of course, at the switch outlet the internal series of digits had to be translated back to the original telephone number.

10.2 Interexchange Control Register

Right after the switch inlet, there was a register translator (control register) that carried out three consecutive steps in this case: (1) information reception, (2) internal control signaling, and (3) information transmission. The translator in this case determines the exchange code (first three dialed digits), the outgoing trunk group, and the type of signals required by the group and the amount of information to be transmitted. This concept added flexibility to the design and operation in multiexchange areas. The interexchange control register concept is shown in Figure 3.10. Outgoing trunk-control registers facilitate the use of tandem exchange operation. Also, incoming trunk calls to an exchange so equipped can be handled directly from a distant exchange if it is similarly equipped.

Interexchange control registers enabled alternative (alternate) routing. When first-attempt routing is blocked, this same technique can be used for second-attempt routing through an exchange by setting up a speech path with switching

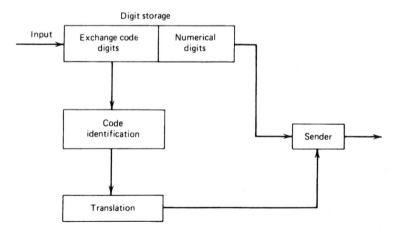

Figure 3.10. Interexchange control register concept.

components different from those of the first setup that failed. The sender in Figure 3.10 generates and transmits signaling information to distant exchanges.

10.3 Common Control (Hard-Wired)

10.3.1 Overview. We call "common control" any control circuitry in a switch that is used for more than one switching device. For purposes of this discussion, common control is defined as providing a means of control of the interconnecting switch network, first identifying the input and output of the terminals of the network that are free and then establishing a path between them. This implies a busy-test of the path before setting the path up. Common control may cover the entire switch or separate control of the originating and terminating halves. Markers are one of the basic elements of common control of a crossbar switch. We discuss markers in the section below.

Source: *Network* is an ambiguous term as used in our text. In this chapter it means an arrangement of components, nodes, and interconnecting branches *inside* a switch. When we refer to the *telecommunication network*, we mean a series of switching nodes interconnected by communication channels. We have also called these channels "trunks."

10.3.2 Common Control Principles. The *marker* sets aside common control systems as we define them from more generalized common control systems previously discussed. On grid circuits (Figure 3.11) the characteristics are such that with specified input and output terminals, various sets of linkage paths exist

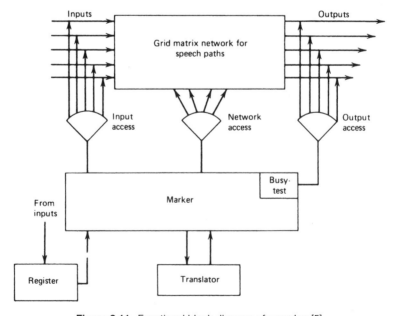

Figure 3.11. Functional block diagram of a marker [5].

that can provide connection between the terminals. It is the marker, with terminal points identified, that locates a path, busy-tests it, and finally sets up a channel through the switch grid network.

A marker always works with one or more registers. A marker is a rapidly operating device serving many calls per minute. It cannot wait for the comparatively slow input information supplied by an incoming line or trunk. Such information, whether dial pulses, touch tones, or interregister signaling information, is stored in the register and released to the marker on demand. The register may receive the entire dialed number and store it before dumping to the marker or may take only the exchange code or area code plus the exchange code (see Section 15 of this chapter for some in-depth information on *numbering*) which is sufficient to identify the trunk group. The register will also identify the location of the call input or setup a control path to the input for the marker. Figure 3.11 illustrates the basic functions of a marker.

10.3.2.1 Translator. Whereas the register provides the dialed exchange number to the marker, the translator provides the marker information on access to the proper trunk group. Of course, the translator may also provide other information, such as the type of signaling required on that trunk group. The reader must bear in mind that modern common-control switches, particularly grid-type switches, use a control code that differs from the numerical dialed code. The dialed code is the directory number (DN) of the called subscriber. The equipment number (EN) is the number the equipment uses and is often a series of five one- or two-digit numbers to indicate location on the grid matrix of the speech-path network. The equipment codes are arbitrary and require changes from time to time to accommodate changes in number assignments. The changes are done on a patch field associated with the translator. One equipment number code is the "two out of five" code, which has ten possibilities to correspond to our decimal base number system and is used on the subscriber dial (or push buttons). As the number implies, combinations of five elements are taken two at a time. This code is shown in Table 3.1.

The number combinations are additive, corresponding to the decimal equivalent except the 4–7 combination that adds to 11 (not to 0). For transmission through the exchange, the numbers are represented by two out of five possible audio frequencies.

TABLE 3.1 Two Out of Five Code

Digit	Two Out of Five 0–1–2–4–7	Digit	Two Out of Five 0–1–2–4–7
1	0–1	6	2–4
2	0–2	7	0–7
3	1–2	8	1–7
4	0–4	9	2–7
5	1–4	0	4–7

On conventional exchanges, one marker cannot serve all incoming calls, particularly during the busy hour (BH). Good marker holding time per call is of the order of half a second. Call attempts may be much greater than two per second, so several markers are required. Each marker must have access from any input grid to any output grid, and thus we can see where markers might compete. This possibility of "double seizure" can cause blockage. Therefore only one control circuit is allowed into a specific grid at any one time.

To reduce marker holding time, fast action switches are required. This is one reason why common (marker) control is more applicable to fine-motion switches such as the crossbar switch. The control circuits themselves must also be fast-acting. Older switches thus used all-relay devices and some vacuum tubes. Newer switches used solid-state control circuitry, which was more reliable and even faster acting.

Figure 3.12 is a simplified functional block diagram of a North American No. 5 crossbar switch. The originating and terminating networks are combined. Two different types of marker are used, dial-tone marker and call-completing marker. A typical No. 5 crossbar has a maximum of four dial-tone markers and eight call-completing markers. The dial-tone marker sets up connections between the calling subscriber line and an originating register. Call-completing markers carry out the remainder of the control functions, such as trunk selection, identification of calling and called line terminals, channel busy-test and selection, junctor group and pattern control, route advance, sender link control, trunk charge control when applicable, overall timing, automatic message accounting (AMA) information, and trouble recorder control.

The marker is also used for exchange troubleshooting and fault location. It is the most convenient check of the switching network because it samples the entire network at frequent intervals. Many markers provide trouble-recording circuits. Trouble on one call setup attempt may have no effect on another. The marker provides a trouble output, returns to service, and reattempts the call. Reattempts

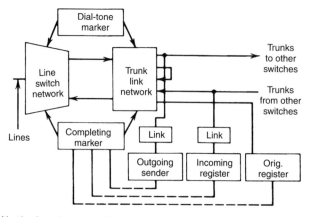

Figure 3.12. North American No. 5 crossbar block diagram showing a typical marker application.

by a marker are usually limited to two so that the marker can return to service other calls.

If a marker finds an "all trunks busy" (ATB) condition on a tandem or transit call, it can ask for instructions to reroute the call. It then handles the call as a first attempt. If again an ATB condition is found and a second alternative route is available, it can make a call attempt on the second rerouting. When all possible routes are busy, the marker returns an ATB signal to the calling subscriber.

11   STORED-PROGRAM CONTROL

11.1   Introduction

Stored-program control (SPC) is a broad term designating switches where common control is carried out to a greater extent or entirely by computerware. Computerware can be a full-scale computer, minimicrocomputer, microprocessors, or other electronic logic circuits. Control functions may be entirely carried out by a central computer in one extreme for centralized processing or partially or wholly by distributed processing utilizing microprocessors. Software may be hard-wired or programmable. Telephone switches are logical candidates for digital computers. A switch is digital in nature,* as it works with discrete values. Most of the control circuitry, such as the marker, works in a binary mode.

The conventional crossbar marker requires about half a second to service a call. Up to 40 expensive markers are required on a large exchange. Strapping points on the marker are available to laboriously reconfigure the exchange for subscriber change, new subscribers, changes in traffic patterns, reconfiguration of existing trunks or their interface, and so on.

Replacing register markers with programmable logic—a computer, if you will—permits one device to carry out the work of 40. A simple input sequence on the keyboard of the computer workstation replaces strapping procedures. System faults are printed out as they occur, and circuit status may be printed out periodically. Due to the speed of the computer, postdial delay is reduced and so on. Computer-controlled exchanges permit numerous new service offerings, such as conference calls, abbreviated dialing, "camp on busy," call forwarding, and incoming-call signal to a busy line [7, 9].

11.2   Basic Functions of Stored-Program Control

There are four basic functional elements of an SPC switching system: (1) switching matrix, (2) call-store (memory), (3) program store (memory), and (4) central processor. These four basic elements are shown in a functional block diagram in Figure 3.13.

---

* This important concept is the basis for the argument set forth in Chapter 9 regarding the rationale for digital switching.

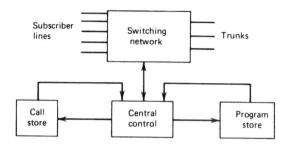

Figure 3.13. A simplified functional block diagram of an SPC exchange.

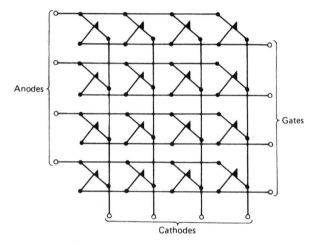

Figure 3.14. An SCR cross-point matrix.

The switching matrix can be made up of electromechanical cross-points, such as in the crossbar switch, reed, correed or ferreed cross-points, or switching semiconductor diodes, often SCR (silicon-controlled rectifier). An SCR matrix is illustrated in Figure 3.14.

The call store is often referred to as the *scratch pad* memory. This is a temporary storage of incoming call information ready for use, on command from the central processor. It also contains availability and status information of lines, trunks and service circuits, and internal switch circuit conditions. Circuit status information is brought to the memory by a method of scanning. All speech circuits are scanned for a busy/idle condition.

The program store provides the basic instructions to the controller (central processor). In many installations, translation information is held in this store, such as DN to EN translation and trunk signaling information.

A typical functional block diagram of a basic (full-up) North American SPC system is shown in Figure 3.15. This is an expansion of Figure 3.13, showing, in addition, scanner circuitry and signal distribution.

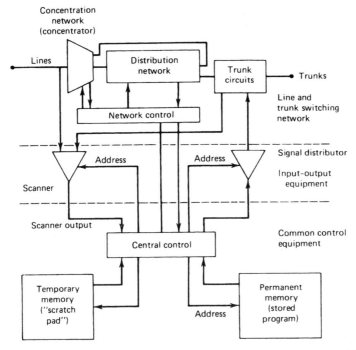

Figure 3.15. Conceptual block diagram of a typical North American semielectronic SPC exchange.

The control network executes the orders given by the central control processor (computer). These orders usually consist of instructions such as "connect" or "release," along with location information on where to carry out the action in the switching network. In one typical (4ESS) system the network control circuits are classified in three major functional categories:

1. Selectors, which set up and release a connection on receipt of location information.
2. Identifiers, which determine the location (called an "address") of a network terminal on one side that is to be connected to a known terminal (address) on the other side of the network.
3. Enablers, which are circuits sequentially enabling junctors to be connected together.

The input–output equipment consists of a line scanner and a signal distributor. Both circuits operate under control of the central control processor. The scanner and distributor carry out the "time-sharing" concept of SPC. The term "time sharing" is the basis where one (or several) computers can control literally thousands of circuits, with each circuit being served serially. The concept of "holding time," important in SPC, is the time taken to serve a circuit, and "delay" is the time each circuit must wait to be served. A computer read–write cycle in a typical SPC is 2–5 μs, with a scanning rate of 2–5 μs per terminal, line scanning during digit reception every 10 μs, and 100-ms supervisory scan. The scanner is an input circuit used for sampling the states (idle or busy) of subscriber lines, trunks, and switch test points to permit monitoring the operation of the system. The signal distributor, on the other hand, is an output circuit directing output signals to various points in the system. In the ESS switch the signal distributor is primarily used on trunk circuits for supervisory and signaling actions.

Common-control equipment, as mentioned previously, is made up of the central processor, call store, and program store. The three units can be considered as making up the control computer, which is capable of transmitting orders to the system as well as detecting signals from the system. The SPC systems with centralized control that we have discussed above have a human interface (I/O = input–output) with the central controller. This, in many instances, is a teleprinter (i.e., keyboard send, printer receive) or keyboard with a visual display unit. The installation adds many advantages and conveniences not found in more conventional switching installations. Several of these are:

- Rerouting and reallocation of trunks
- Traffic statistics
- Renumbering lines
- Changes in subscriber class
- Exchange status
- Fault finding
- Charge records

All these functions can be carried out via the I/O equipment connect to the central processor. Modern switches usually have a dual PC with a printer. [14, 15]

11.3 Evolutionary Stored Program Control and Distributed Processing

11.3.1 Introduction. The presently installed switching plant represents an extremely large investment yet to be fully amortized. The small amount of plant that remains consists of crossbar equipment that is quickly being replaced by digital switches. It was estimated that in 1973, SPC served less than 10% of Bell System subscribers (North America). SPC was just one small step behind a full digital switch. To accelerate SPC implementation, it was common practice to adapt existing plant by modifying crossbar hard-wired control to a computer-controlled installation (SPC) with distributed processing. ITT's Pentaconta 2000 was a step in this direction. (Also see Refs. 8 and 13.)

11.3.2 Modification of Existing Plant. As we mentioned above, the remaining analog plant was crossbar with its inherent distributed control, typically markers, registers, translators, directors, and finders. It was costly to install a large computer in these switches, but cost-effective to replace the markers, registers, and so on, with small microprocessors. This allowed most of the amenities of digital switching without actually being digital. We break these amenities down into two categories: subscriber-related and administration-related. Table 3.2 lists new service features and facilities available to the subscriber, and Table 3.3 lists new service features and facilities administration (telephone company) related.

There were also disadvantages to the implementation of computer control. Electromechanical exchanges were rugged and worked well over wide ranges of temperature, humidity, and vibration. They could withstand high levels of EMI (electromagnetic interference). Not so with solid-state devices. None of the above is valid. Microprocessors, computers, and other solid-state control elements are much more finicky, with heat being a particular problem. In fact it is advisable to keep these devices under air conditioning, vibration, EMI, and humidity control. EMI may turn out to be the worst of them all. Many depend on chopper power supplies which emit high levels of RF noise, often related to the chopping rate.

TABLE 3.2 Service Features and Facilities, Subscriber-Related

Line Classes	Subscriber Facilities
One-party lines	Abbreviated numbering
With dial or push-button-type subscriber sets	One or two digits
With or without special charging categories	For individual lines, or groups
Two-party lines	of lines, or for all lines
With or without privacy, separate ringing, separate	Transfer of terminating calls
charging, revertive call facility	Conversation hold and transfer
Multiparty lines (up to 10 main stations)	Calling party's ring-back
Without privacy	Toll-call offering
With selective or semiselective ringing	Hot line
PABXs	Automatic wake-up
Unlimited number of trunks	Doctor-on-duty service
Uni- or bidirectional trunks	Do-not-disturb service
With or without in-dialing	Absentee service
Coin-box lines	Immediate time and charge
Local traffic	information
Toll traffic	Conference calls
Special applications	Centrex facilities (optional)
Restricted lines	Voice mail
To own exchange	Call-waiting
To urban, regional, national, or toll areas	Call-forwarding
To some specified routes	
Priority lines	
Toll essential	
Essential	
Priority during emergency, overload situations, etc.	

Source: Ref. 9.

TABLE 3.3   Service Features and Facilities: Telephone Company (Administration) Related

Administration Facilities

Interoffice signaling
    Direct-current signaling codes (step-by-step, rotary 7A and 7D, R6 with register or direct
       control, North American dc codes)
    Alternating-current pulse signaling codes
    MF signaling codes for register-controlled exchanges (MFCR 2 code, MF Socotel, North
       American MF codes)
    Direct data transmission over common signaling link between processor-controlled
       exchanges of the time or space division types
Charging
    Control of charge indicators at subscriber premises
    Metering on a single fee or a multifee basis
    Free number service
Numbering
    Full flexibility for equipment number — directory number translation
    For local calls, national toll calls, international toll calls
    Private automatic branch exchanges with direct inward dialing
Routing
    Prefix translation for outgoing or transit calls
    Alternative or overflow routing on route busy or congestion condition
    Resignaling on route busy or congestion condition
    Called side release control
Maintenance
    Plug-in boards
    Automatic fault detection and identification means by diagnostic programs
Operation
    Generalized use of teleprinters; workstations
Possibility of a remote centralized maintenance and operation center

Survivability is another argument against full SPC with centralized control. There is a certain amount of redundancy in the control circuitry of a conventional crossbar switch. Calls can be processed several at a time, each with its own marker, registers, and translator. There is *graceful degradation* in service when one control unit fails. However, let the singular central control fail in a full-up SPC switch, and catastrophic failure results. This is one important reason that later SPC switches and all digital switches have dual processors, some design using the "Ping-Pong" mode. This is where the processors alternate calls, the first call on processor No. 1 and the second on No. 2, and so on. However, if one processor should fail, the other can take the entire load.

One approach to centralization is to have several control processors to be served by one common memory for the switch as a whole. This simplifies the problem of programming new information or modifying existing programs. It also simplifies the I/O such that one PC with printer can serve the entire switch.

Administration and maintenance procedures are often carried out by a separate processor in a modified electromechanical switch. Connection is made to the routing and control processor bus as well as the electromechanical elements. A separate I/O is also provided. (Also see Refs. 14–18.)

12   CONCENTRATORS, OUTSIDE PLANT MODULES, REMOTE SWITCHING, AND SATELLITES

Concentrators or "line concentrators" consolidate subscriber loops, are remotely operated and are a part of or the complete concentration (and expansion) portion of a local switch placed at a remote location. They consolidate subscriber lines but do not switch among them. A typical crossbar type of concentrator is illustrated in Figure 3.16, where 100 subscriber loops are consolidated to 20 trunks plus 2 trunks for control from the nearby "mother" exchange. Concentrators are used in sparsely populated areas that require a long trunk connection to the nearest exchange. In effect, a concentrator extends the serving area of an exchange.

On the trunk side of a concentrator or outside plant module, E1 or T1 carrier techniques are used effectively (Chapter 8). Actually the carrier equipment itself can serve as a concentrator.

An outside plant module is a concentrator with local switching capability. Switching to circuits outside of the local area is carried out back at the "mother" switch. In older texts, this device may be called a *satellite* or satellite switch.

A satellite switch or outside plant module originates and terminates calls from a parent exchange. As we said, it differs from a concentrator in that local calls (i.e., calls originating and terminating inside the outside plant module or satellite exchange serving area) do not have to traverse the parent exchange. A block of numbers is assigned to the satellite or outside plant module serving area and is usually part of the basic number block assigned to the parent exchange. Because of its numbering arrangement, these remote modules can discriminate between local calls and calls to be handled by the parent exchange. An outside plant

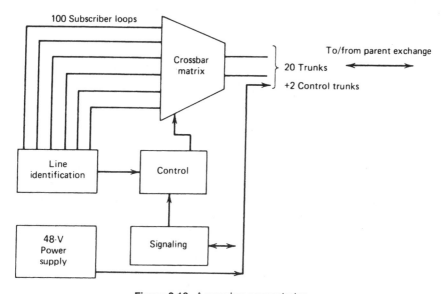

Figure 3.16. A crossbar concentrator.

module can be regarded as a component of the parent exchange that has been dislocated and moved to a distant site. This is still another method of extending the serving area of a main exchange to reduce the cost of serving small groups of subscribers, which, under ordinary circumstances, could only be served by excessively long loops to the parent exchange. Satellites range in size from 300 to 2000 lines. Concentrators are usually more cost-effective than satellites when serving 300 lines or fewer [24, 25].

13   CALL CHARGING: EUROPEAN VERSUS NORTH AMERICAN APPROACHES

In Europe and in countries following European practice, telephone-call charging is simple and straightforward. Each subscriber line is equipped with a meter with a stepping motor at the local exchange. Calls are metered on a time basis. The number of pulses per second actuating the meter are derived from the exchange code or the area code of the dialed number. A local call to a neighbor (same exchange) may be 1 pulse (one step) per minute, a call to a nearby city 3 pulses, or a call to a distant city 10 pulses per minute. International direct-dial calls require checking the digits of the country code that will then set the meter to pulse at an even higher rate. All completed calls are charged; thus the metering circuitry must also sense call supervision to respond to call completion—that is, when the called subscriber goes "off hook" (which starts the meter pulses)—and also to respond to call termination (i.e., when either subscriber goes "on hook") to stop meter pulses. Many administrations and telephone companies refer to this as the *Karlson method* of pulse metering. Some form of number translation is required, in any event, to convert dialed number information to key the metering circuit for the proper number of pulses per unit period.

Pulse metering or "flat rate" billing requires no further record keeping other than periodic meter readings. Charging equipment cost is minimal, and administrative expenses are nearly inconsequential. Detailed billing on toll calls is used in North America (and often flat rate on local calls). Detailed billing is comparatively expensive to install and requires considerable administrative upkeep. Such billing information is recorded by determining first the calling subscriber number and then the called number. Also, of course, there is the requirement of timing call duration.

Automatic message accounting (AMA) is a term used in North American practice. When the accounting is done locally, it is called *local automatic message accounting* (LAMA), and when it is carried out at a central location, it is called *centralized automatic message accounting* (CAMA). Sometimes operator intervention is required, especially when automatic number identification (ANI) is not available or for special lines such as hotels. With automatic identified outward dialing (AIOD), independent data links are sometimes used, or trunk outpulsing may be another alternative.

Only some of the complexities have been described in North American billing practice. One overriding reason why detailed billing is provided on long-distance service is customer satisfaction. In European practice it is difficult to verify a telephone bill. A customer can complain that it is too high but can do little to prove it. Ordinarily, subscriber-dialed long-distance service is lumped with local service, and a bulk amount for the number of meter steps incurred for the intervening billing period is shown on the bill. With detailed billing, individual subscriber-dialed calls can be verified and the bulk local charge is separate. In this case a subscriber has verification for a complaint if it is a valid one.

14 TRANSMISSION FACTORS IN SWITCHING

14.1 Discussion

Here we deal with a family of impairments often overlooked by the telecommunication professional. There are two cases under consideration. The first is an analog switch and the second, of course, is the digital switch.

An analog switch is a source of noise, and to some extent, deteriorates the signal-to-noise ratio. The switch is lossy. This second impairment may even be desirable, because a telephone network, by definition, needs to be lossy. Loss helps us control singing and echo. An analog switch adds loss by its very nature. A digital switch, unless we do something about it, is loss-free. An analog switch distorts attenuation-frequency response of the signal passing through. A digital switch is clear of this impairment.

One more impairment which deserves consideration is return loss at input and output ports. This is especially true where a switch interfaces with subscriber loops. Here is the principal source of echo and singing. A digital switch with an analog input and A/D conversion isolates the line; and return loss, in general, is not really of major importance. In Chapter 9 we will present a major discussion of transmission factors in digital switching.

15 ZERO TEST LEVEL POINT

Signal level in analog telephony (see Chapter 5, Section 2.4) is a most important system parameter. In a telephone network we often deal with relative level measured in dBr, which can be relative to the familiar dBm as follows:

$$dBm = dBm0 + dBr \qquad (3.2)$$

where dBm0 is an absolute unit of power in dBm referred to as the 0 TLP.

The 0 TLP can be located anywhere in the network. North American practice places it at the outgoing switch of the local switching network, whereas CCITT (ITU-T) Rec. G.122 (11) sets levels outgoing from the first international switch in a country at -3.5 dBr and entering the same switch from the international network at -4.0 dBr. The CCITT (ITU-T) then leaves it up to the national telephone company (local administration) to set the 0 TLP, provided it meets the levels just specified [19].

Level certainly is a primary issue in an analog telecommunication network. In a digital network it becomes more of a secondary concern. It is also more difficult to set the 0 TLP in all-digital network [19, 25].

16   NUMBERING CONCEPTS FOR TELEPHONY

16.1   Introduction

Numbering was introduced in Section 2 as a basic element of switching in telephony. This section discusses, in greater detail, numbering as a factor in the design of a telephone network.

16.2   Definitions

There are four elements to an international telephone number. CCITT Rec. E.163 recommends that not more than 12 digits make up an international number. These 12 digits exclude the international prefix, which is that combination of digits used by a calling subscriber to a subscriber in another country to obtain access to the automatic outgoing international equipment; thus we have 12 digits maximum made up of 4 elements. For example, dialing from Madrid to a specific subscriber in Brussels requires only 10 digits (inside the 12-digit maximum).

07	32	2	4561234
International prefix	Country code	National significant number	Subscriber number
		Trunk code (area code)	

Thus the international number is

$$32\ 2\ 456\ 1234$$

According to CCITT international usage (Recs. E.160, E.161, and E.162), we define the following terms.

Numbering Area (Local Numbering Area). This is the area in which any two subscribers use the same dialing procedure to reach another subscriber in the telephone network. Subscribers belonging to the same numbering area may call one another simply by dialing the subscriber number. If they belong to different numbering areas, they must dial the trunk prefix plus the trunk code in front of the subscriber number.

Subscriber Number. This is the number to be dialed or called to reach a subscriber in the same local network or numbering area.

Trunk Prefix (Toll-Access Code). This is a digit or combination of digits to be dialed by a calling subscriber making a call to a subscriber in his own country but outside his own numbering area. The trunk prefix provides access to the automatic outgoing trunk equipment.

Trunk Code (Area Code). This is a digit or combination of digits (not including the trunk prefix) characterizing the called numbering area within a country.

Country Code. This is the combination of one, two, or three digits characterizing the called country.

Local Code. This is a digit or combination of digits for obtaining access to an adjacent numbering area or to an individual exchange (or exchanges) in that area. The national significant number is not used in this situation.

From Madrid, dialing a subscriber in Copenhagen requires 9 digits, Brussels 10, near London (Croydon) 10, Harlow (England) outskirts 11, Harlow center 10, and New York City 11 (not including the international prefix). This raises the concepts of uniform and nonuniform numbering as well as some ambiguity.

The CCITT defines *uniform numbering* as a numbering scheme in which the length of the subscriber numbers is uniform inside a given numbering area. It defines *nonuniform numbering* as a scheme in which the subscriber numbers vary in length within a given numbering area. With uniform numbering, each subscriber, by using the same number of digits, can be reached inside a numbering area and from one numbering area to another inside national boundaries. Theoretically, this is true for North America (north of the Rio Grande River), where a subscriber can dial seven digits and always seven digits for that subscriber to reach any other subscriber inside the calling area. Ten digits is required to reach any subscriber outside the numbering area. Note that in North America the trunk code is called the *area code*. This arrangement is shown in the following formula as it was used at the end of 1973, before the introduction of "interchangeable codes" discussed briefly later in this section.

$$
\begin{array}{cc}
 & \text{Telephone number} \\
\text{Area code} & \text{(subscriber number)} \\
0 & \\
N\!-\!X^* & NNX\!-\!XXXX \\
1 &
\end{array}
$$

where X is any number from 0 to 9, N is any number from 2 to 9, and 0/1 is the number 0 or 1.

Following a plan set up as in this formula, ATT (the major North American telephone administration prior to 1983) provided for an initial arrangement of 152 area codes, each with a capacity of 540 exchange codes. Remember that

---

* Excludes the combination $N11$.

there are 10 digits involved in dialing between areas including all the 50 United States, Canada, Puerto Rico, and parts of Mexico.

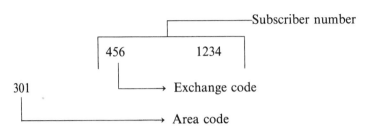

Subsequently, the 540 exchange codes were expanded to 640 when ANC (all number calling) was introduced. This simply means that the use of letters was eliminated and digit combinations 55, 57, 95, and 97 could be utilized where names and resulting letters could not be structured from such combinations, and the code group *NN*0 was also added. This provided needed number relief. This relief was not sufficient to meet telephone growth on the continent, and further code relief had to be provided. This was carried out by realigning code areas, introducing interchangeable codes with code areas where required, and splitting existing code areas and introducing new area codes [23]. The 1995 interchangeable NPA (numbering plan area) codes provide 792 NPA codes (code areas) with its revised *NXX* format.

It should be appreciated that a switch must be able to distinguish between receipt of calls bound both out of and within the calling area. The introduction of "interchangeable" codes precludes the ability of a switch to determine whether a seven-digit number (bound for a subscriber inside the same calling area) or a 10-digit number (for a call bound for a subscriber in another calling area) can be expected. This was previously based on the presence of a "0" or "1" in the second digit position. Now the switch must use either of two different methods to distinguish seven-digit numbers: the "timing method," where the exchange is designed so that the equipment waits for 3–5 s after receiving seven digits to distinguish between seven- and ten-digit calls (if one or more digits are received in the waiting period, the switch then expects a ten-digit call), or the "prefix method," which utilizes the presence of either a "1" or "0" prefix that identifies the call as having a ten-digit format. This, of course, is the addition of an extra digit.

The prefix method described here for ATT is the "trunk prefix" defined previously. Such a trunk prefix is used in Spain, for example. In the automatic service, Spain uses either six or seven digits for the subscriber number. Numbering areas with high population density and high telephone growth, such as Madrid, Barcelona, and Valencia, have seven digits. For numbering areas of low telephone density, six digits are used. However, when dialing between numbering areas, the subscriber always dials nine digits. This is also referred to as a *uniform system*. Inside numbering areas the subscriber number is made up of six or seven digits. Area codes (trunk codes) consist of one or two digits. The ninth digit is used for toll access. To dial Madrid, 91 + 7 digits is used, and to dial Huelva (a small province in southwestern Spain), 955 + 6 digits is used. The United Kingdom presents an example of nonuniform numbering where subscriber numbers in the same numbering area may be five, six, or seven digits. Dialing from one area to another often involves different procedures [10, 24].

16.3 Factors Affecting Numbering

16.3.1 General. In telecommunications network design there are numerous trade-offs between economy and operability. Operability covers a large realm, one aspect of which is the human interface. The subscriber must use his/her telephone, and its use should be easy to understand and apply. Uniform numbering and number length notably improve operability.

Number assignment should leave as large a reserve of numbers as possible for growth. The simplest method to accomplish this tends toward longer numbers and nonuniform numbers. Another goal is to reduce switching costs. One way is to reduce number analysis, that is, the number of digits to be analyzed by a switch for proper routing and charging. We find that an economical analysis becomes increasingly necessary as the network becomes more complex with more and more direct routes.

16.3.2 Routing. Consider the scheme for number analysis in routing illustrated in Figure 3.17. Only one-digit analysis is required to route any call from exchange X to any station in the network. The first digit selects the required outlet. However, if a direct route was established between X and Z_1, two-digit analysis would be required. If a direct route was established to Z_2, thee-digit analysis would be necessary. Note that some freedom of number assignment has been lost because all numbers beginning with digit 2 home on X, 3 on Y, and 4 on Z. With the loss of such freedom, we have gained simplification of switches. We could add more digit analysis capability at each exchange and gain more freedom of digit assignment.

This brings up one additional point, namely, geographical significance and the definition of numbering areas. When dealing with trunk codes, "geographical significance" means that neighboring call areas are assigned digits beginning with the same number. In Spain there is geographical significance; in the United States there is not. In the northwestern corner of Spain, all trunk codes begin with 8. Aside from Barcelona, all codes in northeastern Spain begin with 7. In Andalusia all begin with 5, and around Valencia the codes begin with 6; Alicante is 65 and Murcia is 68. In the United States, New York City is 212 and 718, just to the north is 914, across the Hudson River is 301, and Long Island is 516 and 631.

However, both countries try to use political administrative boundaries to coincide with call-area boundaries. In the United States in many cases we have a whole state or a grouping of counties, but never crossing state boundaries. In Spain we are dealing with provinces, with each province bearing its own call-area number (although some provinces share the same number). Nevertheless, boundaries do coincide with political demarcations. This eases subscriber understanding and simplifies tariffing procedures [10–12].

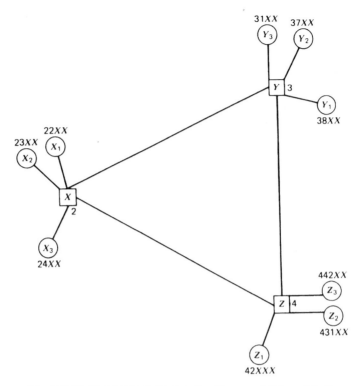

Figure 3.17. Number analysis of routing (X = any digit from 0 to 9).

16.3.3 Tariffs (Charging). The billing of telephone calls is automated in the automatic service. There are two basic methods for charging, bulk billing and detailed billing (see Section 13 of this chapter). Detailed billing is essentially a North American practice. Detailed billing includes in the subscriber's bill a listing of each toll call made, number called, date and time, charge time, and individual charge entry for each toll call. Such a form of billing requires extensive number analysis, which is usually carried out by data-processing equipment in centralized locations.

When charges are determined typically by accumulated meter steps, billing is defined as bulk billing. The subscriber is periodically presented with a bill indicating the total number of steps incurred for the period at a certain money rate per step. Bulk billing is much more economical to install and administer. It does lead to much greater subscriber misunderstanding and, in the long run, to dissatisfaction.

With bulk billing, stepping meters are part of the switching equipment. The switching equipment determines the tariff and call duration and actuates the meters accordingly. For bulk billing the relationship between numbering and billing is much closer than in the case of detailed billing. Ideally, tariff zones should coincide with call areas (routing areas). Care should be taken in the compatibility of charging for tariff zones and the numerical series used in numbering. It follows that the larger the tariff zone, the less numerical analysis required [10].

16.3.4 Size of Numbering Area and Number Length. Size relates not only to geographical dimension but also to the number of telephones encompassed in the area. Areas of very large geographical size present problems in network and switch design. Principally, large areas show a low level of community of interest between towns. In this case, switching requires more digit analysis, and tariff problems may be complicated. If the area has a large telephone population, a larger number of digits may be required, with implied longer dialing times and more postdial delay for analyses. However, large areas are more efficient for number assignment.

With smaller numbering areas, short-length subscriber numbers may be used for those with a higher community of interest. Smaller digit storage would be required for intra-area calls with shorter holding times for switch control units. Smaller areas do offer less flexibility, particularly in the future when uneven growth takes place and forecasts are in error. In the mid-1970s, North America faced a situation requiring reconfiguration and the addition of new areas and also the use of an extra digit or access code.

One report [10] recommends, as a goal, that a numbering area be no greater than $70,000 \text{ km}^2$ nor have less than 100,000 subscribers at the end of a numbering-plan validity period. Subscribers should be able to dial a shorter number when calling other subscribers inside their own numbering area. All other subscribers would be reached by dialing the national number. This is done by adding a simple trunk code in front of the subscriber number. The use of a trunk prefix has advantages in switch design. This should be a single digit, preferably 0. Zero is recommended because few, if any, subscriber numbers start with zero. Thus when a switch receives the initial digit as zero, it is prepared to receive the longer toll number for interarea dialing, whereas if it receives any other digit, it is prepared for receipt of a subscriber number for intra-area dialing. In the United States, the initial digit 1 is often used to indicate a toll call.

For dialed international calls, the ITU-T recommends number length no greater than 12 digits. A little research will show that few international calls require 12 digits. Number length, of course, deals with number of subscribers, code blocks reserved, immediate and future spare capacity, call-area size, and trunk prefix assignment. Uniform and nonuniform numbering are also factors. Other factors are 40-year forecast accuracy, subscriber habits, routing and translation facility availability, switching system capabilities, and existing numbering scheme [10].

16.4 In-Dialing

Numbers, particularly in uniform numbering systems, must be set aside for PABXs with in-dialing capability. This is particularly true when the PABX numbering scheme is built into the national scheme for in-dialing. For example, suppose that a PABX main number is 543-7000, with an extension to that PABX of 678. To dial the PABX extension directly from the outside, we would dial 543-7678, and the 543-7*XXX* block would be lost for use except for 999 extension possibilities for that PABX. If it were a small installation, the numbering block loss could be reduced by 543-7600 (PABX main number), extension 78—543-7678. The PABX would be limited to 99 (or 98) direct in-dialed extensions. If the PABX added digits onto the end of the main number, there would be no impact on national or area numbering. For example:

Main PABX number:	543-7000
PABX extension number:	678
Extension dialed directly from outside:	543-7000678

Here we face the CCITT limitation of 12 digits for international dialing; 10 of the 12 have been used. We must also consider digit storage in switches handling such calls. Of course, the realm of nonuniform numbering has been entered. A third method suggested is shown in the following example:

Main PABX number:	53-87654
PABX extension number:	789
Extension dialed directly from outside:	503-8789

Here the first and third digits locate the local exchange on which the PABX is a subscriber. That local exchange must then analyze the fourth, fifth, and possibly the sixth digits. Disadvantages are that translation is always required and that there is some loss to the numbering system [10–12].

17 TELEPHONE TRAFFIC MEASUREMENT

Statistics on telephone traffic are mandatory both at the individual exchange level and for the network as a whole. Traffic measurements provide the required statistics when carried out in accordance with a well-organized plan. The statistics include traffic intensity (erlangs or CCS) and its distribution by type of subscriber and service on each route and circuit group and its variation daily, weekly, and seasonally. In the process of traffic measurement, congestion is indicated, if present, and switch and network efficiency can be calculated. The most common measurement of "efficiency" is grade of service.

Congestion and its causes tell the systems engineer whether an exchange is overdimensioned, underdimensioned, or of improper traffic balance. Traffic statistics provide a concrete base or starting point from which to forecast growth. They give past traffic evolution of subscriber-generated traffic growth by type of subscriber and class of service. They also provide a prediction of the evolution of local traffic between exchanges of national long-distance (toll) traffic and international traffic. Traffic measurements (plus forecasts) supply the data necessary to dimension new exchanges and for the extension of existing ones. They are especially important when a new exchange is to replace an existing exchange or when a new exchange will replace several existing exchanges.

Traffic measurement involves a number of parameters, including seizures (call attempts), completed calls, traffic intensity (involving holding times), and congestion. The term "seizures" indicates the number of times a switching unit or groups are seized without taking into account holding time. Seizures may be equated to call attempts and give an expression of how much exchange control equipment is being worked. The number of completed calls is of interest in the operation and administration of an exchange. Of real interest are statistics on uncompleted calls that are not attributed to busy lines or lack of answer or that can be attributed to the specified grade of service. However, at a particular exchange a completed call means only that the switch in question has carried out its function, which does not necessarily imply that a connection has been established between two subscribers. The intensity of traffic or traffic volume, a most important parameter, directly provides a measure of usage of a circuit. It is especially useful in those switching units involved directly in the speech network. On the other hand, it is not directly indicative of grade of service. Approximate grade of service should be taken from the traffic tables used to dimension the exchange by using measured traffic intensity as an input. The approximation is most accurate when traffic intensity approaches dimensioned intensity.

Congestion involves three characteristics: "all circuits busy," overflow, and dial-tone delay. "All circuits busy" is an indication of the number of times, and eventually the duration, where all units of a switching group are handling traffic simultaneously. Thus it represents an index of the real grade of service. Its use is particularly effective for those switching groups that are operating near maximum capacity. For an overdimensioned exchange, the "all circuits busy" index is useless and tells us nothing. Overflow is an index of the number of call attempts that have not been able to proceed due to congestion. For networks with alternative routing, overflow tells us the number of offered calls not handled by a specific switching equipment group. Dial-tone delay is directly indicative of overall grade of service that an exchange provides to its subscribers, particularly in the preselection switching stages. Dial-tone delay is normally expressed in the time required in getting dial tone compared to a fixed time, usually 3 s, as a percentage of total calls. [13]

Traffic measurements should be made through the busy hour (BH), which can be determined by reading amperage of exchange battery over the estimated period of occurrence, say, 9:30 A.M. to 12:30 P.M., every 10 min (or by peg count usage devices, microprocessors on lines, equipment or trunk circuits). These measurements should be done daily for at least three weeks. Traffic measurements should be carried out for the work week, one week per month for an entire year. The means of measurement depends on available equipment and exchange type. This may range from simple observation to fully automatic traffic-measuring equipment with recorders on magnetic disk (CD) [21].

18   DIAL-SERVICE OBSERVATION

Dial-service observation provides an index or measurement of the telephone service provided to the customer (see Chapter 1, Section 13). It gives an administration or telephone company a sample measure of maintenance required (or how well it is being carried out), load balance, and adequacy of installed equipment. The number of customer-originated calls sampled per day is of the order of 1–1.5% of the total calls originated; 200 observations per exchange per day is a minimum [20].

Traditionally, service observation has been done manually, requiring the presence of an observer. Service observation positions have some forms of automation, such as automatic recording of a calling number and/or the mark sensing or keypunching to be employed for computer-processing input. Tape recorders may be placed at exchanges to automatically record calls on selected lines. The tapes are then periodically sent to the service observation desk. Tapes usually have a "time hack" to record the time and the duration of calls. A service observation desk usually serves many exchanges or an entire local area. Tandem exchanges are good candidates because they concentrate the service function. The results of service observation are intended to represent average service; thus care should be taken to ensure that subscribers selected represent the average customer. The following is one list of data to be collected per line observed:

- Total call attempts (local and long-distance)
- Completed calls
- Ineffective calls due to calling subscriber: incomplete call, late dialing, unavailable number dialed, call abandoned prematurely
- Ineffective calls due to called subscriber: busy, no answer
- Ineffective calls due to equipment: wrong number, congestion, no ringing, busy tone, no answer, interruptions
- Transmission quality (level, distortion, noise, etc.)
- Wrong number dialed by subscriber
- Dial-tone delay
- Postdial delay

Results of service observation are computerized, and summary tables are published monthly or quarterly. Individual tables are often made for each exchange, relating it to the remainder of the network. Quality of route can be determined since observations identify the called local exchange. A similar set of observations may be made for the long-distance (toll) service exclusively.

There is a tendency to fully automate service observation without any human intervention. The major weakness to this approach is the lack of subjective observation of factors such as type of noise and its effect on a call, subscriber behavior to certain stimuli, and so forth. [26]

REVIEW QUESTIONS

1.   List at least four user requirements with regard to switching.

2.   Why do we use switching in the first place? Explain the rationale of switching versus transmission.

3.   Give two basic functions of a telephone number.

4.   What is a *serving area*?

5.   Without digit blocking, how many distinct entities can be served by a three-digit number? a four-digit number?

6.   What are the two basic functions of a local switch?

7.   In a local area a certain traffic relation is 12 erlangs. What would be the most economical form of routing: direct or tandem?

8.   There are eight basic functions that a conventional generic switch carries out. Name at least six.

9.   What does the term *supervision* mean in switching?

10. In the concentration stage of a local switch, are there more or less inputs than outputs?

11. In the expansion stage of a local switch, the number of outputs equals the number of_____.

12. Define a trunk group using the term *traffic relation*.

13. What does *degeneration* tell us about a particular switch?

14. Differentiate between *limited availability* and *full availability*.

15. Define *grading*.

16. What is the function of latching on a crossbar switching matrix?

17. In a crossbar switch, what unit carries out the common-control function?

18. Why would a marker need more than one register?

19. Why is it advantageous to use *equipment number* (EN) rather than *dial number* (DN) in a switch?

20. SPC simply means that a switch is_____-controlled.

21. Give at least three major advantages of SPC over hard-wired control switches.

22. What is the meaning of *holding time* and *delay* with regard to SPC systems?

23. Give at least two advantages and one disadvantage of using centralized control (versus distributed control).

24. What is the one major difference between a *concentrator* and an outside plant module?

25. Describe the differences between European and North American charging methods. Bring in the acronyms *CAMA* and *LAMA*.

26. Give at least three transmission impairments encountered in an analog switch.

27. Give the ITU-T and North American definitions of 0 TLP.

28. Give the two different characteristic impedances that may be encountered on local exchange VF switch ports.

29. Differentiate between uniform and nonuniform numbering.

30. With switch complexity in mind, discuss the importance of numbering on routing.

31. Give at least three traffic parameters that are measured at a switch that give valuable insight to planners, system design engineers, and telephone company/administration managers.

32. In what period during a weekday is traffic measurement most important?

33. Give at least five parameters to be included in dial-service observation.

REFERENCES

1. *Switching Systems*, American Telephone and Telegraph Company, New York, 1961.
2. M. Hobbs, *Modern Communication Switching Systems*, Tab Books, Blue Ridge Summit, PA, 1974.
3. T. H. Flowers, *Introduction to Exchange Systems*, John Wiley & Sons, New York, 1976.
4. A. F. Joel, "What Is Telecommunication Circuit Switching?" *Proc. IEEE*, **65** (9) (September 1977).
5. *Fundamental Principles of Switching Circuits and Systems*, American Telephone and Telegraph Company, New York, 1963.
6. International Telephone and Telegraph Corporation, *Reference Data for Radio Engineers*, 6th ed., Howard W. Sams, Indianapolis, 1976.
7. J. G. Pearce, *Electronic Switching*, Telephony Publishing Company, Chicago, 1968.
8. J. P. Dartois, "Metaconta L Medium Size Local Exchanges," *Electr. Commun.*, **48** (3) (1973).
9. J. G. Pearce, "The New Possibilities of Telephone Switching," *Proc. IEEE*, **65** (9) (September 1977).
10. "Numbering," Telecommunication Planning, ITT Laboratories (Spain), Madrid, 1973.
11. CCITT, Q Recommendations (Rec. Q), Redbooks, VIII Plenary Assembly, Malaga-Torremolinos, Spain, October 1984.

12. L. F. Goeller, *Design Background for Telephone Switching*, Lees ABC of the Telephone, Geneva, IL, 1977.

13. *Notes on the Network 1980*, American Telephone and Telegraph Company, New York, 1980.

14. B. E. Briley and W. N. Toy, "Telecommunication Processors," *Proc. IEEE*, **65** (9) (September 1977).

15. T. H. Flowers, "Processors and Processing in Telephone Exchanges," *Proc. IEEE*, **119** (3) (March 1972).

16. E. Aro, "Stored Program Control-Assisted Electromechanical Switching—An Overview," *Proc. IEEE*, **65** (9) (September 1977).

17. P. J. Hiner, "TXE4 and the New Technology," *Telecommunications* (January 1977).

18. USITA Symposium, April 1970, Open Questions 18–37.

19. R. L. Freeman, *Telecommunication Transmission Handbook*, 4th ed., John Wiley & Sons, New York, 1998.

20. L. Alvarez Mazo and M. Poza Martinez, "Dial Service Observation," Telecommunication Planning Symposium, STC (SA), Boksburg, South Africa, June 1972.

21. L. Alvarez Mazo, "Network Traffic Measurements," Telecommunication Planning Symposium, STC (SA), Boksburg, South Africa, June 1972.

22. R. Deese, "A History of Switching," *Telecommunications* (magazine), Norwood, MA, February 1984.

23. *Notes on the BOC Intra-Lata Networks—1994*, SR-TSV-002275, Bellcore, Livingston, NJ, April 1994.

24. "Telcordia Notes on the Networks," Telcordia Special Report, SR-2275, Issue 4, Telcordia, Piscataway, NJ, October 2000.

25. *Reference Manual for Telecommunication Engineering*, 3rd ed., John Wiley & Sons, New York, 2002.

26. *Engineering and Operations in the Bell System*, 2nd ed, ATT Bell Laboratories, Murray Hill, NJ, 1983.

4

SIGNALING FOR ANALOG TELEPHONE NETWORKS

1 INTRODUCTION

In a switched telephone network, signaling conveys the intelligence needed for one subscriber to interconnect with any other in that network. Signaling tells the switch that a subscriber desires service and then gives the local switch the data necessary to identify the required distant subscriber and hence to route the call properly. It also provides supervision of the call along its path. Signaling also gives the subscriber certain status information, such as dial tone, busy tone (busy back), and ringing. Metering pulses for call charging may also be considered a form of signaling.

There are several classifications of signaling:

1. General.
   a. Subscriber signaling.
   b. Interswitch signaling.
2. Functional.
   a. Audible–visual (call progress and alerting).
   b. Supervisory.
   c. Address signaling.

Figure 4.1 shows a more detailed breakdown of these functions.

It should he appreciated that on many telephone calls, more than one switch is involved in call routing. Therefore switches must interchange information among switches in fully automatic service. Address information is provided between modern switching machines by interregister signaling, and the supervisory function is provided by line signaling. The audible–visual category of signaling

Telecommunication System Engineering, by Roger L. Freeman
ISBN 0-471-45133-9   Copyright © 2004 Roger L. Freeman

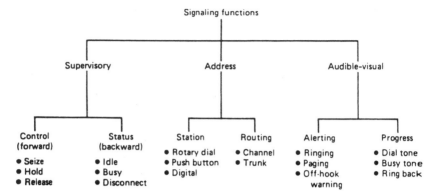

Figure 4.1. A functional breakdown of signaling.

functions inform the calling subscriber regarding call *progress*, as shown in Figure 4.1. The *alerting* function informs the called subscriber of a call waiting or an extended "off-hook" condition of his or her handset. Signaling information can be conveyed by a number of means from subscriber to switch or between (and among) switches. Signaling information can be transmitted by means such as

- Duration of pulses (pulse duration bears a specific meaning)
- Combination of pulses
- Frequency of signal
- Combination of frequencies
- Presence or absence of a signal
- Binary code
- For dc systems, the direction or level of transmitted current

Source: From Ref. 1.

2 SUPERVISORY SIGNALING

Supervisory signaling provides information on line or circuit condition and indicates whether a circuit is in use or idle. It informs the switch and interconnecting trunk circuits whether a calling party is "off hook" or "on hook" or whether a called party is "off hook" or "on hook." The meaning and importance of the terms "on hook" and "off hook" were detailed in Chapter 1, Section 2. The assumption is that a telephone in the network can have one of two states: busy or idle. Idle, of course, is represented by the "on-hook" condition.

The reader must appreciate that supervisory information–status must be maintained end to end on every telephone call. It is necessary to know when a calling subscriber lifts his/her telephone off hook, thereby requesting service. It is equally important that we know when the called subscriber answers (i.e., lifts her telephone off hook), because that is when we may start metering the call to establish charges. It is also important to know when the called and calling subscribers return their telephones to the on-hook condition. Charges stop, and the intervening trunks comprising the talk path as well as the switching points are then rendered idle for use by another pair of subscribers. During the period of occupancy of a talk path end to end, we must know that this particular path is busy (is occupied) so that no other call attempt can seize it.

Dialing of a subscriber line is merely interruption of the subscriber loop's off-hook condition, often called "make and break." The "make" is a current flow condition (or off hook), and the "break" is the no-current condition (or on hook). How do we know the difference between supervisory and dialing? Primarily by duration—the on-hook interval of a dial pulse is relatively short and is distinguishable from an on-hook disconnect signal (subscriber hangs up), which is transmitted in the same direction for a longer duration. Thus the switch is sensitized to duration to distinguish between supervisory and dialing of a subscriber loop. Figure 4.2 is a simplified diagram of a subscriber loop showing its functional signaling elements.

2.1 E and M Signaling

Probably the most common form of trunk supervision is E and M signaling, particularly with multiplex equipment (Chapters 5 and 8). Yet it only becomes true E and M signaling where the trunk interfaces with the switch (see Figure 4.3). E-lead and M-lead signaling systems are semantically derived from historical designation of signaling leads on circuit drawings covering these systems. Historically, the E and M signaling interface provides two leads between the switch and what we may call *trunk-signaling equipment* (signaling interface). One lead is called the "E-lead," which carries signals *to* the switching equipment. Such signal directions are shown in Figure 4.3, where we see that signals from switch A and switch B leave A on the M-lead and are delivered to B on the E-lead. Likewise, from B to A, supervisory information leaves B on the M-lead and is delivered to A on the E-lead.

For conventional E and M signaling (referring to electromechanical exchanges), the following supervisory conditions are valid:

Direction		Condition at A		Condition at B	
Signal A to B	Signal B to A	M-Lead	E-Lead	M-Lead	E-Lead
On hook	On hook	Ground	Open	Ground	Open
Off hook	On hook	Battery	Open	Ground	Ground
On hook	Off hook	Ground	Ground	Battery	Open
Off hook	Off hook	Battery	Ground	Battery	Ground

Source: Ref. 8. Also see Ref. 12.

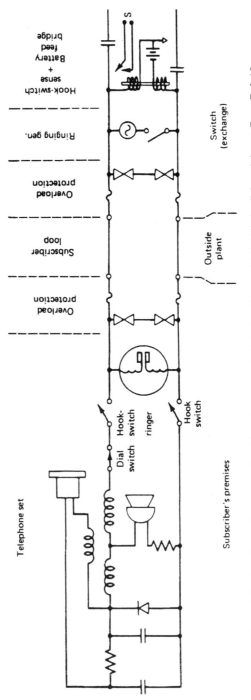

Figure 4.2. Signaling with a conventional telephone subset. Note functions of hook switch, dial, and ringer. From Ref. 16.

Figure 4.3. E and M signaling.

3   AC SIGNALING

3.1   General

Up to this point we have reviewed the most employed means of supervisory trunk signaling (or line signaling). Direct-current signaling, such as reverse-battery signaling, has notable limits on distance because it cannot be applied directly to multiplex systems (Chapters 5 and 8) and is limited on metallic pairs due to the IR drop of the lines involved. Direct-current trunk signaling is addressed in Section 10.

There are many ways to extend these limits, but from a cost-effectiveness standpoint there is a limit that we cannot afford to exceed. On trunks exceeding dc capabilities, some form of ac signaling will be used. Traditionally, ac signaling systems are divided into three categories: low-frequency, in-band, and out-band (out-of-band) systems. Each of these can derive the four E and M signaling states.

3.2   Low-Frequency AC Signaling Systems

An ac signaling system operating below the limits of the conventional voice channel (i.e., <300 Hz) are termed *low frequency*. Low-frequency signaling systems are one-frequency systems, typically 50 Hz, 80 Hz, 135 Hz, or 200 Hz. It is impossible to operate such systems over carrier-derived channels (see Chapter 5) because of the excessive distortion and band limitation introduced. Thus low-frequency signaling is limited to metallic-pair transmission systems. Even on these systems, cumulative distortion limits circuit length. A maximum of two repeaters may be used, and, depending on the type of circuit (open wire, aerial cable, or buried cable) and wire gauge, a rough rule of thumb is a distance limit of 80–100 km.

3.3   In-Band Signaling

In-band signaling refers to signaling systems using an audio tone, or tones inside the conventional voice channel, to convey signaling information. In-band signaling is broken down into three categories: (1) one frequency (SF or single frequency), (2) two frequency (2VF), and (3) multifrequency (MF). As the term implies, in-band signaling is where signaling is carried out directly in the voice channel. As the reader is aware, the conventional voice channel as defined by the CCITT occupies the band of frequencies from 300 Hz to 3400 Hz. Single-frequency and two-frequency signaling systems utilize the 2000- to 3000-Hz portion, where less speech energy is concentrated.

3.3.1 Single-Frequency Signaling.

Single-frequency signaling is used almost exclusively for supervision. In some locations it is used still for interregister signaling, but the practice is diminishing in favor of more versatile methods such as MF signaling. The most commonly used frequency is 2600 Hz, particularly in North America. On two-wire trunks, 2600 Hz is used in one direction and 2400 Hz is used in the other. A diagram showing application of SF signaling on a four-wire trunk is shown in Figure 4.4.

3.3.2 Two-Frequency Signaling.

Two-frequency signaling is used for both supervision (line signaling) and address signaling. We often associate SF and 2VF supervisory signaling systems with carrier (FDM) operation. Of course, when we discuss such types of line signaling (supervision), we know that the term "idle" refers to the on-hook condition while "busy" refers to the off-hook condition. Thus, for such types of line signaling that are governed by audio tones of which SF and 2VF are typical, we have the conditions of "tone on when idle" and "tone on when busy." The discussion holds equally well for in-band and out-of-band signaling methods. However, for in-band signaling, supervision is by necessity tone-on idle; otherwise subscribers would have an annoying 2600-Hz tone on throughout the call.

A major problem with in-band signaling is the possibility of "talk-down," which refers to the premature activation or deactivation of supervisory equipment by an inadvertent sequence of voice tones through the normal use of the channel. Such tones could simulate the SF tone, forcing a channel dropout (i.e., the supervisory equipment would return the channel to the idle state). Chances of

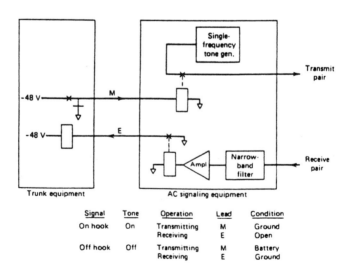

Signal	Tone	Operation	Lead	Condition
On hook	On	Transmitting	M	Ground
		Receiving	E	Open
Off hook	Off	Transmitting	M	Battery
		Receiving	E	Ground

Figure 4.4. Functional block diagram of a single-frequency signaling circuit. (*Note:* Wire pairs, "receive" and "transmit," derive from carrier-equipment "receive" and "transmit" channels.) [2, 6]

simulating a 2VF tone set are much less likely. To avoid the possibility of talk-down on SF circuits, a time-delay circuit or slot filters to bypass signaling tones may be used. Such filters do offer some degradation to speech unless they are switched out during conversation. The tones must be switched out if the circuit is going to be used for data transmission [7].

It becomes apparent why some administrations and telephone companies have turned to the use of 2VF supervision, or out-of-band signaling for that matter. For example, a typical 2VF line signaling arrangement is the CCITT No. 5 code, where f_1 (one of the two VF frequencies) is 2400 Hz and f_2 is 2600 Hz. 2VF signaling is also used widely for address signaling (see Section 4.1 of this chapter) [4].

3.4   Out-of-Band Signaling

With out-of-band signaling, supervisory information is transmitted out of band (i.e., above 3400 Hz). In all cases it is a single-frequency system. Some out-of-band systems use "tone on when idle," indicating the on-hook condition, whereas others use "tone off." The advantage of out-of-band signaling is that either system, tone on or tone off, may be used when idle. Talk-down cannot occur because all supervisory information is passed out of band, away from the speech-information portion of the channel.

The preferred CCITT out-of-band frequency is 3825 Hz, whereas 3700 Hz is commonly used in the United States. It also must be kept in mind that out-of-band signaling is used exclusively on carrier systems, not on wire trunks. On the wire side, inside an exchange, its application is E and M signaling. In other words, out-of-band signaling is one method of extending E and M signaling over a carrier system.

In the short run, out-of-band signaling is attractive in terms of both economy and design. One drawback is that when channel patching is required, signaling leads have to be patched as well. In the long run, the signaling equipment required may indeed make out-of-band signaling even more costly because of the extra supervisory signaling equipment and signaling lead extensions required at each end and at each time that the carrier (FDM) equipment demodulates to voice. The major advantage of out-of-band signaling is that continuous supervision is provided, whether tone on or tone off, during the entire telephone conversation. In-band SF signaling and out-of-band signaling are illustrated in Figure 4.5. An example of out-of-band signaling is the CCITT R-2 System (CCITT Rec. Q.351) (see Table 4.1) [9, 11].

4   ADDRESS SIGNALING: INTRODUCTION

Address signaling originates as dialed digits (or activated push buttons) from a calling subscriber, whose local switch accepts these digits and, using that information, directs the telephone call to the desired distant subscriber. If more than one

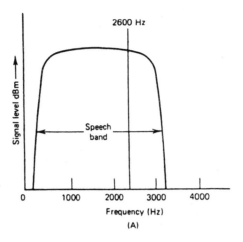

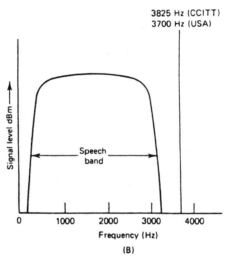

Figure 4.5. Single-frequency signaling: (A) in-band; (B) out-of-band.

TABLE 4.1   R-2 Line Signaling (3825 Hz)

	Direction	
Circuit State	Forward (Go)	Backward (Return)
Idle	Tone on	Tone on
Seized	Tone off	Tone on
Answered	Tone off	Tone off
Clear back	Tone off	Tone on
Release	Tone on	Tone on or off
Blocked	Tone on	Tone off

Source: Ref. 11.

switch is involved in the call setup, signaling is required between switches (both address and supervisory). Address signaling between switches in conventional systems is called *interregister signaling.*

The paragraphs that follow discuss various more popular standard ac signaling techniques such as 2VF, MF pulse, and MF tone. Although interregister signaling is stressed where appropriate, some supervisory techniques are also reviewed. Common-channel signaling is discussed in Chapter 16 where we describe CCITT No. 7 signaling system.

4.1 Two-Frequency Pulse Signaling

Two-frequency signaling is commonly used as an interregister mode of signaling employing the speech band for the transmission of information. It may also be used for line signaling.* There are various methods of using two-voice frequencies to transmit signaling information. For example, CCITT No. 4 uses 2040 Hz and 2400 Hz to represent binary 0 and 1, respectively. It uses a four-element code, permitting 16 different coded characters, as shown in Table 4.2.

TABLE 4.2   CCITT Signal Code System No. 4

		Combination Elements			
Digit	Number	1	2	3	4
1	1	y	y	y	x
2	2	y	y	x	y
3	3	y	y	x	x
4	4	y	x	y	y
5	5	y	x	y	x
6	6	y	x	x	y
7	7	y	x	x	x
8	8	x	y	y	y
9	9	x	y	y	x
0	10	x	y	x	y
Call operator code 11	11	x	y	x	x
Call operator code 12	12	x	x	y	y
Spare code (see CCITT Rec. Q. 104)	13	x	x	y	x
Incoming half-echo suppressor required	14	x	x	x	y
End of pulsing	15	x	x	x	x
Spare code	16	y	y	y	y

Sending duration of binary elements 35 ± 7 ms. Sending duration of blank elements between binary elements 35 ± 7 ms. Element x is 2040 Hz; element y is 2400 Hz.

Source: Ref. 10.

* Supervision (interswitch).

TABLE 4.3   CCITT No. 4 Line Signaling

Forward signals	
Terminal seizing	*Px*
Transit seizing	*Py*
Numerical signals	As in Table 4.2
Clear forward	*Pxx*
Forward transfer	*Pyy*
Backward signals	
Proceed to send	
Terminal	*x*
International transit	*y*
Number received	*p*
Busy flash	*pX*
Answer	*pY*
Clear back	*Px*
Release guard	*Pyy*
Blocking	*Px* (congestion)
Unblocking	*Pyy*

Source: Ref. 10.

With the CCITT No. 4 code both interregister and line signaling utilize the 2VF technique. Interregister signaling in this case is a pulse-type signaling, and line signaling utilizes the combination of the two frequencies and the duration of signal to convey the necessary supervisory information, as shown in Table 4.3.

As we mentioned, line signaling with the CCITT No. 4 code is based on signal duration as well as frequency. The line-signaling format uses both tone frequencies, 2040 Hz and 2400 Hz. Each line signal consists of an initial *prefix (P)* signal followed by a control signal element, called a *suffix*. The *P* signal consists of both frequencies (2VF), and the suffix signal consists of one frequency, where *x* is 2040 Hz and *y* is 2400 Hz (see Table 4.3). Now consider the durations of the following signal elements used for line signaling:

$$P = 150 \pm 30 \text{ ms}$$

$$x, y = 100 \pm 20 \text{ ms (each)}$$

$$xxyy = 350 \pm 70 \text{ ms (each)}$$

This set of values refers to transmitted signal duration—that is, as transmitted by the signaling sender.

Let us see how these signals are used in the line (interswitch supervision) signaling (see Table 4.3).

The supervisory functions "clear forward" and "release guard" do the reverse of "call setup." They take the call down or disconnect, readying the circuit for the next user. "Clear back" is another example.

4.2 Multifrequency Signaling

Multifrequency signaling is used for interregister signaling. It is an in-band method utilizing five or six tone frequencies, two at a time. Multifrequency signaling works equally well over metallic and carrier (FDM) systems. Four commonly used MF signaling systems follow with a short discussion for each [5].

4.2.1 SOCOTEL. SOCOTEL is an interregister signaling system used principally in France, areas of French influence, and Spain with some modifications. The frequency pairs and their digit equivalents are shown in Table 4.4. Line signaling used with SOCOTEL may be dc, 50 Hz, or 2000 Hz. The same frequencies are used in both directions.

4.2.2 Multifrequency Signaling in North America: The R-1 Code. The MF signaling system principally used in the United States and Canada is recognized by the CCITT as the R-1 code. It is a two-out-of-five frequency-pulse system. Additional signals for control functions are provided by combinations using a sixth frequency. Table 4.5 shows digits and other applications and their corresponding frequency combinations as well as a brief explanation of "other applications."

4.2.3 CCITT No. 5 Signaling Code. Interregister signaling with the CCITT No. 5 code is very similar in makeup to the North American R-1 code. Variations

TABLE 4.4 Basic SOCOTEL MF Signaling Code

Tone Frequencies (Hz)[a]	Digit
700 + 900	1
700 + 1100	2
900 + 1100	3
700 + 1300	4
900 + 1300	5
1100 + 1300	6
700 + 1500	7
900 + 1500	8
1100 + 1500	9
1300 + 1500	0

[a] The 1700-Hz frequency is also used for signaling system check and when more code groups are required by a telephone company or national administration. When 1700 Hz is used for coding, 1900 Hz is used for checking the system; 1700 Hz and/or 1900 Hz may also be used for control purposes.

Source: Ref. 3.

TABLE 4.5 The R-1 Code[a] (North American MF)

Digit	Frequency Pair (Hz)
1	700 + 900
2	700 + 1100
3	900 + 1100
4	700 + 1300
5	900 + 1300
6	1100 + 1300
7	700 + 1500
8	900 + 1500
9	1100 + 1500
10(0)	1300 + 1500

Use	Frequency Pair	Explanation
KP	1100 + 1700	Preparatory for digits
ST	1500 + 1700	End-of-pulsing sequence
STP	900 + 1700	
ST2P	1300 + 1100	Used with TSPS (traffic service position system)
ST3P	700 + 1700	
Coin collect	700 + 1100	Coin control
Coin return	1100 + 1700	Coin control
Ring-back	700 + 1700	Coin control
Code 11	700 + 1700	Inward operator (CCITT No. 5)
Code 12	900 + 1700	Delay operator
KP1	1100 + 1700	Terminal call
KP2	1300 + 1700	Transit call

[a] Pulsing of digits is at the rate of about seven digits per second with an interdigital period of 68 ± 7 ms. For intercontinental dialing for CCITT No. 5 code compatibility, the R-1 rate is increased to 10 digits per second. The KP pulse duration is 100 ms [3, 11].
Source: Ref. 13.

with R-1 are shown in Table 4.6. The CCITT No. 5 line-signaling code is shown in Table 4.7.

4.2.4 The R-2 Code. The R-2 code is listed by CCITT (Rec. Q.361) [11] as a European regional signaling code. Taking full advantage of combinations of two-out-of-six tone frequencies, 15 frequency-pair possibilities are available. This number is doubled in each direction by having meaning groups I and II in the forward direction and groups A and B in the backward direction (see Table 4.8).

Groups I and A are said to be of primary meaning, and groups II and B are said to be of secondary meaning. The change from primary to secondary meaning is commanded by the backward signal A-3 or A-5. Secondary meanings can be changed back to primary meanings only when the original change from primary to secondary was made by the use of the A-5 signal. Referring to Table 4.8, the 10 digits to be sent in the forward direction in the R-2 system are in group I and are index numbers 1 through 10 in the table. The index 15 signal (group

TABLE 4.6   CCITT No. 5 Code^a Showing Variations with R-1 Code

Signal	Frequencies (Hz)	Remarks
KP1	1100 + 1700	Terminal traffic
KP2	1300 + 1700	Transit traffic
1	700 + 900	
2	700 + 1100	
3-0	Same as Table 4.5	
ST	1500 + 1700	
Code 11	700 + 1700	Code 11 operator
Code 12	900 + 1700	Code 12 operator

^{a}Line signaling for CCITT No. 5 code is 2VF, with f_1 2400 Hz and f_2 2600 Hz. Line-signaling conditions are shown in Table 4.7.

Source: Ref. 3. Also see Ref. 4.

TABLE 4.7   CCITT No. 5 Line-Signaling Code

Signal	Direction	Frequency	Sending Duration	Recognition Time (ms)
Seizing	→	f_1	Continuous	40 ± 10
Proceed to send	←	f_2	Continuous	40 ± 10
Busy flash	←	f_2	Continuous	125 ± 25
Acknowledgment	→	f_1	Continuous	125 ± 25
Answer	←	f_1	Continuous	125 ± 25
Acknowledgment	→	f_1	Continuous	125 ± 25
Clear back	←	f_2	Continuous	125 ± 25
Acknowledgment	→	f_1	Continuous	125 ± 25
Forward transfer	→	f_2	850 ± 200 ms	125 ± 25
Clear forward	→	$f_1 + f_2$	Continuous	125 ± 25
Release guard	←	$f_1 + f_2$	Continuous	125 ± 25

$f_1 = 2400$ Hz; $f_2 = 2600$ Hz.

Source: Ref. 10.

A) indicates "congestion in an international exchange or at its output." This is a typical backward information signal giving circuit status information. Group B consists of nearly all "backward information" and in particular deals with subscriber status.

The R-2 line-signaling system has two versions: The one used on analog networks is discussed here; the other, on digital (PCM) networks, is briefly covered in Chapter 8. The analog version is an out-of-band tone-on-when-idle system. Table 4.9 shows the line conditions in each direction, forward and backward. Note that the code takes advantage of a signal sequence that has six characteristic operating conditions.

TABLE 4.8 European R-2 System, Address Signaling, DTMF Code

Index No. for Groups I/II and A/B	Frequencies (Hz)						
	1380	1500	1620	1740	1860	1980	Forward Direction I/II
	1140	1020	900	780	660	540	Backward Direction A/B
1	x	x					
2	x		x				
3		x	x				
4	x			x			
5		x		x			
6			x	x			
7	x				x		
8		x			x		
9			x		x		
10				x	x		
11	x					x	
12		x				x	
13			x			x	
14				x		x	
15					x	x	

Source: Ref. 15.

TABLE 4.9 Line Conditions for R-2 Code

Operating Condition of the Circuit	Signaling Conditions	
	Forward	Backward
1. Idle	Tone on	Tone on
2. Seized	Tone off	Tone on
3. Answered	Tone off	Tone off
4. Clear back	Tone off	Tone on
5. Release	Tone on	Tone on or off
6. Blocked	Tone on	Tone off

Source: Ref. 14.

Let us consider several of these conditions.

Seized. The outgoing exchange (call-originating exchange) removes the tone in the forward direction. If seizure is immediately followed by release, removal of the tone must be maintained for at least 100 ms to ensure that it is recognized at the incoming end.

Answered. The incoming end removes the tone in the backward direction. When another link of the connection using tone-on-when-idle continuous signaling precedes the outgoing exchange, the "tone-off" condition must be established on the link as soon as it is recognized in this exchange.

Clear Back. The incoming end restores the tone in the backward direction. When another link of the connection using tone-on-when-idle continuous signaling precedes the outgoing exchange, the "tone-off" condition must be established on this link as soon as it is recognized in this exchange.

Clear Forward. The outgoing end restores the tone in the forward direction.

Blocked. At the outgoing exchange the circuit stays blocked as long as the tone remains off in the backward direction.

4.2.5 Subscriber Tones and Push-Button Codes (North America). Subscriber subsets in many places in the world are either dial or push button. The push-button type is more versatile, and more rapid dialing can be accomplished by a subscriber. Table 4.10 compares digital dialed, dial pulses (breaks), and multifrequency (MF) push-button tones. Table 4.11 shows the audible tones commonly

TABLE 4.10   North American Push-Button Codes

Digit	Dial Pulse (Breaks)	Multifrequency Push-Button Tones
0	10	941,1336 Hz
1	1	697,1209 Hz
2	2	697,1336 Hz
3	3	697,1474 Hz
4	4	770,1209 Hz
5	5	770,1336 Hz
6	6	770,1477 Hz
7	7	852,1209 Hz
8	8	852,1336 Hz
9	9	852,1477 Hz

Source: Refs. 2, 20.

TABLE 4.11   Audible Tones Commonly Used in North America

Tone	Frequencies (Hz)	Cadence
Dial	350 + 440	Continuous
Busy (station)	480 + 620	0.5 s on, 0.5 s off
Busy (network congestion)	480 + 620	0.2 s on, 0.3 s off
Ring return	440 + 480	2 s on, 4 s off
Off-hook alert	Multifrequency howl	1 s on, 1 s off
Recording warning	1400	0.5 s on, 15 s off
Call waiting	440	0.3 s on, 9.7 s off

Source: Refs 2, 20.

used in North America. Functionally, these are the call-progress tones presented to the subscriber.

5   COMPELLED SIGNALING

In many of the signaling systems discussed thus far, signal element duration is an important parameter. For instance, in a call setup an initiating exchange sends a 100-ms seizure signal. Once this signal is received at the distant end, the distant exchange sends a "proceed to send" signal back to the originating exchange; in the case of the R-1 system, this signal is 140 ms or more in duration. Then, on receipt of "proceed to send" the initiating exchange spills all digits forward. In the case of R-1, each digit is an MF pulse of 68-ms duration with 68 ms between each pulse. After the last address digit an ST (end-of-pulsing) signal is sent. In the case of R-1 the incoming (far-end) switch register knows the number of digits to expect. Thus there is an explicit acknowledgment that the call setup has proceeded satisfactorily. Thus R-1 is a good example of noncompelled signaling.

A fully compelled signaling system is one in which each signal continues to be sent until an acknowledgment is received. Thus signal duration is not significant and bears no meaning. The R-2 and SOCOTEL are examples of fully compelled signaling systems. Figure 4.6 shows a fully compelled signaling sequence. Note

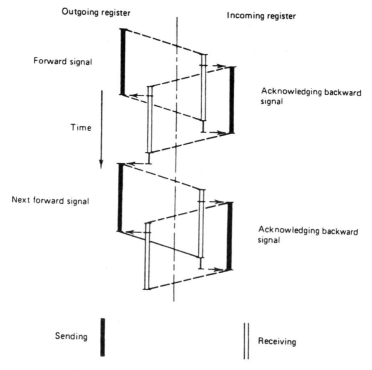

Figure 4.6. Fully compelled signaling procedure.

the small overlap of signals, causing the acknowledging (reverse) signal to start after a fixed time on receipt of the forward signal. This is because of the minimum time required for recognition of the incoming signal. After the initial forward signal, further forward signals are delayed for that short recognition time (see Figure 4.6). Recognition time is normally less than 80 ms.

Fully compelled signaling is advantageous in that signaling receivers do not have to measure duration of each signal, thus making signaling equipment simpler and more economical. Fully compelled signaling adapts automatically to the velocity of propagation, to long circuits, to short circuits, to metallic pairs, or to carrier and is designed to withstand short interruptions in the transmission path. The principal drawback of compelled signaling is its inherent lower speed, thus requiring more time for setup. Setup time over space-satellite circuits with compelled signaling is appreciable and may force the system engineer to seek a compromise signaling system.

There is also a partially compelled type of signaling, where signal duration is fixed in both forward and backward directions according to system specifications; or the forward signal is of indefinite duration and the backward signal is of fixed duration. The forward signal ceases once the backward signal has been received correctly. The CCITT No. 4 is a variation of partially compelled signaling [3].

6 LINK-BY-LINK VERSUS END-TO-END SIGNALING

An important factor to be considered in switching system design that directly affects both signaling and customer satisfaction is postdialing delay. This is the amount of time it takes after the calling subscriber completes dialing until ring-back is received. Ring-back is a backward signal to the calling subscriber telling her that her dialed number is ringing. Postdialing delay must be made as short as possible.

Another important consideration is register occupancy time for call setup as the setup proceeds from originating exchange to terminating exchange. Call-setup equipment, that equipment used to establish a speech path through a switch and to select the proper outgoing trunk, is expensive. By reducing register occupancy per call, we may be able to reduce the number of registers (and markers) per switch, thus saving money.

Link-by-link and end-to-end signaling each affect register occupancy and postdialing delay, each differently. Of course, we are considering calls involving one or more tandem exchanges in a call setup, because this situation usually occurs on long-distance or toll calls. Link-by-link signaling may be defined as a signaling system where *all* interregister address information must be transferred to the subsequent exchange in the call-setup routing. Once this information is received at this exchange, the preceding exchange control unit (register) releases. This same operation is carried on from the originating exchange through each tandem (transit) exchange to the terminating exchange of the call. The R-1 system is an example of link-by-link signaling.

End-to-end signaling abbreviates the process such that tandem (transit) exchanges receive only the minimum information necessary to route the call. For instance, the last four digits of a seven-digit telephone number need be exchanged only between the originating exchange (e.g., the calling subscriber's local exchange or the first toll exchange in the call setup) and the terminating exchange in the call setup. With this type of signaling, fewer digits are required to be sent (and acknowledged) for the overall call-setup sequence. Thus the signaling process may be carried out much more rapidly, decreasing postdialing delay. Intervening exchanges on the call route work much less, handling only the digits necessary to pass the call to the next exchange in the sequence.

The key to end-to-end signaling is the concept of "leading register." This is the register (control unit) in the originating exchange that controls the call routing until a speech path is set up to the terminating exchange before releasing to prepare for another call setup. For example, consider a call from subscriber X to subscriber Y.

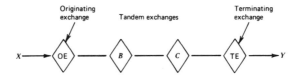

The telephone number of subscriber Y is 345-6789. The sequence of events is as follows using end-to-end signaling:

- A register at exchange OE receives and stores the dialed number 345-6789 from subscriber X.
- Exchange OE analyzes the number and then seizes a trunk (junction) to exchange B. It then receives a "proceed-to-send" signal indicating that the register at B is ready to receive routing information (digits).
- Exchange OE then sends digits 34, which are the minimum necessary to effect correct transit.
- Exchange B analyzes the digits 34 and then seizes a trunk to exchange C. Exchanges OE and C are now in direct contact and exchange B's register releases.
- Exchange OE receives the "proceed-to-send" signal from exchange C and then sends digits 45, those required to effect proper transit at C.
- Exchange C analyzes digits 45 and then seizes a trunk to exchange TE. Direct communication is then established between the leading register for this call at OE and the register at TE being used on this call setup. The register at C then releases.
- Exchange OE receives the "proceed-to-send" signal from exchange TE, to which it sends digits 5678, the subscriber number.
- Exchange TE selects the correct subscriber line and returns to A ringback, line busy, out of order, or other information after which all registers are released.

Thus we see that a signaling path is opened between the leading register and the terminating exchange. To accomplish this, each exchange in the route must "know" its local routing arrangements and request from the leading register those digits it needs to route the call further along its proper course.

Again, the need for backward information becomes evident, and backward signaling capabilities must be nearly as rich as forward signaling capabilities when such a system is implemented.

R-1 is a system inherently requiring little backward information (interregister). The little information that is needed, such as "proceed to send," is sent via line signaling. The R-2 system has major backward information requirements, and backward information and even congestion and busy signals are sent back by interregister signals [3].

7   THE EFFECTS OF NUMBERING ON SIGNALING

Numbering, the assignment and use of telephone numbers, affects signaling as well as switching. It is the number or the translated number, as we found out in Chapter 3, that routes the call. There is "uniform" numbering and "nonuniform" numbering. How does each affect signaling? Uniform numbering can simplify a signaling system. Most uniform systems in the nontoll or local-area case are based on seven digits, although some are based on six. The last four digits identify the subscriber. The first three digits (or the first two in the case of a six-digit system) identify the exchange. Thus the local exchange or transit exchanges know when all digits are received. There are two advantages to this sort of scheme:

1. The switch can proceed with the call once all digits are received because it "knows" when the last digit (either the sixth or seventh) has been received.
2. "Knowing" the number of digits to expect provides inherent error control and makes "time out"* simpler.

For nonuniform numbering, particularly on direct distance dialing in the international service, switches require considerably more intelligence built in. It is the initial digit or digits that will tell how many digits are to follow, at least in theory.

However, in local or national systems with nonuniform numbering, the originating register has no way of knowing whether it has received the last digit, with the exception of receiving the maximum total used in the national system. With nonuniform numbering, an incompletely dialed call can cause a useless call setup across a network up to the terminating exchange, and the call setup is released only after time out has run its course. It is evident that with nonuniform numbering systems, national (and international) networks are better suited to signaling

---

* "Time out" is the resetting of call-setup equipment and return of dial tone to subscriber as a result of incomplete signaling procedure, subset left off hook, and so forth.

systems operating end to end with good features of backward information, such as the R-2 system [3, 20].

8   ASSOCIATED AND DISASSOCIATED CHANNEL SIGNALING

Here we introduce a new concept: disassociated channel signaling. Up to now, we have only considered associated channel signaling. Figure 4.7 is meant to show the difference. The upper drawing in Figure 4.7 shows conventional signaling on an analog circuit. This fits all of the signaling techniques discussed so far. The signaling goes right along with the channel it is associated with, on the same medium. The lower drawing in Figure 4.7 shows "separate channel signaling." Now this signaling may or may not go on that same medium or path. Most often with this type of signaling, that separate channel handles all signaling (especially supervisory signaling) for a group of channels. Typically, the European PCM system called E1 (Chapter 8) uses this type of signaling. One separate digital channel covers all supervisory signaling for 30 traffic channels. If it travels on

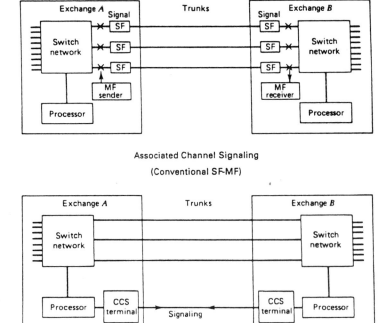

Figure 4.7. Conventional analog associated channel signaling (upper drawing) versus separate channel signaling (lower drawing). *Note:* Signaling on upper drawing accompanies voice paths; signaling on the lower drawing is conveyed on a separate circuit (or time slot). CCS = common channel signaling such as CCITT Signaling System No. 7.

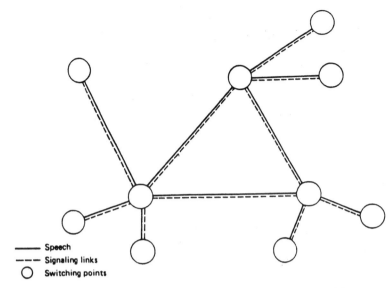

Figure 4.8. Associated channel signaling. As shown, the signaling channel is separate, but associated. With conventional analog signaling, it would be a single solid line, where the signaling is embedded with its associated traffic.

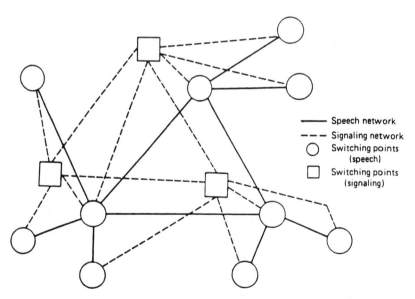

Figure 4.9. Fully disassociated channel signaling. This type of signaling may be used with CCITT Signaling System No. 7, described in Chapter 17.

the same medium and path as its associated traffic channels, it is still associated channel signaling.

We can carry this one step further. That separate channel can follow a different path using, perhaps, different media. CCITT Signaling System No. 7 is always a separate channel, but can be associated or disassociated. Figure 4.8 shows separate channel signaling, but associated. Figure 4.9 shows fully disassociated channel signaling.

9 SIGNALING IN THE SUBSCRIBER LOOP

9.1 Background and Purpose

In Chapter 2 we described loop start signaling, although we didn't call it that. When a subscriber takes a telephone off hook (i.e., out of its cradle), there is a switch closure in the subset (see Figure 4.2 and Section 2 of this chapter); current flows in the loop, alerting the serving exchange that service is desired on that telephone. As a result, dial tone is returned to the subscriber. This is basic supervisory signaling on the subscriber loop.

A problem can arise from this form of signaling. It is called *glare*. Glare is the result of attempting to seize a particular subscriber loop from each direction. In this case it would be an outgoing call and an incoming call nearly simultaneously. There is a much greater probability of glare with a PABX than with an individual subscriber.

Ground-start signaling is the preferred signaling system when lines terminate in a switching system such as a PABX. It operates as follows. When a call is

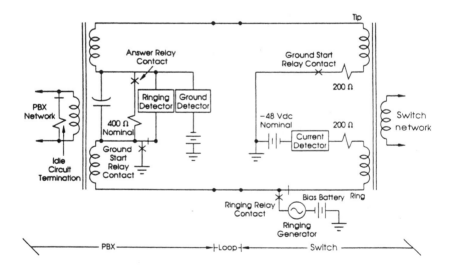

Figure 4.10. Ground start interface block diagram. From Figure 2-7, Ref. 18. Reprinted with permission.

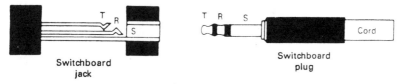

Figure 4.11. Switchboard plug with corresponding jack (R, S, and T are ring, sleeve, and tip, respectively).

from the local serving switch to the PABX, the local switch immediately grounds the conductor tip to seize the line. With some several seconds delay, ringing voltage is applied to the line (where required). The PABX immediately detects the grounded tip conductor and will not allow an outgoing call from the PABX to use this circuit, thus avoiding glare.

In a similar fashion, if a call originates at the PABX and is outgoing to the local serving switch, the PABX grounds the ring conductor to seize the line. The serving switch recognizes this condition and prevents other calls from attempting to terminate on the circuit. The switch then grounds the tip conductor and then returns dial tone after it connects a digit receiver. There can be a rare situation when double seizure occurs, causing glare. Usually one or the other end of the circuit is programmed to back down and allow the other call to proceed. A ground start interface is shown in Figure 4.10.

Terminology in signaling often refers back to manual switchboards or, specifically, to the plug used with these boards and its corresponding jack as shown in Figure 4.11. Thus we have tip (T), ring (R), and sleeve (S). Often, only the tip and ring are used, and the sleeve is grounded and has no real function.

10   METALLIC TRUNK SIGNALING

10.1   Basic Loop Signaling

Many trunks serving the local area are metallic-pair trunks, although in more developed nations they may carry a digital format such as T1 or E1. In other situations they are DC loops much like the subscriber loop. A few of these trunks still use dial pulses for address signaling as well as some sort of supervisory signaling.

Of course these loop-signaling circuits provide two signaling states: one when the circuit is opened and the other when the circuit is closed. A third signaling state is obtained by reversing the direction or by changing the magnitude of the current in the circuit. Combinations of (1) open/close, (2) polarity reversal, and (3) high/low current are used for distinguishing signals intended for one direction of signaling (e.g., dial-pulse signals) from those intended for the opposite direction (e.g., answer signals). We describe the most popular loop method below, namely, reverse-battery signaling.

We have illustrated loop-signaling methods with electromechanical components. These methods are also applicable to SPC (stored-program control) switching systems. SPC switching systems can connect to loop signaling on physical facilities, to analog or digital multiplex systems, or can work directly with a bit stream from a digital carrier (e.g., SLC-96, Chapter 8) [19].

10.2 Reverse-Battery Signaling

Reverse-battery signaling employs basic methods (1) and (2) above, and it takes its name from the fact that battery and ground are reversed on the tip and ring to change the signal toward the calling end from on hook to off hook. Figure 4.12 shows a typical application of reverse-battery signaling in a common-control switch.

In the idle or on-hook condition, all relays are unoperated and the switch (SW) contacts are open. Upon seizure of the outgoing trunk by the calling switch (exchange) (trunk group selection based on the switch or exchange code dialed by the calling subscriber), the following occurs:

- SW1 and SW2 contacts close, thereby closing loop to called office (exchange) and causing the A relay to operate.
- Operation of the A relay signals off-hook (connect) indication to the called switch (exchange).
- Upon completion of pulsing between switches, SW3 contacts close and the called subscriber is alerted. When the called subscriber answers, the S2 relay is operated.
- Operation of the S2 relay operates the T relay, which reverses the voltage polarity on the loop to the calling end.
- The voltage polarity causes the CS relay to operate, transmitting an off-hook (answer) signal to the calling end.

When the calling subscriber hangs up, disconnect timing starts (between 150 and 400 ms). After the timing is completed, SW1 and SW2 contacts are released in the calling switch. This opens the loop to the A relay in the called switch and releases the calling subscriber. The disconnect timing (150–400 ms) is started in the called switch as soon as the A relay releases. When the disconnect timing is completed, the following occurs:

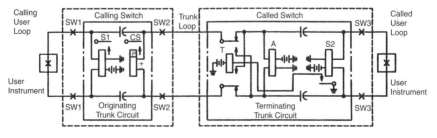

Figure 4.12. Reverse-battery signaling, conceptual drawing.

- If the called subscriber has returned to on hook, SW3 contacts release. The called subscriber is now free to place another call.
- If the called subscriber is still off hook, disconnect timing is started in the called switch. On the completion of the timing interval, SW3 contacts open. The called subscriber is then returned to dial tone. If the circuit is seized again from the calling switch during the disconnect timing, the disconnect timing is terminated and the called subscriber is returned to dial tone. The new call will be completed without interference from the previous call.

When the called subscriber hangs up, the CS relay in the calling switch releases. Then the following occurs:

- If the calling subscriber has also hung up, disconnection takes place as described above.
- If the calling subscriber is still off hook, disconnect timing is started. On the completion of the disconnect timing, SW1 and SW2 contacts are opened. This returns the calling subscriber to dial tone and releases the A relay in the called switch. The calling subscriber is free to place a new call at this time. After the disconnect timing, the SW3 contacts are released, which releases the called subscriber. The called subscriber can place a new call at this time.

Source: Section 10 is based on information taken from Refs. 2, 17, and 19.

REVIEW QUESTIONS

1. Give the three generic signaling functions, and explain the purpose of each.
2. Differentiate between line signaling and interregister signaling.
3. There are seven ways to transmit signaling information, one of which is frequency. Name five others.
4. How does a switch know whether a particular talk path is busy or idle?
5. A most common form of analog line signaling is E and M signaling. Describe how it works in three sentences or less.
6. Compare in-band and out-of-band supervisory signaling regarding tone-on idle/busy, advantages, and disadvantages.
7. What is the most common form of in-band supervisory signaling in North America (analog)?
8. What is the standard out-of-band signaling frequency for North America?
9. Give the principal advantage of 2VF over SF supervisory signaling.
10. Compare CCITT No. 5, R-1, and R-2 supervisory signaling systems.

11. List three types of interregister signaling using MF.

12. Clearly distinguish compelled and noncompelled signaling. Give advantages and disadvantages of each.

13. On connections involving geostationary communication satellites, what would be more attractive regarding postdial delay, compelled or noncompelled signaling?

14. Describe and compare link-by-link and end-to-end signaling. Associate R-1 and R-2 with each.

15. Describe at least four types of backward information.

16. Describe the effects of uniform and nonuniform numbering on signaling/switching. Why is uniform numbering more advantageous to signaling/switching?

17. Distinguish associated channel signaling from separate channel signaling.

18. What is "disassociated channel" signaling? Give some examples.

19. What is the meaning of *glare*?

20. Describe how loop signaling works from the subscriber subset.

21. Reverse-battery signaling is commonly used on two-wire_____. It is useful on these two-wire circuits because it can_____.

REFERENCES

1. C. A. Dahlbom and C. Breen, "Signaling Systems for Control of Telephone Switching," *STJ*, **39** (November 1960).

2. BOC *Notes on the LEC Networks*—1990, BSR-TSV-002275, Issue 1, Bellcore, Piscataway, NJ, March, 1991.

3. "Signaling," from Telecommunication Planning Documents, ITT Laboratories (Spain), Madrid, November 1974.

4. *National Networks for the Automatic Service*, CCITT-ITU, Geneva, 1964.

5. M. Den Hertog, "Inter-register Multifrequency Signaling for Telephone Switching in Europe," *Electr. Commun.*, **38** (1) (1972).

6. *Reference Data for Engineers: Radio, Electronics, Computers and Communications*, 8th ed., Howard W. Sams, Indianapolis, 1993.

7. R. L. Freeman, *Telecommunication Transmission Handbook*, 4th ed., John Wiley & Sons, New York, 1998.

8. "Lenkurt Demodulator" *World-Wide E&M Signaling* (July–August 1977; *A Glossary of Signaling Terms*, April 1974, GTE Lenkurt, San Carlos, CA.

9. *Specifications of Signaling System No. 4, Signal Code*, CCITT Rec. Q.121, ITU, Geneva, 1988.

10. *Specifications of Signaling System No. 5, Signal Code for Line Signaling*, ITU-T Rec. Q.141, ITU, Geneva, March 1993.

11. *Specifications for Signaling System No. 5, Signal Code for Register Signaling*, ITU-T Rec. Q.151, ITU, Geneva, 1993.

12. *Specifications of Signaling System R-1—Line Signaling*, ITU-T Rec. Q.311, Geneva, 1993.

13. *Specifications for Signaling System R-1—Register Signaling*, ITU-T Rec. Q.320, Geneva, 1993.

14. *Specifications of Signaling System R2—Line Signaling, Analog Version*, ITU-T Rec. Q.411, ITU, Geneva, 1993.

15. *Specifications of Signaling System R2—Interregister Signaling*, ITU-T Rec. Q.441, ITU, Geneva, 1993.

16. *Access Area Switching and Signaling Concepts, Issues and Alternatives*, NTIA, Boulder, CO, 1978.

17. C. A. Dahlbom, "Signaling Systems and Technology," *Proc. IEEE*, **65**, 1349–1353 (1977).

18. W. D. Reeve, *Subscriber Loop Signaling and Transmission Handbook, Analog*, IEEE Press, IEEE, New York, 1992.

19. "Telcordia Notes on the Networks," Telcordia Special Report, Issue 4, Telcordia, Piscataway, NJ, October 2000.

20. R. L. Freeman, *Reference Manual for Telecommunication Engineering*, John Wiley & Sons, New York, 2002.

5

INTRODUCTION TO TRANSMISSION FOR TELEPHONY

1 PURPOSE AND SCOPE

The basic building block for transmission is the telephone channel or voice channel. "Voice channel" implies spectral occupancy, whether the voice path is over wire, radio, or coaxial cable or over a fiber-optic system. If a pair of wires of a simple subscriber loop is extended without loading, we can expect to see the spectral content from the average talker with frequencies as low as 20 Hz and as high as 20 kHz if the transducer of the telephone set was at all efficient across this band. Our ear, at least in younger people, is sensitive to frequencies from about 30 Hz to as high as 20 kHz. However, the primary content of a voice signal (energy plus emotion) will occupy a much narrower band of frequencies (approximately 100–4000 Hz). Considering these and other factors, we say that the nominal voice channel occupies the band from 0 to 4 kHz. CCITT defines the voice channel as the band of frequencies between 300 and 3400 Hz. Bell Laboratories [4] states that "the optimum trade-off between economics and quality of transmission occurs when the telephone speech signal is band-limited to the range from about 200 to 3200 Hz."

There are three basic impairments we must deal with regarding the voice channel.

- Attenuation distortion (frequency response)
- Phase distortion
- Noise

Two additional impairments are echo and singing. We will deal with these two later.

Telecommunication System Engineering, by Roger L. Freeman
ISBN 0-471-45133-9   Copyright © 2004 Roger L. Freeman

Level is another important parameter, especially in an analog network. Level must be controlled because it can surely impact quality of service (QoS).

2 THE THREE BASIC IMPAIRMENTS TO VOICE CHANNEL TRANSMISSION

2.1 Attenuation Distortion

A signal transmitted over a voice channel suffers various forms of distortion. That is, the output signal from the channel is distorted in some manner such that it is not an exact replica of the input. One form of distortion is called *attenuation distortion* and is the result of imperfect amplitude-frequency response. Attenuation distortion can be avoided if all frequencies within the passband are subjected to exactly the same loss (or gain). Whatever the transmission medium, however, some frequencies are attenuated more than others. For example, on loaded wire-pair systems, higher frequencies are attenuated more than lower ones. On carrier equipment (see Section 4 of this chapter), band-pass filters are used on channel units, where, by definition, attenuation increases as the band edges are approached. Figure 5.1 is a good example of the attenuation characteristics of a voice channel operating over carrier multiplex equipment.

Attenuation distortion across the voice channel is measured against a reference frequency. The CCITT specifies 800 Hz as a reference, which is universally used in Europe, Africa, and parts of Hispanic America, whereas 1000 Hz is the common reference frequency in North America. Let us look at some ways attenuation distortion may be stated. For example, one European requirement may state that between 600 Hz and 2800 Hz the level will vary no more than -1 to $+2$ dB, where the plus sign means more loss and the minus sign means less loss. Thus if a signal at -10 dBm is placed at the input of the channel, we would expect -10 dBm at the output at 800 Hz (if there were no overall loss

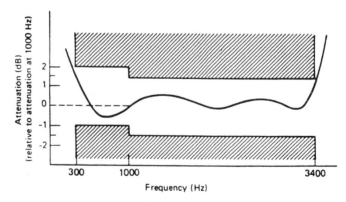

Figure 5.1. Typical attenuation–frequency response (attenuation distortion) for a voice channel. Hatched areas show specified limits.

or gain), but at other frequencies we could expect a variation between -1 and $+2$ dB. For instance, we might measure the level at the output at 2500 Hz at -11.9 dBm and at 1100 Hz at -9 dBm.

2.2 Phase Distortion

A voice channel may be regarded as a band-pass filter. A signal takes a finite time to pass through a telecommunication network. This time is a function of the velocity of propagation, which varies with the media involved.

The velocity of propagation also tends to vary with frequency because of the electrical characteristics associated with the network. Considering the voice channel, therefore, the velocity of propagation tends to increase toward band center and decrease toward band edge. This is illustrated in Figure 5.2, which shows relative delay across the voice channel.

The finite time it takes a signal to pass through the total extension of a voice channel or any network is called *delay*. Absolute delay is the delay a signal experiences while passing through the channel end-to-end at a reference frequency. But we see that the propagation time is different for different frequencies, with the wavefront of one frequency arriving before the wavefront of another in the passband. A modulated signal will not be distorted on passing through the channel if the phase shift changes uniformly with frequency, whereas if the phase shift is nonlinear with respect to frequency, the output signal is distorted compared to the input.

In essence we are dealing with phase linearity of a circuit. If the phase–frequency relationship over a passband is not linear, distortion will occur in the transmitted signal. This phase distortion is often measured by a parameter called *envelope delay distortion* (EDD). Mathematically, envelope delay is the derivative of the phase shift with respect to frequency. The maximum difference in the

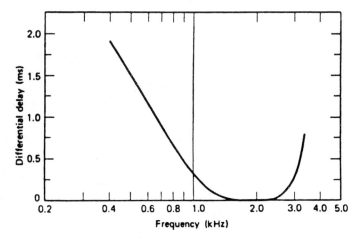

Figure 5.2. Typical differential delay across a voice channel. Frequency-division-multiplex (FDM) equipment back-to-back.

derivative over any frequency interval is called envelope delay distortion. Therefore EDD is always a difference between the envelope delay at one frequency and that at another frequency of interest in the passband. Note that envelope delay is often defined the same as group delay—that is, the ratio of change, with angular frequency, of the phase shift between two points in a network [2].

2.2.1 Notes on Phase Distortion. Absolute delay is minimum around 1700 and 1800 Hz in the voice channel. This is shown in Figure 5.2. The figure also shows that around 1700 or 1800 Hz, envelope delay distortion is flattest. It is for this reason that so many data modems use 1700 or 1800 Hz for the characteristic tone frequency which is modulated by the data.

This brings up the next point. Phase distortion (or EDD) has little effect on speech communications over the telecommunication network. However, for data transmission, phase distortion is the greatest bottleneck for data rate (i.e., number of bits per second that the channel can support). It has probably more effect on limiting data rate than any other parameter [3].

2.3 Noise

2.3.1 General. Noise, in its broadest definition, consists of any undesired signal in a communication circuit. The subject of noise and noise reduction is probably the most important single consideration in analog transmission engineering. It is the major limiting factor in system performance. For the discussion in this text, noise is broken down into four categories:

1. Thermal noise
2. Intermodulation noise
3. Crosstalk
4. Impulse noise

2.3.2 Thermal Noise. Thermal noise occurs in all transmission media and all communication equipment, including passive devices. It arises from random electron motion and is characterized by a uniform distribution of energy over the frequency spectrum with a Gaussian distribution of levels.

Every equipment element and the transmission medium proper contribute thermal noise to a communication system if the temperature of that element or medium is above absolute zero. Thermal noise is the factor that sets the lower limit of sensitivity of a receiving system and is often expressed as a temperature, usually given in units referred to absolute zero. These units are kelvins.

Thermal noise is a general expression referring to noise based on thermal agitations. The term "white noise" refers to the average uniform spectral distribution of noise energy with respect to frequency. Thermal noise is directly proportional to bandwidth and temperature. The amount of thermal noise to be found in 1 Hz of bandwidth in an actual device is

$$P_n = kT \, (\text{W/Hz}) \qquad (5.1)$$

where k is Boltzmann's constant, equal to 1.3803×10^{-23} J/K, and T is the absolute temperature (K) of the circuit (device). At room temperature, $T = 17°C$ or 290 K; thus

$$P_n = 4.00 \times 10^{-21} \text{ W/Hz of bandwidth}$$

$$= -204 \text{ dBW/Hz of bandwidth}$$

$$= -174 \text{ dBm/Hz of bandwidth}$$

For a band-limited system (i.e., a system with a specific bandwidth), $P_n = kTB$ (W), where B refers to the so-called noise bandwidth in hertz. Thus at 0 K we obtain $P_n = -228.6$ dBW/Hz of bandwidth; for a system with a noise bandwidth measured in hertz (B) and whose noise temperature is T we obtain

$$P_n = -228.6 \text{ dBW} + 10 \log T + 10 \log B \qquad (5.2)$$

2.3.3 Intermodulation Noise.

Intermodulation (IM) noise is the result of the presence of intermodulation products. If two signals with frequencies F_1 and F_2 are passed through a nonlinear device or medium, the result will contain IM products that are spurious frequency energy components. These components may be present either inside and/or outside the band of interest for a particular device. IM products may be produced from harmonics of the desired signals in question, either as products between harmonics or as one of the signals and the harmonic of the other(s) or between both signals themselves. The products result when two (or more) signals beat together or "mix." Look at the mixing possibilities when passing F_1 and F_2 through a nonlinear device. The coefficients indicate the first, second, or third harmonics.

- Second-order products $F_1 \pm F_2$
- Third-order products $2F_1 \pm F_2; 2F_2 \pm F_1$
- Fourth-order products $2F_1 \pm 2F_2; 3F_1 \pm F_2 \ldots$

Devices passing multiple signals simultaneously, such as multichannel radio equipment, develop intermodulation products that are so varied that they resemble white noise.

Intermodulation noise may result from a number of causes:

- Improper level setting. If the level of input to a device is too high, the device is driven into its nonlinear operating region (overdrive).
- Improper alignment causing a device to function nonlinearly.
- Nonlinear envelope delay.
- Device malfunction.

To summarize, intermodulation noise results from either a nonlinearity or a malfunction that has the effect of nonlinearity. The cause of intermodulation noise is different from that of thermal noise. However, its detrimental effects and physical nature can be identical with those of thermal noise, particularly in multichannel systems carrying complex signals [12, 14].

2.3.4 Crosstalk. Crosstalk refers to unwanted coupling between signal paths. There are essentially three causes of crosstalk: (1) electrical coupling between transmission media, such as between wire pairs on a voice-frequency (VF) cable system, (2) poor control of frequency response (i.e., defective filters or poor filter design), and (3) nonlinear performance in analog (FDM) multiplex systems. Excessive level may exacerbate crosstalk.

There are two types of crosstalk:

1. *Intelligible*, where at least four words are intelligible to the listener from extraneous conversation(s) in a 7-s period.
2. *Unintelligible*: crosstalk resulting from any other form of disturbing effects of one channel on another.

Intelligible crosstalk presents the greatest impairment because of its distraction to the listener. Distraction is considered to be caused either by fear of loss of privacy or primarily by the user of the primary line consciously or unconsciously trying to understand what is being said on the secondary or interfering circuits; this would be true for any interference that is syllabic in nature.

Received crosstalk varies with the volume of the disturbing talker, the loss from the disturbing talker to the point of crosstalk, the coupling loss between the two circuits under consideration, and the loss from the point of crosstalk to the listener. The most important of these factors for this discussion is the coupling loss between the two circuits under consideration. Talker levels have been discussed elsewhere in this text. Also, we must not lose sight of the fact that the effects of crosstalk are subjective, and other factors have to be considered when crosstalk impairments are to be measured. Among these factors are the type of people who use the channel, the acuity of listeners, traffic patterns, and operating practice [3, 4].

2.3.5 Impulse Noise. Impulse noise is noncontinuous, consisting of irregular pulses or noise "spikes" of short duration, broad spectral density, and relatively high amplitude. In the language of the trade, these spikes are often called "hits." A technician may say that the circuit is getting "hit up." Impulse noise degrades telephony ordinarily only marginally, if at all. However, it may seriously degrade data error performance on data or other digital waveforms. We discuss impulse noise further when covering data transmission in Chapter 10.

2.4 Level

Level is a primary parameter in the analog network. By "primary" we mean very important. With the digital network level it is of secondary importance. In the context of this book, when we use the word *level*, we mean *signal magnitude*. Level could be comparative. The output of an amplifier is 20 dB higher than the input. But more commonly, we mean absolute level, and in telephony it is measured in dBm (decibels referenced to 1 milliwatt) or in milliamperes. In radio (wireless) systems, we will more likely employ dBW (decibels referenced to 12 watts). When dealing with video systems (e.g., television), the unit of measure is voltage. The commonly derived unit is the dBmV, meaning decibels referenced to 1 millivolt.

In the telecommunication network, if levels are too high, amplifiers become overloaded, resulting in intermodulation and other types of distortion such as crosstalk. If levels are too low, customer satisfaction may suffer with a degraded loudness rating.

System levels are important parameters when engineering a telecommunication system. The values are usually taken from a level chart or a reference system drawing made by a planning group or as a part of an engineered job. On the chart a 0 TLP (zero test level point, refer to Chapter 3, Section 15) is established. A test-level point is a location in a circuit or system at which a specified test-tone level is expected during alignment. A 0 TLP is a point at which the test-tone level should be 0 dBm. Here the decibel unit, the dBr,* may enter the discussion. As we briefly discussed in Chapter 3, the dBm can be related to dBr and dBm0 by the following formula:

$$dBm = dBm0 + dBr \tag{5.3}$$

For instance, a value of -32 dBm at a -22-dBr point corresponds to a reference level of -10 dBm0. A -10-dBm0 signal introduced at the 0 dBr point (0 TLP) has an absolute signal level of -10 dBm [4–6].

2.4.1 Typical Levels. Earlier measurement of speech level used the unit of measure VU, standing for volume unit. For 1000-Hz sinusoid signal, 0 VU = 0 dBm. When a VU meter is used to measure the level of a voice signal, it is difficult to equate VU and dBm. One of the problems, of course, is that speech transmission is characterized by spurts of signal. The damping on a VU meter tends to smooth out the spurts of voice signal. We can relate VU to dBm in the following formula:

$$\text{Average power of a telephone talker} \approx VU - 1.4 \text{ (dBm)} \tag{5.4}$$

Today, when discussing speech transmission, we often talk about *equivalent peak level* (EPL) and *long-term conversational level*. The unit of measurement for both is the dBm0 (dBm referenced to the zero test level point or 0 TLP).

EPL can be described approximately as the 95% point on the cumulative probability distribution of instantaneous talker power. Periods of silence are excluded from the distribution.

---

* You may interpret dBr as *decibels "reference."*

In the long-distance network, EPLs can be characterized by a mean of −11.1 dBm0 with a standard deviation of 4.7 dB. Average power for calls in the long-distance network can be characterized by a mean of −25.5 dBm0 and a standard deviation of 5.3 dB. For the local area, the corresponding quantities are a mean EPL of −11.8 dBm0 with a standard deviation of 4.7 dB and a mean average power of −26.5 dBm0 with a standard deviation of 5.4 dB. The distributions of EPL and average power are approximately log-normal (i.e., normal when the independent variable is expressed in dB units) [4, 9, 12, and 13].

2.5 Signal-to-Noise Ratio

When dealing with transmission engineering, signal-to-noise (S/N) ratio is perhaps more frequently used than any other criterion when designing a telecommunication system. S/N ratio expresses in decibels the amount by which a signal level exceeds the noise within a specified bandwidth.

As we review several types of material to be transmitted, each will require a minimum S/N ratio to satisfy the customer or to make the receiving instrument function within certain specified criteria. We might require the following S/N ratios with the corresponding end instruments:

- Voice: 40 dB ⎫
- Voice: 45 dB ⎭ based on customer satisfaction.
- Data: ∼15 dB, based on a specified error rate and modulation type.

In Figure 5.3 a 1000-Hz signal has an S/N ratio of 10 dB, assuming a nominal 4-kHz bandwidth for his example. The level of the noise is +5 dBm, and the signal level is +15 dBm. Thus

$$(S/N)_{dB} = \text{level}_{(\text{signal in dBm})} - \text{level}_{(\text{noise in dBm})} \tag{5.5}$$

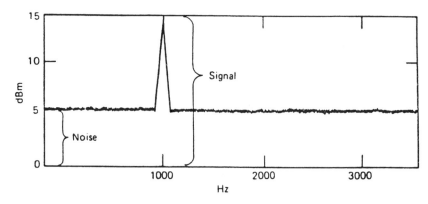

Figure 5.3. Signal-to-noise ratio.

S/N ratio in the PSTN really has limited applicability in characterizing speech transmission because of the "spurtiness" of the human voice. We can appreciate that individual talker signal power can fluctuate widely so that S/N ratio is far from constant from telephone call to telephone call. In lieu of actual voice, we use a test tone to measure level and S/N ratio, which has a constant amplitude and no silent intervals [4].

3 TWO-WIRE AND FOUR-WIRE TRANSMISSION

3.1 Two-Wire Transmission

A telephone conversation inherently requires transmission in both directions. When both directions are carried on the same pair of wires, it is called *two-wire transmission*. The telephones in our homes and offices are connected to a local switching center (exchange) by means of two-wire circuits. A more proper definition for transmitting and switching purposes is that when oppositely directed portions of a single telephone conversation occur over the same electrical transmission channel or path, we call this *two-wire operation*.

3.2 Four-Wire Transmission

Carrier and radio systems require that oppositely directed portions of a single conversation occur over separate transmission channels or paths (or use mutually exclusive time periods). Thus we have two wires for the transmit path and two wires for the receive path, or a total of four wires for a full-duplex (two-way) telephone conversation. For almost all telephone systems, the end instrument (i.e., the telephone subset) is connected to its local serving exchange on a two-wire basis. In other words, the subscriber loop is two-wire.

In fairly well developed nations, the output of the local serving exchange, looking toward the toll network, is four-wire. In many less developed nations the two-wire to four-wire conversion does not take place until the output of the toll-connecting exchange. This is the same point, of course, where the A/D conversion (conversion to PCM, Chapter 8) takes place.

To simplify the explanation, Figure 5.4 illustrates a typical PSTN network showing the two-wire-to-four-wire conversion from the calling-subscriber end, and the converse, conversion from four-wire to two-wire at the called subscriber end. Schematically, the four-wire interconnection is shown as if it were a single-channel wire-line system with amplifiers. However, it would be more likely be a multichannel digital carrier system on cable, fiber-optic lightguide, and/or multiplex over radio. The amplifiers in Figure 5.4 serve to convey the ideas this section considers. As we detail in the figure, conversion from two-wire operation to four-wire is carried out by a *terminating set*, more commonly referred to in the industry as a *term set*, which contains a four-port balanced transformer (a hybrid) or, less commonly, a resistive network.

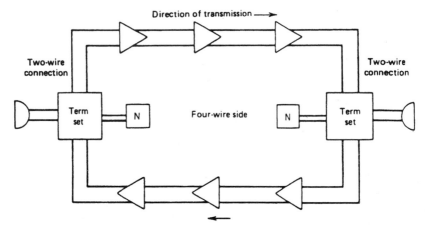

Figure 5.4. A typical long-distance (toll) telephone connection.

3.3 Operation of a Hybrid

A hybrid, in terms of telephony (at voice frequency), is a transformer. For a simplified description, a hybrid may be viewed as a power splitter with four sets of wire-pair connections. A functional block diagram of a hybrid device is shown in Figure 5.5. Two of the wire-pair connections belong to the four-wire path, which consists of a transmit pair and a receive pair. The third pair is the connection to the two-wire link that is eventually connected to the subscriber subset via one or more switches. The last wire pair of the four connects the hybrid to a resistance–capacitance balancing network, which electrically balances the hybrid with the two-wire connection to the subscriber subset over the frequency range of the balancing network. An artificial line may be used for this purpose.

Signal energy entering from the two-wire subset connection divides equally, half of it dissipating in the impedance of the four-wire side receive path and the other half going to the four-wire side transmit path, as shown in Figure 5.5. Here the *ideal* situation is that no energy is to be dissipated by the balancing network (i.e., there is a perfect balance). The balancing network is supposed to display

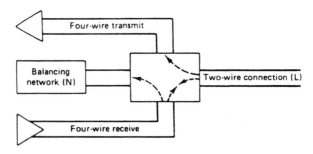

Figure 5.5. Operation of a hybrid transformer.

the characteristic impedance of the two-wire line (subscriber connection) to the hybrid. Signal energy entering from the four-wire side receive path is also split in half in the ideal situation where there is perfect balance. Half of the energy is dissipated by the balancing network (N) and half at the two-wire port (L) (see Figure 5.5).

The reader notes that in the description of the hybrid, in every case, ideally half of the signal energy entering the hybrid is used to advantage and half is dissipated or wasted. Also keep in mind that any passive device inserted in a circuit, such as a hybrid, has an insertion loss. As a rule of thumb, we say that the insertion loss of a hybrid is 0.5 dB. Thus there are two losses here that the reader must not lose sight of:

Hybrid insertion loss	0.5 dB	
Hybrid dissipation loss	3.0 dB	(half of the power)
	3.5 dB	(total)

As far as this section is concerned, any signal passing through a hybrid suffers a 3.5-dB loss. This is a good design number for gross engineering practice. Hybrids used on short loops, however, may have higher losses, as do special resistance-type hybrids.

In Figure 5.5, consider the balancing network (N) and the two-wire side of the hybrid (L). For the older, conventional analog network, the two-wire side of the hybrid connected the subscriber through at least one two-wire switch. Because of the switch, the two-wire side of the hybrid could look into at least 10,000 possible subscriber connections—some short loops, some long loops, and other loops in poor condition. Because of the fixed conditions on the four-wire side, we can pretty much depend on holding a good impedance match. Our concern under these conditions is impedance match on the two-wire side. That is the match between the compromise network (N) and the two-wire side (L).

We measure the capability of impedance match by *return loss*. In this particular case we call it *balance return loss*:

$$\text{Balance return loss}_{dB} = 20 \log_{10} \frac{Z_L + Z_N}{Z_L - Z_N}$$

Let's say, for argument's sake, that we have a perfect match. In other words the impedance of the two-wire subscriber loop side on this particular call was exactly 900 Ω and the balancing network (N) was 900 Ω. Substitute these numbers in the formula above and we get

$$\text{Balance return loss}_{dB} = 20 \log \frac{900 + 900}{900 - 900}$$

Examine the denominator. It is zero. Any number divided by zero is infinity. Thus we have an infinitely high return loss. And this happens when we have a perfect match, an ideal condition. Of course it is seldom realized in real life.

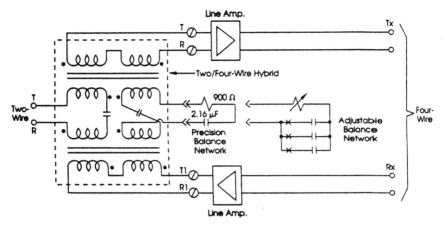

Figure 5.6. Block diagram of two-wire-to-four-wire conversion using a hybrid. From Figure 5-9, page 104, Ref. 11. Reprinted with permission of IEEE Press.

In real life we find that the balance return loss for a large population of hybrids connected in service and serving a large population of two-wire users has a median more of the order of 11 dB with a standard deviation of 3 dB [4]. This is valid for North America. For some other areas of the world, balance return loss median may be lower with a larger standard deviation.

When return loss becomes low (i.e., there is a poor impedance match), there is a reflection of the speech signal. This is echo. We define the cause of echo as any impedance mismatch in the network. Most commonly this mismatch occurs at the hybrid. Echo that is excessive becomes singing. This is caused by high positive feedback on intervening amplifiers. Singing is a highly undesirable impairment. We discuss the control of echo and singing in Chapter 6. Figure 5.6 is a block diagram of a hybrid circuit [10, 14].

3.4   Notes on the Digital Network in the Local Area

Present practice with many telephone companies and administrations is to convert the two-wire subscriber loop, after it terminates in the mainframe, to either a T1 or E1 digital format (see Chapter 8), which, of course is four-wire. Because each subscriber has his or her own hybrid with an incoming A/D (analog-to-digital) conversion, the return losses are generally higher. Add this to the isolation of the A/D, and echo levels are either exceedingly low or nonexistent.

4   MULTIPLEXING

4.1   Definition and Introduction

Reference 1 defines multiplexing as the combining of two or more signals into a single wave from which the signals can be individually recovered. In our case the signals are voice channels and we can literally combine more than 1000 such channels for transmission over a medium. Of course the medium has to be able to accommodate the required bandwidth. On a wire pair we might combine 24, 30, or 48 channels. On LOS microwave we can commonly carry 1800 analog channels or hundreds of digital channels. On fiber optics we can carry thousands of digital channels.

There are essentially two generic ways we can multiplex voice channels:

1. In the frequency domain using frequency division multiplex (FDM)
2. In the time domain using time division multiplex (TDM)

We provide a brief review of FDM in this chapter. TDM, using pulse-code modulation (PCM), is discussed in Chapter 8.

4.2 Frequency Division Multiplex (FDM)

4.2.1 Overview. Frequency division multiplex is a method of allocating a unique band of frequencies in a comparatively wideband frequency spectrum of the transmission medium to each communication channel on a continuous time basis. The communication channel may be the analog voice channel of nominal 4-kHz in bandwidth, a 15-kHz broadcast channel, a 48-kHz data channel, or a 6.0-MHz television channel (concept based on Ref. 7). One positive development just prior to the rapid evolution to a digital network was the final acceptance of a standardized FDM modulation format espoused by CCITT (ITU-T). The digital network has two distinct formats, one which we may call European and the other, North American.

The concept of frequency division multiplexing should be appreciated by the reader. However, its use or application has become obsolete in this day of the digital network.

4.2.2 Impairments. Noise was the principal impairment of the analog network, and a greater portion of this noise derived from the FDM equipment involved. In a digital network, noise is a secondary issue; degradation of bit error rate becomes the primary issue. Insufficient signal-to-noise ratio is just one of several possible causes of this degradation.

5 SHAPING OF A VOICE CHANNEL AND ITS MEANING IN NOISE MEASUREMENT UNITS

This section deals with attenuation distortion or frequency response of a voice channel. Such a channel may be "flat" or "weighted." Weighting means favoring some portion of the frequency spectrum over some other portion or portions. It is simply that we impart more gain or loss for that part of the spectrum which we are favoring. "Flat response" just means that there is no weighting. If, for instance, we measure the attenuation distortion (frequency response) of the voice channel input on multiplex system compared to the response of the voice channel output at the companion demultiplex, we might see a variation of some ±0.5 dB. If we do similar measurements from end-to-end on the PSTN with no weighting involved, the variation may be in the order of ±3 to ±5 dB. It is still called a flat channel.

Now connect a telephone handset transmitter (with appropriate talk battery) to the input of the voice channel modulator and handset receiver to the output of the companion voice channel demodulator and include the acuity of the "average" human ear. Now we see a "shaping" effect. The handset and the ear acuity "shape" the channel when the audio level is compared at various frequencies at the input of the acoustical–electrical transducer (handset transmitter) to the audio output from the electrical–acoustical transducer (the handset receiver).

The frequency response measurement uses a reference frequency located at the point of minimum attenuation in the voice channel. In North America the reference frequency is 1000 Hz, whereas in Europe and many other locations in the world it is 800 Hz. All CCITT recommendations dealing with the voice channel use 800 Hz as reference.

Considering now the shaping effect we just mentioned combining the mouth, handset, and ear, we see that for speech transmission, certain frequencies in the voice channel are attenuated more than others. When transmission systems are tested, *weighting networks* are used to simulate these effects. There are two types of weighting networks now in use:

- C-message, used in North America
- Psophometric, used in the rest of the world and starting to be accepted in North America

Let's look at this weighting from another viewpoint. Admit that noise we hear is annoying. Let the noise be a single-sinusoid tone from an audio signal generator. Pass this tone through the microphone on a standard handset and have someone listen on the earpiece. Adjust the signal generator to 1000 Hz, the reference frequency, and adjust the level to a point where the listener is annoyed. Let's say the level is 0 dBm. Now set the signal generator to 300 Hz and increase the level until the point of same annoyance is reached. It will be noted that we had to increase the level to about +20 dBm for equal annoyance. We can draw a curve for frequencies inside the voice channel which we could call an *annoyance curve*. Such a curve is shown in Figure 5.7. This is the C-message weighting curve (solid line). A similar curve can be drawn for psophometric weighting. It is also shown in Figure 5.7, the dashed line.

Figure 5.8 shows the C-message curve with the *noise advantage*. This is the hatched area between 300 and 3400 Hz. If we are to use a circuit for speech, the noise in the hatched area is not counted against us, but is instead a noise advantage. We can take about 2.5 dB of advantage for psophometric weighting and about 2.0 dB for C-message weighting.

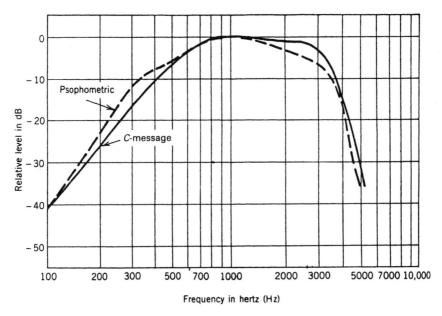

Figure 5.7. Line weighting curves for telephone channel noise [9].

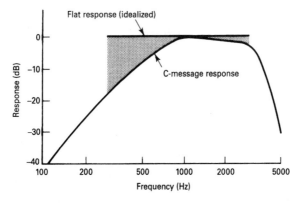

Figure 5.8. C-message weighting curve showing how we achieve about 2.0-dB noise advantage with speech transmission over a voice channel. The hatched area shows the noise advantage [10].

The noise measurement unit for C-message weighting is the dBrnc; when referenced to the 0 TLP, it is the dBrnc0. dBrn may stand for "decibels reference noise." For psophometric weighting, the most common unit for noise measurement is the pWp. It is a linear unit where the dBrnC is a logarithmic unit. pWp stands for "picowatts, psophometrically weighted." Relating pWp to other, more commonly recognized units: pW is 1×10^{-12} watts or 1×10^{-9} milliwatts. If we wished to use decibel units, then we probably would use dBmp or dBm psophometrically weighted. With psophometric weighting the reference frequency is 800 Hz, and with C-message weighting it is 1000 Hz [9].

If we were to use this same circuit (voice channel) for data transmission, the noise advantage goes away, and we must consider the channel as flat as shown in Figure 5.8.

To convert from flat channel noise measurements in pW or dBm, the following excerpt from CCITT Rec. G.223 may be useful:

> If uniform-spectrum random noise is measured in a 3.1-kHz band with flat attenuation frequency characteristic, the noise level must be reduced 2.5 dB to obtain the psophometric power level. For another bandwidth, B, the weighting factor will be equal to:
>
> $$2.5 + 10 \log \left(\frac{B}{3.1} \right) \text{(dB)}$$
>
> When $B = 4$ kHz, for example, this formula gives a weighting factor of 3.6 dB.

The following relationships will be useful for converting from one noise unit to another:

$$-90 \text{ dBm} = -2 \text{ dBrnC and thus} - 92 \text{ dBm} = 0 \text{ dBrnC}$$

$$(\text{white noise loading})$$

$$-92.5 \text{ dBmp} = -90 \text{ dBm (flat, white noise)}$$

$$1 \text{ pWp} = -90 \text{ dBmp}$$

$$\text{value in dBm} = 10 \log(\text{value in pWp} \times 10^{-9}) + 2.5 \text{ dB}$$

$$\text{dBrnC} = 10(\log \text{pWp} \times 10^{-9}) - 0.5 \text{ dB} + 90 \text{ dB}$$

$$\text{value in pW} \times 0.56 = \text{value in pWp}$$

$$\text{value in pWp}/0.56 = \text{value in pW}$$

Source: Material based on Ref. 5.

In telecommunications and in other areas of electronics, as a general rule a passband is defined between 3-dB points. Remember that if power is dropped 3 dB, it is cut in half. The voice channel is the one exception to this rule. The voice channel in this context uses 10-dB points or where the power drops to 1/10th. This definition is valid only in the subscriber plant up to the main frame of the serving exchange [9, 10–13].

REVIEW QUESTIONS

1. There are four basic transmission parameters for a telephone voice channel, three of which are impairments. Name and define each.

2. Define the specified CCITT voice channel.

3. What are the two reference frequencies for the voice channel? One is European and accepted by CCITT and the other is North American.

4. The reference test-tone input to a telephone channel is -16 dBm and the output is $+7$ dBm. What range of values in dBm can be expected at the output if the attenuation distortion is $-1 + 2.5$ dB from 600 to 2600 Hz?

5. What causes phase distortion?

6. What test-tone level should one expect at 0 TLP? A value of -10 dBm at the -12-dBr point corresponds to what dBm0 value?

7. Name the four basic categories of noise.

8. What thermal noise level would one expect with a receiver with 1000 K noise temperature and a bandwidth of 1 kHz?

9. Two frequencies of 1000 Hz and 1200 Hz appear at the input of a nonlinear device. Give values of third-order products.

10. Give at least two causes of intermodulation products.

11. Impulse noise generally does not affect speech telephony. What can it affect seriously?

12. The signal level coming out of a receiver is $+7$ dBm and the thermal noise level is -35 dBm. What is the signal-to-noise ratio in this case?

13. Define two-wire and four-wire transmission. Where would each be applied?

14. What are the two loss components across a model hybrid? In each case explain what causes the loss.

15. What is the problem measuring signal-to-noise ratio for a speech signal? How is the problem conventionally overcome?

16. C-message weighted noise uses what noise measurement units?

17. Express 1 pWp in watts and in milliwatts. Use the powers of 10.

18. What is the power level in dBm of 200 pWp? Give the same level in dBmp.

19. Explain the "annoyance" factor in noise weighting.

20. Why must we use "flat" weighting for data transmission in the voice channel?

REFERENCES

1. *The IEEE Standard Dictionary of Electrical and Electronic Terms*, 6th ed., IEEE Press, New York, 1996.

2. M. E. Van Valkenburg, editor-in-chief, *Reference Data for Engineers: Radio, Electronics, Computer and Communications*, 8th ed., SAMS, Prentice-Hall Computer Publishing, Carmel, IN, 1993.

3. *The Lenkurt Demodulator*, Lenkurt Electric Company, San Carlos, CA, December 1964, June 1965, and September 1965.

4. *Transmission Systems for Communications*, 5th ed., Bell Telephone Laboratories, Holmdel, NJ, 1982.

5. H. H. Smith, *Noise Level Terms in American and International Practice*, ITT Communication Systems, Paramus, NJ, 1964.

6. *The Transmission Plan*, ITU-T Rec. G.101, ITU, Geneva, August 1996.

7. B. D. Holbrook and J. T. Dixon, "Load Rating Theory for Multichannel Amplifiers," *BSTJ*, pages 624–644 (October 1939).

8. "Telcordia Notes on the Networks," Telcordia Special Report, SR-2275, Issue 4, Telcordia, Piscataway, NJ, October 2000.

9. W. Oliver, *White Noise Loading of Multichannel Communication Systems*, Marconi Instruments, St. Albans, Herts, UK, 1976.

10. R. L. Freeman, *Telecommunication Transmission Handbook*, 4th ed., John Wiley & Sons, New York, 1998.

11. W. D. Reeve, *Subscriber Loop Signaling and Transmission Handbook, Analog*, IEEE Press, New York, 1992.

12. *Telecommunication Transmission Engineering*, Vol. 1, 2nd ed., AT&T, New York, 1977.

13. *IEEE Standard Methodologies for Specifying Voicegrade Channel Transmission Parameters and Evaluating Connection Transmission Performance for Speech Telephony*, IEEE Std 823–1989, IEEE, New York, 1989.

14. R. L. Freeman, *Reference Manual for Telecommunication Engineering*, 3rd ed., John Wiley & Sons, New York, 2002.

6

LONG-DISTANCE NETWORKS

1 GENERAL

The design of a long-distance network involves basically three considerations: (1) routing scheme given inlet and outlet points and their traffic intensities, (2) switching scheme and associated signaling, and (3) transmission plan. In the design each criterion will interact with the others. In addition, the system designer must specify type of traffic, lost-call criteria or grade of service, a survivability criterion, forecast growth, and quality of service. The trade-off of all these factors with "economy" is probably the most vital part of initial planning and downstream system design.

Consider transcontinental communications in the United States. Service is now available for people in New York to talk to people in San Francisco. From the history of this service, we have some idea of how many people wish to talk, how often, and for how long. These factors are embodied in traffic intensity and calling rate. There are also other cities on the West Coast to be served and other cities on the East Coast. In addition, there are existing traffic nodes at intermediate points such as Chicago and St. Louis. An obvious approach would be to concentrate all traffic into one transcontinental route with drops and inserts at intermediate points.

Again, we must point out that switching enhances the transmission facilities. From an economic point of view, it would be desirable to make transmission facilities (carrier, radio, and cable systems) adaptive to traffic load. These facilities taken alone are inflexible. The property of adaptivity, even when the transmission potential for it has been predesigned through redundancy, cannot be exercised except through the mechanism of switching in some form. It is switching that makes transmission adaptive.

The following requirements for switching ameliorate the weaknesses of transmission systems: Concentrate light, discretely offered traffic from a multiplicity

Telecommunication System Engineering, by Roger L. Freeman
ISBN 0-471-45133-9   Copyright © 2004 Roger L. Freeman

of sources and thus enhance the utilization factor of transmission trunks; select and make connections to a statistically described distribution of destinations per source; and restore connections interrupted by internal or external disturbances, thus improving reliabilities (and survivability) from the levels on the order of 90% to 99% to levels on the order of 99% to 99.9% or better. Switching cannot carry out this task alone. Constraints have to be iterated or fed back to the transmission systems, even to the local area. The transmission system must not excessively degrade the signal to be transported; it must meet a reliability constraint expressed in MTBF (mean time between failures) and availability and must have an alternative route scheme in case of facility loss, whether switching node or trunk route. This latter may be termed *survivability* and is only partially related to overflow (e.g., alternative routing).

The single transcontinental main traffic route in the United States suggested earlier has the drawback of being highly vulnerable. Its level of survivability is poor. At least one other route would be required. Then why not route that one south to pick up drops and inserts? Reducing the concentration in the one route would result in a savings. Capital, of course, would be required for the second route. We could examine third and fourth routes to improve reliability–survivability and reduce long feeders for concentration at the expense of less centralization. In fact, with overflow, one to the other, dimensioning can be reduced without reduction of overall grade of service.

2   THE DESIGN PROBLEM

The same factors enter into long-distance network design as were discussed for the local area in Chapter 2. The first step is exchange placement. Here we follow North American practice and call an exchange in the long-distance network a "toll exchange."* Rather than base the placement decision on subscriber density and their calling rates, the basic criterion is economy, the most cost-effective optimum. Toll-center placement is discussed in Section 5 of this chapter.

Having selected toll-center locations, the design procedure is to construct the familiar traffic matrix, where cost ratio studies are carried out to determine whether routing will be direct or tandem. The direct routes are called high-usage (HU) routes. The tendency is to use tandem (or "transit" exchanges) working and direct (HU) routes with overflow. The economic decision arises to balance switching against transmission (considering our arguments in Section 1). Compare local versus long-distance networks:

	Switching Cost per Circuit	Transmission Cost per Circuit	Favored Network
Local network	Relatively high	Low	Mesh
Toll network	Relatively low	High	Star

* In European and CCITT terminology it is a transit exchange, and the toll network is often called the *transit network*.

In the past, for the long-distance network we could nearly always assume a hierarchical structure with three, four, or even five levels in the hierarchy. Ideally, the highest levels would be connected in mesh for survivability [1].

Our thinking has changed. We are moving away from the hierarchical concept (albeit slowly) to one using more direct routes. Vestiges of a hierarchy remain in the United States for the general network structure. However, we must also take into account structural considerations with the breakup of the Bell System and the formation of the Regional Bell Operating Companies (RBOCs). The RBOCs with their local exchange carriers (LECs) formed the lower level of this artificial hierarchy, and the interexchange (long-distance) carriers (IXCs) formed an upper level.

In local network design, particularly in metropolitan areas, we could often assume a mesh connection. There might be an exceptional case where tandem working would prove to be economical, where traffic flows were 20 erlangs or less. Because of inherent (comparative) low traffic flows in the long-distance network, a star topology can be assumed at the outset, and we would then proceed to determine cases where direct links may be justified with or without alternative routing.

3 LINK LIMITATION

ITU-T Organization recommends that there be no more than 12 links in tandem on any international connection, except for very large countries where 14 links may be acceptable. On an international connection, the 12 links in tandem are broken down into three groups, each 4 links in tandem as follows:

1. National connection of country originating call
2. International portion
3. National connection of country terminating the call.

This is shown diagrammatically in Figure 6.1, taken from ITU-T Rec. G.101, "Transmission Plan." Also consult ITU-T Rec. E.171 [3].

We define a *link* in this context as the connectivity from one exchange to an adjacent exchange serving the international connection. It does _not_ include that connectivity in the local area from a serving exchange to the calling or called subscriber.

ITU-T places this link limitation in the transmission plan to ensure some minimum transmission quality and to provide efficient operation of signaling, end-to-end. For national network planning we assume that there are no more than four links in tandem; we also assume that there are no more than four links in tandem on the international connectivity.

4 INTERNATIONAL NETWORK

Before 1980 the CCITT routing plan was based on a network of hierarchical structure with descending levels called CT1, CT2, CT3, and CTX. Since 1980

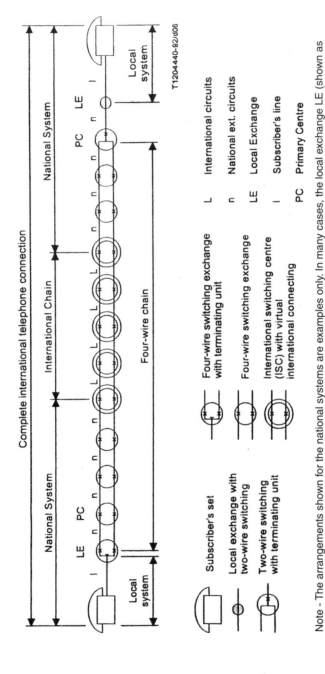

Figure 6.1. An international connection to illustrate the nomenclature adopted and the maximum number of links in tandem for an international connection. From ITU-T Rec.G.101, Figure 6/G. 101, page 12 [2].

Note - The arrangements shown for the national systems are examples only. In many cases, the local exchange LE (shown as analogue in the figure) is digital.

Complete international telephone connection

National System | International Chain | National System

Local system | Four-wire chain | Local system

LE PC   PC LE

Subscriber's set

Local exchange with two-wire switching

Two-wire switching with **terminating unit**

Four-wire switching exchange with **terminating unit**

Four-wire switching exchange

International switching centre (ISC) with virtual international connecting

L — International circuits

n — National ext. circuits

LE — Local Exchange

l — Subscriber's line

PC — Primary Centre

T1204440-92/d06

the CCITT has made a radical change in its international routing plan. The new plan they adopted might be called a *free routing structure*. It assumes that national administrations (telephone companies) will maintain national hierarchical networks. One reason for this change at that time was the universality of earth satellite connections with their long reach. This allowed international high-usage (HU) trunks which could terminate practically anywhere on the terrestrial earth surface.

The CCITT (now ITU-T) International Telephone Routing Plan is contained in ITU-T Rec. E.171 [3]. Some of its highlights are covered below.

- It is not hierarchical.
- Direct traffic should be routed over final (fully provided) or high usage circuit groups.
- No more than four international circuits in tandem should be involved between originating and terminating ISCs (international switching centers).
- Advantage should be taken of the noncoincidence of international traffic by use of alternative routings to effect circuit economies and provide route diversity.
- The routing of transit switched traffic should be so planned to avoid circular routings ("ring-around-the-rosy").
- When a group consists of both terrestrial and satellite circuits, the choice of routing should be governed by:
  - Total delay of connectivity (<400 ms) including both processing delay and propagation delay (see ITU-T Rec. G.114 [7]).
  - The number of satellite circuits in the overall connection. No more than one GEO-link (consists of one up and one down link).
  - Select the circuit that provides the overall better transmission quality.
- Both originating and transit traffic should be routed over the minimum number of international circuits in tandem unless this is in conflict with one of the above-mentioned guidelines.

Source: From Ref. 3.

5   EXCHANGE LOCATION (TOLL/LONG-DISTANCE NETWORK)

5.1   Toll Areas

A country is divided into toll areas. The size of a toll area is impacted by a number of interacting disciplines, such as subscriber density at the end of a forecast period (say 10 years for the argument) for the candidate area; this will impact numbering and tariff areas. Let's say at the outset that tariff areas and toll areas will be the same.

In Chapter 3, Section 16.3.4, we dealt with the size of numbering areas (area codes). A top limit of no greater than 70,000 km^2 nor less than 100,000 subscribers was recommended. That's a very large area implying a low subscriber density. We are also tied to a 3-digit exchange code. In theory, 3 digits allow 1000 exchanges if there were no blocked numbers. With blocked numbers and some reserve, 200 exchanges may be more reasonable. One rough rule of thumb is that a tariff area and thus a toll area have no more than a 50-km diameter. In rural regions, toll areas/tariff areas may be considerably larger. In densely populated urban areas with heavy business concentration, these areas may be smaller.

For the sake of argument, a country is divided into toll areas 50 km in diameter. A toll (long-distance) exchange is tentatively placed in each toll area. Now the system designer should examine adjacent pairs of toll exchanges to determine whether one exchange could serve both areas.

The next step is to examine assignments of toll exchanges regarding numbering. This aspect was covered in discussions in Chapters 3 and 4. Here we saw the impact of numbering on routing a call and on accounting equipment (metering). Numbering may entail consideration of more than one toll exchange in geographically large toll/tariff areas.

Another aspect is the maximum size of a toll exchange. In the following discussion we only consider toll traffic. Of course all this traffic would be connected through the toll-connecting exchange (first toll exchange). Allow 0.003 erlangs per subscriber line; thus a 4000-line toll exchange could serve just under a million subscribers maximum. The exchange capacity should be dimensioned (possibly wired, but not equipped) to the forecast long-distance traffic load 10 years after installation. The exchange location in this case is not very sensitive to traffic. Of course we should make use of the existing infrastructure of telephone plant in the area.

Hierarchy is another essential aspect (see Section 7.2 of this chapter). One important criterion is establishing the number of hierarchical levels in a national network. Today's tendency is to reduce the number of levels. It would seem that we must have at least two levels: local area and toll area. Factors leading to more than two levels are:

- Geographical size
- Telephone density, usually per 100 inhabitants
- Toll traffic trends
- Political factors (e.g., divestiture of Bell System)

The trend toward greater utilization of HU routes may also force the use of less hierarchical levels. Once the number of hierarchical levels has been established, the number of fan-outs must be considered to establish the number of toll exchanges in the network. Fan-outs of six and eight are desirable. Thus with a two-stage hierarchy with a fan-out of six, then to eight, there would be 48 of the lowest-level long-distance (toll) exchanges. A three-level hierarchy, using the same rules, would have $48 \times 8 = 384$, a formidable number.

As can be seen, there are many choices open to the system engineer to establish the route-plan hierarchy. For example, if there are 24 long-distance exchanges in an area, the network will initially be a star connection, either three-stage,

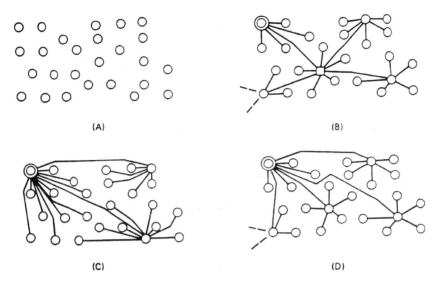

Figure 6.2. Choice of fan-outs: (A) basic pattern; (B) three-stage interconnection; (C) two-stage interconnection with low initial fan-out; (D) two-stage interconnection with high initial fan-out.

two-stage with low (initial) fan-out, or two-stage with high initial fan-out. Here "fan-out" refers to the highest level and works downward. Figure 6.2 illustrates these principles. Figure 6.2A shows an area with 26 exchanges and part of a larger area with perhaps 100 or more exchanges, with the principal city in the upper left-hand corner. Three choices of fan-outs are shown. Figure 6.2B is a three-level hierarchy with a four-to-five fan-out at each stage. For a two-level hierarchy, two possibilities are suggested. Figure 6.2C has low initial fan-out, and Figure 6.2D has a high one. The choice between 6.2C and 6.2D may depend on traffic intensity between nodes or availability of routes. For national networks, the fan-out in Figure 6.2D may be most economical because traffic is brought to a common point more quickly, leaving the individual branches to be least traffic efficient [4]. (Also consult Refs. 9–11.)

6 NETWORK DESIGN PROCEDURES

The attempt to attain a final design of an optimum national network is a major "cut-and-try" process. It lends itself well to computer techniques. Large telephone companies or administrations have such programs "in-house." In other cases, private contractors such as Telcordia should be investigated.

Simple logic demands that the design must first take into account the existing network. Major changes in the network require a large expenditure. The network also represents an existing investment that should be amortized over time. Elements of the system, such as switches, are of varying age; some switches used to remain in service almost 40 years. Not so today. Technology advances are galloping along. Ten years' age of a switch might be the very outside. Even a 5-year-old switch may have to be replaced because of obsolescence.

The good news is that signaling on the national and international networks has been standardized on CCITT Signaling System No. 7 (see Chapter 16). The bad news is that each country or administration has its own national variant of SS No.7.

To simplify the design process, visualize a group of local areas. That is, geographical and demographic areas of interest in which a national network is to be designed is made up of contiguous local areas (Chapter 2). There are now three bases to work from:

1. There are existing local areas, each of which has a toll exchange.
2. There is one or more ISCs placed at the top of the network hierarchy.
3. There will be no more than four links in tandem (Section 3) on any connection to reach an ISC.

Point 1 may be redefined as a toll area made up of a grouping of local areas probably coinciding with a numbering (plan) area. This is illustrated in a very simplified manner in Figure 6.3, where T, in European (CCITT) terminology, is a primary center, or a class 4 exchange in North American terminology. Center T, of course, is a tandem exchange with a fan out of four; these are four local exchanges, *A, B, C,* and *D* homing on T. The entire national geographic area will be made up of small segments, as shown in Figure 6.3, and each may be represented by a single exchange such as T.

The next step is to examine traffic flows to and from (originating and terminating at) each T. This information is organized and tabulated on a traffic matrix. A simplified example is shown in Table 6.1. Care must be taken in the preparation and subsequent use of such a table. The convention used here is that values are read *from* the exchange in the left-hand column *to* the exchange in the top row. For example, traffic from exchange 1 to exchange 5 is 23 erlangs, and traffic from exchange 5 to exchange 1 is 25 erlangs. It is often useful to set up a companion matrix of distances between exchange pairs. The matrix (Table 6.1) immediately offers candidates for high-usage routes. Nonetheless, this step is carried out after a basic hierarchical structure is established.

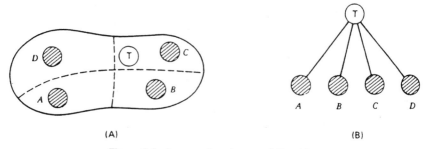

(A) (B)

Figure 6.3. Areas and exchange relationships.

TABLE 6.1   Toll Traffic Matrix (Sample) (in Erlangs)

From Exchange	To Exchange									
	1	2	3	4	5	6	7	8	9	10
1		57	39	73	23	60	17	21	23	5
2	62		19	30	18	26	25	2	9	6
3	42	18		28	17	31	19	8	10	12
4	70	31	23		6	7	5	8	4	3
5	25	19	32	5		22	19	31	13	50
6	62	23	19	8	20		30	27	19	27
7	21	30	17	40	16	32		15	16	17
8	21	5	12	3	25	19	17		18	29
9	25	10	9	1	16	22	18	19		19
10	7	8	7	2	47	25	13	30	17	

We recommend that a hierarchical structure be established at the outset, being fully aware that the structure may be modified or even done away with entirely in the future as dynamic routing disciplines are incorporated (see Section 7). At the top of a country's hierarchy is (are) the international switching center(s). The next level down, as a minimum, would be the long-distance network, thence down to a local network consisting of local serving exchanges and tandem exchanges. The long-distance network might only be divided into two layers.

It will be noted that there is some redundancy between this section and Section 9 of Chapter 1. This chapter dealt with the design of the local area, its network architecture, and some possible routing schemes. This was a "bottom-up" exercise starting with a subscriber subset. In the present chapter, network design is "top-down." The link limit is immediately enforced starting at the international switching center (ISC) and heading downwards to the local serving exchange. No more than four links in tandem are permitted.

It would seem that we are forced to apportion two links to the local network leaving only two more links to reach the international switching center. This certainly is true when a tandem exchange is involved in a connectivity. On the other hand, if a connectivity had a BH flow greater than 20 erlangs from a certain local serving exchange requiring long-distance connectivity, then using the United States as an example, there would be a link from the local serving exchange to the POP (point-of-presence).* In this case, three links would be allocated to the national long-distance network including that connecting to the POP. In other words the POP would serve as the toll-connecting exchange. Then there would be a link from the POP to toll/transit exchange *A*, thence from *A* to toll/transit exchange *B*, and thence from *B* to the ISC. We have one link for the local network plus three links in the toll network to reach an ISC. Using a high-usage (HU) route, perhaps only two links would be required to reach the

---

* POP stands for point of presence, a strictly U.S. phenomenon. This is a point where the local serving area met the long-distance network. The actual interface occurred here.

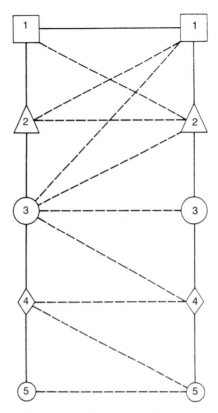

Figure 6.4. A typical hierarchical network. The example illustrated here is the North American network circa 1990. Dashed lines show high-usage trunks. Note how the two highest levels are connected in mesh.

ISC—that is, from local serving exchange to POP, thence directly to the ISC. Again, >20 erlangs of traffic are required.

The earlier ATT network in the United States (see Figure 6.4) was a five-level hierarchy. At first blush, it seems we have broken our four-link rule. But the ATT hierarchy included the final level, from local serving switch to subscriber. The four-link limit specifically did not include that last link. One could say that the present U.S. network consisted of two minihierarchies: from the local subscriber to the POP and from the POP to the ISC [14, 15].

CCITT (ITU-T) defines a routing structure as hierarchical for all traffic streams where all calls offered to a given route, at a specific node, overflow to the same set of routes irrespective of the routes already tested. The routes in the set will always be tested in the same sequence although some routes may not be available for certain call types. The last choice route is the final route in the sense that no traffic streams using this route may overflow further.

A routing structure is nonhierarchical if it violates the above-mentioned definition (e.g., mutual overflow between circuit groups originating at the same exchange).

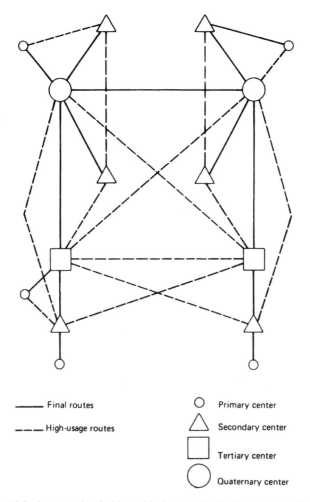

——— Final routes   ○ Primary center

— — — High-usage routes   △ Secondary center

▢ Tertiary center

⬭ Quaternary center

Figure 6.5. An example of a hierarchical network with alternative routing [2].

The outline of another five-level hierarchy is illustrated in Figure 6.5 with HU routes. Note that the lowest level is not shown in the figure, that of the local exchange. HU routes ameliorate the problems of an excessive number of links in tandem on the great majority of completed calls, thereby meeting the intent of CCITT Rec. E.171 or G.101.

Suppose, for example, that a country had four major population centers and could be divided into four areas around each center. Each of the four major population centers would have a tertiary center assigned, one of which would be an ISC. Each tertiary center would have one or several secondary centers homing to it, and a number of primary centers would home on the secondary centers. This procedure is illustrated in Figure 6.5 and is represented schematically in Figure 6.6, thus establishing a hierarchy and setting out the final routes. In this

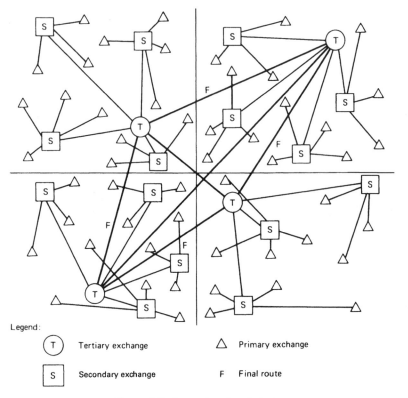

Legend:

(T) Tertiary exchange   △ Primary exchange

[S] Secondary exchange   F Final route

Figure 6.6. A sample network design.

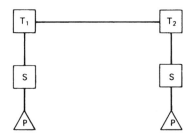

Figure 6.7. Hierarchical representation showing final routes.

case, one of the tertiary exchanges would be an ISC. We define a final route as a route from which no traffic can overflow to an alternative (alternate) route. It is a route that connects an exchange immediately above or below it in the network hierarchy, and there is also connection of the two exchanges at the top level of the network. "Final routes" are said to make up the "backbone" of a network (see Figures 6.7 and 6.8). Calls that are offered to the backbone but cannot be completed are lost calls.

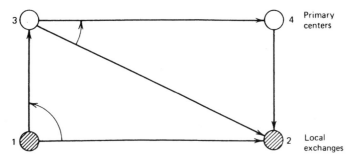

Figure 6.8. Hierarchical network segment.

A *high-usage* (HU) route is defined as any route that is not a final route; it may connect exchanges at a level of the network hierarchy *other than* the top level, such as between tertiary and quaternary centers in Figure 6.5. It may also be a route between exchanges on different hierarchical levels when the lower-level exchange does not home on the higher level. A *direct route* is a special type of HU route connecting exchanges of the lowest rank in the hierarchy. Figure 6.7 illustrates these two definitions. High-usage routes are between exchanges 1 and 2 and exchanges 3 and 2. The direct route is also between exchanges 1 and 2, with exchanges 1 and 2 being the lowest level in the hierarchy.

Before final dimensioning can be carried out, a grade of service (GOS) p must be established, usually no greater than 1% per link on the final route during the busy hour (BH). If the maximum number of links on an international call is established at 12, the very worst GOS would be $12 \times 1\%$, or 12%. However, on most calls, the overall grade of service would be significantly better. Many system planners set the GOS during the BH on a HU route at $p = 0.001$. It is interesting how this ratchets up cost.

Following the $p = 1\%$ rule and with three links in tandem, the connectivity would have a 3% GOS during the busy hour, and with four links in tandem, 4%. The use of HU connections reduces tandem operation and improves overall GOS.

The next step is to lay out HU routes. The conventional method we used was to take data from a traffic matrix, such as that shown in Table 6.1. The concept lends itself to programs that would work well on a PC with just nominal power. One method of dimensioning with overflow (alternative routing) was discussed in Section 8.1 of Chapter 1. The subject is continued in Section 9 of that chapter. Reference should also be made to Section 10 of Chapter 2. Trunks are costly. The exercise is to optimize the number of trunks and maintain a given grade of service.

7   TRAFFIC ROUTING IN THE NATIONAL NETWORK

7.1   Objective of Routing

The objective of routing is to establish a successful connection between any two exchanges in the network. The function of traffic routing is the selection of a particular circuit group, for a given call attempt or traffic stream, at an exchange in the network. The choice of a circuit group may be affected by information on the availability of downstream elements of the network on a quasi-real-time basis.

7.2   Network Topology

A network comprises a number of nodes (i.e., switching centers) interconnected by circuit groups. There may be several direct circuit groups between a pair of nodes and these may be one-way or both-way (two-way).* A simplified illustration of this idea is shown in Figure 6.9.

Remember that a direct route consists of one or more circuit groups connecting adjacent nodes. We define an indirect route as a series of circuit groups connecting two nodes providing an end-to-end connection via other nodes.

7.2.1   Network Architecture.   Under many circumstances with a national network we would develop a hierarchy of switching centers (e.g., local area, regional trunk, and international) with each level of the hierarchy performing different functions. As was mentioned earlier, there is no hierarchy for international switching centers (ISCs). Here telecommunication companies and administrations are free to determine the most suitable utilization of their individual ISCs. (Refer to Section 4, ITU-T Rec. E.171 [3]. Also consult Refs. 6 and 18.)

7.3   Routing Scheme

A routing scheme defines how a set of routes is made available for calls between a pair of nodes. There are *fixed routing schemes* and *dynamic routing schemes.* Of course, for a fixed routing scheme, the routing pattern is always the same. For a dynamic scheme the set of routes in the routing pattern varies.

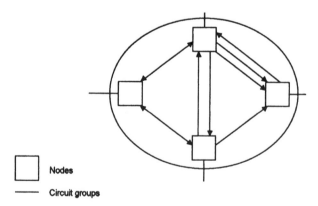

☐ **Nodes**

— **Circuit groups**

Figure 6.9. A simplified network with circuit groups connecting pairs of nodes with one-way and both-way (two-way) working.

---

* One-way and both-way circuits have also been discussed in Chapter 1, Sections 9 and 11.

7.3.1 Fixed Routing Scheme. Routing patterns in a network may be fixed, in that changes in route choices for a given type of call attempt require manual intervention. Changes then represent a "permanent change" to the routing scheme (e.g., the introduction of new routes require a change to a fixed routing scheme).

7.3.2 Dynamic Routing Schemes. Routing schemes may also incorporate frequent automatic variations. Such changes may be time-dependent, state-dependent, and/or event-dependent. The updating of routing patterns may take place periodically or aperiodically, predetermined, depending on the state of the network or depending on whether calls succeed or fail.

7.3.2.1 Time-Dependent Routing. With this type of routing scheme, routing patterns are altered at fixed times during the day (or week) to allow changing traffic demands to be provided for. It is important to note that these changes are preplanned and are implemented consistently over a long time period.

7.3.2.2 State-Dependent Routing. This is a routing scheme where routing patterns vary automatically according to the state of the network. This is *adaptive* routing. In order to support this type of routing, it is necessary to collect information about the status of the network. For example, each exchange compiles records of successful calls or outgoing trunk group occupancies. This information is then distributed through the network to other exchanges or passed to a centralized database. Based on this network status information, routing decisions are made either in each exchange or at a central processor serving all exchanges. This concept is illustrated in Figure 6.10.

7.3.2.3 Event-Dependent Routing. In this case, routing patterns are updated locally on the basis of whether calls succeed or fail on a given route choice. Each exchange has a list of choices, and the updating favors those choices which succeed and discourage those which suffer congestion.

7.4 Route Selection

We define *route selection* as the action to actually select a definite route for a specific call. The selection can be *sequential* or *nonsequential*. Sequential route

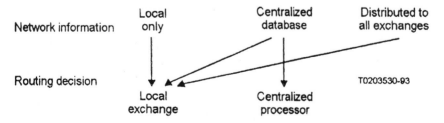

Figure 6.10. Concept of state-dependent routing. From CCITT Rec. E.170, page 3 [5].

selection is where the routes in a set are always tested in sequence and the first available route is selected. For the case of nonsequential routing, the routes in a set are tested in no specific order.

The decision to select a route can be based on the state of the outgoing circuit group or the states of the series of circuit groups in the route. In either case, it can also be based on the incoming path of entry, class of service, or type of call to be routes. One example of the above is selective trunk reservation.

7.5 Call Control Procedures

Call control procedures define the entire set of interactive signals necessary to establish, maintain, and release connection between exchanges. Two types of call control procedures are described below.

7.5.1 Progressive Call Control. Progressive call control uses link-by-link signaling (see Chapter 4) to pass supervisory controls sequentially from one exchange to the next. This type of call control can be reversible or irreversible. In the irreversible case, call control is always passed downstream toward the destination exchange. Call control is reversible when it can be passed backwards (maximum one node), toward the originating exchange, using automatic rerouting or crankback possibilities.

7.5.2 Originating Call Control. Originating call control requires that the originating exchange maintain control of the call setup until a connection between originating and terminating exchanges has been completed.

7.6 Applications

7.6.1 Automatic Alternative Routing. One type of progressive (irreversible) routing is *automatic alternative routing* (AAR). When an exchange has the option of using more than one route to the next exchange, an alternative routing scheme can be used. There are two principal types of this routing available:

1. When there is a choice of direct circuit groups between the two exchanges.
2. When there is a choice of direct and indirect routes between the two exchanges.

Alternative routing takes place when all appropriate circuits in a group are busy. Several circuit groups may be tested sequentially. The circuit order is fixed or is time-dependent.

7.6.2 Automatic Rerouting (Crankback). Automatic rerouting is a routing facility enabling connection of call attempts encountering congestion during the initial call setup phase. Thus, if a signal indicating congestion is received from

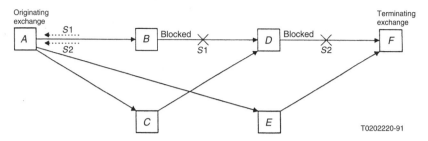

Note - Blocking from *B* to *D* activates signal *S*1 to *A*. Blocking from *D* to *F* activates signal *S*2 to *A*.

Figure 6.11. An example of automatic rerouting. From Figure 4/E.170, page 5, CCITT Rec. E.170 [5].

exchange *B*, subsequent to the seizure of an outgoing trunk from exchange *A*, the call can be rerouted at exchange *A*. See Figure 6.11.

However, it is possible to improve the situation through the use of different signals indicating congestion, *S*1 and *S*2. *S*1 indicates that congestion has occurred on outgoing trunks from exchange *B*. *S*2 indicates that congestion has occurred further downstream—for example, on outgoing trunks from exchange *D*. The action to be taken at exchange *A* is subject to bilateral agreement. In the example given in Figure 6.11, a call from exchange *A* to exchange *D* is routed via exchange *C* because the circuit group *B–D* is congested as shown by the *S*1 indicator. A call from exchange *A* to exchange *F* is routed via exchange *E* because the circuit group *D–F* is congested (*S*2 indicator).

Telephone companies or administrations may wish to consider an increase in signaling load and the number of call setup operations resulting from the use of these signals. If the increase is unacceptable, companies/administrations may restrict the number of reroutings or limit the signaling capability to fewer exchanges. Finally, care should be taken to avoid circular routings which return the call to the point at which blocking previously occurred during the call setup.

7.6.3 Load Sharing. All routing schemes result in the sharing of traffic load between network elements. However, routing schemes can be developed to ensure that call attempts are offered to route choices according to a preplanned distribution.

Figure 6.12 shows this application of load sharing which can be made available as a software function of SPC (stored program control*) exchanges. The system works by distributing the call attempts to a particular destination in a fixed ratio between specified outgoing routing patterns.

7.6.4 Dynamic Routing Examples

7.6.4.1 Example State-Dependent Routing. A centralized routing processor is employed to select optimum routing patterns on the basis of actual occupancy

---

* An SPC exchange is a computer-controlled exchanges. All modern exchanges are SPC exchanges.

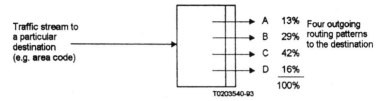

Note - Each outgoing routing pattern (A, B, C, D) may include alternative routing options.

Figure 6.12. An example of a load-sharing application. From Figure 5/E.170, page 5 [5].

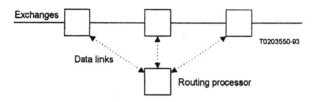

Figure 6.13. State-dependent routing. From Figure 6/E.170, CCITT Rec. E.170, page 6 [5].

levels of the circuit groups and exchanges in the network which are monitored on a periodic basis (e.g., 10 s). See Figure 6.13. In addition, qualitative traffic parameters may also be taken into consideration in the determination of the optimal routing pattern. This routing technique inherently incorporates fundamental principles of network management (Chapter 19) in determining routing patterns. These include:

- Avoiding occupied circuit groups.
- Not using overloaded exchanges for transit.
- In overload circumstances, restriction of routing direction connections.

7.6.4.2 Example of Time-Dependent Routing. For each originating and terminating exchange pair, a particular route pattern is planned depending on the time of day and the day of the week. This concept is illustrated in Figure 6.14. For example, a weekday can be divided into different time periods, with each time period resulting in different route patterns being defined to route traffic streams between the same pair of exchanges.

This type of routing takes advantage of idle circuit capacity in other possible routes between originating and terminating exchanges which may exist due to noncoincident busy hours. Crankback may be used to identify downstream blocking on the second link of each two-link alternative path.

7.6.4.3 Example of Event-Dependent Routing. In a fully connected network, calls between each originating and terminating exchange pair try the direct route with a two-link alternative path selected dynamically. When calls are successfully

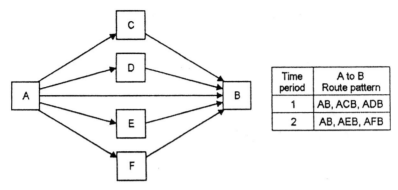

Time period	A to B Route pattern
1	AB, ACB, ADB
2	AB, AEB, AFB

Figure 6.14. Example of time-dependent routing. From Figure 7/E.170, CCITT Rec. E.170, page 6 [5].

routed on a two-link path, that alternative is retained. Otherwise, a new two-link alternative path is selected. This updating, for example, could be at random or weighted by the success of previous calls.

This type of routing scheme routes traffic away from congested links by retaining routing choices where calls are successful. It is simple, adapts quickly to changing traffic patterns, and requires only local information.

Source: Section 7 has been based on CCITT Rec. E.170, 10/92 [5]. Also consult Refs 13–21.

8   TRANSMISSION FACTORS IN LONG-DISTANCE TELEPHONY

8.1   Introduction

The long-distance network is entirely four-wire. As the network is extended, delay becomes more of a problem. Delay has two components:

- Propagation
- Processing time

Not only is delay itself annoying to the listener, it affects the design of data protocols and exacerbates the effects of echo and singing. Total one-way transmission time on a connection is governed by ITU-T Rec. G.114 [7]. See also Refs. 12 & 13.

8.2   Definition of Echo and Singing

8.2.1   *Echo.* As the name implies, echo in telephone systems is the return of a talker's voice. It is most apparent to the talker himself or herself. Secondarily, it can also be an annoyance to the listener. To be an impairment, the returned voice must suffer some noticeable delay. Thus we can say that echo is a reflection of the voice. The cause of echo is impedance mismatches that might be present any place in the electrical telephone connection. Echo is a major annoyance to the telephone user. It affects the talker more than the listener. Two factors determine the degree of annoyance of echo: its loudness and its length of delay.

8.2.2 Singing. Singing is the result of sustained oscillations due to positive feedback in telephone amplifiers or amplifying circuits. Circuits that sing are unusable and promptly overload multiplex equipment, particularly FDM equipment.

Singing may be regarded as echo that is completely out of control. This can occur at the frequency at which the circuit is resonant. Under such conditions the circuit losses at the singing frequency are so low that oscillation will continue, even after cessation of its original pulse.

8.3 Causes of Echo and Singing

Echo and singing can generally be attributed to the mismatch between the balancing network of the hybrid and its two-wire connection associated with the subscriber loop. In older, more conventional networks, two-wire operation was carried to the toll-connecting switch or a tandem exchange just preceding this switch. In this type of network there was at least one two-wire switch between the subscriber and the hybrid. A typical model of the two-wire/four-wire conversion points is shown in Figure 6.15.

We can therefore say that a particular hybrid could have up to one in 10,000 different subscriber lines connected, each with a different impedance—there are long subscriber loops, short subscriber loops, and loops in bad condition electrically.

The mismatch is usually between the two-wire side and the hybrid, where the balancing transformer (B in Figure 6.15) provides the other side of the match. We describe the amount of match and how well we have an impedance match by a term called *return loss*. The higher the return loss value, the better the match.

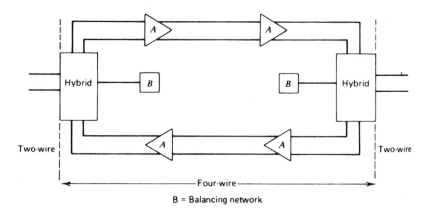

Figure 6.15. Simplified schematic of two-wire/four-wire operation.

We relate return loss, measured in dB, to the impedances of the two-wire line we call N and the balancing network B (in Figure 6.15) by

$$\text{Return loss}_{\text{dB}} = 20 \log_{10}(Z_N + Z_L)/(Z_N - Z_L) \qquad (6.1)$$

If the network perfectly balances (i.e., the impedances at N and L were exactly the same), the $Z_N = Z_L$ and return loss would be infinite.

We use the term *balance return loss* (see ITU-T Rec. G.122 [19]) and classify it as two types:

1. Balance return loss from the point of view of echo.* This is the return loss measured between the frequencies 300 and 3400 Hz.
2. Balance return loss from the point of view of stability. This is the return loss measured between 0 and 4000 Hz.

The band of frequencies most important in terms of echo for the voice channel is that from 300 to 3400 Hz. In the conventional telephone plant where much of the local area is two-wire, the return loss at the hybrid toll–plant interface is usually about 11 dB, with some values of connections dropping to as low as 6 dB. (See ITU-T Recs. G.122 and G.131 [19, 24]). Figure 6.16 shows echo paths in a four-wire circuit.

Echo and singing may be controlled by

- Improved return loss at the term set (hybrid)
- Adding loss on the four-wire side (or on the two-wire side)
- Reducing the gain of the individual four-wire amplifiers

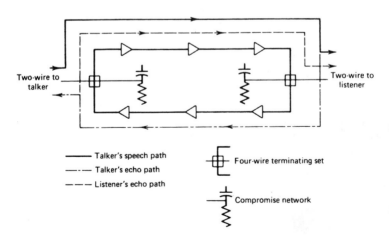

Figure 6.16. Echo paths in a four-wire circuit.

* Called *echo return loss* in North American practice.

The annoyance of echo to a subscriber is very much related to its delay. Delay is a function of the velocity of propagation of the intervening transmission facility. A telephone signal requires considerably more time to traverse 100 km of a voice-pair facility, particularly if it has inductive loading, than it requires to traverse 100 km of a radio facility (i.e., as low as 22,000 km/s for a loaded cable facility and 240,000 km/s for a FDM or TDM facility). Delay is measured in one-way or round-trip propagation time measured in milliseconds. The ITU-T Organization recommends that if the mean round-trip propagation time exceeds 50 ms for a particular circuit, an echo canceler should be used. Practice in North America uses 45 ms as a dividing line. In other words, where echo delay is less than 45 or 50 ms, echo can be controlled by adding loss, as we shall briefly describe below.

An echo canceler generates an echo-canceling signal, which is the mirror image of the echo signal, effectively canceling it out [12, 13].

8.4 Transmission Design to Control Echo and Singing

As stated previously, echo is an annoyance to the subscriber. Figure 6.17 relates echo path delay to echo path loss. The curve in Figure 6.17 traces a group of points at which the average subscriber will tolerate echo as a function of its delay. Remember that the longer the return signal is delayed, the more annoying it is to the telephone talker (i.e., the more the echo signal must be attenuated). For example, if the echo path delay on a particular circuit is 20 ms, an 11-dB loss must be inserted to make echo tolerable to the talker. We should realize that if we insert an 11-dB loss in the circuit, the loudness rating value degrades an equivalent amount. Not a good outcome, we would say.

To control singing, all four-wire paths must have some loss. Once these paths go into a gain condition, and we refer to overall circuit gain, positive feedback

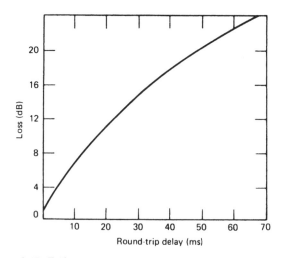

Figure 6.17. Talker echo tolerance for average telephone users [16].

results and the amplifiers begin to oscillate or "sing." North American practice calls for a 4-dB loss on all four-wire circuits to ensure against singing.

8.4.1 Echo Annoyance from the Point of View of the ITU-T Organization.

The degree of annoyance of talker echo depends both on the amount of delay and on the level difference between the voice and the echo signals. ITU-T characterizes this level difference by the measure of *talker echo loudness rating* (TELR). TELR is described in ITU-T Rec. G.122. It should be noted that the delay depends on the physical distance of the connection, the velocity of propagation of the medium (media), and the digital processing time of the equipment that is employed in the connection.

Figure 6.18 shows the minimum requirements on TELR as a function of the mean one-way transmission time T. In general, the "acceptable" curve is the one that ITU-T Rec. G.131 recommends to follow. Only in exceptional cases should values for the "limiting case" in the figure be allowed.

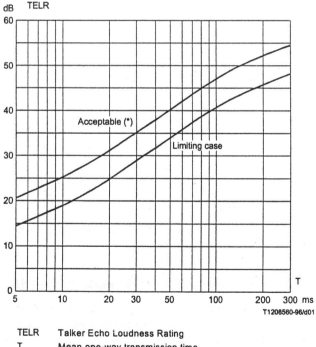

TELR	Talker Echo Loudness Rating
T	Mean one-way transmission time
(*)	The "Acceptable" curve is equivalent to the curve with "1%" probability of encountering objectionable echo.

Figure 6.18. Talker echo tolerance curves. Note that these curves are only valid if sidetone is present in the talker's subset. From Figure 1/G.131, page 2 [24]. With the absence of sidetone, talker echo is more evident.

8.5   Introduction to Transmission-Loss Engineering

One major aspect of transmission system design for a telephone network is to establish a transmission-loss plan. Such a plan, when implemented, is formulated to accomplish three goals:

1. Control singing (stability).
2. Keep echo levels within limits tolerable to the subscriber.
3. Provide an acceptable overall loudness rating value to the subscriber.

Transmission loss engineering of a telephone network is based on the following three guidelines:

1. A certain minimum loss must be maintained in four-wire circuits to ensure against singing.
2. Up to a certain limit of round-trip delay, echo is controlled by loss.
3. It is desirable to limit these losses as much as possible to improve loudness rating values.

These rules were developed for the "all analog" network or for a hybrid network, partially analog, partially digital. In a modern network, which is usually digital from the subscriber input of the local serving exchange of the calling subscriber to the subscriber line output (hybrid or term set) of the called subscriber, there is effective protection against echo and singing brought about by two factors:

- The circuit is four-wire down to the subscriber side of the local serving exchange where each subscriber line (leaving out concentration) has its own hybrid/term set.
- At the subscriber input to the local serving switch, the format is converted from analog to digital.

Because of the effects of these factors, the insertion of loss in the digital network (as just described) has been greatly simplified.

8.6   Loss Plan for the Evolving Digital Networks (United States)

This is the loss plan developed by ANSI in 1998 [22] applicable for the United States, as well as for other countries and administrations who wish to adopt the plan. We will be dealing with added loss values of 0, 3, or 6 dB. Note that the digital network taken in isolation is zero gain/zero loss network.

8.6.1   Overview of the Loss Plan. For digital connections terminated in analog access lines, the required loss values are dependent on the connection architecture:

- For interLATA or interconnecting network connections, the requirement is 6 dB.
- For intraLATA connections involving different LECs (local exchange carriers), 6 dB is the preferred value although 3 dB may apply to connections not involving a tandem switch.
- For intraLATA connections involving the same LEC, the guidelines are:

$$0-6 \text{ dB (typically 0 dB, 3 dB, or 6 dB)}$$

ANSI adds the following guidelines for intraLATA, same exchange carrier operation.

- For intra-DEO/RSU* connections between metallic access lines, 0–6 dB is recommended for DEOs or RSUs with line lengths up to 12 kilofeet, and 0 dB for DEOs or RSUs with line lengths above 12 kilofeet. The selection of appropriate loss values are administered by the exchange carrier and is not required to be an automatic function of the switch.
- For connections involving an RSU/RDT† or multiple RSUs/RDTs, two considerations apply:
  1. The length of the connected metallic loops.
  2. The value of the round-trip delay between the host and remote.

When considering the length of the connected metallic loops, the desired value of network loss is the same as that for intra-DEO/RSU connections between metallic access lines. The following guidelines can be applied for the selection of the network loss based on the round-trip delay between the host and remote:

(a) 0 dB can be used up to 3.2 ms round-trip delay (approximately 80 miles).

(b) 3 dB can be used up to 8 ms round-trip delay (approximately 360 miles).

(c) 6 dB can be used up to 12 ms round-trip delay (approximately 600 miles).

Network loss, when added to the connection, is added only in the receive path.

To facilitate the use of the same network for both analog and digital services, it is desirable to insert loss, where required, as near the end-user terminal as possible. However, for practical reasons, it may be necessary to administer network loss that is dependent upon the type of connection, at the point of switching nearest the end-user terminal. Loss values that are not dependent upon the type of connection can be inserted at the D/A conversion point, which may be at the DEO, RSU, RDT, IDLC (integrated digital loop carrier), or ONU (optical network unit), or at the last point of switching. (From ANSI T1.508–1998 [22].) (Consult also Refs. 18 and 23.)

---

* DEO is digital end office; RSU is remote switching unit.
† RDT is a remote digital terminal.

REVIEW QUESTIONS

1. What are the three basic underlying considerations in the design of a long-distance (toll) network?

2. What is the fallacy of providing just one high-capacity trunk across the United States to serve all major population centers by means of tributaries off that main trunk?

3. How can the utilization factor of trunks be improved?

4. For long-distance (toll) switching centers, what is the principal factor involved in the placement of such exchanges (differing from local exchange placement substantially)—economy, subscriber density, altitude, or what?

5. How are the highest levels of a national hierarchical network connected, and why is this approach used?

6. On a long-distance (toll) connection, why must the number of links in tandem be limited?

7. Describe in one short sentence the current approach to structure of the international network as recommended by ITU-T.

8. What type of routing is used on the majority of international connections?

9. Why do we limit the number of satellites used on an international full-duplex speech telephone connection?

10. Name three principal factors used in deciding how many and where toll (long-distance) exchanges will be located in a given geographic area.

11. Discuss the impact of fan outs on the number of hierarchical levels in a national network.

12. Name the three principal bases required at the outset for the design of a toll (long-distance) network.

13. In the design of a long-distance (toll) network, once the hierarchical levels have been established, what is assembled next?

14. Define a final route.

15. Define high-usage (HU) route and direct route.

16. A grade of service no greater than_____% per link is recommended on a final route.

17. When assembling a traffic matrix, at the ends of what forecast period are the traffic intensities valid for?

18. Define a network hierarchy from the point of view of a routing structure.

19. There are two generic types of routing schemes. What are they?

20. Name the three different types of dynamic routing and explain each type in one sentence.

21. Compare progressive call control procedures with originating call control.

22. What is *crankback*?

23. Give an example of state-dependent routing.

24. One advantage of alternate (alternative) routing is economy. What is the second advantage?

25. Dynamic routing technique (DRT) differs from conventional routing operations in two important areas. What are those two areas?

26. What is the principal cause of echo in the telephone network?

27. What causes singing in the telephone network?

28. Differentiate balance return loss and echo return loss.

29. What is the return loss of a two-wire terminal of a hybrid where the subscriber loop has a measured impedance of 300 Ω and the balancing network is adjusted for a 900-Ω impedance?

30. How can we control echo? Give the two principal ways of doing it.

31. On the loss plan for the United States for the digital network, how much loss is inserted at each end of a 600-mile (960-km) circuit?

REFERENCES

1. *National Networks for the Automatic Service*, ITU, Geneva, 1964.
2. *CCITT Recommendations*, Blue Books, Fascicle VI.1, "General Recommendations on Telephone Switching and Signaling," CCITT Recs. Q.1–Q.118 bis (Study Group XI), IXth Plenary Assembly, Melbourne, 1988.
3. *International Routing Plan*, ITU-T Rec. E.171, ITU, Geneva, 1993.
4. *Telecommunication Planning* (limited circulation), ITT Laboratories, Madrid, 1973 (in particular, Section 2, "Networks").
5. *Traffic Routing*, CCITT Rec. E.170, ITU, Geneva, October 1992.
6. G. R. Ash, "Design and Control of Networks with Dynamic Nonhierarchical Routing," *IEEE Communications Magazine*, IEEE, New York, 1990.
7. *One-Way Transmission Time*, ITU-T Rec. G.114, ITU, Geneva, May 2000.
8. J. E. Flood, *Telecommunication Networks*, IEE Series, London, 1974.
9. *General Network Planning*, CCITT, ITU, Geneva, 1983.
10. "Optimization of Telephone Networks with Hierarchical Structure" (computer program), ITT Laboratories, Madrid, 1973.
11. "Optimization of Telephone Trunking Networks with Alternate Routing" (computer program), ITT Laboratories, Madrid, 1973.

12. R. L. Freeman, *Telecommunication Transmission Handbook*, 4th ed., John Wiley & Sons, New York, 1998.

13. F. T. Andrews and R. W. Hatch, "National Telephone Network Planning in the AT&T," *IEEE Trans. Commun.* (June 1971).

14. R. R. Mina, *Introduction to Teletraffic Engineering*, Telephony Publishing Corporation, Chicago, 1974.

15. *Theory of Telephone Traffic: Tables and Diagrams*, Siemens, Berlin-Munich, Part 1, 1971.

16. M. A. Clement, *Transmission*, reprint from *Telephony* (magazine), Telephony Publishing Corporation, Chicago, 1969.

17. *Transmission Systems for Communications*, 5th ed., Bell Telephone Laboratories, Holmdel, NJ, 1982.

18. "Telcordia Notes on the Networks," Telcordia Special Report, Issue 4, Telcordia, Piscataway, NJ, October 2000.

19. *Influence of National Systems on Stability and Talker Echo in International Connections*, ITU-T Rec. G.122, Helsinki, March, 1993.

20. *Engineering and Operations in the Bell System*, 2nd ed., Bell Telephone Laboratories, Holmdel, NJ, 1984.

21. *ISDN Routing Plan*, CCITT Rec. E.172, October 1992.

22. *Network Performance—Loss Plan for Evolving Digital Networks*, ANSI T1.508–1998 (and supplement), ANSI, New York, 1998.

23. *Telecommunication Transmission Engineering*, 2nd ed., Vols. 1 and 2, AT&T, New York, 1977.

24. *Control of Talker Echo*, ITU-T Rec. G.131, ITU, Geneva, August 1996.

7

THE DESIGN OF LONG-DISTANCE LINKS

1 INTRODUCTION

In Chapter 6 we proposed a methodology for the design of a long-distance network. The network may be defined as a group of switching nodes interconnected by links. We may refer to a link as a transmission highway between switches carrying one or more traffic relations. The link could appear as that in the following diagram, where switches A, B, and C are connected to switches X, Y, and Z over a link as shown. The discussion that follows introduces the essentials of transmission design of such links.

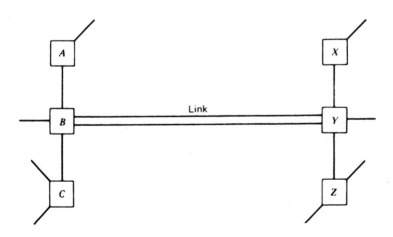

Telecommunication System Engineering, by Roger L. Freeman
ISBN 0-471-45133-9   Copyright © 2004 Roger L. Freeman

2 THE BEARER

British telecommunication engineers are fond of the term "bearer," which is quite descriptive. The bearer is what carries the information signal(s). It could be a pair of wires or two pairs on a four-wire basis, a radio carrier in each direction, a coaxial cable, or a fiber-optic cable. The wire pair could be open wire lines, aerial cable, or buried cable. In the text that follows, it is assumed that the bearer will be transporting some sort of multiplex configuration, probably in a digital format as discussed in Chapter 8.

Modern long-distance links use either radio or fiber-optic cable as the medium of choice. The decision on which one to use is driven by economics more than any other factor. However, capacity can well be another deciding factor. If a requirement for a certain trunk group connectivity exceeds 10,000 equivalent digital voice channels at the end of a forecast period, optical fiber would certainly be the medium of choice at the outset. We do not configure a link for today's traffic requirements. We size the link for probable forecast requirements, or at least engineer a link for that future expected expansion.

Coaxial cable has purposefully been left out of the discussion. It remains, of course, as a transmission medium such as a radio-frequency transmission line for radio systems, and it is very much a contender in cable television systems. Optical fiber is so far superior as transport for telecommunication digital configurations that coaxial cable must be removed from contention. Coaxial cable requires many more active repeaters per unit length than fiber optics. Jitter, a major transmission impairment on digital systems, builds up as a function of the number of repeaters in tandem. Another reason to favor optical fiber is that it needs no equalization, whereas coaxial cable needs equalizers even for modest digital configurations.

3 INTRODUCTION TO RADIO TRANSMISSION

Wire, cable, and fiber are well-behaved transmission media, and they display little variability in performance. The radio medium, on the other hand, displays notable variability in performance. The radio-frequency spectrum is shared with others and requires licensing. Metallic and fiber media need not be shared and do not require licensing (but often require right-of-way).

A major factor in the selection process is information bandwidth. Fiber optics seems to have nearly an infinite bandwidth. Radio systems have very limited information bandwidths. It is for this reason that radio-frequency bands 2 GHz and above are used for PSTN and private network applications. In fact the U.S. Federal Communications Commission requires that users in the 2-GHz band must have systems supporting 96 digital voice channels where bandwidths are still modest. In the 4- and 6-GHz bands, available bandwidths are 500 MHz allocated in 20- and 30-MHz segments for each radio-frequency carrier.

One might ask why use radio in the first place if it has so many drawbacks. Often, it turns out to be less expensive than fiber-optic cable. But there are other factors such as

- No requirement for right-of-way
- Less vulnerable to vandalism
- Not susceptible to "accidental" cutting of the link
- Often more suited to crossing rough terrain
- Often more practical in heavily urbanized areas
- As a backup to fiber-optic cable links

Fiber-optic cable systems provide strong competition with line-of-sight (LOS) microwave, but LOS microwave does have a place and a good market.

Satellite communications is an extension of line-of-sight microwave. It is also feeling the "pinch" of competition from fiber-optic systems. It has two drawbacks. First, of course, is limited information bandwidth. The second is excessive delay when the popular geostationary satellite systems are utilized. It also shares frequency bands with LOS microwave.

One application that continues to show strong growth is very small aperture terminal (VSAT) systems. It is most attractive for private data circuits as an extension of enterprise networks. There are many thousands of these networks now in operation worldwide. All VSAT systems today operate with geostationary orbit satellites (GEO).

Another type of earth-satellite system that seemed to offer great promise was a satellite constellation in a low earth orbit (LEO). One such system was called iridium and was fielded by Motorola. It provided cellular voice operation any place on the globe. The marketing model was fallacious and the system nearly went bankrupt. It operated extremely well and was the first system using active crosslinks. The constellation consisted of 66 satellites. At the date of preparation of this text, it was still operational, under new ownership.

The LEO-type satellite eliminates the notorious delay problem of which GEO satellites suffer. Such systems can be designed for comparatively high elevation angles getting rid of near-horizon conditions where gaseous absorption is maximum. This allows use of millimeter wave frequencies with their broad bandwidths.

4   DESIGN ESSENTIALS FOR LINE-OF-SIGHT MICROWAVE SYSTEMS

4.1   Introduction

Line-of-sight (LOS) microwave provides broadband bearer connectivity over a link or series of links in tandem. We can take advantage of this "line-of-sight" phenomenon at frequencies from 150 MHz and upwards into the millimeter spectrum. Each link can be up to 30 miles (46 km) long or more depending on terrain topology. Some links extend over 100 miles (160 km). A series of LOS links is shown in Figure 7.1. Perhaps the key term here is line-of-sight. It implies that the antenna of the radiolink on one end has to be able to "see" the antenna on

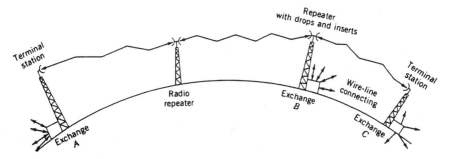

Figure 7.1. A sketch of an LOS microwave radio relay system.

the other end. This may not necessarily be true, but it does give some idea of the problem.

Let us suppose *smooth earth*. This means earth with no mountains or ridges, buildings, or sloping ground of any sort. Here our LOS distance is limited by the horizon. Given an LOS microwave antenna height h_{ft} or h'_{m} above ground surface, the distance d_{mi} or d_{km} to the horizon just where the ray beam will graze the rounded earth surface horizon can be calculated using one of the formulas given below.

To optical horizon ($k = 1$):

$$d = \sqrt{\frac{3h}{2}} \tag{7.1A}$$

and the radio horizon ($k = 4/3$):

$$d = \sqrt{2h} \tag{7.1B}$$

$$d' = 2.9(2h')^{1/2} \tag{7.1C}$$

Equation 7.1A gives the distance to the optical horizon (i.e., the ray travels a straight line) in miles (d) and feet (h), equation 7.1B gives the distance to the radio horizon in miles and h is in feet, and equation 7.1C gives the distance to the radio horizon in kilometers and h is in meters. The concept of optical and radio horizon is shown in Figure 7.2.

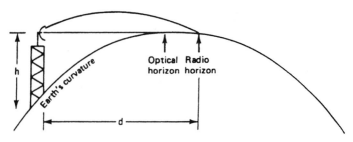

Figure 7.2. Radio and optical horizon (smooth earth).

The distance to the radio horizon varies with the index of refraction of the intervening space. Some designers say it is 4/3 the distance to the radio horizon, with the microwave ray beam being bent toward the earth. However, this generalization may be overly optimistic under certain circumstances.

The design of a microwave LOS link involves five basic steps:

1. Setting performance requirements.
2. Site selection and preparation of a path profile to determine antenna tower heights.
3. Carrying out a path analysis, also called a *link budget*.
4. Running a path/site survey.
5. Test of the system prior to cutover to traffic.

In the following paragraphs we review each of the five steps.

4.2 Setting Performance Requirements

Often a microwave link is part of an extensive system of multiple links in tandem. Thus we must first set system requirements based on the output of the far-end receiver of the several or many links. If the system were analog, the specification would be given for noise in a voice channel; if it were video, a signal-to-noise ratio specification would be provided. In the case we will emphasize here, it will be a bit error rate on a digital bit stream. This is an electrical signal of "1s" and "0s," probably in one of the serial formats described in Chapter 8.

The specification should be based on an existing standard either directly or modified to meet more stringent requirements (e.g, superior error performance per link). One such family of standard may be found in "Telcordia Notes on the Networks" [1], or we may turn to standards issue by ITU-T and/or ITU-R Organizations such as ITU-T Rec. G.821 [2]. ITU-T Rec. G.826 [3], or ITU-R Rec. 594-4 [4]. Note that there is a tendency to tightening standards and express performance in errored seconds (ES), errored second ratio (ESR), and severely errored second ratio (SESR). A bit error rate (BER) on a single link may have a 1×10^{-12} requirement during unfaded conditions. For many digital links, a threshold floor of no worse than 1×10^{-3} is set. This value is related to supervisory signaling where, if further degraded, supervisory signaling is lost and the link drops out (i.e., dial-tone is returned to the subscriber).

4.3 Site Selection and Preparation of a Path Profile

4.3.1 Site Selection. In this step we will select operational sites where we will install and operate radio equipment. After site selection, we will prepare a path profile of each link to determine the heights of radio towers to achieve "line of sight." Sites are selected using large topographical maps. If we are dealing with a long system crossing a distance of hundreds of miles or kilometers, we should minimize the number of sites involved. There will be two terminal sites, where the system begins and ends. Along the way, repeater sites will be required. At some repeater sites, we may have need to drop and insert traffic. Other sites will just be repeaters. This concept is shown in Figure 7.3. The figure shows the drops and inserts of traffic at telephone exchanges. These drop and insert points may just as well be buildings or other facilities in a private/corporate network. There is considerable iteration between site selection and path profile preparation to optimize the route.

In essence, the sites selected for drops and inserts will be points of traffic concentration. There are several trade-offs to be considered:

1. Bringing traffic in by wire or cable rather than adding drop and insert (add–drop) capabilities at relay points.
2. Siting based on propagation advantages (or constraints) only versus colocation with exchange (or corporate facility) (saving money for land and buildings).
3. Method of feeding (feeders)*: by light-route radio, fiber-optic cable, and wire-pair cable.

In gross system design, exchange location (or corporate facility location), particularly with tandem/transit exchanges, must be considered in light of probable radio and cable routes. Another consideration is electromagnetic compatibility (EMC). Midcity repeater-relay or terminal sites have the following advantages:

- Colocation with a local or toll exchange.
- Use of tall buildings as natural towers.

And they have the following disadvantages:

- Wave reflections (multipath) off buildings.
- Electromagnetic compatibility (EMC) problems, particularly from other nearby emitters and industrial emission.
- Low-grade labor market.

Sitings in the country have fewer EMC problems and usually a better labor force (for operators, other operational personnel, and technicians), and right-of-way for cable is easier.

We are now led to propagation constraints. Terminal sites will be in or near heavily populated areas and preferably collocated with a toll exchange. The tops of modern large office buildings, if properly selected, are natural towers. Relay sites are heavily influenced by intermediate terrain. Accessible hilltops or mountain tops are good prospective locations. Draw a line along the path of the desired

---

* Here the word *feeders* refers to feeding the mainline trunk radio system. Feeders could be called *spurs*.

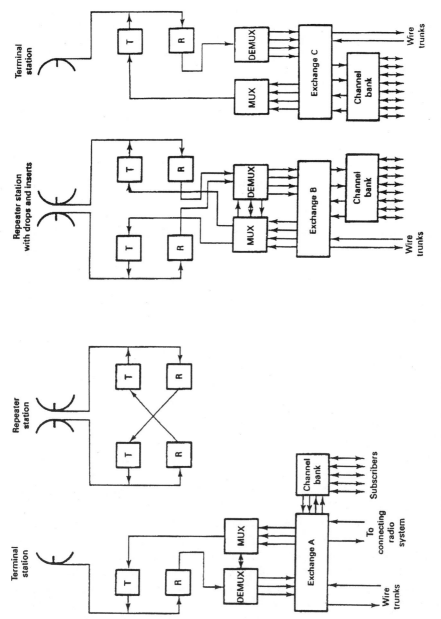

Figure 7.3. Simplified functional block diagram of the LOS microwave system shown in Figure 7.1.

191

route. Sites would zigzag along the line with optical or "radio" separation distances. If tower costs were $300 per foot ($900 per meter), 300-ft or 100-m towers might be the height limit for economic reasons. If hilltop or mountain-top sites are well selected, towers that high may never have to be considered. High towers are the rule over flat country. The higher the tower, the longer the line-of-sight distance. Thus, on a given link, fewer repeaters would be required if towers could be higher. Hence there is a trade-off between tower height and number of repeaters.

The system planner should put a cap on tower height. In my seminars, I tell the attendees that a rule-of-thumb for maximum tower height is 300 ft. Certainly, towers can be built higher. For example, there are broadcast towers in excess of 1000 ft. As a tower goes above 300 ft, the cost of maintaining twist and sway requirements begins to escalate [19]. Tower stiffening is expensive. To further exacerbate the problem, solid back reflector antennas mounted along the tower structure act as sails applying still more twist and sway pressure on the structure.

4.3.2 Calculation of Tower Heights. Assume now that sites along a microwave radio relay route have been carefully selected. The next step in engineering is the determination of tower heights. The objective is to keep the tower height as low as possible and still maintain effective communication. The towers must just be high enough for the radio beam to surmount obstacles in the path. As the discussion proceeds, the term "high *enough*" is carefully defined. What obstacles might there be in the path? To name some, there are terrains such as mountains, ridges, hills, and earth curvature—which is highest at midpath—and buildings, towers, grain elevators, and so on.

All obstacles along the path must be scaled on graph paper in an exercise called *path profiling*. Good topographical maps are required of the region. Ideally, such maps should be 1 : 24,000, although 1 : 62,500 maps are acceptable. A straight line is drawn between the sites in question and then on linear graph paper scaled to 1 in. for 2 miles on the horizontal or 1 cm for 1 or 2 km. Vertical scales depend on the rate of change of elevation along the path. An ideal scaling is 100 ft/in. or 1 cm for 10 m and over hilly country, 1 in. equivalent to 200 ft or 1 cm equivalent to 20 m. In mountainous country the vertical scale may have to be as much as 1 in. equivalent to 1000 ft or 1 cm equivalent to 100 m. Each obstacle encountered must be identified with a letter or number on the horizontal scale. The next step is to establish a point directly on top of each obstacle, giving altitude above mean sea level. The bottom of the chart need not be mean sea level; it may be mean sea level plus so many meters. Once the reference altitude has been established, we must give several additional clearances. If the obstacle is terrain with vegetation, especially trees, a clearance for trees and growth must be established. If no other values are available, use 40 ft (12 m) and 10 ft (3 m), respectively.

To the altitude or height of each obstacle must be added "earth bulge," the number of feet or meters an obstacle is raised higher in elevation (into the path) as a result of earth curvature (EC) or "earth bulge." The amount of earth bulge at any point in the path may be calculated by the formula(s)

$$h = 0.677d_1d_2 \qquad (h \text{ in feet}; d \text{ in miles}) \qquad (7.2A)$$

$$h = 0.078d_1d_2 \qquad (h \text{ in meters}; d \text{ in km}) \qquad (7.2B)$$

where d_1 is the distance from the near end of the hop to the obstacle in question and d_2 is the distance from the far end of the hop to the obstacle in question. Equation 7.2 is for a ray beam that is a straight line (i.e., no bending). Atmospheric refraction may cause the beam to be bent either toward or away from the earth. This bending effect is handled by adding the factor K to equation 7.2, where

$$K = \frac{\text{Effective earth radius}}{\text{True earth radius}}$$

such that

$$h_{\text{ft}} = \frac{0.667d_1d_2}{K} \qquad (d \text{ in miles}) \qquad (7.3A)$$

$$h_{\text{m}} = \frac{0.078d_1d_2}{K} \qquad (d \text{ in km}) \qquad (7.3B)$$

If the factor K is greater than 1, the ray beam is bent toward the earth and the radio horizon is greater than the optical horizon. If K is less than 1, the radio horizon is less than the optical horizon. For general system planning purposes, $K = \frac{4}{3}$ may be used. However, for specific path engineering, K must be selected with care. The value of h or earth curvature corrected for K from equation 7.3 must be added to obstacle height in the path-profile exercise for each obstacle.

Still another factor must be added to obstacle height, namely, Fresnel zone clearance. This factor derives from the electromagnetic wave theory that a wavefront, which our ray beam is, has expanding properties as it travels through space. These expanding properties result in reflections and phase transitions as the wave passes over an obstacle. The outcome is an increase or a decrease in received signal level. The amount of additional clearance over obstacles that must be allowed to avoid problems of the Fresnel phenomenon (diffraction) is expressed in Fresnel zones. The first Fresnel zone radius may be calculated from the following formula:

$$R_{\text{ft}} = 72.1\sqrt{\frac{d_1d_2}{FD}} \qquad (7.4A)$$

where F is the frequency in gigahertz, d_1 is the distance from transmit antenna to obstacle (statute miles), d_2 is the distance from path obstacle to receive antenna (statute miles), and $D = d_1 + d_2$. For metric units,

$$R_{\text{m}} = 17.3\sqrt{\frac{d_1d_2}{FD}} \qquad (7.4B)$$

where F is the frequency in gigahertz and d_1, d_2, and D are the same as in equation 7.4A, but d and D are in kilometers and R in meters.

Previously, a clearance of 0.6 Fresnel zone (0.6 is the value of R in equation 7.4) was considered sufficient. A new rule of thumb is evolving, namely, when $K = \frac{2}{3}$, at least 0.3 Fresnel zone clearance is required, and 1.0 Fresnel zone clearance must be allowed when $K = \frac{4}{3}$. At points near the ends of a path, Fresnel zone clearances should be at least 6 m or 20 ft [5].

The three basic increment factors that must be added to obstacle heights are now available: vegetation height and its growth, earth bulge corrected for K factor, and Fresnel zone clearance. These are marked as indicated previously on our path-profile chart. A straight line is drawn from right to left, just clearing the obstacle points as corrected for the three factors. Another line is then drawn from left to right. A sample profile is shown in Figure 7.4. Some balance is desirable so at one extreme we have a very tall tower and at the other extreme we have a little stubby tower. This is true but for one exception: when a reflection point exists at an inconvenient spot along the path.

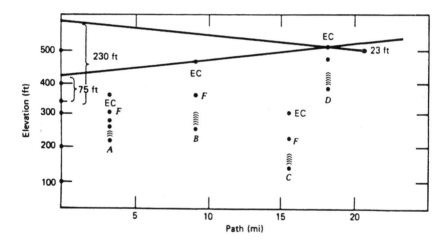

	Obstacle	d_1	d_2	Basic Height (ft)	F (Fresnel) (ft)	EC (ft)	T and G (ft)	Adjusted Total Height (ft)
Tree conditions: 40 + 10 ft growth (T and G)	A	3.5	19.0	220	30	49	50	349
Frequency band: 6 GHz	B	10	12.5	270	41	91.7	50	453.7
Midpath Fresnel (0.6)	C	17	5.5	160	36	68.6	50	314.6
= 42 ft	D	20	2.5	390	25.2	36.6	50	501.8

Figure 7.4. Practice path profile (x in miles, y in feet; assume that $K = 0.9$). *Note:* EC = earth curvature of each bulge.

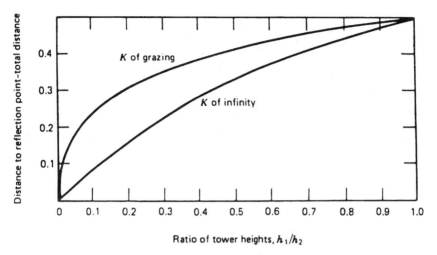

Figure 7.5. Calculation of reflection points.

4.3.3   Reflection Point.   Possible reflection points may be obtained from the profile. The objective is to adjust tower heights such that the reflection point is adjusted to fall on land area where the reflected energy will be broken up and scattered. Bodies of water and other smooth surfaces cause reflections that are undesirable. Figure 7.5 can facilitate calculations for the adjustment of the reflection point. It uses a ratio of tower heights, h_1/h_2, and the shorter tower height is always h_1. The reflection area lies between a K factor of grazing ($K = 1$) and a K factor of infinity. The distance expressed is always from h_1, the shorter tower. The reflection point can be moved by adjusting the ratio h_1/h_2.

For a path that is highly reflective for much of its length, space-diversity operation may minimize the effects of multipath reception.

4.4   Path Analysis or Link Budget

4.4.1   Introduction.   The path analysis (or link budget) is carried out to dimension the link. What is meant here is to establish operating parameters such as transmitter power output, parabolic antenna aperture (diameter), and receiver noise figure, among others. The link is assumed to be digital. Digital formats are described in Chapter 8. The type of modulation and modulation rate (number of transitions per second) are also important parameters.

Table 7.1 shows basic parameters in two columns. The first we call "normal" and would be the most economic; the second column is titled "special," giving improved performance parameters, but at an increased price.

Diversity reception is another option that may wish to be considered. It entails greater expense. The options in Table 7.1 and diversity reception will be addressed further on.

TABLE 7.1 LOS Microwave Basic Equipment Parameters

Parameter	Normal	Special	Comments
Transmitter power	1 W	10 W	500 mW common above 10 GHz
Receiver noise figure	8–12 dB	Down to 1.2 dB	Use of LNA (low noise amplifier)
Antenna	Parabolic 2–12 ft	Same	Antennas over 12 ft not recommended
Modulation	64–128 QAM	Up to 512 QAM	Based on bandwidth/bit rate constraints

4.4.2 Approach.

We can directly relate the desired performance to the receive signal level (RSL) at the first active stage of the far-end receiver and the receiver's noise characteristics. A reference RSL is established.

Next, we calculate the free-space loss between the transmit antenna and receive antenna. This is a function of distance and frequency (i.e., the microwave transmitter operational frequency). We then calculate the EIRP (effective isotropically radiated power) at the transmit antenna. The EIRP is the sum of the transmitter power output, minus transmission line losses plus the antenna gain, all in decibel units.

When we add the EIRP to the free-space loss (in dB), the result is the isotropic* receive level (IRL). When we add the receive antenna gain to the IRL and subtract the receive transmission line losses, we get the receive signal level (RSL).

This relationship of path gains and losses is shown in Figure 7.6.

Path Loss. For all intents and purposes, path loss up to about 10 GHz can be considered as only "free-space loss." To introduce the reader to the problem, consider an isotropic antenna—that is, an antenna that radiates uniformly in all directions. If the isotropic radiator is fed by a transmitted power P_t, it radiates $P_t/4\pi d^2$ (W/m^2) at a distance d, and if a radiator has a gain G_t, the power flow is enhanced by the factor G_t. Finally, the power intercepted by an antenna of effective cross section A (related to the gain by $G_r = 4\pi A/\lambda^2$) is $P_t G_t G_r(\lambda/4\pi d)^2$. The term $(\lambda/4\pi d)^2$ is known as the free-space loss and represents the steady decrease of power flow (in W/m^2) as the wave propagates. From this we can derive the more common formula of free-space path loss, which reduces to

$$L = 96.6 + 20 \log_{10} F + 20 \log_{10} D \tag{7.5}$$

where L is the free-space attenuation between isotropic antennas in dB, F is the frequency in GHz, and D is the path distance in statute miles. In the metric

---

* An isotropic antenna is an antenna that is uniformly omnidirectional with 0 dB gain. It is an imaginary reference antenna. The isotropic receive level is the power level we would expect to achieve at that point using an isotropic antenna.

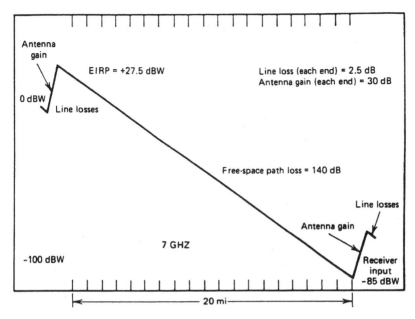

Figure 7.6. LOS microwave link gains and losses (simplified). Transmitter output is 1 watt or 0 dBW.

system we obtain

$$L_{dB} = 92.4 + 20 \log_{10} F_{GHz} + 20 \log_{10} D_{km} \qquad (7.6)$$

Consider the problem from a different aspect. It requires 22 dB to launch a wave to just 1 wavelength (1λ) distant from an antenna. Thus for an antenna emitting +10 dBW, we could expect the signal one wavelength away to be 22 dB down, or −12 dBW. Whenever we double the distance, we incur an additional 6 dB of loss. Hence at 2λ from the +10-dBW radiator, we would find −18 dBW; at 4λ, −24 dBW; 8λ, −30 dBW; and so on. Now suppose that we have an emitter where $F = 1$ GHz. What is the path loss at 1 statute mile?

$$L = 96.6 + 20 \log_{10} 1 + 20 \log_{10} 1 = 96.6 \text{ dB}$$

From rough calculations, the 6-dB relationship is worthwhile and also gives insight in that if we have a 20-mile path and shorten or lengthen it by a mile, our signal level will be affected little.

Calculation of EIRP. Effective isotropically radiated power is calculated by adding decibel units: the transmitter power output (in dBm or dBW), the transmission line losses in dB (a negative value because it is a loss), and the antenna gain in dBi.* Figure 7.7 shows this graphically.

---

* dBi = decibels referenced to an isotropic (antenna).

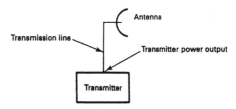

EIRP = Trans. output (dBW) – Trans. line loss (dB) + Ant. gain (dB) (7.6)

Figure 7.7. Elements in the calculation of EIRP.

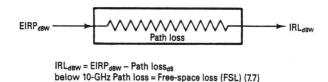

$IRL_{dBW} = EIRP_{dBW} -$ Path loss$_{dB}$
below 10-GHz Path loss = Free-space loss (FSL) (7.7)

Figure 7.8. Calculation of isotropic receive level.

Example: If a microwave transmitter has a 1-watt (0-dBW) power output, the waveguide loss is 3 dB, and the antenna gain is 34 dBi, what is the EIRP in dBW?

$$EIRP_{dBW} = 0 \text{ dBW} - 3 \text{ dB} + 34 \text{ dBi}$$
$$= +31 \text{ dBW}$$

Calculation of Isotropic Receive Level (IRL). The IRL is the RF power level impinging on the receive antenna. It would be the power we would measure at the base of an isotropic receive antenna. The calculation is shown graphically in Figure 7.8.

Calculation of Receive Signal Level (RSL). The receive signal level (RSL) is the power level entering the first active stage of the receiver:

$$RSL_{dBW} = IRL_{dBW} + \text{Rec. ant. gain (dB)} - \text{Rec. trans. line losses (dB)} \quad (7.8)$$

(*Note*: Power levels can be in dBm as well, but we must be consistent.)

Example: Suppose the isotropic receive level (IRL) was -121 dBW, the receive antenna gain was 31 dB, and the line losses were 5.6 dB. What would the RSL be?

$$RSL = -121 \text{ dBW} + 31 \text{ dB} - 5.6 \text{ dB}$$
$$= -95.6 \text{ dBW}$$

Calculation of Receiver Noise Level. The thermal noise level of a receiver is a function of the receiver noise figure and its bandwidth. For analog radio systems, receiver thermal noise level is calculated using the bandwidth of the intermediate frequency (IF). For digital systems, the noise level of interest is in only 1 Hz of bandwidth using the notation N_0, the noise level in a 1-Hz bandwidth.

The noise that a device self-generates is given by its noise figure (dB) or a noise temperature value. Any device, even passive devices, above absolute zero generates thermal noise. We know the thermal noise power level in a 1-Hz bandwidth of a perfect receiver operating at absolute zero. It is

$$P_n = -228.6 \text{ dBW/Hz}$$

where P_n is the noise power level. Many will recognize this as Boltzmann's constant expressed in dBW.

We can calculate the thermal noise level of a perfect receiver operating at room temperature using the following formula:

$$P_n = -228.6 \text{ dBW/Hz} + 10 \log 290 \text{ (K)}$$
$$P_n = -204 \text{ dBW/Hz}$$

(7.9)

The value, 290 K, is room temperature, or about 17°C or 68°F.

Noise figure simply tells us how much noise has been added to a signal while passing through a device in question. Noise figure (dB) is the difference in signal-to-noise ratio between the input to the device and the output of that same device.

We can convert noise figure to noise temperature in kelvins with the following formula:

$$\text{NF}_{dB} = 10 \log(1 + T_e/290) \tag{7.10}$$

where T_e is the effective noise temperature of a device. Suppose the noise figure of a device is 3 dB. What is the noise temperature? 290 K.

$$3 \text{ dB} = 10 \log(1 + T_e/290)$$
$$0.3 = \log(1 + T_e/290)$$
$$1.995 = 1 + T_e/290$$

We round 1.995 to 2; thus

$$2 - 1 = T_e/290$$
$$T_e = 290 \text{ K}$$

The thermal noise power level of a device operating at room temperature is

$$P_n = -204 \text{ dBW/Hz} + \text{NF}_{dB} + 10 \log \text{BW}_{Hz} \tag{7.11}$$

where BW is the bandwidth of the device in Hz.

Example: A microwave receiver has a noise figure of 8 dB and its bandwidth is 10 MHz. What is the thermal noise level (sometimes called the thermal noise threshold)?

$$P_n = -204 \text{ dBW/Hz} + 8 \text{ dB} + 10 \log(10 \times 10^6)$$

$$= -204 \text{ dBW/Hz} + 8 \text{ dB} + 70 \text{ dB}$$

$$= -126 \text{ dBW}$$

If the receiver in the above example was operating in a digital regime, we'd want to calculate N_0.

$$N_0 = -204 \text{ dBW/Hz} + NF_{dB}$$

$$= -196 \text{ dBW/Hz}$$

Calculation of E_b/N_0 in Digital Radio Systems. Many readers are familiar with signal-to-noise ratio (S/N). It was introduced in Chapter 5. In digital systems we use E_b/N_0, meaning energy per bit per noise spectral density ratio. We can relate E_b/N_0 to bit error rate (BER) given the modulation type in question.

We defined N_0 above. E_b is the energy per bit. Suppose the RSL was 1 watt and we were receiving 1000 bits per second. How much energy is imparted to 1 bit? It is 1 mW. We simply divided 1 watt by 1000 bits per second. In radio work it is easier to do the division logarithmically because we work with decibels. E_b can be stated as follows:

$$E_b = RSL - 10 \log(\text{bit rate}) \tag{7.12}$$

Example: A certain radio system receives 1.544 Mbps and the RSL is -108 dBW. What is the energy per bit (E_b)?

$$E_b = -108 \text{ dBW} - 10 \log(1.544 \times 10^6)$$

$$= -108 \text{ dBW} - 61.88 \text{ dB}$$

$$= -169.88 \text{ dBW}$$

We can now develop a formula for E_b/N_0:

$$E_b/N_0 = RSL_{dBW} - 10 \log(\text{bit rate}) - (-204 \text{ dBW} + NF_{dB}) \tag{7.13}$$

Simplifying, we obtain

$$E_b/N_0 = RSL_{dBW} - 10 \log(\text{bit rate}) + 204 \text{ dBW} - NF_{dB} \tag{7.14}$$

Some Notes on E_b/N_0 and Its Use. E_b/N_0, for a given BER, will be different for different types of modulation (e.g., FSK, PSK, QAM, etc.).

When working with E_b, we divide RSL by the bit rate, not the symbol rate nor the baud rate.

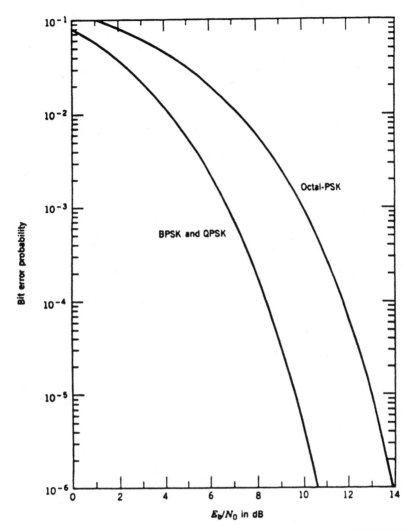

Figure 7.9. Bit error probability (BER) versus E_b/N_0 performance of coherent BPSK/QPSK and 8-ary PSK (octal PSK).

There is a theoretical E_b/N_0 and a practical E_b/N_0. The practical is always a greater value than the theoretical, greater by the *modulation implementation loss* in dB.

Figure 7.9 is an example where we can relate BER to E_b/N_0. There are two curves in the figure. The first is from the left if for BPSK/QPSK (binary phase shift keying/quadrature phase shift keying), and the second is for 8-ary PSK (an 8-level PSK modulation). The values are for coherent detection. Coherent detection means that the receiver has a phase reference as a basis to make its binary decisions.

Digital Modulation of LOS Microwave Radios. Digital systems, typically standard PCM as discussed in Chapter 8, are notoriously wasteful of bandwidth compared to their analog counterparts. For example, the analog voice channel is nominally 4 kHz, whereas the digital voice channel, assuming one bit per hertz of bandwidth, is 64 kHz. This is a 16-to-1 difference in required bandwidth. Thus various national regulatory authorities, such as the U.S. FCC, require that digital systems be bandwidth conservative. One term that is used is *bit packing*. This means packing more bits into a hertz of bandwidth. Roughly, the FCC requires about 4.5 bits per hertz of bandwidth. To meet these requirements, digital LOS microwave utilizes some form of quadrature amplitude modulation (QAM), and as a minimum at the 64-QAM level, or often 128-QAM or 256-QAM. 64-QAM has a theoretical bit packing capability of 6 bits/Hz, 128-QAM at 7 bits/Hz, and 256-QAM at 8 bits/Hz. Figure 7.10 compares bit error rate performance versus E_b/N_0 for various QAM schemes [12]. (Also consult Ref. 18.)

To sum up this section on digital LOS microwave, we will work an example problem. A digital link operates in the 7-GHz band with a link 37 km long. The bit rate is 155 Mbps and the modulation is 64-QAM. The specified BER for the link is 1×10^{-7} and the modulation implementation loss is 2 dB. The receiver noise figure is 8 dB. The antennas have 35-dB gain at each end, and transmission line losses are 1.8 dB at each end. What link margin can be expected?

First turn to Figure 7.10 and derive the required E_b/N_0. This is 19.5 dB; add to this the modulation implementation loss of 2 dB and the result is that the required value for E_b/N_0 is 21.5 dB.

The next step is to calculate a candidate RSL value. We know that E_b must be 21.5 dB above N_0. We can calculate N_0 because we have the receiver noise figure.

$$N_0 = -204 \text{ dBW} + 8 \text{ dB}$$
$$= -196 \text{ dBW} \quad \text{and}$$
$$E_b = -196 \text{ dBW} + 21.15 \text{ dB}$$
$$= -174.5 \text{ dBW}$$

Thus RSL, in this case, is $10 \log(1.544 \times 10^6)$ greater than E_b.

$$\text{RSL}_{\text{dBW}} = E_b + 10 \log(1.544 \times 10^6) \quad \text{(from equation 7.8)}$$
$$= -174.5 \text{ dBW} + 61.88 \text{ dB}$$
$$= -112.61 \text{ dBW}$$

We will hold this minimum RSL value for future reference, and now turn to the transmit side of the link. Assume the transmitter has a 1-watt output or 0 dBW. Calculate EIRP in dBW.

$$\text{EIRP}_{\text{dBW}} = 0 \text{ dBW} - 1.8 \text{ dB} + 35 \text{ dB} \quad \text{(equation 7.6)}$$
$$= +33.2 \text{ dBW}$$

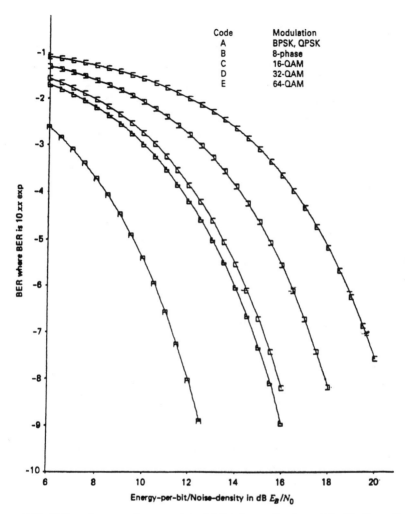

Figure 7.10. BER performance for several modulation types. Courtesy of the Raytheon Company [5].

Calculate the free-space loss (path loss):

$$\text{FSL}_{dB} = 92.4 + 20\log 37 + 20\log 7 \qquad \text{(equation 7.5B)}$$
$$= 92.4 + 31.36 + 16.90$$
$$= 140.66 \text{ dB}$$

Calculate the IRL:

$$\text{IRL}_{dBW} = +33.2 \text{ dBW} - 140.66 \text{ dB} \qquad \text{(equation 7.7)}$$
$$= -107.46 \text{ dBW}$$

Calculate RSL:

$$RSL = -107.46 \text{ dBW} + 35 \text{ dB} - 1.8 \text{ dB} \qquad \text{(from equation 7.8)}$$

$$= -74.26 \text{ dBW}$$

Calculate the margin:

$$\text{Margin} = -74.26 \text{ dBW} - (-112.61 \text{ dBW})$$

$$= 38.35 \text{ dB}$$

Often we are faced with the problem of "What antenna gain will provide the margin or provide the gain necessary to meet performance objectives?"

Parabolic Antenna Gain. At a given frequency the gain of a parabolic antenna is a function of its effective area and may be expressed by the formula

$$G = 10 \log_{10}(4\pi A\eta/\lambda^2) \qquad (7.15)$$

where G is the gain in decibels relative to an isotropic antenna, A is the area of antenna aperture, η is the aperture efficiency, and λ is the wavelength at the operating frequency. Commercially available parabolic antennas with a conventional horn feed at their focus usually display a 55% efficiency or somewhat better. With such an efficiency, gain (G, in decibels) is then

$$G = 20 \log_{10} D + 20 \log_{10} F + 7.5 \qquad (7.16)$$

where F is the frequency in gigahertz and D is the parabolic diameter in feet. In metric units, we have

$$G = 20 \log_{10} D + 20 \log_{10} F + 17.8 \qquad (7.17)$$

where D is measured in meters and F in gigahertz.

What size antenna would be required in the preceding example? Let $G = 35$ dB and $F = 6$ GHz.

$$35 \text{ dB} = 20 \log_{10} D + 20 \log_{10} 7 + 7.5$$

$$20 \log D = 35 - 20 \times 0.8451 - 7.5$$

$$= 10.598/20$$

$$= 3.38 \text{ feet}$$

Parabolic dish antennas, with waveguide (horn) feeds (see Figure 7.11), are probably the most economic antennas for radiolinks operating from 3 GHz upward. From 50 MHz to about 3 GHz, coaxial feeds are used, and often the antennas are

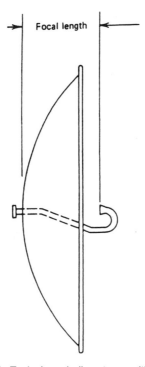

Figure 7.11. Typical parabolic antenna with front feed.

Yagi's. Coaxial cable transmission lines deliver the RF energy from/to transmitter/receiver to the antenna in this range. Above 3 GHz, coaxial cable becomes too lossy and waveguide is more practical.

Other types of antennas may also be used, such as the "cornucopia," horn, and spiral. Besides cost and gain, other features are front-to-back ratio, side lobes, and efficiency. For instance the "cornucopia," called such because it looks like "the horn of plenty," has efficiencies in excess of 60% and improved side-lobe discrimination but is more costly.

4.5 Running a Path/Site Survey

This can turn out to be the most important step in the design of an LOS microwave link (or hop). We have found through experience that mountains move (i.e., map error), buildings grow, grain elevators appear where none were before, east of Madrid a whole high-rise community goes up, and so forth.

Another point from experience: If someone says "line-of-sight" conditions exist on a certain path, **don't believe it!** Line of sight must be precisely defined. We reiterate that for each obstacle in the LOS microwave path, earth curvature with proper K-factor must be added to obstacle height, 0.6 of the first Fresnel zone must be added on top of that,* and then 50 ft for trees and 10 ft more for growth must be added if in a vegetated area (to avoid foliage loss penalties).

Much of the survey is to verify findings and conclusions of the path profile. Of course each site must be visited to determine the location of the radio equipment shelter, the location of the tower, whether site improvement is required, the nearest prime power lines, and site access, among other items to be investigated.

Site/path survey personnel must personally inspect the sites in question, walking/driving the path or flying in a helicopter, or a combination thereof. From our experience the use of reliable GPS receivers is mandatory. The positions must then be periodically cross-checked by a true-north reading on a known location.

GPS (Global Positioning System) receivers are available from hundred to thousands of dollars. There are differential GPSs that can bring position reading accuracy to less than a meter in three dimensions.

4.6   System Test Prior to Cutover

A series of tests should be carried out to verify if the link (or system) meets the performance requirements established in Section 4.2. The first is the measurement of receive signal level (RSL) at a link's far-end receiver. The second test is the bit error rate test (BERT). Ideally, the tests should be done over time. Here we mean to run the BERT continuously for at least 12 or 24 hours or more to capture the effects of fading.

4.7   Fades, Fading, and Fade Margins

In Section 4.4.2 we showed how path loss (free-space loss) can be calculated. This was a fixed loss which can be simulated in the laboratory with an attenuator. On very short radio paths below about 10 GHz, the signal level impinging on the distant end receiving antenna, assuming full LOS conditions, can be calculated to less than 1 dB. If the transmitter continues to give the same output, the receive signal level will remain uniformly the same over long periods of time, for years. As the path is extended, the measured RSL will vary around a median. The signal level may remain at that median for minutes or hours, and then suddenly drop and then return to the median again. In other periods and/or on other links, this level variation can be continuous for periods of time. Drops in level can be as much as 30 dB or more. This phenomenon is called *fading*. The system and link design must take fading into account when sizing or dimensioning the system/link.

As the RSL drops in level, so does the E_b/N_0. As the E_b/N_0 decreases, there is a deterioration in error performance; the BER degrades. Fades vary in depth, duration, and frequency (i.e., number of fade events per unit of time). We cannot eliminate the fades, but we can mitigate their effects. The primary tool we have is to overbuild each link by increasing the margin.

---

*Often it is advisable to add 10 ft (or 3 m) of safety factor on top of the 0.6 first Fresnel zone clearance to avoid any diffraction loss penalties.

Link margin is the number of dB we have as a surplus in the link design. We could design an LOS microwave link so we just achieve the RSL at the distant receiver to satisfy the E_b/N_0 (and BER) requirements using free-space loss as the only factor in link attenuation (besides transmission line loss). Unfortunately we will only meet our specified requirements about 50% of the time. So we must add margin to compensate for the fading.

We have to determine what percentage of the time the link meets BER performance requirements. We call this *time availability*.* If a link meets its performance requirements 99% of the time, then it does not meet performance requirements 1% of the time. We call this latter factor *unavailability*.

To improve time availability, we must increase the link margin, often called the *fade margin*. How many additional dB are necessary? There are several approaches to the calculation of a required fade margin. One of the simplest and most straightforward approaches is to assume that the fading follows a Rayleigh distribution, often considered worst-case fading. If we base our premise on a Rayleigh distribution, then the following fade margins can be used:

Time Availability (%)	Required Fade Margin (dB)
90	8
99	18
99.9	28
99.99	38
99.999	48

More often than not, LOS microwave systems consist of multiple hops. Here our primary interest is the time availability at the far-end receiver in the system after the signal has progressed across all of the hops. From this time availability value we will want to assign an availability value for each hop or link.

Suppose a system has nine hops and the system time availability specified is 99.95%, and we want to calculate the time availability per hop or link. The first step is to calculate the system time unavailability. This is simply $1.0000 - 0.9995 = 0.0005$. We now divide this value by 9 (i.e., there are nine hops or links):

$$0.0005/9 = 0.0000555$$

Now we convert this value to time availability:

$$\text{Per-hop time availability} = 1.0000000 - 0.0000555$$

$$= 0.99994 \text{ or } 99.994\%$$

The most common cause of fading is multipath conditions. Refer to Figure 7.12. As the term implies, signal energy follows multiple paths from the transmit

---

* Other texts call this "reliability." The use of this term should be deprecated because it is ambiguous and confusing. In our opinion, reliability should relate to equipment failure rate, not propagation performance.

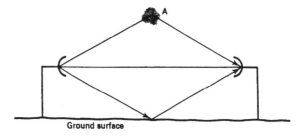

A = layers of different refractive index

Figure 7.12. Multipath is the most common cause of fading.

antenna to the receive antenna. Two additional paths, besides the main ray beam, are shown in Figure 7.12. Most of the time the delayed signal energy (from the reflected/refracted paths) will be out of phase with the principal ray beam which causes fading. In digital systems, there is the additional impairment of dispersion caused by multipath. Of course, the delay energy arrives later, spilling into the next bit or binary symbol position, increasing the probability that that bit decision will be in error.

Probably the most economic way to overbuild a link is to increase the antenna aperture. Every time we double the aperture (i.e., in this case, doubling the diameter of the parabolic dish), we increase the gain by 6 dB (see equations 7.16 and 7.17). We recommend that apertures for LOS microwave antennas not exceed 12 ft (3.7 m). Not only does the cost of the antenna get notably greater as aperture increases over 8 ft (2.5 m), but the equivalent sail area of the dish starts to have an impact on system design. Wind pressure on large dishes increases tower twist and sway, resulting in movement out of the capture area of the ray beam at the receive antenna. This forces us to stiffen the tower, which could dramatically increase system cost. Also, as antenna aperture increases, gain increases and beamwidth decreases.

Other measures we can take to overbuild a link are:

- Insert a low-noise amplifier (LNA) in front of the receiver–mixer. Improvement: 6–12 dB.
- Use an HPA (high-power amplifier). Usually a traveling-wave tube (TWT) amplifier; 10 watts output. Improvement: 10 dB.
- Implement FEC (forward error correction). Improvement: 1–5 dB. Involves adding a printed circuit board at each end. It will affect link bandwidth. See Ref. 17 for description of FEC.
- Implement some form of diversity. Space diversity is preferable in many countries. Can be a fairly expensive measure. Improvement: 5–20 dB or more. Diversity is described below.

It should be appreciated that fading varies with path length, frequency, climate, and terrain. The rougher the terrain, the more reflections are broken up. Flat terrain, and especially paths over water, tends to increase the incidence of fading. For example, in dry, windy, mountainous areas the multipath fading phenomenon may be nonexistent. In hot, humid coastal regions a very high incidence of fading may be expected.

4.8 Diversity and Hot-Standby Operation

Diversity reception means the simultaneous reception of the same radio signal over two or more paths. Each "path" is handled by a separate receiver chain and then combined by predetection or postdetection combiners in the radio equipment so that effects of fading are mitigated. The separate diversity paths can be based on space, frequency, and/or time diversity. The simplest form of diversity is space diversity. Such a configuration is shown in Figure 7.13.

The two diversity paths in space diversity are derived at the receiver end from two separate receivers with a combined output. Each receiver is connected to its own antenna, separated vertically on the same tower. The separation distance should be at least 70 wavelengths and preferably 100 wavelengths. In theory, fading will not occur on both paths simultaneously.

Frequency diversity is more complex and more costly than space diversity. It has advantages as well as disadvantages. Frequency diversity requires two transmitters at the near end of the link. The transmitters are modulated simultaneously by the same signal but transmit on different frequencies. Frequency separation must be at least 2%, but 5% is preferable. Figure 7.14 is an example of a frequency-diversity configuration. The two diversity paths are derived in the frequency domain. When a fade occurs on one frequency, it will probably

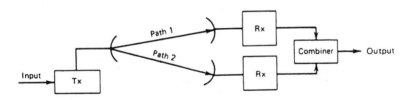

Figure 7.13. A space-diversity configuration.

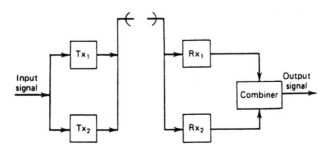

Figure 7.14. A frequency-diversity configuration.

not occur on the other frequency. The more one frequency is separated from the other, the less chance there is that fades will occur simultaneously on each path.

Frequency diversity is more expensive, but there is greater assurance of path reliability. It provides full and simple equipment redundancy and has the great operational advantage of two complete end-to-end electrical paths. In this case, failure of one transmitter or one receiver will not interrupt service, and a transmitter and/or a receiver can be taken out of service for maintenance. The primary disadvantage of frequency diversity is that it doubles the amount of frequency spectrum required in this day and age when spectrum is at a premium. In many cases it is prohibited by national licensing authorities. For example, the U.S. Federal Communications Commission (FCC) does not permit frequency diversity for industrial users. It also should be appreciated that it will be difficult to get the desired frequency spacing.

The full equipment redundancy aspect is very attractive to the system designer. Another approach to achieve diversity improvement in propagation plus reliability improvement by fully redundant equipment is to resort to the "hot-standby" technique. On the receive end of the path, a space-diversity configuration is used. On the transmit end a second transmitter is installed as in Figure 7.14, but the second transmitter is on "hot standby." This means that the second transmitter is on but its signal is not radiated by the antenna. On a one-for-one basis the second transmitter is on the same frequency as the first transmitter. On failure of transmitter 1, transmitter 2 is switched on automatically.

One-for-N hot standby is utilized on large radiolink systems employing several radio carriers, where the cost for duplicate equipment for each channel may be prohibitive. In this case, one full set of spare equipment in the "on" condition serves to replace one of several operational channels, and the spare equipment is assigned its own frequency. On the receive side there is just one extra receiver. Relay cut-over must be provided, and no space-diversity improvement is afforded. Likewise, there is no paralleling of inputs on the transmit side; thus the switching from the operational pair to the standby pair is much more complex. On such multi-RF channel arrangements it is customary to assign some sort of priority arrangement. Often the priority channel enjoys the advantage of frequency diversity, whereas the other RF channels do not. On failure of one of the other channels, the diversity improvement is lost on the priority channel, with the diversity pair switched to carry the traffic on the failed pair. In another arrangement the standby channel carries low-priority traffic and does not operate in a frequency-diversity arrangement, while providing protection for perhaps two or three other channels carrying the higher-priority traffic. On failure of a high-priority channel, the lower-priority channel drops its traffic, replacing one of the RF channels carrying the more important traffic flow. Once this occurs, the remaining channels operate without standby equipment protection.

Diversity Improvement. Propagation reliability improvement can be exemplified as follows. If a 30-mile path required a 51-dB fade margin to achieve a 99.999% reliability on 6.7 GHz without diversity, with space diversity on the same path, only a 33-dB fade margin would be required for the same propagation reliability, namely, 99.999% (Vigants, *IEEE Trans. Commun.*, December 1968 and Ref. 6). For frequency diversity in the nondiversity condition, assuming Rayleigh fading, a 30-dB fade margin would display something better than a 99.9% path reliability. But under the same circumstances with frequency diversity, with only a 1% frequency separation, propagation reliability on the same path would be improved to 99.995% [6].

4.9 LOS Microwave Repeaters

Digital LOS microwave repeaters completely demodulate the incoming signal to baseband (i.e., to the raw electrical signal of "1s" and "0s."). This full demodulation causes regeneration of the "1s" and "0s." The outgoing signal is "squared up" and retimed. Regeneration is explained in Chapter 8. The regenerated baseband signal is remodulated, upconverted, and retransmitted on a different frequency. RF and IF repeaters are not recommended for digital microwave systems because such repeaters do not have the vital regeneration stage.

4.10 Frequency Planning and Frequency Assignment

4.10.1 General. To derive optimum performance from an LOS microwave system, the design engineer must set out a frequency-usage plan that may or may not have to be approved by the national regulatory organization.

The problem has many aspects. First, the useful RF spectrum is limited from above dc to about 150 GHz. The upper limit is technology-restricted. To some extent it is also propagation-restricted. The frequency ranges for this discussion cover the bands in Table 7.2. Those frequencies above 10 GHz could be called rainfall-restricted, because at about 10 GHz is where excess attenuation due to rainfall can become an important design factor.

Then there is the problem of congestion. Around urban and built-up areas, frequency assignments below 10 GHz are hard to obtain from national regulatory authorities. If we plan properly for excess rainfall attenuation, nearly equal performance is available at those higher frequencies.

4.10.2 Radio-Frequency Interference (RFI). There are three facets to RFI in this context. (1) Own microwave can interfere with other LOS microwave and satellite communication earth stations nearby, (2) nearby LOS microwave

TABLE 7.2 LOS Microwave Frequency Bands

2110–2130 MHz	18,920–19,160 MHz
3700–4200 MHz	19,260–19,700 MHz
5925–6425 MHz	21,200–23,600 MHz
6525–6875 MHz	27,500–29,500 MHz
10,700–11,700 MHz	31,000–31,300 MHz
17,700–18,820 MHz	38,600–40,000 MHz

and satellite communication facilities can interfere with own microwave, and (3) own microwave can interfere with itself. To avoid self-interference (No. 3), it is advisable to use frequency plans of CCIR (ITU-R organization) as set forth in the RF Series (Fixed Service). Advantage is taken of proper frequency separation, transmit and receive, and polarization isolation. CCIR also provides methods for interference analysis (coordination contour), also in the RF Series. Another alternative is specialist companies that provide a service of electromagnetic compatibility (EMC) analysis.

5 SATELLITE COMMUNICATIONS

5.1 Introduction

Satellite communications is an extension of LOS microwave technology covered in Section 4. The satellite must be within LOS of each participating earth terminal. We are more concerned about noise in satellite communication links than we were with LOS microwave. In most cases, received signals will be of a lower level. On satellite systems operating below 10 GHz, very little link margin is required; there is essentially no fading, as experienced in LOS microwave.

5.2 Application

Satellite communication is another method of extending the digital network (Chapters 8 and 9). These digital trunks may be used as any other digital trunks for telephony, data, facsimile, and video. Satellite links may prove optimum for a variety of applications, including the following:

1. On international high-usage trunks country to country.
2. On national trunks, between switching nodes that are fairly well separated in distance [i.e., >200 miles (320 km)] in highly developed countries. Again, the tendency is to use satellite links for direct high-usage connectivity. It may serve as an adjunct to LOS microwave and fiber optics.
3. In areas under development where satellite links replace HF radio and a high growth is expected to be eventually supplemented by radiolink and fiber-optic cable.
4. In sparsely populated, highly rural, "out-back" areas where it may be the only form of communication. Northern Canada and Alaska are good examples.
5. On final routes for overflow on a demand-assignment basis. Route length again is a major consideration.
6. In many cases, on international connections reducing such connections to one link.
7. On private and industrial networks including VSAT (very small aperture terminal) networks.

8. On specialized common carriers.

9. On thin-line communications and tracking systems.

5.3 Definition

A number of world bodies, including the ITU-R and the U.S. Federal Communications Commission (FCC) have adopted the term *earth station* as a radio frequency facility located on the earth's surface that communicates with satellites. A *terrestrial station* is a radio facility on the earth's surface that communicates with other similar facilities on the earth's surface. Section 4 of this chapter dealt with one form of terrestrial station. The term "earth station" as used in the current literature has come more to mean a radio station operating with other stations on the earth via an orbiting satellite relay.

The preponderance of commercial satellites are geostationary. Such satellites orbit the earth in a 24-h period. Thus they appear stationary over a particular geographic location on earth. For a 24-h synchronous orbit the altitude of a geostationary satellite is 22,300 statute miles or 35,900 km above the equator of the earth.

The reader should leave his/her mind open to the concept of LEO (low earth orbit) satellites. One such constellation is operational providing cellular radio telephone service over the entire earth's surface. The system was called *Iridium*. It was conceptualized and developed by Motorola and soon after its initial operation it was sold off to a holding company. A LEO satellite has an altitude of from 500 to 800 km above the earth's surface. Its principal advantage is the short transmission delay interval when compared to a GEO satellite, where the one-way delay is about 125 ms. The following discussion is valid for GEO satellite systems.

5.4 The Satellite

Most of the presently employed communication satellites are RF repeaters. A typical RF repeater used in a communication satellite is shown in Figure 7.15. The tendency today is to call these types of satellite "bent pipe" satellites as opposed to processing satellites. A processing satellite, as a minimum, regenerates the received digital signal. It may decode and recode a digital bit stream. It also may have some bulk switching capability, switching to crosslinks connecting to other satellites. Theoretically, as mentioned earlier, three such satellites placed correctly in equatorial geostationary orbit could provide communication from one earth station to any other located anywhere on the earth surface (see Figure 7.16). However, high latitude service is marginal and nil north of 80°N and south of 80°S.

5.5 Three Basic Technical Problems

As the reader can appreciate, satellite communication is nothing more than radiolink (microwave LOS) communication using one or two RF repeaters located at

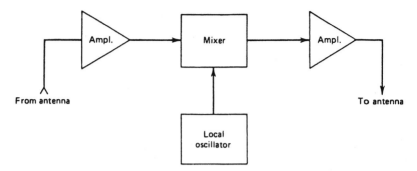

Figure 7.15. Simplified functional block diagram of one transponder of a typical communication satellite.

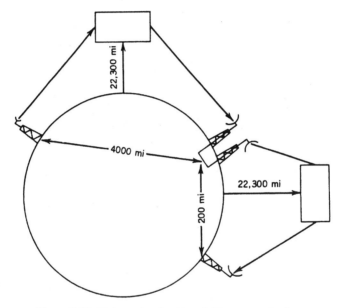

Figure 7.16. Distance involved in satellite communications.

great distances from the terminal earth stations, as shown in Figure 7.16. Because of the distance involved, consider the slant range from earth antenna to satellite to be the same as the satellite altitude. This would be true if the antenna were pointing at zenith to the satellite. Distance increases as the pointing angle to the satellite decreases (elevation angles).

We thus are dealing with very long distances. The time required to traverse these distances—namely, earth station to satellite to another earth station—is on the order of 250 ms. Round-trip delay will be 2×250 or 500 ms. These propagation times are much greater than those encountered on conventional terrestrial systems. So one major problem is propagation time and resulting echo on telephone circuits. It influences certain data circuits in delay to reply for block or packet transmission systems and requires careful selection of telephone signaling systems, or call-setup time may become excessive.

Naturally, there are far greater losses. For LOS microwave we encounter free-space losses possibly as high as 145 dB. In the case of a satellite with a range of 22,300 miles operating on 4.2 GHz, the free-space loss is 196 dB and at 6 GHz, 199 dB. At 14 GHz the loss is about 207 dB. This presents no insurmountable problem from earth to satellite, where comparatively high power transmitters and very high gain antennas may be used. On the contrary, from satellite to earth the link is power-limited for two reasons: (1) in bands shared with terrestrial services such as the popular 4-GHz band to ensure noninterference with those services and (2) in the satellite itself, which can derive power only from solar cells. It takes a great number of solar cells to produce the RF power necessary; thus the down-link, from satellite to earth, is critical, and received signal levels will be much lower than on comparative radiolinks, as low as -150 dBW. A third problem is crowding. The equatorial orbit is filling with geostationary satellites. Radio-frequency interference from one satellite system to another is increasing. This is particularly true for systems employing smaller antennas at earth stations with their inherent wider beamwidths. It all boils down to a frequency congestion of emitters.

It should be noted that by the year 2000, we can expect to see several low earth-orbit satellite systems in operation. These satellites typically orbit some 500 km above the earth.

5.6 Frequency Bands: Desirable and Available

The most desirable frequency bands for commercial satellite communication are in the spectrum 1000–10,000 MHz. These bands are:

3700–4200 MHz (satellite-to-earth or downlink)
5925–6425 MHz (earth-to-satellite or uplink)
7250–7750 MHz* (downlink)
7900–8400 MHz* (uplink)

These bands are preferred by design engineers for the following primary reasons:

- Less atmospheric absorption than higher frequencies.
- Rainfall loss not a concern.
- Less noise, both galactic and man-made.
- A well-developed technology.
- Less free-space loss compared to the higher frequencies.

* These two bands are intended mainly for military application.

There are two factors contraindicating application of these bands and pushing for the use of higher frequencies:

- The bands are shared with terrestrial services.
- There is orbital crowding (discussed earlier).

Higher-frequency bands for commercial satellite service are:

10.95–11.2 GHz (downlink)
11.45–12.2 GHz (downlink)
14.0–14.5 GHz (uplink)
17.7–20.2 GHz (downlink)
27.5–30.0 GHz (uplink)

Above 10 GHz, rainfall attenuation and scattering and other moisture and gaseous absorption must be taken into account. The satellite link must meet a BER of 1×10^{-6} at least 99.9% of the time. One solution is a space-diversity scheme where we can be fairly well assured that one of the two antenna installations will not be seriously affected by the heavy rainfall cell affecting the other installation. Antenna separations of 4–10 km are being employed. Another advantage with the higher frequencies is that requirements for downlink interference are less; thus satellites may radiate more power. This is often carried out on the satellite using spot-beam antennas rather than general-coverage antennas.

5.7   Multiple Access of a Satellite

Multiple access is defined as the ability of a number of earth stations to interconnect their respective communication links through a common satellite. Satellite access is classified (1) by assignment, whether quasi-permanent or temporary, namely, (a) preassigned multiple access or (b) demand-assigned multiple access (DAMA), and (2) according to whether the assignment is in the frequency domain or the time domain, namely, (a) frequency-division multiple access (FDMA) or (b) time-division multiple access (TDMA). On comparatively heavy routes (≥ 10 erlangs), preassigned multiple access may become economical. Other factors, of course, must be considered, such as whether the earth station is "INTELSAT" standard as well as the space-segment charge that is levied for use of the satellite. In telephone terminology, "preassigned" means dedicated circuits. Demand-assigned multiple access is useful for low-traffic multipoint routes where it becomes interesting from an economic standpoint. Also, an earth station may resort to DAMA as a remedy to overflow for its FDMA circuits.

5.7.1   *Frequency-Division Multiple Access.* Historically, FDMA has the highest usage and application of the various access techniques. The several RF bands assigned by international treaty (Section 5.6) have 500-MHz bandwidths.

There is a notable exception for the 6/4-GHz frequency pair where the bandwidth has been expanded to 575 MHz.

INTELSAT (International Telecommunication Satellite [consortium]) has the largest constellation of GEO satellites in the world. Satellite series INTELSAT I/II date back to the early 1960s. INTELSAT IX is the latest design series providing both 6/4 GHz with 72 equivalent 36-MHz transponders and 14/11-GHz service with 22 transponders. Frequency reuse allows this dense packing of transponder bandwidths inside the operational bands.

With FDMA a user is assigned a frequency in the 6-GHz uplink band with an associated bandwidth. This bandwidth is a function of the bit rate. INTELSAT commonly employs quadrature phase shift keying (QPSK). The satellite receives the uplink carrier on a spot beam (often a parabolic antenna). It then mixes the uplink frequency with a 2225-MHz local oscillator frequency source; the lower sideband is selected, producing an output in the 4 GHz band. The output is fed to a desired spot beam with coverage at the indicated destination. Any earth station in the spot beam coverage area can receive this downlink carrier.

Suppose the uplink were 6.0 GHz. This carrier is then mixed with a 2225-MHz source and the difference frequency is selected:

6000 MHz − 2225 MHz = 3775 MHz, the equivalent downlink frequency.

This mixing operation is shown in the simplified block diagram of Figure 7.15. Table 7.3 shows the information rate for INTELSAT Intermediate Data Rate (IDR) QPSK service. The BER for this service is 1×10^{-10} for better than 95.90% of the time clear sky conditions.

5.7.2 Time-Division Multiple Access.

Time-division multiple access (TDMA) operates in the time domain, whereas FDMA operates in the frequency domain. TDMA may only be used for digital network connectivity. Use of a satellite transponder is on a time-sharing basis. Individual time slots are assigned to earth stations in a sequential order. Each earth station has full and exclusive use of the transponder bandwidth during its time-assigned segment. Depending on the bandwidth of the transponder, bit rates from 10 Mbits/s (or below) up to 100 Mbits/s are used.

With TDMA operation, earth stations use digital modulation and transmit with bursts of information. The duration of a burst lasts for the time period of the slot assigned. (Note that a time slot is slightly longer than the period of a burst to allow some guard time to compensate for timing errors.) Of course, timing and synchronization are major considerations.

A frame, in digital format, may be defined as a repeating cycle of events. It occurs in a time period containing a single burst from each accessing earth station. There are guard periods or guard times between bursts as mentioned above. A sample frame is illustrated in Figure 7.17 for earth stations 1, 2, and 3 to N. Typical frame periods are 750 μs for INTELSAT and 250 μs for the Canadian Telesat.

TABLE 7.3   INTELSAT QPSK/IDR Information Rates and Associated Overhead

Number of 64 kbits/s Bearer Channels (n)		Information Rate (n × 64 kbits/s)		Type of Overhead		
				No Overhead (1)	With 96 kbits/s ESC Overhead (2)	With 6.7% IBS Overhead (2, 3)
1		64		X		X
2		128				X
3		192		X		
4		256				X
6		384		X		X
8		512				X
12		768				X
16		1024				X
24		1536				X
24		1544	(4)		X	
30	(31)	2048	(4)		X	
90	(93)	6312	(4)		X	
120	(124)	8448	(4)		X	
480		32064	(4)		X	
480	(496)	34368	(4)		X	
630	(651) or	44736	(4)		X	
672						

X = Recommended rate corresponding to the type of overhead.
Notes:

(1) For rates less than 1544 kbits/s, it is possible to use any $n \times 64$ kbits/s information rate without overhead, but the only INTELSAT-recommended rates are 64 kbits/s, 192 kbits/s, and 384 kbits/s. The use of the optional Reed–Solomon outer coding is not defined for any information rate less than 1.544 Mbits/s, which does not use overhead.

(2) The optional Reed–Solomon outer coding can be used with the information rates shown with an "X."

(3) The carriers in this column are small QPSK/IDR carriers some of which can be used with the circuit multiplication concept described in Appendix B of reference document. [For a definition of the IBS overhead framing, see IESS–309 (IBS).]

(4) These are standard ITU–T hierarchical bit rates. Other $n \times 64$ kbits/s information rates above 1.544 Mbits/s are also possible and must have an ESC overhead of at least 96 kbits/s (the overhead framing will be defined on a case-by-base basis by INTELSAT).

Source: From INTELSAT IESS-308 (Rev. 10), Table 1, page 36, Ref. 10
Note: also see Table 7.4.

As we mentioned, timing is a major consideration in TDMA systems; it is crucial to their effective operation. The greater N becomes (i.e., the more stations operating in the frame period), the more the clock timing affects the system. The secret lies in the "carrier and clock (timing) recovery pattern" as shown in Figure 7.17. One way to ensure that all accessing stations synchronize to a master clock is to place a sync burst and the first element in the format frame. INTELSAT does just this. The burst carries 44 bits, starting with 30 bits for carrier and bit timing recovery, 10 bits for the "unique word," and 4 bits for the station identification code.

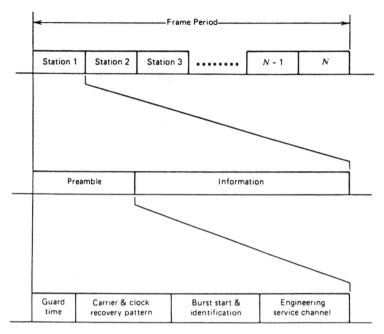

Figure 7.17. Example of TDMA burst format.

Why use TDMA in the first place? It lies in a major detraction of FDMA. Satellites use traveling-wave tubes (TWTs) in their transmitter final amplifiers. A TWT has the undesirable property of nonlinearity in its input–output characteristics when operated at full power. When there is more than one carrier accessing the transponder simultaneously, high levels of intermodulation (IM) products are produced, thus increasing noise and crosstalk. When a transponder is operated at full power output, such noise can be excessive and intolerable. Thus input must be backed off (i.e., level reduced) by ≥ 3 dB. This, of course, reduces the EIRP and results in reduced efficiency and reduced information capacity. Consequently, each earth station's uplink power must be carefully coordinated to ensure proper loading of the satellite. The complexity of the problem increases when a large number of earth stations access a transponder, each with varying traffic loads.

On the other hand, TDMA allows the transponder's TWT to operate at full power because only one earth-station carrier is providing input to the satellite transponder at any one instant.

To summarize, consider the following advantages and disadvantages of FDMA and TDMA. The major advantages of FDMA are as follows:

- No network timing is required.
- Channel assignment is simple and straightforward.

The major disadvantages of FDMA are as follows:

- Uplink power levels must be closely coordinated to obtain efficient use of transponder RF output power.
- Intermodulation difficulties require power back-off as the number of RF carriers increases with inherent loss of efficiency.

The major advantages of TDMA are as follows:

- There is no power sharing and IM product problems do not occur.
- The system is flexible with respect to user differences in uplink EIRP and data rates.
- Accesses can be reconfigured for traffic load in almost real time.

The major disadvantages of TDMA are as follows:

- Accurate network timing is required.
- There is some loss of throughput due to guard times and preambles.
- Large buffer storage may be required if frame lengths are long.

5.7.3 INTELSAT Demand-Assignment Multiple Access System (DAMA).

The basic communication services that are supported by this system include information rates* of 8 kbits/s† for voice service using CS-ACELP (conjugate-structure algebraic code excited linear predictive coding), 16 kbits/s using LD-CELP (low-delay code excited linear predictive coding) for voice, facsimile, and voice-band data, and 64 kbits/s for switched digital data. Higher information of $n \times 64$ kbits/s up to 8.448 Mbits/s can also be supported subject to capacity availability.

The DAMA system is a multinode network where circuits are established between two nodes in a mesh configuration upon demand. Call setup and termination are accomplished via control messages exchanged between the NMCC (network management control center) and traffic earth stations. Upon receipt of a call request message from a traffic earth station, the NMCC assigns connection details to the calling and called terminal, subject to satellite circuit availability within permitted connectivities. The connection assignment information contained in the NMCC control messages is utilized by the DAMA terminal equipment to automatically tune to the assigned operating frequencies and to automatically set the transmit EIRP. The control messages also convey system management information.

When a call is terminated, the connection is released and the associated resources returned to a common resource pool.

---

* INTELSAT defines the *information rate* as the bit rate entering the modem/FEC (forward error correction) subsystem, prior to the application of FEC.
† Offered in the hemispheric beam to support domestic rural telephony and is not to be used for international PSTN.

DAMA Carrier Characteristics. Three types of RF carriers are transmitted by an INTELSAT DAMA traffic earth station: traffic carrier, control carrier, and engineering service channel (ESC). All earth stations in the DAMA network are capable of transmitting and receiving traffic and control carriers. All INTELSAT standard A, B, and F-3 DAMA earth stations are equipped for transmitting and receiving ESC carriers.

Equivalent Isotropically Radiated Power (EIRP). DAMA earth stations are sized to meet the maximum EIRP values shown in Table 7.4. A DAMA earth station may operate with a variety of receive earth station standards from Standard A to Standard H-2. The maximum required EIRP is determined by the maximum information rate and/or smallest receive earth station anticipated in the communication connectivities. In addition, the sizing of earth station HPA should take into account the number and types of DAMA carriers as well as other INTELSAT approved carriers that may need to be transmitted simultaneously.

Carrier Performance Objectives, INTELSAT DAMA System. Table 7.5 shows the carrier performance objectives for the INTELSAT DAMA system.

5.7.4 INTELSAT QPSK/IDR Application Block Diagram. Figure 7.18 shows an application block diagram for the INTELSAT QPSK/IDR system.

5.7.5 INTELSAT SCPC (Single Channel per Carrier)/FM Diagram. Figure 7.19 shows SCPC/Companded FM system block diagram with satellite communication application. The baseband interface is analog.

TABLE 7.4 Example EIRP Requirements for Operation with INTELSAT IX (Beam Edge, dBW)[a,b]

Beam	Information Rate (kbit/s)	Receive Earth Station							
		A	B	F-3	F-2	F-1	H-4	H-3	H-2
Global	16	46.4	47.1	48.2	49.5	52.4	52.9	56.3	N/A[c]
	64[d]	51.1	51.8	52.9	54.2	57.1	57.6	61.0	N/A
Hemi	8	36.9	37.2	37.7	38.6	40.5	40.8	43.6	46.2
	16	42.7	43.0	43.5	44.4	46.3	46.6	49.4	52.0
	64[d]	47.4	47.7	48.2	49.1	51.0	51.3	54.1	56.7

[a] These maximum EIRP values have been computed using a beam edge saturation flux density of −81.0 dBW/m² for Global beam and −83.0 dBW/m² for Hemi beam.
[b] The maximum EIRPs shown in the above table assume transmit and receive earth stations are located at beam edge.
[c] NA, not available.
[d] EIRP requirements for $n \times 64$ kbits/s ($n = 2$ to 132) carrier will be defined in the future.
Source: Table 4, page 18, IESS-311 (Rev. 1) [20].

TABLE 7.5   Carrier Performance Objectives for the INTELSAT DAMA System

Parameter				Units
1. Information rate	8	16	64	kbits/s
2. FEC rate	3/4	3/4	1/2	
3. Modulation	BPSK	QPSK	QPSK	
4. Occupied BW	10.7	10.7	64.0	kHz
5. Allocated BW/Channel spacing[a]	17.5	17.5	90.0	kHz
6. Threshold (minimum performance)				
BER	10^{-3}	10^{-6}	10^{-6}	dB
E_b/N_0	6.1	8.9	7.5	dBK/K
C/T	-183.5	-177.7	-173.0	dBW/K
7. Margin to threshold	2.0	2.0	2.0	dB
8. Clear sky performance				
BER	2×10^{-6}	10^{-9}	10^{-11}	
C/T	-181.5	-175.7	-171.0	dBW/K
9. Link availability (w.r.t. threshold)	99.9	99.9	99.9	% of year

[a] These are nominal carrier channel spacing values. INTELSAT reserves the right to change these carrier channel spacing values as operational requirement dictates.
Source: Table 1, page 15, IESS-311 [20].

5.8   Earth Station Link Engineering

5.8.1   Introduction. Up to this point we have discussed basic satellite communication concepts such as access and coverage. This section briefly covers link engineering methods. Our approach introduces the reader to essential path engineering, expanding on the basic principles previously discussed in Section 4 of this chapter. Links to an from satellites are nothing more than specialized line-of-sight microwave links. As we saw in Section 5.3, a communication satellite, as we discuss here, is nothing more than a distant RF repeater. By international treaty, the downlink power from a satellite must be limited so as not to interfere with terrestrial services sharing the same frequency band. A companion uplink does not have these limitations.

5.8.2   Earth Station Receiving System Figure of Merit, G/T. The figure of merit of an earth station receiving system, G/T, has been introduced into the technology to describe the capability of an earth station or a satellite to "*receive*" a signal. It is also a convenient tool in the link budget analysis. A link budget is used by the system engineer to size components of earth stations and satellites, such as RF output power, antenna gain and directivity, and receiver front-end characteristics.

G/T can be written as a mathematical identity:

$$G/T = G_{dB} - 10 \log T_{sys} \qquad (7.18)$$

where G is the net antenna gain up to an arbitrary reference point or reference plane in the downlink receive chain (for an earth station). Conventionally, in commercial practice the reference plane is taken at the input of the low-noise

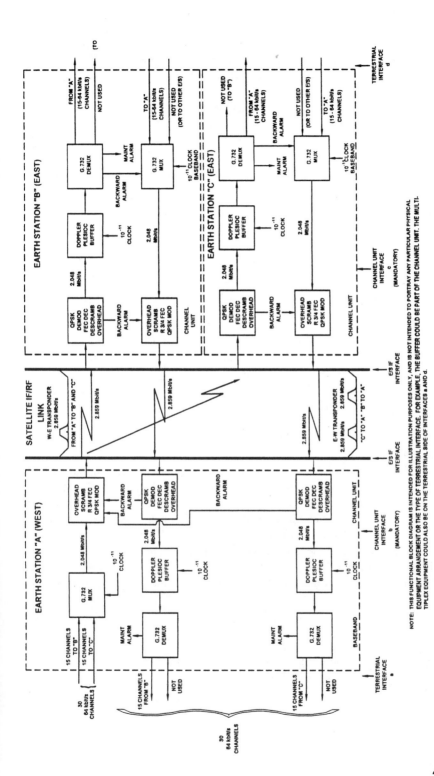

Figure 7.18. Illustration of a typical INTELSAT multidestinational QPSK/IDR application. From Figure 20, page 66, INTELSAT IESS-308 (Rev. 10), [10].

223

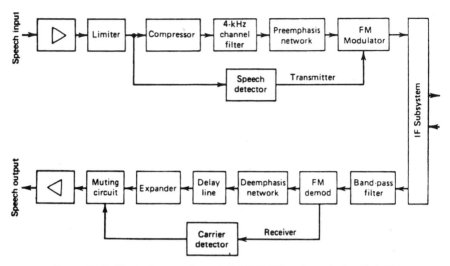

Figure 7.19. Block diagram of a typical SCPC FM station unit. See Ref. 22.

amplifier (LNA). Thus G is simply the gross gain of the antenna minus all losses up to the LNA. These losses include feed loss, waveguide loss, bandpass filter loss, and, where applicable, directional coupler loss, waveguide switch insertion loss, radome loss, and transition losses.

T_{sys} is the effective noise temperature of the receiving system and

$$T_{sys} = T_{ant} + T_{recvr} \tag{7.19}$$

T_{ant} or the antenna noise temperature includes all noise-generating components up to the reference plane. The components include sky noise (T_{sky}) plus the thermal noise generated by ohmic losses created by all devices inserted into the system up to the reference plane, including the radome. A typical earth station receiving system is shown in Figure 7.20 for an 11-GHz downlink. For a 4-GHz downlink the minimum elevation angle would be 5°. The elevation angle is that angle measured from the horizon (0°) to the antenna main beam when pointed at the satellite. Antenna noise (T_{ant}) is calculated by the following formula:

$$T_{ant} = \frac{(l_a - 1)290 + T_{sky}}{l_a} \tag{7.20}$$

where l_a is the numeric equivalent of the sum of the ohmic losses up to the reference plane. l_a is calculated by

$$l_a = \log_{10}^{-1} \frac{L_a}{10} \tag{7.21}$$

where L_a is the sum of the losses in decibels.

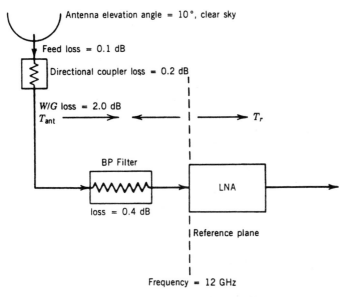

Antenna elevation angle = 10°, clear sky

Feed loss = 0.1 dB

Directional coupler loss = 0.2 dB

W/G loss = 2.0 dB
T_{ant} ⟶ ⟵   ⟶ T_r

BP Filter

LNA

loss = 0.4 dB

Reference plane

Frequency = 12 GHz

Figure 7.20. Example of an earth station receiving system.

Sky noise varies with frequency and elevation angle. Typical values of sky noise are (from CCIR Rep. 720) [8]:

Frequency (GHz)	Elevation Angle	Sky Noise (K)
4.0	5°	25
7.5	5°	35
11.7	10°	30
20.0	10°	110

An earth station operating at 12 GHz with a 10° elevation angle will typically display an antenna noise temperature of 150 K [9].

Using values given in Figure 7.20 and T_{sky} of 30 K, we can calculate T_{ant} for a typical earth station where the downlink operating frequency is 12 GHz.

Sum the losses:	Feed	0.1 dB
	Directional coupler	0.2 dB
	Waveguide	2.0
	Bandpass filter	0.4 dB
	Total	2.7 dB = L_a

Using equation 7.21 we obtain

$$l_a = \log_{10}^{-1}(2.7/10)$$
$$= 1.86$$

Now use the value for l_a and T_{sky} in equation 7.20:

$$T_{ant} = [(1.86 - 1)290 + 30]/1.86$$
$$= 150 \text{ K}$$

The noise figure for a typical commercial LNA or down-converter is 4 dB at 12 GHz. Convert the 4-dB value to equivalent noise temperature. Noise figure can be related to noise temperature (T_e) by the following formula:

$$NF_{dB} = 10 \log(1 + T_e/290) \qquad (7.22)$$

For the sample problem, $T_e = T_{recvr}$:

$$4 \text{ dB} = 10 \log(1 + T_e/290)$$
$$0.4 = \log(1 + T_e/290)$$

Take the antilog of 0.4, and

$$2.51 = 1 + T_e/290$$
$$T_e = 438 \text{ K} = T_{recvr}$$
$$T_{sys} = T_{ant} + T_{recvr}$$
$$= 150 + 438$$
$$= 588 \text{ K}$$

Suppose that the antenna in Figure 7.20 had a 47-dB gross gain. What, then, is the G/T of the receiving system at a $10°$ elevation angle? Calculate net antenna gain.

$$G_{net} = 47 - 2.7$$
$$= 44.3 \text{ dB}$$
$$G/T = 44.3 - 10 \log T_{sys}$$
$$= 44.3 - 10 \log 588$$
$$= 44.3 - 27.7$$
$$= +16.7 \text{ dB/K}$$

5.8.3 *Station Margin.* One major consideration in the design of LOS micro-wave systems is the fade margin, which is the additional signal level added in the system calculations to allow for fading. This value is often on the order of 20–50 dB. In other words, the receive signal level system in question was over-built to provide 20–50 dB above threshold to overcome most fading conditions or to ensure that noise would not exceed a certain norm for a fixed time frame.

As we saw in Section 4, fading is caused by anomalies in the intervening medium between stations or by the reflected signal, thus causing interference to the direct ray signal. There would be no fading phenomenon on a radio signal being transmitted through a vacuum well above the earth's surface. Thus satellite earth station signals are subject to fade only during the time they traverse the atmosphere. For this case, most fades, if any, may be attributed to rainfall or very low elevation angle refractive anomalies.

Margin or station margin is an additional design advantage that compensates for deteriorated propagation conditions or fading. The margin designed into an LOS system is large and is achieved by increasing antenna size, improving the receiver noise figure, or increasing transmitter output power. The station margin of a satellite earth station in comparison is small, on the order of 4–6 dB. Typical rainfall attenuation exceeding 0.01% of a year may be from 1 dB to 2 dB (in the 4-GHz band) without a radome on the antenna and when the antenna is at 5° elevation angle. As the antenna elevation increases to zenith, the rain attenuation notably decreases because the signal passes through less atmosphere. The addition of a radome could increase the attenuation to 6 dB or greater during precipitation. Receive station margin for an earth station, those extra decibels on the downlink, is sometimes achieved by use of threshold-extension demodulation techniques for FM systems and forward error correction coding on digital systems. Uplink margin is provided by using larger transmitters and by increasing power output when necessary. A G/T ratio in excess of the minimum required for clear sky conditions at the 5° elevation angle will also provide margin but may prove expensive to provide.

5.8.4 Typical Downlink Power Budget.

A link budget is a tabular method of calculating space communication system parameters. The approach is similar to that used on line-of-sight microwave links (radiolinks) (see Section 4.4). We start with the EIRP of the satellite for the downlink or the EIRP of the earth station for the uplink. The bottom line is C/N_0 and link margin (in decibels). C/N_0 is the carrier-to-noise ratio in 1 Hz of bandwidth at the input of the LNA. (*Note:* RSL or receive signal level and C are synonymous.) Expressed as an equation:

$$\frac{C}{N_0} = \text{EIRP} - \text{FSL}_{\text{dB}} - (\text{other losses}) + G/T_{\text{dB/K}} - k \qquad (7.23)$$

where FSL* is the free-space loss to the satellite for the frequency of interest and k is Boltzmann's constant expressed in decibel-watts. "Other losses" may include (where applicable):

- Polarization loss (0.5 dB).
- Pointing losses, terminal and satellite (0.5 dB each).
- Off-contour loss (depends on satellite antenna characteristics).

---

* Remember that geostationary satellite range varies with elevation angle and is minimum at zenith.

- Gaseous absorption loss (varies with frequency, altitude, and elevation angle).
- Excess attenuation due to rainfall (for systems operating above 10 GHz).

The loss values in parentheses are conservative estimates and should be used only if no definitive information is available.

The off-contour loss refers to spacecraft antennas that provide a spot or zone beam with a footprint on a specific geographical coverage area. There are usually two contours, one for G/T (uplink) and the other for EIRP (downlink). Remember that these contours are looking from the satellite down to the earth's surface. Naturally, an off-contour loss would be invoked only for earth stations located outside of the contour line. This must be distinguished from satellite pointing loss, which is a loss value to take into account that satellite pointing is not perfect. The contour lines are drawn as if the satellite pointing were "perfect."

Gaseous absorption loss (or atmospheric absorption) varies with frequency, elevation angle, and altitude of the earth station. As one would expect, the higher the altitude, the less dense the air and thus the less loss. Gaseous absorption losses vary with frequency and inversely with elevation angle. Often, for systems operating below 10 GHz, such losses are neglected. Reference 17 suggests a 1-dB loss at 7.25 GHz for elevation angles under $10°$ and for 4 GHz, 0.5 dB below $8°$ elevation angle.

Example of a Link Budget: Assume the following: a 4-GHz downlink, $5°$ elevation angle, EIRP is $+30$ dBW; satellite range is 25,573 statute miles (sm), and the terminal G/T is $+20.0$ dB/K. Calculate the downlink C/N_0.

First calculate the free-space loss. Use equation 7.5:

$$L_{dB} = 96.6 + 20\log F_{GHz} + 20\log D_{sm}$$
$$= 96.6 + 20\log 4.0 + 20\log 25{,}573$$
$$= 96.6 + 12.04 + 88.16$$
$$= 196.8 \text{ dB}$$

Example Link Budget: Downlink

EIRP of satellite	$+30$ dBW
Free-space loss	-196.8 dB
Satellite pointing loss	-0.5 dB
Off-contour loss	0.0 dB
Excess attenuation rainfall	0.0 dB
Gaseous absorption loss	-0.5 dB
Polarization loss	-0.5 dB
Terminal pointing loss	-0.5 dB
Isotropic receive level	-168.8 dBW
Terminal G/T	$+20.0$ dB/K
Sum	-148.8 dBW
Boltzmann's constant (dBW)	$-(-228.6$ dBW)
C/N_0	79.8 dB

On repeatered satellite systems, sometimes called "bent-pipe satellite systems" (those that we are dealing with here), the link budget is carried out only as far as C/N_0, as we did above. It is calculated for the uplink and for the downlink separately. We then calculate an equivalent C/N_0 for the system (i.e., uplink and downlink combined). Use the following formula to carry out this calculation:

$$\left(\frac{C}{N_0}\right)_{(s)} = \frac{1}{1/(C/N_0)_{(u)} + 1/(C/N_0)_{(d)}} \tag{7.24}$$

Example: Suppose that an uplink has a C/N_0 of 82.2 dB and its companion downlink has a C/N_0 of 79.8 dB. Calculate the C/N_0 for the system $(C/N_0)_s$. First calculate the equivalent numeric value (NV) for each C/N_0 value.

$$NV(1) = \log^{-1}(79.8/10) = 95.5 \times 10^6$$
$$NV(2) = \log^{-1}(82.2/10) = 166 \times 10^6$$
$$C/N_0 = 1/[(10^{-6}/95.5) + (10^{-6}/166)]$$
$$= 1/(0.016 \times 10^{-6}) = 62.5 \times 10^6 = 77.96 \text{ dB}$$

This is the carrier-to-noise ratio in 1 Hz of bandwidth. To derive C/N for a particular RF bandwidth, use the following formula:

$$C/N = C/N_0 - 10\log BW_{\text{Hz}} \tag{7.25}$$

Suppose the example system had a 1.2-MHz bandwidth with the C/N_0 of 77.96 dB. What is the C/N?

$$C/N = 77.96 \text{ dB} - 10\log(1.2 \times 10^6)$$
$$= 77.96 - 60.79$$
$$= 17.17 \text{ dB}$$

5.8.5   Uplink Considerations. A typical specification for INTELSAT states that the EIRP per voice channel must be $+61$ dBW (example); thus to determine the EIRP for a specific number of voice channels to be transmitted on a carrier, we take the required output per voice channel in dBW (the above) and add logarithmically $10 \log N$, where N is the number of voice channels to be transmitted.

For example, consider the case for an uplink transmitting 60 voice channels; thus

$$+61 \text{ dBW} + 10\log 60 = 61 + 17.78 = +78.78 \text{ dBW}$$

If the nominal 50-ft (15-m) antenna has a gain of 57 dB (at 6 GHz) and losses typically of 3 dB, the transmitter output power, P_t, required is

$$\text{EIRP}_{\text{dBW}} = P_t + G_{\text{ant}} - \text{line losses}_{\text{dB}} \qquad (7.26)$$

where P_t is the output power of the transmitter (in decibel-watts) and G_{ant} is the antenna gain (in decibels) (uplink). Then in the example we have

$$+78.78 \text{ dBW} = P_t + 53 - 3$$

$$P_t = +24.78 \text{ dBW}$$

$$= 300.1 \text{ W}$$

5.9 Digital Communication by Satellite

In commercial telecommunications there are three methods of handling digital communication by satellite: TDMA, FDMA, and over a VSAT network. TDMA was covered in Section 5.7.2, and VSAT networks are discussed in Section 5.10. DAMA (demand assignment multiple access) is a more specialized digital access and modulation system covered in Section 5.7.3. In this section we describe a typical digital FDMA system, the INTELSAT International Business System (IBS). Like other FDMA systems, several users can share a common transponder and there are power backoff rules for downlink EIRP to meet international regulations and to limit the level of intermodulation (IM) products produced. If not controlled, these products can degrade error performance.

Because of the restrictions of EIRP levels on the satellite downlink, one approach to meet minimum receive signal level (RSLs) is to employ FEC (forward error correction). INTELSAT allows its correspondents to use either $R = \frac{1}{2}$ or $R = \frac{3}{4}$ coding rates where the coding rate R equals information bit rate/coded symbol rate. Where BER performance objectives typically on INTELSAT links were 1×10^{-7}, they now are better than 1×10^{-10} under clear sky conditions. That is three orders of magnitude improvement. To illustrate a practical example of a fairly widely used digital FDMA system, we provide a brief overview of INTELSAT's IDR (Intermediate Data Rate) system.

5.9.1 INTELSAT IDR.
INTELSAT's IDR (system) is a satellite FMDA system using quadrature phase-shift keying (QPSK) modulation. It accepts standard data rates in the range of 64 kbit/s to 44.736 Mbit/s. It operates in both the 6/4-GHz and 14/11-GHz bands. Table 7.5 illustrates the QPSK characteristics and Transmission Parameters for the IDR system. Table 7.6 shows RF/baseband transmission parameters such as bandwidth occupancy.

5.10 Very Small Aperture Terminal (VSAT) Networks

5.10.1 Definition and Rationale.
VSATs are defined by their antenna aperture (diameter) which can vary from 0.5 m (1.6 ft) to 2 m (6.5 ft). A VSAT

TABLE 7.6 QPSK Characteristics and Transmission Parameters for IDR Carriers

Parameter	Requirement
1. Information rate (IR)	64 kbits/s to 44.736 Mbits/s
2. Overhead data rate for carriers with IR $\geq$ 1.544 Mbits/s	96 kbits/s
3. Forward error correction encoding	Rate 3/4 convolutional encoding/Viterbi decoding
4. Energy dispersal (scrambling)	As per ITU-R524-4
5. Modulation	Four-phase Coherent PSK
6. Ambiguity resolution	Combination of differential encoding (180°) and FEC (90°)
7. Clock recovery	Clock timing must be recovered from the received data stream
8. Minimum carrier bandwidth (allocated)	0.7 R Hz or [0.933(IR + Overhead)]
9. Noise bandwidth (and occupied bandwidth)	0.6 R Hz or [0.8(IR + Overhead)]
10. E_b/N_0 at BER (Rate 3/4 FEC)	10^{-3} 10^{-7} 10^{-8}
a. Modems back-to-back	5.3 dB 8.3 dB 8.8 dB
b. Through satellite channel	5.7 dB 8.7 dB 9.2 dB
11. C/T at nominal operating point	$-219.9 + 10\log_{10}$(IR + OH), dBW/K
12. C/N in noise bandwidth at nominal operating point (BER $\leq 10^{-7}$)	9.7 dB
13. Nominal bit error rate at operating point	1×10^{-7}
14. C/T at threshold (BER = 1×10^{-3})	$-222.9 + 10\log_{10}$(IR + OH), dBW/K
15. C/N in noise bandwidth at threshold (BER = 1×10^{-3})	6.7 dB
16. Threshold bit error rate	1×10^{-3}

Notes:

(1) IR is the information rate in bits per second.

(2) R is the transmission rate in bits per second and equals (IR + OH) times 4/3 for carriers employing Rate 3/4 FEC.

(3) The allocated bandwidth will be equal to 0.7 times the transmission rate, rounded up to the next highest odd integer multiple of 22.5-kHz increment (for information rates less than or equal to 10 Mbits/s) or 125-kHz increment (for information rates greater than 10 Mbits/s).

(4) Rate 3/4 FEC is mandatory for all IDR carriers.

(5) OH = overhead.

(6) Also see Table 7.2. See also Ref. 7.

Source: IESS-308 (Rev. 7), which can be considered as part of Rev. 10 [10]. Courtesy of INTELSAT.

network consists of one comparatively large hub earth station and remote VSAT terminals. Some networks in North America have over 3000 remote VSAT terminals. Many such networks exist, often large retailers such as gas stations or grocery stores. There are three underlying reasons for the use of VSAT networks:

1. An economic alternative to establish a data network, particularly if traffic flow is to/from a central facility, usually a corporate headquarters to/from outlying remotes.

2. To bypass telephone companies with a completely private network.

TABLE 7.7 Transmission Parameters for IDR Carriers with Rate 3/4 Coding

Information Rate (bits/s)	Overhead Rate (kbits/s)	Data Rate, (IR + OH) (bits/s)	Transmission Rate (bits/s)	Occupied Bandwidth (Hz)	Allocated Bandwidth (Hz)	C/T (dBW/K)	C/N₀ (dB − Hz)	C/N (dB)
64 k	0	64 k	85.33 k	51.2 k	67.5 k	−171.8	56.8	9.7
192 k	0	192 k	256.00 k	153.6 k	202.5 k	−167.1	61.5	9.7
384 k	0	384 k	512.00 k	307.2 k	382.5 k	−164.1	64.5	9.7
1.544 M	96	1.640 M	2.187 M	1.31 M	1552.5 k	−157.8	70.8	9.7
2.048 M	96	2.144 M	2.859 M	1.72 M	2002.5 k	−156.6	72.0	9.7
6.312 M	96	6.408 M	8.544 M	5.13 M	6007.5 k	−151.8	76.8	9.7
8.448 M	96	8.544 M	11.392 M	6.84 M	7987.5 k	−150.6	78.0	9.7
32.064 M	96	32.160 M	42.880 M	25.73 M	30125.0 k	−144.8	83.8	9.7
34.368 M	96	34.464 M	45.952 M	27.57 M	32250.0 k	−144.5	84.1	9.7
44.736 M	96	44.832 M	59.776 M	35.87 M	41875.0 k	−143.4	85.2	9.7

Notes:

(1) The above table illustrates parameters for recommended carrier sizes. However, any other information rate between 64 kbits/s and 44.736 Mbits/s can be used.

(2) C/T, C/N_0, and C/N values have been calculated for a 10^{-7} BER and assume the use of Rate 3/4 FEC.

(3) For carrier information rates of 10 Mbits/s and below, carrier frequency spacings will be odd multiples of 22.5 kHz. For greater rates, they will be on multiples of 125 kHz.

(4) Rate 3/4 FEC is mandatory for all IDR carriers.

(5) IR, information rate; OH, overhead.

Source: IESS-308 (Rev. 7), which can be considered as part of Rev. 10 [10]. Courtesy of INTELSAT. See also Ref. 21.

3. To provide quality telecommunication connectivity where other means are substandard or nonexistent.

Regarding reason 3, the author is aware of one emerging nation where 124 bank branches had no electrical communication whatsoever with the headquarters institution in the capital city.

5.10.2 Characteristics of a VSAT Network.

On conventional VSAT networks, the hub is designed to compensate for the VSAT handicap (i.e., its small size). For example, a hub antenna aperture is 5–11 m (16–50 ft) [11]. High-power amplifiers (HPAs) run from 100 to 600 watts output power. Low-noise amplifiers, typically at 12 GHz, display (a) noise figures from 0.5 to 1.0 dB and (b) low-noise downconverters in the range of 1.5-dB noise figure. Hub G/T values range from +29 to +34 dB/K.

VSAT terminals have transmitter output powers ranging from 1 to 50 watts, depending on service characteristics. Receiver noise performance using a low-noise downconverter is about 1.5 dB, otherwise 1 dB with an LNA. G/T values for 12.5-GHz downlinks are between +14 and +22 dB/K, depending greatly on antenna aperture. The idea is to make a VSAT terminal as inexpensive as possible.

Figure 7.21 shows the hub/VSAT concept of a star network with the hub at the center.

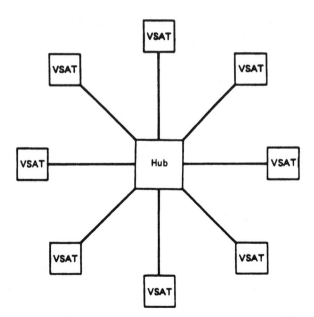

Figure 7.21. VSAT network topology. Note the star configuration and that the outlying VSAT remotes can number in the thousands.

5.10.3 Access Techniques

Inbound and Outbound. Inbound refers to traffic from VSAT(s) to hub, and *outbound* refers to traffic from hub to VSAT(s). The outbound link is commonly a time-division multiplex (TDM) serial bit stream, often 56 kbps, and some high-capacity systems reach 1.544 or 2.048 Mbits/s. The inbound links can take on any one of a number of flavors, typically 9600 bits/s.

More frequently VSAT systems support interactive data transactions, which are very short in duration. Thus, we can expect bursty operation from a remote VSAT terminal. One application is to deliver, in near real time, point-of-sale (POS) information, forwarding it to headquarters where the VSAT hub is located. Efficiency of bandwidth use is not a primary motivating factor in system design. Thus, for the interactive VSAT data network environment, low delay, simplicity of implementation, and robust operation are generally of greater importance than the bandwidth efficiency achieved.

Message access on any shared system can be of three types: fixed assigned, contention (random access), or reservation (controlled access). There are hybrid schemes between contention and reservation.

In the fixed assigned multiple access, VSAT protocols are SCPC*/FDMA, CDMA (a spread spectrum technique), and TDMA. All three are comparatively inefficient in the bursty environment with hundreds or thousands of potential users.

With the contention/random access category, there is the famous ALOHA protocol and selective reject (SREJ) ALOHA. Both have to have some form of collision resolution. In ALOHA, stations transmit new messages on the channel as they are generated. Collision resolution is achieved simply by retransmitting colliding packets with random delay. ALOHA has low efficiency of channel use but this is offset by low access delay, the ability to handle variable-length packets, robust operation, and minimal equipment complexity.

SREJ ALOHA employs subpacketization of messages in conjunction with a selective reject strategy and is able to achieve notably higher random access throughout than pure ALOHA. Here data are formed into a contiguous sequence of independently detectable fixed-length subpackets, each with its own header and acquisition preamble. The idea here is that most collisions in an asynchronous channel result in partial overlap of contending packets, so that only the smaller subpackets encounter conflict and must be retransmitted. SREJ ALOHA is an efficient access protocol.

Slotted ALOHA is based on the principle of reducing the vulnerability of a packet by constraining transmission to fixed-length packets which begin and end in TDMA-like slot boundaries. Thus maximum throughput is increased and the excellent delay characteristics of pure ALOHA are maintained.

Another variant is the tree collision resolution algorithm (CRA) random access system, which uses fixed-length packets. Here packets involved in collision participate in a systematic partitioning procedure for collision resolution, during

---

* SCPC stands for single channel per carrier.

which time new messages are not allowed to access the channel. The tree-type protocols have excellent capacity capabilities but can suffer deadlocks due to incorrect channel observation.

The third group of protocols is the reservation/controlled access type. These are based on demand assignment multiple access (DAMA) schemes and are useful variable length data frames. As in our previous discussion in Section 5.7.3, DAMA is a two-step process. There was a call-setup stage before the actual call went into operation. In this case, there is phase 1, where short reservation packets are transmitted from the VSAT requesting service and giving information regarding the station's demand. The second phase is the actual passing of data.

Reservation can be made by either a contention process or a fixed assigned process, both described above. One major advantage of a DAMA protocol is that data messages can be scheduled in a conflict-free manner. Well-designed DAMA systems have advantages with variable-length message traffic with relatively high overall channel throughput. However, this higher throughput is accompanied by a relatively large minimum latency (>0.500 ms) delay due to the reservation mechanism.

A subset of DAMA is DAMA with TDMA reservations. In this case, channel frames are made up into request and packet transmission intervals. Slots are available in each frame for VSATs to request data slot allocation. Of course, the allocation assignment is returned to the VSAT in the outbound TDM bit stream. Due to satellite propagation delays and TDMA frame structure, such protocols are characterized by a relatively high latency delay and are limited regarding the maximum number of accesses.

Another form of DAMA is DAMA with slotted ALOHA reservations. This access protocol can be used to support large VSAT populations because it uses contention access rather than fixed assigned TDMA-type access. DAMA with slotted ALOHA access provides good overall performance and can handle mixed interactive file-transfer traffic [11].

5.10.4 VSAT Transponder Operation.
A total VSAT system may occupy no more than 1 MHz of transponder space. Other, larger VSAT systems may require more transponder bandwidth. A typical 1-MHz frequency assignment may be apportioned as shown in Figure 7.22.

TDMA inbound carriers to the hub, as shown in Figure 7.22, can be configured for one user per carrier (more similar to an FDMA configuration) or multiple users per carrier using one of several TDMA access techniques, or a DAMA discipline. There are many subnetworking possibilities as well. For example, each TDMA carrier may be assigned to a family of users. Forward error correction (FEC) is commonly used and the employment of ARQ* is almost universal.

Carrier modulation techniques for VSAT systems is commonly BPSK, often favored over QPSK because it is more robust.

---

* ARQ stands for automatic repeat request. See Chapter 11 for a discussion of ARQ.

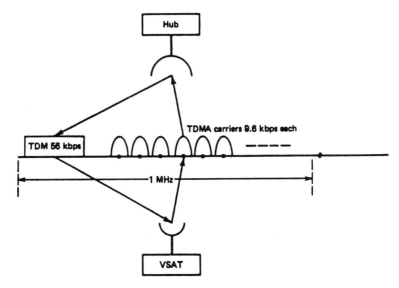

Figure 7.22. VSAT system operation with a 1-MHz allocation on a satellite transponder [12].

6 FIBER-OPTIC COMMUNICATION LINKS

Bandwidth capacity of a fiber-optic cable is virtually unlimited. It has excellent attenuation properties, as low as 0.25 dB/km or even slightly lower. A major advantage fiber has over coaxial cable as a transmission medium is that no equalization is necessary. Also, repeater separation is on the order of 10–100 times that of coaxial cable for equal transmission bandwidths. Other advantages are:

- Electromagnetic immunity
- Ground loop elimination
- Security
- Small size and lightweight
- Expansion capabilities requiring change out of electronics, in most cases
- No licensing required

Fiber has analog transmission application, particularly for video/TV. However, for this discussion we will be considering only digital applications, principally as a PCM highway or "bearer."

Fiber-optic transmission is used for links under 1 ft in length all the way up to and including transoceanic undersea cable. In fact, all transoceanic cables presently being installed and planned for the future are based on fiber optics.

Fiber-optic technology was developed by physicists, and, following the convention of optics, wavelength rather than frequency is used to denote the position of light emission in the electromagnetic spectrum. The fiber optics of today uses

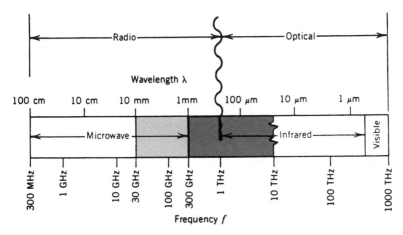

Figure 7.23. Frequency spectrum above 300 MHz. The usable wavelengths are just above and below 1 μm.

three wavelength bands: around 800 nm, 1300 nm, and 1600 nm or near-visible infrared. This is shown in Figure 7.23.

The application and further development of wavelength division multiplexing (WDM) have further advance the bandwidth capabilities. We expect 40-Gbit/s carriers on fiber to be deployed at least by 2005. WDM with 100–200 carriers per fiber strand just in the 1550-nm band. Simple multiplication illustrates an eventual capacity in gigabits per second per fiber of 40×200 or 8000 Gbits/s.

6.1 Scope

This section covers an overview of how a fiber-optics link works, including types of optical fiber, a discussion of sources, detectors, amplifiers, WDM technology, connectors, and splices as well as link design using the familiar link budget as applied to optical fiber. Included is a review of impairment peculiar to optical fiber transmission.

6.2 Introduction to Optical Fiber as a Transmission Medium

Optical fiber consists of a core and a cladding as shown in Figure 7.24. At present the most efficient core material is silica SiO_2.

The practical propagation of light through an optical fiber may best be explained using ray theory and Snell's law. Simply stated, we can say that when light passes from a medium of higher refractive index (n_1) into a medium of lower refractive index (n_2), the refractive ray is bent away from the normal. For instance, a ray traveling in water and passing into an air region is bent away from the normal to the interface between the two regions. As the angle of incidence becomes more oblique, the refracted ray is bent more until finally the refracted energy emerges at an angle of 90° with respect to the normal and just grazes the surface. Figure 7.25 shows

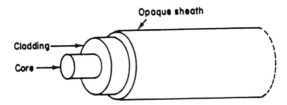

Figure 7.24. Structure of optical fiber consisting of a central core and a peripheral transparent cladding surrounded by protective packaging.

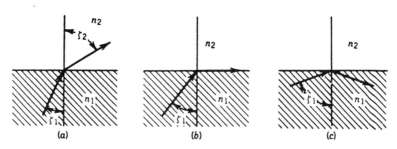

Figure 7.25. Ray paths for several angles of incidence ($n_1 > n_2$).

the various incidence angles. Figure 7.25b illustrates what is called the *critical angle*, where the refracted ray just grazes the surface. Figure 7.25c is an example of total internal reflection. This occurs when the angle of incidence exceeds the critical angle. A glass fiber, for the effective transmission of light, requires total internal reflection.

Another property of the fiber for a given wavelength λ is the normalized frequency V; then

$$V = \frac{2\pi a}{\lambda}\sqrt{n_1^2 - n_2^2} \tag{7.27}$$

where a is the core radius, n_1 is the index of refraction of the core, and n_2 is the index of refraction of the cladding. n_2 of unclad fiber equals 1 (air). In equation 7.27 the term $\sqrt{n_1^2 - n_2^2}$ is called the *numerical aperture* (NA).

In essence the numerical aperture is used to describe the light-gathering ability of fiber. In fact, the amount of optical power accepted by a fiber varies as the square of the numerical aperture. It is also interesting to note that the numerical aperture is independent of any physical dimension of the fiber [12].

As shown in Figure 7.26, there are three basic elements in an optical-fiber transmission system: the optical source, the fiber link, and the optical detector. Regarding the fiber link itself, there are two basic impairments that can limit the length of such a link without resorting to repeaters or that can limit the distance between repeaters. Optical amplifiers can notably extend that distance. These impairments are loss, usually expressed in decibels per kilometer, and dispersion, usually expressed as bandwidth per unit length, such as megahertz

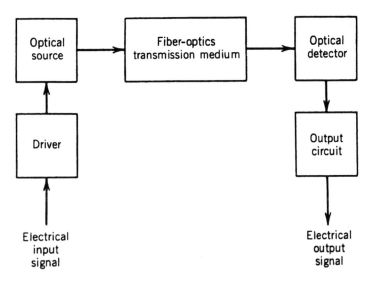

Figure 7.26. Typical fiber-optic communication link.

per kilometer. A particular fiber-optic link may be *power-limited* or *dispersion-limited*. A number of other impairments begin to appear as the number of WDM channels rise and the power on the link is increased to compensate for the losses of passive elements required to form WDM configurations.

Dispersion comes in several flavors manifesting itself in intersymbol interference on the receive side. It can be brought about by a number of factors. There is material dispersion, modal dispersion and chromatic dispersion. Material dispersion derives when the emission spectral line from the fiber source is wide, typically from an LED (light emitting diode). Certain frequencies inside the emission line travel faster than others, causing some transmitted energy from a light pulse to arrive later than other energy. This delayed energy causes intersymbol interference. Modal dispersion occurs when several different modes are launched. To reach the far end, some of the modes have more reflections than others, thus, again, causing some energy from higher-order modes to be delayed compared to lower-order modes.

One way of limiting the number of modes N that a fiber can support is by applying equation 7.27. The modes propagated can be reduced by reducing V, the normalized frequency, and by keeping the ratio n_1/n_2 as small as practical, often 1.01 or less. V can be reduced by reducing a, the radius of the core. When the radius is about 4 μm, $V = 2.405$ and only the HE_{11} mode will propagate. This very thin glass fiber is called *monomode fiber*, and using this type of fiber with some of the longer light wavelengths puts us well on the way to reducing or nearly eliminating dispersion [13].

We should only be concerned with chromatic dispersion on monomode fiber carrying a transmission data rate greater than 1 Gbit/s. We will handle chromatic dispersion later on.

6.3 Types of Optical Fiber

There are three categories of optical fiber as distinguished by their modal and physical properties:

- Step index (multimode)
- Graded index (multimode)
- Single mode (also called monomode)

Step-index fiber is characterized by an abrupt change in refractive index, and graded index is characterized by a continuous and smooth change in refractive index (i.e., from n_1 to n_2). Figure 7.27 shows the fiber construction and refractive index profile for step-index fiber (7.27a) and graded-index fiber (7.27b). Both step-index and graded-index fibers are characterized as "multimode" because more than one mode can propagate. Graded index has a superior bandwidth–distance product compared to that of step index. In other words, it can carry a higher bit rate further than step index. It is also more expensive.

Single-mode fiber is designed such that only one mode can propagate. To do this, $V \leq 2.405$. Such fiber exhibits no modal dispersion at all (theoretically). Typically we might encounter a fiber with indices of refraction of $n_1 = 1.48$ and $n_2 = 1.46$. If the optical source wavelength is 1.2 μm, for single-mode operation we'd find the core radius to be about 4 μm [15].

There are two additional factors that the fiber-optic communication system designer must take into account, namely, minimum bending radius and fiber

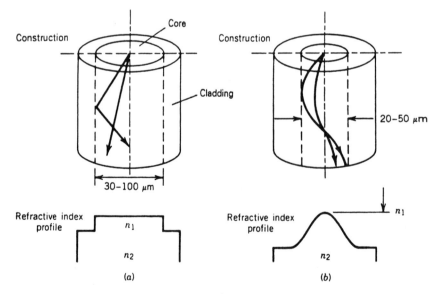

Figure 7.27. Construction and refractive index properties for (a) step-index fiber and (b) graded-index fiber.

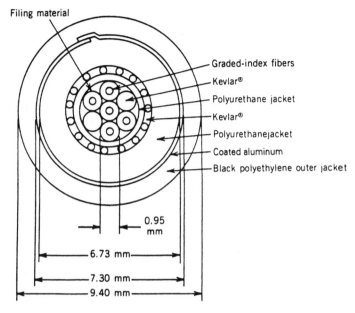

Figure 7.28. Direct-burial optical fiber cable.

strength. Radiation losses at fiber waveguide bends are usually quite small and may be neglected in system design unless the bending radius is smaller than that specified by the fiber manufacturer. Minimum bending radii vary from about 2 cm to 10 cm, depending on the cable characteristics, or, as a rule of thumb, about 10 times the cable diameter. Fiber cable strength is also specified by the manufacturer. For example, one manufacturer for a specific cable type specifies a maximum pulling tension of 1780 N (400 lb) at 20°C, a maximum permissible compression load of 655 N/cm (375 lb/in.) flat plate, and a maximum permissible impact force of 280 N/cm (160 lb/in.).

Figure 7.28 shows a typical five-fiber cable for direct burial.

6.4 Splices and Connectors

Optical fiber cable is commonly available in 1-km sections; it is also available in longer sections, in some types up to 10 km or more. In any case there must be some way of connecting the fiber to the source and to the detector as well as connecting the reels of cable together, whether in 1 km or more lengths, as required. There are two methods of connection, namely, splicing or using connectors. The objective in either case is to transfer as much light as possible through the coupling. A good splice couples more light than the best connectors.

A good splice can have an insertion loss as low as 0.09 dB, whereas the best connector loss can be as low as 0.3 dB.

An optical fiber splice requires highly accurate alignment and an excellent end finish to the fibers. There are three causes of loss at a splice:

1. Lateral displacement of fiber axes
2. Fiber end separation
3. Angular misalignment

Splice loss also varies directly with the numerical aperture of the fiber in question.

There are two types of splice now available, the mechanical splice and the fusion splice. With a mechanical splice an optical matching substance is used to reduce splicing losses. The matching substance must have a refractive index close to the index of the fiber core. A cement with similar properties is also used, serving the dual purpose of refractive index matching and fiber bonding. The fusion splice, also called a *hot splice*, is where the fibers are fused together. The fibers to be spliced are butted together and heated with a flame or electric arc until softening and fusion occur.

Splices require special splicing equipment and trained technicians. Thus it can be seen that splices are generally hard to handle in a field environment such as a cable manhole. Connectors are much more amenable to field connecting. However, connectors are lossier and can be expensive. Repeated mating of a connector may also be a problem, particularly if dirt or dust deposits occur in the area where the fiber mating takes place.

However, it should be pointed out that splicing equipment is becoming more economic, more foolproof, and more user-friendly. Technician training is also becoming less of a burden.

Connectors are nearly universally used at the source and at the detector to connect the main fiber to these units. This makes easier change-out of the detector and source when they fail or have degraded operation.

6.5 Light Sources

A light source, perhaps more properly called a *photon source*, has the fundamental function in a fiber-optics communication system to convert efficiently electrical energy (current) into optical energy (light) in a manner that permits the light output to be effectively launched into the optical fiber. The light signal so generated must also accurately track the input electrical signal so that noise and distortion are minimized.

The link designer can select from three different light sources for fiber-optic communication systems listed in increasing output power capability:

- LED—light-emitting diode
- VCSEL—vertical cavity surface emitting laser
- Laser diodes (LD)

All three devices are fabricated from the same basic semiconductor compounds and have similar heterojunction structures. They do differ in the way they emit light and in their performance characteristics.

An LED is a forward-biased *p–n* junction that emits light through sponta-neous emission, a phenomenon referred to as *electroluminescence*. LDs emit light through stimulated emission. LEDs are less efficient than LDs but are con-siderably more economical. They also have a longer operational life. The emitted light of an LED is incoherent with a relatively wide spectral line width (from 30 to 60 nm) and a relatively large angular spread, about 100°. On the other hand, a semiconductor laser emits a comparatively narrow line width (from <2 to 4 nm). Figure 7.29a shows the spectral line for an LED, and Figure 7.29b shows the spectral line for a semiconductor laser.

With present technology the LED is capable of launching about 100 μW(−10 dBm) or less of optical power into the core of a fiber with a numerical aperture of 0.2 or better. A semiconductor laser with the same input power can couple up to 7 mW (+8.5 dBm) into the same cable. The coupling efficiency of an LED is on the order of 2%, whereas the coupling efficiency of an LD (semiconductor laser) is better than 50%.

Methods of coupling a source into an optical fiber vary, as do coupling efficien-cies. To avoid ambiguous specifications on source output powers, such powers should be stated at the *pigtail*. A pigtail is a short piece of optical fiber coupled to the source at the factory and, as such, is an integral part of the source. Of course, the pigtail should be the same type of fiber as that specified for the link.

Component lifetimes for LEDs are on the order of 100,000 h (MTBF) with up to a million hours reported in the literature. Many manufacturers guarantee a semiconductor laser for 20,000 h or more. About 150,000 h can be expected from semiconductor lasers after stressing and culling of unstable units. These hi-rel (high reliability) lasers are used in undersea cables such as TAT-8/9/10/11 (transatlantic), TPC-4/5 (transpacific), and Columbus 3 connecting Europe with the Americas and the Americas with Asia.

It should be noted that the life expectancy of a semiconductor laser is reduced when it is overdriven to derive more coupled power, such as more than 7 mW.

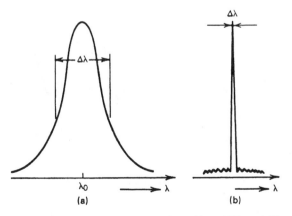

Figure 7.29. Spectral distribution of the emission from (a) an LED and (b) a semiconductor laser (LD). λ is the optical wavelength.

Often these devices are operated at 1 mW (0 dBm). LEDs also have a drawback. Output power tends to drop off with age.

The semiconductor laser is a temperature-dependent device. Its threshold current increases nonlinearly with temperature. Rather than attempt to control the device's temperature, a negative feedback circuit is used whereby a portion of the emitted light is sampled, detected, and fed back to control the drive current. Such circuits are similar to the familiar AGC circuits used on radio receivers.

The third type of light source, which is just reaching maturity, is the VCSEL (vertical-cavity surface emitting laser). It costs less than a laser diode and more than an LED. It is easier to couple light to a fiber from a VCSEL than from either an LED or a laser diode because the output of the VCSEL is circular more matching the circular fiber rather than the elliptical of the LED and LD. VCSELs may be arrayed for use with WDM systems, otherwise, they are employed in fiber-optic LANs and local loops.

Fiber-optic communication systems operate in the nominal wavelength regions of 820 nm, 1330 nm, and 1550 nm. If we examine the attenuation versus wavelength curve in Figure 7.30, we see that for the 820-nm region the lowest attenuation we can expect is 3 dB/km. As wavelength increases, going to the right on the curve, we see another valley around 1330 nm, where the loss per unit length drops to 0.5 dB/km, and there is still another valley at 1550 nm, where the loss is only 0.25 dB/km or a little less. There is mature technology available for all three wavelengths.

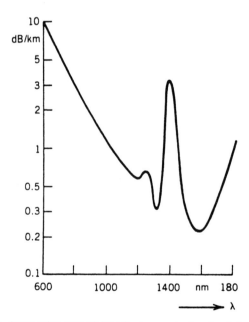

Figure 7.30. Attenuation per unit length versus wavelength of glass fiber [12].

6.6 Light Detectors

The most commonly used detectors (receivers) for fiber-optic communication systems are photodiodes, either PIN or APD. The terminology *PIN* derives from the semiconductor construction of the device where an intrinsic (I) material is used between the p–n junction of the diode.

A photodiode can be considered a photon counter. The photon energy E is a function of frequency and is given by

$$E = h\nu \tag{7.28}$$

where h is Planck's constant (W/s^2) and ν is the frequency in hertz. E is measured in watt-seconds or kilowatt-hours.

The receiver power in the optical domain can be measured by counting, in quantum steps, the number of photons received by a detector per second. The power in watts may be derived by multiplying this count by the photon energy, as given in equation 7.28.

The efficiency of the optical-to-electrical power conversion is defined by a photodiode's *quantum efficiency*, which is the average number of electrons released by each incident photon. A highly efficient photodiode would have a quantum efficiency of 1, and decreasing from 1 indicates progressively poorer efficiencies. The quantum efficiency, in general, varies with wavelength and temperature.

For the fiber-optic communication system engineer, *responsivity* is a most important parameter when dealing with photodiode detectors. Responsivity is expressed in amperes per watt or volts per watt and is sometimes called *sensitivity*. Responsivity is the ratio of the root mean square (rms) value of the output current or voltage of a photodetector to the rms value of the incident optical power. In other words, responsivity is a measure of the amount of electrical power we can expect at the output of a photodiode, given a certain incident light power signal input. For a photodiode the responsivity R is related to the wavelength λ of the light flux and to the quantum efficiency η, the fraction of the incident photons that produce a hole–electron pair. Thus

$$R = \frac{\eta\lambda}{1234} \left(\frac{A}{W} \right) \tag{7.29}$$

with λ measured in nanometers.

The avalanche photodiode (APD) is a gain device displaying gains on the order of 15–20 dB. The PIN diode is not a gain device. Table 7.8 summarizes detector sensitivities with the standard BER of 1×10^{-9} for some common bit rates. Noise equivalent power (NEP) is often used as the figure of merit of a photodiode. NEP is defined as the rms value of optical power required to produce a unit signal-to-noise ratio (i.e., signal-to-noise ratio = 1) at the output of a light-detecting device. NEPs vary for specific diode detectors between 1×10^{-13} W/Hz$^{1/2}$ and 1×10^{-14} W/Hz$^{1/2}$.

TABLE 7.8 Receive Levels, BER Values, and Bit Rates for PIN Diode and APD Light Detectors

Bit Rate	BER	Level (dBm)	Comments
155 Mbits/s	1×10^{-10}	−33	Alcatel, PIN
2.5 Gbits/s	1×10^{-10}	−26	Alcatel, APD
622 Mbits/s	1×10^{-10}	−27	Alcatel, PIN
155 Mbits/s	1×10^{-10}	−35	Alcatel, PIN
622 Mbits/s	1×10^{-10}	−28	ITU-T G.957
2.5 Gbits/s	1×10^{-10}	−23	Lucent, PIN
2.5 Gbits/s	1×10^{-10}	−32	Lucent, APD
155 Mbits/s	1×10^{-10}	−38	Fujitsu, PIN
10 Gbits/s	1×10^{-10}	−16.3	Discovery PIN
10 Gbits/s	1×10^{-10}	−26	Epitaxx APD

Source: Proprietary sources and the ITU.

Of the two types of photodiodes discussed here, the PIN is more economical and requires less complex circuitry than does its APD counterpart. The PIN diode has peak responsivity from about 800 nm to 900 nm for silicon devices. These responsivities range from 300 μA/mW to 600 μA/mW. The overall response time for the PIN diode is good for about 90% of the transient but sluggish for the remaining 10%, which is a "tail." The power response of the tail portion of a pulse may limit the net bit rate on digital systems.

The PIN detector does not display gain, whereas the APD does. The response time of the APD is far better than that of the PIN diode, but the APD displays certain temperature instabilities where responsivity can change significantly with temperature. Compensation for temperature is usually required in APD detectors and is often accomplished by a feedback control of bias voltage. It should be noted that bias voltages for APDs are much higher than for PIN diodes, and some APDs require bias voltages as high as 200 V. Both the temperature problem and the high-voltage bias supply complicate repeater design.

6.7 Optical Fiber Amplifiers

Optical amplifiers amplify incident light through stimulated emission, the same mechanism as used with lasers. These amplifiers are the same as lasers without feedback. Optical gain is achieved when the amplifier is pumped either electrically or optically to realize population inversion.

There are semiconductor laser amplifiers, Raman amplifiers, Brillouin amplifiers, and erbium-doped fiber amplifiers (EDFAs). Certainly the EDFA shows the widest acceptance. One reason is that they operate near the 1.55-μm wavelength region where fiber loss is at a minimum. Reference 15 states that it is possible to achieve high amplifier gains in the range of 30–40 dB with only a few milliwatts of pump power when EDFAs are pumped by using 0.980-μm or 1.480-μm semiconductor lasers.

Figure 7.31 shows erbium-doped fiber core geometry and Figure 7.32 illustrates a typical block diagram of a modern, low-noise EDFA.

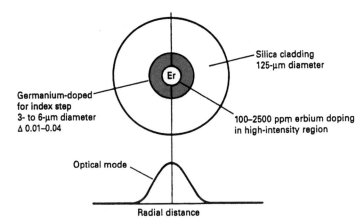

Figure 7.31. Erbium-doped fiber core geometry. From Ref. 14. Courtesy of Hewlett-Packard.

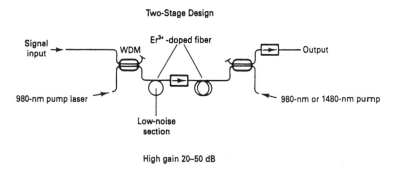

Figure 7.32. EFDA block diagram. From Ref. 14.

In Figure 7.32, optical pumping is provided by fiber pigtailed semiconductor lasers with typically 100 mW of power. Low-loss wavelength division multiplexers efficiently combine pump and signal powers and can also be used to provide a pump power bypass around the internal isolator. The EDFA has an input stage that is codirectionally pumped and an output stage counterdirectionally pumped. Such multistage EDFA designs have simultaneously achieved a low noise figure of 3.1 dB and a high gain of 54 dB [14, 15].

The loops of fiber should be noted in Figure 7.32. These are lengths of fiber with the dopant. The length of erbium-doped fiber required for a particular amplifier application depends on the available pump power, doping concentration, the design topology, and gain and noise requirements. Generally, low-noise amplifiers will have shorter active fiber lengths (e.g., 6–8 m at 300 ppm erbium doping where the optimum gain length may be 10 m). For a constant pump power, reducing the erbium-doped fiber length increases the fractional population of erbium ions in the metastable state. This reduces the amount of spontaneous emission generated by the amplifier. Since the noise figure and gain characteristics vary

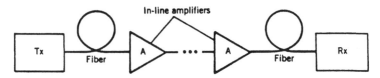

Figure 7.33. In-line amplifier application of EFDAs.

slowly about the optimum values, great precision is not generally required in the active fiber length.

Erbium-doped fiber amplifiers are often installed directly after a semiconductor laser source (transmitter) and/or directly before the PIN or APD receiver at the distant end. Figure 7.33 illustrates this concept. This can extend the length of a fiber-optic link without repeaters about 100–200 km, or it can extend the distance between repeaters by a similar amount.

6.8   Fiber-Optic Link Design

Certainly the strongest argument in favor of fiber optics as a transmission medium is its information carrying capacity. Only radiated light competes with it in some few circumstances. The required capacity should be estimated at some five or ten years. Fiber-optic cable may be installed in an aerial configuration or as buried cable. One question requiring an economic trade-off is whether we are to depend on wavelength division multiplexing to expand capacity in the future or add fiber strands now to support capacity projections.

Fiber-optic communication links have wide application. It has broad use in the cable television (CATV) industry using analog transmission techniques; there is also growing interest in a digital application for CATV. It is used for low-level signal transmission in radio systems such as for long runs of IF, and even for RF. However, in this text we stress digital applications, some of which are listed below.

- On-premises data bus (Ref. EIA/TIA-568 [23] and BICSI [24])
- LANs, more and more use as the transmission medium
- High-level PCM or CVSD* configurations; SONET or SDH transport
- Radar data links; digital video
- Sensor data from remotes where higher data rates are required
- Conventional data links where bit/symbol rates exceed 56/64 kbit/s

Cost erosion continues of fiber-optic cable and light components. Fiber-optic repeaters are considerably more expensive than their PCM metallic wire and cable counterparts. The powering of repeaters is more involved and complex.

---

* CVSD stands for continuous variable slope delta (modulation), a form of digital modulation where the coding is one bit at a time. It is very popular with the armed forces.

For one thing, OA&M* monitoring information has to be formatted and reinserted back into the cable bit stream for display at the NMCC. The powering of repeaters/amplifiers can be more complex. One reason is that reliability requirements are so demanding. The most common approach to providing power for the repeaters is to take power from the cable itself. This means that the cable has a metallic element to conduct the power from a terminal end to downstream repeaters and/or amplifiers. The presence of metal in the cable makes us lose a primary advantage of fiber optics. When there is no conductive path, ground loops are eliminated; the propagation of EMI† tends to be stopped at the source. Another approach is supply power locally with a small floating battery supply at each installation.

A key advantage of fiber when compared to metallic cable is *less repeaters* per unit length. In Chapter 8 we will shows that regenerative repeaters in tandem are the principal cause of jitter, a major impairment to digital systems such as PCM. Reducing the number of repeaters reduces jitter accordingly. In fact, fiber-optic cable systems require a small fraction of the number of regenerative repeaters compared to a metallic PCM system of the same length, whether wirepair or coaxial cable.

6.8.1 Design Approach. The first step in designing a fiber-optic communication system is to establish the basic system parameters. Among these we would wish to know at the outset:

- Type of signal to be transmitted (e.g., CATV analog, PCM, or CVSD); bit rate and format (e.g., SONET, SDH or 8-bit data, digital TV, compression type).
- System length, fiber portion, end-to-end.
- Growth requirements (additional circuits, increased bit rates). This could mean total number of fibers, use of WDM, or both.
- Tolerable signal impairment level stated as signal-to-noise ratio or BER at the electrical output of the terminal-end detector.

The link BER should be established based on end-to-end requirements. We recommend using one of the standards accepted worldwide such as Telcordia TSGR [25], or a proprietary standard. For example, U.S. Sprint set 1×10^{-12} as a link BER requirement. The following ITU-T standards are relevant: ITU-T Rec. G.826 [3] and G.957 [26]. We'll set the BER for practice link budgets at 1×10^{-10}. (Also consult Ref. 25.)

Throughout the design procedure, when working with trade-offs, the system engineer establishes whether he/she is working in the power-limited domain or dispersion-limited domain. At the lower bit rates, say from 622 Mbit/s and below, expect to work in the power-limited domain under all circumstances.

---

* OA&M stands for operation, administration, and maintenance.
† EMI stands for electromagnetic interference. Also called RFI, radio frequency interference.

Modern system design can eliminate a number of major causes of dispersion. Let's simply look at dispersion as a delay. We have a stream of bits. The first bit in the stream does its job, but there is some power from that bit that is delayed which slips into the time slot of bit two. If there is sufficient power from bit one in time slot for bit two, the receiver is confused and may make an incorrect decision whether it is a binary 1 or a binary 0. As the link bit rate is increased, pulse widths get shorter and the problem of dispersion becomes more acute.

One example of eliminating the cause of dispersion deals with the type of fiber we select. To eliminate multimode dispersion, use monomode fiber. We then can turn to using the zero dispersion wavelength which is at approximately 1310 nm for production silicon fiber. By doing this we remove the opportunity to use the low loss band at about 1550 nm. To overcome this shortcoming, we spend more on fiber and buy dispersion-shifted fiber. That moves the zero dispersion wavelength to the 1550 band by changing the fiber geometry.

The designer is now left with chromatic dispersion. This is the phenomenon where even with the narrow line width of a laser diode, different frequencies appearing in that spectral line travel at different velocities.

There are more factors to consider. The designer must select the most economic alternatives among the following:

- Fiber parameters: single mode or multimode; if multimode, step index or graded index; number of fibers, cable makeup, strength.
- Transmission wavelength: 820 nm, 1330 nm, or 1550 nm.
- Source type: LED or semiconductor laser; there are subsets to each source type.
- Detector type: PIN or APD.
- Use of EDFA (amplifiers).
- Repeaters, if required, and how they will be powered.
- Modulation will probably be intensity modulation (IM), but the electrical waveform entering the source is important; possibly consider Manchester coding.

There is the splice and connector trade-off as well as the type of splice and type of connector. Permanently installed systems would opt for splices because of lower insertion losses. Temporarily installed systems, such as used by the military in the tactical environment, may prefer connectors because of ease and speed of mating and demating.

6.8.2 Loss Design. As a first step, assume that the system is power-limited. Probably a majority of systems being installed today can stay in the power-limited regime if monomode fiber is used with semiconductor lasers.

With systems operating at such high data rates such that chromatic dispersion may be a problem, the selection of the laser and the design of the transmitter itself become important. It is highly desirable to minimize the line width, and we can achieve very narrow widths by using a distributed feedback (DFB) laser and an external modulator. It also may be wise to select dispersion-shifted fiber, where the zero dispersion line is shifted to the low-loss 1550-nm band. We now get the best of both worlds for extra cost.

The next step is to develop a link budget, which in format is very similar to the link budgets in LOS microwave (Section 4) and satellite communications (Section 5) of this chapter. It is a tabular format where the first entry is the transmitter output. If the transmitter initially selected is a laser diode, a 0-dBm output is a good starting point. For shorter links with lower bit rates, the LED transmitter should be a candidate because of cost, lower complexity, and longer life. Then all the losses/gains of the link are entered, enumerating and identifying each. Among these losses we would expect to find the following items:

1. *Connector Loss.* There are two connectors, one at the output of the transmitter pigtail and one at the input of the receiver pigtail. Budget 0.5 dB for each. Enter minus signs for losses, plus signs for gains.

   A *pigtail* is a short length of fiber coupled to the transmitter output on one end, and usually with a connector at the other. Likewise, at the receiver there is a pigtail with a connector which connects the input of the receiver top with the main line fiber. Pigtails are factory-installed.

2. *Fiber Loss.* The fiber selected for the link will have a loss specified by the manufacturer given in dB/km for a particular wavelength of interest. Multiply that value by the length of the link plus 5% for slack.

3. *Splice Loss.* Assuming the link is more than 1 or 2 km long, there will be a splice to connect the fiber from one reel to the fiber of the next reel. Good fusion splices have a very low insertion loss. Budget 0.1 dB each. Multiply this value by the number of splices in the link.

4. *Amplifier Gain.* Links longer than 30 to 50 km, in some cases over 100 km (depending on design), will use an amplifier, as shown in Figure 7.33. Budget +30 dB for each in-line amplifier. There is a trade-off where the amplifier is installed. Another candidate location is at the input to the receiver to extend the receiver threshold; still another is at the output of the transmitter to increase output value. Installing with either (or) transmitter or receiver removes the onus of remote power.

5. *Dispersion Compensation Loss.* Budget 1.0 dB for this value.

6. *Link Margin Reserve.* This is a cost without an immediate return on investment. Many designers reduce the value as much as possible. ITU-T recommends 3.0 dB. (We believe this value is low). This dB loss value is set aside for the following contingencies:

   - Cable reel loss variability.
   - Future added splices due to cable repair and their insertion loss.

- Component degradation over the life of the system. This is particularly pronounced for LED output.

One major difference between a link budget for LOS radio and for fiber optics is that there is no fading on a fiber-optic link. Another consideration is the use of an optical power attenuator on very short links so as not to overload the receiver.

The next step is to sum the losses and gains with the output value of the transmitter. The summed value should be stated in dBm. This value helps the system designer to select the type and make of the receiver, whether a PIN diode or an APD. The receiver threshold in dBm is established for the bit rate and BER desired. Turn to Table 7.8 for example values at 1×10^{-10} BER for some of the more popular SONET/SDH bit rates (see Chapter 8 and Ref. 26).

Another consideration is an equivalent loss, a penalty, stemming from the laser diode *extinction ratio*. One imagines that with intensity modulation, a binary 1 is represented by the "on" condition and a binary 0 by an "off" condition. This is not true in the case of the "off" condition. The laser is not completely off; it still conducts a small amount. There is a low output level. The extinction ratio is then defined as

$$r_{ex} = P_0/P_1$$

where P_0 is the laser diode output power for the binary 0 condition and P_1 is the output power for the binary 1 condition. We would want the value of r_{ex} to be as small as possible. In many practical cases it is less than 0.05. This derives a power penalty of <0.4 dB. The value has been derived from Figure 7.34.

The following link budget example will help high-light some of the major points of this section on fiber optics. Some of the material was derived from

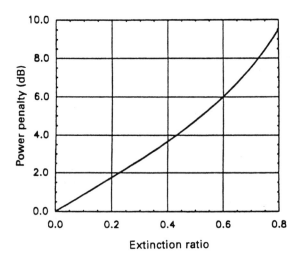

Figure 7.34. Extinction ratio (r_{ex}) versus power penalty (dB). From Ref. 15. Reprinted with permission.

TABLE 7.9   Example Fiber-Optic Link Budget

Item or Parameter	Value	Comments
Laser output	0 dBm	
Extinction ratio penalty	0.4 dB	
Connector losses	1.0 dB	2 connectors
Dispersion power loss	1.0 dB	
Fiber loss 315 km @ 0.25 dB/km	78.75 dB	
Splice losses @0.1 dB/splice	31.4 dB	314×0.1
Margin	4.0 dB	A good judgment estimate
Three EDFA in-line amplifiers	+90.0 dB	
Sum	26.55 dB	
Level at receiver input	−26.55 dBm	−23 dBm required for PIN diode. Shortfall = 3.55 dB

Ref. 27. The link operates at 2.5 Gbps and is 300 km long. The characteristics of the fiber selected for the link are based on ITU-T Rec. G.654. The link operates in the 1550-nm band, and the loss (rounded off) is 0.25 dB/km. The fiber comes on 5-km reels. The total fiber necessary is 1.05×300 km or 315 km. This includes the necessary slack. There will be 63 fiber segments (i.e., 315/5) requiring 314 fusion splices per strand and two connectors. The light source is a distributed feedback laser diode with an extinction ratio penalty of 0.4 dB. It is externally modulated. The link BER is 1×10^{-10}. The candidate light detector is a PIN diode and with the specified BER, the receiver threshold is −23 dBm. Table 7.8 is then constructed.

To compensate for the shortfall of 3.55 dB, the following are possible measures that can be taken:

1. Shorten link by 15 km.
2. Increase gain of each amplifier by 1.2 dB, assuming that EDFAs were operating at less than full gain.
3. Increase output power of laser diode transmitter by 3.55 dB. This may shorten the life of the device.
4. Reduce margin accordingly. Highly undesirable.
5. Turn to using an APD rather than a PIN diode. The threshold is found to be −32 dBm, or a 9-dB improvement. The APD has a shorter life than a PIN diode; its reliability is lower; it is more expensive and it is temperature- and humidity-sensitive. On the positive side, the margin has been increased by 5.45 dB.

6.9   Wavelength-Division Multiplexing (WDM)

6.9.1   Deriving a Definition.
To add operational capacity to a fiber-optic link, we can physically add fibers, utilize dark fibers if there are any, or take advantage of existing capacity of a fiber strand. In this section we deal with the last item, taking more advantage of existing capacity.

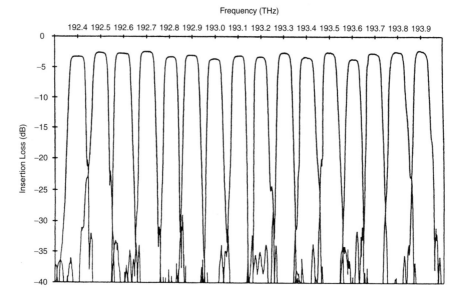

Figure 7.35. Spectrum showing a WDM formation at the output a 16-channel multiplexer. Courtesy of ADC Communications.

In early systems, there was just one bit stream on a fiber. If this bit stream was on a light carrier whose wavelength was in the 1310-nm band, and additional wavelength was added in the 1550-nm band. As a result, the operational capacity of the fiber was doubled. This was WDM in its most primitive form.

Once practical EDFAs were fielded, adding light carriers at different wavelengths in the desirable 1550-nm band became practical. WDM multiplexers and demultiplexers are assembled using passive components such as filters, signal splitters, signal couplers, gratings, lenses, fiber Bragg grating circulator, and so on. These optical components tend to be lossy, and the EDFA(s) compensate for these losses. Figure 7.35 shows an oscilloscope snapshot of a 16-channel WDM configuration. Table 7.10 shows a portion of the ITU WDM grid with 50- and 100-GHz spacing between light carriers. The ITU has established 193.10 THz as the reference frequency for the 1550-nm band. If there were 50-GHz spacing, there would be 80 carrier wavelengths for the ITU grid, part of which is shown in Table 7.10. When it becomes feasible to reduce the spacing to 25 GHz, the number of light carriers could be doubled to 160.

For an analysis of dispersion and system bandwidth on a fiber-optic link, consult Refs. 12, 27, and 28. The manufacturers of sources, detectors, and EDFAs should be able to provide the appropriate rise times of the devices that are purchased from them.

TABLE 7.10   A Portion of the 1550-nm-Band ITU WDM Signal Grid

Nominal Central Frequencies (THz) for Spacings of 50 GHz	Nominal Central Frequencies (THz) for Spacings of 100 GHz and Above	Nominal Central Wavelengths (nm)
193.50	193.50	1549.32
193.45	—	1549.72
193.40	193.40	1550.12
193.35	—	1550.52
193.30	193.30	1550.92
193.25	—	1551.32
193.20	193.20	1551.72
193.15	—	1552.12
193.10	193.10	1552.52
193.05	—	1552.93
193.00	193.00	1553.33
192.95	—	1553.73
192.90	192.90	1554.13
192.85	—	1554.54
192.80	192.80	1554.94
192.75	—	1555.34
192.70	192.70	1555.75
192.65	—	1556.15
192.60	192.60	1556.55
192.55	—	1556.96
192.50	192.50	1557.36
192.45	—	1557.77
192.40	192.40	1558.17
192.35	—	1558.58
192.30	192.30	1558.98
192.25	—	1559.39
192.20	192.20	1559.79
192.15	—	1560.20
192.10	192.10	1560.61

Note: The endpoints of this table are illustrative only. Future evolutions of multichannel systems are anticipated to include frequencies beyond those limits.

Source: Table A.1/G.692, ITU-T Rec. G.692 [6].

REVIEW QUESTIONS

1. What transmission medium should be considered as a principal candidate for a digital waveform that would carry more than 10,000 full-duplex telephone circuits at the end of the forecast period?

2. What are the advantages of using the RF bands from 2 GHz to 10 GHz for trunk telephony? Name at least two.

3. Frequencies above 10 GHz have an additional major impairment that must be taken into account in radiolink design. What is the cause of this impairment?

4. Discuss the problem of delay in speech telephone circuits traversing a satellite. Can you think of any problems in telephone signaling or in data transmission?

5. Give three of the five basic procedure steps in the design of line-of-sight radiolinks.

6. Name three basic planning considerations in siting radiolinks.

7. Name two advantages and at least two disadvantages to a radiolink terminal sited in the middle of an urban area.

8. Why do we limit the heights of radio towers serving radiolink (line-of-sight) systems?

9. Describe how earth bulge varies from one end of a radiolink hop to the other. Where is earth bulge maximum?

10. When a K factor of $\frac{4}{3}$ is used, does the microwave ray beam bend toward or away from the earth?

11. In the path profile, what are the three basic increment factors that must be added to obstacle height?

12. Name at least five of the basic factors that a radiolink design engineer must deal with when carrying out the path calculation (analysis) stage of the design effort.

13. Calculate the free-space loss in decibels of a radiolink hop 24 statute miles long operating at 4100 MHz.

14. Express Boltzmann's constant in decibel-watts.

15. If the signal-to-noise ratio into a device is 31 dB and the signal-to-noise ratio at the device output is 28 dB, what is the device noise figure?

16. A receiving system operates at room temperature; its noise figure is 11 dB, and its bandwidth is 2000 kHz. What is the thermal noise threshold of the system?

17. A receive signal level is -120 dBW and the bit rate is 2.048 Mbits/s. What is E_b?

18. The receiver on a digital LOS microwave link has a noise figure of 7 dB. What is N_0 for this receiver?

19. What is modulation implementation loss? Refer it to E_b/N_0.

20. Why is the selection of modulation type so important on digital LOS microwave?

21. The gain of a parabolic dish antenna is a function of which three variables?

22. What is the standard efficiency (%) used with LOS microwave parabolic dish antennas?

23. Calculate the EIRP of an LOS microwave transmitting system where the transmitter has 1.0-W output, 125 ft of waveguide with a loss of 0.015 dB/ft, and an antenna gain of 35 dB.

24. What is the most common type of fading encountered on an LOS microwave link? Discuss its causes and techniques for mitigation.

25. On a 20-mile (32-km) LOS microwave hop, the desired path propagation reliability is 99.995%. Based on a Rayleigh fading criterion, what fade margin (in dB) is required?

26. Why should "line of sight" be precisely defined?

27. A picowatt equals how many watts?

28. There are two types of diversity which may be employed on line-of-sight microwave links. What are they? Discuss pros and cons of each.

29. Why is frequency diversity not advisable to use?

30. An LOS microwave link does not meet error rate specifications. What steps can be taken, in ascending order of cost, to assure the system meets specifications.

31. What are the two ways of configuring hot-standby equipment?

32. Satellite communications is an extension of line-of-sight microwave; thus a satellite must be in _____ _____ _____ of an earth station.

33. For speech telephony circuits, why would we want to limit GEO satellite links to just one in such a connectivity?

34. How does range (free-space loss) vary with elevation angle for a satellite communication system?

35. There is one overriding reason why downlink power is limited on a satellite system. What is that reason? What is a second important reason?

36. Give at least three reasons why frequencies between 1 and 10 GHz are preferred for satellite communications.

37. Give the two popular methods of satellite multiple access. Define each in no more than two sentences.

38. What are two of the major advantages of TDMA?

39. Define G/T. What are the two components of receiving system noise temperature for a satellite receiving system?

40. Excess attenuation due to rainfall for a satellite downlink varies with at least four major factors. Give three of them.

41. Give two major reasons why a corporation might opt for a VSAT system.

42. Discuss three generic protocols that may be used for VSAT access.

43. What is the principal overriding advantage of fiber-optic cable?

44. There are three wavelength bands currently used on fiber-optic systems. Identify these bands by wavelength and compare each with loss per unit length of fiber.

45. A glass fiber consists of a _____ and a covering called a _____. Relate these two with indices of refraction.

46. Name the three basic components of a fiber-optic link.

47. How many modes propagate in monomode fiber?

48. There are two basic types of light sources and two basic types of light detectors. Name each and compare.

49. What is the standard BER for a fiber-optic link?

50. What is a pigtail?

51. Give the two common locations where we would install fiber-optic amplifiers.

52. Give pros and cons for having a metallic pair in a fiber-optic cable.

53. A fiber-optic link can be either _____ limited or _____ limited.

54. In the link budget for a fiber-optic link, we have a loss item called *link margin*. What are the three factors we should consider when assigning a dB value to link margin?

55. A fiber-optic link with a bit rate of 622 Mbits/s has a laser diode with an output of 0 dBm and an APD-based receiver with a threshold of −40 dBm. Design a link around these parameters. What is the maximum link length achievable without repeaters? Show rationale.

REFERENCES

1. "Telcordia Notes on the Networks," Telcordia Special Report, Issue 4, Telcordia, Piscataway, NJ, October 2000.

2. *Error Performance of an International Digital Connection Operating at a Bit Rate below the Primary Rate and Forming Part of an Integrated Services Digital Network*, ITU-T Rec. G.821, ITU, Geneva, August 1996.

3. *Error Performance Parameters and Objectives for International Constant Bit Rate Digital Path at or above the Primary Rate*, ITU-T Rec. G.826, ITU, Geneva, February 1999.

4. *Allowable Bit Error Rates at the Output of the Hypothetical Reference Digital Path for Radio-Relay Systems which May Form Part of an Integrated Services Digital Network*, ITU-R Rec. 594-4, F Series, Fixed Service, ITU, Geneva, September 30, 1997.

5. *Principles of Digital Transmission*, Raytheon Company, ER79-4307, under U.S. Government contract mDA-904-79-C-0470, Sudbury, MA 1979 (limited circulation).

6. *Engineering Considerations for Microwave Communications Systems*, GTE-Lenkurt, San Carlos, CA, 1975.

7. *INTELSAT VIII Satellite Characteristics*, IESS-417, Rev. 1A, INTELSAT, Washington, DC, November 30, 1998.

8. *Radio Emission from Natural Sources in the Frequency Range above about 50 MHz*, CCIR Report 720, Annex to Vol. V, XVIIth Plenary Assembly, Dusseldorf, 1990.

9. *Satellite Communication Reference Data Handbook*, Defense Communication Agency (NTIS), Washington, DC, 1972.

10. INTELSAT Earth Station Standards (IESS), *Performance Characteristics for Intermediate Data Rate Digital Carriers Using Convolutional Encoding/Viterbi Encoding and QPSK Modulation (QPSK/IDR)*, IESS-308, Rev. 10, INTELSAT Washington, DC, February 10, 2000.

11. J. Everett, *VSATs Very Small Aperture Terminals*, IEE/Peter Peregrinus, Stevenage, Herts, UK, 1992.

12. R. L. Freeman, *Telecommunication Transmission Handbook*, 4th ed., John Wiley & Sons, New York, 1998.

13. *Optical Fibres System Planning Guide*, CCITT/ITU, Geneva, 1989.

14. *1993 Lightwave Symposium*, Hewlett-Packard, Burlington, MA, March 23, 1993.

15. G. P. Agraval, *Fiber-Optic Communication Systems*, 3rd ed., John Wiley & Sons, New York, 2001.

16. *Telecommunication Transmission Engineering*, 3rd ed., Vol. 2, Bellcore, Piscataway, 1991.

17. R. L. Freeman, *Radio System Design for Telecommunications*, 2nd ed., John Wiley & Sons, New York, 1997.

18. *Digital Radio Theory and Measurements*, H-P Application Note 355A, Hewlett-Packard, San Carlos, CA, 1992.

19. *Structural Standards for Steel Antenna Towers and Antenna Supporting Structures*, EIA/TIA-222-E, Telecommunication Industry Association, Alexandria, VA, March 1991.

20. *Performance Characteristics for Demand Assignment Multiple Access Digital Carriers*, INTELSAT Earth Station Standard-311 (IESS-311), Rev. 1, INTELSAT, Washington, DC, May 11, 2000.

21. INTELSAT Earth Station Standards, *INTELSAT IX Satellite Characteristics* (IESS-422), INTELSAT Washington, DC, August 10, 1999.

22. *SCPC/QPSK and SCPC/PCM/QPSK System Specifications*, IESS-303, Rev. 4B, INTELSAT Washington, DC, February 10, 2000.

23. *Commercial Building Telecommunications Cabling Standard—Part 1: General*, TIA/EIA-568-B.1, Telecommunication Industry Association, Alexandria, VA April 2001.

24. *Methods Manual*, Building Industry Consulting Services International (BICSI), 8610 Hidden River Pkwy, Tampa, FL 33637-1000, January 3, 2001.

25. *Transport Systems Generic Requirements: Common Requirements*, Bellcore GR-499-CORE, Issue 2, Piscataway, NJ, December 1998.

26. *Optical Interfaces for Equipments and Systems Relating to the Synchronous Digital Hierarchy*, ITU-T Rec. G.957, ITU, Geneva, June 1999.

27. R. L. Freeman, *Fiber-Optic Systems for Telecommunications*, John Wiley & Sons, New York, 2002.

28. R. L. Freeman, *Reference Manual for Telecommunication Engineering*, 3rd ed., John Wiley & Sons, New York, January 2002.

29. *Optical Interfaces for Multichannel Systems with Optical Amplifiers*, ITU-T Rec. G.692, ITU Geneva 1998.

8

DIGITAL TRANSMISSION SYSTEMS

1 DIGITAL VERSUS ANALOG TRANSMISSION

The IEEE dictionary [22] contrasts analog and digital transmission as follows:

An analog signal implies *continuity*, as contrasted to a digital signal that is concerned with *discrete* states. Often the means of carrying information is the distinguishing feature between analog and digital. The information content of an analog signal is conveyed by the value or magnitude of some characteristics of the signal such as phase, amplitude, frequency of a voltage, the amplitude or duration of a pulse, and so on. To extract the information, it is necessary to compare the value or magnitude of the signal to a standard. The information content of a digital signal is concerned with discrete states of the signal, such as the presence or absence of a voltage, a contact in the open or closed position, or a hole or no hole in certain positions on a card. The digital signal is given meaning by assigning numerical values or other information to the various possible combinations of the discrete states of the signal.

There are three notable advantages to digital transmission that make it extremely attractive to the telecommunication system engineer when compared to its analog counterpart. Dealing in generalities, we can say:

1. Noise does not accumulate on a digital system as it does on an analog system. Noise accumulation stops at each regenerative repeater where the digital signal is fully regenerated. Noise accumulation was the primary concern in analog network design.
2. The digital format lends itself ideally to solid-state technology and, in particular, to integrated circuits.
3. It is theoretically compatible with digital data, telephone signaling, and computers.

Telecommunication System Engineering, by Roger L. Freeman
ISBN 0-471-45133-9   Copyright © 2004 Roger L. Freeman

At the time of preparation of this fourth edition, only about 50% of the traffic carried on the North American PSTN is voice traffic which is initially analog. These analog signals must be converted to a digital format compatible with the digital network. The remaining 50% of the PSTN traffic is digital data, a great portion of which is Internet-related.

The digital waveform on the PSTN is based on *pulse-code modulation* (PCM). There are two varieties of PCM: the North American, which is popularly called T1, and the European variety, which is popularly called E1.

The PSTN in the United States is 100% digital. This means that not only the transmission facilities are digital, but also the switches are digital. Nearly all major switch vendors are marketing fourth-generation digital switches. Digital switching is covered in Chapter 9.

2   BASIS OF PULSE-CODE MODULATION

Pulse-code modulation is a method of modulation in which a continuous analog wave is transmitted in an equivalent digital mode. The cornerstone of an explanation of the functioning of PCM is the Nyquist sampling theorem, which states [1, Section 21]:

> If a band-limited signal is sampled at regular intervals of time and at a rate equal to or higher than twice the highest significant signal frequency, then the sample contains all the information of the original signal. The original signal may then be reconstructed by use of a low-pass filter.

As an example of the sampling theorem, the nominal 4-kHz voice channel would be sampled at a rate of 8000 samples per second (i.e., 4000×2). A high-fidelity 15-kHz program channel* would be sampled at 30,000 times per second (i.e., $15,000 \times 2$).

To develop a PCM signal from an analog signal, three processing steps are required: *sampling, quantization,* and *coding.* The result is a serial binary signal or bit stream,[†] which may or may not be applied to the line without additional modulation or conditioning steps. One major advantage of digital transmission is that signals may be regenerated at intermediate points along a transmission path as well as at the end points. One price for this advantage is the increased bandwidth required for PCM. Conventional PCM systems as found in our digital network require 16 times the bandwidth of their analog counterpart (i.e., a 4-kHz analog voice channel requires 16×4 or 64 kHz when transmitted by PCM), assuming 1 bit per hertz of bandwidth. Regeneration of a digital signal is simplified and particularly effective when the transmitted line signal is binary, whether neutral, polar, or bipolar. An example of a bipolar bit stream is shown in Figure 8.1.

---

* A program channel is a communications channel that carries broadcast material such as music and commentary. This is a facility provided to broadcasters and cable TV operators.

[†] A bit stream is a continuous series of 1s and 0s.

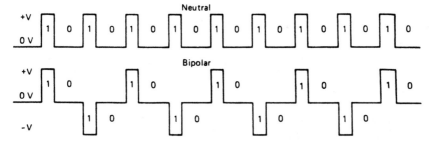

Figure 8.1. Neutral versus bipolar bit streams. The upper diagram illustrates alternating 1s and 0s transmitted in the neutral mode; the lower diagram illustrates the equivalent in a bipolar mode, which is also called AMI or alternate mark inversion.

Binary transmission tolerates considerably higher noise levels (i.e., degraded signal-to-noise ratio), when compared to its analog counterpart (i.e., FDM in Chapter 5). This fact, in addition to the regeneration capability, is a great step forward in transmission engineering. The regeneration that takes place at each repeater and switch by definition re-creates a new signal; therefore noise, as we know it, does not accumulate.

Error performance is another important factor in the design of PCM systems. Intelligibility is maintained even with an error rate as poor as 1 bit error in 100 bits (BER $= 1 \times 10^{-2}$). However, because of line or supervisory signaling concerns, digital networks maintain error rates no worse than 1×10^{-3}. Because the PSTN carries such a large percentage of data in the total traffic load, much more stringent error performance is required. These new requirements are set out in ITU-T Recs. G.826 [25] and G.828 [26]. We make the following assumptions based on G.826. For a bit stream of from 55 to 160 Mbits/s, 20,000 bits per block, the errored second ratio (ESR) is 0.16. ESR is the ratio of errored seconds to total seconds of available time. An errored second (ES) is a 1-s period where at least one error or event occurred. This is end-to-end performance. To achieve these performance numbers, link error performance is often set at 1×10^{-10} to 1×10^{-12} or better.

Another important factor in the design of PCM cable installations is crosstalk, which can degrade error performance. This is crosstalk spilling from one PCM system into another or into the same system from the send path to the receive path inside the same cable sheath.

3 DEVELOPMENT OF A PULSE-CODE MODULATION SIGNAL

3.1 Sampling

Consider the sampling theorem given previously. If we now sample the standard CCITT voice channel, 300–3400 Hz (a bandwidth of 3100 Hz), at a rate of 8000 samples per second, we will have complied with the Nyquist sampling theorem and can expect to recover all the information in the original analog signal. Therefore a sample is taken every 1/8000 s, or every 125 μs. These are key parameters for our future argument.

Another example may be a 15-kHz program channel. Here the sampling rate would be 30,000 times per second. Samples would be taken at 1/30,000-s intervals, or at 33.3 μs.

3.1.1 The Pulse Amplitude Modulation Wave.

With the exception of several specialized applications, practical PCM systems involve time-division multiplexing. Sampling in these cases does not involve just one voice channel but several. In practice, one system (i.e., U.S. T1) samples 24 voice channels in sequence, and another (E1) samples 30 voice channels. The result of the multiple sampling is a pulse amplitude modulation (PAM) wave. A basic PAM wave is illustrated in Figure 8.2, in this case a single sinusoid. A simplified diagram of the processing involved to derive a multiplexed PAM wave is shown in Figure 8.3.

If the nominal 4-kHz voice channel must be sampled 8000 times per second and a group of 24 such voice channels are to be sampled sequentially to interleave them, forming a PAM multiplexed wave, this could be done by gating. The gate should be open for 5.2 μs (125/24) for each voice channel to be sampled successively from channels 1 through 24. This full sequence must be done in a 125-μs period $\frac{1}{8000}$. We call this 125-μs period a frame, and inside the frame all 24 channels are successively sampled once. This 24-channel system is popularly called T1, but we will call it DS1.

Another system widely used outside of the United States and Canada is E1, which is a 30-voice-channel system plus an additional two service channels for a total of 32 channels. By definition, this system must sample 8000 times per second because it is also optimized for voice operation, and thus its frame period is 125 μs. To accommodate the 32 channels, the gate is open 125/32 or about 3.906 μs.

3.2 Quantization

Our goal is to assign a binary sequence to each voltage sample. For argument's sake, we will contain the maximum excursion of the PAM wave to within +1 to −1 V. In the PAM waveform there could be an infinite number of different values of voltage between +1 and −1 V. For instance, one value could be

Figure 8.2. A PAM wave as a result of sampling a single sinusoid.

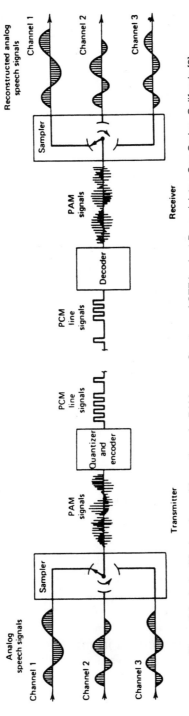

Figure 8.3. A simplified analogy of formation of a PAM wave. Courtesy of GTE Lenkurt Demodulator, San Carlos, California [2].

−0.3875631 V. To assign a different binary sequence to each voltage value, we would have to construct a code of infinite length. So we must limit the number of voltage values between +1 and −1 V, and the values must be discrete. For example, we could set 20 discrete values between +1 and −1 V, each value at a 0.1-V increment.

Because we are working in the binary domain, we select the total number of discrete values to be a binary number multiple (i.e., 2, 4, 8, 16, 32, 64, 128, etc.). This facilitates binary coding. For instance, if there were four values, they would be as follows: 00, 01, 10, and 11. This is a 2-bit code. A 3-bit code would yield eight different binary numbers. We find, then, that the number of total possible different binary combinations given a code of n binary symbols (bits) is 2^n. A 7-bit code has 128 different binary combinations (i.e., $2^7 = 128$).

For the quantization process, we want to present to the coder a discrete voltage value. Suppose our quantization steps were on 0.1-V increments and our voltage measure for one sample was 0.37 V. That would have to be rounded off to 0.4 V, the nearest discrete value. Note here that there is a 0.03-V error, the difference between 0.37 and 0.40 V.

Figure 8.4 shows one cycle of the PAM wave of Figure 8.2 where we use a 4-bit code. In the figure a 4-bit code is used which allows 16 different binary-coded possibilities or levels between +1 and −1 V. Thus we can assign eight possibilities above the origin and eight possibilities below the origin. These 16 quantum steps are coded as follows:

Step Number	Code	Step Number	Code
0	0000	8	1000
1	0001	9	1001
2	0010	10	1010
3	0011	11	1011
4	0100	12	1100
5	0101	13	1101
6	0110	14	1110
7	0111	15	1111

Examination of Figure 8.4 shows that step 12 is used twice. Neither time it is used is it the true value of the impinging sinusoid. It is a rounded-off value. These rounded-off values are shown with the dashed line in Figure 8.4, which follows the general outline of the sinusoid. The horizontal dashed lines show the point where the quantum changes to the next higher or next lower level if the sinusoid curve is above or below that value. Take step 14 in the curve, for example. The curve, dropping from its maximum, is given two values of 14 consecutively. For the first, the curve is above 14, and for the second, below. That error, in the case of 14, from the quantum value to the true value, is called *quantizing distortion*. This distortion is the major source of imperfection in PCM systems.

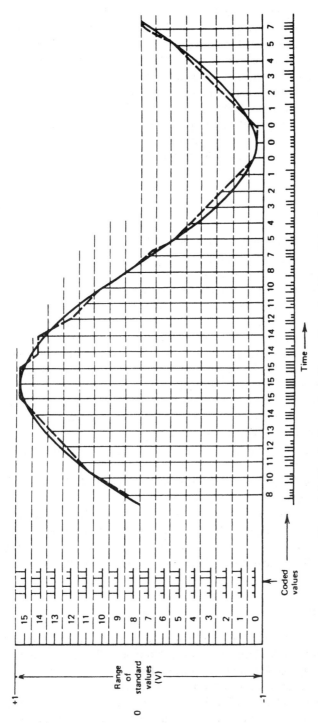

Figure 8.4. Quantization and resulting coding using 16 quantizing steps.

In Figure 8.4, maintaining the -1, 0, $+1$ V relationship, let us double the number of quantum steps from 16 to 32. What improvement would we achieve in quantization distortion? First determine the step increment in millivolts in each case. In the first case the total range of 2000 mV would be divided into 16 steps, or 125 mV/step. The second case would have 2000/32 or 62.5 mV/step. For the 16-step case, the worst quantizing error (distortion) would occur when an input to be quantized was at the half-step level or, in this case, 125/2 or 62.5 mV above or below the nearest quantizing step. For the 32-step case, the worst quantizing error (distortion) would again be at the half-step level, or 62.5/2 or 31.25 mV. Thus the improvement in decibels for doubling the number of quantizing steps is

$$20 \log \frac{62.5}{31.25} = 20 \log 2 \text{ or 6 dB (approximately)}$$

This is valid for linear quantization only (see Section 3.2 of this chapter). Thus increasing the number of quantizing steps for a fixed range of input values reduces quantizing distortion accordingly.

Voice transmission presents a problem. It has a wide dynamic range, on the order of 50 dB. That is the level range from the loudest syllable of the loudest talker to lowest-level syllable of the quietest talker. Using linear quantization, we find it would require 2048 discrete steps to provide any fidelity at all; 2048 is 2^{11}. This means we would need an 11-bit code. Such a code sampled 8000 times per second leads to 88,000-bit/s equivalent voice channel and an 88-kHz bandwidth, assuming 1 bit/Hz. Designers felt this was too great a bit rate/bandwidth.

They turned to an old analog technique of companding. *Companding* stands for two words: compression–expansion. Compression takes place on the transmit side of the circuit; expansion on the receive side. Compression reduces the dynamic range with little loss of fidelity, and expansion returns the signal to its normal condition.

This is done by favoring low-level speech over higher-level speech. In other words, more code segments are assigned to speech bursts at low level than at the higher levels, progressively more as level goes down. This is shown graphically in Figure 8.5, where eight coded sequences are assigned to each level grouping. The smallest range rises only 0.0666 V from the origin (0 V). The largest range extends over 0.5 V, and it is assigned only eight coded sequences [3].

3.3 Coding

Older PCM systems used a 7-bit code, and modern systems use an 8-bit code with its improved quantizing distortion performance. The companding and coding are carried out together, simultaneously. The compression and later expansion functions are logarithmic. A pseudologarithmic curve made up of linear segments imparts finer granularity to low-level signals and less granularity to the higher-level signals. The logarithmic curve follows one of two laws, the A-law and the

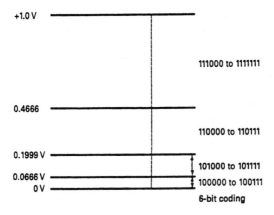

Figure 8.5. A simple graphic representation of compression. Six-bit coding, eight six-bit sequences per segment.

μ-law. The curve for the A-law may be plotted from the formula

$$F_A(x) = \left(\frac{A|x|}{1 + \ln(A)}\right), \qquad 0 \le |x| \le \frac{1}{A}$$

$$F_A(x) = \left(\frac{1 + \ln|Ax|}{1 + \ln(A)}\right), \qquad \frac{1}{A} \le |x| \le 1$$

where $A = 87.6$. The curve for the μ-law may be plotted from the formula

$$F_\mu(x) = \frac{\ln(1 + \mu|x|)}{\ln(1 + \mu)}$$

where x is the signal input amplitude and $\mu = 100$ for the original North American T1 system (now outdated) and 255 for later North American (DS1) systems and the CCITT 24-channel system (CCITT Rec. G.733). Note the use of the natural logarithms (ln) in these formulas [4].

A common expression used in dealing with the "quality" of a PCM signal is *signal-to-distortion* ratio (expressed in dB). Parameters A and μ, for the respective companding laws, determine the range over which the signal-to-distortion ratio is comparatively constant, about 26 dB. For A-law companding, an S/D $= 37.5$ dB can be expected ($A = 87.6$) and for μ-law companding, an S/D $= 37$ dB($\mu = 225$) [5].

Turn now to Figure 8.6, which shows the companding curve and resulting coding for the European E1 system. Note that the curve consists of linear piecewise segments, seven above and seven below the origin. The segment just above and the segment just below the origin each consists of two linear elements. Counting the collinear elements by the origin, there are 16 segments. Each segment has 16 8-bit PCM codewords assigned. These are the codewords that identify the voltage level of a sample at some moment in time. Each codeword, often called a PCM

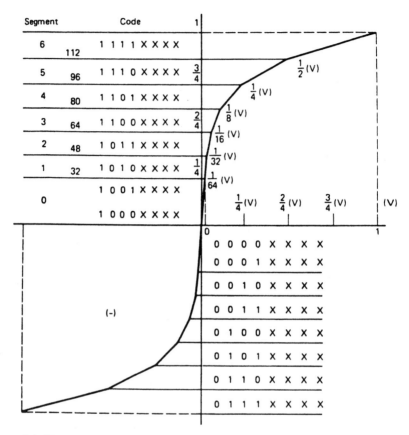

Figure 8.6. The 13-segment approximation of the *A*-law curve used with E1 PCM equipment.

"word," consists of 8 bits. The first bit (most significant bit) tells the distant-end receiver if that sample is a positive or a negative voltage. Note that all the PCM words above the origin start with a binary 1, and those below the origin start with a binary 0. The next three bits identify the segment. There are 8 segments (or collinear equivalents) above the origin and 8 below ($2^3 = 8$). The last 4 bits, shown in the figure as XXXX, identify where in the segment that voltage line is located.

Suppose the distant end received the binary sequence 11010100 in an E1 system. The first bit indicates that the voltage is positive (i.e., above the origin in Figure 8.6). The next three bits, 101, indicate that the sample is in segment 4. The last 4 bits, 0100, tell us where it is in that segment as illustrated in Figure 8.7. Note that the 16 steps inside the segment are linear.

Figure 8.8 shows an equivalent logarithmic curve for the North American DS1 system. It uses a 15-segment approximation of the logarithmic μ-law curve ($\mu = 255$). The segments cutting the origin are collinear and are counted as one. So, again, we have a total of 16 segments.

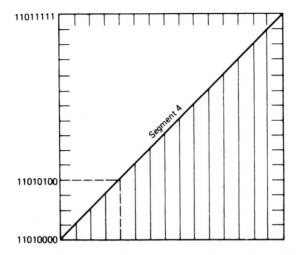

Figure 8.7. The European E1 system, coding of segment 4 (positive).

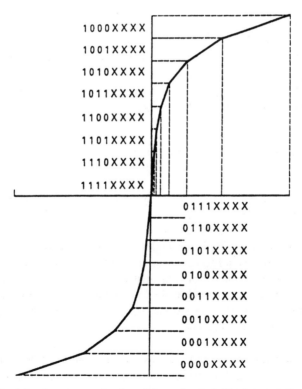

Figure 8.8. Piecewise linear approximation of the μ-law logarithmic curve. Coding based on ITU-T Rec. G.711 [6].

Code Level		Digit Number							
		1	2	3	4	5	6	7	8
255	(Peak positive level)	1	0	0	0	0	0	0	0
239		1	0	0	1	0	0	0	0
223		1	0	1	0	0	0	0	0
207		1	0	1	1	0	0	0	0
191		1	1	0	0	0	0	0	0
175		1	1	0	1	0	0	0	0
159		1	1	1	0	0	0	0	0
143		1	1	1	1	0	0	0	0
127	(Center levels)	1	1	1	1	1	1	1	1
126	(Nominal zero)	0	1	1	1	1	1	1	1
111		0	1	1	1	0	0	0	0
95		0	1	1	0	0	0	0	0
79		0	1	0	1	0	0	0	0
63		0	1	0	0	0	0	0	0
47		0	0	1	1	0	0	0	0
31		0	0	1	0	0	0	0	0
15		0	0	0	1	0	0	0	0
2		0	0	0	0	0	0	1	1
1		0	0	0	0	0	0	1	0
0	(Peak negative level)	0	0	0	0	0	0	1[a]	0

[a]One digit is added to ensure that the timing content of the transmitted pattern is maintained.

Figure 8.9. Eight-level coding of the North American DS1 PCM system. Note that there are actually only 255 quantizing steps because steps 0 and 1 use the same bit sequence, thus avoiding a code sequence with no transitions (i.e., all 0s).

The coding process in PCM systems utilizes straightforward binary codes. Examples of such codes are shown in Figure 8.6, which is expanded in Figure 8.7 and in Figure 8.8.

The North American DS1 (T1) PCM system uses a 15-segment approximation of the logarithmic μ-law ($\mu = 255$), shown in Figure 8.8. The segments cutting the origin are collinear and are counted as one. Figure 8.9 shows some sample DS1 code sequences for particular code levels, for both positive and negative voltages. As can be seen in Figure 8.8, similar to Figure 8.6, the first code element (bit), whether a 1 or a 0, indicates to the distant end whether the sample voltage is positive or negative, above or below the horizontal axis. The next three elements (bits) identify the segment, and the last four elements (bits) identify the actual quantum level inside the segment.

3.3.1 The Concept of Frame. As is shown in Figure 8.3, PCM multiplexing is carried out in the sampling process, sampling the analog sources sequentially. These sources may be the nominal 4-kHz voice channels or other information sources that have a 4-kHz bandwidth, such as data or freeze-frame video. The final result of the sampling and subsequent quantization and coding is a series of electrical pulses, a serial bit stream of 1s and 0s that requires some identification or indication of the beginning of a scanning sequence. This identification is necessary so that the far-end receiver knows exactly when the sampling sequence starts. The receiver knows a priori (in the case of DS1) that 24 eight-bit slots follow. It synchronizes the receiver. Such identification is carried out by a *framing bit,* and one full sequence or cycle of samples is called a *frame* in PCM terminology.

Consider the framing structure of the two widely implemented PCM systems: the North American DS1 and the European E1.* The North American DS1 system is a 24-channel PCM system using 8-level coding (e.g., $2^8 = 256$ quantizing steps or distinct PCM code words). Supervisory signaling is "in-band" where bit 8 of every sixth frame[†] is "robbed" for supervisory signaling. The DS1 signal format, shown in Figure 8.10, has one bit added as a framing bit called an "S" bit. The DS1 frame thus consists of

$$(8 \times 24) + 1 = 193 \text{ bits}$$

making up a full sequence or frame. By definition, 8000 frames are transmitted per second (i.e., 4000×2, the Nyquist sampling rate), so the bit rate for DS1 (T1) is

$$193 \times 8000 = 1,544,000 \text{ bits/s or } 1.544 \text{ Mbits/s}$$

The DS1 frame structure is further clarified in Figure 8.11.

The E1 European PCM system is a 32-channel system. Of the 32 channels, 30 transmit speech (or data) derived from incoming telephone trunks and the

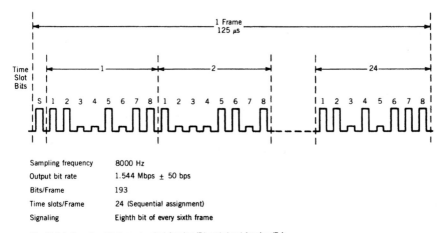

Sampling frequency	8000 Hz
Output bit rate	1.544 Mbps ± 50 bps
Bits/Frame	193
Time slots/Frame	24 (Sequential assignment)
Signaling	Eighth bit of every sixth frame

The S-bit is time-shared between terminal framing (F_t) and signal framing (F_S).

Figure 8.10. DS1 signal format.

* Previously, the European system was called CEPT30 + 2, where CEPT stands for Conference European Post & Telegraph. The 30 + 2 means that it has 30 traffic channels and two service channels.
[†] Note that on each frame that has bit 8 "robbed," 7-bit coding is used versus 8-bit coding used on the other five frames.

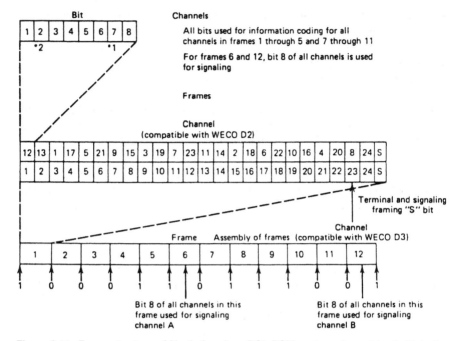

Figure 8.11. Frame structure of North American DS1 PCM system channel bank. Note the "bit robbing" technique used on each sixth frame to provide supervisory signaling information. *Notes:* (1) If bits 1 to 6 and 8 are 0, then bit 7 is transmitted as binary 1; (2) bit 2 is transmitted as binary 0 on all channels for transmission of end-to-end alarm; (3) composite pattern 000110111001, etc. [7].

remaining 2 channels transmit synchronization-alignment and signaling information. Each channel is allotted an 8-bit time slot (TS), and we tabulate TS 0 through 31 as follows:

TS	Type of Information
0	Synchronizing (framing)
1–15	Speech
16	Signaling
17–31	Speech

In TS 0 a synchronizing code or word is transmitted every second frame, occupying digits 2 through 8 as follows:

$$0011011$$

In those frames without the synchronizing word, the second bit of TS 0 is frozen at a 1 so that in these frames the synchronizing word cannot be imitated. The remaining bits of time slot 0 can be used for the transmission of supervisory information signals [8].

As we said, E1 in its primary rate format transmits 32 channels of 8-bit time slots. An E1 frame therefore has $8 \times 32 = 256$ bits. There is no framing bit. Framing alignment is carried out in TS 0. The E1 bit rate to the line is

$$256 \times 8000 = 2,048,000 \text{ bits/s or } 2.048 \text{ Mbits/s}$$

Framing and basic timing should be distinguished. "Framing" ensures that the PCM receiver is aligned regarding the beginning (and end) of a bit sequence or frame; "timing" refers to the synchronization of the receiver clock, specifically, that it is in step with its companion far-end transmit clock. Timing at the receiver is corrected via the incoming "1"-to-"0" and "0"-to-"1" transitions.* It is important that periods of no transitions do not occur. This point is discussed later in reference to line codes and digit inversion.

3.3.2 Quantization Distortion.

Quantizing distortion has been defined as the difference between the signal waveform as presented to the PCM multiplex (codec) and its equivalent quantized value. For a linear codec with n binary digits per sample, the ratio of the full-load sine wave power to quantizing distortion power (S/D) is [25]

$$\frac{S}{D} = 6n + 1.8 \text{ dB}$$

where n is the number of bits per PCM word, the word expressing the sample. For instance, the older ATT D1 system used a 7-bit word to express a sample (level), and the $30 + 2$ and DS1 systems use essentially 8 bits. If we had a 7-bit word and uniform quantizing, S/D would be 43.8 dB. Each binary digit added to the PCM code word increases the S/D ratio 6 dB for linear quantization. Practical S/D values range in the order of 33–38 dB, depending largely on the talker levels (using 8-bit words).

4 PULSE-CODE MODULATION SYSTEM OPERATION

Pulse-code modulation (PCM) equipment operates on a four-wire basis. Voice-channel inputs and outputs to and from a PCM multiplex channel bank are four-wire, or must be converted to four-wire in the channel bank. The term "codec" is a contraction of the word group *coder–decoder* even though the equipment carries out more functions than just coding and decoding. A block diagram of a typical codec (PCM channel bank) is shown in Figure 8.12.

A codec accepts 24 or 30 voice channels, depending on the system used; digitizes and multiplexes the information; and delivers a serial bit stream to the line of 1.544 Mbits/s or 2.048 Mbits/s. It accepts a serial bit stream at one or

---

*A transition in this context is a change of electrical state. We often use the term "mark" for a binary 1 and "space" for a binary 0. The terms mark and space come from old-time telegraphy and have been passed on through the data world to the parlance of digital communications technology.

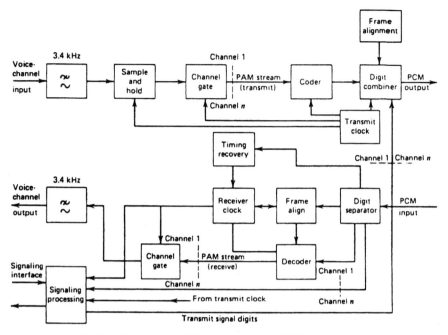

Figure 8.12. Simplified functional block diagram of a PCM codec or channel bank.

the other modulation rate, demultiplexes the digital information, and performs digital-to-analog conversion. Output to the analog telephone network is the 24 or 30 nominal 4-kHz voice channels. Figure 8.12 illustrates the processing of a single analog voice channel through a codec. The voice channel to be transmitted is passed through a 3.4-kHz low-pass filter. The output of the filter is fed to a sampling circuit. The sample of each channel of a set of n channels (n usually equals 24 or 30) is released in turn to the pulse amplitude modulation (PAM) highway. The release of samples is under control of a channel gating pulse derived from the transmit clock. The input to the coder is the PAM highway. The coder accepts a sample of each channel in sequence and then generates the appropriate 8-bit signal character corresponding to each sample presented. The coder output is the basic PCM signal that is fed to the digit combiner where framing-alignment signals are inserted in the appropriate time slots, as well as the necessary supervisory signaling digits corresponding to each channel (European approach), and are placed on a common signaling highway that makes up one equivalent channel of the multiplex serial bit stream transmitted to the line. In North American practice, supervisory signaling is carried out somewhat differently by "bit robbing," such as bit 8 in frame 6 and bit 8 in frame 12. Thus each equivalent voice channel carries its own signaling (see Figure 8.11).

On the receive side the codec accepts the serial PCM bit stream, inputting the digit separator where the signal is regenerated and split, delivering the PCM signal to four locations to carry out the following processing functions:

(1) timing recovery, (2) decoding, (3) frame alignment, and (4) signaling (supervisory). Timing recovery keeps the receive clock in synchronism with the far-end transmit clock. The receive clock provides the necessary gating pulses for the receive side of the PCM codec. The frame-alignment circuit senses the presence of the frame-alignment signal at the correct time interval, thus providing the receive terminal with frame alignment. The decoder, under control of the receive clock, decodes the code character signals corresponding to each channel. The output of the decoder is the reconstituted pulses making up a PAM highway. The channel gate accepts the PAM highway, gating the n-channel PAM highway in sequence under control of the receive clock. The output of the channel gate is fed in turn to each channel filter, thus enabling the reconstituted analog voice signal to reach the appropriate voice path. Gating pulses extract signaling information in the signaling processor and apply this information to each of the reconstituted voice channels with the supervisory signaling interface as required by the analog telephone system in question.

5   PRACTICAL APPLICATIONS

5.1   General

In an early use, PCM was widely employed in expanding interoffice trunks (junctions) that have reached or will reach exhaust* in the near future. An interoffice trunk is one pair of a circuit group that connects two switching points (exchanges). Figure 8.13 sketches the interoffice trunk concept. Depending on the particular application, at some point where distance d is exceeded it will be more economical to install PCM on existing cable plant than to rip up streets and add more VF cable pairs. For distances less than d, additional VF cable pairs are installed for expanding plant. When the length of the trunk cable exceeds d, two voice pairs are taken out of service and either an E1 or T1 configuration is placed in service. So for the loss of two VF channels, we gain either 30 or 24 equivalent channels [19].

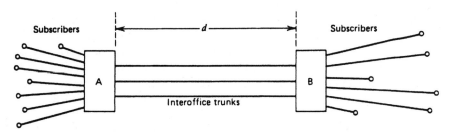

Figure 8.13. Simplified application drawing of PCM as applied to interexchange plant in the local area. A and B are two local serving switches separated by some distance d.

Exhaust is an outside plant term meaning that the useful pairs of a cable have been used up (assigned) from a planning point of view.

In the 1980s many toll switches (long-distance switches) were converted to digital operation. This conversion was accelerated because much of the transmission plant had been converted. The final element in the conversion to digital operation was the local switch. The only vestige remaining of analog plant is much the subscriber pair. We will find that from a transmission quality perspective, it is highly desirable to limit the number of A/D and D/A conversions. In Chapter 9, we will show that ideally there should only be one A/D and one D/A conversion in any telephone connection.

6 PCM LINE CODES

The original design of a PCM system was for application to a wire-pair cable medium. The line code is bipolar in this case as shown in Figure 8.1. The marks or binary 1s have only a 50% duty cycle. There are several advantages to this mode of transmission:

- No dc return is required; thus transformer coupling can be used on the line.
- The power spectrum of the transmitted signal is centered at a frequency equivalent to half the bit rate.

It will be noted in bipolar transmission that the 0s are coded as absence of pulses and 1s are alternately coded as positive and negative pulses, with the alternation taking place at every occurrence of a 1. This mode of transmission is also called *alternate mark inversion* (AMI).

One drawback to straightforward AMI transmission is that when a long string of 0s is transmitted (e.g., no transitions), a timing problem may arise because repeaters and decoders have no way of extracting timing without transitions. The problem can be alleviated by forbidding long strings of 0s. Codes have been developed that are bipolar but with N 0s substitution; they are called "BNZS" codes. For instance, a B6ZS code substitutes a particular signal for a string of six 0s. B8ZS is used on subscriber loop carrier.

Another such code is the HDB3 code (high-density binary 3), where the 3 indicates substitution for binary sequences with more than three consecutive 0s. With HDB3, the second and third 0s of the string are transmitted unchanged. The fourth 0 is transmitted to the line with the same polarity as the previous mark sent, which is a "violation" of the AMI concept. The first 0 may or may not be modified to a 1 to ensure that the successive violations are of opposite polarity. HDB3 is used with European E series PCM systems and is similar to B3ZS.

7 REGENERATIVE REPEATERS

As the reader is probably aware, pulses passing down a digital transmission line suffer attenuation and are badly distorted by the frequency characteristic of the

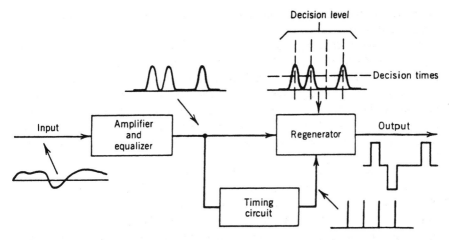

Figure 8.14. Simplified functional block diagram of a regenerative repeater for use on PCM cable systems [3].

line. A regenerative repeater amplifies and reconstructs such a badly distorted digital signal and develops a nearly perfect replica of the original at its output. Regenerative repeaters are an essential key to digital transmission in that we could say that the "noise stops at the repeater."

Figure 8.14 is a simplified block diagram of a regenerative repeater and shows typical waveforms corresponding to each functional stage of signal processing. As shown in the figure, the first stage of signal processing is amplification and equalization. Equalization is often a two-step process. The first is a fixed equalizer that compensates for the attenuation–frequency characteristic of the nominal section, which is the standard length of transmission line between repeaters (often 6000 ft). The second equalizer is variable and compensates for departures between nominal repeater section length and the actual length and loss variations due to temperature. The adjustable equalizer uses automatic line build-out (ALBO) networks that are automatically adjusted according to characteristics of the received signal.

The signal output of the repeater must be accurately timed to maintain accurate pulse width and space between the pulses. The timing is derived from the incoming bit stream. The incoming signal is rectified and clipped, producing square waves that are applied to the timing extractor, which is a circuit tuned to the timing frequency. The output of the circuit controls a clock-pulse generator that produces an output of narrow pulses that are alternately positive and negative at the zero crossings of the square wave input.

The narrow positive clock pulses gate the incoming pulses of the regenerator, and the negative pulses are used to run off the regenerator. Thus the combination is used to control the width of the regenerated pulses.

Regenerative repeaters are the major source of timing jitter in a digital transmission system. Jitter is one of the principal impairments in a digital network, giving rise to pulse distortion and intersymbol interference. Jitter is discussed in more detail in Section 11.2.

Most regenerative repeaters transmit a bipolar (AMI) waveform (see Figure 8.1). Such signals can have one of three possible states in any instant in time, positive, zero, or negative and are often designated +, 0, −. The threshold circuits are gated to admit the signal at the middle of the pulse interval. For example, if the signal is positive and exceeds a positive threshold, it is recognized as a positive pulse. If it is negative and exceeds a negative threshold, it is recognized as a negative pulse. If it has a value between the positive and negative thresholds, it is recognized as a 0 (no pulse).

When either threshold is exceeded, the regenerator is triggered to generate a pulse of the appropriate duration, polarity, and amplitude. In this manner the distorted input signal is reconstructed as a new output signal for transmission to the next repeater. It should be kept in mind that the regenerative function is not unique to a device called a regenerative repeater. PCM switches, LOS radio receivers, satellite terminal receivers, and other similar equipment also regenerate digital signals.

8 SIGNAL-TO-GAUSSIAN-NOISE RATIO ON PULSE-CODE MODULATION REPEATED LINES

As we have mentioned earlier, noise accumulation on PCM systems is not a primary consideration. However, this does not mean that Gaussian noise* or crosstalk (or impulse noise) is unimportant. Indeed, it may affect error performance expressed as error rate. Errors are cumulative, as is the error rate. A decision in error, whether 1 or 0, made anywhere in the digital system is not recoverable. Thus such an incorrect decision made by one regenerative repeater adds to the existing error rate on the line, and errors taking place in subsequent repeaters further down the line add in a cumulative manner, thus tending to deteriorate the received signal.

In a purely binary transmission system, if there is a 22-dB signal-to-noise ratio, the system operates nearly error-free. In this respect, consider Table 8.1.

As we discussed in Section 6, in practice on wire-pair cable, PCM is transmitted in the alternate mark inversion format. The marks have a 50% duty cycle,

TABLE 8.1 Error Rate of a Binary Transmission System Versus Signal-to-RMS-Noise Ratio

Error Rate	S/N (dB)	Error Rate	S/N (dB)
10^{-2}	13.5	10^{-7}	20.3
10^{-3}	16.0	10^{-8}	21.0
10^{-4}	17.5	10^{-9}	21.6
10^{-5}	18.7	10^{-10}	22.0
10^{-6}	19.6	10^{-11}	22.2

* Same as thermal noise.

permitting energy concentration at a frequency equivalent to half the transmitted bit rate. Thus it is advisable to add 1 or 2 dB to the values shown in Table 8.1 to achieve the desired error performance in a practical system.

9 PCM SYSTEM ENHANCEMENTS

9.1 North American DS1

9.1.1 Frame and Superframe. In Section 3.3.1 of this chapter we defined a frame. In that section the major difference between North American DS1 framing strategy and European E1 framing was pointed out. The North American system inserts one S-bit in each frame.

A superframe (a multiframe in ITU parlance) consists of 12 consecutive frames. As a result we have developed 12 S-bits and they all are used for frame alignment/synchronization. Thus, there is a 12-bit sequence, one S-bit frame from each frame. This 12-bit sequence is subdivided into two sequences. The frame alignment pattern is 101010 and is located in the odd-numbered frames. The superframe-alignment pattern is 001110 and is located in the even-numbered frames. The superframe pattern is shown in Table 8.2. (Note that the ITU uses the term multiframe, which is "superframe" in North American terminology.)

9.1.2 Extended Superframe. Table 8.3 shows the extended superframe (ESF). With modern processing technology, it is not necessary to tell the distant-end PCM receiver 8000 times a second where a frame started. Thus, the extended

TABLE 8.2 Multiframe (Superframe) Structure

Frame Number	Frame-Alignment Signal[a]	Multiframe-Alignment Signal (S-bit)	Bit Number(s) in Each Channel Time Slot		Signaling Channel Designation[b]
			For Character Signal	For Signaling	
1	1	—	1–8	—	
2	—	0	1–8	—	
3	0	—	1–8	—	
4	—	0	1–8	—	
5	1	—	1–8	—	
6	—	1	1–7	8	A
7	0	—	1–8	—	
8	—	1	1–8	—	
9	1	—	1–8	—	
10	—	1	1–8	—	
11	0	—	1–8	—	
12	—	0	1–7	8	B

[a] When the S-bit is modified to signal the alarm indications to the remote end, the S-bit in frame 12 is changed from state 0 to 1.
[b] Channel-associated signaling provides two independent 667-bit/s signaling channels designated A and B or one 1333-bit/s signaling channel.
Source: Table 10/ITU-T G.704, page 23, ITU-T Rec. G.704, ITU, Geneva, October 1998 [9].

TABLE 8.3   Multiframe Structure for the 24-Frame Multiframe (Extended Superframe)

Frame Number Within Multiframe	Bit Number Within Multiframe	F-bit Assignments			Bit Number(s) in Each Channel Time Slot		Signaling Channel Designation[d]
		FAS[a]	DL[b]	CRC[c]	For Character Signal[d]	For Signaling[d]	
1	1	—	m	—	1–8	—	
2	194	—	—	e_1	1–8	—	
3	387	—	m	—	1–8	—	
4	580	0	—	—	1–8	—	
5	773	—	m	—	1–8	—	
6	966	—	—	e_2	1–7	8	A
7	1159	—	m	—	1–8	—	
8	1352	0	—	—	1–8	—	
9	1545	—	m	—	1–8	—	
10	1738	—	—	e_3	1–8	—	
11	1931	—	m	—	1–8	—	
12	2124	1	—	—	1–7	8	B
13	2317	—	m	—	1–8	—	
14	2510	—	—	e_4	1–8	—	
15	2703	—	m	—	1–8	—	
16	2896	0	—	—	1–8	—	
17	3089	—	m	—	1–8	—	
18	3282	—	—	e_5	1–7	8	C
19	3475	—	m	—	1–8	—	
20	3668	1	—	—	1–8	—	
21	3861	—	m	—	1–8	—	
22	4054	—	—	e_6	1–8	—	
23	4247	—	m	—	1–8	—	
24	4440	1	—	—	1–7	8	D

[a] FAS Frame-alignment signal (. . . 001011 . . .).
[b] DL 4-kbit/s data link (message bits m).
[c] CRC CRC-6 block check field (check bits $e_1 \ldots e_6$).
[d] Only applicable in the case of channel-associated signaling.
Source: Table 1/G.704, page 2, ITU-T Rec. G.704, October 1998 [9].

superframe was developed (called 24-frame multiframe in ITU-T G.704). It consists of 24 sequential frames with 24 available S-bit positions (see Figure 8.11). Every fourth position in the ESF was dedicated to the frame-alignment signal (FAS). The remaining 18 bit positions (Table 8.3) in the 24-frame ESF were put to good use.

Six of the 18 bits developed a cyclic redundancy check (CRC) pattern permitting system operators to monitor gross link error performance in quasi-real time. The generating polynomial for the pattern is $X^6 + X + 1$ based on CRC-6. A complete description of how CRC works is given in Chapter 10.

The remaining 12 bits are used to form a 4-kbit/s data link. This data link provides a communication path between primary hierarchical level terminals and contains data, n idle data link sequence or a loss of frame alignment alarm sequence. A loss of frame alignment alarm sequence is used when a loss of frame alignment (LFA) condition has been detected. After a loss of frame alignment condition is detected at local end A, one 16-bit LFA sequence of eight 1s and eight 0s (1111111100000000) is transmitted in the m-bits of the 4-kbit/s data link continuously to remote end B [7, 9].

9.2 Enhancements to E1

Allocation of bits 1 to 8 of the E1 frame, time slot (channel 0) is shown in Table 8.4. Table 8.5 shows the complete CRC-4 multiframe structure of E1, also TS 0.

Each CRC-4 multiframe, which is composed of 16 frames, numbered 0 through 15, is divided into two 8-frame sub-multiframes (SMF), designated SMF I and SMF II, which signifies their respective order of occurrence within the CRC-4 multiframe structure. The SMF is the cyclic redundancy check-4 better known

TABLE 8.4   Allocation of Bits 1 to 8 in TS0 of the E1 Frame

Bit Number	1	2	3	4	5	6	7	8
Alternate Frames								
Frame containing the frame alignment signal	S_i	0	0	1	1	0	1	1
	Note 1	Frame alignment signal						
Frame not containing the frame alignment signal	S_i	1	A	S_{a4}	S_{a5}	S_{a6}	S_{a7}	S_{a8}
	Note 1	Note 2	Note 3	Note 4				

Note 1: S_i = Bits reserved for international use. One specific use is described in 2.3.3 of the reference document. Other possible uses may be defined at a later stage. If no use is realized, these bits should be fixed at 1 on digital paths crossing an international border. However, they may be used nationally if the digital path does not cross a border.

Note 2: The bit is fixed at 1 to assist in avoiding simulations of the frame alignment signal.

Note 3: A = remote alarm indication. In undisturbed operation, set to 0; in alarm condition, set to 1.

Note 4: S_{a4} to S_{a8} = additional spare bits whose use may be as follows:

   (i) Bits S_{a4} to S_{a8} may be recommended by ITU-T for use in specific point-to-point applications (e.g., transcoder equipments conforming to Recommendation G.761).

   (ii) Bit S_{a4} may be used as a message-based data link to be recommended by ITU-T for operations, maintenance and performance monitoring. If the data link is accessed at intermediate points with consequent alterations to the S_{a4} bit, the CRC-4 bits must be updated so as to retain the correct end-to-end path termination functions associated with the CRC-4 procedure. The data-link protocol and messages are for further study.

   (iii) Bits S_{a5} to S_{a7} are for national usage where there is no demand on them for specific point-to-point applications [see (i) above].

   (iv) One of the bits S_{a4} to S_{a8} may be used in a synchronization interface to convey synchronization status messages.

   Bits S_{a4} to S_{a8} (where these are not used) should be set to 1 on links crossing an international border.

Source: Table 5A/G.704, page 9, ITU-T Rec. G.704, October 1998 [9].

TABLE 8.5 CRC-4 Multiframe Structure

Sub-multiframe (SMF)	Frame Number	Bits 1 to 8 of the Frame							
		1	2	3	4	5	6	7	8
Multiframe I	0	C_1	0	0	1	1	0	1	1
	1	0	1	A	S_{a4}	S_{a5}	S_{a6}	S_{a7}	S_{a8}
	2	C_2	0	0	1	1	0	1	1
	3	0	1	A	S_{a4}	S_{a5}	S_{a6}	S_{a7}	S_{a8}
	4	C_3	0	0	1	1	0	1	1
	5	1	1	A	S_{a4}	S_{a5}	S_{a6}	S_{a7}	S_{a8}
	6	C_4	0	0	1	1	0	1	1
	7	0	1	A	S_{a4}	S_{a5}	S_{a6}	S_{a7}	S_{a8}
II	8	C_1	0	0	1	1	0	1	1
	9	1	1	A	S_{a4}	S_{a5}	S_{a6}	S_{a7}	S_{a8}
	10	C_2	0	0	1	1	0	1	1
	11	1	1	A	S_{a4}	S_{a5}	S_{a6}	S_{a7}	S_{a8}
	12	C_3	0	0	1	1	0	1	1
	13	E	1	A	S_{a4}	S_{a5}	S_{a6}	S_{a7}	S_{a8}
	14	C_4	0	0	1	1	0	1	1
	15	E	1	A	S_{a4}	S_{a5}	S_{a6}	S_{a7}	S_{a8}

Note 1: E = CRC-4 error indication bits (see 2.3.3.4 of the reference document).
Note 2: S_{a4} to S_{a8} = spare bits (see Note 4 to Table 8.4).
Note 3: C_1 to C_4 = cyclic redundancy check 4 (CRC-4) bits.
Note 4: A = remote alarm indication (see Table 8.4).
Source: Table 5B/G.704, page 11, ITU-T Rec. G.704, October 1998 [9].

as CRC-4 block size (i.e., 2048 bits). In those frames containing the frame-alignment signal, bit 1 is used to transmit the CRC-4 bits. There are four CRC-4 bits, designated C_1, C_2, C_3, and C_4 in each SMF [9].

In those frames not containing the frame-alignment signal as defined above, bit 1 is used to transmit the 6-bit CRC-4 multiframe-alignment signal and two CRC-4 error indication bits (E). The CRC-4 multiframe-alignment signal has the form 001011.

The E-bits are used to indicate received errored sub-multiframes by setting the binary state of one E-bit from 1 to 0 for each errored sub-multiframe. Any delay between detection of an errored sub-multiframe and the setting of the E-bit that indicates the error state must be less than 1 s. The E-bits are always taken into account even if the SMF which contains them is found to be errored, since there is little likelihood that the E-bits themselves will be errored. The generating polynomial for CRC-4 is $X^4 + X + 1$ [9].

10 HIGHER-ORDER PCM MULTIPLEX SYSTEMS

10.1 Introduction

Higher-order PCM multiplex is developed out of several primary multiplex sources. Primary multiplex is typically DS1 in North America and E1 in Europe;

some countries have standardized on E1, such as most of Hispanic America. Not only are E1 and DS1 incompatible, the higher-order multiplexes, as one might imagine, are also incompatible. First we introduce *stuffing*, describe some North American higher-level multiplex, and then discuss European multiplexes based on the E1 system.

10.2 Stuffing and Justification

Stuffing (justification) is common to all higher-level multiplexers that we describe below. Consider the DS2 higher-level multiplex. It derives from an M12 multiplexer, taking inputs from four 24-channel channel banks. The clocks in these channel banks are free running. The transmission rate output of each channel bank is *nominally* 1,544,000 bits/s. However, there is a tolerance of ±50 ppm (±77 bits/s). Suppose all four DS1 inputs were operating on the high side of the tolerance or at 1,544,077 bit/s. The input to the M12 multiplexer is a buffer. It has a finite capacity. Unless bits are read out of the buffer faster than they are coming in, at some time the buffer will overflow. This is highly undesirable. Thus we have bit stuffing.

Stuffing in the output aggregate bit stream means adding extra bits. It allows us to read out of a buffer faster than we write into it.

In Ref. 24 the IEEE defines *stuffing bits* as "bits inserted into a frame to compensate for timing differences in constituent lower rate signals." CCITT uses the term *justification*.

Figure 8.15 illustrates the stuffing concept.

10.3 North American Higher-Level Multiplex

The North American PCM digital hierarchy is shown in Figure 8.16. The higher-level multiplexers are type-coded in such a way that we know the DS levels which are being combined. For example, an M34 has inputs from level 3 (DS3) and the output is at level 4 (DS4). We describe the operation of the M12 multiplexer because it is typical of this series.

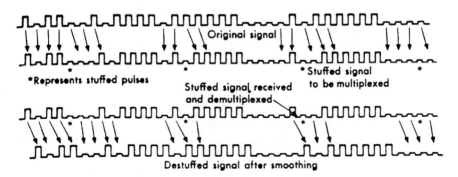

Figure 8.15. Concept of stuffing and justification. Based on Ref. 3.

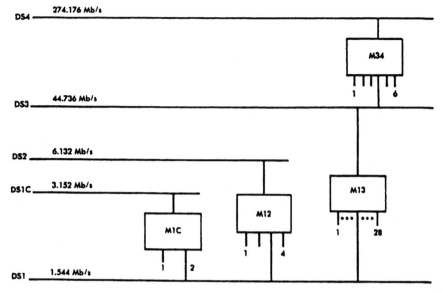

Figure 8.16. The North American digital hierarchy [3].

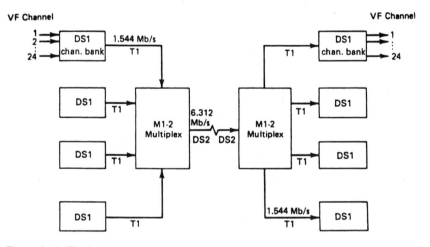

Figure 8.17. The formation of the DS2 signal from four DS1 signal in a M12 multiplexer [3].

The formation of the second-level North American multiplex, DS2, operating from four DS1 inputs is shown in Figure 8.17. There are four inputs, each operating at a nominal 1.544 Mbits/s. The output bit rate is 6.312 Mbits/s. Multiply 1.544 by 4 and we get 6.176 Mbits/s. In other words, the output of the M12 multiplexer is operating 136 kbits/s faster than the aggregate of the four inputs. Some of these extra bits are overhead bits and the remainder are stuff bits. Figure 8.18 shows the makeup of a DS2 frame.

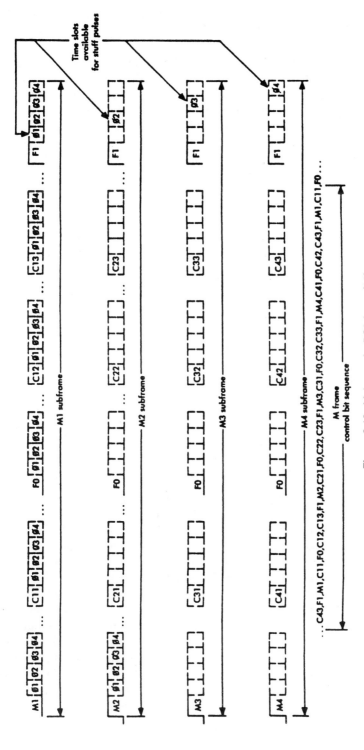

Figure 8.18. Makeup of a DS2 frame [10].

The M12 multiplex frame consists of 1176 bits. The frame is divided into four 294-bit subframes as illustrated in Figure 8.17. There is a control bit word that is distributed throughout the frame and that begins with an M bit. Thus each subframe begins with an M bit. There are four M bits forming the series 011X, where the fourth bit (X), which may be a 1 or a 0, may be used as an alarm indicator bit. When transmitted as a 1, no alarm condition exists. When it is transmitted as a 0, an alarm is present. The 011 sequence for the first three M bits is used in the receiving circuits to identify the frame.

It is noted in Figure 8.18 that each subframe is made up of six 49-bit blocks. Each block starts with a control bit which is followed by a 48-bit block of information. Of these 48 information bits, 12 bits are taken from each of the four DS1 signals. These are interleaved sequentially in the 48-bit block. The first bit in the third and sixth block is designated an F bit. The F bits are a 0101... sequence used to identify the location of the control bit sequence and the start of each block of information bits.

The stuff-control bits are transmitted at the beginning of each of the 48-bit blocks numbered 2, 4, and 5 within each subframe. When these control bits, designated C, are 000, no stuff pulse is present; when the C bits are 111, a stuff pulse is added in the stuff position.

The stuff bit positions are all assigned to the sixth 48-bit block in each subframe. In subframe No. 1, the stuff bit is the first bit after the F1 bit. In subframe No. 2, the stuff bit is the second bit after the F1 bit, and so on through the fourth subframe. The nominal stuffing rate is 1796 bits/s for each DS1 input signal. The maximum is 5367 bits/s.

Prior to multiplexing at the M12 multiplex unit, input signals 2 and 4 are logically inverted. This is done to improve the statistical properties of the output DS2 signal.

10.4   The European E1 Digital Hierarchy

The E1 hierarchy is identified in a similar manner as DS1. E1 (30 voice channels) is the primary multiplex; E2 is the second level and is derived from four E1s. Thus E2 contains 120 equivalent voice channels. E3 is the third level and is derived from four E2 inputs and contains 480 equivalent voice channels. E4 derives from four E3 formations and contains the equivalent of 1920 voice channels. International digital hierarchies are compared in Table 8.6.

Table 8.7 gives the basic parameters of the formation of the E2 level in the European digital hierarchy. ITU-T Rec. G.745 recommends cyclic interleaving in the tributary (i.e., E1 inputs) numbering order and positive/zero/negative justification with two-command control. The justification control signal is distributed and the C_{jn} bits ($n = 1, 2, 3$; see Table 8.7) are used for justification control.

Positive justification is indicated by the signal 111, transmitted in each of two consecutive frames. Negative justification is indicated by the signal 000, also transmitted in each of two consecutive frames. No justification is indicated by the signal 1111 in one frame and 000 in the next frame. Bits 5, 6, 7, and 8 in

TABLE 8.6 Higher-Level PCM Multiplex Comparison

System Type	Level				
	1	2	3	4	5
North American T/D type	1	2	3	4	
Number of voice channels	24	96	672	4032	
Line bit rate (Mbit/s)	1.544	6.312	44.736	274.176	
Japan					
Number of voice channels	24	96	480	1440	5760
Line bit rate (Mbit/s)	1.544	6.312	32.064	97.728	400.352
Europe					
Number of voice channels	30	120	480	1920	
Line bit rate (Mbit/s)	2.048	8.448	34.368	139.264	

Source: Ref. 11.

TABLE 8.7 8448-kbit/s Digital Multiplexing Frame Structure Using Positive/Zero/ Negative Justification

Tributary bit rate (kbits/s)	2048
Number of tributaries	4
Frame Structure	**Bit Number**
	Set I
Frame-alignment signal (11100110)	1 to 8
Bits from tributaries	9 to 264
	Set II
Justification control bits C_{j1} (see Note)	1 to 4
Bits for service functions	5 to 8
Bits from tributaries	9 to 264
	Set III
Justification control bits C_{j2} (see Note)	1 to 4
Spare bits	5 to 8
Bits from tributaries	9 to 264
	Set IV
Justification control bits C_{j3} (see Note)	1 to 4
Bits from tributaries available for negative justification	5 to 8
Bits from tributaries available for positive justification	9 to 12
Bits from tributaries	12 to 264
Frame length	1056 bits
Frame direction	125 μs
Bits per tributary	256 bits
Maximum justification rate per tributary	8 kbit/s

Note: C_{jn} indicates *n*th justification control bit of the *j*th tributary.
Source: Table 1/G.745, ITU-T G.745, page 3, 1993 [12].

Set IV (Table 8.7) are used for negative justification of tributaries 1, 2, 3, and 4, respectively, and bits 9 to 12 for positive justification of the same tributaries.

Besides, when information from tributaries 1, 2, 3, and 4 is not transmitted, bits 5, 6, 7, and 8 in Set IV are available for transmitting information concerning the type of justification (positive or negative) in frames containing commands of positive justification and intermediate amount of jitter in frames containing commands of negative justification. The maximum amount of justification rate per tributary is shown in Table 8.7 [12].

11 LONG-DISTANCE PCM TRANSMISSION

11.1 Transmission Limitations

Digital waveforms lend themselves to transmission by wire pair, coaxial cable, fiber-optic cable, and wideband radio media. The PCM multiplex format was first applied to wire-pair cable as described in Section 5.1. As time progressed, its employment has become nearly universal in the long-distance telephone plant.

Each medium has transmission limitations brought about by impairments. In one way or another each limitation is a function of length and transmission rate (bit rate). We have discussed loss, for example. As loss increases (i.e., between regenerative repeaters), signal-to-noise ratio suffers, directly impacting bit error performance. Dispersion is another impairment that limits circuit length over a particular medium, especially as transmission rate increases. The following transmission impairments to PCM transmission are covered: jitter, distortion, noise, crosstalk, and echo.

11.2 Jitter

In the context of digital transmission, *jitter* is defined as short-term variation of the sampling instant from its intended position in time or phase. Longer-term variation of the sampling instant is called *wander*. Jitter can cause transmission impairments such as:

- Displacement of the ideal sampling instant. This leads to a degradation in system error performance.
- Slips in timing recovery circuits manifesting itself in degraded error performance.
- Distortion of the resulting analog signal after decoding at the receive end of the circuit.

The random phase modulation, or *phase jitter*, introduced at each repeater accumulates in a repeater chain and may lead to crosstalk and distortion in the reconstructed analog signal. In digital switching systems, jitter on the incoming lines is a potential source of slips.* The sources of timing jitter may be classified

---

* Slips will be discussed in Chapter 9 as a principal impairment in digital networks.

as systematic or nonsystematic according to whether or not they are related to the pulse pattern. Systematic jitter sources lead to jitter which degrades the bit stream in the same way at each repeater in the chain. Systematic sources include intersymbol interference, finite pulse width, and clock threshold effects. Nonsystematic jitter sources such as mistuning and crosstalk result in timing degradations which are random from repeater to repeater. Thermal and impulse noise are not serious contributors to timing jitter. That is, if the total noise at the regenerator input is low enough to permit the regenerator to operate with an acceptably low error rate, the noise passed by the narrow-band timing extractor is orders of magnitude less and therefore negligible. In a long repeater chain, the total accumulated jitter is dominated by components produced by systematic sources [3].

Jitter accumulation is a function of the number of regenerative repeaters in tandem. Keep in mind that switches, fiber-optic receivers, and radios are also regenerative repeaters. The mean square value of jitter in a long chain of repeaters increases with N (the number of repeaters), and the rms value of jitter increases with $(N)^{1/2}$. Jitter is also proportional to the timing filter bandwidth, which leads to the conclusion that higher-Q tuned circuits in the repeater reduce jitter.

Certainly by reducing the number of regenerative repeaters in tandem, we reduce jitter accordingly. Wire-pair systems have repeaters every 6000 ft, and coaxial cable has repeaters approximately every mile. If we are to reduce jitter, these transmission media are not good candidates on long circuits. Fiber-optic systems, depending on design and bit rate, have repeaters every 40–200 miles. This is another reason why fiber-optic systems are favored for digital transmission. Microwave radio, strictly for budgeting purposes, may have repeaters every 30 miles, so it too is a candidate for long systems. Satellite links have the least repeaters, at least one in a long circuit (say 4000 miles).

The principal effect of jitter on the resulting analog signal after decoding is to distort the signal. The analog signal derives from a PAM pulse train, which is then passed through a low-pass filter. Jitter displaces the PAM pulses from their proper location, showing up as undesired pulse-position modulation (PPM) [3].

11.3 Distortion

On metallic transmission links, such as coaxial cable and wire-pair cable, line characteristics distort and attenuate the digital signal as it traverses the medium. There are three cable characteristics that create this distortion: loss, amplitude distortion (amplitude–frequency response), and delay distortion. Thus the regenerative repeater must provide amplification and equalization of the incoming digital signal before regeneration. There are also trade-offs between loss and distortion on the one hand and repeater characteristics and repeater section length on the other.

11.4 Thermal Noise

As in any electrical communication system, thermal noise, impulse noise, and crosstalk affect system design. Because of the nature of a digital system, these impairments need only be considered on a per-repeater-section basis because noise does not accumulate due to the regenerative process carried out at repeaters and nodes. Bit errors do accumulate, and this impairment family is one of several that create these errors. One way to limit error accumulation is to specify a stringent BER for each repeater section. Repeater sections are often specified with a median BER between 1×10^{-10} to 1×10^{-12}.

It is interesting to note that PCM provides reasonable voice performance for a BER as poor as 1 in 10^2. However, the worst tolerable BER is 1 in 10^3 at system end points. This value is required to ensure the correct operation of supervisory signaling. The reader should appreciate that such degraded BER values are completely unsuitable for data transmission.

11.5 Crosstalk

Crosstalk is a major impairment in PCM wire-pair systems, particularly when "go" and "return" channels are carried in the same cable sheath. The major offender of single-cable operation is near-end crosstalk (NEXT). When the two directions of transmission are carried in separate cables or use shielded pairs in a common cable, far-end crosstalk (FEXT) becomes dominant.

One characteristic has been found to be a major contributor to poor crosstalk coupling loss. This is the capacitance imbalance between wire pairs.

Stringent quality control during cable manufacture is one measure taken to ensure minimum balance values are met.

11.6 Echo

Echo is caused by impedance discontinuities in the transmission line, including repeaters and terminations (MDFs, codecs, switch ports). Good impedance match across the entire system eliminates the cause of echo or reduces its level. On a PCM transmission system there are many causes of echo, such as gas plugs and splices.

Gas plugs are used on cable systems to allow gas to be applied under pressure to the cable to prevent moisture buildup. The plug tends to add capacitance to the line. To compensate for this, repeater sections that incorporate plugs are made short to accommodate the added capacitance.

Other sources of echo are where gauge and insulation changes take place along the cable run. Bridged taps are still another potential source of mismatch.

12 DIGITAL LOOP CARRIER

Digital subscriber loop carrier (DLC) is one method of extending the metallic subscriber plant by using one or more DS1 configurations. As an example, the SLC-96 employs four DS1 configurations to derive an equivalent 96 voice channels.

The digital transmission facility used by a DLC system may be repeated wire-pair cable, optical fibers, either or both combined with digital multiplexers, or other appropriate media. In Bellcore (Telcordia) terminology, the central office termination (COT) is the digital terminal collocated with the local serving switch. The RT is the remote terminal. The RT must provide all of the features to a subscriber loop that the local serving switch normally does, such as battery, supervision, ringing, address signaling, both dial pulse and touch tone, and so on [13].

13   SONET AND SDH

13.1   Introduction

SONET is an acronym for *Synchronous Optical Network*. SONET provides digital formats extending to 9953.28 Mbits/s. It is a North American development. The equivalent European format is called SDH or *Synchronous Digital Hierarchy*. The two are very similar. Either one can accommodate the standard DS1 family (i.e., 1.544 Mbits/s, etc.) and E1 family (i.e., 2.048 Mbits/s, etc.) of line rates [21].

13.2   SONET

SONET's higher-level digital format was originally intended for transmission over optical fiber facilities. It can, however, be accommodated on any transmission medium that meets the bandwidth requirements. Figure 8.19 is a functional diagram depicting SONET section, line, and path for the purpose of definition.

13.2.1   SONET Rates and Formats. SONET was designed to have a synchronous hierarchy that has sufficient flexibility to carry many different capacity signals such as DS1, DS3, E1, and ATM. This is realized by defining a basic module with a bit rate of 51.840 Mbits/s, as well as defining a byte-interleaved multiplex scheme that results in a family of signals with N times 51.840 Mbits/s, where N is an integer [15].*

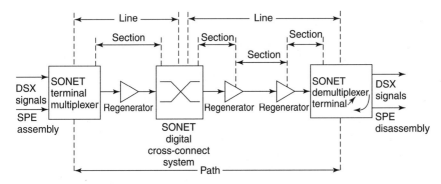

Figure 8.19. SONET: section, line and path definitions [14, 16].

* Note the difference with plesiochronous digital hierarchy (PDH), what the industry calls the DS1 multiplex hierarchy (DS1C, DS2, etc.). These are *not* integer multiples of DS1 [22]

The basic SONET module is divided into a portion assigned to overhead and a portion that carries the payload. The payload can be used to transport DS3 signals or a variety of sub-DS3 signals. Because some signals requiring transport have rates greater than the basic rate (e.g., broadband-ISDN), a technique of linking several basic modules together to build a transport signal of increased capacity is provided. To maintain a consistent payload structure while providing for the transport of a variety of lower-rate payloads (e.g., DS1, DS1C, and DS2 signals), a structure called a *virtual tributary* (VT) is defined. Payloads below the DS3 rate are transported with a VT structure.

13.2.1.1 Synchronous Hierarchical Rates. The synchronous transport signal level 1 (STS-1) is the basic SONET module. It has a bit rate of 51.840 Mbits/s. The optical counterpart of STS-1 is optical carrier-level 1 signal (OC-1).

Higher-level SONET signals are obtained by synchronous multiplexing lower-level modules. When these lower-level modules are multiplexed, the result is denoted STS-N, where N is an integer. STS-N can be converted OC-N or STS-N electrical signal. The popular SONET line rates are shown in Table 8.8. All these higher-level signals are integer multiples of 51.84 Mbits/s (STS-1).

An STS-1 signal is a specific sequence of 810 bytes (6480 bits), which includes various overhead bytes and an envelope capacity for transporting payloads. An STS-1 frame structure is illustrated in Figure 8.20. It has a 90-column by 9-row structure. The frame duration is 125 μs (i.e., 8000 frames per second), deriving a bit rate of 51.840 Mbits/s. The order of transmission of bytes in Figure 8.21 is row by row, from left top right [14].

Transport overhead occupies the first three columns of the STS-1 frame for a total of 27 bytes. The remaining 87 columns of the STS-1 frame, a total of 783 bytes, are allocated to the synchronous payload envelope (SPE) signal. This provides a channel capacity of 50.11 Mbit/s in the STS-1 signal structure for carrying tributary payloads intact across the synchronous network.

It should be noted that at 8000 frames per second, each byte within the SONET signal structure represents a channel bandwidth of 64 kbits/s (i.e., 8 bits/byte × 8000 bytes/second = 64 kbits/s). This is the same bit rate of a PCM voice channel or a DS0/E0 time slot.

Figure 8.21 shows the SPE providing 87 columns and 9 rows of payload. The figure gives the appearance that an SPE is wholly contained in one STS-1 frame.

TABLE 8.8 Line Rates for Standard SONET Interface Signals

OC-N Level	STS-N Electrical Level	Line Rate (Mbits/s)
OC-1	STS-1 electrical	51.84
OC-3	STS-3 electrical	155.52
OC-12	STS-12 electrical	622.08
OC-24	STS-24 electrical	1244.16
OC-48	STS-48 electrical	2488.32
OC-192	STS-192 electrical	9953.28

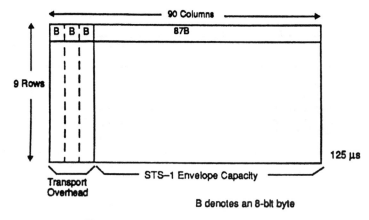

Figure 8.20. SONET STS-1 frame structure.

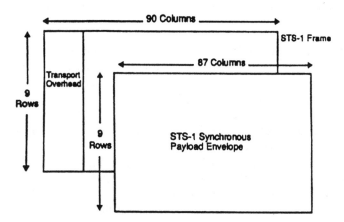

Figure 8.21. STS-1 synchronous payload envelope (SPE) [14].

This is not necessarily true. In fact the STS-1 SPE can begin anywhere in the STS-1 envelope capacity, as shown in Figure 8.22. Of course, if it begins in one STS-1 frame and is not wholly contained in that frame, the remainder of the SPE appears in the following contiguous frame. Allowing the SPE to begin anywhere in the STS-1 frame facilitates efficient multiplexing (especially add–drop multiplexing) and cross-connection of signals in the synchronous network.

When an SPE is assembled into the transport frame, additional bytes, referred to as the *payload pointer*, are made available in the transport overhead. These bytes contain a pointer value which indicates the location of the first byte (J1) of the STS-1 SPE. The SPE is allowed to float freely within the space made available for it in the transport frame so that timing phase adjustments can be made as required between the SPE and the transport frame. The payload pointer identifies the first byte location of the SPE [14].

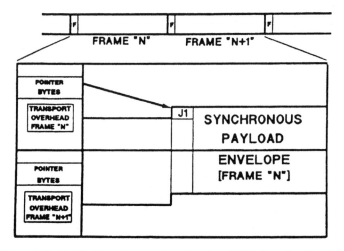

Figure 8.22. Link between transport overhead and SPE. Based on Ref. [16].

An STS path overhead (POH) associated with each payload is used to communicate various information from the point where a payload is mapped into the STS-1 SPE to where it is delivered. The POH is contained in the first column of the SPE and thus consists of 9 bytes. This signal capacity provides such facilities as alarm and performance monitoring required to support and maintain the transport of the SPE between *path terminations*. A path termination is where the SPE is either assembled or disassembled.

The frame structure of the STS-N is shown in Figure 8.23. It consists of N × 810 bytes. The STS-N is formed by byte-interleaving STS-1 and STS-M (M < N) modules. The transport overhead of the individual STS-1 and STS-M modules are frame-aligned before interleaving, but the associated STS SPEs are

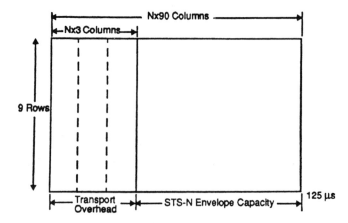

Figure 8.23. STS-N frame.

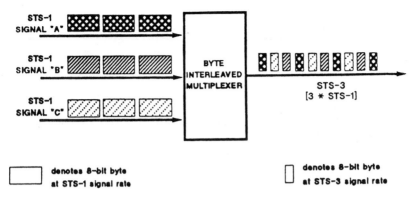

Figure 8.24. The concept of byte-interleaved multiplexing. Based on Ref. 16.

not required to be aligned because each STS-1 has a payload pointer to indicate the location of the SPE. The concept of byte interleaving is shown in Figure 8.24.

13.2.1.2   Virtual Tributaries. The virtual tributary (VT) structure is designed for transport and switching of sub-STS payloads. There are four sizes of VTs: VT1.5 (1.728 Mbits/s), VT21 (2.304 Mbits/s), VT3 (3.456 Mbits/s), and VT6 (6.912 Mbits/s); they are illustrated in Figure 8.25.

A VT1.5 packed in an STS-1 SPE is shown in Figure 8.26.

13.2.1.3   SPE Assembly/Disassembly Process. Fundamental to the SONET format of transmission is the concept of a tributary signal (such as a DS3) being

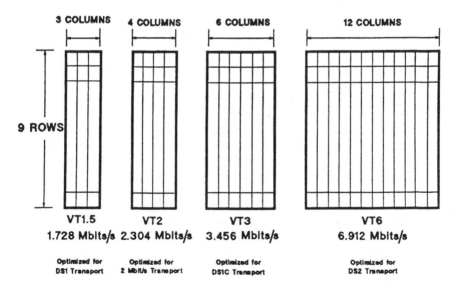

Figure 8.25. The four different VT frames. Based on Refs. [13, 16].

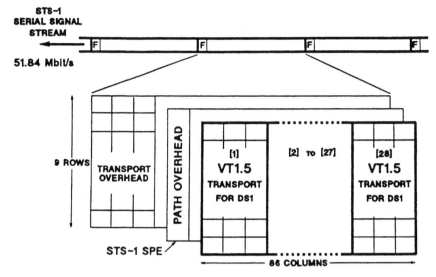

Figure 8.26. VT1.5 packaged in an STS-1 SPE. Based on Ref. [16].

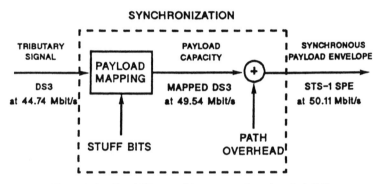

Figure 8.27. The SPE assembly process. Based on Ref. [16].

assembled into an SPE to be transported end-to-end across the synchronous network. The assembly process is called *payload mapping*. The payload capacity provided for each tributary signal is always slightly greater than that required by the tributary signal. This provides uniformity across all SONET transport facilities. The mapping process synchronizes the tributary signal with the payload capacity. This is achieved by stuffing bits to the signal stream as part of the mapping process. The assembly process is illustrated in Figure 8.27. In this example, a DS3 tributary enters at its nominal 44.736 Mbits/s and need to be synchronized with the payload capacity of 49.54 Mbits/s provided by the STS-1 SPE. The bit rate is increased to 50.11 Mbits/s by addition of the path overhead (POH).

At the point of exit from the synchronous network, the payload tributary signal that has been transported over the network needs to be recovered from the SPE.

DESYNCHRONIZATION

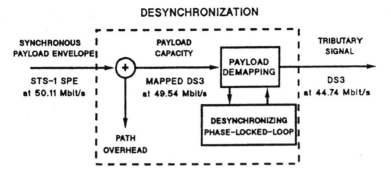

Figure 8.28. The SPE disassembly process. Based on Ref. [16].

The process of disassembling the tributary signal from the SPE is referred to as *payload demapping*. This disassembly process is illustrated in Figure 8.28 [16].

13.2.1.4 Super Rate Payloads. Multiple STS-1 SPEs are required to transport super rate payloads as might be encountered with broadband-ISDN/ATM. To accommodate such a payload, an STS-Nc module is formed by linking N constituent STS-1s together in a fixed phase alignment. The super rate payload is then mapped into the resulting STS-Nc SPE for transport. The STS-Nc can be carried by an OC-N, STS-N electrical signal or higher. Concatenation indicators contained in the second through Nth STS payload pointers are used to show that the STS-1s of an STS-Nc are linked together. The STS-Nc SPE is shown in Figure 8.29.

The STS-Nc consists of NX783 bytes and can be deposited as an N × 97 column by 9-row structure. Only one set of STS POH is required in the STS-Nc SPE. The STS-Nc SPE is carried within the STS-Nc so that the STS POH will always appear in the first of the N STS-1s that make up the STS-Nc.

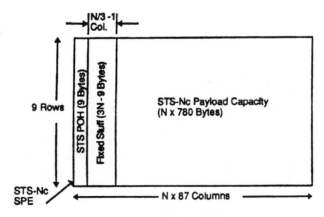

Figure 8.29. STS-Nc SPE.

In all the super rate payload mappings, the first $(N/3) - 1$ columns of the STS-Nc SPE following the STS POH are not used for payload, but are designated fixed stuff columns (i.e., columns of undefined bytes). Only mappings into STS-3c and STS-12c have been defined in Bellcore/Telcordia GR-253 [14]. Other mappings may be defined in the future.

13.2.2 The Three Overhead Levels in SONET. The three embedded overhead levels of SONET are

- Path (POH)
- Line (LOH)
- Section (SOH)

These overhead levels, represented as spans, are illustrated in Figure 8.19. One important function of overhead is to support OA&M (operations, administration, and maintenance). Another function is the payload pointer, described in Section 13.2.3. Figure 8.30 outlines the overhead on a byte for byte basis for an STS-1 in a synchronous transport frame.

The path overhead (POH) is the column of 9 bytes on the far right of Figure 8.30. It is created and included in the SPE as part of the SPE assembly process. The POH provides the facilities to support and maintain the transport of the SPE between path terminations, where the SPE is assembled and disassembled.

The path overhead functions include:

- A fixed length string of 64 bytes (J1) that is transmitted one byte per SPE. This string may contain any alphanumeric message that can associated with the path. Continuity to the unique path source may be verified at any receiving terminal by monitoring the message string. An example is the Common Language Location Identifier (CLLI) code terminated with a carriage return, line feed, and null characters.
- *Bit-Interleaved Parity (BIP-8)* check (B3) is calculated over all bits of the previous SPE and placed in the current SPE (performance monitoring for the SPE).

	Transport Overhead			Path Overhead
Section overhead	Framing (A1)	Framing (A2)	STS-1ID (C1)	Path Trace (J1)
	BIP-8 (B1)	Orderwire (E1)	User (F1)	BIP-8 (B3)
	Data Com (D1)	Data Com (D2)	Data Com (D3)	Signal Level (C2)
Line overhead	Pointer H1	Pointer H2	Pointer H3	Path Status (G1)
	BIP-8 (B2)	APS (K1)	APS (K2)	User Channel (F2)
	Data Com (D4)	Data Com (D5)	Data Com (D6)	Multiframe (H4)
	Data Com (D7)	Data Com (D8)	Data Com (D9)	Growth (Z3)
	Data Com (D10)	Data Com (D11)	Data Com (D12)	Growth (Z4)
	Sync (S1/Z1)	FEBE (M0/M1/Z2)	Orderwire (E2)	Tandem (Z5)

Figure 8.30. STS-1 overhead [28].

- A signal label (C2) or an 8-bit code value that specifies the SPE structure. There are 256 possible structures such as the status of mapped payloads (e.g., all 0s are unequipped and no signals).
- The path status byte (G1) communicates path alarm and path performance information back to the transmitting NE and allows in-service testing (e.g., a count of BIP-8 errors, path *Remote Defect Indication (RDI-P)*.
- A user channel (F2) for proprietary network operator communications between path and terminating NEs.
- A multiform indicator (H4) that may be used by the VT structure payloads. For example, the VT multiframe indicator in the path overhead identifies the phase of the VT multiframe being carried by that SPE. The first STS-1 of an STS-N is xxxxxx01, the second is xxxxxx10, etc.
- The tandem connection (Z5) byte allows for tandem connection maintenance per ANSI T1.105 [27].

The transport overhead is contained in the left three columns of Figure 8.30. The upper three rows, 34 columns wide, contain the section overhead (SOH) and the lower six rows, also three columns wide, contain the line overhead (LOH).

The section overhead functions include:

- Two frame alignment bytes (A1 and A2) that carry the repeating pattern "11110110 0010100" to recognize framing. These bytes are in all STS-1s of an STS-N.
- An STS-1 order of appearance byte (C1) that contains a binary number representing the sequence of the STS-1 in a byte interleaved STS-N frame. This byte may be used in the deinterleaving process and is in all STS-1s of an STS-N. The first STS-1 is given the count 00000001. Future uses of this byte for section trace and section growth may redefine this byte.
- A parity byte B1 for section errors (performance monitoring for the STS-N). This byte uses a *Bit Interleaved Parity (BIP-8)* code with even parity. The calculation is on the previous STS-N prior to scrambling.
- A local orderwire (E1) for voice communications between regenerators, hubs, and remote terminal locations. This byte is defined in the first STS-1 of an STS-N.
- A user byte (F1) for proprietary communications between section termination equipment. This byte is in the first STS-1 of an STS-N.
- A data communication channel (bytes D1, D2, and D3) for administration monitoring maintenance and alarm at 192 kbits/s between section terminating equipment. These bytes are defined in the first STS-1 in an STS-N frame.

The line overhead functions include:

- Three bytes (H1, H2, H3) that facilitate the operation of the STS-1 payload pointer. These bytes are present in all STS-1s in an STS-N frame.

- A parity byte (B2) that is calculated from the line overhead and SPE of the previous STS-1 frame (performance monitoring for the STS-1). This *Bit Interleaved Parity* (BIP-8) byte is computed for the bits of the previous STS-1 frame and placed in the current frame prior to scrambling. These bytes are present in all STS-1s in an STS-N.
  - *Automatic Protection Switching* (APS) bytes (K1 and K2) provide communication between the line terminating units. These bytes are defined in the first STS-1 in an STS-N. Line level switching is present in linear APS and bidirectional line-switched ring architectures. Byte K2 may also be used to detect *Alarm Indication Signal (AIS)*.
- A data communications channel (bytes D4 to D12) of 576 kbits/s for systems administration, monitoring, maintenance, and alarms. These bytes communicate between line terminating equipment and are defined in the first STS-1 of an STS-N.
- A synchronization byte (S1) conveys synchronization status in bits 5 to 8. The other bits are undefined. The byte is in the first STS-1 of an STS-N. The corresponding bytes in the second through Nth STS-1s of an STS-N are undefined growth bits (Z1).
- A *Far-End Block Error (FEBE)* byte (MM0) conveys the line FEBE status in bits 5 to 8 as an error count back to the transmitting NE. The other bits are undefined. This byte is in an *Optical Carrier (OC)-1* or an STS-1. The OC-N and STS-N use FEBE byte (M1) that is in the third STS-1 of an STS-N. The corresponding bytes in the other STS-1s of an STS-N are undefined growth bits (Z2).
- An orderwire byte (E2) for a 2-way voice communication channel between line terminating equipment. The byte is defined in the first STS-1 of an STS-N signal.

13.2.3 The SONET Payload Pointer.

The STS payload pointer provides a method of allowing flexible and dynamic alignment of the STS SPE within the STS envelope capacity, independent of the actual contents of the SPE.

SONET, be definition, was designed to operate as a synchronous network. It derives its timing, as discussed in Chapter 9, from the underlying network delivering signals slaved to the network master clock or from independent stratum 1 sources.

Modern digital networks must make provision for more than one master clock. Examples in North America are the several interexchange carriers which interface with local exchange carriers (LECs), each with their own master clock. Each master clock operates independently. Each independent master clock has excellent stability (i.e., better than 1×10^{-11}/month), yet there may be a small variance in time among the clocks. Likewise, SONET must take into account loss of master clock or a segment of its timing delivery system. In this case, switches fall back on lower-stability internal clocks. This situation must also be handled by SONET.

Therefore synchronous transport must be able to operate effectively under these conditions where network nodes are operating at slightly different rates.

To accommodate these clock offsets, the SPE can be moved (justified) in the positive or negative direction one octet at a time with respect to the transport frame. This is accomplished by recalculating or updating the payload pointer at each SONET network node. In addition to clock offsets, updating the payload pointer also accommodates any other timing reference at the SONET node.

This is what is meant by dynamic alignment where the STS SPE is allowed to float within the STS envelope capacity.

The payload pointer is contained in the H1 and H2 octets in the line overhead and designates the location of the octet where the STS SPE begins. These two octets are viewed as one word, as illustrated in Figure 8.31. Bits 1 through 4 carry the *new data flag* (NDF), and bits 7 through 16 carry the pointer value. Bits 5 and 6 are undefined.

The pointer value is a binary number with a range of 0 to 782. It indicates the offset of the pointer word and the first octet of the STS SPE (i.e., the J1 octet). The transport overhead octets are not counted in the offset. For example, a point value of 0 indicates that the STS SPE starts in the octet location that immediately follows the H3 octet, whereas an offset value of 87 indicates that it starts immediately after the K2 octet location.

Payload pointer processing introduces a signal impairment known as *payload pointer jitter*. This impairment appears on a received tributary signal after recovery from an SPE that has been subjected to payload pointer changes. The operation of the network equipment processing the tributary signal immediately downstream is influenced by this excessive jitter. By careful design of the timing

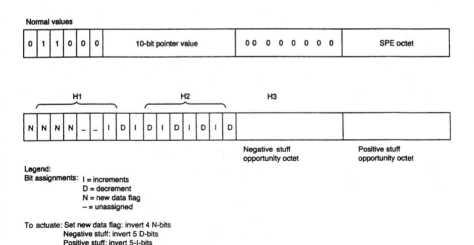

Figure 8.31. STS payload pointer (H1, H2) coding.

distribution for the synchronous network, payload pointer adjustments can be minimized, thus reducing the level of tributary jitter that can be accumulated through synchronous transport.

13.3   Synchronous Digital Hierarchy (SDH)

13.3.1   Introduction. Synchronous digital hierarchy (SDH) was a European development, whereas SONET was a North American development. They are very similar. The principal difference is that SONET's basic rate, 51.84 Mbits/s, has no SDH equivalent. The SDH basic rate is equivalent to SONET STS-3, 155.520 Mbits/s. Some texts point out that SONET appears to have been designed for private networks whereas SDH was designed more with the flavor of a public network. There are also differences in the payload point and OA&M overhead.

13.3.2   SDH Standard Bit Rates. Table 8.9 shows the standard SDH bit rates. ITU-T Rec. G.707 [17] states "—that the first level of the digital hierarchy shall be 155,520 kbits/s—and—that higher synchronous digital hierarchy bit rates shall be obtained as integer multiples of the first level bit rate."

13.3.3   Interface and Frame Structure of SDH. Figure 8.32 illustrates the relationship between various multiplexing elements that are given below and shows generic multiplexing structures. Figures 8.33 to 8.35 show specific derived multiplexing methods.

Definitions

Synchronous Transport Module (STM). An STM is the information structure used to support section layer connections in the SDH. It consists of information payload and section overhead (SOH) fields organized in a block frame structure which repeats every 125 μs. The information is suitably conditioned for serial transmission on selected media at a rate which is synchronized with the network. A basic STM is defined at 155,520 kbits/s. This is termed STM-1. Higher capacity STMs are formed at rates equivalent to N times this basic rate. STM capacities for $N = 4, N = 16$, and $N = 64$ are defined. Higher rates are under consideration by the ITU.

TABLE 8.9   SDH Bit Rates with SONET Equivalents

Synchronous Digital Hierarchy Level	Hierarchical Bit Rate (kbits/s)	SONET Equivalent Line Rate
1	155,520	STS-3/OC-3
4	622,080	STS-12/OC-12
16	2,488,320	STS-48/OC-48
64	9,953,280	STS-192/OC-192

Note: The ITU states that the specification for levels higher than 64 requires further study.

Source: Table 1/G.707, March 1996 [17].

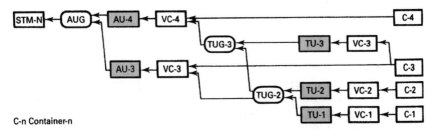

Figure 8.32. Generalized SDH multiplexing structure. From Figure 2-1/G.708, page 3, March 1993 [18].

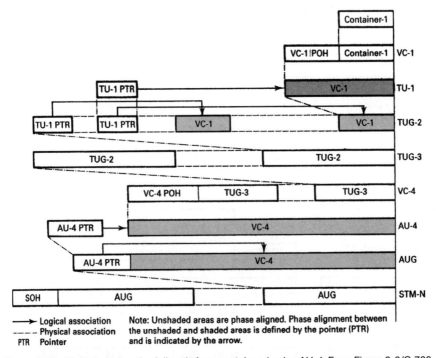

Figure 8.33. Multiplexing method directly from container-1 using AU-4. From Figure 2-2/G.708, page 3, ITU-T Rec. G.708, March 1993 [18].

The STM-1 comprises a single administrative unit group (AUG) together with the SOH. The STN-N contains N AUGs together with SOH. The STM-N hierarchical rates are given in Table 8.9.

Virtual Container-n (VC-n). A virtual container is the information structure used to support path layer connections in the SDH. It consists of information payload and path overhead (POH) information fields organized in a block frame structure, which repeats every 125 or 500 μs. Alignment information to identify VC-n frame start is provided by the server network layer.

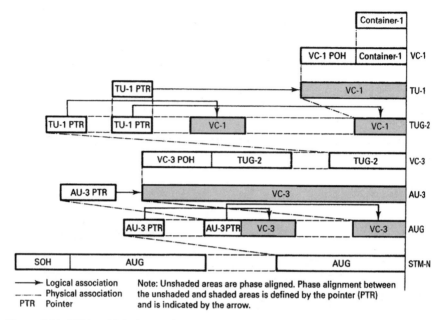

Figure 8.34. SDH multiplexing method directly from container-1 using AU-3. From Figure 2-3/G.708, page 4, ITU-T Rec. G.708, March 1993 [18].

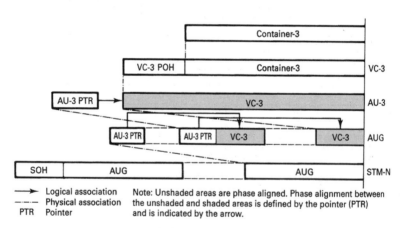

Figure 8.35. Multiplexing method directly from container-3 using AU-3. From Figure 2-4/G.708, page 5, ITU-T Rec. G.708, March 1993 [18].

Two types of virtual containers have been identified.

- Lower-order virtual container-n: VC-n ($n = 1, 2, 3$)
  This element comprises a single container-n ($n = 1, 2, 3$) plus the lower-order virtual container POH appropriate to that level.

- Higher-order virtual container-n: VC-n ($n = 3, 4$)
This element comprises a single container-n ($n = 3, 4$) or an assembly of tributary unit groups (TUG-2s or TUG-3s) together with virtual container POH appropriate to that level.

Administrative Unit-n (AU-n): An administrative unit is the information structure which provides adaptation between higher-order path layer and multiplex section layer. It consists of an information payload (the higher-order virtual container) and an administrative unit pointer which indicates the offset of the payload frame start relative to the multiplex section start.

Two administrative units are defined. The AU-4 consists of a VC-4 plus an Administrative Unit pointer which indicates the phase alignment of the VC-4 with respect to the STM-N frame. The AU-3 consists of a VC-3 plus the Administrative Unit pointer which indicates the phase alignment of the VC-32 with respect to the STM-N frame. In each case the administrative unit pointer location is fixed with respect to the STM-N frame.

One or more administrative units occupying fixed, defined positions in an STM payload are called an administrative unit group (AUG). An AUG consists of a homogeneous assembly of AU-3s or an AU-4.

Tributary Unit-n (TU-n). A tributary unit is an information structure which provides adaptation between the lower-order path layer and the higher-order path layer. It consists of an information payload (the lower-order virtual container) and a tributary unit pointer which indicates the offset of the payload frame start relative to the higher-order virtual container frame start.

The TU-n ($n = 1, 2, 3$) consists of a VC-n together with a tributary unit pointer.

One or more tributary units, occupying fixed, defined positions in a higher-order VC-n payload is termed a tributary unit group (TUG). TUGs are defined in such a way that mixed capacity payloads made up of different size tributary units can be constructed to increase flexibility of the transport network.

A TUG-2 consists of a homogeneous assembly of identical TU-1s or a TU-2. A TUG-3 consists of a homogeneous assembly of TUG-2s or a TU-3.

Container-n ($n = 1$–4). A container is the information structure which forms the network synchronous information payload for a virtual container. For each of the defined virtual containers, there is a corresponding container. Adaptation functions have been defined for many common network rates into a limited number of standard containers. These include those rates already defined in ITU-T Rec. G.702. Further adaptation functions will be defined by the ITU in the future for new broadband rates.

Concatenation. A procedure whereby a multiplicity of virtual containers is associated with another with the result that their combined capacity can be used as a single container across which bit sequence integrity is maintained.

SDH Aligning. A procedure by which the frame offset information is incorporated into the tributary unit or the administrative unit when adapting to the frame reference of the supporting layer.

13.3.3.1 Frame Structure. The basic frame structure, STM-N, is shown in Figure 8.36. The three main areas of the STM-1 frame are section overhead, AU pointers, and STM-1 payload.

Section Overhead. Section overhead is shown in rows 1–3 and 5–9 of columns 1–9 × *N* of the STM-N in Figure 8.34.

Administrative Unit (AU) Pointers. Row 4 of column 9 × *N* in Figure 8.36 is available for AU pointers. The positions of the pointers of the AUs for different organizations of the STM-1 payload are shown in Table 8.10. See ITU-T Rec. G.709 for application of pointers and their detailed specifications. The rules for interpreting the AU-*n* pointers are summarized below [20].

1. During normal operation, the pointer locates the start of the VC-*n* within the AU-*n* frame.
2. Any variation from the current pointer value is ignored unless a consistent new value is received three times consecutively or it is preceded by one of the rules 3, 4, or 5. Any consistent new value received three consecutive overrides (i.e., takes priority over) rules 3 and 4.
3. If the majority of the I-bits of the pointer word are inverted, a positive justification operation is indicated. Subsequent pointer values shall be incremented by one.
4. If the majority of the D-bits of the pointer word are inverted, a negative justification operation is indicated. Subsequent pointer values shall be decremented by one.
5. If the NDF (new data flag) is set to "1001," then the coincident pointer value shall replace the current one at the offset indicated by the new pointer value unless the receiver is in a state that corresponds to a loss pointer.

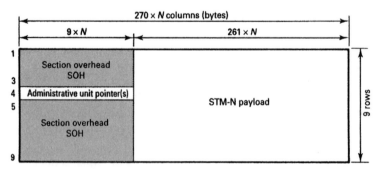

Figure 8.36. STM-N frame structure. From Figure 3-1/G.708, page 7, ITU-T Rec. G.708, March 1993 [18].

TABLE 8.10   AU-n/TU-3 Pointer (H1, H2, H3) Coding

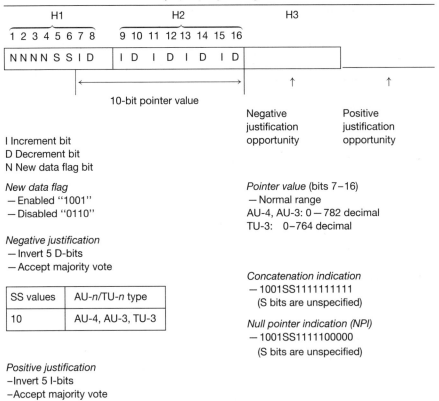

I Increment bit
D Decrement bit
N New data flag bit

New data flag
— Enabled "1001"
— Disabled "0110"

Negative justification
— Invert 5 D-bits
— Accept majority vote

SS values	AU-n/TU-n type
10	AU-4, AU-3, TU-3

Positive justification
–Invert 5 I-bits
–Accept majority vote

Negative justification opportunity

Positive justification opportunity

Pointer value (bits 7–16)
— Normal range
AU-4, AU-3: 0 — 782 decimal
TU-3:   0–764 decimal

Concatenation indication
— 1001SS1111111111
    (S bits are unspecified)

Null pointer indication (NPI)
— 1001SS1111100000
    (S bits are unspecified)

Notes:

1. NPI value applies only to TU-3 pointers.

2. The pointer is set to all "1"s when an AIS occurs.

Source: Figure 8-3/G.707/Y.1322.

Administrative units in the STM-N. The STM-N payload can support N AUGs where each AUG may consist of one AU-4 or three AU-3s. The VC-n associated with each AU-n does not have a fixed phase with respect to the STM-N frame. The location of the first byte of the VC-n is indicated by the AU-n pointer. The AU-n pointer is in a fixed location in the STM-N frame. This is shown in Figures 8.33 to 8.38. The AU pointer bytes H1 and H2 with bit meaning are shown in Table 8.10.

The AU-4 may be used to carry, via the VC-4, a number of TU-ns ($n = 1, 2, 3$) forming a two-stage multiplex. An example of this arrangement is illustrated in Figures 8.33 and 8.37a. The VC-n associated with each TU does not have a fixed-phase relationship with respect to the start of the VC-4. The TU-n pointer is in a fixed location in the VC-4, and the location of the first byte of the VC-n is indicated by the TU-n pointer.

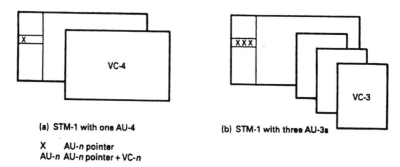

(a) STM-1 with one AU-4   (b) STM-1 with three AU-3s

X AU-*n* pointer
AU-*n* AU-*n* pointer + VC-*n*

Figure 8.37. Administrative units in the STM-1 frame. From Figure 3-2/G.708, page 8, ITU-T Rec. G.708, March 1993 [18].

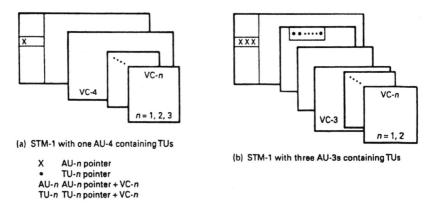

(a) STM-1 with one AU-4 containing TUs

X   AU-*n* pointer
•   TU-*n* pointer
AU-*n* AU-*n* pointer + VC-*n*
TU-*n* TU-*n* pointer + VC-*n*

(b) STM-1 with three AU-3s containing TUs

Figure 8.38. Two-stage multiplex. Based on Figure 3-3/G.708, page 9, ITU-T Rec. G.708, March 1993 [18].

The AU-3 may be used to carry, via the VC-3, a number of TU-*n*s ($n = 1, 2$) forming a two-stage multiplex. An example of this arrangement is illustrated in Figures 8.34 and 8.38b. The VC-*n* associated with each TU-*n* does not have a fixed-phase relationship with respect to the start of the VC-3. The TU-*n* pointer is in a fixed location in the VC-3, and the location of the first byte of the VC-*n* is indicated by the TU-*n* pointer [18].

13.3.4 Interconnection of STM-1s. SDH has been designed to be universal, allowing transport of a large variety of signals including those specified in ITU-T Rec. G.703, such as the North American 1.544 Mbits/s and European 2.048 Mbits/s regimes. However, different structures can be used for the transport of virtual containers. The following interconnection rules are used:

1. The rule for interconnecting two AUGs based upon two different types of administrative unit, namely AU-4 and AU-3, is to use the AU-4 structure.

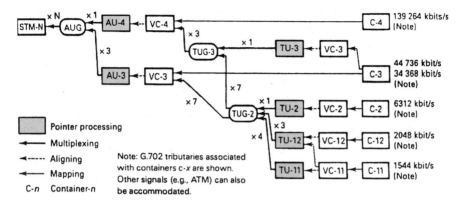

Figure 8.39. SDH multiplexing structure. Based on Figure 6-1/G.707.1.1322, page 18, ITU-T Rec. G.709, February 2001 [20].

Therefore, the AUG based upon AU-3 is demultiplexed to the TUG-2 or VC-3 level according to the type of the payload and is remultiplexed within an AUG via the TUG-3/VC-4/AU-4 route.

2. The rule for interconnecting VC-11s transported via different types of tributary unit, namely TU-11 and TU-12, is to use the TU-11 structure. VC-11, TU-11, and TU-12 are described in ITU-T Rec. G.709 [20].

13.3.5 Basic SDH Multiplexing Structure. The basic SDH multiplexing structure is shown in Figure 8.39. This is an expansion in detail of Figure 8.32.

14   SUMMARY OF ADVANTAGES AND DISADVANTAGES OF DIGITAL TRANSMISSION

The advantages of digital transmission tend to far outweigh the disadvantages. This is being borne out with the rapid conversion of the trunk plant to all digital. Because of the large investment in the subscriber loop plant, conversion to all digital may take more time. However, this time may be shortened if ISDN (Chapter 12) implementation is accelerated.

Some advantages and disadvantages of digital transmission are listed below.

Advantages

1. System noise is controlled by the design of the terminal (quantization noise) and is essentially independent of the length of the system or of line noise and distortion. This assumes that the full length of the system is digital.

2. The signal-to-distortion performance of the system increases linearly with the number of bits per sample, giving a more efficient noise/bandwidth trade-off than other *bandwidth expansion* techniques such as FM. This

efficiency, combined with the ruggedness of digital transmission, gives better utilization of noisy media such as wire-pair cable.

3. Increases in device speed (i.e., ICs, VLSI) allow common circuit components to be shared by many channels, thus lowering the per-channel cost of a PCM terminal. This is one major factor that makes the per-channel cost of PCM more cost-effective than FDM.

4. Digital systems are insensitive to traffic loading up to their full capacity. FDM is highly sensitive to traffic loading.

5. Likewise, digital time-division multiplex treats all channels alike in contrast to FDM regarding phase and amplitude distortion (and resulting noise degradation) suffered by channels at band edge.

6. There is no appreciable degradation incurred in multiplexing/demultiplexing so that facility arrangements do not need to take into account the number of previous multiplex/demultiplex operations.

7. Digital transmission gives complete freedom to multiplex digital data, voice, video, facsimile, etc., on the same facility, whereas analog transmission does not.

8. Digital systems are more efficient than analog in the transmission of digital data in that fewer voice channels must be displaced to obtain a given digital capacity.

9. Digital transmission provides the most economically possible interface to digital switching systems. Analog systems, on the other hand, require full demultiplexing/remultiplexing at switching nodes. Digital switches have inlets/outlets at the digital multiplex rates.

10. Fiber-optic transmission systems tend to favor digital transmission in that its attenuation is relatively independent of frequency. This makes bandwidth expansion transmission techniques to reduce baseband noise particularly attractive since the added bandwidth on fiber-optic systems is almost "free," and PCM is one of the most efficient of such schemes. In addition, light sources used in optical transmission exhibit nonlinearities that make them better suited to nonlinear modulation techniques such as PCM than to linear modulation methods such as AM.

11. Signaling is digital. Signaling on analog systems has to be converted to something compatible, such as a tone or multitone format. On digital systems, only a bit has to be changed in state for supervisory signaling and a bit sequence for address signaling (SSN No. 7). A digital system is also compatible with digital processors used in SPC switches.

Disadvantages

1. Bit errors accumulate across a digital system. These are not recoverable unless we resort to an error-correction system such as ARQ or FEC, both of which require still additional bandwidth.

2. System timing is a major issue and is discussed in the following chapter.

3. Although digital terminals tend to be less expensive than their analog counterparts on a per-channel basis, digital transmission lines (metallic media) tend to be more expensive than their analog counterparts. With the cost of fiber-optic systems dropping, digital transmission on fiber is less expensive than on metallic media [24].

REVIEW QUESTIONS

1. Give the three principal advantages of digital transmission.

2. What are the three basic steps in the development of a PCM signal from an analog source, typically voice?

3. Following the Nyquist sampling theorem, what is the sampling rate of a 4-kHz voice channel? Of a 7.5-kHz program channel? Of a 4.2-MHz video channel?

4. What is the polarity of a mark (1) and of a space (0) in AMI (bipolar) transmission?

5. For a 24-channel PCM system, calculate the period of one frame.

6. Define *quantization distortion.*

7. If the number of quantization steps is doubled in a particular PCM design using linear quantization, what improvement in quantization distortion (noise) is achieved? Express the answer in decibels.

8. Why is it desirable to reduce the size (number of bits or elements) of a PCM code word as much as possible and yet maintain reasonable voice quality?

9. There are only 16 quantization steps in a particular PCM system. What minimum length (bits) code word is required to accommodate these 16 discrete levels?

10. Identify the two distinct logarithmic companding laws used in modern PCM systems.

11. What key piece of information does the first significant bit in a PCM code word tell us?

12. Derive the value 1.544 Mbit/s in accordance with DS1 format.

13. Name at least five differences between European and North American PCM systems.

14. How is supervisory signaling carried out in the North American PCM system? In the European system? Argue pros and cons of each approach.

15. In the earliest implementations of PCM, where was it applied and why? (This does not imply that today it is not applied for the same reason in various situations.)

16. In the North American T1/DS1 system, what is the repeater spacing in feet? Going out from a switching node, why is there always a half-section rather than a full repeater section? For extra credit, what additional benefit can we get from that particular repeater spacing distance? (*Hint*: Turn back to subscriber/trunk loop design.)

17. Give at least two reasons why AMI (bipolar) waveform is used on T1/T1C (DS1/DS1C) cable transmission systems.

18. In a simplified functional block diagram of a regenerative repeater there are three basic functional blocks. Name and describe the function of each.

19. Describe BNZS. Why do we use it? What would B3ZS be?

20. How do we derive "enhancements" to a DS1 PCM system? Talk about framing bits. (I tell my classes: "Do you mean that I have to tell you 8000 times a second where a frame begins?")

21. In the DS1 extended superframe, what are the two basic enhancements (besides frame alignment, which we have had right along)?

22. Of what use is stuffing (justification) in higher-order PCM multiplex systems?

23. Jitter accumulation, a major impairment on long-distance digital transmission, is a function of what?

24. Besides jitter, give at least two other transmission impairments for digital transmission.

25. SONET, synchronous optical network, implies transmission on optical fiber. Can the SONET format be transmitted on other media? What would be the principal constraint, say, for transmitting SONET format on digital LOS radio?

26. Compare the first level digital line rate of SONET and SDH. The only concern in this question is the transmission rate in bps.

27. What is the payload capacity of STS-1?

28. What is the primary purpose of the payload pointer? Describe another benefit we get from the payload pointer.

29. How do we derive other line rates from STM-1?

30. Comment on the superiority of DS1 over E1, or the superiority of E1 over DS1.

31. If PCM is performance limited by the number of regenerative repeaters in tandem, then how can we get around this problem?

32. If the fact that noise does not accumulate on a digital network, and we see that as the principal advantage, then what is the principal disadvantage of, say, PCM transmission? (*Hint*: It deals with accumulation.)

REFERENCES

1. *Reference Data for Radio Engineers*, 6th ed., ITT/Howard W. Sams, Indianapolis, 1976.

2. GTE Lenkurt Demodulator, *PCM Update, Parts 1 and 2*, GTE-Lenkurt Electric Company, San Carlos, CA, February 1975.

3. *Transmission Systems for Communications*, 5th ed., Bell Telephone Laboratories, Holmdel, NJ, 1982.

4. J. Bellamy, *Digital Telephony*, 3rd ed., John Wiley & Sons, New York, 2000.

5. D. R. Smith, *Digital Transmission Systems*, 2nd ed., Van Nostrand Reinhold, New York, 1993.

6. *Pulse Code Modulation (PCM) of Voice Frequencies*, ITU-T Rec. G.711, ITU, Geneva, 1993.

7. *Digital Channel Bank—Requirements and Objectives*, Bell System Technical Reference Publication 43801, American Telephone & Telegraph Co., Basking Ridge, NJ, 1982.

8. *Physical and Electrical Characteristics of Hierarchical Digital Interfaces*, ITU-T Rec. G.703, ITU, Geneva, 1988.

9. *Synchronous Frame Structures Used at Primary and Secondary Hierarchical Levels*, ITU-T Rec. G.704, ITU, Geneva, October 1998.

10. R. L. Freeman, *Reference Manual for Telecommunications Engineering*, 3rd ed., John Wiley & Sons, New York, 2002.

11. R. L. Freeman, *Telecommunication Transmission Handbook*, 4th ed., John Wiley & Sons, New York, 1998.

12. *Second Order Digital Multiplex Equipment Operating at 8448 kbps Using Positive/Zero/Negative Justification*, ITU-T Rec. G.745, ITU, Geneva, 1993.

13. *Functional Criteria for Digital Loop Carrier Systems*, Bellcore Technical Reference TR-NWWT-000057, Issue 2, Bellcore (Telcordia), Piscataway, NJ, 1993.

14. *Synchronous Optical Network (SONET) Transport Systems: Common Generic Criteria*, Bellcore GR-253-CORE, Issue 1, Bellcore (Telcordia), Piscataway, NJ, December 1994.

15. *Telcordia Notes on the Networks*, Special Report SR-2275, Issue 4, Telcordia, Piscataway, NJ, October 2000.

16. *Introduction to SONET*, A Hewlett-Packard Seminar, Burlington, MA, 1993.

17. *Synchronous Digital Hierarchy Bit Rates*, ITU-T Rec. G.707, ITU, Helsinki, March 1993.

18. *Network Node Interface for Synchronous Digital Hierarchy*, ITU-T Rec. G.708, ITU, Geneva, March 1993.

19. *Digital Hierarchy Bit Rates*, CCITT Rec. G.702, ITU, Geneva, 1988.

20. *Network Node Interface for the Synchronous Digital Hierarchy*, ITU-T Rec. G.707, ITU, Geneva, Oct. 2000.

21. R. L. Freeman, "An Overview of Digital Transmission and Multiplexing," a tutorial presentation to the MITRE Institute, Bedford, MA, October 1987.

22. *IEEE Standard Dictionary of Electrical and Electronic Terms*, 6th ed., IEEE Std. 100-1996, IEEE Press, New York, 1996.

23. K. W. Catermole, *Principles of Pulse Code Modulation*, Illiffe, London, 1969.

24. *Telecommunications Transmission Engineering* (3 volumes), ATT-Western Electric, Winston-Salem, NC, 1977.

25. *Error Performance Parameters and Objectives for International Constant Bit Rate Paths at or above the Primary Rate*, ITU-T Rec. G.826, ITU, Geneva, February 1999.

26. *Error Performance Parameters and Objectives for International Constant Bit Rate Paths*, ITU-T Rec. G.828, ITU, Geneva, March 2000.

27. *Synchronous Optical Network (SONET) Basic Description Including Multiplex Structure, Rates and Formats*, ANSI T1.105-1995, ANSI, New York, 1995.

28. "Service Description and Interface Requirements for Pacific Bell's SONET Services," Pub L-780046-PB/NB, Issue 1, May 1995.

29. *Interfaces for the Optical Transport Network*, ITU-T Rec. G.709, ITU, Geneva, March 2001.

9

DIGITAL SWITCHING
AND NETWORKS

1 INTRODUCTION

In Chapter 3 we dealt with analog space-division switching in which a metallic path is set up between calling and called subscriber. "Space division" in this context refers to the fact that speech paths are physically separated (in space). Figure 9.1A illustrates this concept by showing a representative crosspoint matrix. Time-division switching (Figure 9.1B) permits a single common metallic path to be used by many calls separated one from the other in the time domain. In this context, with time-division switching the speech or other information to be switched is digital in nature, either PCM or delta modulation (DM). Samples of each telephone call are assigned time slots, as described in Chapter 8. PCM or DM switching involves the distribution of these slots in sequence to the desired destination port(s) of the switch. Internal functional connectivities in the switch are carried out by digital "highways." A highway consists of sequential speech path time slots.

This chapter describes PCM switching in a simplified step-by-step fashion. It covers generic switch architectures and explains basic functional operations. This chapter then discusses a move toward modernization that higher internal bit rates allow these switches to become smaller and more adaptable.

This is followed by an overview of digital networks and their topologies, where the switch is probably the most important functional element. The discussion of digital networks includes network synchronization and timing, network error performance, and digital network impairments such as jitter, wander, and slips. Our technical vocabulary is based on ITU-T Rec. G.701 [13] and the *IEEE Standard Dictionary*, 6th ed. [24].

Telecommunication System Engineering, by Roger L. Freeman
ISBN 0-471-45133-9   Copyright © 2004 Roger L. Freeman

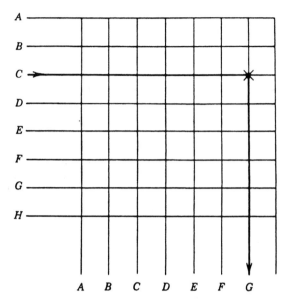

Figure 9.1A. A space-division switch showing connectivity from user C to user G [1].

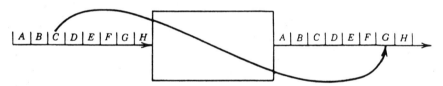

Figure 9.1B. A time-division switch which is a time-slot interchanger (TSI). Connectivity is from user C (in incoming times slot C) to user G (in outgoing time slot G) [1].

1.1 Radical New Directions

If we move ahead to Chapters 10 and 11 where we discuss data communications and data networks, we find some very different philosophies, particularly in the area of signaling. It could be said that each data frame repeats address signaling over and over again. Digital telephony also uses a frame concept, but address information is not repeated after the first frame. It is sent just once to set up a circuit. However, some form of supervisory signaling is required to maintain that circuit so set up in a "busy condition." It remains this way until one or the other end of the connectivity goes "on hook" (i.e., hangs up).

In the case of data communications, where the service often used is "connectionless." Here the frame or packet is an independent entity. It is delivered to the network and it is on its own to find its way to the destination. In a data network, a *router* is the key device, much like a switch is the key device in a PCM (digital) network. It examines the header of a data frame or packet where the address and control information may be found. Based on the destination address in the header, it routes the message directly to its destination or via one or more routers thence to the destination.

The radical new direction of a digital telecommunication network is to have just one service, the data network. Digital voice samples are placed in the payload of a data packet as any other form of data. We see the final elimination of a PCM voice network as we know it today. There will be just one, singular network handling voice and data as though they were just one form or another of information. This new approach is commonly referred to as *voice over IP* or *voice over packet*. In our new network a router will carry out the function of the PCM switch. "Voice over IP" is described in detail in Chapter 12.

Another element of this chapter is the digital cross-connect (DXC). A DXC can be likened to a digital switch, but the circuit connectivity is via a form of patch-panel (manual) or one that is controlled by a local processor with keypad inputs. The connection may last for minutes, hours, days, months, or years.

The final discussion on digital switching deals with the programmable switch controlled by a host computer. This switch will handle voice, data, and video (image) traffic in one integrated approach.

The last portion of the chapter discusses the digital network per se dealing with new requirements not found in its analog counterpart. There are also new and different impairments such as jitter and slips that will be new to many readers.

2 ADVANTAGES AND ISSUES OF PCM SWITCHING WHEN COMPARED TO ITS ANALOG COUNTERPART

There are both economic and technical advantages to digital switching; in this context we refer to PCM switching (or course, most of these same arguments hold for delta/CVSD switching as well). The economic advantages of time-division PCM switching include the following:

- There are notably fewer equivalent cross-points for a given number of lines and trunks than in a space-division switch.
- A PCM switch is of considerably smaller size.
- It has more common circuitry (i.e., common modules).
- It is easier to achieve full availability within economic constraints.

The technical advantages include the following:

- It is regenerative (i.e., the switch does not distort the signal; in fact, the output signal is "cleaner" than the input).
- It is noise-resistant.
- It is computer-based and thus incorporates all the advantages of SPC.
- The binary message format is compatible with digital computers. It is also compatible with signaling.

- A digital exchange is lossless. There is no insertion loss as a result of a switch inserted in the network.
- It exploits the continuing cost erosion of digital logic and memory; LSI, VLSI, and VHSIC (very high speed integrated circuit) insertion.

Two technical issues may be listed as disadvantages:

- A digital switch deteriorates error performance of the system. A well-designed switch may only impact network error performance minimally, but it still does it.
- Switch and network synchronization and the reduction of wander and jitter can be gating issues in system design.

Source: Ref. 1.

3 APPROACHES TO PCM SWITCHING

3.1 General

A classical digital switch is made up of two functional elements: a time switch called "T" and a space-switch abbreviated "S." The architecture of a digital switch is described in sequences of Ts and Ss. For example, the AT&T (now Lucent) 4ESS is a TSSSST switch. In other words, the input stage is a time switch, followed by four space switches in sequence and the last stage is a time stage. The Northern Telecom DMS-100 and its family of derivatives is a TSTS switch that is folded back on itself.

One thing in common with these switches is that they had multiple space (S) stages. This has now changed. Many of the new switches or enhanced versions of the switches just mentioned have very large capacities (e.g., 100,000 lines) and are simply TST or STS switches.

We will describe a simple time switch, a space switch, and methods of making up an architecture combining T and S stages. We will show that designing a switch with fairly high line and trunk capacity requires multiple stages. Then we will discuss the "new look" at the time stage.

3.2 Time Switch

In a most simplified way, Figure 9.1B is a time-switch or time-slot interchanger (TSI). From Chapter 8 we know that a time slot in conventional PCM contains 8 sequential bits, and a basic frame is 125 μs in duration. For the North American DS1 format the basic frame contains 24 time slots, and for the European E1 it has 32 time slots. The time duration of an 8-bit time slot in each case is $125/24 = 5.2083$ μs for DS1 and $125/32 = 3.906$ μs for E1. Time slot interchanging involves moving the data contained in each time slot from the incoming bit stream to an outgoing bit stream but with a different time-slot arrangement in accordance with the destination of each time slot. What is done, of course, is to generate a new frame for transmission at the appropriate switch outlet.

Obviously, to accomplish this, at least one time slot must be stored in memory (write) and then called out of memory in a changed position (read). The operations must be controlled in some manner, and some of these control actions must be kept in memory together with the software managing such actions. Typical control functions are time-slot "idle" or "busy." Now we can identify three of the basic functional blocks of a time switch:

1. Memory for speech
2. Memory for control
3. Time-slot counter or processor

These three blocks are shown in Figure 9.2. There are two choices in handling the time switch: (1) sequential write, random read as shown in Figure 9.2A and (2) the reverse, namely, random write, sequential read. In the first case, sequential write, the time slots are written into the speech memory as they appear in the *incoming* bit stream. For the second case, random write, the incoming time slots are written into memory in the order of appearance in the *outgoing* bit stream. This means that the incoming time slots are written into memory in the desired *output* order. The writing of incoming time slots into the speech memory can be controlled by a simple time-slot counter and can be sequential (e.g., in the order in which they appear in the incoming bit stream, Figure 9.2A). The readout of the speech memory is controlled by the control memory. In this case the readout is random where the time slots are read out in the desired output order. The memory has as many cells as there are time slots. For the DSI example, there would be 24 cells. This time switch, as shown, works well for a single inlet–outlet switch. With just 24 cells it could handle 23 stations besides the calling subscriber, not an auspicious number.

How can we increase a switch's capacity? Enter the space switch (S). Figure 9.3 affords a simple illustration of this concept. For example, time slot B_1 on the B trunk is moved to the Z trunk into time slot Z_1, and time slot C_n is moved to trunk W into time slot W_n. However, we see that there is no change in time-slot position.

3.3 Space Switch

A typical time-division space switch is shown in Figure 9.4. It consists of a cross-point matrix made up of logic gates that allow the switching of time slots in the spatial domain. These PCM time-slot bit streams are organized by the switch into a pattern determined by the required network connectivity. The matrix consists of a number of input horizontals and a number of output verticals with a logic gate at each cross-point. The array, as shown in the figure, has M horizontals and N verticals, and we call it an $M \times N$ array. If $M = N$, the switch is nonblocking. If $M > N$, the switch concentrates; and if $N > M$, the switch expands.

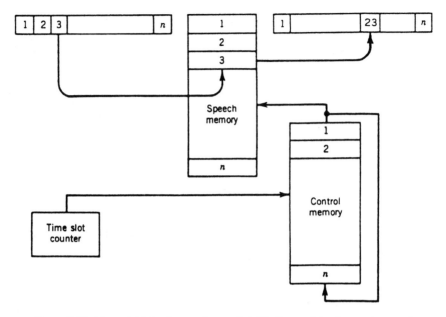

Figure 9.2A. Time-slot interchange: time switch (T). Sequential write, random read.

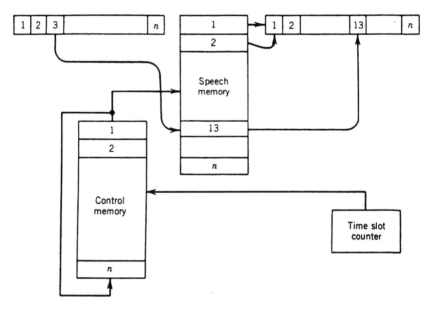

Figure 9.2B. Time-switch, time-slot interchange (T). Random write, sequential read.

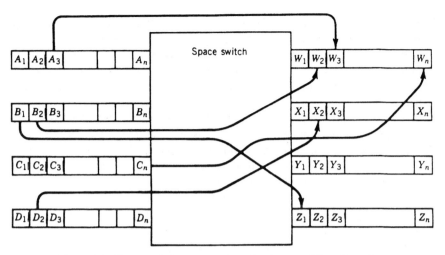

Figure 9.3. Space switch connects time slots in a spatial configuration.

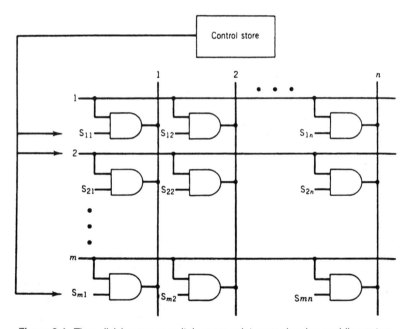

Figure 9.4. Time-division space switch cross-point array showing enabling gates.

Return to Figure 9.4. The array consists of a number of (M) input horizontals and (N) output verticals. For a given time slot, the appropriate logic gate is enabled and the time slot passes from the input horizontal to the desired output vertical. The other horizontals, each serving a different serial stream of time slots, can have the same time slot (e.g., a time slot from time slots number 1–24,

1–30, or 1–n; for instance, time slot 7 on each stream) switched into other verticals enabling their gates. In the next time-slot position (e.g., time slot 8), a completely different path configuration could occur, again allowing time slots from horizontals to be switched to selected verticals. The selection, of course, is a function of how the traffic is to be routed at that moment for calls in progress or being set up.

The space array (cross-point matrix) does not switch time slots as does a time switch (time-slot interchanger). This is because the occurrences of time slots are identical on the horizontal and on the vertical. It switches in the space domain, not in the time domain. The control memory in Figure 9.4 enables gates in accordance with its stored information.

If an array has M inputs and N outputs, M and N may be equal or unequal depending on the function of the switch on that portion of the switch. For a tandem or transit switch we would expect $M = N$. For a local switch requiring concentration and expansion, M and N would be unequal.

If, in Figure 9.4, it is desired to transmit a signal from input 1 (horizontal) to output 2 (vertical), the gate at the intersection would be activated by placing an enable signal on S_{12} during the desired time-slot period. Then the eight bits of that time slot would pass through the logic gate onto the vertical. In the same time slot, an enable signal on S_{M1} on the Mth horizontal would permit that particular time slot to pass to vertical 1. From this we can see that the maximum capacity of the array during any one time-slot interval measured in simultaneous call connections is the smaller value of M or N. For example, if the array is 20×20 and a time-slot interchanger is placed on each input (horizontal) line and the interchanger handles 30 time slots, the array then can serve $20 \times 30 = 600$ different time slots. The reader should note how the TSI (time-slot interchanger) multiplies the call-handling capability of the array when compared to its analog counterpart [5, 6].

3.4   Time–Space–Time Switch

Digital switches are composed of time and space switches in any order or in time switches only. We use the letter T to designate a time-switching stage and use S to designate a space-switching stage. For instance, a switch that consists of a sequence of a time-switching stage, a space-switching stage, and a time-switching stage is called a TST switch. A switch consisting of a space-switching stage, a time-switching stage, and a space-switching stage is designated an STS switch. There are other combinations of T and S. The ATT No. 4 ESS switch is an example. It is a TSSSST switch.

Figure 9.5 illustrates the time–space–time (TST) concept. The first stage of the switch is the time-slot interchanger (TSI) or time stages that interchange time slots (in the time domain) between external incoming digital channels and the subsequent space stage. The space stage provides connectivity between time stages at the input and output. It is a multiplier of call-handling capacity. The multiplier is either the value for M or value for N, whichever is smaller. We also

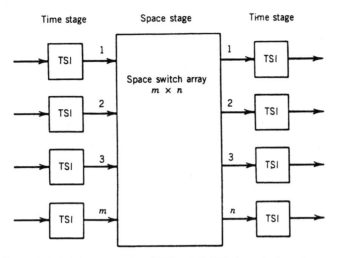

Time stage   Space stage   Time stage

Figure 9.5. A time–space–time (TST) switch. TSI, time-slot interchanger.

saw earlier that space-stage time slots need not have any relation to either external incoming or outgoing time slots regarding number, numbering, or position. For instance, incoming time slot 4 can be connected to outgoing time slot 19 via space network time slot 8.

If the space stage of a TST switch is nonblocking, blocking in the overall switch occurs if there is no internal space-stage time slot during which the link from the inlet time stage and the link to the outlet time stage are both idle. The blocking probability can be minimized if the number of space-stage time slots is large. A TST switch is strictly nonblocking if

$$l = 2c - 1 \tag{9.1}$$

where l is the number of space-stage time slots and c is the number of external TDM time slots [3].

3.5 Space–Time–Space Switch

A space–time–space (STS) switch reverses the architecture of a TST switch. The STS switch consists of a space cross-point matrix at the input followed by an array of time-slot interchangers whose ports feed another cross-point matrix at the output. Such a switch is shown in Figure 9.6. Consider this operational example with an STS. Suppose that an incoming time slot 5 on port No. 1 must be connected to an output slot 12 at outgoing port 4. This can be accomplished by time-slot interchanger No. 1 which would switch it to time slot 12; then the outgoing space stage would place that on outgoing trunk No. 4. Alternatively, time slot 5 could be placed at the input of TSI No. 4 by the incoming space switch where it would be switched to time slot 12, thence out port No. 4.

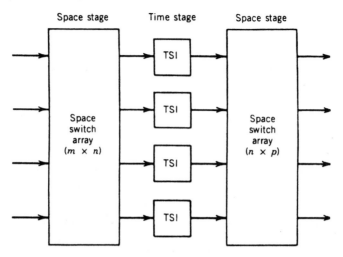

Figure 9.6. A space–time–space (STS) switch.

3.6   TST Compared to STS

Both TST and STS switches can be designed with identical call-carrying capacities and blocking probabilities. It can be shown that a direct one-to-one mapping exists between time-division and space-division networks [2].

The architecture of TST switching is more complex than STS switching with space concentration. The TST switch becomes more cost-effective because time expansion can be achieved at less cost than space expansion. Such expansion is required as link utilization increases because less concentration is acceptable as utilization increases.

It would follow, then, that TST switches have a distinct implementation advantage over STS switches when a large amount of traffic must be handled. Bellamy [3] states that for small switches STS is favored due to reduced implementation complexities. The choice of a particular switch architecture may be more dependent on such factors as modularity, testability, and expandability.

One consideration that generally favors an STS implementation is the relatively simpler control requirements. However, for large switches with heavy traffic loads, the implementation advantage of the TST switch and its derivative is dominant. A typical large switch is the Lucent (previously AT&T) 4ESS, which has a TSSSST architecture and has the capability of terminating 107,520 trunks with a blocking probability of 0.5% and channel occupancy of 0.7.

4   DIGITAL SWITCHING CONCEPTS – BACKGROUND

4.1   Early Implementations

In Section 3 of this chapter, the reader was probably led to believe that the elemental time-switching stage, the time-slot interchanger (TSI), would have 24 or

30 time-slot capacity to match the North American DS1 rate or the European E1 rate, respectively. That means a competitive manufacturer would have to make two distinct switches, one to satisfy the North American market and another for the European market. For more cost-effective production, a typical switch manufacturer made just one switch with peripheral gear to interface either regime and a common internal switching network, consisting of time and space arrays that we just discussed. For one thing, they could map 5 DS1 groups into 4 E1 groups, the common denominator being 120 DS0/E0 (64 kbits/s channels) channels. Those same peripheral modules also cleaned up any signaling disparities that might have existed.

All digital switches have a common internal digital format and bit rate. Take the Lucent (previously AT&T) 4ESS, for example. It uses the number "120." It maps 120 8-bit time slots into 128 time slots. The 8 time slots of the remainder are used for diagnostic and maintenance purposes.

That common internal digital format of a switch might or might not use 8-bit time slots, even though the outside world (e.g., DS1 or E1) required an 8-bit octet interface and frame of 125-μs duration. The Northern Telecom DMS-100 maps the external 8-bit time slot into an internal 10-bit time slot as illustrated in Figure 9.7. The example used in the figure is the DS1.

Note in Figure 9.7 that one bit is a parity bit (bit 0) and the other appended bit (bit 1) carries the supervisory signaling information: the line is idle or busy. Bits 2 through 9 are the bits contained in the original 8-bit time slot. Because Northern Telecom, in their DMS-100 series switch wanted a switch that was simple to convert from E1 to TS1, they built their internal bit rate to 2.560 Mbits/s as follows: 10 bits per time slot, 32 time slots × 8000 (the frame rate) or 2.560 Mbits/s.

Lucent's 5ESS maps each 8-bit time slot into a 16-bit internal PCM word. It actually appends 8 additional bits onto the 8-bit PCM word as shown in Figure 9.8.

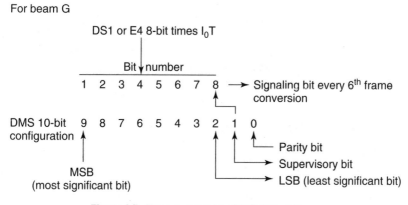

Figure 9.7. Bit mapping in the DMS-100, -200 ---.

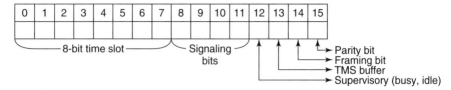

Figure 9.8. The make-up of the 16-bit internal time slot Lucent 5ESS.

4.2 Higher-Level Multiplex Structures Internal to a Digital Switch

We pictured a simple time-slot interchanger switch with 24 8-bit time slots to satisfy DS1 requirements. It would meet the needs of a population of 24 subscribers without blocking. There is no reason why we couldn't build a TSI with greater capacity. Suppose that instead of the DS1 rate, we built the TSI with a DS3 rate. The basic TSI then could handle 672 subscribers (i.e., 672 time slots). If we built a concentrator in front of it for 4 : 1 concentration, then it could handle 4×672 or 2688 subscribers. But note an important aspect, the gating time per channel. In a DS3 frame there are 672 time slots available for traffic. Each time slot is visited once for each frame, and thus the time of dwell is 125 μs/672 = 186 ns. This is still inside the state of the art. But as we increase the bit rate, that gating time per time slot can be very small indeed. For example, a 50,000 time-slot switch would only have a 2.5-ns gating time.

The Lucent (previously AT&T) 5ESS is a TST switch. It has the capacity of 100,000 or more lines. They are able to accomplish this simpler architecture by using larger-capacity time-slot interchangers (TSIs) and higher bit rates on the space stage. A 5ESS TSI handles 512 time slots.* However, each TSI port has an incoming/outgoing time-slot rate of 256 time slots. Two ports are required (in one direction) to handle the 512 time slots: one for odd-numbered channels and one for even-numbered channels. Thus the bit rate at a TSI port is $256 \times 16 \times 8000 =$ 32.768 Mbits/s. This odd-channel, even-channel arrangement carries through the entire switching fabric with each port handling 256 time slots or 32.768 Mbits/s.

Another example of a widely implemented digital switch is the Northern Telecom DMS-100 with supernode/ENET[†]. They modified the older DMS100 conventional switch, which had a TSTS-folded architecture. Like the 5ESS, they also moved into the 2048-time-slot domain in the ENET (extended network). But their time slot is 10 bits, and the ENET uses a 10-bit parallel format, so each line (i.e., there are 10 lines) has 2048×8000 or 16.384 Mbits/s.

4.3 Remote Switching Capabilities

4.3.1 Remote Switching Defined. A remote switch is a module taken from the principal switch and displaced to a remote location. This location may be

---

* Remember that a time slot here has 16 bits. See Figure 9.8 for a time-slot layout.
[†] ENET is nortel term for "enhanced network

just hundreds of feet (meters) or miles (kilometers) from that of the principal or "mother" switch. The functions carried out in the remote module may very considerably with the manufacturer and even model number. As a minimum, the remote module must interface with a subscriber. This interface may be a conventional analog interface of a single channel digital interface.

Among the minimum function set, we expect to find:

- battery supply, often −48 volts DC
- signaling: supervisory and address signaling
- alerting the subscriber, some form of "ring-down"

On the other side of the module there must be some way of communicating with the principal switch or "mother." Among the most common methods we find and E1 or T1 configuration on one or better yet, two wire pairs. Depending on the type of signaling used, there may be one or two time-slot voice channels dedicated to signaling. [12]

Remote switching provides many advantages to the telecommunication system designer.

- It can serve as a community dial office (CDO) where a full-blown switch would not be justified.
- It can dramatically extend the operational area of a switch
- It can serve as an A/D and D/A point of conversion providing analog interface with a subscriber and the digital interface with the network.
- A remote switch can serve as a concentrator. In this case it may provide a capability of switching calls inside its own serving area (or it may not). The most simple configuration is a device that just concentrates. To carry its complexity one step further: it can concentrate and provide A/D conversion. When we say "concentration," we mean a device that serves, say, 120 subscribers and has trunk connectivity with the mother switch with only T1 capacity. Therefore it has a concentration capacity of 120-to-24. (See Chapter 3, Section 12).

4.4   Digital Cross-Connects

4.4.1   Scale and Scope of Digital Cross-Connects.

The Digital Cross-connect (DCS) has been with us virtually since the advent of the digital network. The DCS has been taking on much greater importance as a network element to meet the requirements of exploding traffic volumes and of different traffic types. A digital cross-connect is a device that handles the connections between two or more telecommunication transmission facilities. The types of network cross-connects handled by a DCS can range from nearly terabit data rates of fiber-optic cable to relatively low-speed data rates of copper pairs used to provide access to a group of residences.

A common question arises as to the difference between a DCS and a PSTN switch (typically what has been described in previous sections of this chapter). A digital switch, whether serving the local area, tandem, or toll, sets up a short-term virtual circuit where a connection may last just seconds, minutes, or several hours. A DCS has more permanency where the duration of a connection may be minutes, hours, days, weeks, or years. It is usually a simpler installation.

In the past there was usually a centralized DCS network architecture because of the high cost of grooming and switching traffic. It was more economical to backhaul all the traffic in a region or in a metropolitan network to a tandem switch or hub facility than to distribute the bit rate capacity management capability closer to the network edge.

A new generation of products has emerged over the last several years which are referred to as multiservice provisioning platforms (MSPPs) and a new generation of add–drop multiplexers (ADM). These systems may be deployed at the network edge and are capable of grooming and switching traffic more economically.

4.4.2 Distributed and Centralized DCS Strategies.

There are two strategies for DCS: centralized and distributed. For traffic that both originates and terminates in a metropolitan area network, the distributed DCS strategy makes sense as it eliminates the need to backhaul the traffic to and from a tandem switch or large metropolitan hub site. This will save both bit rate capacity and equipment costs in the form of DCS ports and ADM equipment. However, where traffic originates in a metropolitan network and terminates in some other network, the distributed model does not work so well. This traffic is usually a mix of PSTN voice, data, and other long-distance services. Such traffic must first be passed through a gateway at a tandem switch or metro core site in order to be compatibly routed to other service provider networks whether metropolitan or long-haul.

The volume and intensity of traffic routed through a metropolitan core gateway often greatly exceeds the volume of traffic that both originates and terminates within a metropolitan network itself. A distributed architecture may require capacities in the range of 50–90 Gbits/s with SONET STS-1 switching capability and wideband capacities limited to no more than 10,000 protected circuits. A distributed wideband DCS architecture may have switching requirements within 1000 STS-1s equivalent capacity.

A centralized architecture finds favor at a tandem switch because traffic at these back-end locations often exceeds 200 Gbits/s and will continue to grow. The centralized platform at tandem switch/metropolitan core sites allows for large, scalable grooming. This centralized grooming platform is required to operate as a bit rate capacity gateway to the carrier's services, long-haul networks, and other metropolitan networks.

We have described DCS platforms as systems that perform the switching and grooming of circuits. However, aside from bit-rate capacity management, these systems are critical for other service management operations that include performance monitoring, service assurance, and test access. This is particularly true for wideband systems that are responsible for service management of DS1/E1 circuits—as much as 80% of current service provider revenues are generated from DS1 (E1)-based services.

Source: Converge Network Digest, www.convergedigest.com/columns/0210polaris/g-sgosa14.1ap, 2/24/03, Sections 4.6, 4.6.1, and 4.6.2 [29].

4.4.3 Six Types of DCS Devices

Integrated Access Device (IAD). This device is located on the customer premises. It interfaces with both voice and several data protocols. It aggregates and cross-connects these to an enhanced copper wire loop.

Customer Service Node. This is similar to the IAD but is used on customer premises where bit rate capacities are larger and where fiber-optic cable is required.

Digital Loop Carrier. This is an enhanced DLC which aggregates multiple user protocols and is connected to the transmission network through a SONET/SDH ring.

Access Manager. This device is located at a switching center and is equipped to handle enhanced speed copper wire loops with multiple protocols. It segregates and cross-connects these to appropriate links in a metro network.

Ring Manager. This DCS device is also located at a switching center. It is equipped to manage and cross-connect channels from multiple SONET rings. Such rings can be found in both an access network and in a metro or interswitch network.

Trunk Manager. This DCS device will be found at a hub or tandem switching center. It is equipped to manage and cross-connect channels from multiple fiber-optic trunks.

Note the use of the term "manager" to denote the ever-growing importance of DCS devices to a network management function. A DCS is applied only to transmission management, where, as in most telecommunication services based on switch control, it is an inherent element of the service provided to customers. DCS control is an engineering and provisioning process. A DCS also enhances survivability.

Source: Section 4.4.3 is based on Ref. 30, www.insight-corp.com/dcs.html.

4.5 A New Direction – Programmable Switching

The architecture of programmable switching is based on the open and distributed client–server computing model so common in the computer industry. Previously, telecommunication switches (e.g., 5ESS) were based on large, proprietary solutions. Now switches are based on open, scalable platforms. The importance of

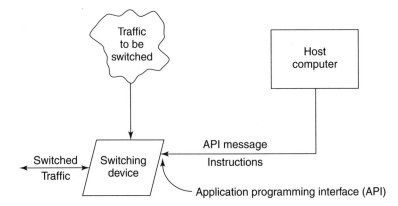

Figure 9.9. A typical programmable switch in a client–server configuration with an API.

standards are emphasized as well as the interoperability between hardware platforms and software applications. We also are seeing the concept of efficient clients and servers where system resources can be distributed geographically where and when required.

Another advantage of programmable switching is that it allows developers to implement new and unique services. Some manufacturers design switches with open strategies allowing developers access to the software environment at one or more levels. These open strategies involve application programming interfaces (APIs), industry-standard interconnection devices, and high-level host-resident development tools. These programmable switches can be designed to meet unique market needs of a particular situation.

Figure 9.9 shows typical programmable-switching architecture. From the figure we can identify three major elements: a switching system, its host, and real-time API messaging.

4.5.1 The Host Computer.

Figure 9.9 shows the host computer and the programmable switch in a client–server relationship. The function of the host is to store and execute varying amounts of call control application software that manages the switching matrix. The host computer controls the matrix switch with API structures and other subroutines. The open architecture between the switch and the host allows developers and users to tailor the switch to meet local market requirements. In the past, switches have had a type of firmware to perform a set of specific duties such as local area or tandem switching.

4.5.2 The Application–Programming Interface (API).

As shown in Figure 9.9, the switch matrix or switching device is controlled by a host computer via the API. Host instructions through the API are transmitted to the switch via messaging. The larger, richer, and more open the message set of the API, the greater the flexibility in programming the switch. The key here is open programming even down to the functional elements of the switch such as DSPs (digital signal processors), common channel signaling modules (e.g., SSN#7), and network data protocols.

4.5.3 Programmable Switching Device. The switch consists of a combination and space switch matrix and time-slot interchangers, line interface units, and interconnects with one or more networks. Modern switches have a high level of redundancy to improve availability and survivability. Overall switch MTBF can often be measured in tens of years.

There are a great many different types of lines and trunks with which a switch may require an interface. They may be analog and/or digital, single channel or multichannel—for example, E0s and/or E1s. These connectivities are usually routed through interface cards. There should also be provision for ample growth. There may be frame relay interfaces, ATM and Ethernet interfaces, and EIA/TIA-530 or 232, loop start, wink start, and others.

We would also expect to find service interfaces, especially signaling, ISDN basic and primary rate, subrate switching, and so on.

API switches are scalable where the user can start out "small" and grow to meet demand. A fitting term is "wired" but not equipped. Another descriptive term is "hot insertion and removable." Here we mean that new modules can be inserted and older ones removed while the switch is in operation, without interruption in service.

4.5.4 Switch Features. Not only should a switch with its incumbent host computer be reliable, it should also be easily maintainable. This measures the ease and speed with which operation can be restored after an interruption in service due to a failure. Maintainability is measured in MTTR (mean time to repair) which should be in several minutes, seconds, or even milliseconds. In fault-tolerant switches, the time should be zero. Failed components should be quickly isolated and their status or condition read out on a PC or similar device.

Redundancy is particularly important in two areas: prime power and its conversion devices (e.g., ac–dc converters, dc–dc converters) and synchronization. The leading edges of pulses on certain T1 or E1 bit streams serve as sync sources. If these sources are lost, the master switch clock takes over with only a very small increase in slip rate.

The new API switches will be uni-service devices, handling voice, data, and image indistinguishably. The forerunner of this is VoIP described in Chapter 12. It will find equal applicability whether in the PSTN or with an enterprise network.

5   THE DIGITAL NETWORK

5.1   Introduction

The North American and European public switched telecommunication networks are 100% digital. Some industrially evolving countries still have rural components which are analog. The international interconnecting network is nearly 100%

digital. The transport circuits, whether national or international, are either PDH using T1 (DS1) or E1 regimes. However, more and more are turning to SONET and SDH because of the greater available capacity. The reader, nevertheless, should keep in mind that the network structure itself changes slowly, even though integrating time gets shorter. Structure change involves two factors:

1. Political
2. Technological (laws of physics and economics of new and old technologies)

In the United States, certainly divestiture of the Bell System affected network structure with the formation of LECs (local exchange carriers) and IXC (interexchange carriers). Then came the CLECs (competitive local exchange carriers). Outside of North America, the movement toward privatization of government telecommunication monopolies has, in one way or another, affected network structure. Another political aspect has been that national regulating authorities have allowed a greater acceptance of new and developing technology. One must also take into consideration the political strength of industrial segments of our society such as the broadcast industry. The more democratic the society, the greater the influence. Broadcaster emitters occupy the major element of the most desirable spectrum. The emitters should be shut down, and the services should be folded into cable television. Then the spectrum allotted to cellular/mobile users could be tripled; public service facilities would not be so crowded, eliminating many interference problems.

Technology and its advances certainly may be equally or even more important than political causes. For example, satellite communications, we believe, brought about the move by CCITT (now ITU-T) away from any sort of international network hierarchy. International high-usage and direct routes became practical. We should not lose sight of the fact that every digital exchange has, at its heart, a powerful computer, permitting millisecond routing decisions for each call. This was greatly aided by the implementation of CCITT Signaling System No. 7 (Chapter 17). This same computer power has allowed PSTN carriers to offer a multifold increase of enhanced services to their customers effectively changing the day-to-day life of many people. Another evident factor is the pervasiveness of fiber-optic cable on many, if not most, terrestrial routes and all undersea trunk routes. The bit-rate capacity on these routes has multiplied a thousandfold or more. The universal format, ATM (asynchronous transfer mode, Chapter 16), seems to be turning out to be a disappointment, similar to ISDN.

Cellular radio (Chapter 18), known in Europe simply as mobile, has revolutionized most of our lives, particularly with its ubiquity. The low earth orbit (LEO) satellite (e.g., "Iridium") has turned out to be a disappointment, but still may surprise us. The "last mile" connectivity, or so-called broadband (Chapter 19), is truly beginning to make its mark, particularly for the internet.

All of the above successes have affected or will affect network structure including topology. However, idealism is tempered by the reality of current investment in existing plants yet to be amortized. This forces the macroplanner to employ gradual conversion rather than to apply rapid, radical change.

Our first consideration in this section is digital extension to the subscriber or user. We then cover some of the design issues of an all-digital PSTN. Among these issues are:

- Change in profile of services that digital brings about
- New technology, specifically SONET and SDH
- Digital network performance and performance requirements
- International interface (European versus North American PCM standards)
- Signaling: CCITT Signaling System No. 7 (covered in Chapter 17)
- Universal digital format. IP, OSI, SONET/SDH, 802.3, ATM, and others

5.2 Digital Extension to the Subscriber

The argument in this section from the third edition of this book has been bypassed, overrun. Much of these bypassing technologies are described in Chapter 19 on last mile connectivity. The great majority of customers, at least in North America, have a CATV cable running near their residence or business. Besides entertainment TV, the CATV cable system can and will provide digital telephony and megabit downstream internet. LMDS (see Chapter 19) and some satellite systems (direct broadcast satellite) offer similar services. DSL, particularly ADSL, and MMDS can also provide the end-user with digital telephony and internet.

5.3 Change of Profile of Services

This certainly has occurred, but not exactly as we saw it in the third edition. The powerful SPC computer in the digital switch has brought about one group of changes and new offerings, of which we will mention just a few:

- Caller ID
- Voice mail
- Call waiting
- Call forwarding—etc.

The internet can be brought into a residence or business by various methods (see Chapter 19) where the downstream bit rate is a major sales point. By "downstream," we mean the direction of traffic from the local serving switch or headed toward the end-user. CATV and LMDS can provide links in the megabit range. Small and mid-size businesses can get DS1/E1 circuit groups for internet, data (VPN), IP, frame relay, conference TV, security systems, and other services. Facsimile, even in residences, has become universal.

5.4 Digital Transmission Network Models — ITU-T Organization (CCITT)

Digital transmission network models are hypothetical entities of defined length and composition for use in the study of digital transmission impairments such as

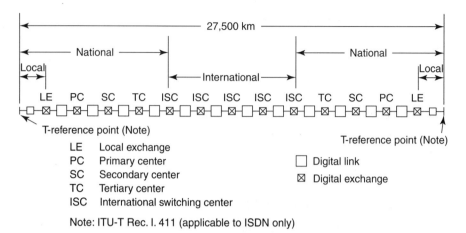

Figure 9.10. The standard CCITT (ITU-T) hypothetical reference connection (HRX), longest length. From ITU-T Rec. G.801, Figure 1/G.801, page 5, 1988 [14].

bit and block errors, jitter, wander, transmission delay, and slips. CCITT (ITU-T) provides a *standard hypothetical reference connection* (HRX) based on an all-digital 64-kbit/s circuit. The standard HRX is shown in Figure 9.10.

The implementation of the standard HRX for a particular application must be tailored for that application. ITU-T has two additional models for shorter connections and where a local exchange connects directly to an international switching center (ISC).

The HRX is further broken down into hypothetical reference digital links (HRDLs) 2500 km in length. A digital link is defined in ITU-T Rec. G.701 [13] as "the whole of the means of digital transmission of a digital signal of a specified rate between two digital distribution frames (or equivalent)." The HRDL is made up of hypothetical reference digital sections with lengths of 50 and 280 km.

5.5 Digital Network Synchronization

5.5.1 Need for Synchronization. When a PCM bit stream is transmitted over a telecommunication link, there must be synchronization at three different levels: bit, time slot, and frame. Bit synchronization refers to the need for the transmitter (coder) and receiver (decoder) to operate at the same bit rate. It also refers to the requirement that the receiver decision point be exactly at the mid-position of the incoming bit. Bit synchronization assures that the bits will not be misread by the receiver.

Obviously a digital receiver must also know where a time slot begins and ends. If we can synchronize a frame, time-slot synchronization can be assured. Frame synchronization assumes that bit synchronization has been achieved. We know where a frame begins (and ends) by some kind of marking device. With DS1 it is the framing bit. In some frames it appears as a 1 and in others it appears as a 0. If the 12-frame superframe is adopted, it has 12 framing bits, one in each of the 12 frames. This provides the 000111 framing pattern [3]. In the case of the 24-frame extended superframe, the repeating pattern is 001011, and the framing bit occurs only once in four frames.

E1, as we remember in Chapter 8, has a separate framing and synchronization channel, namely channel 0. In this case the receiver looks in channel 0 for the framing sequence in bits 2 through 8 (bit 1 is reserved) of every other frame. The framing sequence is 0011011. Once the framing sequence is acquired, the receiver knows exactly where frame boundaries are. It is also time-slot-aligned.

All digital switches have a master clock. Outgoing bit streams from a switch are slaved to the switch's master clock. Incoming bit streams to a switch derive timing from bit transitions of that incoming bit stream. It is mandatory that each and every switch in a digital network generate outgoing bit streams whose bit rate is extremely close to the nominal bit rate. To achieve this, network synchronization is necessary. Network synchronization can be accomplished by synchronizing all switch (node) master clocks so that transmissions from these nodes have the same average line bit rate. Buffer storage devices are judiciously placed at various transmission interfaces to absorb differences between the actual line bit rate and the average rate. Without this network-wide synchronization, *slips* will occur. Slips are a major impairment in digital networks. Slip performance requirements are discussed in Section 5.6.5. A properly synchronized network will not have slips (assuming negligible phase wander and jitter). In the next paragraph we explain the fundamental cause of slips.

As we mentioned above, timing of an outgoing bit stream is governed by the switch clock. Suppose a switch is receiving a bit stream from a distant source and expects this bit stream to have a transmission rate of $F(0)$ in Mbits/s. Of course this switch has a buffer of finite storage capacity into which it is streaming these incoming bits. Let's further suppose that this incoming bit stream is arriving at a rate slightly greater than $F(0)$, yet the switch is draining the buffer at exactly $F(0)$. Obviously, at some time, sooner or later, that buffer must overflow. That overflow is a *slip*. Now consider the contrary condition: The incoming bit stream has a bit rate slightly less than $F(0)$. Now we will have an underflow condition. The buffer has been emptied and for a moment in time there are no further bits to be streamed out. This must be compensated for by the insertion of idle bits, false bits, or frame. However, it is more common just to repeat the previous frame. This is also a slip. We may remember the discussion of stuffing in Chapter 8 in the description of higher-order multiplexers. Stuffing allowed some variance of incoming bit rates without causing slips.

When a slip occurs at a switch port buffer, it can be controlled to occur at frame boundaries. This is much more desirable than to have an uncontrolled slip that can occur anywhere. Slips occur for two basic reasons:

1. Lack of frequency synchronization among clocks at various network nodes
2. Phase wander and jitter on the digital bit streams

Thus, even if all the network nodes are operating in the synchronous mode and synchronized to the network master clock, slips can still occur due to transmission impairments. An example of environmental effects that can produce phase wander of bit streams is the daily ambient temperature variation affecting the electrical length of a digital transmission line.

Consider this example. A 1000-km coaxial cable carrying 300 Mbits/s (3×10^8 bits/s) will have about 1 million bits in transit at any given time, each bit occupying about 1 meter of the cable. A 0.01% increase in propagation velocity, as would be produced by a 1°F decrease in temperature, will result in 100 fewer bits in the cable; these bits must be absorbed to the switch's incoming elastic store buffer. This may end up causing an underflow problem forcing a controlled slip. Because it is underflow, the slip will be manifested by a frame repeat; usually the last frame just before the slip occurs.

In speech telephony, a slip only causes a click in the received speech. For the data user, the problem is far more serious. At least one data frame or packet will be corrupted.

Slips due to wander and jitter can be prevented by adequate buffering. Therefore adequate buffer size at the digital line interfaces and synchronization of the network node clocks are the basic means by which to achieve the network slip rate objective [15].

5.5.2 Methods of Synchronization.

There are a number of methods that can be employed to synchronize a digital network. Six such methods are shown graphically in Figure 9.11.

Figure 9.11a illustrates plesiochronous operation. In this case each switch clock is free running (i.e., it is *not* synchronized to a network master clock).

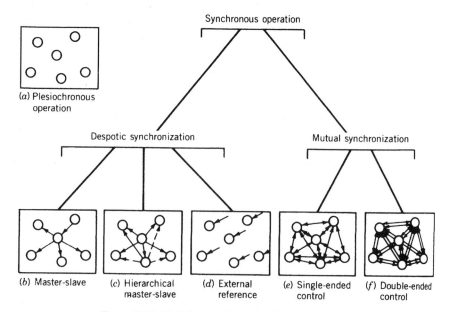

Figure 9.11. Digital network synchronization methods.

Each network nodal switch has identical high-stability clocks operating at the same nominal rate. When we say high-stability, we mean a stability range from 1×10^{-11} to about 5×10^{-13} per month. This implies an atomic clock, rubidium or cesium. The accuracy and stability of each clock are such that there is almost complete coincidence in time-keeping, and the phase drift among many clocks is, in theory, avoided or the slip rate between network nodes is acceptably low. This requires that all switching nodes, no matter how small, have such high-precision clocks. For commercial networks, this causes a high cost burden. However, for military networks it is very attractive for survivability because there is no mutual network synchronization. Thus the loss of a node and its clock does not affect the rest of the network timing. CCITT (ITU-T) recommends plesiochronous operation on transnational connectivities (i.e., at international switching centers). (Refer to ITU-T Rec. G.811 [16]).

Another general synchronization scheme is mutual synchronization, which is shown in Figures 9.11e and 9.11f. Here all nodes in the network exchange frequency references, thereby establishing a common network clock frequency. Each node averages the incoming references and uses the result to correct its local transmitted clock. After an initialization period, the network aggregate clock normally converges to a single stable frequency.

A number of military systems as well as a fair number of private entities (e.g., SBC Communications) use external synchronization. This is illustrated in Figure 9.11d. Switch master clocks use *disciplined* oscillators slave to an external radio source. The source of choice today is GPS (Global Positioning System), which disseminates coordinated universal time called UTC, an acronym deriving from the French. It is a multiple satellite system in polar orbits. Its orbits and parameters are such that there are always three or four satellites in view at once anywhere on the earth's surface. Its time transfer capability is in the 10- to 100-ns range from UTC.

5.5.3 North American Synchronization Plan as Specified by ANSI/Telcordia.

The North American network uses a hierarchical timing distribution system conceptually illustrated in Figure 9.11c. It is based on a four-level hierarchy as shown in Figure 9.12. These levels are called strata (stratum singular). Table 9.1 shows strata timing/synchronization specifications.

The stratum levels for synchronized clocks are based on three parameters:

1. *Free-Run Accuracy.* This is the maximum fractional frequency offset that a clock may have when it has never had a reference or has been in holdover for an extended period, greater than several days or weeks.

2. *Holdover Stability.* This is the amount of frequency offset that a clock experiences after it has lost its synchronization reference. Holdover is specified for stratum 2. The stratum 3 holdover extends beyond one day and it breaks the requirement up into components for initial offset, drift, and temperature. See Bellcore document TR-1244 [17].

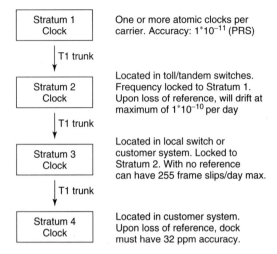

Figure 9.12. North American hierarchical network synchronization. From Ref. 32. Also see Ref. 17. Courtesy of Cirrus Logic; Austin, TX, January 20, 2004, Ref. 33.

TABLE 9.1 Stratum Level Specifications

Stratum Level	Free-Run Accuracy	Holdover Stability	Pull-in/Hold-in
1	$\pm 1.0 \times 10^{-11}$	NA	NA
2	$\pm 1.6 \times 10^{-8}$	$\pm 1. \times 10^{-10}$ per day	$\pm 1.6 \times 10^{-8}$
3E	$\pm 4.6 \times 10^{-6}$	$\pm 1. \times 10^{-8}$ day 1	$\pm 4.6 \times 10^{-6}$
3	$\pm 4.6 \times 10^{-6}$	< 255 slips during first day of holdover	$\pm 4.6 \times 10^{-6}$
4	$\pm 32. \times 10^{-6}$	No holdover	$\pm 32. \times 10^{-6}$

NA, not applicable.

Source: Table 132.1, page 169, November 2001 [31].

3. *Pull-in/Hold-in.* This is a clock's ability to achieve or maintain synchronization with a reference that may be off-frequency. A clock is required to have a pull-in/hold-in range at least as wide as its free-run accuracy. This ensures that a clock of a given stratum level can achieve and maintain synchronization with the clock of the same or higher stratum level.

A stratum 1 clock, which is the PRS (primary reference source), is required to have a timing signal whose long-term accuracy is maintained at 1×10^{-11} or better and completely autonomous of other references. Currently, cesium beam atomic references are the only type clocks that are true stratum 1 references in that they are autonomous and achieve the desired stability. Stratum 1 clocks must be capable of being verified with Universal Coordinated Time (UTC). Alternatively, the PRS source may not be a completely autonomous implementation, in which case it may employ direct control from UTC-derived frequency and time dissemination services such as Loran C or GPS (mentioned above).

Stratum 2 clocks are typically based on either double-oven crystal oscillators or rubidium oscillators. To take advantage of their stable oscillators and provide the best holdover estimate possible, stratum 2 clocks usually have long time constants for averaging their input frequency reference. Stratum 2 clocks are historically deployed as part of tandem and transit switches. To improve reliability, stratum 2 clocks often use two input references with automatic protection switching.

Stratum 3 is a clock with reduced capability in terms of holdover performance. These clocks are typically based on temperature-compensated crystal oscillators (TCXO). Stratum 3 clocks are commonly used by local serving switches. A stratum 3E clock filters its reference timing input to clean up the large amounts of wander to create a timing signal output with low levels of wander. 3E clocks provide significantly better holdover performance.

Stratum 4 clocks are commonly used for digital channel banks. Digital PABXs use a special stratum 4E clock which has two input references [15].

5.5.3.1 Building Integrated Timing Supply (BITS) Plan. In this plan each switching center has one master clock called the BITS. All other synchronized clocks in the switching center are timed from the BITS clock. The BITS clock is the same or higher stratum level than all other clocks in the switching center and is the only clock that has an external reference.

With the introduction of DS0 dataports, all channel banks must derive timing from the BITS clock.

The recommended BITS implementation is to have one network element (NE) which is a TSG (timing signal generator). It provides DS1 and CC (composite clock) timing to all synchronized clocks in the switching center. The TSG is timed by two DS1 signals (primary and backup) using any one of the following methods:

- Bridging off DS1 traffic streams coming into the switching center.
- Terminating a DS1 dedicated to synchronization distribution.
- Having SONET network elements derive DS1s as described in Telcordia GR-253 [18].
- Receiving a DS1 from a co-located PRS. This may derive from a disciplined oscillator slaved to GPS or LORAN C.

All equipment in the switching center takes timing (DS1 or CC) from the TSG. See Figure 9.13. A typical composite clock (CC) waveform is illustrated in Figure 9.14. [19]

5.5.3.2 Holdover and Slip Performance. When a BITS clock loses its references, it enters holdover and drifts off frequency. The magnitude of this frequency drift determines the average slip rate experienced by equipment that depends on that clock timing source. Table 9.2 shows the number of slips expected after 1 day and 1 week of holdover given limited ambient temperature variations of $\pm 1°F$ in the switching center. The table shows the difference between stratum

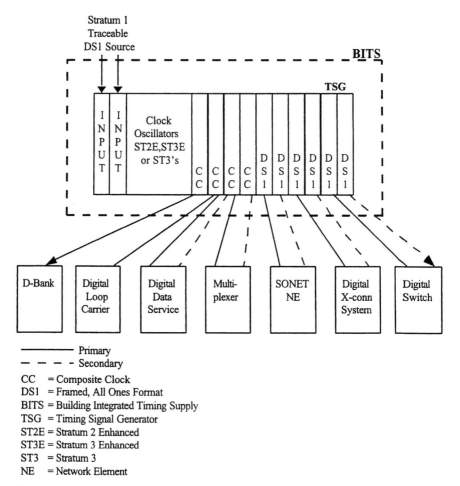

Figure 9.13. A typical BITS arrangement. Courtesy of Qwest, Technical Publication 77386, Figure 13-1, November 2001 [31].

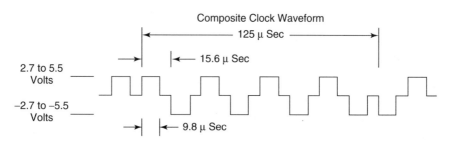

Figure 9.14. A composite clock waveform. Courtesy of Qwest, Technical Publication 77386, Figure 13-3, Issue G, November 2001 [31].

TABLE 9.2   Expected Slip Performance in Holdover

Stratum Level	Slips in Day 1	Slips in Week 1
2	1 or less	2
3E	1	22
3	48	919

Source: Table 13-2, Qwest [31].

levels for performance during holdover. If maintenance actions are prompt when that unusual holdover occurs and we base a network on stratum 2 or 3E clocks, a virtually slip-free network can be expected [15].

5.5.3.3 Timing Distribution. The synchronization network has historically been tied to the switching network, using traffic-carrying DS1s (T1s) between digital switches for synchronization distribution. The switching network includes local serving exchanges and tandem switches in the local network, as well as transit exchanges in the long-distance network). The tandem switches (and transit switches), which are connected to many downstream local switches, provide a convenient point to distribute synchronization. Synchronization has now evolved to become SONET-based. SONET facility hubs have become synchronization distribution hubs. Thus Telcordia recommends that synchronization hubs that distribute timing to several downstream switches have stratum 2 BITS clocks, and consideration should be given to making them PRS sites.

Figure 9.15 shows an example at a local exchange where multiple timing sources feed a system synchronizer and how system clock is derived.

5.5.4 Synchronization between Autonomous Networks.
When two disparate networks must interface, we are up against what some call the *two-clock problem*. In the United States, typically this is the interface between a local

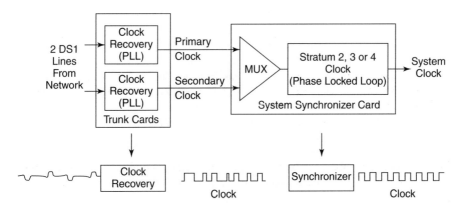

Figure 9.15. Multiple timing sources feed a system synchronizer. From Crystal 62411, Figure 3 [32]. Courtesy of Cirrus Logic, Austin, TX, Ref. 33.

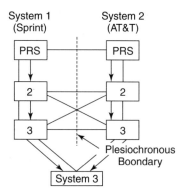

Figure 9.16. Plesiochronous operation where two autonomous systems terminate in a third system (typically a LEC in the United States). Arrowed lines show synchronization path.

exchange carrier (LEC) and one or several interexchange carriers (IXCs). Each network is autonomous for timing synchronization, and each has a clock traceable to some sort of PRS. This can also be an interface between a private digital network and a public network. When two disparate networks have traceability each to their own PRS, it is called *plesiochronous operation*. The concept is illustrated in Figure 9.16.

The effect of frequency offset due to separate stratum 1 traceability is a low slip rate. The worst case when each PRS is operating at stratum 1 requirement extremes of 1×10^{-11}, the expected slip rate between the networks is less than 1 slip in 72 days.

5.5.5   *ITU-T (CCITT) Synchronization Plans.*   ITU-T Rec. G.811 [16] deals with synchronization of international links. Plesiochronous operation is preferred (see Section 5.5.2). The recommendation states the problem at the outset:

> International digital links will be required to interconnect a variety of national and international networks. These networks may be of the following form:
>
> (a) a wholly synchronized network in which the timing is controlled by a single reference clock.
> (b) a set of synchronized subnetworks in which the timing of each is controlled by a reference clock but with plesiochronous operation between the subnetworks.
> (c) a wholly plesiochronous network (i.e., a network where the timing of each node is controlled by a separate reference clock).

Plesiochronous operation is the only type of synchronization that can be compatible with all three types listed. Such operation requires high-stability clocks. Thus Rec. G.811 states that all clocks at network nodes that terminate international links will have a long-term frequency departure of not greater than 1×10^{-11}. This is further described in what follows.

The theoretical long-term mean rate of occurrence of controlled frame or octet (time slot) slips under ideal conditions in any 64-kbit/s channel is consequently not greater than *1 in 70 days* per international digital link.

Any phase discontinuity due to the network clock or within the network node should result only in the lengthening or shortening of a time signal interval and should not cause a phase discontinuity in excess of one-eighth of a unit interval on the outgoing digital signal from the network node.

Rec. G.811 states that when plesiochronous and synchronous operation coexist within the international network, the nodes will be required to provide both types of operation. It is therefore important that the synchronization controls do not cause short-term frequency departure of clocks, which is unacceptable for plesiochronous operation. The magnitude of the short-term frequency departure should meet the requirements specified in Section 5.5.5.1.

5.5.5.1 Time Interval Error and Frequency Departure. Time interval error (TIE) is based on the variation of ΔT, which is the time delay of a given timing signal with respect to an ideal timing signal, such as UTC (universal coordinated time). The TIE over a period of S seconds is defined to be the magnitude of difference between time delay values measured at the end and at the beginning of the period:

$$\text{TIE}(S) = |\Delta T(t + S) - \Delta T(t)| \qquad (9.2)$$

This is shown diagrammatically in Figure 9.17. The corresponding normalized frequency departure $\Delta f / f$ is the TIE divided by the duration of the period (i.e., S seconds).

The TIE at the output of a reference clock is specified in CCITT Rec. G.811 for three values of frequency as follows:

The TIE over a period of S seconds shall not exceed the following limits:

(a) $(100S)$ ns $+ 1/8$ unit interval. Applicable to S less than 5. These limits may be exceeded during periods of internal clock testing and rearrangements. In such cases the following conditions should be met: TIE over

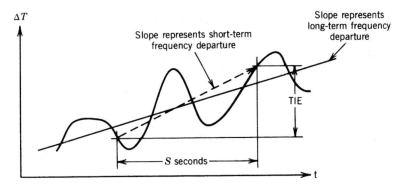

Figure 9.17. Definition of time interval error (TIE). From CCITT Rec. G.811, page 298, 1988 [16].

TABLE 9.3 Maximum Permissible Degradation of Timing at a Network Node

Performance Category[a]	Frequency Departure $\left|\dfrac{\Delta f}{f}\right|$ of Node Timing[b]		Proportion of Time During Which Degradation May Occur, Referred to Total Time[d]	
	Local[c]	Transit	Local[c]	Transit
Nominal	See Section 5.5.5.1	See Section 5.5.5.1	$\geq 98.89\%$	$\geq 99.945\%$
(a)	$10^{-11} < \left\|\dfrac{\Delta f}{f}\right\| \leq 10^{-8}$	$10^{-11} < \left\|\dfrac{\Delta f}{f}\right\| \leq 2.0 \times 10^{-9}$	$\leq 1\%$	$\leq 0.05\%$
(b)	$10^{-8} < \left\|\dfrac{\Delta f}{f}\right\| \leq 10^{-6}$	$2.0 \times 10^{-9} < \left\|\dfrac{\Delta f}{f}\right\| \leq 5.0 \times 10^{-7}$	$\leq 0.1\%$	$\leq 0.005\%$
(c)	$\left\|\dfrac{\Delta f}{f}\right\| > 10^{-6}$	$\left\|\dfrac{\Delta f}{f}\right\| > 5.0 \times 10^{-7}$	$\leq 0.01\%$	$\leq 0.0005\%$

[a] The performance categories (b) and (c) correspond to (b) and (c) in Rec. G.822 while category (a) in Rec. G.822 corresponds to "Nominal" and (a) in Rec. G.811, combined.
[b] All values are provisional.
[c] The values for local nodes are given for guidance only, and administrations are free to adopt other performance levels provided the overall controlled slip performance objective of Rec. G.822 are met.
[d] These values are more stringent than would be strictly required by Rec. G.822 for a 64-kbits/s connection, to allow for the future introduction of services at higher bit rates that may require a better slip performance. They also allow a margin for possible network effects.
Source: CCITT Rec. G.811, page 298, 1988 [16].

any period up to 2 UI (unit intervals) should not exceed 1/8 of a UI. For periods greater than 2 UI, the phase variation for each interval of 2 UI should not exceed 1/8 UI up to a total maximum TIE of 500 ns.

(b) $(5S + 500)$ ns for values of S between 5 and 500.
(c) $(10^{-2S} + 3000)$ ns for values of S greater than 500.

The allowance in (c) of 3000 ns is for component aging and environmental effects.

For clarification, CCITT defines UI in Rec. G.701 [13] as the nominal difference in time between consecutive significant instants of an isochronous signal. With NRZ coding we can think of a unit interval (UI) as the duration of 1 bit or the bit period. The bit period in NRZ is the inverse of the bit rate. In the case of AMI coding, the mark and space are usually not of equal duration.

Table 9.3 shows the permissible degradation of the timing of a network node. The performance category refers to slip rate performance objectives given in Section 5.6.3.

5.6 Digital Network Performance Requirements

5.6.1 Blocking Probability.

Telcordia recommended blocking probability is $B = 0.01$. This is a quality of service objective. With judicious use of alternative routing, a blocking probability of $B = 0.005$ might be expected.

5.6.2   Error Performance — Telcordia Perspective

Definitions

BER. The BER is the ratio of the number of bits in error to the total number of bits transmitted during the measurement period.

Errored Seconds (ES). An errored second is any one-second interval containing at least one bit error.

Burst Errored Seconds. A burst errored second is any second containing at least 100 bit errors.

Severely Errored Seconds (SES). These are a measure of the burst character of error streams. These are defined at the DS0 and DS1 rates as 1-s intervals in which the BER exceeds 1×10^{-3}.

Telcordia TSGR Error Performance Requirements. Note that the measurement period consists of a series of 1-s intervals. The values below are all for one-way system options.

1. The BER at the interface levels DSX-1, DSX-1C, DSX-2, and DSX-3 and shall be less than 2×10^{-10}, excluding all burst errored seconds in the measurement period. During a burst errored second, neither the number of bit errors nor the number of bits is counted.

2. The frequency of burst errored seconds, other than those caused by protection switching induced by hard equipment failures, shall average no more than four per day at each of the interface levels DSX-1, DSX-1C, DSX-2, and DSX-3.

3. For systems interfacing at the DS1 level, the long-term percentage of errored seconds (measured at the DS1 rate) shall not exceed 0.04%. This is equivalent to 99.96% error-free seconds. It is equivalent to no more than 10 errored seconds during a 7-h, one-way loopback test.

4. For systems interfacing at the DS3 level, the long-term percentage of errored seconds (measured at the DS3 rate) shall not exceed 0.4%. This is equivalent to 99.6% error-free seconds. It is equivalent to no more than 29 errored seconds during a 2-h, one-way loopback test.

5. For systems interfacing at the DS3 level, the long-term percentage of errored seconds (measured at the DS3 rate) shall not exceed 0.2%. This is equivalent to no more than 15 errored seconds during a 2-h, one-way loopback test.

Telcordia specifies that errored second performance shall be evaluated separately from BER performance.

Source: Telcordia TSGR, December 1998 [20].

5.6.3   Error Performance — ITU-T Perspective.

ITU-T Rec. G.821 [21] provides digital network error performance objectives on ISDN circuits based

TABLE 9.4 Error Performance Objectives for International ISDN Connections

Performance Classification	Objective (Notes 1, 2)
Severely errored second ratio	<0.002
Errored second ratio	<0.08

Note 1: The ratios are calculated over the available time. The observation time has not been specified since the period may depend upon the application. A period of the order of any one month is suggested as a reference.
Note 2: Annex B of the reference document illustrates how the overall performance should be assessed.
Source: Table 1/G.821, page 3 [21].

on an "errored-second ratio (ESR)" and "severely errored second ratio (SESR)." These two parameters are defined below. The definition of *available time* is also given.

Errored-Second Ratio (ESR). The ratio of ES (errored seconds) to total seconds in available time during a fixed measurement interval.

Severely Errored Second Ratio (SESR). The ratio of SES (severely errored seconds) to total seconds in available time during a fixed measurement interval.

Available Time (Available State). A period of *unavailable time* begins when the bit error ratio (BER) in each second is worse than 1×10^{-3} for a period of ten consecutive seconds. These ten seconds are considered unavailable time. A new period of *available time* begins with the first second of a period of ten consecutive seconds each of which has a BER better than 1×10^{-3}.

These error performance objectives from ITU Rec. G.821 apply to $N \times 64$ kbits/s switched connection and are detailed in Table 9.4.

5.6.3.1 Network Error Performance Based on ITU-T Rec. G.828. Rec. G.828 provides error performance objectives both for SDH and ATM paths.

TERMS AND DEFINITIONS FROM ITU-T REC. G.828

Hypothetical Reference Path. A hypothetical reference path (HRP) is defined as the whole means of digital transmission of a digital signal of a specified rate, including path overhead, between equipment at which the signal originates and terminates. An end-to-end HRP spans a distance of 27,500 km.

SDH Digital Path. An SDH digital path is a trial carrying an SDH payload and associated overhead through the layered transport network between path terminating equipment. A digital path may be bidirectional or unidirectional and may comprise both customer-owned portions and network-operator-owned portions.

Generic Definition of the Block. The reference ITU-T Recommendation is based upon the error performance measurement of blocks consistent with an SDH frame. This paragraph defines the term "block" as follows: A block is a set of consecutive bits associated with the path; each bit belongs to one and only one block. Consecutive bits may not be contiguous in time.

Errored Block (EB). A block in which one or more bits are in error.

Errored Second (ES). A one-second period with one or more errored blocks or at least one defect. A defect is a change in performance state. It is further defined in ITU-T Recs. G.707 and G.783.

Severely Errored Second (SES). A one-second period which contains $\geq 30\%$ errored blocks or at least one defect. SES is a subset of ES.

Background Block Error (BBE). An errored block not occurring as part of an SES.

Errored-Second Ratio (ESR). The ratio of ES in available time to total seconds in available time during a fixed measurement interval.

Severely Errored Second Ratio (SESR). The ratio of SES in available time to total seconds in available time during a fixed measurement interval.

Background Block Error Ratio (BBER). The ratio of BBE in available time to total blocks in available time during a fixed measurement period.

Severely Errored Period (SEP). A sequence of between 3 and 9 consecutive SES. The sequence is terminated by a second which is not an SES.

Severely Errored Period Intensity (SEPI). The number of SEP events in available time, divided by the total available time in seconds.

ITU-T REC. G.828 PERFORMANCE OBJECTIVES. Table 9.5 specifies the end-to-end objectives for a 27,500-km HRP in terms of parameters defined above. The actual objectives applicable to a real path may be derived from Table 9.5 using the allocation principles given below. Each direction of the path will independently satisfy the allocated objectives for all parameters. In other words, a path fails to satisfy Rec. G.828 if any parameter exceeds the allocated objective in either direction at the end of the given evaluation period. The objectives given herein are understood to be long-term objectives to be met over an evaluation period of typically 30 consecutive days (1 month). (See note 1, Table 9.5.)

Note that Table 9.5 specifically refers to SDH type paths. SDH is covered in Chapter 8, Section 13.3.

Rec. G.828 continues telling us that synchronous digital paths operating at bits rates covered herein are carried by transmission systems (digital sections) operating at higher bit rates. Such systems must meet their allocations of the end-to-end objectives for the highest bit-rate paths which are foreseen to be carried. Meeting the allocated objectives for this highest bit-rate path should be sufficient to ensure that all paths through the system are achieving their objective. For example, in SDH, an STM-1 section may carry a VC-4 path and therefore the STM-1 section should be designed such that it will ensure that the objectives as specified herein for the bit rate corresponding to a VC-4 path are met.

TABLE 9.5   End-to-End Error Performance Objectives for a 27,500-km International Synchronous Digital HRP

Bit Rate (kbits/s)	Path Type	Blocks/s	ESR	SESR	BBER	SEPI
1,664	VC-11, TC-11	2,000	0.01	0.002	5×10^{-5}	(Note 3)
2,240	VC-12, TC-12	2,000	0.01	0.002	5×10^{-5}	(Note 3)
6,848	VC-2, TC-2	2,000	0.01	0.002	5×10^{-5}	(Note 3)
48,960	VC-3, TC-3	8,000	0.02	0.002	5×10^{-5}	(Note 3)
150,336	VC-4, TC-4	8,000	0.04	0.002	1×10^{-4}	(Note 3)
601,344	VC-4-4c, TC-4-4c	8,000	(Note 1)	0.002	1×10^{-4}	(Note 3)
2,405,376	VC-4-16c, TC-4-16c	8,000	(Note 1)	0.002	1×10^{-4}	(Note 3)
9,621,504	VC-4-64c, TC-4-64c	8,000	(Note 1)	0.002	1×10^{-3} (Note 2)	(Note 3)

Note 1: ESR objectives tend to lose significance for applications at high bit rates and are therefore not specified for paths operating at bit rates above 160 Mbits/s. Nevertheless, it is recognized that the observed performance of synchronous digital paths is error-free for long periods of time even at Gbits/s rates; and that significant ESR indicates a degraded transmission system. Therefore, for maintenance purposes ES monitoring should be implemented within any error performance measuring devices operating at these rates.
Note 2: This BBER objective corresponds to an equivalent bit error ratio of 8.3×10^{-10}, an improvement over the bit error ratio of 5.3×10^{-9} for the VC-4 rate. Equivalent bit error ratio is valuable as a rate-independent indication of error performance, as BBER objectives cannot remain constant as block sizes increase.
Note 3: SEPI objectives require further study.
Source: Table 1/G.828, page 7, ITU-T Rec. G.828, March 2000 [22].

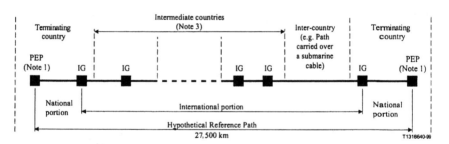

NOTE 1 – If a path is considered to terminate at the IG, only the international portion allocation applies.
NOTE 2 – One or two International Gateways (entry or exit) may be defined per intermediate country.
NOTE 3 – Four intermediate countries are assumed.

Figure 9.18. A model of the hypothetical reference path. From Figure 3/G.828, page 8, March 2000 [22].

APPORTIONMENT OF OBJECTIVES. Figure 9.18 is a model of the HRP under consideration. We will wish to apportion performance objectives for the national and international segments of the HRP. Let the boundary between the national and international portions be defined as an international gateway (IG). The exact location will reside in a cross-connect, a higher-order multiplexer, or a switch which may be N-ISDN or B-ISDN. G.828 states that IGs are always terrestrially based equipment physically resident in the terminating (or intermediate) country. Higher-order paths (relative to the HRP under consideration) may be used between IGs. Such paths receive only the allocation corresponding to the international portion between the IGs. In intermediate countries, the IGs are only located in order to calculate the overall length of the international portion of the path in order to deduce the overall allocation.

Allocation to the National Portion of the End-to-End Path. Each national portion (there are two national portions) is allocated a fixed block allowance of 17.5% of the end-to-end objective. Furthermore, a distance-based allocation is added to the block allowance. The actual route length between the PEP (path endpoint) and the IG should also be determined and multiplied by an appropriate routing factor. This routing factor is specified as follows:

- If the air route distance is <1000 km, the routing factor is 1.5.
- If the air route distance is >1000 km and <1200 km, the calculated route length is taken to be 1500 km.
- If the route distance is ≥1200, the routing actor is 1.25.

When both actual and calculated route lengths are known, the smaller value is retained. The distance should be rounded up to the nearest 100 km. An allocation of 0.2% per 100 km is then applied to the resulting distance. The two national portions are allocated a minimum of 500 km (i.e., 1%) each.

Suballocation of the 17.5% national portion allowance is left to national standardization bodies. It is called to the readers attention that there is empirical evidence that most of the error impairments can occur in that part of the path nearest to its endpoint.

When a national portion includes a satellite hop, a total allowance of 42% end-to-end objectives in Table 9.5 is allocated to the national portion. The 42% allowance completely replaces both the distance-based allowance and the 17.5% block allowance otherwise given to national portions.

Allocation to the International Portion of the End-to-End Path. The international portion is allocated a block allowance of 2% per intermediate country plus 1% for each terminating country. Furthermore, a distance-based allocation is added to the block allowance. In that the international path may pass through intermediate countries, the actual route length through consecutive IGs (one or two for each intermediate country) should be added to calculate the overall length of the international portion. The air route distance between consecutive IGs should also be determined and multiplied by an appropriate routing factor. This routing factor has been specified by Rec. G.828 as follows for each element between IGs:

- If the air route distance between IGs is <1000 km, the routing factor is 1.5.
- If the air route distance is ≥1000 km and <1200 km, the calculated route length is taken to be 1500 km.
- If the air route distance between two IGs is ≥1200 km, the routing factor is 1.25.

When both actual and calculated route lengths are known, the smaller value is retained for each element between IGs for the calculation of the overall length of the international portion. This overall distance should be rounded up to the nearest 100 km but should not exceed 26,500 km. An allocation of 0.2% per 100 km is then applied to the resulting distance. (Also see Ref. 23.)

CONVERTING BIP MEASUREMENTS INTO ERRORED BLOCKS. Since Rec. G.828 defines a block as consecutive bits associated with a path, each BIP-*n* (Bit Interleaved Parity, order *n*) in the SDH path overhead pertains to a single defined block. For the purpose of this subsection, a BIP-*n* corresponds to a G.828 block. BIP-*n* is *not* interpreted as checking *n* separate interleaved parity check blocks. If any of the *n* separate parity checks fails, the block is assumed to be in error.

Block Size for Monitoring SDH Paths. The number of bits per block for in-service performance monitoring of SDH paths, as specified in ITU-T Rec. G.707 is given in Table 9.6. Paths operating at VC-11, VC-12, or VC-2 rates use 500-μs measurement blocks (i.e., 2000 blocks per second) [23].

5.6.4 Jitter.
Jitter was discussed in Chapter 8, where we stated it to be a major digital transmission system impairment. We also said that systematic jitter magnitude was a function of the number of regenerative repeaters that are in tandem. In this section we look at jitter as an overall digital network impairment.

Definition of Timing Jitter. Timing jitter is the short-term variations of a digital signal's significant instants (e.g., optimum sampling instants) from their ideal positions in time. Short-term variations are phase oscillations of frequency greater than a demarcation point that is specified for each interface rate (e.g., DS1, phase modulation that, after demodulation, passes through a high-pass filter with a cut-off frequency of 10 Hz and a 20-dB/decade roll-off). Tables 9.7A and 9.7B show maximum permissible jitter at hierarchical interfaces. Table 9.7A provides jitter

TABLE 9.6 Block Sizes for Synchronous Digital Path Performance Monitoring

Bit Rate (kbits/s)	Path Type	SDH Block Size Used in G.828	EDC
1,664	VC-11, TC-11	832 bits	BIP-2
2,240	VC-12, TC-12	1,120 bits	BIP-2
6,848	VC-2, TC-2	3,424 bits	BIP-2
48,960	VC-3, TC-3	6,120 bits	BIP-8
150,336	VC-4, TC-4	18,792 bits	BIP-8
601,344	VC-4-4c, TC-4-4c	75,168 bits	BIP-8
2,405,376	VC-4-16c, TC-4-16c	300,672 bits	BIP-8
9,621,504	VC-4-64c, TC-4-64c	1,202,688 bits	BIP-8

Source: Table B.1/G.828, page 12, ITU-T Rec. G.828, March 2000 [22].

TABLE 9.7A Maximum Permissible Jitter at 2.048-Mbits/s Traffic Interfaces

Interface	Measurement Bandwidth, −3-dB Frequencies (Hz)	Peak-to-Peak Amplitude (UIpp) (Note 3)
64 kbits/s	20 to 20 k	0.25
(Note 1)	3 k to 20 k	0.05
2048 kbits/s	20 to 100 k	1.5
	18 k to 100 k (Note 2)	0.2
8448 kbits/s	20 to 400 k	1.5
	3 k to 400 k (Note 2)	0.2
34,368 kbits/s	100 to 800 k	1.5
	10 k to 800 k	0.15
139,264 kbits/s	200 to 3.5 M	1.5
	10 k to 3.5 M	0.075

Note 1: For the codirectional interface only.
Note 2: For 2048-kbit/s and 8448-kbit/s interfaces within the network of an operator, the high-pass cutoff frequency may be specified to be 700 Hz (instead of 18 kHz) and 80 kHz (instead of 3 kHz), respectively. However, at interfaces between different operator networks, the values in the table apply, unless involved parties agree otherwise.
Note 3:

64 kbits/s	1 UI = 15.6 μs
2048 kbits/s	1 UI = 488 ns
8448 kbits/s	1 UI = 118 ns
34,368 kbits/s	1 UI = 29.1 ns
139,264 kbits/s	1 UI = 7.18 ns

Note 4: UI = unit interval which is the duration of one bit with NRZ operation.
Source: Table 1/G.823, page 5, ITU-T Rec. G.823, March 2000 [25].

information for the E1 hierarchy and Table 9.7B for the DS1 hierarchy. A test setup for measuring input jitter at a digital interface is shown in Figure 9.20.

Wander. When jitter occurs at a 10-Hz or lower rate, it is arbitrarily defined as wander. Jitter and wander can be present simultaneously. Figure 9.19 shows high-frequency jitter superimposed upon low-frequency wander.

Due to characteristics of phase-lock loop technology used in system trunk cards and system synchronizers (see Figure 9.15), jitter can be more readily filtered than can wander. Normally, most wander which is input on a timing reference source will be relayed to all downstream systems.

TABLE 9.7B   Maximum Permissible Jitter at 1.544-Mbits/s Hierarchy
Traffic Interfaces

Digital Rate (kbits/s)	Measurement Filter Bandwidth, −3-dB Frequencies (Hz)	Peak-to-Peak Amplitude (UIpp)
1,544	10 to 40 k	5.0
	8 to 40 k	0.1
6,312	10 to 60 k	5.0
	3 to 60 k	0.1 (Note 1)
32,064	10 to 400 k	5.0
	8 to 400 k	0.1 (Note 1)
44,736	10 to 400 k	5.0
	30 to 400 k	0.1
97,728	10 to 1000 k	5.0
	240 to 1000 k	0.1

Note 1: This value requires further study.
Note 2:

1544 kbits/s	1 UI = 647 ns
6312 kbits/s	1 UI = 158 ns
32,064 kbits/s	1 UI = 31.1 ns
44,736 kbits/s	1 UI = 22.3 ns
97,728 kbits/s	1 UI = 10.2 ns

Source: Table 1/G.824, page 4, ITU-T Rec. G.824, March 2000 [26].

A primary source of wander is frequency instability in a system synchronizer during a switch between timing reference sources due to upstream clock rearrangement. While acquiring lock to the new source, wander can be created and sent to all downstream timing sources. Stratum 2 clocks can take hours/minutes to reacquire lock after a switch-over. Stratum 4 clocks take seconds to reacquire lock. This wander can be amplified by each of the downstream synchronizers, causing ever-larger wave-of-wander to spread out through the network.

The actual measurement setup in Figure 9.20 is determined by the following considerations:

System Clock.   The equipment under test clock can be externally synchronized (if a reference input is available) or be synchronized from the interface under test.

Constraints on Δf.   The clock generator can be used to generate a fixed frequency offset Δf upon which the jitter and wander is modulated. The

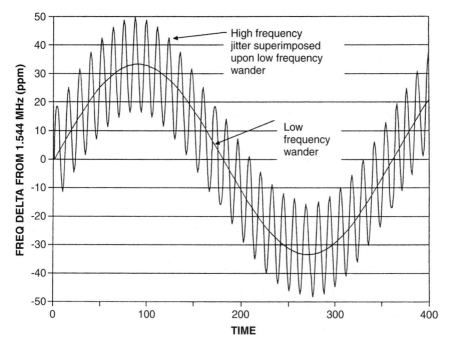

Figure 9.19. Jitter superimposed upon wander. From Crystal document AN12, Rev.2, June 1994 [32]. Courtesy Cirrus Logic, Austin, TX, Ref. 33.

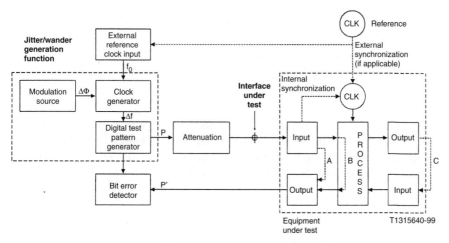

Figure 9.20. Generic measurement setup for input jitter and wander testing. From Figure III.1/GG.823, page 38, ITU-T Rec. G.823, March 2000 [25].

frequency offset must be limited to the values applicable to the interface or equipment under test. The frequency offset should be held constant during a stabilization period and the subsequent measurement. The allowed frequency offset can be dependent on the path that the measurement signal takes through the system and the way that the equipment under test clock is synchronized.

Constraints on $\Delta\Phi$. The modulation source is used to superpose a jitter or wander effect $\Delta\Phi$ on top of the clock signal, which can also have a fixed frequency offset Δf. These jitter and wander phase perturbations usually have a sinusoidal, triangular or noisy (PRBS-generated) characteristic. The exact perturbations are prescribed in the applicable jitter and wander tolerance requirements.

Choice of Test-Pattern (P and P'). The test-pattern P must match the bit rate of the particular interface that is being subjected to the jitter and wander tolerance test. Pattern P' is not necessarily the same as pattern P, but it is of importance that a part of pattern P is present in P'. This part, let us call it Q, is passed transparently through the equipment under test. The bit error detector can only search for errors in this common part Q.

Routing the Signal Through the Equipment Under Test. Depending upon which parts of the system are actually to be tested and the capabilities of the equipment under test, the signal can be looped back in different configurations. For example:

(a) directly behind the input (path A), to test the tolerance of the receiving circuitry;

(b) in the routing functionality (path B), which could test, in addition, buffer hysteresis, stuffing mechanisms, etc.; or

(c) externally through some other inputs and outputs of the system (path C).

The choice of the actual path can influence the selection of test pattern P' and the part Q, over which errors can be monitored.

Attenuation. The attenuation function is needed for optical interfaces to be able to determine the 1-dB sensitivity penalty (in terms of optical power) at a certain bit error ratio. For electrical interfaces the (frequency-dependent) attenuation should represent the worst-case cable length.

Source: Ref. 26.

WANDER LIMITS (ITU)

Synchronous 1544-kbits/s Network Interface. At the network interface, the wander of a 1544-kbit/s network signal shall not exceed an MTIE (τ) of 28 UI (18 μs) for $\tau = 24$ h, nor shall it exceed an MTIE (τ) of 13 UI (8.4 μs) for $\tau = 15$ min. See Tables 9.8 and 9.9.

TABLE 9.8 Synchronous Network Interface for 1544-kbit/s Rate

Observation Interval τ (s)	MTIE (μs)
$\tau \leq 900$	8.4
$900 < \tau \leq 86{,}400$	18.0

MTIE, maximum time interval error.

Source: Table 2/G.824, page 5, March 2000 [26].

TABLE 9.9 Wander Limit for 44,736-kbit/s Network Interface

Observation Interval τ(s)	MRTIE, τ (μs)
$0.1 < \tau \leq 0.195$	$7700\,\tau$
$0.195 < \tau \leq 5200$	$1400 + 230\,\tau^{0.5}$
$5200 < \tau$	18,000

MRTIE, maximum relative time interval error.

Source: Table 3/G.824, page 5, March 2000 [26].

TABLE 9.10 2048-kbit/s Output Interface Wander Limit

Observation Interval, τ (s)	MRTIE Requirement (μs)
$0.05 < \tau \leq 0.2$	$46\,\tau$
$0.2 < \tau \leq 32$	9
$32 < \tau \leq 64$	$0.28\,\tau$
$64 < \tau \leq 1000$ (Note)	18

Note: For the asynchronous configuration (refer to Figure B.1 in the reference document), the maximum observation interval to be considered is 80 s.

Source: Table 2/G.823, page 6, ITU-T Rec. G.823, March 2000 [25].

2048-kbit/s Interface Output Wander Limit. The maximum level of wander that may exist at a 2048-kbit/s network interface, expressed in MRTIE, should not exceed the limit given in Table 9.10.

5.6.5 Slips. Slips are a major impairment in digital networks. The cause of slips was explained in Section 5.5.1 of this chapter.

North American Perspective. Based on data taken from Telcordia TSGR [20], when stratum 3 conditions are trouble-free, the nominal clock slip rate is 0. If there is trouble with the primary reference, a maximum of one slip on any trunk will result from a switched-reference or any other rearrangement. If there is a loss of all references, the maximum slip rate is 255 slips for the first day on any trunk. This occurs when the stratum-3 clocks drift a maximum 0.37 parts per million from their reference frequency.

ITU-T Perspective. With plesiochronous operation, the number of slips on international links will be governed by the sizes of buffer stores and inaccuracies and stabilities of the interconnecting national clocks. The end-to-end performance should satisfy the service requirements for telephone and nontelephone services on a 64-kbit/s digital connection in an ISDN.

The slip rate objectives for an international end-to-end connection are stated with reference to the standard hypothetical reference connection (HRX) (see Section 5.4) of 27,500 km in length.

The theoretical slip rate is one slip in 70 days per plesiochronous interexchange link assuming clocks with specified accuracies (see Section 5.5.1) and provided that the performance of the transmission and switching requirements remain within their design limits.

In the case where the international connection includes all of the 13 nodes identified in the HRX and those nodes are all operating together in a plesiochronous mode, the nominal slip performance of a connection could be 1 in 70/12 days (12 links in tandem) or 1 in 5.8 days. In practice, however, some nodes in such a connection would be part of the same synchronized network. Therefore, a better nominal slip performance can be expected (e.g., where the national networks at each end are synchronized). The nominal slip performance of the connection would be 1 in 70/4 or 1 in 17.5 days. Note that these calculations assume a maximum of four international links.

The performance objectives for the rate of *octet* slips on an international connection of 27,500 km in length of a corresponding bearer channel are given in Table 9.8. CCITT (ITU-T) [28] adds that further study is required to confirm that these values are compatible with other objectives such as error performance given in Section 5.6.3 of this chapter.

Allocation of Slips. Because the impact of slips occurring in different parts of a connection varies in importance depending on the type of service and level

TABLE 9.11 Controlled Slip Performance on a 64-kbit/s International Bearer Channel

Performance Category	Mean Slip Rate	Proportion of Time[a]
(a)[b]	≤5 slips in 24 h	>98.9%
(b)	> 5 slips in 24 h and ≤ 30 slips in 1 h	<1.0%
(c)	> 30 slips in 1 h	<0.1%

[a]Total time ≥ 1 year.
[b]The nominal slip performance due to plesiochronous operation alone is not expected to exceed 1 slip in 5.8 days.
Source: Table 1/G.822, page 37, CCITT Rec. G.822, 1988 [28].

TABLE 9.12   Allocation of Controlled Slip Performance

Portion of HRX Derived from Figure 9.26[a]	Allocated Proportion of Each Objective in Table 9.10[b]	Objectives as Proportion of Total Time[c]	
		(b)	(c)
International transit portion	8.0%	0.08%	0.008%
Each national transit portion[d]	6.0%	0.06%	0.006%
Each local portion[d]	40.0%	0.4%	0.04%

[a] The portions of the HRX are defined in Figure 9.27. They are derived from, but not identical to, Rec. G.801.
[b] Performance levels are defined in Table 9.11.
[c] Total time ≥ 1 year.
[d] The allocation between national transit portion and local portion is given for guidance only. Administrations are free to adopt a different apportionment provided the total for each national portion (local plus transit) does not exceed 46%.

Source: Table 2/G.822, page 38, CCITT Rec. G.822, 1988 [28].

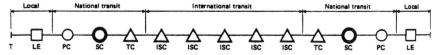

Figure 9.21. Subdivision of the HRX for the purpose of allocation of slip performance objectives. From Figure 1/G.822, page 38, CCITT Rec. G.822, 1988 [28].

of traffic affected, the allocation process includes placing higher limits on slips detected at international and national transmit exchanges and less stringent on slips detected on small local exchanges. The allocation process is based on subdividing the percentage of time objective for performance categories (b) and (c) in Table 9.11. Table 9.12 shows these allocations assigned to the various portions of the HRX (see Section 5.4 of this chapter). CCITT (ITU-T) states that these allocations are provisional. Figure 9.21 shows the subdivision of the HRX for slip allocation. [27]

5.7   *A*-Law Conversion to μ-Law; Digital Loss

Chapter 8 discussed two quite different PCM systems. In North America, there is the DS1 or T1* system; in Europe and in those countries under European hegemony there is the E1 system, previously called CEPT30 + 2. The logarithmic commanding curve used for the development of the DS1 signal is the μ-law, where $\mu = 255$. E1 uses the *A*-law, where $A = 87.6$. An international switching center is probably the most convenient location to convert from one to the other. To carry out such conversion, a processor is required with a look-up table. The processor forces some small changes in the binary value of time-slot samples based on look-up table values. As we mentioned earlier, to match E1 to DS1 channel-by-channel, five DS1s are mapped into four E1s, and during this mapping

---

* T1 and DS1 are synonymous.

the commanding conversion can also take place. The conversion equipment must also handle the different signaling and framing strategies used by each system. This can be carried out in fairly straightforward software.

Like its analog counterpart, the digital network must be lossy to control echo and singing. Specific loss plans were described in Chapter 6. In North America, 3 dB is a common value for loss on one end of a long-distance connection. It was a fairly simple matter to switch in a 3-dB pad (attenuator) in an analog switch. It is not so straightforward if we wish to insert such a loss on the digital side. Remember that loss equates to reducing level. To do this on the digital side, we again can resort to the use of a processor and a look-up table.

REVIEW QUESTIONS

1. Give a simple definition of time-division switching. Compare space-division switching with time-division switching.

2. Give at least five advantages of time-division switching compared to space-division switching.

3. Give two technical issues relating to time-division switching and digital networks. (*Hint*: These may be listed as disadvantages in relation to question 2.)

4. What are the two principal functional elements of digital switching? These elements carry out the actual switching of digital connections (e.g., not control).

5. Define the terms *read* and *write* as used in this chapter. Differentiate between sequential read–random write and random read–sequential write.

6. What are the three basic building blocks of a time switch?

7. What are the limitations of a time (T) switch? How are these solved by the addition of space (S) switching capability?

8. A space array has M horizontals and N verticals in its matrix. Relate M to N for nonblocking tandem switching, for local switch expansion, and for local switch concentration.

9. We have a 30 × 30 space array with a time-slot interchange (TSI) at each input. The TSI is designed for DS1 operation. How many total time slots can the array handle?

10. Why is a TST switch more desirable than an STS switch? Where would an STS switch have application?

11. How can blocking probability be reduced in a TST switch?

12. What is the function of "junctors" in a switch (e.g., a DMS-100)?

13. Internal bit rates of switches differ from the external interface of 8-bit time slots. Northern Telecom DMS-100 maps the 8-bit PCM "word" into a _____-bit time slot. What is/are the function(s) of the extra two bits?

14. To make a switch "universal"—that is, easy to change over from DS1 format to E1 format and vice versa—often internal bit rates were also based on an X/Y ratio between DS1 and E1. What are the values of X and Y?

15. The AT&T 5ESS has 100,000-line capacity. How does it accomplish this capacity with only a TST architecture (i.e., only one space stage)?

16. Suppose a digital switch were based solely on a time stage consisting of an E3 format. How many lines could it handle?

17. Why would fiber-optic transmission links find application internally to these third-generation PCM switches?

18. Give at least two applications of remote switching, particularly in light of Chapter 2.

19. What are the two principal factors that change network structure?

20. In your opinion, what is delaying "fiber to the home"?

21. What is the effect on the subscriber when his/her telephone subset operates without sidetone?

22. What is the meaning of each letter in BORSCHT?

23. Of what use to the telecommunication system engineer is a hypothetical reference circuit or hypothetical reference connection?

24. In the digital network, digital bit streams require synchronization at three levels. What are they?

25. Why is it imperative that every outgoing bit stream from a digital switch have its bit rate *very* close to the nominal? If not, what impairment results?

26. What is the cause of slips?

27. Which type of user of the digital network suffers the most when a slip occurs?

28. What is the most common boundary for a controlled slip? Why there?

29. The velocity of propagation for optical fiber is 2×10^8 m/s. A data source is transmitting at 1000 Mbits/s. How many bits will there be on 1000 km of fiber-optic cable?

30. Six methods of synchronizing a digital network were given in the text. Name four of them.

31. Give a value of frequency stability when we mean *high* stability regarding synchronization of the digital network.

32. What is the most common way for a digital network to disseminate timing?

33. We commonly meet what is termed the "two-clock problem." An international switch is an example, where it is more of a multiclock problem. How can a switch handle such a problem?

34. Define plesiochronous operation.

35. What is a unit interval (UI)? Define it.

36. What is the end-to-end BER specified by CCITT for an ISDN?

37. Why did CCITT come up with a new way of specifying error performance on the digital network?

38. What is the basic performance measure of a digital network?

39. What primary performance parameter does timing jitter affect?

40. If network synchronization is working properly, what slip rate can be expected?

41. What is the most common method of converting A-law to μ-law and vice versa? (We face this when E1 interfaces with DS1 and vice versa.)

42. To control echo, we can insert loss in a voice channel. How can loss be inserted on the digital side?

REFERENCES

1. J. P. Ronadyne, *Introduction to Digital Communications Switching*, Howard W. Sams & Co., Indianapolis, IN, 1986.

2. J. C. McDonald, ed., *Fundamentals of Digital Switching*, 2nd ed., Plenum Press, New York, 1990.

3. J. C. Bellamy, *Digital Telephony*, 3rd ed., John Wiley & Sons, New York, 2000.

4. *Planning Guide: DMS-100/200 Family*, Northern Telecom, Research Triangle Park, NC 1985.

5. *Network Frame*, General Specification, Northern Telecom, Research Triangle, NC, 1983/1988.

6. R. L. Freeman, *Reference Manual for Telecommunication Engineering*, 3rd ed., John Wiley & Sons, New York, 2000.

7. *5ESS Switch and 5ESS-2000 Switch System Description*, AT&T 235-100-125, Issue 7.00, Winston-Salem, NC, November 1994.

8. *Enhanced Network*, a brief description of ENET, Northern Telecom, Research Triangle, NC, Issue 2, August 1992.

9. *DMS-100 Advantage*, Issue 2, Northern Telecom, Research Triangle, NC, November 1993.

10. Northern Telecom U.S. Patent 4,470,139, issued September 4, 1984.

11. Private communication with Ernst Munter, Northern Telecom, Ottawa, March 25, 1995.

12. W. D. Reeve, *Subscriber Loop Signaling and Transmission Handbook, Digital*, IEEE Press, New York, 1995.

13. *Vocabulary of Digital Transmission and Muliplexing and Pulse Code Modulation (PCM) Terms*, ITU-T Rec. G.701, ITU, Geneva, March 1993.

14. *Digital Transmission Models*, ITU-T Rec. G.801, ITU, Geneva, 1988.

15. *Digital Network Synchronization Plan*, Bellcore Generic Requirements, GR-436-CORE, Bellcore, Piscataway, NJ, June 1994.

16. *Timing Requirements at the Outputs of Reference Clocks and Network Nodes Suitable for Plesiochronous Operation of International Digital Links*, CCITT Rec. G.811, Fascicle III.5, page 298, IXth Plenary Assembly, Melbourne, 1988.

17. *Clocks for the Synchronized Network: Common Generic Criteria*, TR-NWT-001244, Issue 1, Bellcore, Piscataway, NJ, June 1993.

18. *Synchronous Optical Network (SONET) Transport Systems: Common Generic Criteria*, GR-253-CORE, Bellcore, Piscataway, NJ, December 1994.

19. *Telcordia Notes on the Networks*, Special Report, SR-2275, Issue 4, Telcordia, Piscataway, NJ, October 2000.

20. *Transport Systems Generic Requirements (TSGR): Common Requirements*, GR-499-CORE, Issue 2, Bellcore (now Telcordia), Piscataway, NJ, December 1998.

21. *Error Performance of an International Digital Connection Operating at a Bit Rate below the Primary Rate and Forming Part of an ISDN*, ITU-T Rec. G.821, ITU, Geneva, August 1996.

22. *Error Performance Parameters and Objectives for International Constant Bit Rate Synchronous Digital Paths*, ITU-T Rec. G.828, Geneva, March 2000.

23. M. Shafi and P. J. Smith, "The Impact of G.826," *IEEE Commun. Mag.* (September 1993).

24. *The New IEEE Standard Dictionary of Electrical and Electronics Terms*, 6th ed., IEEE Std 100-1996, IEEE, New York, 1996.

25. *The Control of Jitter and Wander within Digital Networks which Are Based on the 2048 kbps Hierarchy*, ITU-T Rec. G.823, ITU, Geneva, March 2000.

26. *The Control of Jitter and Wander within Digital Networks which Are Based on the 1544 kbps Hierarchy*, ITU-T Rec. G.824, ITU, Geneva, March 2000.

27. *BOC Notes on the LEC Networks—1990*, Issue 1, Bellcore, Piscataway, NJ, 1990.

28. *Controlled Slip Rate Objectives on an International Digital Connection*, CCITT Rec. G.822, Fascicle III.5, IXth Plenary Assembly, Melbourne, 1988.

29. Converge Network Digest, from the Web, www.convergedigest.com/columns/0210polaris/g-sgosa14.1ap, 2/24/03, Sections 4.6, 4.6.1, and 4.6.2.

30. From the web. www.insight-corp.com/dcs.htrml 03/03/03.

31. *Interconnection and Collocation for Transport and Switched Unbundled Network Elements and Finished Services*, Table 13.1, Qwest, Technical Publication 77386, Nov. 2001, Issue G (from the Web).

32. *AT&T 62411 Design Considerations*, AN12, Rev. 2, Crystal (Inc.), Austin, TX, 1994 (from the Web).

33. Private Communication, Cirrus Logic, Austin, TX, January 20, 2004, (Kristin Frye).

10

INTRODUCTION TO DATA COMMUNICATIONS

1 OVERVIEW

Data communications emerged as a technology slightly delayed from the development of the computer. Today it is probably the most rapidly growing branch of telecommunications. My initial experience with data communications was in 1961 with DoD's LogComNet, which became AutoDiN. Many of its concepts, technology, and certainly language derived from the much older telegraph transmission.

The IEEE [1] defines *data communications* (data transmission) as "The movement of encoded information by means of communications techniques." In this chapter we will discuss the *encoding of data* and then the *techniques of its communication*. In Chapter 11, we will address data networks and their operation. This discussion will be a general treatment of wide area (data) networks (WANs). Chapter 12 delves into the combined data-voice technology, voice-over IP (VoIP). Local area networks (LANs) are covered in Chapter 13. Chapter 14 deals with integrated services digital networks (ISDN), and Chapter 15 covers broadband data communication technologies such as frame relay. Chapter 16, dealing with the asynchronous transfer mode, is the final chapter in the data grouping.

The fact that this book covers data in 6 out of 18 chapters shows its significance in the world of telecommunication systems. The present North American public switched telecommunication network (PSTN) is optimized for voice communication. It is estimated that less than 50% of the traffic on that network is voice traffic. On other networks in the world, the percentage is higher.

This begs the question, What is the other 50%? It is a mix of data, facsimile, and broadcast and conference television. However, facsimile as transmitted on

---

Telecommunication System Engineering, by Roger L. Freeman
ISBN 0-471-45133-9 Copyright © 2004 Roger L. Freeman

the network is data, and conference television is transmitted in a digital format. If we said that the remaining 50% was data, we would be pretty much correct. Today, though, much of that 50% or more is taken up by the internet, which is based on data communication technologies, pure and simple.

In this chapter we address data transmission on wide area networks. Facsimile will be considered as just another data transmission circuit. Wide area networks often use the underlying PSTN for transport. But not always. Private networks that have nothing to do with the PSTN* are becoming ever more prevalent. VSAT networks (Chapter 7) are just one example. Often such networks are 100% dedicated to data communication. Other private networks carry a mix of voice, data, and possibly conference television.

2   THE BIT

The *bit* is often called the most elemental unit of information. The IEEE [1] calls it a contraction of *binary digit*, a unit of information represented by either a zero or a one. These are the same bits we were dealing with in Chapters 8 and 9. In those chapters the primary purpose of those bits was to signal the distant end of the voltage level of an analog voice channel at some moment in time. Here we will be assembling bit groupings which will represent the letters of the alphabet, numerical digits 0 through 9, punctuation, graphic symbols, or just operational sequences that have meaning to a machine but have no ostensible outward meaning to us.

From old-time telegraphy, much of the terminology used has migrated to data communications. A *mark* is a binary 1 and a *space* is a binary 0. A space or 0 is represented by a positive-going voltage, and a mark or 1 is represented by a negative-going voltage. Now I am getting confused. When I was growing up in the industry, a 1 or mark was a positive-going voltage (and so forth).

3   REMOVING AMBIGUITY — BINARY CONVENTION

To remove ambiguity of the various ways, we can express a 1 and a 0. CCITT in Recommendation V.1 states clearly how to represent a 1 and a 0. This is summarized in Table 10.1 with several additions from other sources.

Table 10.1 defines the *sense* of transmission so that the mark and space, the 1 and 0, respectively, will not be inverted. Inversion can take place by just changing the voltage polarity. We call it reversing the sense. Some data engineers often refer to such a table as a "table of mark-space convention."

---

* Some private networks lease circuits from or overlay on the PSTN. That is why the statement is made as it is.

TABLE 10.1 Equivalent Binary Designations: Summary of Equivalence

Symbol 1	Symbol 0
Mark or marking	Space or spacing
Current on	Current off
Negative voltage	Positive voltage
Hole (in paper tape)	No hole (in paper tape)
Condition Z	Condition A
Tone on (amplitude modulation)	Tone off
Low frequency (frequency shift keying)	High frequency
Inversion of phase	No phase inversion (differential phase shift keying)
Reference phase	Opposite to reference phase

Source: CCITT Recs. V.1 [2], V.10, and V.11.

4 CODING

4.1 Introduction to Binary Coding Techniques

Written information must be coded before it can be transmitted over a data network. One bit carries very little information. There are only those two possibilities: the 1 and the 0. It serves good use for supervisory signaling where a telephone line could only be in one of two states. It is either idle or busy. As a minimum, we would like to transmit every letter of the alphabet and the 10 basic decimal digits plus some control characters such as a space and carriage return, along with some punctuation.

Suppose we stick two bits together for transmission. There are four possibilities*:

$$00 \quad 10$$
$$01 \quad 11$$

or four pieces of information. Suppose 3 bits are transmitted in sequence. Now there are eight possibilities:

$$000 \quad 100$$
$$001 \quad 101$$
$$010 \quad 110$$
$$011 \quad 111$$

We can now see that for a binary code, the number of distinct information characters available is equal to two raised to a power equal to the number of elements or bits per character. For instance, the last example was based on a three-element code giving eight possibilities or information characters, or 2^3.

---

* To a degree, this is a review of an argument in Chapter 8.

Another, more practical example is the CCITT ITA No. 2 teleprinter code (Figure 10.1), which has 5 bits or information elements per character. Therefore the number of different graphics* and characters available is $2^5 = 32$. The American Standard Code for Information Interchange (ASCII) has seven information elements per character, or $2^7 = 128$; thus it has 128 distinct combinations of marks and spaces that are available for assignment as characters or graphics.

The number of distinct characters for a specific code may be extended by establishing a bit sequence (a special character assignment) to shift the system

Characters				Code Elements [a]						
Letters Case	Communications	Weather	CCITT #2[b]	START	1	2	3	4	5	STOP
A	-	↑			X	X				X
B	?	⊕			X			X	X	X
C	:	O				X	X	X		X
D	$	↗	WRU		X			X		X
E	3	3			X					X
F	1	→	Unassigned		X		X	X		X
G	&	↖	Unassigned			X		X	X	X
H	STOP[c]	↓	Unassigned				X		X	X
I	8	8				X	X			X
J	'	↗	Audible signal		X	X		X		X
K	(	←			X	X	X	X		X
L	)	↘				X			X	X
M	.	.					X	X	X	X
N	,	⦶					X	X		X
O	9	9						X	X	X
P	0	0				X	X		X	X
Q	1	1			X	X	X		X	X
R	4	4				X		X		X
S	BELL	BELL	'		X		X			X
T	5	5							X	X
U	7	7			X	X	X			X
V	;	⦶	=			X	X	X	X	X
W	2	2			X	X			X	X
X	/	/			X		X	X	X	X
Y	6	6			X		X		X	X
Z	"	+	+		X				X	X
BLANK		-								X
SPACE							X			X
CAR. RET.								X		X
LINE FEED						X				X
FIGURE					X	X		X	X	X
LETTERS					X	X	X	X	X	X

[a] Blank, spacing element; crosshatched, marking element.

[b] This column shows only those characters that differ from the American "communications" version

[c] Figures case H(COMM) may be STOP or +.

Figure 10.1. Communication and weather codes. CCITT Alphabet No. 2 (ITA#2 Code).

* In this context a graphic is a printing character other than a letter or number. Typical graphics are asterisks, punctuation, parentheses, dollar signs, and so forth.

or machine to uppercase (as is done with a conventional typewriter). Uppercase is a new character grouping. A second distinct bit sequence is then assigned to revert to lowercase. For example, the CCITT ITA No. 2 code (Figure 10.1) is a five-unit code with 58 letters, numbers, graphics, and operator sequences. The additional characters and graphics (additional above $2^5 = 32$) originate from the use of uppercase. Operator sequences appear on a keyboard as "space" (spacing bar), "figures" (uppercase), "letters" (lowercase), "carriage return," "line feed" (spacing vertically), and so on. When we refer to a 5-unit, 6-unit, or 12-unit code, we refer to the number of information units or elements that make up a single character or symbol. That is, we refer to those elements assigned to each character that carry information and that distinguish it from all other characters or symbols of the code.

4.2 Specific Binary Codes for Information Interchange

The most commonly used binary source code for data communications is the ASCII code. ASCII stands for American Standard Code for Information Interchange. It is used worldwide. ASCII is a 7-unit or 7-level or 7-bit code and is illustrated in Figure 10.2. Almost universally ASCII has a parity bit appended making it an 8-level code. Table 10.2 gives a description of ASCII control characters. Figure 10.3 shows EBCDIC (extended binary coded decimal interchange code) developed by IBM. It is a true 8-level code with 2^8 or 256 coded character

b_7 b_6 b_5 →				Column Row	0 0 0	0 0 1	0 1 0	0 1 1	1 0 0	1 0 1	1 1 0	1 1 1
Bits b_4	b_3	b_2	b_1		0	1	2	3	4	5	6	7
0	0	0	0	0	NUL	DLE	SP	0	@	P	`	p
0	0	0	1	1	SOH	DCI	!	1	A	Q	a	q
0	0	1	0	2	STX	DC2	"	2	B	R	b	r
0	0	1	1	3	ETX	DC3	#	3	C	S	c	s
0	1	0	0	4	EOT	DC4	$	4	D	T	d	t
0	1	0	1	5	ENQ	NAK	%	5	E	U	e	u
0	1	1	0	6	ACK	SYN	&	6	F	V	f	v
0	1	1	1	7	BEL	ETB	'	7	G	W	g	w
1	0	0	0	8	BS	CAN	(	8	H	X	h	x
1	0	0	1	9	HT	EM	)	9	I	Y	i	y
1	0	1	0	10	LF	SUB	*	:	J	Z	j	z
1	0	1	1	11	VT	ESC	+	;	K	[	k	{
1	1	0	0	12	FF	FS	,	<	L	\	l	¦
1	1	0	1	13	CR	GS	-	=	M	]	m	}
1	1	1	0	14	SO	RS	.	>	n	^	n	~
1	1	1	1	15	SI	US	/	?	O	_	o	DEL

Figure 10.2. American Standard Code for Information Interchange. From Mil-Std-188-100 [31] and from ANSI with updates from ANSI X3.4 1986 [3]. See Table 10.2 for definitions.

TABLE 10.2   Definitions for Figure 10.2, ASCII Control Characters, and Format Effectors

ACK (Acknowledge). A transmission control character transmitted by a receiver as an affirmative response to the sender.

BEL (Bell). A control character that is used when there is a need to call for attention; it may control alarm or attention devices.

BS (Backspace). A format effector that causes the active position to move one character position backwards.

CAN (Cancel). A character, or the first character of a sequence, indicating that the data preceding it is in error. As a result, this data is to be ignored. The specific meaning of this character shall be defined for each application and/or defined between sender and recipient.

CR (Carriage Return). A format effector that causes the active position to move to the first character position on the same line.

DC1 (Device Control One). A device control character that is primarily intended for turning on or starting an ancillary device. If it is not required for this purpose, it may be used to restore a device to the basic mode of operation (see also DC2 and DC3), or for any other device control function not provided by other DCs.

DC2 (Device Control Two). A device control character that is primarily intended for turning on or starting an ancillary device. If it is not required for this purpose, it may be used to set a device to a special mode of operation (in which case DC1 is used to restore the device to the basic mode), or for any other device control function not provided by other DCs.

DC3 (Device Control Three). A device control character that is primarily intended for turning off or stopping an ancillary device. This function may be a secondary level stop — for example, wall, pause, standby, or halt (in which case DC1 is used to restore normal operation). If it is not required for this purpose, it may be used for any other ancillary device control function not provided by other DCs.

DC4 (Device Control Four). A device control character that is primarily intended for turning off, stopping, or interrupting an ancillary device. If it is not required for this purpose, it

may be used for any other device control function not provided by other DCs.

DEL (Delete). A character used primarily to erase or obliterate an erroneous or unwanted character in punched tape. DEL characters may also serve to accomplish media-fill or time-fill. They may be inserted into or removed from a stream of data without affecting the information content of that stream, but such action may affect the information layout and/or the control of equipment. If media-fill or time-fill is required, it is preferred that the NUL character be used.

DLE (Data Link Escape). A transmission control character that changes the meaning of a limited number of contiguously following bit combinations. It is used exclusively to provide supplementary transmission control functions. Only graphic characters and transmission control characters may be used in DLE sequences. Appropriate sequences are defined in ANSI X3.28-1976.

EM (End of Medium). A control character that may be used to identify the physical end of a medium, the end of the used portion of a medium, or the end of the wanted portion of data recorded on a medium. The position of this character does not necessarily correspond to the physical end of the medium.

ENQ (Enquiry). A transmission control character used as a request for a response from a remote station — the response may include station identification and/or station status. When a "Who are you" function is required on a switched transmission network, the first use of ENQ after the connection is established shall have the meaning "Who are you" (station identification). Subsequent use of ENQ may or may not include the function "Who are you," as determined by agreement.

EOT (End of Transmission). A transmission control character used to indicate the conclusion of the transmission of one or more texts.

ESC (Escape). A control character that is used to provide additional characters (code extension). It alters the meaning of a limited number of contiguously following bit combinations. The use of this character is specified in ANSI X3.41-1974.

TABLE 10.2 (*continued*)

ETB (End of Transmission Block). A transmission control character used to indicate the end of a transmission block of data where data are divided into such blocks for transmission purposes.

EXT (End of Text). A transmission control character that terminates a text.

FF (Form Feed). A format effector that causes the active position to advance to the corresponding character position on a predetermined line of the next form or page.

FS (File Separator) (Information Separator Four). A control character used to separate and qualify data logically; its specific meaning has to be defined for each application. If this character is used in hierarchical order, as specified in the general definition of the information separators, it delimits a data item called a "file."

GS (Group Separator) (Information Separator Three). A control character used to separate and qualify data logically; its specific meaning has to be defined for each application. If this character is used in hierarchical order, as specified in the general definition of the information separators, it delimits a data item called a "group."

HT (Horizontal Tabulation). A format effector that causes the active position to advance to the next predetermined character position.

LF (Line Feed). A format effector that causes the active position to advance to the corresponding character position of the next line.

NAK (Negative Acknowledge). A transmission control character transmitted by a receiver as a negative response to the sender.

NUL (Null). A control character used to accomplish media-fill or time-fill. NUL characters may be inserted into or removed from a stream of data without affecting the information content of that stream, but such action may affect the information layout and/or the control of equipment.

RS (Record Separator) (Information Separator Two). A control character used to separate and quality data logically; its specific

meaning has to be defined for each application. If this character is used in hierarchical order, as specified in the general definition of the information separators, it delimits a data item called a "record."

SI (Shift-In). A control character that is used in conjunction with SO and ESC to extend the graphic character set of the code. It may reinstate the standard meanings of the bit combinations that follow it. The effect of this character is described in ANSI X3.41-1974.

SO (Shift-Out). A control character that is used in conjunction with SI and ESC to extend the graphic character set of the code. It may alter the meaning of the bit combinations that follow it until an SI character is reached. The effect of this character is described in ANSI X3.41-1974.

SOH (Start of Heading). A transmission control character used as the first character of a heading of an information message.

STX (Start of Text). A transmission control character that precedes a text and that is used to terminate a heading.

SUB (Substitute Character). A control character used in the place of a character that has been found to be invalid or in error. SUB is intended to be introduced by automatic means, as, for example, when a transmission error is detected.

SYN (Synchronous Idle). A transmission control character used by a synchronous transmission system in the absence of any other character (Idle condition) to provide a signal from which synchronism may be achieved or retained between data terminal equipment.

US (Unit Separator) (Information Separator One). A control character used to separate and quality data logically; its specific meaning has to be defined for each application. If this character is used in hierarchical order, as specified in the general definition of the information separators, it delimits a data item called a "unit."

VT (Vertical Tabulation). A format effector that causes the active position to advance to the corresponding character position on the next predetermined line.

B I T S 8765 \ 4321	0000	0001	0010	0011	0100	0101	0110	0111	1000	1001	1010	1011	1100	1101	1110	1111
0000	NUL				PF	HT	LC	DEL								
0001					RES	NL	BS	IL								
0010					BYP	LF	EOB	PRE			SM					
0011					PN	RS	UC	EOT								
0100	SP										¢	.	<	(	+	|
0101	&										!	$	*	)	;	¬
0110	-	/									^	,	%	—	>	?
0111											⁄:	#	@	'	=	"
1000		a	b	c	d	e	f	g	h	i						
1001		j	k	l	m	n	o	p	q	r						
1010			s	t	u	v	w	x	y	z						
1011																
1100		A	B	C	D	E	F	G	H	I						
1101		J	K	L	M	N	O	P	Q	R						
1110			S	T	U	V	W	X	Y	Z						
1111	0	1	2	3	4	5	6	7	8	9						¤

PF — Punch Off   RES — Restore   BYP — Bypass
HT — Horiz. Tab   NL — New Line   LF — Line Feed
LC — Lower Case   BS — Backspace   EOB — End of Block
DEL — Delete   IL — Idle   PRE — Prefix
SP — Space   PN — Punch On   RS — Reader Stop
UC — Upper Case   EOT — End of Transmission SM — Start Message

Figure 10.3. Extended binary-coded decimal interchange code (EBCDIC).

possibilities. As shown in Figure 10.3, some of the bit sequences have their meaning left blank (i.e., no meaning has been assigned).

5   ERRORS IN DATA TRANSMISSION

5.1   Introduction

In data transmission one of the most important design goals is to minimize the error rate. *Error rate* (or error ratio) may be defined as the ratio of the number of bits incorrectly received to the total number of bits transmitted. Up to several years ago, an error rate of one error in 1×10^6 bits (often expressed as 1×10^{-6}) was acceptable. Today one should expect end-to-end performance of 2×10^{-10}. On a single link a design objective often is 1×10^{-12}. Such values will be found in Telcordia TSGR [32].

One method of optimizing error performance would be to provide a "perfect" transmission channel, one that introduces no errors in the transmitted information at the output of the receiver. However, that perfect channel can never be achieved. "It is against the laws of nature."

Besides improvement of the channel transmission parameters themselves, error performance can be improved by selecting an optimum modulation scheme and by employing some form of systematic redundancy. In old-time Morse code, on a bad circuit words were often sent twice; this is redundancy in its simplest form. Of course, it took twice as long to send a message. Such a practice is not very economical if the number of useful words per minute received is compared to channel occupancy.

This illustrates the trade-off between redundancy and channel efficiency. Redundancy can be increased such that the error rate could approach zero. Meanwhile, the information transfer across the channel would also approach zero. Thus unsystematic redundancy is wasteful and merely lowers the rate of useful communication. On the other hand, maximum efficiency could be obtained in a digital transmission system if all redundancy and other code elements, such as "start" and "stop" elements, parity bits, and other "overhead" bits, were removed from the transmitted bit stream. In other words, the channel would be 100% efficient if all bits transmitted were information bits. Obviously, there is a trade-off of cost and benefits somewhere between maximum efficiency on a data circuit and systematically added redundancy (see Chapter 11, Section 5).

5.2 Throughput

Throughput of a data channel is the expression of how much data are put through. In other words, throughput is an expression of channel efficiency. The term gives a measure of *useful* data put through the communication link. These data are directly useful to the computer or DTE (data-terminal equipment).

Therefore, on a specific circuit, throughput varies with the raw-data rate, is related to the error rate and the type of error encountered (whether burst or random), and varies according to the type of error detection and correction system used, the message-handling time, and the block length from which we must subtract overhead bits such as parity, flags, and cyclic redundancy checks. Throughput and the operational features of data circuits are described in detail in Chapter 11.

5.3 The Nature of Errors

In binary transmission an error is a bit that is incorrectly received. For instance, suppose a 1 is transmitted in a particular bit location and at the receiver the bit in that same location is interpreted as a 0. Bit errors occur either as single random errors or as bursts of errors.

Random errors occur when the signal-to-noise ratio deteriorates. This assumes, of course, that the noise is thermal noise. In this case, noise peaks, at certain moments of time, are of sufficient level as to confuse the receiver's decision, whether a 1 or a 0.

Burst errors are commonly caused by fading on radio circuits. Impulse noise can also cause error bursts. Impulse noise can derive from lightning, car ignitions, electrical machinery, and certain electronic power supplies, to name a few.

5.3.1 Error Performance. We discussed error performance in Section 5.1 of this chapter as well as in Sections 5.6.2 and 5.6.3 of Chapter 9. Today, a great portion of the wide area network (WAN) data is transported on digital circuits on the public network. We can then safely assume that the error performance on these data circuits will be the same as the underlying digital network whether of the PDH (typically DS1 or E1) or SDH varieties. Rather than use powers of 10 to express BER, error performance is described in terms of *errored seconds* (ES) and *severely errored seconds* (SES).

To review, an errored second is a 1-s period in which one or more bits are in error. A severely errored second is a 1-s period which has a bit error ratio $\geq 1 \times 10^{-3}$. An errored second ratio (ESR) is the ratio of ES to total seconds of available time during a fixed measurement interval. A severely errored second ratio is the ratio of SES to total seconds of available time during a fixed measurement interval. "Available time" is defined in Chapter 9, Section 5.6.3.

ITU-T Rec. G.821 [4], which deals with error performance on ISDN circuits, states that the ESR should be <0.08 and the SESR <0.002.

5.4 Error Detection and Error Correction

Error detection just identifies that a bit (or bits) has been received in error. Error correction corrects errors at a far-end receiver. Both require a certain amount of redundancy to carry out the respective function. Redundancy, in this context, means those added bits or symbols that carry out no other function than as an aid in the error detection or error correction process.

One of the earliest methods of error detection was the *parity check*. With the 7-bit ASCII code, a bit was added for parity, making it an 8-bit code. This is character parity. It is also referred to as *vertical redundancy checking* (VRC).

We speak of *even parity* and *odd parity*. One system or the other is used. Either system is based on the number of marks or 1s in a 7-bit character and the eighth bit is appended accordingly, either a 0 or a 1. Let's assume even parity and we transmit the ASCII bit sequence 1010010. There are three 1s, an odd number. Thus a 1 is appended as the eighth bit to make it an even number.

Suppose we use odd parity and transmit the same character. There is an odd number of 1s (marks) so we append a 0 to leave the total number of 1s an odd number. With odd parity, try 1000111. If you added a 1 as the eighth bit, you'd be correct.

Character parity has the weakness that a lot of errors can go undetected. Suppose two bits are changed in various combinations and locations. Suppose a 10 became a 01; a 0 became a 1 and a 1 became a 0; two sequential 1s became two sequential 0s. All would get by the system undetected.

To strengthen this type of parity checking, the *longitudinal redundancy check* (LRC) was included as well as the VRC. This is a summing of the 1s in a vertical column of all characters, including the 1s in the 8th bit location. The sum is now appended at the end of a message block or frame which is a special field for error detection, which we can call the block check count or BCC. Later, we will call that field the *frame check sequence* (FCS). At the distant end receiver, the same addition is carried out and if the sum agrees with the BCC value received, the block is accepted as error-free. If not, it contains at least one error, and a request is sent to the transmit end to re-transmit the block.*

Even with the addition of LRC, errors can get through. In fact, no error detection system is completely foolproof. There is another method, though, that has excellent error detection properties. This is the *cyclic redundancy check (CRC)*. It comes in a number of varieties.

5.4.1 Cyclic Redundancy Check (CRC). In very simple terms the CRC error detection technique works as follows. A data block or frame is placed in storage. We can call it a k-bit sequence and can be represented by a polynomial which is called $G(x)$. Various modulo-2 arithmetic† operations are carried out on $G(x)$, and the result is divided by a known generator polynomial called $P(x)$. This results in a quotient $Q(x)$ and a remainder $R(x)$. The remainder is appended to the frame as an FCS (frame check sequence), and the total frame with FCS is transmitted to the distant end receiver where the frame is stored, then divided by the same generating polynomial $P(x)$. The calculated remainder is compared to the received remainder (i.e., the FCS). If the values are the same, the frame is error-free. If they are not, there is at least one bit in error in the frame.

For many WAN applications the FCS is 16 bits long, and on LANs it is often 32 bits long; in Chapter 8 we dealt with FCS of length 4 and 6 bits. Generally speaking, the greater the number of bits, the more powerful the CRC is for catching errors.

The following are two common generating polynomials:

- ANSI CRC-16: $X^{16} + X^{15} + X^2 + 1$
- CRC-CCITT: $X^{16} + X^{12} + X^5 + 1$

producing a 16-bit FCS.

CRC-16 provides error detection of error bursts up to 16 bits in length. Additionally, 99.955% of error bursts greater than 16 bits can be detected [31].

5.5 Forward-Acting Error Correction (FEC)

Forward-acting error correction (FEC) uses certain binary codes that are designed to be self-correcting for errors introduced by the intervening transmission media. In this form of error correction the receiving station has the ability to reconstitute messages containing errors.

The codes used in FEC can be divided into two broad classes: block codes and convolutional codes. In block codes information bits are taken k at a time, and c

---

* A *block* is a group of bits, bytes, or octets transmitted as a unit over which an error control procedure is applied. We may also use the terms packet or frame synonymously for block. The more popular term today is frame.

† Modulo-2 arithmetic is the same as binary arithmetic but without carries or borrows.

parity bits are added, checking combinations of the k information bits. A block consists of $n = k + c$ digits. When used for the transmission of data, block codes may be systematic. A systematic code is one in which the information bits occupy the first k positions in a block and are followed by the $(n - k)$ check digits.

Still another block code is the group code, where the modulo-2 sum of any two n-bit code words is another code word. Modulo-2 addition is denoted by the symbol $\oplus$. It is a binary addition without the "carry" or $1 + 1 = 0$, and we do not carry the 1. Summing 10011 and 11001 in modulo-2, we get 01010.

The minimum Hamming distance is a measure of the error detection and correction capability of a code. This "distance" is the minimum number of digits in which two encoded words differ. For example, to detect E digits in error, a code of a minimum Hamming distance of $(E + 1)$ is required. To correct E errors, a code must display a minimum Hamming distance of $(2E + 1)$. A code with a minimum Hamming distance of 4 can correct a single error *and* detect two digits in error.

A convolution(al) code is another form of coding used for error correction. As the word "convolution" implies, this is one code wrapped around or convoluted on another. It is the convolution of an input-data stream and the response function of an encoder. The encoder is usually made up of shift registers. Modulo-2 adders are used to form check digits, each of which is a binary function of a particular subset of the information digits in the shift register.

Error performance can also be improved by the assistance of a microprocessor in the decoder. Mark or space decisions made by a demodulator are not hard or irrevocable decisions; rather, these are called "soft" decisions. In this case a tag of 3 bits is attached to each received digit to indicate the confidence level of the decision in the demodulator before processing. After processing, when errors are indicated, the bits with the lowest confidence level are changed from 0 to 1 or 1 to 0, as the case may be.

FEC as is works on random errors only. It does not help with burst errors. But we can fool the system by using an interleaver–deinterleaver. This concept is shown in Figure 10.4. An interleaver first stores a stream of x number of serial bits. It then shuffles the bits into a pseudorandom sequence and releases those x number of bits for transmission. The deinterleaver operates at the receive end using the same randomizing polynomial, but de-randomizes, placing the bits back in their original order before they entered the interleaver. Of course the deinterleaver has to be time-synchronized with the interleaver. The value of x is

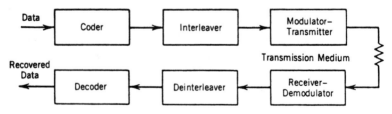

Figure 10.4. FEC scheme for a channel with burst errors.

important and x can be related to time. If we transmit 1200 bits/s and we wish time to be 1 s, then x must be 1200 bits; x must be greater than the expected duration of a burst. If we can meet these conditions, burst errors will be treated as though they are random errors.

5.6   Error Correction with Feedback Channel

Two-way or feedback error correction is used widely today on data and some telegraph circuits. Such a form of error correction is called ARQ. The letter sequence ARQ derives from the old Morse and telegraph signal, "automatic repeat request."

There are three varieties of ARQ:

- Stop-and-wait ARQ
- Selective or continuous ARQ
- Go-back-n ARQ

Stop-and-wait ARQ is simple to implement and may be the most economic in the short run. It works on a frame-by-frame basis. A frame is generated; it goes through CRC processing and an FCS is appended. It is transmitted to the distant end where the frame runs through CRC processing. If no errors are found, an acknowledgment (ACK) signal is sent to the transmitter, which now proceeds to send the next frame, and so forth. If a bit error is found, a negative acknowledgment (NACK) signal is sent to the transmitter, which then proceeds to repeat that frame. It is the waiting time of the transmitter as it waits for either ACK or NACK signals. Many point to this wait time as wasted time. It could be costly on high-speed circuits. However, the control software is simple and the storage requirements are minimal (i.e., only one frame).

Selective ARQ, sometimes called *continuous ARQ*, eliminates the waiting. The transmit side pours out a continuous stream of contiguous frames. The receive side stores and CRC processes as before, but it is processing a continuous stream of frames. When a frame is found in error, it informs the transmit side on the return channel. The transmit side then picks that frame out of storage and places it in the transmission queue. Several points become obvious to the reader. First, there must be some way to identify frames. Second, there must be a better way to acknowledge or "negative-acknowledge." The two problems are combined and solved by the use of send sequence numbers and receive sequence numbers. The header of a frame has bit positions for a send sequence number and a receive sequence number. The send sequence number is inserted by the transmit side; the receive sequence number by the receive side. The receive sequence numbers forwarded back to the transmit side are the send sequence numbers acknowledged by the receive side. Of course the receive side has to insert the corrected frame in its proper sequence, before passing the data message to the end user. *Continuous* or *selective ARQ* is more costly in the short run compared to stop-and-wait ARQ. It requires more complex software and notably more storage on both sides of the link. However, there are no gaps in transmission and no time is wasted waiting for the ACK or NACK.

Go-back-n ARQ is a compromise. In this case the receiver does not have to insert the corrected frame in its proper sequence, thus less storage is required. It works this way. When a frame is received in error, the receiver informs the transmitter to "go-back-*n*," *n* being the number of frames back to where the errored frame was. The transmitter then repeats all *n* frames, from the errored frame forward. Meanwhile, the receiver has thrown out all frames from the errored frame forward. It replaces this group with the new set of *n* frames it received, all in proper order.

6   THE DC NATURE OF DATA TRANSMISSION

6.1   Loops

Binary data are transmitted on a dc loop. More correctly, the binary data end instrument delivers to the line and receives from the line one or several dc loops. In its most basic form a dc loop consists of a switch, a dc voltage, and a termination. A pair of wires interconnects the switch and termination. The voltage source in data and telegraph work is called the *battery*, although the device is usually electronic, deriving the dc voltage from an ac power line source. The battery is placed in the line to provide voltage(s) consistent with the type of transmission desired. A simplified dc loop is shown in Figure 10.5.

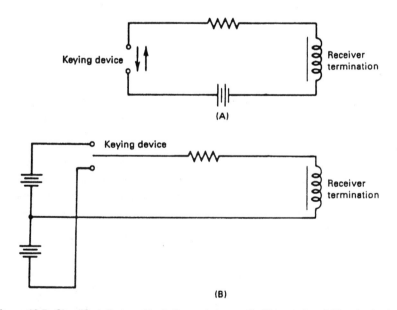

Figure 10.5. Simplified diagram illustrating a dc loop with (A) neutral and (B) polar keying.

6.2 Neutral and Polar DC Transmission Systems

Older telegraph and data systems operated in the *neutral* mode. Nearly all present data transmission systems operated in some form of *polar* mode. The words "neutral" and "polar" describe the manner in which battery is applied to the dc loop. On a "neutral" loop, following the convention of Table 10.1, battery is applied during spacing (0) conditions and is switched off during marking (1). Current therefore flows in the loop when a space is sent and the loop is closed. Marking is indicated on the loop by a condition of no current. Thus we have two conditions for binary transmission, an open loop (no current flowing) and a closed loop (current flowing). Keep in mind that we could reverse this, namely, change the convention and assign marking to a condition of current flowing or closed loop and spacing to a condition of no current or an open loop.* As we mentioned, this is called "changing the sense." Either way, a neutral loop is a dc loop circuit where one binary condition is represented by the presence of voltage and the flow of current, and the other condition is represented by the absence of voltage and current. Figure 10.5A illustrates a neutral loop.

Polar transmission approaches the problem differently. Two battery sources are provided, one "negative" and the other "positive." Following the convention in Table 10.1, during a condition of spacing (binary 0), a positive battery (i.e., positive voltage) is applied to the loop, and a negative battery is applied during marking (binary 1). In a polar loop, current is always flowing. For a mark or binary "1" it flows in one direction, and for a space or binary "0" it flows in the opposite direction. Figure 10.5B shows a simplified polar loop. Notice that the switch used to select the voltage we called a *keying device.* Figure 10.6 shows two electrical waveforms.

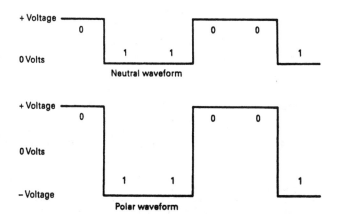

Figure 10.6. Neutral and polar electrical waveforms.

* In fact, this was the older convention, about prior to 1960.

7 BINARY TRANSMISSION AND THE CONCEPT OF TIME

7.1 Introduction

Time and timing are most important factors in digital transmission. For this discussion, consider a binary end instrument sending out in series a continuous run of marks and spaces. Those readers who have some familiarity with the Morse code will recall that the spaces between dots and dashes told the operator where letters ended and where words ended. With the sending device or transmitter delivering a continuous series of characters to the line, each consisting of five, six, seven, eight, or nine elements (bits) per character, a receiving device that starts its print cycle when the transmitter starts sending and subsequently is perfectly in step with the transmitter can be expected to provide good printed copy and few, if any, errors at the receiving end.

It is obvious that when signals are generated by one machine and received by another, the speed of the receiving machine must be the same or very close to that of the transmitting machine. When the receiver is a motor-driven device, timing stability and accuracy are dependent on the accuracy and stability of the speed of rotation of the motors used. Most simple data–telegraph receivers sample at the presumed center of the signal element. It follows, therefore, that whenever a receiving device accumulates timing error of more than 50% of the period of one bit, it will print in error.

The need for some sort of synchronization is illustrated in Figure 10.7. A five-unit code is employed, and three characters transmitted sequentially are shown. Sampling points are shown in Figure 10.7 as vertical arrows. Receiving timing begins when the first pulse is received. If there is a 5% timing difference between the transmitter and receiver, the first sampling at the receiver will be 5% away from the center of the transmitted pulse. At the end of the tenth pulse or signal element the receiver may sample in error. The eleventh signal element will, indeed, be sampled in error, and all subsequent elements will be errors. If the timing error between transmitting machine and receiving machine is 2%, the cumulative error in timing would cause the receiving device to print all characters in error after the 25th bit.

7.2 Asynchronous and Synchronous Transmission

In the earlier days of printing telegraphy, "start–stop" transmission, or asynchronous operation, was developed to overcome the problem of synchronism.

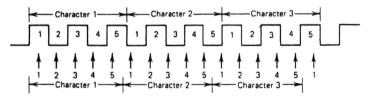

Figure 10.7. Five-unit synchronous bit stream with timing error.

Here timing starts at the beginning of a character and stops at the end. Two signal elements are added to each character to signal the receiving device that a character has begun and ended.

For example, consider a five-element code such as CCITT No. 2 (see Figure 10.1). In the front of a character an element called a "start space" is added, and a stop mark is inserted at the end of each character. To send the letter Y in Figure 10.1, the receiving device starts its timing sequence on the first signal element, which is a space or 0, followed by 10101, which is the code sequence for the character Y, followed by a stop mark, which terminates the timing sequence, as shown in Figure 10.8. In such an operation, timing errors can accumulate only inside each character. Suppose the receiving device is again 5% slower or faster than its transmitting counterpart; now the fifth information element will be no more than 30% displaced in time from the transmitted pulse and well inside the 50% or halfway point for correct sampling to take place.

In start–stop transmission, information signal elements are each of the same duration, which is the duration or pulse width of the start element. The stop element has an indefinite length or pulse width beyond a certain minimum. If a steady series of characters is sent, the stop element is always of the same width or has the same number of unit intervals. Consider the transmission of two Y's, 0101011010101111 → 11111. The start space (0) starts the timing sequence for six additional elements, which are the five code elements in the letter Y and the stop mark. Timing starts again on the mark-to-space transition between the stop mark of the first Y and the start of the second. Sampling is carried out at pulse center for most asynchronous systems. Note that a continuous series of marks is sent at the end of the second Y; thus the signal is a continuation of the stop element or just a continuous mark. It is the mark-to-space transition of the start element that tells the receiving device to start timing a character.

Minimum lengths of stop elements vary. The preceding example above shows a stop element of one-unit interval duration (1 bit). Some are 1.42-unit intervals, others are of 1.5- and 2-unit interval duration. The proper semantics of data–telegraph transmission would describe the code of the previous paragraph as a five-unit start–stop code with a one-unit stop element.

A primary objective in the design of data systems is to minimize errors received or to minimize the error rate. Two of the prime causes of errors are

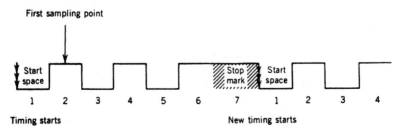

Figure 10.8. Five-unit start–stop stream of bits with a 1.5-unit stop element.

noise and improper timing relationships. With start–stop systems a character begins with a mark-to-space transition at the beginning of the start space. Then 1.5-unit intervals later, the timing causes the receiving device to sample the first information element, which simply is a mark or space decision. The receiver continues to sample at one-bit intervals until the stop mark is received. In start–stop systems the last information bit is most susceptible to cumulative timing errors. Figure 10.8 is an example of a five-unit start–stop bit stream with a 1.5-unit stop element.

Another problem in start–stop systems is the mutilation of the start element. Once this happens, the receiver starts a timing sequence on the next mark-to-space transition it sees and then continues to print in error until, by chance, it cycles back properly on a proper start element.

Synchronous data systems do not have start and stop elements, but consist of a continuous serial stream of information elements or bits such as shown in Figure 10.7. With start–stop systems, timing error could only accumulate inside a character, those five or eight bits of character length. This is not so for synchronous systems. Timing error can accumulate for the entire length of a frame.

With start–stop systems, the receiving device knows when a character starts by the mark-to-space transition at the start space. In a synchronous transmission system, some marker must be provided to tell the receiver when a frame starts. This "marker" is the *unique field*. Every data frame starts with a unique field. A generic data frame is shown in Figure 10.9. View the frame from left to right. The first field is the unique field or flag, and it generally consists of the binary sequence 01111110. A frame always starts with this field and ends with the same field. If one frame follows another contiguously, the unique field ending frame No. 1 is the unique field starting frame No. 2, and so forth. Once a frame knows where it starts, it will know a priori where the following fields begin and end by simple bit/octet counting. However, with some data link protocols, the

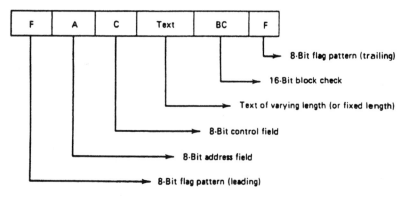

Figure 10.9. A generic data frame. Flag pattern = unique field. Text often called "info" field or information field.

information field is of variable length. In this case, the information field length will appear as a subfield in the control field.

The unique field *aligns* the frame. Suppose the unique field binary sequence (01111110) occurred inside the frame. The receiver, of course, would misalign, and the entire frame would be in error. We don't want our hands tied in the use of a particular binary sequence. A frame should be *transparent*, meaning that we can use any 8-bit sequence we desire, even 01111110. Thus care must be taken with sequences of contiguous 1s. To avoid interpreting that unique field (or flag), *bit stuffing*, sometimes called *zero insertion*, is used.

The rule for bit stuffing is to insert (stuff) a 0 into the data stream of the frame proper after each successive appearance of five 1s. This concept is shown in Figure 10.10. Thus, the frame, after stuffing, never contains more than five consecutive 1s, and the unique field (flag) at the end of the frame is singularly recognizable. At the receiving end of the link, the first 0 after each string of five consecutive 1s is deleted. If, however, a string of five 1s is followed by a 1, the frame is declared to be finished.

Bit stuffing is used for other purposes beyond eliminating flags within the frame. Most data link protocols have an abort capability in which a frame can be aborted by transmitting seven or more 1s in a row. In addition, a link is regarded as idle if 15 or more 1s in a row are received.

Synchronous data transmission systems not only require frame alignment, but must be bit-aligned as well. This was an easy matter on start–stop systems because we could let the receive clock run freely inside the 5 or 8 bits of a character. There is no such freedom with synchronous systems. Suppose we assume a free-running receive clock. However, if there was a timing error of 1% between the transmit and receive clocks, not more than 100 bits could be transmitted until the synchronous receiving device would be off in timing by the duration of 1 bit from the transmitter, and all bits received thereafter would be in error. Even if the timing accuracy of one relative to the other was improved to 0.05%, the correct timing relationship between transmitter and receiver would exist for only the first 2000 bits transmitted. It follows, therefore, that no timing error whatsoever can be permitted to accumulate since anything but absolute accuracy in timing would cause eventual malfunctioning. In practice, the receiver is provided with an accurate clock that is corrected by small adjustments based on the transitions of the received bit stream, as explained in Section 7.3.

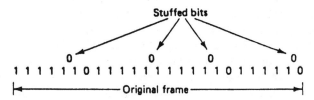

Figure 10.10. The concept of bit stuffing (zero insertion). A 0 bit is stuffed after each consecutive five 1s inside the original frame. The unique field (flag) has no stuffing.

7.3 Timing

All currently used data transmission systems are synchronized in phase and symbol rate in some manner. Start–stop synchronization has already been discussed. All fully synchronous transmission systems have timing generators or clocks to maintain stability. The transmitting device and its companion receiver at the far end of the circuit must maintain a timing system. In normal practice, the transmitter is the master clock of the system. The receiver also has a clock that in every case is corrected by some means to its transmitter's master clock equivalent at the far end.

Another important timing factor is the time it takes a signal to travel from the transmitter to the receiver. This is called *propagation time*. With velocities of propagation as low as 20,000 miles/s, consider a circuit 200 miles in length. The propagation time would then be 200/20,000 s or 10 ms. Ten milliseconds is the time duration of 1 bit at a data rate of 100 bit/s; thus the receiver in this case must delay its clock by 10 ms to be in step with its incoming signal. Temperature and other variations in the medium may also affect this delay, as well as variations in the transmitter master clock.

There are basically three methods of overcoming these problems. One is to provide a separate synchronizing circuit to slave the receiver to the transmitter's master clock. However, this wastes bandwidth by expending a voice channel or subcarrier just for timing. A second method, which was quite widely used until several years ago, was to add a special synchronizing pulse for groupings of information pulses, usually for each character. This method was similar to start–stop synchronization and lost its appeal largely because of the wasted information capacity for synchronizing. The most prevalent system in use today is one that uses transition timing, where the receiving device is automatically adjusted to the signaling rate of the transmitter by sampling the transitions of the incoming pulses. This type of timing offers many advantages, particularly automatic compensation for variations in propagation time. With this type of synchronization the receiver determines the average repetition rate and phase of the incoming signal transition and adjusts its own clock accordingly by means of a phase-locked loop.

In digital transmission the concept of a transition is very important. The transition is what really carries the information. In binary systems the space-to-mark and mark-to-space transitions (or lack of transitions) placed in a time reference contain the information. In sophisticated systems, decision circuits regenerate and retime the pulses on the occurrence of a transition. Unlike decision circuits, timing circuits that reshape a pulse when a transition takes place must have a memory in case a long series of marks or spaces is received. Although such periods have no transitions, they carry meaningful information. Likewise, the memory must maintain timing for reasonable periods in case of circuit outage. Note that synchronism pertains to both frequency and phase and that the usual error in high-stability systems is a phase error (i.e., the leading edges of the received pulses are slightly advanced or retarded from the equivalent clock pulses of the receiving device). Once synchronized, high-stability systems need only a small amount of correction in timing (phase). Modem internal timing systems may have a long-term stability of 1×10^{-8} or better at both the transmitter and receiver. At 2400 bits/s, before a significant timing error can build up, the accumulated time difference between transmitter and receiver must exceed approximately 2×10^{-4} s. Whenever the circuit of a synchronized transmitter and receiver is shut down, their clocks must differ by at least 2×10^{-4} s before significant errors take place once the clocks start back up again. This means that the leading edge of the receiver-clock equivalent timing pulse is 2×10^{-4} in advance or retarded from the leading edge of the pulse received from the distant end. Often an idling signal is sent on synchronous data circuits during periods of no traffic to maintain the timing. Some high-stability systems need resynchronization only once a day.

Note that thus far in our discussion we have considered dedicated data circuits only. With switched (dial-up) synchronous circuits, the following problems exist:

- No two master clocks are in perfect phase synchronization.
- The propagation time on any two paths may not be the same.

Thus such circuits will need a time interval for synchronization for each call setup before traffic can be passed.

To summarize, synchronous data systems use high-stability clocks and the clock at the receiving device is undergoing constant but minuscule corrections to maintain an in-step condition with the received pulse train from the distant transmitter, which is accomplished by responding to mark-to-space and space-to-mark transitions. The important considerations of digital network timing were also discussed in Chapter 9.

7.4 Distortion

It has been shown that the key factor in data transmission is timing. Although the signal must be either a mark or space, this alone is not sufficient. The marks and spaces (or 1s and 0s) must be in a meaningful sequence based on a time reference.

In the broadest sense, distortion may be defined as any deviation of a signal in any parameter, such as time, amplitude, or wave shape, from that of the ideal signal. For binary data transmission, distortion is defined as a displacement in time of a signal transition from the time that the receiver expects to be correct. In other words, the receiving device must make a decision as to whether a received signal element is a mark or a space. It makes the decision during the sampling interval, which is usually at the center of where the received pulse or bit should be; thus it is necessary for the transitions to occur between sampling times and preferably halfway between them. Any displacement of the transition instants is called "distortion." The degree of distortion suffered by a data signal as it traverses the transmission medium is a major contributor in determining the error rate that can be realized.

7.5 Bits, Bauds, and Symbols

There is much confusion among professionals in the telecommunication industry over terminology, especially in differentiating bits, bauds, and symbols. The bit, a binary digit, has been defined previously.

The baud is a unit of transmission rate or modulation rate. It is a measure of transitions per second. A transition is a change of state. In *binary* systems, bauds and bits per second (bits/s) are synonymous. In higher-level systems, typically *m*-ary systems, bits and bauds have different meanings. For example, we will be talking about a type of modulation called QPSK. In this case, every transition carries two bits. Thus the modulation rate in bauds is half the bit rate.

The industry often uses symbols per second and bauds interchangeably. It would be preferable, in our opinion, to use "symbols" for the output of a coder or other conditioning device. For the case of a channel coder (or encoder), bits go in and symbols come out. There are more symbols per second in the output than bits per second in the input. They differ by the coding rate. For example, a 1/2 rate coder (used in FEC) may have 4800 bits/s at the input and then would have 9600 symbols per second at the output.

7.5.1 The Period of a Bit, Symbol, or Baud. The period of a bit is the time duration of a bit pulse. When we use NRZ coding (discussed in Section 7.6), the period of a bit, baud, or symbol is simply 1/(bit rate) or 1/(symbol rate). For example, if we are transmitting 2400 bits/s, what is the period of a bit? It is 1/2400 or 416 μs; for 19.2 kbits/s it is $1/19,200 = 52.08$ μs and for 1.544 Mbits/s it is 647 ns.

7.6 Digital Data Waveforms

Digital symbols may be represented in many different ways by electrical signals to facilitate data transmission. All these methods for representing (or coding) digital symbols assign electrical parameter values to the digital symbols. In binary coding, of course, these digital symbols are restricted to two states, space (0) and mark (1). The electrical parameters used to code digital signals are levels (or amplitudes), transitions between different levels, phases (normally $0°$ and $180°$ for binary coding), pulse duration, and frequencies or a combination of these parameters. There is a variety of coding techniques for different areas of application, and no particular technique has been found to be optimum for all applications, considering such factors as implementing the coding technique in hardware, type of transmission technique employed, decoding methods at the data sink or receiver, and timing and synchronization requirements.

In this section we discuss several basic concepts of *electrical coding* of binary signals. In the discussion, reference is made to Figure 10.11, which graphically illustrates several line coding techniques.

Figure 10.11A shows what is still called by many today "neutral transmission." This was the principal method of transmitting telegraph signals until about 1960. In many parts of the world, neutral transmission is still used. First, this waveform

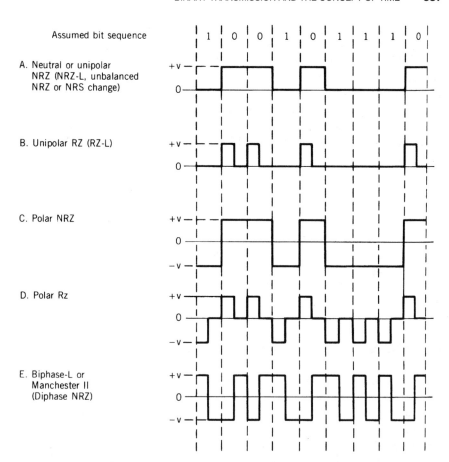

Figure 10.11. Digital data transmission waveforms.

is a non-return-to-zero (NRZ) format in its simplest form. "Non-return-to-zero" basically means that if a string of 1s (marks) is transmitted, the signal remains in the mark state with no transitions. Likewise, if a string of 0s is transmitted, there are no transitions and the signal remains in the 0 state until a 1 is transmitted. As we can now see, with NRZ transmission, we can transmit information without transitions.

Figures 10.11B and 10.11D show the typical "return-to-zero" (RZ) waveform, where, when a continuous string of marks (or spaces) is transmitted, the signal element (i.e., amplitude) returns to the zero voltage condition at each element or bit. Obviously, RZ transmission is much richer in transitions than NRZ.

In Section 6.2 of this chapter we discussed neutral and polar dc transmission systems. Figure 10.11A shows a typical neutral waveform where the two state conditions are 0 V for the mark or 1 condition and some positive voltage for the space or 0 condition. On the other hand, with polar transmission, as shown in Figures 10.11C and 10.11D, a positive voltage represents a space and a negative voltage, a mark. With NRZ transmission, the pulse width is the same duration as the duration of a unit interval or bit. Not so with RZ transmission, where the pulse width is less than the duration of a unit interval (i.e., signal element). This is because we have to allow time for the pulse to return to the zero voltage condition.

Bi-phase or Manchester coding (Figure 10.11E) is a code format that is being used ever more widely on digital systems such as wire pair and fiber optics. Here the binary information is carried in the transition. By convention a logic 0 is defined as a positive-going transition while a logic 1 is defined as a negative-going transition. It should be noted that Manchester coding has a signal transition in the middle of each unit interval (bit or signal element). Manchester coding is a form of phase coding.

The reader should be cognizant of and be able to differentiate between two sets or ways of classifying binary digital waveforms. The first set is *neutral* and *polar*. The second set is *NRZ* and *RZ*. Manchester coding is still another way to represent binary digital data where the transition takes place in the middle of the unit interval. In Chapter 8 one more class of waveform was introduced: alternate mark inversion (AMI).

8   DATA INTERFACE – THE PHYSICAL LAYER

When we wish to transmit data over a conventional analog network, the electrical representation of the data signal is essentially direct current. As such it is incompatible with that network that accepts information channels in the band 300 to 3400 Hz.* A data modem is a device that brings about this compatibility. It translates the electrical data signal into a modulated frequency tone in the range of 300 to 3400 Hz, often 1800 Hz. Also, more often than not, the digital network (described in Chapters 8 and 9) extensions are analog and require the same type(s) of modem. If the digital network extends to the user's premise, a digital conditioning device (CSU/DSU) is required for bit rate compatibility.

For this discussion of data interface, we will call the modem or digital conditioning device *data communication equipment (DCE)*. This equipment has two interfaces, one on each side as shown in Figure 10.12. The first, which is discussed

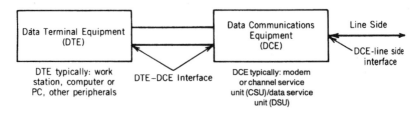

Figure 10.12. Data circuit interfaces, physical layer.

---

* This is the traditional analog voice channel.

in this section, is on the user side, which is called *data terminal equipment (DTE)*, and the applicable interface is the DTE–DCE interface. The second interface is on the line side, which is covered in Section 9. It should be noted that the DTE–DCE interface is well-defined.

The most well-known DTE–DCE standard was developed by the (U.S.) Electronics Industries Association (EIA) and is called EIA-232F. It is essentially equivalent to international standards covered by CCITT Recs. V.24 and V.28 and ISO IS2110.

EIA-232F and most of the other standards discussed are applicable to the DTE–DCE interface employing serial binary data interchange. It defines signal characteristics, mechanical interface characteristics, and functional descriptions of the interchange circuits. EIA-232F is applicable for data transmission rates up to 20,000 bits/s and for synchronous/asynchronous serial binary data systems.

Section 2.1.3 is quoted from EIA-232F. It is crucial to the understanding of signal state conventional and level:

> For data interchange circuits, the signal shall be considered in the marking condition when the voltage (V_1) on the interchange circuit, measured at the interface point [Figure 10.13], is more negative than minus three volts with respect to circuit AB (signal ground), The signal shall be considered in the spacing condition when the voltage V_1 is more positive than plus three volts with respect to circuit AB. . . . The region between plus three volts and minus three volts is defined as the transition region. The signal state is not uniquely defined when the voltage (V_1) is in this transition region.

During the transmission of data, the marking condition is used to denote the binary state *ONE* and the spacing condition is used to denote the binary state *ZERO*.

Figure 10.13 shows the interchange equivalent circuit.

Besides EIA-232, there are many other interface standards issued by EIA, CCITT, U.S. federal standards, U.S. military standards, and ISO. Each define the DTE–DCE interface. Several of the more current standards are briefly described below.

EIA-530* [6] is a comparatively recent standard developed by the EIA. It provides for all data rates below 2.1 Mbits/s and it is intended for all applications requiring a balanced electrical interface. It can also be used for unbalanced operation.

Let us digress for a moment. An unbalanced electrical interface is where one of the signal leads is grounded; for a balanced electrical interface, no ground is used.

EIA-530 applies for both synchronous and nonsynchronous (i.e., start–stop) operation. It uses a standard 25-pin connector; alternatively, it can use a 26-pin connector. A list of interchange circuits showing circuit mnemonic, circuit name, circuit direction, and circuit type is presented in Table 10.3.

---

* More properly called ANSI/EIA/TIA-530-A.

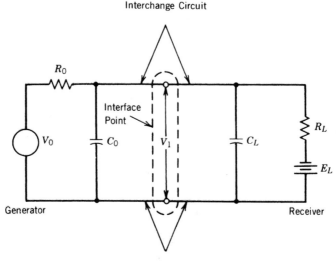

Circuit AB, Signal Ground

Figure 10.13. Interchange equivalent circuit. Courtesy of Telecommunication Industry Association (TIA). From EIA/TIA-232F, July 1996 [5].

V_0 is the open-circuit generator voltage.
R_0 is the generator internal dc resistance.
C_0 is the total effective capacitance associated with the generator, measured at the interface point and including any cable to the interface point.
V_1 is the voltage at the interface point.
C_L is the total effective capacitance associated with the receiver, measured at the interface point and including any cable to the interface point.
R_L is the receiver load dc resistance.
E_L is the open-circuit receiver voltage (bias).

The electrical characteristics of EIA-530 are described in EIA/TIA-422 and EIA/TIA-423 [7, 8].

EIA-422 deals with a balanced electrical interface, and EIA-423 deals with an unbalanced electrical interface. The state transition region for these two standards is between +2 V and −2 V at the generator side. The sense is the same as that for EIA-232.

The applicable U.S. military standard is MIL-STD-114. This provides a potpourri of EIA-232 and EIA-422/423 parameters. One significant difference reflected in MIL-STD-188-114 [9] is that the receiver has a balanced input even though it is used with an unbalanced generator. The choice of a balanced receiver was done deliberately for the following reasons:

1. Noise immunity and reducing problems of ground potential differences between generator and receiver(s).
2. Convenience of inverting mark and space signaling sense.
3. Uniformity of receiver design for economic advantages in mass production.

TABLE 10.3   EIA-530 Interchange Circuits

Circuit Mnemonic	CCITT Number	Circuit Name	Circuit Direction	Circuit Type
AB	102	Signal Common		Common
AC	102B	Signal Common		
BA	103	Transmitted Data	To DCE	Data
BB	104	Received Data	From DCE	
CA	105	Request to Send	To DCE	
CB	106	Clear to Send	From DCE	
CF	109	Received Line Signal Detector	From DCE	
CJ	133	Ready for Receiving	To DCE	
CE	125	Ring Indicator	From DCE	Control
CC	107	DCE Ready	From DCE	
CD	108/1, /2	DTE Ready	To DCE	
DA	113	Transmit Signal Element Timing (DTE Source)	To DCE	
DB	114	Transmit Signal Element Timing (DCE Source)	From DCE	Timing
DD	115	Receiver Signal Element Timing (DCE Source)	From DCE	
LL	141	Local Loopback	To DCE	
RL	140	Remote Loopback	To DCE	
TM	142	Test Mode	From DCE	

Source: ANSI/EIA/TIA-530-A, June 1992 [6].

CCITT has issued a number of recommendations for the DTE–DCE interface. The equivalent of EIA-232 is CCITT Rec. V.24 [12]. The following are other pertinent CCITT Recommendations:

Rec. V.10, *Electrical Characteristics for Unbalanced Double-Current Interchange Circuits for General Use with Integrated Circuit Equipment in the Field of Data Communications.* The term *double-current* is synonymous with polar transmission discussed in Sections 6.2 and 7.6 [10].

Rec. V.11, *Electrical Characteristics for Balanced Double-Current Interchange Circuits for General Use with Integrated Circuit Equipment in the Field of Data Communications* [11].

Rec. V.28, *Electrical Characteristics for Unbalanced Double-Current Interchange Circuits* [13].

Rec. V.31, V.31 bis, *Electrical Characteristics for Single-Current Interchange Circuits Controlled by Contact Closure.* Single-current refers to *neutral transmission* [14].

8.1   TIA/EIA-644 Low-Voltage Differential Signaling (LVDS)

LVDS has been designed for point-to-point data circuits where the signaling rates exceed 100 Mbits/s. LVDS power supplies are available at 5 V, 3.3 V, and 2.5 V.

Signaling is carried out with voltage swings to about 300 mV. Noise margins and noise immunity are increased even further by the use of differential data transmission. Such differential signals are immune to common mode noise, which is the primary source of system noise. LVDS changes signal voltage levels without a fast slew rate. The slow transition rate decreases the radiated field strength. The rule-of-thumb design guide states that pulse rise and fall times should be no more than two-thirds of bit width; then signals with 333-ps transitions can operate at data rate greater than 1 Gbits/s with margin.

LVDS has excellent EMC characteristics. Its generated EMI is of a low level because of its low voltage swings, slow edge rates, odd-mode differential signals, and minimal I_{cc} spikes from constant current drivers. LVDS susceptibility is minimal because of its differential signal paths. Balanced differential lines have equal but opposite currents, called odd-mode signals. The differential signals also have the advantage of tolerating interference from outside sources such as inductive radiation from electric motors or crosstalk from nearby transmission lines.

Source: International Engineering Consortium (IEC) tutorial www.iec.org and TIA/EIA-644-1995.

9 DIGITAL TRANSMISSION ON AN ANALOG CHANNEL

9.1 Introduction

Although digital connectivity in telecommunications is more and more reaching end-to-end, we are of the opinion that the reader should have a good grasp of transmitting digital data on an analog facility. Usually these facilities have been designed primarily for voice traffic and are not amenable to the transmission of dc binary digit or bit streams of data. To permit the transmission of data over voice facilities (i.e., the analog telephone network), it is necessary to convert the dc data into a signal within the voice-frequency range. The equipment that performs the necessary conversion to the signal is generally called a *modem*, an acronym for *mo*dulator–*dem*odulator.

9.2 Modulation–Demodulation Schemes

A modem modulates and demodulates a carrier signal with digital data signals. The types of modulation used by present-day modems may be one or a combination of the following:

Amplitude modulation, double sideband (DSB).
Amplitude modulation, vestigial sideband (VSB).
Frequency shift modulation, commonly called frequency shift keying (FSK).
Phase shift modulation, commonly called phase shift keying (PSK).

9.2.1 Amplitude Modulation: Double Sideband. With the double-sideband (DSB) modulation technique, binary states are represented by the presence or

absence of an audio tone or carrier. More often it is referred to as "on–off telegraphy." For data rates up to 1200 bits/s, one such system uses a carrier frequency centered at 1600 Hz. For binary transmission, amplitude modulation has significant disadvantages, which include (1) susceptibility to sudden gain change and (2) inefficiency in modulation and spectrum utilization, particularly at higher modulation rates (see CCITT Rec. R.70).

9.2.2 Amplitude Modulation: Vestigial Sideband.

An improvement in the amplitude modulation double-sideband (DSB) technique results from the removal of one of the information-carrying sidebands. Since the essential information is present in each of the sidebands, there is no loss of content in the process. The carrier frequency must be preserved to recover the dc component of the information envelope. Therefore digital systems of this type use VSB modulation in which one sideband, a portion of the carrier, and a "vestige" of the other sideband are retained. This is accomplished by producing a DSB signal and filtering out the unwanted sideband components. As a result, the signal takes only about 75% of the bandwidth required for a DSB system. Typical VSB data modems are operable up to 2400 bits/s in a telephone channel. Data rates up to 4800 bits/s are achieved using multilevel (*M*-ary) techniques. The carrier frequency is usually located between 2200 Hz and 2700 Hz.

9.2.3 Frequency Shift Modulation.

Many data transmission systems utilize frequency shift modulation (FSK*). The two binary states are represented by two different frequencies and are detected by using two frequency-tuned sections, one tuned to each of the 2-bit frequencies. The demodulated signal is then integrated over the duration of 1 bit, and a binary decision is based on the result.

Digital transmission using FSK modulation has the following advantages: (1) The implementation is not much more complex than an AM system; and (2) since the received signals can be amplified and limited at the receiver, a simple limiting amplifier can be used, whereas the AM system requires sophisticated automatic gain control for operation over a wide level range. Another advantage is that FSK can show a 3- to 4-dB improvement over AM in most types of noise environment, particularly at distortion threshold (i.e., at the point where the distortion is such that good printing is about to cease). As the frequency shift becomes greater (i.e., a greater frequency separation between the mark and space frequencies), the advantage over AM improves in a noisy environment.

Another advantage of FSK is its immunity from the effects of nonselective level variations, even when they occur extremely rapidly. Thus a major application is on worldwide high-frequency radio transmission where rapid fades are a common occurrence. In the United States, FSK has nearly universal application for the transmission of data at the lower data rates (i.e., ≤1200 bits/s).

9.2.4 Phase Shift Modulation.

For systems using higher data rates, phase modulation becomes more attractive. Various forms are used, such as two-phase,

---

* FSK meaning frequency shift keying. "Keying" derives from old-time telegraph transmission.

relative phase, and quadrature phase. A two-phase system uses one phase of the carrier frequency for one binary state and the other phase for the other binary state. The two phases are ideally 180° apart and are detected by a synchronous detector using a reference signal at the receiver that is of known phase with respect to the incoming signal. This known signal operates at the same frequency as the incoming signal carrier and is arranged to be in phase with one of the binary signals. In the relative-phase system a binary 1 is represented by sending a signal burst of the same phase as that of the previous signal burst sent. A binary 0 is represented by a signal burst of a phase opposite to that of the previous signal transmitted. The signals are demodulated at the receiver by integrating and storing each signal burst of 1-bit period for comparison in phase with the next signal burst. In the quadrature-phase system (QPSK), two binary channels (2 bits) are phase multiplexed onto one tone by placing them in phase quadrature, as shown in the following sketch. An extension of this technique places two binary channels on each of several tones spaced across the voice channel of a typical telephone circuit. Almost all modems operating at 2400 bits/s use QPSK. The baud rate in the case is one-half of the bit rate.

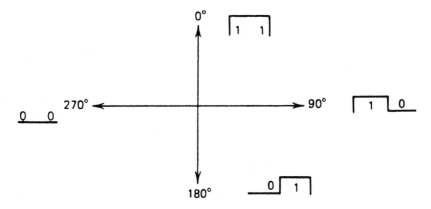

Some of the advantages of phase modulation are as follows:

1. All available power is utilized for intelligence conveyance.
2. The demodulation scheme has good noise-rejection capability.
3. The system yields a smaller noise bandwidth.

A disadvantage of such a system is the complexity of equipment required, compared to FSK systems. (For further information, see References 34–39.)

9.3 Critical Parameters

The effect of the various telephone-circuit impairments on the capability of a circuit to transmit data is a most important consideration. These impairments are

- Phase distortion
- Amplitude distortion
- Noise

In Chapter 5 there was a detailed discussion of these parameters.

The following is a rule of thumb for phase distortion (group delay):

The group delay distortion (envelope delay distortion) should be kept under the period of the bit (or baud) of the bit stream of interest. For the analog voice channel, the frequency band of interest is often stated as 1000–2600 Hz. If the circuit is carrying 1000 bits/s (binary), then the envelope delay distortion should be under a millisecond.

Envelope delay distortion is additive on circuits in tandem, end-to-end. Envelope delay is minimum in band center (around 1700 or 1800 Hz) and increases toward band edge. The increase is nonlinear, often parabolic. For this reason the modulated frequency of a modem is usually at 1700 or 1800 Hz. We generally consider delay distortion as the greatest bottleneck toward bit rate.

Amplitude distortion (frequency response) is less of a headache for data transmission than delay distortion. It also is additive on circuits end-to-end [15].

Noise is a major impairment to a data bit stream, whether in the analog or digital environment. We consider two types of noise for this discussion:

1. Thermal noise
2. Impulse noise

Both were treated in some fair detail in Chapter 5, Section 2.3. In this section we approach the problems with some different perspective, namely, how these two types of noise affect the BER performance of a data bit stream [18,19 and 20].

Thermal Noise. Thermal noise is our most common, "garden variety" type of noise. It is often called "resistance noise," "white noise," or "Johnson noise," and it is of a Gaussian nature or completely random. Any system or circuit operating at a temperature above absolute zero inherently will display thermal noise. The noise is caused by the random motions of discrete electrons in the condition path.

Random noise, when measured by a typical transmission measurement set, appears to have a relatively constant value. However, the instantaneous value of the noise fluctuates over a wide range of amplitude values. If the instantaneous noise voltage is of the same magnitude as the received signal, the receiving detection equipment may yield an improper interpretation of the received signal, and an error or errors will occur. Thus we need some way of predicting the behavior of data transmission in the presence of noise. We can expect this noise on either an analog channel or a digital channel. Being that thermal noise is of a Gaussian nature, we can indeed make some statistical predictions. From the probability distribution curve for Gaussian noise shown in Figure 10.14, we can examine noise voltage peaks. For example, there is a probability of 1×10^{-5} that noise peaks will have an amplitude of 12.5 dB above the rms (root mean square)

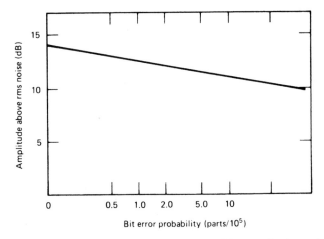

Figure 10.14. Probability of bit error in Gaussian noise, binary polar transmission with a Nyquist bandwidth.

voltage level value. Hence, if we wish to ensure an error rate in a particular system using binary polar modulation, the rms noise should be at least 12.5 dB below the signal level [17, p. 114]. This simple analysis is only valid for the type of modulation used (i.e., binary polar baseband modulation), assuming that no other degrading factors are present and that a cosine-shaped receiving filter is used. If we were to interject distortion such as EDD into the system, we could translate the degradation into an equivalent signal-to-noise ratio improvement necessary to restore the desired error rate. For example, if the delay distortion were the equivalent of one pulse width, the signal-to-noise ratio improvement required for the same error rate would now be 17.5 dB. (For Standardized signal level, consult references 16 and 21.)

Impulse Noise. Unlike random noise, which is measured by its rms value when we measure level, impulse noise is measured by the number of "hits" or "spikes" per interval of time above a certain threshold. In other words, it is a measurement of the recurrence rate of noise peaks over a specified level. The word "rate" should not mislead the reader. The recurrence is not uniform per unit time, as the word "rate" may indicate, but we can consider a sampling and convert it to an average.

However, impulse noise is not usually a limiting impairment for voice transmission. Not so in the case of data transmission. It will cause random hits deteriorating BER on a data bit stream.

Bellcore (Telcordia) [33] is quoted:

Impulse noise is any burst of noise, usually less than 5 but up to 10 ms in duration, that produces a voltage exceeding the rms (root means square) noise voltage (i.e., the mean noise as measured with an AT&T Technologies 3A-type noise measuring set using C-message weighting or its equivalent) by a given magnitude. The impulse noise threshold is nominally 12 dB above the rms value for 3-kHz bandwidth systems, but in some systems, particularly microwave systems, it may increase to 16 dB. Impulse noise objectives and requirements are normally stated in terms of threshold noise setting and the number, or "count," of times impulse noise exceeds the threshold during a given time interval.

The performance objective is that there should be no more than five counts in 5 minutes on at least 50 percent of the switching system connections of each connection type. The objectives apply to a working (installed EO*) environment and over the life of the office.[†] These objectives are associated with the following noise thresholds:

- 47 dBrnC0 for analog and digital offices on connections between voice frequency interfaces including remote LI[‡]-to-remote LI (intra-RSU[§]), but excluding remote LI to host.
- 60 dBrnC0 with a −23 dBm0 holding tone of 1004 Hz and 66 dBrnC0 with a −13 dBm0 holding tone at 1004 Hz for remote LI-to-host voice frequency connections. The requirement is that 99 percent or more of the connections of each connection type between voice frequency interfaces (excluding the carrier facility used in RSU-host connections) should have zero counts in 5 minutes above 54 dBrnC0 for any possible combination of laboratory-simulated switching system operations. It is desirable that there be no counts in 5 minutes above 47 dBrnC0 under these same conditions.

CCITT Rec. Q.45 states that "in any four-wire international exchange the busy hour impulsive noise counts should not exceed 5 counts in 5 minutes at a threshold level of −35 dBm0."

Remember that random noise has a Gaussian distribution and will produce peaks at 12.5 dB over the rms value (unweighted) 0.001% of the time on a data-bit stream for an equivalent error rate of 1×10^{-5}. It should be noted that some references use 12 dB, some use 12.5 dB, and others use 13 dB. The 12.5 dB above the rms random noise floor should establish the impulse noise threshold for measurement purposes. We should assume that in a well-designated data transmission system traversing the telephone network, the signal-to-noise ratio of the data signal will be well in excess of 12.5 dB. Thus impulse noise may well be the major contributor to degradation of the error rate.

When an unduly high error rate has been traced to impulse noise, there are some methods for improving conditions. Noisy areas may be bypassed, repeaters may be added near the noise source to improve signal-to-impulse-noise ratio, or in special cases pulse smearing techniques may be used. This latter approach uses two delay-distortion networks that complement each other such that the net delay distortion is zero. By installing the networks at opposite ends of the circuit, impulse noise passes through only one network[¶] and hence is smeared

---

* EO = end office, meaning a local switch.
[†] "Office" in North America means a switching center.
[‡] LI = line interface.
[§] RSU = remote switching unit.
[¶] This assumes that impulse noise enters the circuit at some point beyond the first network.

because of the delay distortion. The signal is unaffected because it passes through both networks.

Signal-to-noise ratio may be traded for the implementation of forward-acting error correction (FEC). This trade-off may be economically feasible or even mandatory, such as in certain digital satellite circuits, to reduce power output of a transmitter or reduce level out of a modem. Reducing power by 3 dB (Figure 10.14) on a circuit displaying a bit error rate of 1×10^{-5} would deteriorate error rate to some 5×10^{-2}. The error rate could be recovered by selecting the proper method of FEC, decoding algorithm and possibly implementing soft mark–space decisions. Soft decision decoding can improve error performance by an equivalent 2 dB. For additional discussion of FEC, see Section 5 of this chapter.

9.4 Channel Capacity

A leased or switched voice channel represents a financial investment. Therefore one goal of the system engineer is to derive as much benefit as possible from the money invested. For the case of digital transmission, this is done by maximizing the information transfer across the system. This section discusses how much information in bits can be transmitted, relating information to bandwidth, signal-to-noise ratio, and error rate. These matters are discussed empirically in Section 9.5.

First, looking at very basic information theory, Shannon stated in his classic paper [23] that if input information rate to a band-limited channel is less than C (bits/s), a code exists for which the error rate approaches zero as the message length becomes infinite. Conversely, if the input rate exceeds C, the error rate cannot be reduced below some finite positive number.

The usual voice channel is approximated by a Gaussian band-limited channel (GBLC) with additive Gaussian noise. For such a channel, consider a signal wave of mean power of S watts applied at the input of an ideal low-pass filter that has a bandwidth of W (Hz) and contains an internal source of mean Gaussian noise with a mean power of N watts uniformly distributed over the passband. The capacity in bits per second is given by

$$C = W \log_2 \left(1 + \frac{S}{N}\right)$$

Applying Shannon's "capacity" formula to an ordinary voice channel (GBLC) of bandwidth (W) 3000 Hz and a signal-to-noise (S/N) ratio of 1023, the capacity of the channel is 30,000 bits/s. (Remember that bits per second and bauds are interchangeable in binary systems.) Neither S/N nor W is an unreasonable value. Seldom, however, can we achieve a modulation rate greater than 3000 bauds. The big question in advanced design is how to increase the data rate and keep the error rate reasonable.

One important item not accounted for in Shannon's formula is intersymbol interference (ISI). A major problem of a pulse in a band-limited channel is that

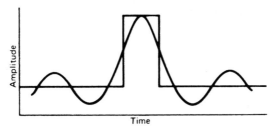

Figure 10.15. Pulse response through a Gaussian band-limited channel (GBLC).

the pulse tends not to die out immediately, and a subsequent pulse is interfered with by "tails" from the preceding pulse. See Figure 10.15.

Nyquist [24] provided another approach to the data rate problem, this time using intersymbol interference (the tails in Figure 10.15) as a limit. This resulted in the definition of the so-called Nyquist rate $= 2\ W$ symbols/s, where W is the bandwidth (Hz) of a band-limited channel. In binary transmission we are limited to 2W bits/s, where a symbol is 1 bit. If we let $W = 3000$ Hz, the maximum data rate attainable is 6000 bits/s. Some refer to this as "the Nyquist 2-bit rule."

The key here is that we have restricted ourselves to binary transmission and are limited to $2W$ bits/s no matter how much we increase the signal-to-noise ratio. The Shannon GBLC equation indicates that we should be able to increase the information rate indefinitely by increasing the signal-to-noise ratio. The way to attain a higher C value is to replace the binary transmission system with a multilevel system, often termed an M-ary transmission system, with $M > 2$. An M-ary channel can pass $2W \log_2 M$ bits/s with an acceptable error rate. This is done at the expense of signal-to-noise ratio. As M increases (as the number of levels increases), so must S/N increase to maintain a fixed error rate [17].

Table 10.4 shows some typical (possibly historical) modem bit rates, corresponding modulation rates in bauds, and the required bandwidth in hertz.

9.5 Equalization

Of the critical circuit parameters mentioned in Section 9.3, two that have severely deleterious effects on data transmission can be reduced to tolerable limits by *equalization*. These two are amplitude–frequency response (amplitude distortion) and EDD (delay distortion).

The most common method of performing equalization is the use of several networks in tandem. Such networks tend to flatten response and, in the case of amplitude response, add attenuation increasingly toward channel center and less toward its edges. The overall effect is one of making the amplitude response flatter. The delay equalizer operates in a similar manner. Delay increases toward channel edges parabolically from the center. To compensate, delay is added in the center much like an inverted parabola, with less and less delay added as the band edge is approached. Thus the delay response is flattened at some small cost to absolute delay, which has no effect in most data systems. However, care

TABLE 10.4   Medium- and High-Rate Modems for a Telephone Voice Channel

Data Rate (bit/s)	Modulation Rate (Bauds)	Modulation	Bits per Hertz[a]	Bandwidth Required (Hz)[a]
1. 2400 Synchronous (e.g., Rec. V.26)	1200	Differential four-phase	2	1200
2. 4800 Synchronous (e.g., Rec. V.27)	1600	Differential eight-phase	3	1600
3. 3600 Synchronous	1200	Differential four-phase, two-level (combined PSK-AM)	3	1200
4. 2400 Synchronous	800	Differential eight-phase	3	800
5. 9600 Synchronous (e.g., Rec. V.29)	2400	Differential four-phase, two-level	4	2400[b]
6. 14,400 Synchronous (e.g., Rec. V.33)	2400	Differential four-phase, six-level, trellis coding	6	2400

[a] Theoretical values.
[b] Uses automatic equalizer.

must be taken with the effect of a delay equalizer on an amplitude equalizer and, conversely, of an amplitude equalizer on the delay equalizer. Their design and adjustment must be such that the flattening of the channel for one parameter does not entirely distort the channel for the other.

Another type of equalizer is the transversal type of filter, which is useful where it is necessary to select among, or to adjust, several attenuation (amplitude) and phase characteristics. The basis of the filter is a tapped delay line to which the input is presented. The output is taken from a summing network that adds or sums the outputs of the taps. Such a filter is adjusted to the desired response (equalization of both phase and amplitude) by adjusting the contributions for each tap.

If the characteristics of the line are known, one could equalize using predistortion of the output signal of the data set. Some equalizers use a shift register and a summing network. If the equalization needs to be varied, a feedback circuit would be required from the receiver to the transmitter to control the shift register. This type of dynamic predistortion is practical for binary transmission only.

A major drawback of all the equalizers discussed (with the exception of the latter one with the feedback circuit) is that they are useful only on dedicated or leased circuits where the circuit characteristics are known and remain fixed. Obviously, a switched circuit would require a variable automatic equalizer, or conditioning would be required on every circuit in the switched system that would be transmitting data.

Circuits are usually equalized on the receiving end. This is called *post-equalization*. An equalizer must be balanced and must present the proper impedance to the line. Administrations may choose to condition trunks and attempt to eliminate the need to equalize station lines; the economy of considerably fewer equalizers is obvious. In addition, each circuit that would possibly carry high-speed data in the system would have to be equalized, and the equalization must be sufficient for any possible combination to meet the overall requirements. If equalization requirements become greater (i.e., parameters more stringent), the maximum number of circuits (trunks) in tandem may have to be restricted still further (i.e., <12).

Equalization to meet amplitude–frequency response requirements is less exacting in the overall system than is envelope delay.* Equalization for envelope delay and its associated measurements are time consuming and expensive. In general, envelope delay is arithmetically cumulative. If there is a requirement of overall envelope delay distortion of 1 ms for a circuit between 1000 Hz and 2600 Hz, then for 3 links in tandem, each link must be better than 333 μs between the same frequency limits. For 4 links in tandem, each link would have to be at least 250 μs. In practice, accumulation of delay distortion is not entirely arithmetical, because it results in a reduction of requirements by about 10%. Delay distortion tends to be inversely proportional to the velocity of propagation. Loaded cables display greater delay distortion than do nonloaded cables. Likewise, with sharp filters a greater delay is experienced for frequencies approaching band edge than for filters with a more gradual cutoff.

Automatic equalization for both amplitude and delay are effective, particularly for switched data systems. Such devices are self-adaptive and require a short adaptation period after switching, on the order of <1 s [34]. This can be carried out during synchronization. Not only is the modem clock being "averaged" for the new circuit on transmission of a synchronous idle signal, but the self-adaptive equalizers adjust for optimum equalization as well. The major drawback of adaptive equalizers is cost.

Equalization, especially automatic equalization, is more in the realm of analog data modems. As we move into the all-digital network, equalization becomes a secondary issue or no issue at all.

9.6   Data Transmission on the Digital Network

9.6.1   The Problem.
Chapters 8 and 9 described the digital network, whether based on DS1 or E1; the digital voice channel is either 56 kbits/s or 64 kbits/s. These transmission rates, which are incompatible with CCITT (ITU-T) recommendations V.5 [35] or V.6 [36], are 600, 1200, 2400, 4800, 9600, or 14,400 bits/s. Other acceptable data rates are 3000, 6000, 7200, 12,000, and 28,800 bits/s.

In this section, two methods are described to interface standard data rates with a 56/64-kbit/s digital voice channel. The first is AT&T's Digital Data System (DDS), and the second is based on ITU-T Rec. V.110.

9.6.2   The AT&T Digital Data System (DDS).
The AT&T digital data system (DDS) provides duplex point-to-point and multipoint private line digital data

---

* Remember "envelope delay distortion" (EDD) is a measure of phase distortion.

transmission at a number of synchronous data rates. This system is based on the standard 1.544-Mbit/s DS1 PCM line rate, where individual bit streams have data rates that are submultiples of that line rate (i.e., based on 64 kbits/s). However, pulse slots are reserved for identification in the demultiplexing of individual user bit streams as well as for certain status and control signals and to ensure that sufficient line pulses are transmitted for receive clock recovery and pulse regeneration. The maximum data rate available to a subscriber is 56 kbits/s, some 87.5% of the 64-kbit/s theoretical maximum.

The 1.544-Mbit/s line signal as applied to DDS service consists of 24 sequential 8-bit words (i.e., channel time slots) plus one additional framing bit. This entire sequence is repeated 8000 times per second. Note that again we have $(192 + 1)8000 = 1.544$ Mbits/s, where the value 192 is 8×24 (see Chapter 8). Thus the line rate of a DDS facility is compatible with the DS1 (T1) PCM line rate and offers the advantage of allowing a mix of voice (PCM) and data where the full dedication of a DS1 facility to data transmission would be inefficient in most cases.

AT&T calls the basic 8-bit word a *byte*. One bit of each 8-bit word is reserved for network control and for stuffing to meet nominal line bit rate requirements. This control bit is called a *C*-bit. With the *C*-bit removed we see where the standard channel bit rate is derived, namely, 56 kbits/s or 8000×7. Three subrates or submultiple data rates are also available: 2.4, 4.8, and 9.6 kbits/s. However, when these rates are implemented, an additional bit must be robbed from the basic byte to establish flag patterns to route each subrate channel to its proper demultiplexer port. This allows only 48 kbits/s out of the original 64 kbits/s for the transmission of user data. The 48-kbits/s composite total may be divided down to five 9.6-kbit/s channels, or ten 4.8-kbit/s channels, or twenty 2.4-kbit/s channels [37, 38]. The subhierarchy of DDS signals is illustrated in Figure 10.16.

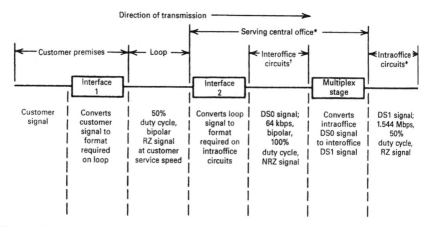

Figure 10.16. Subhierarchy of DDS signals. *Note:* Inverse processing must be provided for the opposite direction of transmission. Four-wire transmission is used throughout. *Exchange. †PCM trunk.

Often in practical implementations of DDS, the link connecting the local digital serving switch to the customer premise is via digital loop carrier such as the SLC-96, described in Chapter 8.

9.6.3 Transmitting Data on the Digital Network Based on ITU-T Rec. V.110.
This ITU-T recommendation covers data rate adaptation for 48 and 56 kbits/s to 64 kbits/s E0 or clear DS0 channel.

In the two-step case, the first conversion is to take the incoming data rate and convert it to an appropriate intermediate rate expressed by $2^k \times 8$ kbits/s, where $k = 0$, 1, or 2. The second conversion takes the intermediate rate and converts it to 64 kbits/s.

Simple division of 64,000 bits/s by standard data rates shows that as a minimum, a lot of bit stuffing would be required. These would essentially be wasted bits. CCITT makes use of these bits to provide framing overhead, status information, and control information. A frame (step 1 conversion) consists of 10 octets ($10 \times 8 = 80$ bits). Six octets carry user data; the first octet is all 0s for frame alignment, and bit 1 of the remaining 9 octets is set to 1. There are 15 overhead bits. Nine different frames accommodate the various data rates, and each frame has 80 bits (10 octets).

The recommendation also covers conversion of start–stop data rates including 50, 75, 110, 150, and 300 bits/s, and the standard rates up through 19.2 kbits/s [39]. [Further information may be found in Refs 22–30.]

REVIEW QUESTIONS

1. What is the basic element of information in a binary data system? How much information does it contain? (*Hint*: How many distinct states may it have?)

2. How does one extend the information content of that basic information element (question 1), for example, to construct a code that represents our alphabet?

3. There are many ways we can express the binary 1 and the binary 0. Also, we do not want to confuse the 1 and the 0 (i.e., reverse them). Give at least four ways we can express a 1 and a 0, and show that we are in common agreement on how to do this such that all ambiguity is removed.

4. How many distinct characters or symbols can be represented by a 4-unit binary code? a 7-unit binary code? an 8-unit binary code? (*Hint*: For this argument, consider that a unit is a bit.)

5. Name at least three nonprinting characters that might be encountered in a practical binary code.

6. How many information elements (bits) are in an ASCII character?

7. What are some of the more common causes of burst errors?

8. An ASCII character is represented as 1001011 with odd parity assumed. Give the value of the eighth bit. In another situation even parity is assumed and the ASCII character is 0101110. What is the value of the eighth bit?

9. Would you imagine that throughput is at all sensitive to the underlying error rate of a circuit? Explain your answer.

10. Why the insistence on careful definition of throughput?

11. Which is the more powerful error detection scheme: vertical redundancy check, longitudinal redundancy check, or cyclic redundancy check (CRC)?

12. Describe the two common methods of correcting errors that occur on data links.

13. A frame check sequence (FCS) contains either a 16-bit or 32-bit sequence of bits. What is it and how is it derived?

14. When FEC (forward error correction) is implemented, what types of errors can be corrected? How can one use FEC to correct burst errors?

15. Define "Hamming distance."

16. Name the three different types of ARQ. Which is the simplest to implement? Which is the most efficient for maximum "throughput."

17. Differentiate between neutral and polar transmission.

18. A data link transmits 2400 bits/s synchronously. There is a time base stability difference between transmitter and distant-end receiver of 20 ppm (parts per million). If the link can be considered "error free" except for timing, how many bits will the receiver receive correctly before printing in error? Assume, of course, that the receiver started copying the first bit out of the transmitter and that their two clocks started exactly in synchronism.

19. On a start–stop circuit, where does the receiver start counting information bits? At which signal element?

20. There are three major causes of errors in a data system. Name two of them.

21. The mark-to-space transition for the start space element tells a data receiver when a character is to begin for start–stop systems. How does a synchronous data receiver know when things start so it can begin a bit/byte count?

22. Show and explain the function of each of the 5/6 fields of a generic data frame.

23. How does a synchronous data receiver keep in synchronization with an incoming bit stream?

24. Explain timing distortion and how it can affect error performance.

25. A serial bit stream with an NRZ waveform has a bit rate of 19.2 kbits/s. What is the period of one bit?

26. Why should we be careful on the use of terms such as baud and bit?

27. What is notably richer in transitions per second, RZ or NRZ coding? Why would that be important anyway?

28. Keeping strictly in the electrical domain (i.e., the physical layer), draw a simple functional block diagram of a data circuit identifying the DTE and the DCE, and show two major interface points.

29. On the diagram prepared for question 28, show the TIA/EIA-232 interface point.

30. Distinguish between a binary 1 and a binary 0 based on TIA/EIA-232 using voltage values.

31. Name at least two advantages using a balanced interface between DTE and DCE.

32. What are the three basic impairments to data transmission?

33. Name the two types of noise that impairs data transmission (There are four all told). Discuss how each impacts a stream of bits.

34. Phase distortion, in general, has little effect on speech communication. What can we say about it for data transmission, or any stream of bits, for that matter?

35. Shannon's formula for capacity (in bits/s) for a particular bandwidth was based on only one other parameter. What was it?

36. We commonly equalize two voice channel impairments. What are they and what does equalization do?

37. Why do higher speed analog voice channel modems use a center tone frequency around 1700–1800 Hz?

38. Although we said in Chapter 8 that the digital network was compatible with conventional data bit streams, what seems to be the real problem? Discuss three ways we can transmit binary digital data on the digital network.

REFERENCES

1. *The IEEE Standard Dictionary of Electrical and Electronics Terms*, 6th ed., IEEE Std 100-1996, IEEE, New York, 1996.

2. *Equivalence Between Binary Notation Symbols and the Significant Conditions of a Two-Condition Code*, CCITT Rec. V1, Fascicle VIII.1, IXth Plenary Assembly, Melbourne, 1988.

3. *Coded Character Sets—7-Bit American National Standard Code for Information Interchange (7-bit ASCII)*, ANSI X3.4-1986, ANSI, New York, 1986.

4. *Error Performance on an International Digital Connection Forming Part of an Integrated Services Digital Network*, ITU-T Rec. G.821, Fascicle III.5, IXth Plenary Assembly, Melbourne, 1996.

5. *Interface between Data Terminal Equipment and Data Circuit-Terminating Equipment Employing Serial Binary Data Interchange*, EIA/TIA-232F, Telecommunications Industry Association, Washington, DC, July 1996.

6. *High Speed 25-Position Interface for Data Terminal Equipment and Data Circuit-Terminating Equipment Including Alternative 26-Position Connector*, EIA/TIA-530-A, Electronics Industries Association, Washington, DC, June 1992.

7. *Electrical Characteristics of Balanced Voltage Digital Interface Circuits*, EIA-422-A, Electronic Industries Association, Washington, DC, December 1978.

8. *Electrical Characteristics of Unbalanced Voltage Digital Interface Circuits*, EIA-423-A, Electronics Industries Association, Washington, DC, December 1978.

9. *Electrical Characteristics of Digital Interface Circuits*, MIL-STD-188-114A, U.S. Department of Defense, Washington, DC, September 1985.

10. *Electrical Characteristics for Unbalanced Double-Current Interchange Circuits for General Use with Integrated Circuit Equipment in the Field of Data Communications*, ITU-T Rec. V.10, ITU, Geneva, March 1993.

11. *Electrical Characteristics of Balanced Double-Current Interchange Circuits for General Use with Integrated Circuit Equipment in the Field of Data Communications*, ITU-T Rec. V.11, ITU, Geneva, October 1996.

12. *List of Definitions of Interchange Circuits Between Data Terminal Equipment (DTE) and Data Circuit-Terminating Equipment (DCE)*, ITU-T Rec. V.24, ITU, Geneva, February 2000.

13. *Electrical Characteristics of Unbalanced Double-Current Interchange Circuits*, ITU-T Rec. V.28, ITU, Geneva, March 1993.

14. *Electrical Characteristics of Single-Current Interchange Circuits Controlled by Contact Closure*, CCITT Rec. V.31, Fascicle VIII.1, IXth Plenary Assembly, Melbourne, 1988.

15. *Attenuation Distortion*, CCITT Rec. Q.44, Extract from the Blue Book, 1988.

16. *Reference Data for Engineers: Radio, Electronics, Computer and Communications*, 8th ed., Howard W. Sams, Indianapolis, IN, 1993.

17. W. R. Bennett and J. R. Davey, *Data Transmission*, McGraw-Hill, New York, 1965.

18. *Automatic Level Control Devices*, CCITT Rec. G.169, Fascicle III.1, IXth Plenary Assembly, Melbourne, 1988.

19. *LATA Switching Systems Generic Requirements, Transmission*, Bellcore TR-TSY-000507, Issue 3, Piscataway, NJ, 1989.

20. *CCITT Recommendations, Blue Books*, Volume VI, Fascicle VI.1, IXth Plenary Assembly, Melbourne, 1988.

21. *Power Levels for Data Transmission over Telephone Lines*, CCITT Rec. V.2, Fascicle VIII.1, IXth Plenary Assembly, Melbourne, 1988.

22. *Error on the Reconstituted Frequency*, CCITT Rec. G.135, Fascicle III.1, IXth Plenary Assembly, Melbourne, 1988.

23. C. E. Shannon, "A Mathematical Theory of Communications," BSTJ 27, Bell Telephone Laboratories, Holmdel, NJ, 1948.

24. H. Nyquist, "Certain Topics in Telegraph Transmission Theory," BSTJ, 6-7-644, Bell Telephone Laboratories, Holmdel, NJ, April 1928.

25. R. L. Freeman, *Reference Manual for Telecommunications Engineering*, John Wiley & Sons, New York, 2002.

26. *Synchronous Signaling Rates for Data Transmission*, EIA-269A, Electronics Industries Association, Washington, DC, May 1968.

27. *Support of Data Terminal Equipments with V-Series Type Interfaces by an Integrated Services Digital Network*, CCITT Rec. V.110, ITU, Geneva, September 1992.

28. *Digital Data System Data Service Unit Interface Specification*, Bell System Technical Reference Publication 41450, AT&T, New York, 1981.

29. *Digital Data System Channel Interface Specification*, Bell System Technical Reference Publication 62310, AT&T, New York, 1983.

30. R. L. Freeman, *Telecommunication Transmission Handbook*, 4th ed., John Wiley & Sons, New York, 1998.

31. *Common Long-Haul and Tactical Communication System Technical Standard*, MIL-STD-188-100, U.S. Department of Defense, Washington, DC, November 1972.

32. *Transport Systems Generic Requirements (TSGR): Common Requirements*, GR-499-Core Issue 2, Telcordia, Piscataway, NJ, December 1998.

33. "LATA Switching System Generic Requirements (LSSGR)," FR-64, Bellcore, Telcordia, Piscataway, NJ, 1998.

34. K. Pahlavan and J. L. Holsinger, "Voice-Band Communication Modems: A Historical Review, 1919–1988," *IEEE Commun. Mag.* **26**(1), January 1988.

35. *Standardization of Data Signaling Rates for Synchronous Data Transmission in the General Switched Telephone Network*, CCITT Rec. V.5, Fascicle VIII.1, IXth Plenary Assembly, Melbourne, 1988.

36. *Standardization of Data Signaling Rates for Synchronous Data Transmission on Leased Telephone-Type Circuits*, CCITT Rec. V.6, Fascicle VIII.1, IXth Plenary Assembly, Melbourne, 1988.

37. *Digital Data System Data Service Unit Interface Specification*, Bell System Technical Reference Publication 41450, AT&T, New York, 1981.

38. *Digital Data System Channel Interface Specification*, Bell System Technical Reference Publication 62310, AT&T, New York, 1983.

39. *Support of Data Terminal Equipments with V-Series Type Interfaces by an Integrated Services Digital Network*, CCITT Rec. V.110, ITU, Geneva, September 1992.

11

DATA NETWORKS AND THEIR OPERATION

1 INTRODUCTION

Data networking deals with the rapid transporting of information. The transport system should be efficient and cost-effective. There are two generic data network regimes that have developed and evolved over the past 40 years. These are the *wide area network* (WAN) and the *local area network* (LAN). There is also the *metropolitan area network* (MAN), which is written about in the literature but has turned out to be a late bloomer.

The WAN appeared on the scene first, providing connectivity among computer and computer-related sources and destinations that are geographically widely dispersed. Data circuits were set up using the ubiquitous telephone network connections either by dial-up or by leased lines. This means of establishing data circuits is still very prevalent today. Entire networks are built up from leased telephone network facilities. These could be private networks dedicated completely to a single business enterprise, or public subscription networks such as Tymnet and Telenet, both of which offer packet data services. An example of an early large WAN is ARPANET shown in Figure 11.1, which went on line in 1969. ARPANET evolved into what we know today as the internet.

Local area networks (LANs) came along later, particularly due to the widespread use of the PC in the enterprise environment. It is interesting to note that the first LAN protocol, Ethernet, was formulated in 1972, some three years before the PC was introduced. Ethernet still remains the most popular LAN access protocol. LANs are discussed in Chapter 13.

In the design of a WAN, what sort of topology will be used? In this context we will be dealing with how to connect up these data assets, be they switches, routers, PCs, printers, servers, mass storage devices, and so on. Topology is

---

Telecommunication System Engineering, by Roger L. Freeman
ISBN 0-471-45133-9 Copyright © 2004 Roger L. Freeman

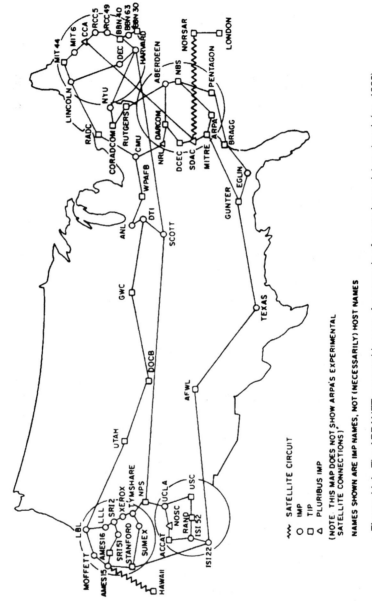

SATELLITE CIRCUIT
○ IMP
□ TIP
△ PLURIBUS IMP

(NOTE: THIS MAP DOES NOT SHOW ARPA'S EXPERIMENTAL SATELLITE CONNECTIONS.)

NAMES SHOWN ARE IMP NAMES, NOT (NECESSARILY) HOST NAMES

Figure 11.1. The ARPANET geographic map. An example of a very large data network (ca. 1989).

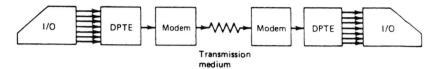

Figure 11.2. Elemental point-to-point data connectivity. The DTE, described in Chapter 10, consists of the DPTE and I/O combined. The DCE is the modem. DPTE, data processing terminal equipment; I/O, input/output device.

defined by Rosenberg [26] as the physical or logical placement of nodes in a computer network. One very common approach is to use point-to-point connectivity as shown in Figure 11.2.

Cost, quantity of data, urgency, and reliability would drive the designer to other types of topology such as mesh and star, described in Chapters 2 and 6. We will also describe multipoint and ring networks, which have broad application in the enterprise network world.

There is a disparity between LANs and WANs. LANs operate at comparatively high megabit data rates and WANs operate at kilobit rates and lower megabit data rates, and they operate with entirely different protocol regimes. Shortly after LANs came into existence, the problem arose interconnecting them over comparatively long distances. Enter frame relay and ATM, described in Chapters 15 and 16. VSATs, discussed in Chapter 7, brought a whole new dimension to data networking. They completely bypassed the telephone network.

Around 1994 the internet really began to take hold. It became a common highway of communication with email. It is a worldwide source of knowledge, information, and publicity—some for better, some for worse. The internet is an unparalleled source for research of any kind. Access is free to the user. It is a wide-access WAN with backbone routes carried on dedicated circuits using PDH (e.g., DS1/E1), SONET, and SDH data rates and formats.

1.1 Applications

Data and data networks enter every facet of our society, particularly in the business or "enterprise" world. The list of applications is extremely broad, although we will emphasize those of business and government. Certainly, local area networks (LANs) evolved from the enterprise world. Virtual private networks (VPNs) became a reality as a result, developing into a method of cost-effective connectivity of disparate LANs.

The basic building block of the digital network is the 64-kbit/s voice channel. We have described two ways where we can transport data on these digital voice channels. There is, of course, the common approach of superimposing a 64-kbit/s data bit stream directly on the standard 64-kbit/s digital bit stream. Many enterprises desire a greater bit rate where they will lease or build their own DS1, removing the framing bit. Here the bit rate will be 1.536 Mbits/s rather than the 1.544 Mbits/s we are accustomed to. Some users go even further using a DS3 or E3 bit stream—and in some cases, a SONET or SDH bit stream.

Voice communication (i.e., telephony) is marrying with data communications to provide a singular service: voice-over IP (VoIP). Here packets carrying voice can mix and match with IP packets carrying conventional data. VoIP is described in Chapter 12.

2 INITIAL DESIGN CONSIDERATIONS

2.1 General

Data network architecture can vary from a complex, widely distributed network shown in Figure 11.1 to the elemental point-to-point network shown in Figure 11.2. Between these extremes there is a large variety of data networks regarding size, configuration, and capability. Some of the more important considerations entering into the network design are:

- Data sources and destinations: locations and geographic dispersal
- Applicability of LANs
- Traffic profile among all connectivities
- Network organization, topology: existing and planned
- Private network, use of PSTN, hybrid schemes (i.e., partially private network, partial use of PSTN)
- For PSTN usage, tariffs, and tariff structures
- The applicability and extension of VPNs
- The applicability, cost-effectiveness of VSAT networks; of frame relay and ATM networks
- Data security requirements
- Type of data communication now in use or planned; compatibility for upgrade
- Requirements for data perishability (e.g., inquiry-response), reducing latency

Modern data networks all have distributed processing capability. Certainly the ubiquitous PC was the major instigator for distributed processing. One or more mainframe computers and minicomputers are also connected to most networks. They are just represented as another node, and they will probably be associated with a LAN. The LAN usefulness as a data aggregator, a data traffic concentrator, has turned into a boon for the data network designer.

2.2 Data Terminals, Workstations, PCs, and Servers

A key element of a data network, whether LAN or WAN, is the data terminal or workstation. It may be no more than a PC, or it could be a powerful Sun workstation. Generally, a network consists of workstations and other computer assets, such as CPUs,* servers, high-speed printers, and in some cases mass

---

* CPU stands for central processing unit, a computer.

storage devices. Our interest in this section is how data terminals and workstations can drive network design.

A workstation has a human operator. It is the principal input/output device for the network. As a minimum, it consists of a display, a keyboard or keypad input device, and a processor. To interface a data network, it must also have a communication interface device, which may or may not be physically a part of its processor, such as a plug-in card. It may also have an associated printer. For planning the system, design engineers should determine the following:

- The type of protocol(s) supported
- Physical interface (OSI Layer 1): EIA/TIA-232, EIA/TIA-530, MIL-STD-188-114C, etc.
- Limitations of the underlying transmission media
- LAN interface and NIC (network interface card)
- Operating system
- Security devices, software, limitations

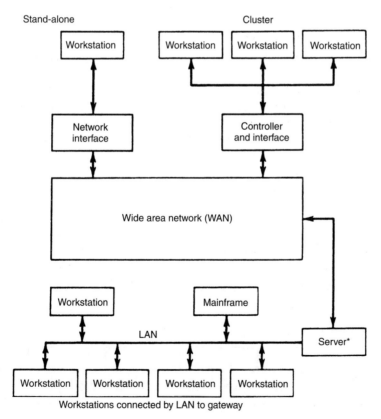

Figure 11.3. Graphic representation of three methods of connecting terminals or workstations to a WAN.

A *server* is a network device that provides service to the network users by managing shared resources. Note that the term is often used in the context of a client–server architecture for a local area network. Typical specialized servers are printer server and file server.

Figure 11.3 illustrates the incorporation of workstations, servers, and mainframe into a WAN.

3 NETWORK TOPOLOGIES AND CONFIGURATIONS

For the wide area network, there are seven ways of configuring data assets which are illustrated in Figures 11.4A to 11.4G:

1. Point-to-point (Figure 11.4A)
2. Multipoint or multidrop (Figure 11.4B)
3. Bus/tree (Figure 11.4C)
4. Star with host computer at hub (Figure 11.4D)
5. Multistar/hierarchical (Figure 11.4E)
6. Ring network (Figure 11.4F)
7. Grid network, packet-switched or circuit-switched (Figure 11.4G)

There are advantages and disadvantages of a certain topology. An advantage of a particular type of topology in one situation may turn out to be a disadvantage in another. For instance, a multidrop (multipoint, Figure 11.4B) is simple and economical, but should only be used when each access has a low traffic

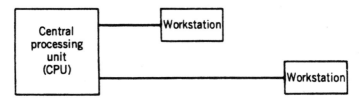

Figure 11.4A. A simplified diagram of a point-to-point network.

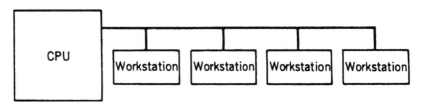

Figure 11.4B. A simplified multipoint or multidrop network.

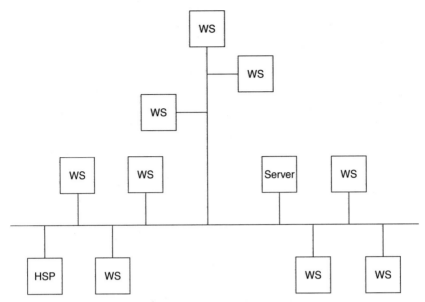

Figure 11.4C. Bus network. A tree network is a simple modification of a bus network. CATV networks generally have a tree topology. HSP, high-speed printer; WS, workstation (or PC).

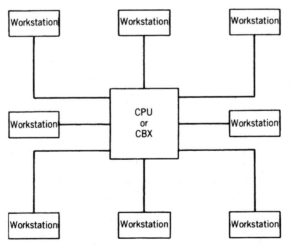

Figure 11.4D. A simplified diagram of a star network with a CPU at the center or hub of the star.

intensity. For the high traffic intensity and occupancy case, a point-to-point network (Figure 11.4A) is more attractive and its expense is a function of geographic dispersal of data terminals.

A star network (Figure 11.4D) has a processing and switching capability at its hub. The hub is usually a computer. The star network gives the appearance

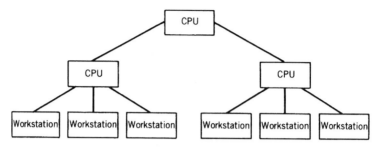

Figure 11.4E. A simplified diagram of a multistar or hierarchical network.

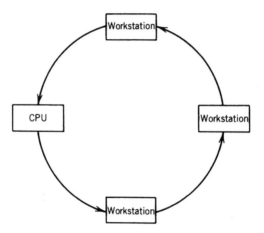

Figure 11.4F. A simplified diagram of a ring network.

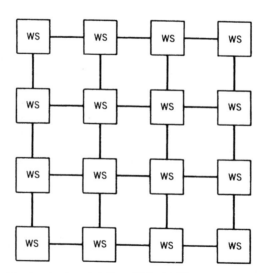

Figure 11.4G. A grid network of workstations (WS). A CPU can replace a workstation at any location or be colocated with a workstation.

of an extension of the point-to-point concept. The exception is that workstations not only can access the hub computer, but they can access each other through the hub. VSAT networks are star networks, but only in special VSAT networks can remote terminals access each other. The star network has poor grades for survivability.

Figure 11.4E expands the star network concept shown in Figure 11.4D. Our present telephone network is a modified hierarchical network (Chapters 1 and 6). One application for data communications is to place workstations at the lowest levels in the hierarchy. These workstations access nearby computers, which carry on the brunt of the processing and also provide a concentration capability for data communication to a large mainframe, centrally located computer (at the top of the figure). Some of the same techniques employed in the telephone network can be applied to the hierarchical data network. One is the application of high-usage (HU) and/or direct routes for high-traffic-intensity data relations.

Figure 11.4F illustrates a ring network. We note in the figure the unidirectionality of data flow. In the LAN arena, this is typical of a token ring. If a second ring is added, overlayed on the first, but providing traffic flow in the other direction, survivability and availability are improved. If there is a failure or cut across both rings, traffic can be routed in opposite directions avoiding catastrophic breakdown. If one ring fails, the other still operates. Of course, there is the added cost of the additional ring.

In large networks, survivability and availability* can be enhanced still further with a grid network, shown in Figure 11.4G. This is because of the large number of alternative routing possibilities available with a grid network. Whereas the ring network can withstand one failure, the grid network can withstand multiple failures. A grid network also lends itself well to packet communications, where each node has at least three inlets and outlets. Packets can travel on a variety of routes between any two nodes.

4 OVERVIEW OF DATA SWITCHING

4.1 Introduction

Data switching is the heart of data communications. In years past we thought of three varieties of data switching:

- Circuit switching
- Message switching
- Packet switching

Plain old telephone service, affectionately called POTS, is based on circuit switching. A data switch can be set up in a similar manner. A PVC (permanent virtual

---

* Availability is related to reliability and expresses the percentage of time that a network is up and fully operational.

circuit, such as found in frame relay) is a good example. Previously we discussed a dial-up connection to exchange data. This uses the POTS circuit-switching capability of the telephone network. It is fairly simple and straightforward. Circuit switching of data is well-suited for long-lasting connections where initial circuit establishment time cost is balanced by low forwarding time cost.

Message switching, often called store-and-forward switching, accepts messages from originators, stores the messages, and then forwards each message to the next node or destination when circuits become available. There are two important points we wish to make when comparing circuit switching and message switching. Circuit switching provides end-to-end connectivity in near real time. Message switching does not. There is usually some delay as a message makes its way through the system to its destination. The second point deals with efficient use of expensive transmission links. A well-designed message switching system keeps a uniform load through the working day and even into the night. Circuit-switching systems, on the other hand, are designed for busy-hour loading, and the system tends to loaf after-hours. One could say they are notoriously inefficient.

Packet switching uses some of the advantages of message switching and circuit switching, and it mitigates some of the disadvantages of both. A data *packet* is a comparatively short block of message data of fixed length. Complete data messages are broken down into short packages, each with a header. These packets are sent on diverse routes to their eventual destination, and each packet is governed by an ARQ error-correction protocol. Because packets travel on diverse routes, they may not arrive at the far-end receiving node in sequential order. Thus the far-end node must have the capability to store incoming packets and rearrange them in sequential order. The destination node then reformats the message as it was sent by the originator and forwards it to the final destination.

4.2   Traffic Engineering — A Modified Meaning

A modified meaning of *traffic engineering* (TE) has evolved because of the advancement of packet networks (e.g., IP and the internet), varying (to some degree) with our definition in Chapter 1, Section 5. Traffic engineering consists in optimizing resource utilization in a (data) network by choosing appropriate paths followed by flows of data, according to certain static and dynamic constraints. Here the main goal is to balance the load in the network (i.e., to avoid congestion and blockage on links in a network while other links are underutilized). To achieve these goals, traffic engineering methods vary from offline capacity planning algorithms to automatic dynamic changes. In the case of circuit switching of data, a fixed path is allocated for each flow and circuits can be established according to traffic engineering algorithms.

4.3   Packet Networks and Packet Switching

We will discuss two types of packet switching: datagram packet switching and virtual circuit packet switching.

4.3.1 Datagram Packet Switching. In circuit switching, resources are allocated during circuit setup. To carry out this function the network must have resources available so a circuit can be setup when required, no waiting. Once the communication is completed, the circuit is torn down and returned to the pool of idle circuits. These circuits are wasted when not carrying data.

Datagram service is the conventional packet handling service as we think of it. The switching is done by a router which uses a look-up table or *routing table* for each incoming packet. A routing table contains a "mapping" of routes to the final destination(s) and identifies the outgoing port of a path to the destination(s). Routing tables can be very large because they are indexed by all possible destinations in the network. To do this, they make look-ups and routing decisions that are computationally expensive. Here the full forwarding process is comparatively slow when compared to circuit switching. In a datagram packet switching network, each datagram (packet) must carry the address of the destination host(s) and use the destination address(es) to make forwarding decisions. As a consequence, routers do not have to modify the destination addresses of packets when forwarding packets.

Routing tables can be dynamically changed "on the fly." This almost assures different routing of two contiguous packets of the same message if the change is made at an instant between the two packets. Secondly we must consider the case where there is a dynamic network routing topology change due to link failure or congestion. As a result, the routing protocol will automatically recompute the routing tables so as to take the new topology into account to avoid the failed or congested link. As opposed to circuit switching, no additional traffic engineering algorithm is required to reroute traffic.

Consider then that routers make routing decisions locally for each packet, independent of the data flow to which a packet belongs. Thus, traffic engineering (as defined above), which controls the route(s) of traffic, is more difficult to implement for datagram packet switching than for circuit switching.

4.3.2 Virtual Circuit Packet Switching. Virtual circuit (VC) packet switching is a packet switching technique which combines datagram packet switching with circuit switching to take advantage of the attributes of both. VC packet switching is a variation of datagram packet switching for which no physical resources (i.e., frequency slots or time slots) are allocated. Each packet carries a circuit identifier which is local to a link and is updated at each switch in the route path of the packet from its source to its destination. Let's define a *virtual circuit* as a sequence of mappings between a link taken by packets that the circuit identifier packets carry on this link. This sequence is set up at connection establishment time, and these identifiers are returned to an idle pool at circuit termination.

One of the trade-offs a system planner should consider is that between connection establishment and forwarding time costs, which exists in circuit switching and datagram packet switching. In VC packet switching, routing is performed at circuit establishment time to keep packet forwarding fast. Among some other advantages of VC packet switching are that it has (a) the traffic engineering

TABLE 11.1 Summary of Data Switching Methods

Switching Method	Advantages	Disadvantages
Circuit switching	Mature technology Near real-time connectivity Excellent for inquiry and response Leased service attractive SVC (switched virtual circuit) example	Comparatively high cost of switch. Lower system utilization, particularly link utilization. Privately owned service can only be justified with high traffic volume.
Message switching	Efficient trunk utilization Cost-effective for low-volume leased service	Delivery delay may be a problem. Not viable for inquiry and response. Survivability problematical. Requires considerable storage.
Packet switching	Efficiency Routers are cost-effective Highly reliable, survivable IP and internet good examples Two varieties: datagram and virtual circuit packet switching	For full survivability, multiple node and route network can be expensive. Traffic volume can justify private ownership. Datagram has dynamic reconfiguration capability; virtual circuit requires traffic engineering.

capability of circuit switching and (b) the resources-usage capability of datagram packet switching. Needless-to-say, a principal issue of VC packet-switched networks is the behavior of topology change. As opposed to datagram packet switched networks, where traffic routing tables are automatically recomputed on a topology change such as link failure, in VC packet-switching all virtual circuits that pass through a failed link are interrupted. Thus, rerouting in VC packet-switching relies on traffic engineering techniques.

Major implementations of VC-packet switching techniques are X.25, ATM, and MPLS networks.

Table 11.1 reviews advantages and disadvantages of several data switching techniques.

4.4 Interior Gateway Routing Protocol (IGRP)

There are a number of routing protocols. One of particular interest to us is IGRP (interior gateway routing protocol), which is proprietary to Cisco Systems. It is in the family of distance-vector routing protocols, which means that each router sends all or a portion of its routing table in a routing message update at regular intervals to each of its neighboring routers. Unlike many other routing protocols that just include one or two metrics for routing decision to determine the best path,

IGRP uses five criteria to determine the best path. These are: link's speed, delay, packet size, loading, and reliability. Network managers can set the weighting factors for each of these metrics.

Source: Ref. 30.

5 CIRCUIT OPTIMIZATION

Data circuits have sources and destinations with varying requirements regarding *quantity* of data and its urgency. For example, an analog voice-grade line can support data rates often up to 56 kbits/s. An E0 or DS0 has 64 kbits/s, and groups of these digital channels can be aggregated such as two channels at 128 kbits/s, four with 256 kbits/s up to E1 at 2.048 Mbits/s and DS1 at 1.536 Mbits/s. The next step graduates to E3 or DS3 (34.368 Mbits/s or 44.736 Mbits/s). These will turn out to be expensive leases, but the traffic level may justify the expense.

It would be uneconomical to underutilize such an expensive facility. Let's consider a number of scenarios. The first is where urgency is not an underlying consideration, and the data rate then is selected at the most economical full-rate scenario or 28.8 kbits/s. The circuit can share in the time domain with other users. Then we consider a 64-kbit/s circuit E0 or DS0 channel), which can also be shared. Table 11.2 expresses these results for time periods from 10 s to 24 h. In the table, data flow is treated as continuous with no breaks and no repeats. This would be unlikely in the real world. The table does give a good idea of data quantity, although on the liberal side.

One could say that these are *throughput* values in Table 11.2. Of course it depends on how we would define throughput. Let's look at a simple case of the ASCII 7-bit code. In this example, it is transmitted in a start–stop format with a 2-bit stop element. Thus there is a start element (bit) and a stop element (2 bits). These 3 bits sole purpose is to facilitate timing and synchronization; they carry no useful information. If Table 11.2 is based on such a start–stop format, only 72% of the bits are useful. There is also a parity bit which is usually appended as the eighth bit. It carries no useful information. It goes along for the ride, making each character 8 bits long, an octet. This is useful for the organization of 8-bit, 16-bit, 32-bit, and so on, processors. It did, at one time, serve an error detection

TABLE 11.2 Circuit Capacity in kbits/s Versus Bit Rate Versus Time Period

Time Unit	28.8 kbits/s	64 kbits/s	768 kbits/s	2.048 Mbits/s
10 s	288 kbits/s	640 kbits/s	7680 kbits/s	20.480 Mbits/s
1 min	1728 kbits/s	3840 kbits/s	46,080 kbits/s	122.880 Mbits/s
10 min	17.280 Mbits/s	38.4 Mbits/s	460.8 Mbits/s	1228.8 Mbits/s
1 h	103.68 Mbits/s	232.8 Mbits/s	2764.8 Mbits/s	7372.8 Mbits/s
8 h	829.44 Mbits/s	1862.4 Mbits/s	22,118.4 Mbits/s	59,982.4 Mbits/s
24 h	2488.32 Mbits/s	5587.2 Mbits/s	66,355.2 Mbits/s	176,947.2 Mbits/s

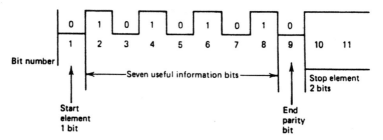

Figure 11.5. An ASCII start–stop bit sequence with a 2-bit stop element and a parity bit.

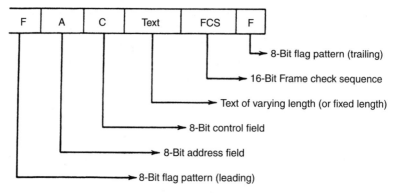

Figure 11.6. A typical data frame, block, or packet.

function, but seldom does now. It is overhead, bearing no useful information. We are now reduced to a 63% efficiency. In other words, based on our start–stop assumption, only 63% of the bits in Table 11.2 are useful to the end-user. This stop–start format is shown in Figure 11.5.

We now turn to synchronous transmission, where the data traffic must be sent in frames, blocks, packets, or cells. A typical frame is shown in Figure 11.6. In this case it is the "text" field, often called the "info" field, that contains the "useful" data for the destination user. Based on the figure, we must amortize this "text" over 48 bits of overhead (or more). Obviously, the longer the text field is, the better we amortize the overhead bits, and our apparent efficiency goes up. Suppose the text had 48 bits and there were 48 bits of overhead; we don't count the second flag pattern. There is a total of 96 bits in the frame, of which 48 were overhead. Thus our efficiency is 48/96, or 50%. If the number of bits now are doubled for the text, we have 96/144, or 67% efficiency. One can see that if we had a very long text and maintained the overhead at 48 bits, the efficiency becomes very high.

A high noise level, or even just a few noise spikes, tends to ruin our good record. This is because of repeats required of errored messages. The longer the frame, the more time is spent on a repeat through an ARQ regime. There is a general rule that on noisy channels, short frames lead to greater throughput.

Some links are so overburdened with inefficiencies that, for example, a 56-kbit/s data link may only afford a net 28 kbits/s of effective, useful bits put through. This poor performance is attributed to large overhead and frame repeats due to errors. As we mentioned, just one noise spike of short duration, a few microseconds, could cause just one bit in error, requiring that the whole frame be repeated.

5.1 Throughput from Another Perspective

The preceding discussion leads to the question of throughput. Throughput means different things to different people. We define throughput as the net *useful* bits put through per unit time. "Useful" is the key word. Useful to whom? The start and stop bits are useful to make the transmission system work. They are not useful to the data user who couldn't care less how the system works. The same argument holds for the synchronous overhead bits (not that start and stop systems do not have additional overhead).

Turning back to the frame format in Figure 11.6, the text field contains the "useful" bits. As we will see later, even bits in this field are required for control overhead, which can be argued as *not useful* bits (noninformation characters). We also must try to partition the data communication system in some way to determine where the telecommunication responsibility leaves off and the data user responsibility begins. This point will be discussed in Section 6.

5.2 Cost-Effective Options to Meet "Throughput" Requirements

In this section we treat point-to-point service exclusively for off-premises connectivity. In general, cost is a function of the product of bit rate and distance. This assumes a tariff structure that is based on circuit length. Table 11.3 gives a number of options available. In many cases, the smallest segment available is a full-period digital voice channel (56/64 kbits/s). This will prove more efficient than the use of a V.34 modem, the first row in Table 11.3.

Cost can be shared among users by breaking up a less expensive (per channel) digital configuration such as DS1 with its twenty-four 64-kbit/s channels. Such an approach is not very common.

A major consideration is whether to lease or set up a private network. The trouble with a private network is that "you own it." You own maintenance and replacement responsibilities. Under the leased situation, the lessor is responsible.

Often an overall circuit cost can be reduced on a per 64-kbit/s circuit basis by turning to fractional DS1 or E1. For example, the 384-kbit/s option is 1/4 of a DS1 configuration, assuming the availability of clear 64-kbit/s channels.* In the case of E1, 64-kbit/s channels are available. Here six contiguous E0 or special DS0 channels make up the 384-kbit/s configuration. It is a popular bit rate for

---

* We will recall from Chapter 8 that the North American DS1 commonly provides the user with a 56-kbit/s channel because of embedded signaling bits. However, clear 64-kbit/s channels can be provided in many instances at extra cost.

TABLE 11.3   Some Available Data Transmission Options in Order of Increasing Cost

Data Rate	Modulation	Cost Basis and Notes
28.8 kbits/s	V.34 modem	Cost of modem + line usage.
56 kbits/s	V.90 modem	Cost of modem + line usage. Nearly double throughput for same price is first item.
64 kbits/s	Digital E0, DS0	Additional throughput for less cost than 2nd item.
768 kbits/s	12 E0 or DS0 aggregate	Consider lease E1 or T1 accept extra capacity for growth.
DS1	Full T1	Room for negotiation. Mix voice and data.
E1	30 64-kbit/s channels	Separate signaling channel. In one configuration, 31 channels.
DS3	28 DS1 configurations	Dedicated 44.736-Mbit/s channel.
E3	Four 8448-kbit/s channels	$64 \times 4 \times 32 = 8192$ kbits/s or 128 E0 channel + admin/stuff bits.

television conferencing and for frame relay. Table 11.3 shows one-half of a DS1 configuration, namely 768 kbits/s.

The final two entries in Table 11.3 are full configurations for North America DS3 or E3 for countries using the ITU-T Organization E1 family of configurations. The DS3 may be used for the full 44.736 Mbits/s or broken up into 28 DS1s channels. Channels for E1. In most cases, telephone companies or telecommunication administrations offer the option of "mix and match." In some cases, such as for frame relay, the entire aggregate can be used for data. In other situations, say a 384-kbit/s group can be used for frame relay or other data connectivity, and other channels can be used for voice and/or data. In North America, leasing of a DS3 configuration is available providing 28 DS1s for the user.

In the United States, where private networks are prevalent, organizations, typically power companies, have excess capacity in their networks and lease circuits to other users. In many cases, such lease revenues not only pay for the private network, but also turn a profit on top of that.

6   DATA NETWORK OPERATION

6.1   Introduction

When data networks involve more than two terminals or nodes, some form of circuit discipline should be invoked. The first case is multidrop or multipoint illustrated in Figure 11.4B. Here several terminals or nodes share a common transmission medium. They could operate on a "first come, first served" basis. This is known as *contention*, where terminals or nodes compete for access to the medium. There is no discipline. As more terminals are added, such networks become unwieldy.

One simple form of discipline in this situation is *polling*. In this situation, one of the terminals is assigned as master station and polls the others periodically, querying them whether they have traffic for transmission. If so, the traffic is transmitted in frames and received by all stations on the network. It is only copied, though, by those stations identified in the frame header.

There are wide area and local area networks with thousands of users. To further complicate the matter, these networks consist of terminals, nodes, servers, routers, and CPUs from different vendors loaded with different software. Thus there is a problem offering equitable access to a network, but there is also an interface problem at many levels. We have already covered the electrical interface, such as TIA/EIA-232, which is more formally called the *physical* interface or *physical layer*.

In the late 1960s a term came into popular use, the *protocol*. Here we will slightly amplify the IEEE definition of protocol [1]: "*A formal set of rules and conventions governing the format and relative timing of message exchange among two or more communication terminals.*"

There are literally many dozens of protocols. Some have been developed by standardization bodies such as the ISO, IEEE, the U.S. Department of Defense, ITU-T Organization, and, to some lesser extent, ANSI and EIA/TIA. Others are proprietary, having been developed by various interests of the large population of data communication and processing equipment manufacturers such as IBM, Cisco, and Extreme.

In the next sections we will introduce protocols and discuss the ISO (International Standards Organization) Open Systems Interconnection (OSI), several protocols fitting into the OSI environment, and the U.S. DoD TCP/IP.

6.2 Protocols

6.2.1 Basic Protocol Functions.
Stallings [2] lists protocol functions in the following categories:

- Segmentation and reassembly (SAR)
- Encapsulation
- Connection control
- Ordered delivery
- Flow control
- Error control
- Multiplexing

A short description of each functional category is provided below.

Segmentation and Reassembly. Segmentation refers to breaking up the data into blocks with some bounded size. Depending on the semantics or system, these blocks may be called frames or packets. Reassembly is the counterpart of segmentation—that is, putting the blocks or packets back into their original order. Another name used for a data block is *protocol data unit* (PDU).

Encapsulation. Encapsulation is the adding of control information on either side of the data *text* of a block. Typical control information is the *header*, which contains address information and sequence numbers. An error control field is appended at the end of a block.

Connection Control. There are three stages of connection control:

1. Connection establishment
2. Data transfer
3. Connection termination

Some of the more sophisticated protocols also provide connection interrupt and recovery capabilities to cope with errors and other sorts of interruptions.

Ordered Delivery. PDUs are assigned sequence numbers to ensure an ordered delivery of the data at the destination. In a large network, especially if it operates in the packet mode, PDUs (packets) can arrive at the destination out of order. With a unique PDU numbering plan using a simple numbering sequence, it is a rather simple task for a long data file to be reassembled at the destination in its original order.

Error Control. Error control is a technique that permits recovery of lost or errored PDUs. There are three possible functions involved in error control:

1. Acknowledgment of each PDU or string of PDUs
2. Sequence numbering of PDUs (e.g., missing numbers)
3. Error detection (see Chapter 10)

Acknowledgment may be carried out by returning to the source the source sequence number of a PDU. This ensures delivery of all PDUs to the destination. Error detection initiates retransmission of errored PDUs.

Flow Control. Flow control refers to the management of data flow from source to destination such that buffers do not overflow but maintain full capacity of all facility components involved in the data transfer. Flow control must operate at several peer layers of a protocol, as will be discussed later.

Multiplexing. We will be talking about a vertical layered architecture in the next section. With this in mind, multiplexing can be used in one of two directions, upwards and downwards. *Upward multiplexing* can be used when multiple higher-level (layer) connections are multiplexed on a single lower-layer connection. This could be done to make more efficient use of the lower-layer service. This, typically, may be the multiplexing of several transport connections on a single network connection. *Downward multiplexing*, sometimes called *splitting*, is the building up of a single high-layer connection on top of multiple lower-layer connections. Reasons for downward multiplexing are to improve performance, reliability, and/or efficiency.

6.2.2 Open System Interconnection

6.2.2.1 Rationale. Data communication systems can be very diverse and complex. These systems involve very elaborate software which must run on equipment having an ever-increasing processing capacity. Under these conditions, it is desirable to ensure maximum independence between the various software and hardware elements of a system for several reasons:

- To facilitate intercommunication among disparate elements
- Tend to eliminate the "ripple effect" when there is a modification to one software element that may affect all elements

The International Standards Organization set about to make this data intercommunication problem more manageable. It developed its famous Open Systems Interconnection (OSI) reference model. Instead of trying to solve the global dilemma, it decomposed the problem into more manageable parts. This provided standard-setting agencies with an architecture that defines communication tasks. The OSI model provides the basis for connecting open systems for distributed applications processing. The term *open* denotes the ability of any two systems conforming to the reference model and associated standards to interconnect. OSI thus provides a common groundwork for the development of families of standards permitting data assets to communicate.

ISO broke data communications down into seven areas or layers arranged vertically starting at the bottom with layer 1, the input/output ports of a data device. The OSI reference model is shown in Figure 11.7. It takes at least two to communicate. Thus we consider the model in twos, one entity to the left in the figure and one to the right. ISO and the ITU-T Organization use the term *peers*. Peers are corresponding entities on either side of Figure 11.7. A peer on one side (system A) communicates with its peer on the other side (system B) by means of a common protocol. For example, the transport layer of system A communicates with its peer transport layer at system B. It is important to note that there is no *direct* communication between peer layers except at the physical layer (layer 1). That is, above the physical layer, each protocol entity sends data down to the next lower layer, and so on to the physical layer, then across and up to its peer on the other side. Even the physical layer may not be directly connected to its peer on the other side of the "connection" such as in packet communications. This we call *connectionless service*. However, peer layers must share a common protocol in order to communicate.

There are seven OSI layers, as shown in Figure 11.7. Any layer may be referred to an N layer. Within a particular system there are one or more active entities in each layer. An example of an entity is a process in a multiprocessing system. It could simply be a subroutine. Each entity communicates with entities above and below it across an interface. The interface is at a service access point

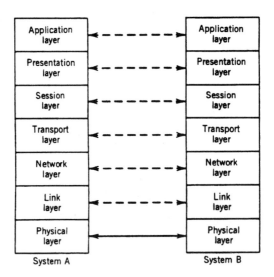

Figure 11.7. The OSI reference model.

(SAP). An $(N - 1)$ entity provides services to an N entity by use of primitives. A primitive [1, 2] specifies the function to be performed and is used to pass data and control information.

ITU-T Rec. X.200 [3] describes four types of primitive used to define the interaction between adjacent layers of the OSI architecture. A brief description of each of these primitives is given below.

Request. A primitive issued by a service user to invoke some procedure and to pass parameters needed to fully specify the service.

Indication. A primitive issued by a service provider either to invoke some procedure or to indicate that a procedure has been invoked by a service user at the peer service access point.

Response. A primitive issued by a service user to complete at a particular SAP some procedure invoked by an *indication* at that SAP.

Confirm. A primitive issued by a service provider to complete at a particular SAP some procedure previously invoked by a request at that SAP. ITU-T Rec. X.200 [3] adds this note: Confirms and responses can be positive or negative depending on the circumstances.

The data that pass between entities are a bit grouping called a *data unit*. We discussed protocol data units (PDUs) earlier. Data units are passed downward from a peer entity to the next OSI layer, called the $(N - 1)$ layer. The lower layer calls the PDU a *service data unit* (SDU). The $(N - 1)$ layer adds control information, transforming the SDU into one or more PDUs. However, the identity of the SDU is preserved to the corresponding layer at the other end of the connection. This concept is shown in Figure 11.8.

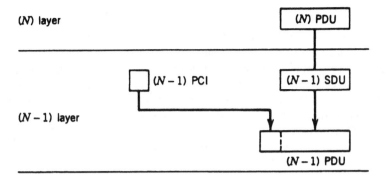

PCI = protocol control information
PDU = protocol data unit
SDU = service data unit

Figure 11.8. An illustration of mapping between data units in adjacent layers.

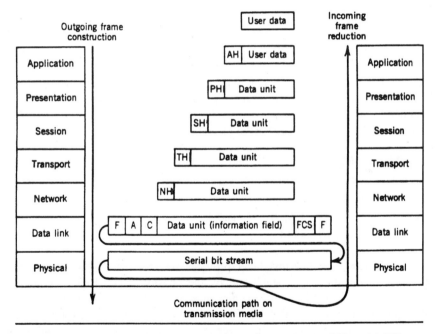

Figure 11.9. Buildup and breakdown of a data message following the OSI model. OSI encapsulates at every layer, except layer 1, adding significant overhead.

When we discussed throughput in Section 5.1, it became apparent that throughput must be viewed from the eyes of the user. With OSI some form of encapsulation takes place at every layer above the physical layer. To a greater or lesser extent OSI is used on every and all data connectivities. The concept of encapsulation, the adding of overhead, from layers 2 through 7 is shown in Figure 11.9.

6.2.2.2 Functions of OSI Layers

PHYSICAL LAYER. The physical layer is layer 1, the lowest OSI layer. It provides the physical connectivity between two data terminals who wish to communicate. The services it provides to the data-link layer (layer 2) are those required to connect, maintain the connection, and disconnect the physical circuits that form the physical connectivity. The physical layer represents the traditional interface between data terminal equipment (DTE) and data communication equipment (DCE). This was described in Chapter 10, Section 8.

The physical layer has four important characteristics:

1. Mechanical
2. Electrical
3. Functional
4. Procedural

The mechanical aspects include the actual cabling and connectors necessary to connect the communications equipment to the media. Electrical characteristics cover voltage and impedance, balanced and unbalanced. Functional characteristics include connector pin assignments at the interface and the precise meaning and interpretation of the various interface signals and data set controls. Procedures cover sequencing rules that govern the control functions necessary to provide higher-layer services such as establishing a connectivity across a switched network. (See also Refs. 11 and 14.)

Some applicable standards for the physical layer are:

- EIA-232, EIA-422, EIA-423, and EIA-530
- ITU-T Recs. V.10, V.11, V.24, V.28, X.21, and X.21 bis
- ISO 2110, 2593, 4902, and 4903
- U.S. Fed, Stds, 1020A, 1030A, and 1031
- U.S. MIL-STD-188-114

DATA-LINK LAYER. The data-link layer provides services for reliable interchange of data across a data link established by the physical layer. Link-layer protocols manage the establishment, maintenance, and release of data-link connections. These protocols control the flow of data and supervise error recovery. A most important function of this layer is recovery from abnormal conditions. The data-link layer services the network layer or logical link control (LLC; in the case of LANs) and inserts a data unit into the INFO portion of the data frame or block. A generic data frame generated by the link layer is shown in Figure 11.10.

Some of the more common data-link layer protocols are:

- ISO HDLC, ISO 3309, 4375
- CCITT LAP-B and LAP-D

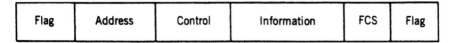

Flag	Address	Control	Information	FCS	Flag

Figure 11.10. Generalized data-link layer frame.

- IBM BSC, SDLC
- DEC DDCMP
- ANSI ADCCP (also a U.S. government standard)

NETWORK LAYER. The network layer moves data through the network. At relay and switching nodes along the traffic route, layering concatenates. In other words, the higher layers (above layer 3) are not required and are utilized only at user endpoints.

The concept of relay open system is shown in Figure 11.11. At the relay or switching point, only the first three layers of OSI are required.

The network layer carries out the functions of switching and routing, sequencing, logical channel control, flow control, and error recovery functions. We note the duplication of error recovery in the data-link layer. However, in the network layer error recovery is network-wide, whereas on the data-link layer error recovery is concerned only with the data link involved.

The network layer also provides and manages logical channel connections between points in a network such as virtual circuits across the public switched network (PSN). It will be appreciated that the network layer concerns itself with the network switching and routing function. On simpler data connectivities, where a large network is not involved, the network layer is not required and can be eliminated. Typical of such connectivities are point-to-point circuits, multipoint circuits, and LANs. A packet-switched network is a typical example where the network layer is required.

The best-known layer 3 standard is ITU-T Rec. X.25.

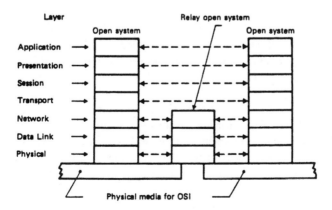

Figure 11.11. Only the first three layers of OSI are required at a relay (switching) point.

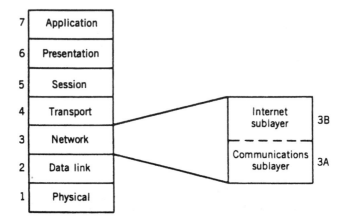

Figure 11.12. Sublayering of OSI layer 3 to achieve internetworking.

LAYER 3.5: INTERNETWORK PROTOCOLS. Layer 3.5 is a suggested sublayer of the OSI network layer.* It carries out the functions of internetworking, that is, the interconnection of two disparate networks. This sublayer is shown in Figure 11.12. The internet function is carried out by routers or gateways.

There are two applicable protocols for internetworking: ITU-T Rec. X.75 and may be considered a subset of X.25, and IP (internet protocol), which is often accompanied by TCP, the transmission control protocol. IP and TCP are members of a fairly large family of protocols developed by the U.S. Department of Defense DARPA (Defense Advanced Research Projects Agency). There is an equivalent standard developed by ISO (ISO 8473).

The ITU-T X.75 protocol is a companion of X.25 and assumes that all networks involved are based on the latter protocol. We will discuss X.25 in Section 6.3 of this chapter. A brief description of IP and its companion TCP is provided in Section 7.

TRANSPORT LAYER. The transport layer (layer 4) is the highest layer of the services associated with the provider of communication services. One can say that layers 1–4 are the responsibility of the communication system engineer. Layers 5, 6, and 7 are the responsibility of the data end-user. However, we believe that the telecommunication system engineer should have a working knowledge of all seven layers.

The transport layer has the ultimate responsibility for providing a reliable end-to-end data delivery service for higher-layer users. It is defined as an end system function, located in the equipment using network service or services. In this way its operations are independent of the characteristics of all the networks that are involved. Services that a transport layer provides are as follows:

---

* Layer 3.5 is not included in ITU-T Rec. X.200. It has been suggested by Ref. 2 and bears a lot of merit.

- *Connection Management.* This includes establishing and terminating connections between transport users. It identifies each connection and negotiates values of all needed parameters.

- *Data Transfer.* This involves the reliable delivery of transparent data between the users. All data are delivered in sequence with no duplication or missing parts.

- *Flow Control.* This is provided on a connection basis to ensure that data are not delivered at a rate faster than the user's resources can accommodate.

The TCP (transmission control protocol) was the first working version of a transport protocol and was created by DARPA for DARPANET (also known as ARPANET). All the features in TCP have been adopted in the ISO version. TCP is often lumped with the internet protocol and referred to as TCP/IP.

The ISO transport protocol messages are called *transport protocol data units* (TPDUs). There are connection management TPDUs and data transfer TPDUs. The applicable ISO references are ISO 8073 OSI (*Transport Protocol Specification*) and ISO 8072 OSI (*Transport Service Definition*).

SESSION LAYER. The purpose of the session layer is to provide the means for cooperating presentation entities to organize and synchronize their dialogue and to manage the data exchange. The session protocol implements the services that are required for users of the session layer. It provides the following services for users:

1. The establishment of session connection with negotiation of connection parameters between users.
2. The orderly release of connection when traffic exchanges are completed.
3. Dialogue control to manage the exchange of session user data.
4. A means to define activities between users in a way that is transparent to the session layer.
5. Mechanisms to establish synchronization points in the dialogue and, in case of error, resume from a specified point.
6. Interrupt a dialogue and resume it later at a specified point, possibly on a different session connection.

Session protocol messages are called session protocol data units (SPDUs). The session protocol uses the transport layer services to carry out its function. A session connection is assigned to a transport connection. A transport connection can be reused for another session connection if desired. Transport connections have a maximum TPDU size. The SPDU cannot exceed this size. More than one SPDU can be placed on a TPDU for transmission to the remote session layer.

Reference standards for the session layer are ISO 8327 [*Session Protocol Definition* (CCITT Rec. X.225)] and ISO 8326 [*Session Services Definition* (CCITT Rec. X.215)].

PRESENTATION LAYER. The presentation layer services are concerned with data transformation, data formatting, and data syntax. These functions are required to adapt the information handling characteristics of one application process to those of another application process.

The presentation layer services allow an application to interpret properly the data being transferred. For example, there are often three syntactic versions of the information to be exchanged between end-users A and B as follows:

- Syntax used by the originating application entity A.
- Syntax used by the receiving application entity B.
- Syntax used between presentation entities. This is called the *transfer syntax* [5].

Of course, it is possible that all three or any two of these may be identical. The presentation layer is responsible for translating the representation of information between the transfer syntax and each of the other two syntaxes as required.

The following standards apply to the presentation layer:

- ISO 8822 Connection-Oriented Presentation Service Definition
- ISO 8823 Connection-Oriented Presentation Service Specification
- ISO 8824 Specification of Abstract Syntax Notation One
- ISO 8824 Specification of Basic Encoding Rules for Abstract Syntax Notation One
- CCITT Rec. X.409 Message Handling Systems: Presentation Transfer Syntax and Notation

APPLICATION LAYER. The application layer is the highest layer of the OSI architecture. It provides services to the application processes. It is important to note that the applications do not reside in the application layer. Rather, the layer serves as a window through which the application gains access to the communication services provided by the model.

This highest OSI layer provides to a particular application all services related to communication in such a format that easily interfaces with the user application and is expressed in concrete quantitative terms. These include identifying cooperating peer partners, determining the availability of resources, establishing the authority to communicate, and authenticating the communication. The application layer also establishes requirements for data syntax and is responsible for overall management of the transaction.

Of course, the application itself may be executed by a machine, such as a CPU in the form of a program, or by a human operator at a workstation.

The following standards apply to the application layer:

- ISO 8449/3 Definition of Common Application Service Elements
- ISO 8650 Specification of Protocols for Common Application Service Elements [2–6].

6.2.3 High-Level Data-Link Control—A Typical Link-Layer Protocol.

High-Level Data-Link Control (HDLC) was developed by the International Standards Organization (ISO). It has spawned many related or nearly identical protocols. Among these are ANSI ADCCP, CCITT LAPB and LAPD, and IBM SDLC [7–9].

HDLC DEFINITIONS. Stations, configurations, and three modes of operation.

Primary Station. A logical primary station is an entity that has primary link control responsibility. It assumes responsibility for organization of data flow and for link level error recovery. Frames issued by the primary station are called *commands*.

Secondary Station. A logical secondary station operates under control of a primary station. It has no direct responsibility for control of the link but instead responds to primary station control. Frames issued by a secondary station are called *responses*.

Combined Station. A combined station combines the features of primary and secondary stations. It may issue both commands and responses.

Unbalanced Configuration. An unbalanced configuration consists of a primary station and one or more secondary stations. It supports full-duplex and half-duplex operation, point-to-point and multipoint circuits. An unbalanced configuration is shown in Figure 11.13a.

Balanced Configuration. A balanced configuration consists of two combined stations in which each station has equal and complementary responsibility of the data link. A balanced configuration operates only in the point-to-point mode and supports full-duplex and half-duplex operation. Figure 11.13b shows a balanced configuration.

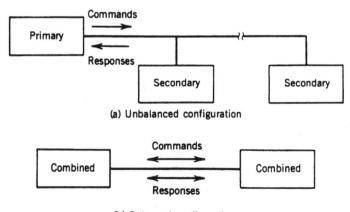

(a) Unbalanced configuration

(b) Balanced configuration

Figure 11.13. HDLC link configurations.

Modes of Operation. With *normal response mode (NRM)* a primary station initiates data transfer to a secondary station. A secondary station transmits data only in response to a poll from the primary station. This mode of operation applies to an unbalanced configuration. With *asynchronous response mode (ARM)* a secondary station may initiate transmission without receiving a poll from a primary station. It is useful on a circuit where there is only one active secondary station. The overhead of continuous polling is thus eliminated. *Asynchronous balanced mode (ABM)* is a balanced mode that provides symmetric data transfer capability between combined stations. Each station operates as if it were a primary station, can initiate data transfer, and is responsible for error recovery. One application of this mode is hub polling, where a secondary station needs to initiate transmission.

6.2.3.1   The HDLC Frame. Figure 11.14 shows the HDLC frame format. It has a similar format of generic data frames shown previously in this chapter and in Chapter 10. Moving from left to right in the figure, we have the flag field (*F*) which delimits the frame at both ends with the unique bit pattern 01111110. If frames are sent sequentially, the closing flag of the first frame is the opening flag of the next frame.

We called the flag sequence unique. Receiving stations constantly search for the flag bit sequence to mark the beginning and end of a frame. A data frame should be transparent to any 8-bit sequence of bits. Suppose, then, that the sequence 01111110 appeared in the middle of a frame. This would incorrectly tell a receiving station that the frame has ended and a new frame begun. Of course, this corrupts the whole frame and probably some subsequent frames. To avoid the problem, no sequence of six consecutive 1s bracketed by 0s is permitted. This is accomplished by what some call *bit stuffing* and others call *zero insertion*. With the exception of the flags, a transmitting station inserts an extra 0 bit after each occurrence of five 1s in the frame.

After detecting the opening flag, a receiver monitors the bit stream for a pattern of five contiguous 1s. If such a pattern occurs, the sixth bit is examined.

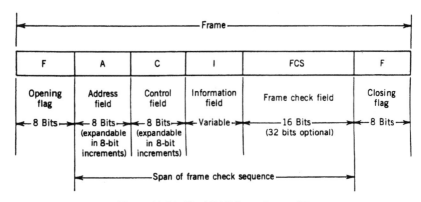

Figure 11.14. The HDLC frame format [7].

If that bit is a 0, it is deleted. If the sixth bit is a 1 and the seventh bit is a 0, the combination is accepted as a flag. If bits six and seven are both 1s, the transmitting station is assumed to be sending an *abort condition* [7].

The *address field* (A) immediately follows the opening flag of a frame and precedes the control field (C). Each station in the network normally has an individual address and a group address. A group address identifies a family of stations. It is used when data messages must be accepted from or destined to more than one user. Normally the address is 8 bits long, providing 256 bit combinations or addresses ($2^8 = 256$). In HDLC (and ADCCP) the address field can be extended in increments of 8 bits. When this is implemented, the least significant bit is used as an extension indicator. When that bit is 0, the following octet is an extension of the address field. The address field is terminated when the least significant bit of an octet is 1. Thus we can see that the address field can be extended indefinitely.

The *control field* (C) immediately follows the address field (A) and precedes the information field (I). The control field conveys commands, responses, and sequence numbers to control the data link. The basic control field is 8 bits long and uses modulo 8 sequence numbering. There are three types of control field: (1) I frame (information frame), (2) S frame (supervisory frame), and (3) U frame (unnumbered frame). The three control field formats are shown in Figure 11.15.

Consider the basic 8-bit format as shown in Figure 11.15. The information flows from left to night. If the frame in Figure 11.14 has a 0 appear as the first bit in the control field, the frame is an I frame (see Figure 11.15a). If the bit is a 1, the frame is an S or a U frame, as shown in Figures 11.15b, c. If that first bit is a 1 followed by a 0, it is an S frame, and if the bit is a 1 followed by a 1, it is a U frame. These bits are called format identifiers.

Turning now to the information (I) frame (Figure 11.15a), its purpose is to carry user data. Bits 2, 3, and 4 of the control field in this case carry the *send* sequence count of transmitted messages (i.e., I frames).

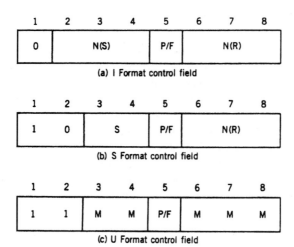

Figure 11.15. The three control field formats of HDLC.

We now digress to describe a *window* of frames. One generally thinks of a receiver on a point-to-point link with one buffer. This works well with stop-and-wait ARQ. Station X sends a message to station Y, which stores the message in a single buffer. On receipt of the entire bit string making up the message, the receiver processes the FCS for errors. If the frame is error free, the message or frame is acknowledged. We should note that stop-and-wait ARQ can be slow, tedious, and inefficient.

Suppose now that receiver Y has seven buffers (the number was picked arbitrarily). Thus Y can accept seven frames (or messages) and X is allowed to send seven frames without acknowledgment. To keep track of which frames have been acknowledged, each is labeled with a sequence number 0–7 (modulo 8). Station Y acknowledges a frame by sending the next sequence number expected. For instance, if Y sends a sequence number 3, this acknowledges frame number 2 and is awaiting frame number 3. Such a scheme can be used to acknowledge multiple frames. As an example, Y could receive frames 2, 3, and 4 and withhold all acknowledgments until frame 4 arrives. By sending sequence number 5, it acknowledges the receipt of frames 2, 3, and 4 all at once. Station X maintains a list of sequence numbers that it is allowed to send, and Y maintains a list of sequence numbers it is prepared to receive. These lists are thought of as a *window of frames*.

HDLC allows a maximum window size of 7, or 127 frames. In other words, a maximum number of 7, or 127 unacknowledged frames, can be sent or one less than the modulus 8 or 128. $N(S)$ is the sequence number of the next frame to be transmitted and $N(R)$ is the sequence number of the frame to be received.

Each frame carries a poll/final (P/F) bit. It is bit 5 in each of the three different types of control fields shown in Figure 11.15. The bit serves a function in both command and response frames. In a command frame it is referred to as a poll (P) bit; in a response frame as a final (F) bit. In both cases the bit is sent as a 1.

The P bit is used to solicit a response or sequence of responses from a secondary or balanced station. On a data link only one frame with a P bit set to 1 can be outstanding at any given time. Before a primary or balanced station can issue another frame with a P bit set to 1, it must receive a response frame from a secondary or balanced station with the F bit set to 1. In the NRM mode, the P bit is set to 1 in command frames to solicit response frames from the secondary station. In this mode of operation the secondary station may not transmit until it receives a command frame with the P bit set to 1.

Of course, the F bit is used to acknowledge an incoming P bit. A station may not send a final frame without prior receipt of a poll frame. As can be seen, P and F bits are exchanged on a one-for-one basis. Thus only one P bit can be outstanding at a time. As a result the $N(R)$ count of a frame containing a P or F bit set to 1 can be used to detect sequence errors. This capability is called *check pointing*. It can be used not only to detect sequence errors but to indicate the frame sequence number to begin retransmission when required.

Supervisory (S) frames, shown in Figure 11.15b, are used for flow and error control. Both go-back-*n* and selective ARQ can be accommodated. There are four types of supervisory or S frames:

1. Receive ready (RR): 1000 P/F $N(R)$
2. Receive not ready (RNR): 1001 P/F $N(R)$
3. Reject (Rej): 1010 P/F $N(R)$
4. Selective reject (SRej): 1011 P/F $N(R)$

The RR frame is used by a station to indicate that it is ready to receive information and acknowledge frames up to and including $N(R) - 1$. Also a primary station may use the RR frame as a command with the poll (P) bit set to 1.

The RNR frame tells a transmitting station that it is not ready to receive additional incoming I frames. It does acknowledge receipt of frames up to and including sequence number $N(R) - 1$. I frames with sequence number $N(R)$ and subsequent frames, if any, are not acknowledged. The Rej frame is used with go-back-n ARQ to request retransmission of I frames with frame sequence number $N(R)$, and $N(R) - 1$ frames and below are acknowledged.

Unnumbered frames are used for a variety of control functions. They do not carry sequence numbers, as the name indicates, and do not alter the flow or sequencing of I frames. Unnumbered frames can be grouped into the following four categories:

1. Mode-setting commands and responses
2. Information transfer commands and responses
3. Recovery commands and responses
4. Miscellaneous commands and responses

The information field follows the control field (Figure 11.14) and precedes the FCS field. The I field is present only in information (I) frames and some unnumbered (U) frames. The I field may contain any number of bits in any code, related to character structure or not. Its length is not specified in the standard (ISO 3309 [7]). Specific system implementations, however, usually place an upper limit on I field size. Some implementations require that the I field contain an integral number of octets.

Frame Check Sequence (FCS). Each frame includes a frame check sequence (FCS). The FCS immediately follows the I field, or the C field if there is no I field, and precedes the closing flag (F). The FCS field detects errors due to transmission. The FCS field contains 16 bits, which are the result of a mathematical computation on the digital value of all bits excluding the inserted zeros (zero insertion) in the frame and including the address, control, and information fields.

It should be noted that previously we had called the FCS field the BCC or block check count. To most of us, FCS implies the use of CRC for error detection. BCC, to many of us, can have a wider implication. Namely, it may mean only some form of parity check including CRC.

With most HDLC implementations, the FCS field is 16 bits long using the ITU-T-recommended CRC (see Chapter 10, Section 5.4.1). Reference should be made to CCITT Rec. V.41 [10]. In some situations that require stringent undetected error conditions and/or because of frame length, a 32-bit FCS may be used. This 32-bit CRC is similar to the one used with 802 series LANs. It is described in Chapter 13. For more information on FCS, see Ref. 11.

6.3   X.25: A Packet-Switched Network Access Standard

6.3.1   Introduction to CCITT Rec. X.25. CCITT Rec. X.25 defines the procedures necessary for a packet mode data terminal to access the services provided by a packet-switched public data network (PDN). The original CCITT recommendation was approved in 1976 and subsequently has undergone a number of modifications. Figure 11.16 shows the X.25 concept of accessing the PDN. An example of the PDN is a public ISDN discussed in Chapter 14.

Data terminals defined by X.25 operate in a synchronous full-duplex mode with a data rate of 2400, 4800, 9600, and 14,400 bits/s; 48, 64, 128, 192, 256, 384, 512, 1024, 1536, and 1920 kbits/s.

6.3.2   X.25 Architecture and Its Relationship to OSI. X.25 spans the lowest three layers of the OSI reference models. Figure 11.17 illustrates the architecture and its relationship to OSI. It can be seen that X.25 is compatible

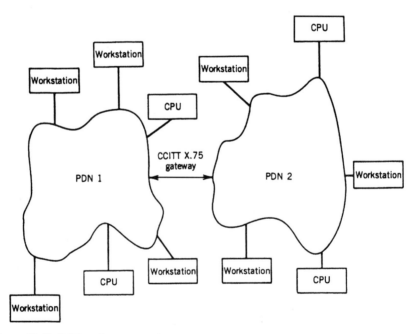

Figure 11.16. X.25 packet communications operates with the public (switched) data network (PDN). CPU, central processing unit or host computer.

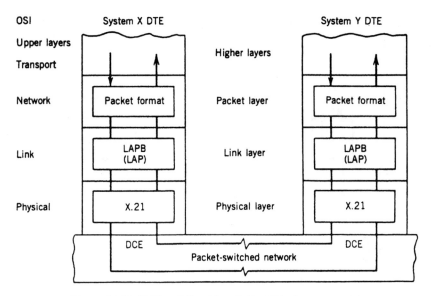

Figure 11.17. X.25's relationship with the OSI reference model.

with OSI up to the network layer. In this context there are differences at the network/transport layer boundary. CCITT leans toward the view that the network and transport layer services are identical and that these are provided by X.25 virtual circuits.

6.3.2.1 User Terminal Relationship to the PDN. CCITT Rec. X.25 calls the user terminal the DTE, and the DCE resides at the related PDN node. The entire recommendation deals with this DTE–DCE interface, not just the physical layer interface. For instance, a node (DCE) may connect to a related user (DTE) with one digital link, which is covered by SLP (single-link procedure) or several links covered by MLP (multilink procedure). Multiple links from a node to a DTE are usually multiplexed on one transmission facility.

The user (DTE) to user (DTE) connectivity through the PDN based on OSI is shown in Figure 11.18. A three-node connection is illustrated in this example. Of course, OSI layers 1–3 are Rec. X.25-specific. Note that the DTE protocol peers for these lower three layers are located in the PDN nodes and not in the distant DTE. The first DTE layer operating end to end is layer 4, the transport layer. Furthermore, the Rec. X.25 protocol operates only at the interface between the DTE and its related PDN node and does not govern internodal network procedures.

6.3.3 The Three Layers of X.25

6.3.3.1 The Physical Layer. The physical layer is layer 1 where the requirements are defined for the functional, mechanical, procedural, and electrical interface between the DTE and DCE. CCITT Rec. X.21 or X.21 bis is the applicable

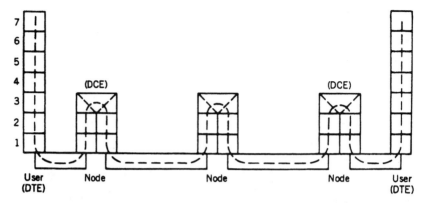

Figure 11.18. X.25 user (DTE) connects through the PDN to a distant user (DTE).

standard for the interface as called out in CCITT Rec. X.25. X.21 bis is similar to TIA/EIA-232.

CCITT Rec. X.21 [13] specifies a 15-pin DTE–DCE interface connector (refer to ISO 4903). The electrical characteristics for this interface are the same as CCITT Recs. V.10 and V.11, depending on whether electrically balanced or unbalanced operation is desired.

6.3.3.2   The Link Layer. The Rec. X.25 link layer specifies LAPB (link access protocol B). Earlier versions of Rec. X.25 permitted the use of LAP, which was based on the ISO ARM protocol. LAPB is preferred, but LAP is still permitted. LAPB is fully compatible with HDLC link-layer access protocol, balanced asynchronous class (see Section 6.2.3). The information field in the LAPB frame carries the user data, in this case the layer 3 packet.

LAPB provides several options for link operation. These include modulo-8 or modulo-128 control fields. It also supports MLP or multilink procedures. MLP allows a group of links to be managed as a single transmission facility. It carries out the function of resequencing packets in the proper order at the desired destination. When MLP is implemented, an MLP control field of two octets in length is inserted as the first 16 bits of the information field. It contains a multilink sequence number and four control bits. See Figure 11.19.

6.3.3.3   Datagrams, Virtual Circuits, and Logical Connections. There are three approaches used with Rec. X.25 operation to manage the transfer and routing of packet streams: datagrams, virtual connections (VCs), and permanent virtual connections (PVCs). Datagram service uses optimal routing on a packet-by-packet basis, usually over diverse routes. In the virtual circuit approach, there are two operational modes: virtual connection and permanent virtual connection. These are analogous to a dial-up telephone connection and a leased line connection, respectively. With the virtual connection a logical connection is established before any packets are sent. The packet originator sends a call request to its serving node, which sets up a route in advance to the desired destination. All

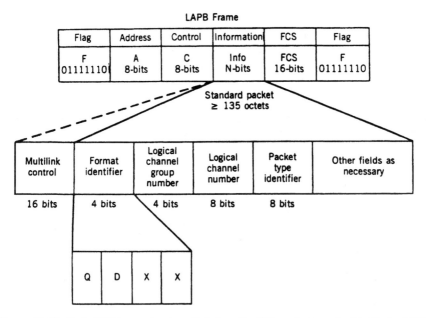

Figure 11.19. Basic X.25 frame structure. Note how the X.25 packet is embedded in the LAPB frame. For the extended LAPB structure the control will have 16 bits. There is also an extended modulo-128 structure for the X.25 packet. For a basic data packet, XX = 01. For an extended (modulo 128) data packet, XX = 10. Q = qualifier bit; D = delivery confirmation bit.

packets of a particular message traverse this route, and each packet of the message contains a virtual circuit identifier (logical channel number) and the packet data. At any one time each station can have more than one virtual circuit to any other station and can have virtual circuits to more than one station. With virtual circuits routing decisions are made in advance. With the datagram approach ad hoc decisions are made for each packet at each node. There is no call-setup phase with datagrams; there is with virtual connections. Virtual connections are advantageous for high community-of-interest connectivities, datagram for low community-of-interest relations.

Datagram service is more reliable because traffic can be alternately routed around network congestion points. Virtual circuits are fixed-routed for a particular call. Call-setup time at each node is eliminated on a packet basis with the virtual connection technique. Rec. X.25 also allows the possibility of setting up permanent virtual connections and is network assigned. This latter alternative is economically viable only for very high traffic relations; otherwise these permanently assigned logical channels will have long dormant periods.

6.3.4 X.25 Frame Structure: Layer 3, The Packet Layer.

The basic data-link layer (LAPB) frame structure is given in Figure 11.19. Its similarity to the HDLC frame structure (Figure 11.14) is apparent. For the X.25 case the packet is embedded in the LAPB information field, as mentioned. When applicable, the other part of the information field contains the MLP, which is appended in front of the X.25 packet and is the first subfield in the information (I) field. Actually the MLP is not part of the actual packet and is governed by the layer 2 LAPB protocol.

6.3.4.1 Structure Common to All Packets. Table 11.4 shows 17 packet types involved in X.25. Every packet transferred across the X.25 DTE–DCE interface consists of at least three octets.* These three octets contain a general format

TABLE 11.4   Packet Type Identifier

Packet Type		Octet 3							
					Bit				
From DCE to DTE	From DTE to DCE	8	7	6	5	4	3	2	1
	Call Setup and Cleaning								
Incoming call	Call request	0	0	0	0	1	0	1	1
Call connected	Call accepted	0	0	0	0	1	1	1	1
Clear indication	Clear request	0	0	0	1	0	0	1	1
DCE clear confirmation	DTE clear confirmation	0	0	0	1	0	1	1	1
	Data and Interrupt								
DCE data	DTE data	X	X	X	X	X	X	X	0
DCE interrupt	DTE interrupt	0	0	1	0	0	0	1	1
DCE interrupt confirmation	DTE interrupt confirmation	0	0	1	0	0	1	1	1
	Flow Control and Reset								
DCE RR (modulo 8)	DTE RR (modulo 8)	X	X	X	0	0	0	0	1
DCE RR (modulo 128)[a]	DTE RR (modulo 128)[a]	0	0	0	0	0	0	0	1
DCE RNR (modulo 8)	DTE RNR (modulo 8)	X	X	X	0	0	1	0	1
DCE RNR (modulo 128)[a]	DTE RNR (modulo 128)[a]	0	0	0	0	0	1	0	1
	DTE REJ (modulo 8)[a]	X	X	X	0	1	0	0	1
	DTE REJ (modulo 128)[a]	0	0	0	0	1	0	0	1
Reset indication	Reset request	0	0	0	1	1	0	1	1
DCE reset confirmation	DTE reset confirmation	0	0	0	1	1	1	1	1
	Restart								
Restart indication	Restart request	1	1	1	1	1	0	1	1
DCE restart confirmation	DTE restart confirmation	1	1	1	1	1	1	1	1
	Diagnostic								
Diagnostic[a]		1	1	1	1	0	0	0	1
	Registration[a]								
	Registration request	1	1	1	1	0	0	1	1
Registration confirmation		1	1	1	1	0	1	1	1

[a]Not necessarily available on every network.
Note: A bit that is indicated as X may be set to either 0 or 1.
Source: ITU-T Rec. X.25, Table 5-2/X.25, page 52 [12].

---

* An octet is an 8-bit sequence. There is a growing tendency to use octet rather than byte when describing such a sequence. It removes ambiguity on the definition of byte.

identifier, a logical channel identifier, and a packet type identifier. Other packets are appended as required. This is shown in Figure 11.19.

Now consider the general format identifier in Figure 11.19. This is a 4-bit sequence. Bit 8 is the qualifier bit only in data packets. In call setup and clearing packets it is the A-bit, and in all other packets it is set to 0. The D-bit, when set to 1, specifies end-to-end delivery confirmation. This confirmation is provided through the packet receive number $[P(R)]$. When set to 01 the XX bits indicate a basic packet (i.e., modulo-8), and when set to 10 they indicate an extended modulo-128 packet. The extension involves sequence number lengths.

Logical channel assignment is shown in Figure 11.20. The logical channel group and logical channel number subfields identify logical channels with the capability of identifying up to 4096 channels (2^{12}). This permits a DTE to establish up to 4095 simultaneous virtual circuits through its DCE to other DTEs. As we mentioned, this is usually done by multiplexing these circuits over a single transmission facility.

Permanent virtual circuits (PVCs) have permanently assigned logical channels, whereas those for virtual calls are assigned channels only for the duration of a call, as shown in Figure 11.20. Channel 0 is reserved for restart and diagnostic functions. To avoid collisions, the DCE starts assigning logical channels at the lowest number end and the DTE from the highest number end. There are one-way

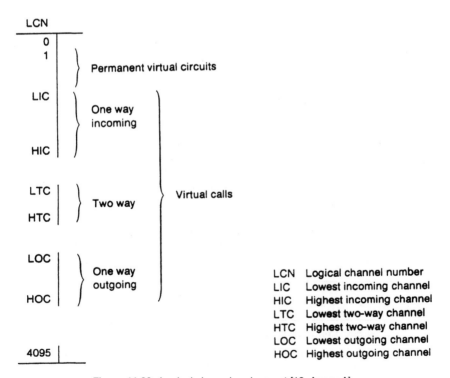

Figure 11.20. Logical channel assignment [12, Annex A].

and both-way (two-way) circuits. The both-way circuits are reserved for overflow to avoid chances of double seizure (i.e., both ends seize the same circuit).

Octet 3 in Figure 11.19 is the packet-type identifier subfield. The packet type and its corresponding coding are shown in Table 11.4. We can see in the table that packet types are identified in associated pairs carrying the same packet identifier (bit sequence). A packet from the calling terminal (DTE) to the network (DCE) is identified by one name. The associated packet delivered by the network to the called terminal (DTE) is referred to by another associated name.

6.3.4.2 Several Typical Packets

CALL REQUEST AND INCOMING CALL PACKET. This type of packet sets up the call for the virtual circuit. The format of the call request and incoming packet is shown in Figure 11.21. Octets 1–3 have been described in the previous subsection. Octet 4 consists of the address length field indicators for the called and calling DTE addresses. Each address length indicator is binary coded, and bit 1 or 5 is the low-order bit of the indicator. Octet 5 and the following octets consist of the called DTE address, when present, and then the calling DTE address, when present.

The facilities length field (one octet) indicates the length of the facilities field that follows. The facility field is present only when the DTE is using an optional user facility requiring some indication in the call request and incoming call packets. The field must contain an integral number of octets with a maximum length of 109 octets.

Optional user facilities are listed in CCITT Rec. X.2. There are 45 listed. Several examples are listed to give some idea of what is meant by *facilities* in CCITT Rec. X.25:

- Nonstandard default window
- Flow control parameter negotiation

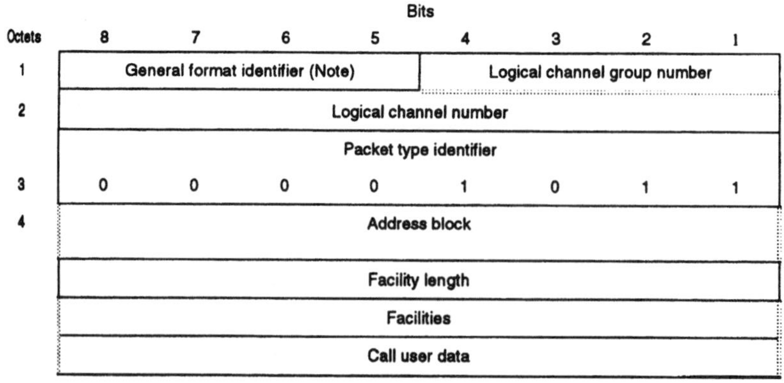

Octets	Bits 8	7	6	5	4	3	2	1
1	General format identifier (Note)				Logical channel group number			
2	Logical channel number							
	Packet type identifier							
3	0	0	0	0	1	0	1	1
4	Address block							
	Facility length							
	Facilities							
	Call user data							

Note: Coded XX01 (modulo 8) or XX10 (modulo 128).

Figure 11.21. Call request and incoming call packet. The general format identifier is coded 0X01 (modulo-8) and 0X10 (modulo-128). From ITU-T Rec. X.25, Figure 5-3/X.25, page 58, ITU-T Organization, Helsinki, March 1993 [12].

- Throughput class negotiation
- Incoming calls barred
- Outgoing calls barred
- Closed user group (CUG)
- Reverse charging acceptance
- Fast select

DTE AND DCE DATA PACKET. Of the 17 packet types listed in Table 11.4, only one truly carries user information, the DTE and DCE data packet. Figure 11.22 illustrates this packet format.

Octets 1 and 2 have been described. Bits 6, 7, and 8 of octet 3 or bits 2–8 of octet 4, when extended, are used for indicating the packet received sequence number $P(R)$. It is binary coded and bit 6, or bit 2 when extended, is the low-order bit.

In Figure 11.22, M, which is bit 5 in octet 3 or bit 1 in octet 4 when extended, is used for more data (M bit). It is coded 0 for "no more data" and 1 for "more data to follow."

Bits 2, 3, and 4 of octet 3, or bits 2–8 of octet 3 when extended, are used for indicating the packet send sequence number $P(S)$. Bits following octet 3, or octet 4 when extended, contain the user data.

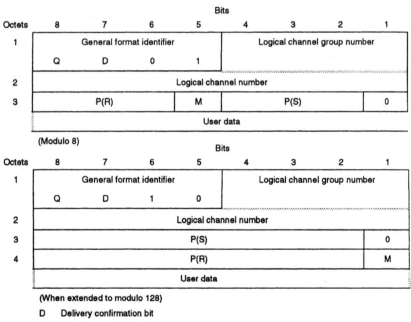

Figure 11.22. DTE and DCE packet format. From ITU-T Rec. X.25, Figure 5-7/X.25, page 64, ITU-T Organization, Helsinki, March 1993 [12].

The standard maximum user data field length is 128 octets. CCITT Rec. X.25 (paragraph 4.3.2) states: "In addition, other maximum user data field lengths may be offered by Administrations from the following list: 16, 32, 64, 256, 512, 1024, 2048, and 4096 octets .... Negotiation of maximum user data field lengths on a per call basis may be made with the flow parameter negotiation facility."

6.3.5 *Tracing the Life of a Virtual Call.* A call is initiated by a DTE by the transfer to the network of a call request packet. It identifies the logical channel number selected by the originating DTE, the address of the called DTE (destination), and optional facility information and can contain up to 16 octets of user information. The facility and user fields are optional at the discretion of the source DTE. The receipt by the network of the call request packet initiates the call-setup sequence. This same call request packet is delivered to the destination DTE as an incoming call packet. The destination DTE in return sends a "call accepted" packet, and the source DTE receives a "call confirmation" packet. This completes the call-setup phase, and the data transfer can begin.

Data packets (Figure 11.22) carry the user data to be transferred. There may be one or a sequence of packets transferred during a virtual call. It is the M bit that tells the destination that the next packet is a logical continuation of the previous packet(s). Sequence numbers verify correct packet order and are the packet acknowledgment tools.

The last phase in the life of a virtual call is the call clearing (takedown). Either the DTE or the network can clear a virtual call. The "clear" applies only to the logical channel that was used for that call. Three different packet types are involved in the call-clearing phase. The clear request packet is issued by the DTE initiating the clear. The remote DTE receives it as a clear indication packet. Both the DTE and DCE then issue clear confirmation packets to acknowledge receipt of the clear packets.

Flow control is called out by the following packet types: receive ready (RR), receive not ready (RNR), and reject (Rej). Each of these packets is normally three octets long or four octets long using modulo-128 numbering. Sequence numbering also assists flow control.

7 TCP/IP AND RELATED PROTOCOLS

7.1 Background and Scope

The transmission control protocol/internet protocol (TCP/IP) family was developed for the ARPANET (Advanced Research Projects Agency Network; see Figure 11.1). ARPANET was one of the first large advanced packet-switched networks. It was initially designed and operated to interconnect the very large university and industrial defense research community to share research resources. It dates back to 1968 and was well into existence before ISO and CCITT took interest in layered protocols.

OSI	TCP/IP AND RELATED PROTOCOLS		
Application	File transfer	Electronic mail	Terminal emulation
Presentation	File transfer protocol (FTP)	Simple mail transfer protocol (SMTP)	Telnet protocol
Session			
Transport	Transmission control protocol (TCP)		User datagram protocol (UDP)
Network	Address resolution protocol (ARP)	Internet protocol (IP)	Internet control message protocol (ICMP)
Data link	———— Network interface cards ———— CSMA/CD (Ethernet), token ring, ARCNet, StarLan		
Physical	———— Transmission media ———— Wire pair, fiber optics, coaxial cable, radio		

Figure 11.23. How TCP/IP and associated protocols relate to OSI.

The TCP/IP suite of protocols [16–24] has wide acceptance today, especially in the commercial and industrial community worldwide. These protocols are used on both LANs and WANs. They are particularly attractive for their internetworking capabilities. The internet protocol (IP) [16] competes with ITU-T Rec. X.75 protocol [15], but is notably more versatile and has a much wider application.

The architectural model of the IP [16] uses terminology that differs from the OSI reference model.* Figure 11.23 shows the relationship between TCP/IP and related DoD protocols and the OSI reference model. Tracing data traffic from an originating host, which runs an applications program, to another host in another network is shown in Figure 11.24. This may be a LAN-to-WAN-to-LAN connectivity as shown in the figure. It may also be a LAN-to-LAN or it may be a WAN-to-WAN connectivity. The host would enter its own network by means of a network access protocol such as HDLC or an IEEE 802 series protocol (Chapter 13).

A LAN connects via a router (or gateway) to another network. Typically a router (or gateway) is loaded with three protocols. Two of these protocols connect to each of the attached networks (e.g., LAN and WAN), and the third protocol is the IP which provides the network-to-network interface.

Hosts typically are equipped with four protocols. To communicate with routers or gateways, a network access protocol and internet protocol are required. A transport layer protocol assures reliable communication between hosts because end-to-end capability is not provided in either the network access or internet protocols. Hosts also must have application protocols such as e-mail or file transfer protocols (FTPs).

7.1.1 A New Version of IP. What we describe in Section 7 is Internet Protocol, Version 4, better known in the Industry as IPV.4. IPV.4 pervades the

* IP predates OSI.

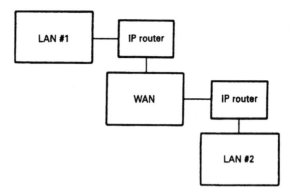

Figure 11.24. Connecting one LAN to another LAN via a WAN with routers equipped with IP.

telecommunications environment. To meet some addressing disparities, a version 6 of IP has been developed. However, there seems to be some reticence in accepting IPV.6. We think it is beginning to catch on in the community. In Section 7.5 a very brief description of IPV.6 is presented.

7.2   TCP/IP and Data-Link Layers

TCP/IP is transparent to the type of data-link layer involved, and it is also transparent whether it is operating in a LAN or WAN domain or among them. However, there is document support for Ethernet, IEEE 802 series, ARCNET LANs, and X.25 for WANs [18, 19].

Figure 11.25 shows how upper OSI layers are encapsulated with TCP and IP header information and then incorporated into the data-link layer frame.

For the case of IEEE 802 series LAN protocols, advantage is taken of the LLC common to all 802 protocols. The LLC extended header contains the SNAP (subnetwork access protocol) such that we have three octets for the LLC header and

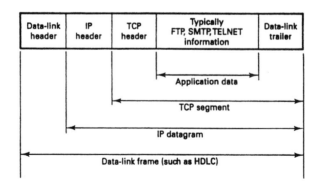

Figure 11.25. The incorporation of upper-layer PDUs into a data-link layer frame showing the relationship with TCP and IP

five octets in the SNAP. The LLC header has its fields fixed as follows (LLC is discussed in Chapter 13):

DSAP = 10101010   (destination service access point)

SSAP = 10101010   (source service access point)

Control = 00000011   [For unnumbered information (UI frame)]

The five octets in the SNAP have three assigned for protocol ID or organizational code and two octets for "EtherType." EtherType assignments are shown in Table 11.5. EtherType refers to the general class of LANs based on CSMA/CD. (See Chapter 13 for a discussion of CMSA/CD.)

Figure 11.26 shows the OSI relationships with TCP/IP working with the IEEE 802 LAN protocol group. Figure 11.27 illustrates an IEEE 802 frame incorporating TCP, IP, and LLC (logical link control, Chapter 13).

Oftentimes, addressing formats are incompatible from one protocol in one network to another protocol in another network. A good example is the mapping of the 32-bit internet address into a 48-bit IEEE 802 address. In IPV.4 this problem is resolved with ARP (Address Resolution Protocol). Another interface problem is the limited IP datagram length of 576 octets, where with the 802

TABLE 11.5   EtherType Assignments

Ethernet Decimal	Hex	Description
512	0200	XEROX PUP
513	0201	PUP address translation
1536	0600	XEROX NS IDP
2048	0800	DOD internet protocol (IP)
2049	0801	X.75 internet
2050	0802	NBS internet
2051	0803	ECMA internet
2052	0804	Chaosnet
2053	0805	X.25 level 3
2054	0806	Address resolution protocol (ARP)
2055	0807	XNS compatibility
4096	1000	Berkeley trailer
21000	5208	BBN Simnet
24577	6001	DEC MOP dump/load
24578	6002	DEC MOP remote control
24579	6003	DEC DECnet phase IV
24580	6004	DEC LAT
24582	6005	DEC
24583	6006	DEC
32773	8005	HP probe
32784	8010	Excelan
32821	8035	Reverse ARP
32824	8038	DEC LANBridge
32823	8098	Appletalk

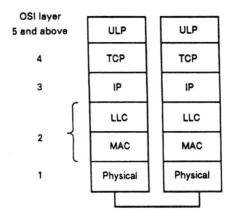

Figure 11.26. How TCP/IP working with IEEE 802 series of LAN protocols relates to OSI.

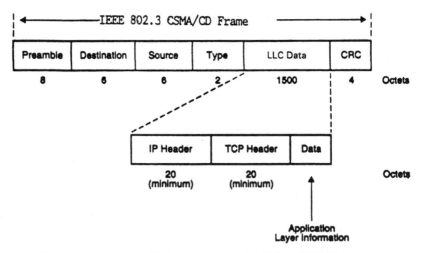

Figure 11.27. A typical IEEE 802 frame showing LLC and TCP/IP functions.

series, frames have considerably larger length limits. IPV.6 resolves both of these dilemmas as discussed in Section 7.5.

7.3 The IP Routing Function

In OSI the network layer functions include routing and switching of a datagram through the telecommunications subnetwork. The IP provides this essential function. It forwards the datagram based upon the network address contained within the IP header. Each datagram is independent and has no relationship with other datagrams. There is no guaranteed delivery of the datagram from the standpoint of the internet protocol. However, the next higher layer, the TCP layer, provides for the reliability that the IP lacks. It also carries out segmentation and reassembly functions of a datagram to match frame sizes of data-link layer protocols.

Addresses determine routing, and at the far end, equipment (hardware). Actual routing derives from the IP address, and equipment addresses derive from the data-link layer header (typically the 48-bit Ethernet address) [19, 20].

User data from upper-layer protocols are passed to the IP layer. The IP layer examines the network address (IP address) for a particular datagram and determines if the destination node is on its own local area network or some other network. If it is on the same network, the datagram is forwarded directly to the destination host. If it is on some other network, it is forwarded to the local IP router (gateway). The router, in turn, examines the IP address and forwards the datagram as appropriate. Routing is based on a look-up table residing in each router or gateway.

7.3.1 *Detailed IP Operation.*
The IP provides connectionless service, meaning that there is no call-setup phase prior to exchange of traffic. There are no flow control or error control capabilities incorporated in IP. These are left to the next higher layer, the transmission control protocol (TCP). The IP is transparent to subnetworks connecting at lower layers; thus different types of networks can attach to an IP gateway.

Whereas prior to this discussion we have used the term *segmentation* to mean breaking up a data file into manageable segments, frames, packets, blocks, and so on, the IP specifications refer to this as *fragmentation*. IP messages are called *datagrams*. The minimum datagram length is 576 octets and the maximum length is 65,535 octets. Fragmentation resolves PDU (protocol data unit) sizes of the different networks with which IP carries out an interface function. For example, X.25 packets typically have data fields of 128 octets; Ethernet limits the size of a PDU to 1500 octets, and so forth. Of course, IP does a reassembly of the "fragments" at the opposite end of its circuit [16].

7.3.1.1 *Description of the IP Datagram.*
The IP datagram format is shown in Figure 11.28. The datagram format should be taken in context with Figures 11.26 and 11.27 showing how the IP datagram relates to TCP and the data-link layer. In Figure 11.28, we move from left to right and top down in our description.

The *version* field (4 bits) gives the release number of the IP version for which a particular gateway or router is equipped.

The *header length* (4 bits) is measured in units of 32 bits. A header without certain options, such as QoS, typically has 20 octets. Thus its length is 5 (32 bits or 4 octets per unit; $5 \times 4 = 20$).

The *type of service* (TOS) field (8 bits) is used to identify several QoS parameters provided by IP. The field has eight bits broken down into four active groupings, and the last two bits are reserved. The first is *precedence*, which consists of three bits as follows:

000	routine	001	priority
010	immediate	011	flash
100	flash override	101	CEITIC/ECP
110	internetwork control	111	network control

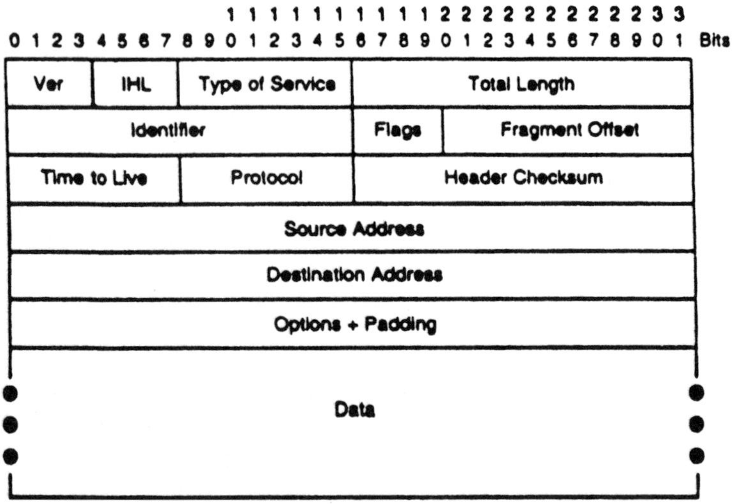

```
                        1 1 1 1 1 1 1 1 1 1 2 2 2 2 2 2 2 2 2 2 3 3
        0 1 2 3 4 5 6 7 8 9 0 1 2 3 4 5 6 7 8 9 0 1 2 3 4 5 6 7 8 9 0 1  Bits
```

Ver	IHL	Type of Service	Total Length

Identifier	Flags	Fragment Offset

Time to Live	Protocol	Header Checksum

Source Address

Destination Address

Options + Padding

Data

Figure 11.28. The IP datagram format [16].

Delay, 1 bit:	$0 =$ normal, $1 =$ low
Throughput, 1 bit:	$0 =$ normal, $1 =$ high
Reliability, 1 bit:	$0 =$ normal, $1 =$ high

The *total length* field specifies the total length of an IP datagram in question. The unit of measure is the octet, and it includes the length of the header and data fields. The maximum possible length of a datagram is $2^{16} - 1$; the minimum length is 576 octets.

Segmentation (fragmentation) and reassembly are controlled by three fields in the header. These are *identifier* (16 bits), *flags* (3 bits), and *fragment offset* (3 bits). The term *fragment* suggests part of a whole; thus the identifier field identifies a fragment as part of a complete datagram, along with the source address. The flag bits determine if a datagram can be fragmented. When it can be fragmented, one of the flag bits shows whether the fragment is the last fragment of a datagram. The fragmentation offset field gives the relative position of the fragment regarding the original datagram. It is initially set at 0 and then set to the proper number by the fragmenting gateway.

The *time-to-live* (TTL) field's basic purpose is to prevent routing loops. In this we mean a routing that eventually routes back on itself. In telephony, it is sometimes called *ring-around-the-rosy*. It is used to measure the time a datagram has been in the internet. Each internet gateway checks this number and will discard that datagram if the TTL equals zero. There are numerous ways to implement TTL, some being vendor-specific. It is sometimes used for diagnostics by network management features such as SNMP.*

---

* SNMP is discussed in Chapter 19.

The *protocol* field (8 bits) identifies the next-higher-layer protocol the datagram expects at the destination host. Table 11.6, taken from RFC 1060 [17] and Ref. 18, shows the protocol decimal numbering scheme and the corresponding protocol for each number.

The next field is the *header checksum* field (16 bits). This provides error detection on the header.

Source address and *destination address* fields are each 32 bits long. Of course, the source address is the address of the originating host and the destination address is the address of the destination host.

The IP address structure is shown in Figure 11.29. It shows four address formats in which the lengths of the two component fields making up the address field change with each of the different formats. These component fields are the "network address" and the "local address" fields. The first bits of the address field specify the format or "class," there being classes A, B, C, or D. The "local address" is often called "host address."

The *class A* addressing is for very large networks, such as what was ARPANET. The field starts with binary 0, indicating that it is a class A format. In this case the local or host address component field is 24 bits long and has an address capacity of 2^{24}. *Class B* addressing is for medium-sized networks,

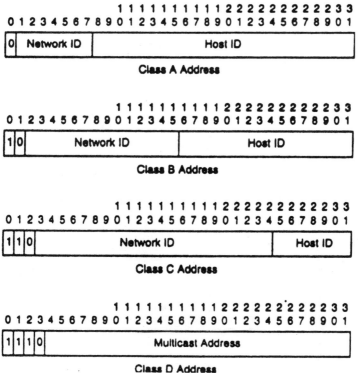

Figure 11.29. Internet protocol address formats [16].

TABLE 11.6 IP Protocol Field Numbering (Assigned Internet Protocol Numbers)

Decimal	Key Word	Protocol
0		Reserved
1	ICMP	Internet control message protocol
2	IGMP	Internet group management protocol
3	GGP	Gateway-to-gateway protocol
4		Unassigned
5	ST	Stream
6	TCP	Transmission control protocol
7	UCL	UCL
8	EGP	Exterior gateway protocol
9	IGP	Interior gateway protocol
10	BBN-MON	BBN-RCC monitoring
11	NVP-II	Network voice protocol
12	PUP	PUP
13	ARGUS	ARGUS
14	EMCON	EMCON
15	XNET	Cross net debugger
16	CHAOS	Chaos
17	UDP	User datagram protocol
18	MUX	Multiplexing
19	DCN-MEAS	DCN measurement subsystems
20	HMP	Host monitoring protocol
21	PRM	Packet radio monitoring
22	XNS-IDP	XEROX NS IDP
23	TRUNK-1	Trunk-1
24	TRUNK-2	Trunk-2
25	LEAF-1	Leaf-1
26	LEAF-2	Leaf-2
27	RDP	Reliable data protocol
28	IRTP	Internet reliable TP
29	ISO-TP4	ISO transport class 4
30	NETBLT	Bulk data transfer
31	MFE-NSP	MFE network services
32	MERIT-INP	MERIT Internodal protocol
33	SEP	Sequential exchange
34–60		Unassigned
61		Any host internal protocol
62	CFTP	CFTP
63		Any local network
64	SAT-EXPAK	SATNET and backroom EXPAK
65	MIT-SUBN	MIT subnet support
66	RVD	MIT remote virtual disk
67	IPPC	Internet plur. packet core
68		Any distributed file system
69	SAT-MON	SATNET monitoring
70		Unassigned
71	IPCV	Packet core utility
72–75		Unassigned
76	BRSAT-MON	Backroom SATNET monitoring
77		Unassigned
78	WB-MON	Wideband monitoring
79	WB-EXPAK	Wideband EXPAK
80–254		Unassigned
255		Reserved

Source: RFC 1060 [17] and *Internet Protocol Transition Workbook* [18].

such as campus networks. The field begins with 10 to indicate that it is a class B format; the network component field is 14 bits long, and the host or local address component field is 16 bits in length. The *class C* format is for small networks with a very large network ID field with an addressing capacity of 2^{24} and a considerably small host ID field of only 8 bits (2^8 addressing capacity). The address field in this case starts with 110. The *class D* format is for multicasting, a form of broadcasting. Its first four bits are the sequence 1110.

7.3.1.2 IP Routing.

A gateway (router) needs only the network ID portion of the address to perform its routing function. Each router or gateway has a routing table which consists of: destination network addresses and specified next-hop gateway.

Three types of routing are performed by the routing table:

1. Direct routing to locally attached devices
2. Routing to networks that are reached via one or more gateways
3. Default routing to destination network in case the first types of routing are unsuccessful

Suppose a datagram (or datagrams) is (are) directed to a host which is not in the routing table resident in a particular gateway. Likewise, there is a possibility that the network address for that host is also unknown. These problems may be resolved with the *address resolution protocol* (ARP) [20].

First the ARP searches a mapping table which relates IP addresses with corresponding physical addresses. If the address is found, it returns the correct address to the requester. If it cannot be found, the ARP broadcasts a *request* containing the IP target address in question. If a device recognizes the address, it will reply to the request where it will update its ARP *cache* with that information. The ARP cache contains the mapping tables maintained by the ARP module.

There is also a *reverse address resolution protocol* (RARP) [21]. It works in a fashion similar to that of the ARP, but in reverse order. RARP provides an IP address to a device when the device returns its own hardware address. This is particularly useful when certain devices are booted and only know their own hardware address.

Routing with IP involves a term called *hop*. A hop is defined as a link connecting adjacent nodes (gateways) in a connectivity involving IP. A *hop count* indicates how many gateways (nodes) must be traversed between source and destination.

One part of an IP routing algorithm can be *source routing*. Here an upper-layer protocol (ULP) determines how an IP datagram is to be routed. One option is that the ULP passes a listing of internet addresses to the IP layer. In this case information is provided on the intermediate nodes required for transit of a datagram in question to its final destination.

Each gateway makes its routing decision based on a resident routing list or routing table. If a destination resides in another network, a routing decision is required by the IP gateway to implement a route to that other network. In many cases, multiple hops are involved and each gateway must carry out routing decisions based on its own routing table.

A routing table can be static or dynamic. The table contains IP addressing information for each reachable network and closest gateway for the network, and it is based on the concept of shortest routing, thus routing through the closest gateway.

Involved in IP shortest routing is the *distance metric*, which is a value expressing minimum number of hops between a gateway and a datagram's destination. An IP gateway tries to match the destination network address contained in the header of a datagram with a network address entry contained in its routing table. If no match is found, the gateway discards the datagram and sends an ICMP message back to the datagram source.

7.3.1.3 Internet Control Message Protocol (ICMP). ICMP [22] is used as an adjunct to IP when there is an error in datagram processing. ICMP uses the basic support of IP as if it were a higher-level protocol; however, ICMP is actually an integral part of IP and is implemented by every IP module.

ICMP messages are sent in several situations: for example, when a datagram cannot reach its destination, when a gateway does not have the buffering capacity to forward a datagram, and when the gateway can direct the host to send traffic on a shorter route.

ICMP messages typically report errors in the processing of datagrams. To void the possibility of infinite regress of messages about messages, and so on, no ICMP messages are sent about ICMP messages. Also ICMP messages are only sent about errors in handling fragment zero of fragmented datagrams. (*Note*: Fragment zero has the fragment offset equal to zero.)

MESSAGE FORMATS. ICMP messages are sent using the basic IP header (see Figure 11.28). The first octet of the data portion of the datagram is an ICMP type field. The "data portion" is the last field at the bottom of Figure 11.28. The ICMP type field determines the format of the remaining data. Any field labeled *unused* is reserved for later extensions and is fixed at zero when sent, but receivers should not use these fields (except to include them in the checksum). Unless otherwise noted under individual format descriptions, the values of the internet header fields are as follows:

- Version: 4.
- IHL (internet header length): length in 32-bit words.
- Type of service: 0.
- Total length: length of internet header and data in octets.
- Identification, flags, and fragment offset: used in fragmentation, as in basic IP protocol described above.
- Time to live: in seconds; as this field is decremented at each machine in which the datagram is processed, the value in this field should be at least as great as the number of gateways which this datagram will traverse.

- Protocol: ICMP = 1.
- Header checksum: the 16-bit one's complement of the one's complement sum of all 16-bit words in the header. For computing the checksum, the checksum field should be zero. The reference RFC (RFC 792) states that this checksum may be replaced in the future.
- Source address: the address of the gateway or host that composes the ICMP message. Unless otherwise noted, this is any of a gateway's addresses.
- Destination address: the address of the gateway or host to which the message should be sent.

There are eight distinct ICMP messages covered in RFC 792:

1. Destination unreachable message
2. Time exceeded message
3. Parameter problem message
4. Source quench message
5. Redirect message
6. Echo or echo reply message
7. Timestamp or timestamp reply message
8. Information request or information reply message

EXAMPLE: DESTINATION UNREACHABLE MESSAGE. The ICMP fields in this case are shown in Figure 11.30.

IP Fields

- Destination address: the source network and address from the original datagram's data.

ICMP Fields

- Type 3
- Code: 0 = net unreachable
  1 = host unreachable
  2 = protocol unreachable
  3 = port unreachable

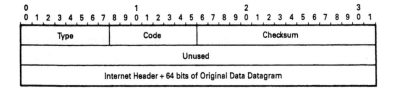

Figure 11.30. A typical ICMP message format; destination unreachable message [22].

4 = fragmentation needed and DF set

5 = source route failed.

- Checksum: as above
- Internet header +64 bits of data datagram

The internet header plus the first 64 bits of the original datagram's data is used by the host to match the message to the appropriate process. If a higher-level protocol uses port numbers, they are assumed to be in the first 64 data bits of the original datagram's data.

DESCRIPTION. If, according to the information in the gateway's routing tables, the network specified in the internet destination field of a datagram is unreachable (e.g., the distance to the network is infinity), the gateway may send a destination unreachable message to the internet source host of the datagram. In addition, in some networks, the gateway may be able to determine if the internet destination host is unreachable. Gateways in these networks may send destination unreachable messages to the source host when the destination host is unreachable.

If, in the destination host, the IP module cannot deliver the datagram because the indicated protocol module or process port is not active, the destination host may send a destination unreachable message to the source host.

Another case is when a datagram must be fragmented to be forwarded by a gateway yet the "Don't Fragment" flag is on. In this case the gateway must discard the datagram and may return a destination unreachable message.

It should be noted that codes 0, 1, 4, and 5 may be received from a gateway; codes 2 and 3 may be received from a host (RFC 792 [22]).

7.4 The Transmission Control Protocol (TCP)

7.4.1 TCP Defined. TCP [23, 24] was designed to provide reliable communication between pairs of processes in logically distinct hosts on networks and sets of interconnected networks. TCP operates successfully in an environment where the loss, damage, duplication or misorder of data, and network congestion can occur. This robustness in spite of unreliable communications media makes TCP well-suited to support commercial, military, and government applications. TCP appears at the transport layer of the protocol hierarchy. Here, TCP provides connection-oriented data transfer that is reliable, ordered, full duplex, and flow controlled. TCP is designed to support a wide range of upper-layer protocols (ULPs). The ULP can channel continuous streams of data through TCP for delivery to peer ULPs. The TCP breaks the streams into portions which are encapsulated together with appropriate addressing and control information to form a segment—the unit of exchange between TCPs. In turn, the TCP passes the segments to the network layer for transmission through the communication system to the peer TCP.

As shown in Figure 11.31, the layer below the TCP in the protocol hierarchy is commonly the internet protocol (IP) layer. The IP layer provides a way for

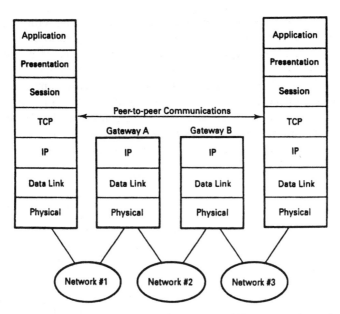

Figure 11.31. Protocol layers showing the relationship of TCP with other layered protocols.

the TCP to send and receive variable-length segments of information enclosed in internet datagram "envelopes." The internet datagram provides a means for addressing source and destination TCPs in different networks. The IP also deals with fragmentation or reassembly of TCP segments required to achieve transport and delivery through the multiple networks and interconnecting gateways. The IP also carries information on the precedence, security classification, and compartmentation of the TCP segments, so this information can be communicated end-to-end across multiple networks.

7.4.2 TCP Mechanisms. TCP builds its services on top of the network layer's potentially unreliable services with mechanisms such as error detection, positive acknowledgments, sequence numbers, and flow control. These mechanisms require certain addressing and control information to be initialized and maintained during data transfer. This collection of information is called a *TCP connection.* The following paragraphs describe the purpose and operation of the major TCP mechanisms.

PAR MECHANISM. TCP uses a positive acknowledgment with retransmission (PAR) mechanism to recover from the loss of a segment by the lower layers. The strategy with PAR is for a sending TCP to retransmit a segment at timed intervals until a positive acknowledgment is returned. The choice of retransmission interval affects efficiency. An interval that is too long reduces data throughput while one that is too short floods the transmission media with superfluous segments. In TCP, the timeout is expected to be dynamically adjusted to approximate the segment round-trip time plus a factor for internal processing; otherwise performance degradation may occur. TCP uses a simple checksum to detect segments damaged in transit. Such segments are discarded without being acknowledged. Hence, damaged segments are treated identically to lost segments and are compensated for by the PAR mechanism. TCP assigns sequence numbers to identify each octet of the data stream. These enable a receiving TCP to detect duplicate and out-of-order segments. Sequence numbers are also used to extend the PAR mechanism by allowing a single acknowledgment to cover many segments' worth of data. Thus, a sending TCP can still send new data although previous data have not been acknowledged.

FLOW CONTROL MECHANISM. TCP's flow control mechanism enables a receiving TCP to govern the amount of data dispatched by a sending TCP. The mechanism is based on a *window* which defines a contiguous interval of acceptable sequence-numbered data. As data are accepted, TCP slides the window upward in the sequence number space. This window is carried in every segment, enabling peer TCPs to maintain up-to-date window information.

MULTIPLEXING MECHANISM. TCP employs a multiplexing mechanism to allow multiple ULPs within a single host and multiple processes in a ULP to use TCP simultaneously. This mechanism associates identifiers, called *ports*, to ULP processes accessing TCP services. A ULP connection is uniquely identified with a *socket*, the concatenation of a port and an internet address. Each connection is uniquely named with a socket pair. This naming scheme allows a single ULP to support connections to multiple remote ULPs. ULPs which provide popular resources are assigned permanent sockets, called *well-known sockets*.

7.4.3 ULP Synchronization.
When two ULPs wish to communicate, they instruct their TCPs to initialize and synchronize the mechanism information on each to open the connection. However, the potentially unreliable network layer (i.e., IP layer) can complicate the process of synchronization. Delayed or duplicate segments from previous connection attempts might be mistaken for new ones. A handshake procedure with clock-based sequence numbers is used in connection opening to reduce the possibility of such false connections. In the simplest handshake, the TCP pair synchronizes sequence numbers by exchanging three segments, thus the name *three-way handshake*.

7.4.4 ULP Modes.
A ULP can open a connection in one of two modes, passive or active. With a passive open, a ULP instructs its TCP to be *receptive* to connections with other ULPs. With an active open, a ULP instructs its TCP to actually initiate a three-way handshake to connect to another ULP. Usually an active open is targeted to a passive open. This active/passive model supports server-oriented applications where a permanent resource, such as a database management process, can always be accessed by remote users. However, the three-way handshake also coordinates two simultaneous active opens to open

a connection. Over an open connection, the ULP pair can exchange a continuous stream of data in both directions. Normally, TCP groups the data into TCP segments for transmission at its own convenience. However, a ULP can exercise a *push* service to force TCP to package and send data passed up to that point without waiting for additional data. This mechanism is intended to prevent possible deadlock situations where a ULP waits for data internally buffered by TCP. For example, an interactive editor might wait forever for a single input line from a terminal. A push will force data through the TCPs to the awaiting process. A TCP also provides the means for a sending ULP to indicate to a receiving ULP that "urgent" data appear in the upcoming data stream. This urgent mechanism can support, for example, interrupts or breaks. When a data exchange is complete, the connection can be closed by either ULP to free TCP resources for other connections. Connection closing can happen in two ways. The first, called a *graceful close*, is based on the three-way handshake procedure to complete data exchange and coordinate closure between the TCPs. The second, called an *abort*, does not allow coordination and may result in the loss of unacknowledged data.

7.4.5 Scenario. The following scenario provides a walk-through of a connection opening, data exchange, and connection closing as might occur between the database management process and the user mentioned above. The scenario focuses more on (a) the three-way handshake mechanism in connection with opening and closing and (b) the positive acknowledgment with retransmission mechanism supporting reliable data transfer. Although not pictured, the network layer transfers the information between TCPs. For the purpose of this scenario, the network layer is assumed not to damage, lose, duplicate, or change the order of data unless explicitly noted. The scenario is organized into three parts:

a. A simple connection opening (steps 1–7) (Figure 11.32)
b. Two-way data transfer (steps 8–17) (Figure 11.33)
c. A graceful connection close (steps 18–25) (Figure 11.34)

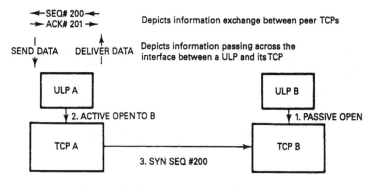

Figure 11.32A. A simple connection opening.

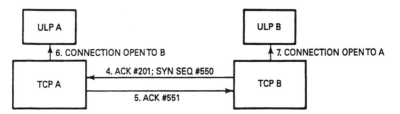

Figure 11.32B. A simple connection opening.

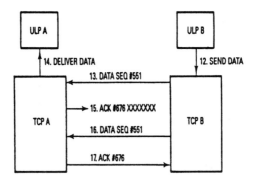

Figure 11.33A. Two-way data transfer.

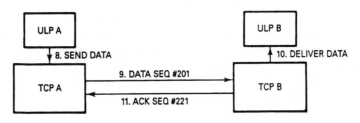

Figure 11.33B. Two-way data transfer.

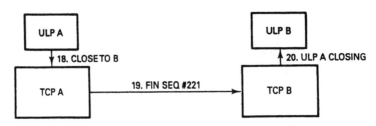

Figure 11.34A. A graceful connection close.

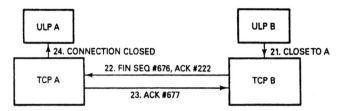

Figure 11.34B. A graceful connection close.

SCENARIO NOTATION. The following notation is used in the diagrams.

1. ULP B (the DB manager) issues a PASSIVE OPEN to TCP B to prepare for connection attempts from other ULPs in the system.
2. ULP A (the user) issues an ACTIVE OPEN to open a connection to ULP B.
3. TCP A sends a segment to TCP B with an OPEN control flag, called a SYN, carrying the first sequence number (shown as SEQ #200) it will use for data sent to B.
4. TCP B responds to the SYN by sending a positive acknowledgment, or ACK, marked with the next sequence number expected from TCP A. In the same segment, TCP B sends its own SYN with the first sequence number for its data (SEQ #550).
5. TCP A responds to TCP B's SYN with an ACK showing the next sequence number expected from B.
6. TCP A now informs ULP A that a connection is open to ULP B.
7. Upon receiving the ACK, TCP B informs ULP B that a connection has been opened to ULP A.
8. ULP A passes 20 octets of data to TCP A for transfer across the open connection to ULP B.
9. TCP A packages the data in a segment marked with current "A" sequence number.
10. After validating the sequence number, TCP B accepts the data and delivers it to ULP B.
11. TCP 3 acknowledges all 20 octets of data with the ACK set to the sequence number of the next data octet expected.
12. ULP B passes 125 bytes of data to TCP B for transfer to ULP A.
13. TCP B packages the data in a segment marked with the "B" sequence number.
14. TCP A accepts the segment and delivers the data to ULP A.
15. TCP A returns an ACK of the received data marked with the number of the next expected data octet. However, the segment is lost by the network and never arrives at TCP B.

16. TCP B times out waiting for the lost ACK and retransmits the segment. TCP A receives the retransmitted segment, but discards it because the data from the original segment has already been accepted. However, TCP A re-sends the ACK.

17. TCP B gets the second ACK.

18. ULP A closes its half of the connection by issuing a CLOSE to TCP A.

19. TCP A sends a segment marked with a CLOSE control flag, called a FIN, to inform TCP B that ULP A will send no more data.

20. TCP B gets the FIN and informs ULP B that ULP A is closing.

21. ULP B completes its data transfer and closes its half of the connection.

22. TCP B sends an ACK of the first FIN and its own FIN to TCP A to show ULP B's closing.

23. TCP A gets the FIN and the ACK, then responds with an ACK to TCP B.

24. TCP A informs ULP A that the connection is closed.

25. (Not pictured) TCP B receives the ACK from TCP A and informs ULP B that the connection is closed.

Source: Ref. 24.

7.4.6 *TCP Header Format.* The TCP header format is shown in Figure 11.35. It should be noted that TCP works with 32-bit segments.

Source Port. The "port" represents the source ULP initiating the exchange. The field is 16 bits long.

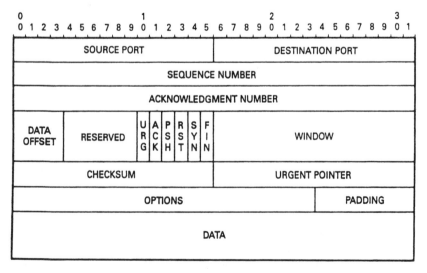

Figure 11.35. TCP header format [23].

Destination Port. This is the destination ULP at the other end of the connection. This field is also 16 bits long.

Sequence Number. Usually, this value represents the sequence number of the first data octet of a segment. However, if an SYN is present, the sequence number is the initial sequence number (ISN) covering the SYN; the first data octet is then numbered SYN + 1. The "SYN" is the *synchronize control flag.* It is the opening segment of a TCP connection. SYNs are exchanged from either end. When a connection is to be closed there is a similar "FIN" sequence exchange.

Acknowledgment Number. If the ACK control bit* is set (bit 2 of the 6-bit control field), this field contains the value of the next sequence number that the sender of the segment is expecting to receive. This field is 32 bits long.

Data Offset. This field indicates the number of 32-bit words in the TCP header. From this value the beginning of the data can be computed. The TCP header is an integral number of 32 bits long. The field size is 4 bits.

Reserved. This is a field of 6 bits set aside for future assignment. It is set to zero.

Control Flags. The field size is six bits covering six items (1 bit per item):

(a) URG: Urgent pointer field significant
(b) ACK: Acknowledgment field significant
(c) PSH: Push function
(d) RST: Reset the connection
(e) SYN: Synchronize sequence numbers
(f) FIN: No more data from sender

Window. The number of data octets beginning with the one indicated in the acknowledgment field which the sender of this segment is willing to accept. The field is two octets in length.

Checksum. The checksum is the 16-bit one's complement of the one's complement sum of all 16-bit words in the header and text. The checksum also covers a 96-bit pseudo-header conceptually prefixed to the TCP header. This pseudo-header contains the source address, the destination address, the protocol, and TCP segment length.

Urgent Pointer. This field indicates the current value of the urgent pointer as a positive offset from the sequence number in this segment. The urgent pointer points to the sequence number of the octet following the urgent data. This field is only to be interpreted in segments with the URG control bit set. The urgent pointer field is two octets long.

Options. This field is variable in size; and, if present, options occupy space at the end of the TCP header and are a multiple of 8 bits in length. All

---

* ACK: A control bit (acknowledge) occupying no sequence space, which indicates that the acknowledgment field of this segment specifies the next sequence number the sender of this segment is expecting to receive, hence acknowledging receipt of all previous sequence numbers.

options are included in the checksum. An option may begin on any octet boundary. There are two cases of an option:

(a) Single octet of option-kind
(b) An octet of option-kind, an octet of option length, and the actual option data octets

Options include "end of option list," "no-operation," and "maximum segment size."

Padding. The field size is variable. The padding is used to ensure that the TCP header ends and data begins on a 32-bit boundary. The padding is composed of zeros.

7.4.6.1 TCP Entity State Diagram. Figure 11.36 summarizes TCP operation with a TCP entity state diagram.

Source: Section 7.4 is based on RFC 793 [23] and MIL-STD-1778 [24].

7.5 Brief Overview of Internet Protocol Version 6 (IPV6)

RFC 1883, which is the principal reference for IPV6, states that this is a new version of the Internet Protocol, designed as a successor to IPV4 (RFC-791). The changes from IPV4 to IPV6 fall primarily into the following categories:

- *Expanded Addressing Capabilities.* IPV6 increases the IP address size from 32 bits to 128 bits, to support more levels of addressing hierarchy, a much greater number of addressable nodes, and simpler auto-configuration of addresses. The scalability of multicast routing is improved by adding a "scope" field to multicast addresses. IPV6 also defines a new type of address called an "anycast address" and is used to send a packet to any one of a group of nodes.
- *Header Format Simplification.* Some IPV4 header fields have been dropped or made optional, to reduce the common-case processing cost of packet handling to limit the bit rate capacity cost of the IPV6 header.
- *Improved Support for Extensions and Options.* This involves changes in the way IP header options are encoded and allows for more efficient forwarding, less stringent limits on the length of options, and greater flexibility for introducing new options in the future.
- *Flow Labeling Capability.* A new capability has been added to enable the labeling of packets belonging to particular traffic "flows" for which the sender requests special handling, such as nondefault quality of service or "real-time" service.

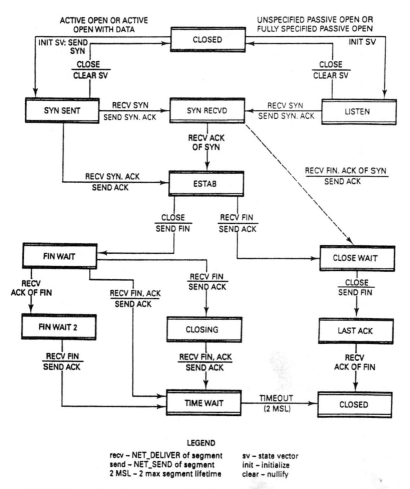

ACTIVE OPEN OR ACTIVE
OPEN WITH DATA

UNSPECIFIED PASSIVE OPEN OR
FULLY SPECIFIED PASSIVE OPEN

CLOSED

INIT SV: SEND
SYN

INIT SV

CLOSE
CLEAR SV

CLOSE
CLEAR SV

SYN SENT

RECV SYN
SEND SYN. ACK

SYN RECVD

RECV SYN
SEND SYN. ACK

LISTEN

RECV ACK
OF SYN

RECV FIN. ACK OF SYN
SEND ACK

RECV SYN. ACK
SEND ACK

ESTAB

CLOSE
SEND FIN

RECV FIN
SEND ACK

FIN WAIT

CLOSE WAIT

RECV
ACK OF FIN

RECV FIN
SEND ACK

CLOSE
SEND FIN

RECV FIN, ACK
SEND ACK

FIN WAIT 2

CLOSING

LAST ACK

RECV FIN
SEND ACK

RECV FIN, ACK
SEND ACK

RECV
ACK OF FIN

TIME WAIT

TIMEOUT
(2 MSL)

CLOSED

LEGEND

recv – NET_DELIVER of segment   sv – state vector
send – NET_SEND of segment   init – initialize
2 MSL – 2 max segment lifetime   clear – nullify

Figure 11.36. TCP entity state summary [24]. Note that this figure is intended only as a summary and does not supersede the formal definitions that precede.

- *Authentication and Privacy Capabilities.* IPV6 provides extensions to support authentication, data integrity, and (optional) data confidentiality.

8 MULTIPROTOCOL LABEL SWITCHING (MPLS)

8.1 Overview

Routing a packet through a conventional connectionless packet network (e.g., IP) is complex. Each router in the packet's path makes its own routing decision for the next hop based on its analysis of the packet address. To carry out its switching function, a router (e.g., an IP router) has to apply a complex algorithm and consult a look-up table or database.

MPLS simplifies and enhances these operations. It is essentially a labeling system design to accommodate multiple protocols. Another term for a "label" is a virtual path identifier.

8.2 Acronyms and Definitions

The following define some of the terminology and acronyms used when dealing with MPLS.

- *LSR (Label Switch Router).* A core router in an MPLS network. It participates in the establishment of label switch paths (LSPs) using the label signaling protocol. It performs high-speed switching of labeled traffic on established LSPs.
- *Label.* A short, fixed-length, physically continuous identifier which is used to identify an FEC. It usually is of local significance.
- *FEC (Forward Equivalence Class).* The FEC represents a group of packets that share the same requirements for their transport. All packets in an FEC travel the same LSP.
- *LER (Label Edge Router) (also called Ingress and Egress LSRs).* An LER is a device at the edge of an MPLS network. It assigns a packet to an FEC. The packets in an FEC are then assigned to an LSP based on traffic criteria. The egress LSR removes labels from traffic coming in from an incoming LSP.
- *LIB (Label Information Base).* Database in each MPLS switch that relates an incoming label and interface to outgoing label and interfaces.
- *LSP (Label Switch Path).* This is the logical path that an MPLS packet travels the network.
- *FEC (Forward Equivalence Class) (also called Functional Equivalent Class).* The FEC represents a group of packets that share the same requirements for their transport. All packets in an FEC travel the same LSP.
- *Label Swapping.* A forwarding paradigm allowing streamlined forwarding of data by using labels to identify classes of data packets which are treated indistinguishably when forwarded.

8.3 MPLS Description

A packet is labeled at the network entrance, more properly called the ingress. This function is carried out by the LER (label edge router) which converts IP packets into MPLS packets. There is a special "label information base" in the LER which matches the destination address of the packet to the label. The LER then attaches a *shim header* to the packet and after that the packet traverses the network.

Shim is an appropriate adjective because the shim header is slipped in between OSI layers 3 and 2 as if it were a shim. The shim header is independent, neither

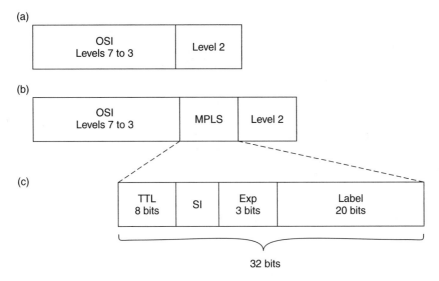

Figure 11.37. Construction of a shim header. In (a) and (b), reference is made to OSI layers; in (c), TTL is "time to live" from IP; S is the stack function (1 bit), and exp (3 bits) is experimental, assigned QoS function. The label has 20 bits.

a part of layer 3 nor a part of layer 2. It does, though, relate information dealing with both layers 2 and 3. Figure 11.37 shows the construction of a shim header.

A shim header (MPLS header) is 32 bits long. It contains the label, which is 20 bits long. Other fields are the *exp* bits (experimental bits) which are allocated to handle eight QoS classes. It should be noted here that MPLS allows hierarchical domain nesting. Consider that when a packet enters an MPLS domain which is contained in another MPLS domain, a new label is appended to the packet, which is already carrying one label. This is called *label stacking* and is illustrated in Figure 11.38. Figure 11.38 shows two nested MPLS domains. A *bottom of stack* (bos) bit (not shown) indicates whether a label is the last in the stack or not. Finally, each LER/LSR decrements an 8-bit time-to-live (TTL) field and discards packets when the value of this field initially set by the ingress LER reaches zero. The reason for this mechanism is to avoid packets from indefinitely looping around in case a circular virtual circuit is mistakenly created.

Figure 11.39 shows a network cloud with LERs at the ingress and egress and LSRs inside the network. An LSR is a router that bases its routing decision on the label value in an MPLS packet. In general, an LSR performs a label swapping function. It removes the label of an incoming packet, replacing it with one of its own.

MPLS handles labels just like all other virtual circuit identifiers are handled in other virtual circuit switching regimes. Consider Figure 11.39, where an IP packet is transmitted from host A to host B. The packet is forwarded through an MPLS network, called a domain in the MPLS RFC [27], between A and B. When a packet arrives at the first MPLS router (an ingress LER) of the MPLS domain,

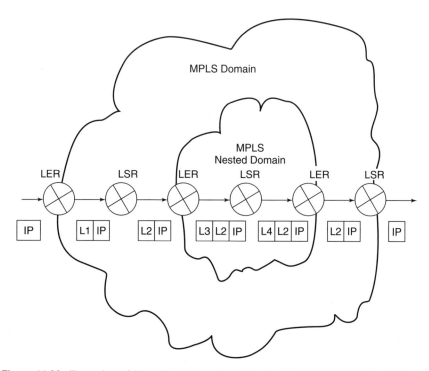

Figure 11.38. Illustration of hierarchical domain nesting in MPLS. Note that two labels are stacked in the inner MPLS domain.

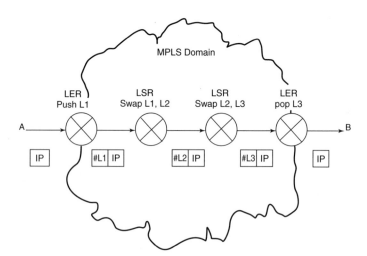

Figure 11.39. An MPLS network showing label edge routers (LERs) and label switch routers (LSRs). FEC determined by ingress LER for packets from A to B and "pushes" a label on packets. LSRs swap labels. At the egress the LER "pops" the label, and packet leaves the MPLS domain for destination B with no labels.

the source and destination addresses are analyzed and the packet is then placed in a *forwarding equivalence class* (FEC). The first rule in MPLS is that all packets in the same FEC use the same virtual circuit. This circuit is called a *label switched path* or LSP. Let's suppose a virtual circuit has already been established for the FEC by the packet sent by A to B. Then the ingress LER inserts or *pushes* an MPLS header (i.e., shim) on the packet (L1 in Figure 11.39). Subsequent routers (LSRs) of the MPLS domain update the MPLS header by swapping the label (i.e., L1 against L2 and then L2 against L3). Finally, the last router, the egress LER, removes or *pops* the MPLS header (L3 in the figure) so that the packet can be handled by subsequent MPLS (IP routers unaware of previous MPLS actions).

All MPLS routers have a table called a *forwarding information base* (FIB) that is characteristic for each MPLS router. There are three possible types of entries in an FIB.

- A *Next Hop Label Forwarding Entry* (NHLFE), which contains the information necessary to forward a packet for which a label has already been assigned. An NHLFE supplies two pieces of information: the packet's next hop address, and whether the MPLS header (the shim) must be swapped or popped. In the case where the MPLS header or shim must be swapped, then the NHLFE also will contain the new label for the packet.
- The *Incoming Label Map* (ILM) has the mappings between labels carried by incoming packets and NHLFE entries.
- The *FEC-to-NHLFE* (FTN) contains the mappings between incoming packet FECs and NHLFE entries.

MPLS routers use their FIBs as follows. Let's suppose a packet with no label arrives at an MPLS router. The router first determines the FEC for the packet. It then looks up in the FIB for the FTN that matches the FEC of the packet. The FTN has a label and an NHLFE which in turn contains the routing information for the next hop of the packet. The MPLS router pushes the shim (MPLS header) that contains the label read in the FTN. It then forwards the packet according to information contained in the NHLFE.

Let's now look at the case there a labeled packet arrives at a MPLS router. The router searches in the FIB for an ILM that matches the label of the packet and reads the associated NHLFE. The NHLFE can either indicate that the MPLS header must be swapped against a new label, or popped. In the case of the former, the MPLS router swaps the MPLS header and forwards the packet to the next hop that was specified in the NHLFE. In the case of the latter, the MPLS router pops the label and forwards the packet as instructed by the NHLFE.

8.4 Notes on FEC

In a traffic engineering context, an important point on FECs is that they support aggregation. All packets from different sources but that enter the MPLS domain through the same LER and that are bound for the same egress LER can be assigned to the same FEC and thus to the same virtual circuit. This is a great advantage in that there is no need to establish a new virtual circuit for each source-destination pair read in the headers of incoming packets. As we know, the ingress LER determines the FEC of a packet. Once this is done, it assigns a virtual circuit to the packet via a label number. Also, the definition of an FEC can take into account IP packet sources as well as destinations. Then, of course, there is the case of *load balancing*. Let's say that two packets enter the MPLS domain through the same LER and are going to the same destination. These packets can be assigned to travel different links to achieve load balancing. This is done to distribute the traffic load evenly across the network. An FEC can also depend on additional parameters such as the Type-of-Service bits of the IP header to provide differentiated services to IP traffic.

MPLS depends on a *signaling protocol* to establish virtual circuits and map-pings or *bindings* between FECs and labels and building the FIB at each MPLS router. The MPLS architecture (RFC 3031 [27]), does not impose the use of any specific signaling protocol. RFC-3031 imposes the requirement that on a given link and for a given LSP, labels are assigned by the downstream node and adver-tised by the upstream node. MPLS architects have devised two refinements from this requirement:

1. In *downstream unsolicited label distribution* mode, a node recognizes that it is a downstream node for an FEC and sends a label binding message to the upstream node for that FEC. The downstream node decides by itself to send the binding message without any trigger from its corresponding upstream node.

2. Conversely, in *downstream on demand* mode, the upstream node identifies that it needs a label binding message for a particular FEC and requests that the downstream node sends this label binding message.

As we mentioned, no label signaling protocol is imposed by the MPLS standards, and only two signaling protocols have been developed for MPLS. The first is RSVP-TE, which is based on the resource reservation protocol for the internet, to which it adds the capability to advertise LSPs. The second protocol is the label distribution protocol (LDP). It has been defined from scratch as part of the MPLS design effort. There is an extension to LDP called constraint routing LDP or CR-LDP. This adds important features to LDP with respect to traffic engineering. Another important improvement to CR-LDP is to support explicit routing. This is where a single node or an offline server which precomputes paths can fully define and advertise LSPs.

9   VIRTUAL PRIVATE NETWORKS (VPNs)

9.1   Why VPNs?

A VPN provides a highly economic alternative to leased lines, frame relay, or an ATM service. An internet-based VPN uses the open, distributed infrastructure of the internet to interexchange data between/among corporate sites. An enterprise using an internet-based VPN set up connections to the local connection points. These are called points-of-presence (POPs) of their internet service provider (ISP). The ISP, then, ensures that the data are forwarded to the appropriate destinations via the internet. The rest of the connectivity details are left to the ISP's network and the internet infrastructure. Because the internet is a public network with open transmission of most data, internet-based VPNs include measures for encrypting data passed between VPN sites. This protects the data against eavesdropping and tampering by unauthorized parties.

VPNs can also connect to mobile installations. This can be done by simple dialing into their ISP/POP. A big advantage here is the reduction of needs for large banks of modems and outlays for long-distance charges.

9.2 Two Major Requirements

When deploying a VPN over the public internet, there are two major concerns: security and transmission performance. When the internet was developed and first deployed by DARPA performance and security were not really issues. For one thing, the internet at that time was a closed network, not open to public correspondence. Transmission performance was good but not truly guaranteed. The network was based on the transmission control protocol (TCP) and the internet protocol (IP).

There are four critical security functions required of a VPN. These are:

1. *Data Integrity*. Ensuring that no one tampers with the data as it travels across the internet.

2. *Confidentiality*. Preventing anyone from reading or copying the data as it moves across the internet.

3. *Access Control*. Restricting unauthorized user from gaining access to the network.

4. *Authentication*. To ensure that the data originates from the source that it claims to originate from.

Several methods of authentication have been offered to solve the problem. One method is the "challenge handshake authentication protocol" (CHAP). Another is "remote authentication dial-in user service" (RADIUS). Hardware-based tokens and digital certificates can also be used to authenticate users on a VPN and to control access to network resources. Encryption of the data protects its privacy as it travels across the network.

The internet is a public network shared by many users over common connections. Other parties may glean information on a shared connectivity such as routing, source, and estimation addressees. To avoid these shortcomings, a number of protocols have been developed to create tunnels. Tunneling allows senders to encapsulate their data in IP packets that hide the underlying routing and switching information of the internet of both source and destination(s).

The term *virtual* in VPN tells us that connectivity is made as needed and taken down when a communication is completed, much like a telephone call. It is a virtual circuit.

There are two possible endpoints for a virtual circuit: as computer or a LAN with a security gateway. This latter may be a router or some sort of firewall. Connectivities may be LAN-to-LAN tunneling. There will be a security gateway at each end which serves as an interface between the tunnel and the private LAN.

The second type of endpoint is the mobile user connecting to a corporate LAN. The mobile user is the client that establishes a tunnel to exchange data traffic with the corporate network. Special software is required at the mobile end to communicate with the gateway protecting the destination LAN.

9.3   Specialized VPN Internet Protocols

For creating a VPN over the internet, one of four specialized protocols should be considered. These are:

1. Point-to-point tunneling protocol (PPTP)
2. Layer-2 forwarding (L2F)
3. Layer-2 tunneling protocol (L2TP)
4. IP security protocol (IPSec)

One group of protocols is designed for traffic traveling in secure tunnels via the internet between protected LANs. This is IPSec's main focus. For the dial-up VPN mobile user, we would apply one of three: PPTP, L2F, or LSTP.

PPTP was developed by Microsoft for Windows 95 and 98 operating systems. It has not been endorsed by standardization bodies such as the Internet Engineering Task Force (IETF).

For remote access of the internet, probably the most popular protocol is the point-to-point protocol (PPP). PPTP is a derivative of the PPP for remote access using tunneling to a destination site. In the present version, PPTP provides remote access by tunneling through the internet to a destination site. This is done by encapsulating PPP packets based on a modified version of the generic routing encapsulation (GRE) protocol. This gives PPTP the flexibility to handle protocols other than IP, such as internet package exchange (IPX).

PPTP runs on OSI layer 2, whereas other protocols such as IPSec run on layer 3. This allows PPTP to transmit other protocols besides IP through its tunnels.

IPSec is the latest of the VPN protocols. Unlike other predecessor VPN protocols, IPSec is being developed by the Internet Engineering Task Force. It can be used by both IPv4 and IPv6 versions. One of the major challenges in its development was security management and key exchanges. The problem has been largely solved by the implementation of the "internet key exchange" or IKE.

Two security operations are carried out by IPSec: authentication and/or encryption of each IP packet. Separating the application of packet authentication and encryption has led to two different methods of using IPSec. These methods are called *modes*. There is the transport mode and the tunnel mode. When just the transport-layer segment of an IP packet is authenticated or encrypted, it is the transport mode. When the entire IP packet is authenticated and encrypted, it is in the tunnel mode. This latter mode is the most secure against certain attacks and traffic level monitoring which might be encountered on the internet. Key exchange and management for encryption can be carried out in one of two ways: manual keying and IKE for automatic key management. Manual keying is suitable for small VPNs. For large networks, automatic key management (IKE) is recommended.

The best solution for a VPN is considered to be IPSec, but it only works with IP. Either PPTP or L2TP should be considered in a non-IP, multiprotocol environment.

9.4 Principal Components of a VPN Based on the Internet

There are four main components of a VPN that is based on the internet. These are: (1) the internet itself, (2) security gateways, (3) certificate authorities, and (4) policy servers. The basic transport medium is the internet. The essential interface between the public internet and a private data device or LAN is the security gateway. It prevents unauthorized intrusion into a private network via a VPN. There are four categories of security gateways. In that regard we may encounter one of the following: router, firewall, integrated VPN hardware, or VPN software.

Generally, it is most practical to incorporate packet encryption in a router. It may be a separate card or be included in the router software.

Firewalls often include a tunneling capability. These devices, like IP routers, pass all traffic based on filters defined for the firewall. It has been found that firewalls used for tunnel are ill-suited for large networks with a great deal of traffic. Firewalls used for tunneling are recommended for smaller networks with less traffic volume.

Specialized hardware is another VPN solution. The hardware is designed for tasks such as tunneling, encryption, and user authentication. This specialty hardware can be viewed as an encryption bridge placed between the router and the connecting WAN. The most popular application for these bridges is LAN-to-LAN connectivity; some, however, may be used for mobile client-to-LAN application.

A good low-cost choice for creating and managing tunnels is to use VPN software. It may be applied between gateways or between a remote client and a security gateway. The software solution can run on existing servers and can share resources with them. Also consider VPN software for mobile client-to-LAN interconnections.

The third important component of a VPN is security policy server. This device maintains access-control lists and other information about users that is used to determine which traffic is authorized on a VPN. One type of policy server is called the *RADIUS server*.

Certificate authorities verify keys shared between VPN sites. They can also be used to verify individuals using digital certificates. Some enterprises maintain their own database of digital certificates by setting up a corporate certificate server. With an extranet, an outside certificate authority may have to be used to verify users that are among the business partners.

REVIEW QUESTIONS

1. Define *topology* regarding data networks.

2. Give at least four major differences between WANs and LANs.

3. Give six of the seven considerations listed in the text which should be quantified or qualified before data network design begins. We can call them "inputs," if need be.

4. Give the three approaches to data switching. Provide a short description of each.

5. What is the efficiency of start–stop data transmission only from the point of view of useful bits above the physical layer? In this case the stop element is 2 bits long. Assume an ASCII format with one-bit parity.

6. Which is one very prevalent way power utilities make money with their telecommunication network? Carry this thought just one step further!

7. Define a protocol.

8. Describe two early forms of data network access. Let's define a network as some form of interconnectivity with more than two users.

9. Define SAR (segmentation and reassembly). Define encapsulation.

10. In one or two sentences, give the rationale why the OSI 7-layer model was developed.

11. Give the remaining six layers of OSI in proper order starting with the layer just above the physical layer.

12. What are the basic responsibilities of the data-link layer (OSI layer 2)?

13. Both the data-link layer and the network layer have provisions for error control. How can they both do it? Differentiate the responsibilities of each regarding error recovery.

14. What are the four basic types of primitives recommended in OSI for one layer to communicate with an adjoining layer?

15. If we didn't do something about it, what would happen if the bit sequence 01111110 (the flag sequence) appeared in the middle of an HDLC frame?

16. Define the term *transparent*.

17. What are the three generic functions of the control field in the HDLC frame?

18. If an address field is 8 bits long, how many distinct addresses can it indicate?

19. Describe a *window of frames*.

20. X.25 deals primarily with which OSI layer? LAPB, its access protocol, resides in layer 2.

21. LAPB is almost identical to and derives from which data-link layer protocol?

22. Differentiate permanent virtual circuits (PVCs) and virtual calls.

23. X.25 was designed as the basic protocol for a packet network. However, experience has shown that X.25 generally uses permanent virtual circuits. Discuss this with respect to full packet operation.

24. What is the primary purpose of IP? Use the word *interface* in the answer.

25. Even though TCP/IP predates OSI, in what OSI layers would we expect to find IP and TCP?

26. What is the purpose of the *time-to-live field* in IP?

27. What is the purpose of the ARP (address resolution protocol) used in conjunction with IP?

28. How does IP use *source routing*?

29. In what way does ICMP help IP?

30. Give four of the ICMP type messages of the eight listed.

31. Describe the purpose of TCP, especially how it works with IP.

32. Describe the PAR mechanism incorporated with TCP and in support of IP.

33. What is the purpose of the *three-way handshake*?

34. TCP operates on what kind of boundaries or segments (express in bits)?

35. IPV6 has augmented addressing compared to IPV4. Give the size of an IPV4 address and of IPV6 address fields.

36. How is IPV6 more efficient than IPV4? Treat "efficiency" as we have defined it.

37. In MPLS, what is a "shim" and what is its function?

38. Where is the BASIC application of MPLS? What packet protocol does it serve?

39. In MPLS, differentiate an LER from and LSR. Incorporate the words push, swap, and pop.

40. What is the function of LDP?

41. What are the four basic elements of a VPN?

42. What is the basic transport mechanism of a most economic VPN?

43. What is a certificate authority?

44. What are the two defining issues of a VPN?

45. What are the two basic applications of a VPN?

REFERENCES

1. *The New IEEE Standard Dictionary of Electrical and Electronics Terms*, 6th ed., IEEE Std 100–1996, IEEE, New York, 1996.
2. W. Stallings, *Handbook of Computer Communications Standards*, Vol. 1, Macmillan, New York, 1987.
3. *Reference Model of Open Systems Interconnection for CCITT Applications*, ITU-T Rec. X.200, ITU, Geneva, July 1994.
4. *Information Processing Systems: Open Systems Interconnection—Basic Reference Model*, ISO 7498, Geneva, 1984.
5. *Open Systems Interconnection Layer Service Definition Conventions*, CCITT Rec. X.210, ITU, Geneva, November 1993.
6. D. Bertsekas and R. Gallager, *Data Networks*, 2nd ed., Prentice-Hall, Englewood Cliffs, NJ, 1987.
7. *High-Level Data Link Control Procedures—Frame Structure*, ISO 3309, International Standards Organization, Geneva, 1979.
8. *Advanced Data Communication Control Procedures*, X3.66 ANSI, New York, 1979.
9. *High-Level Data Link Control Procedures—Consolidation of Elements of Procedures*, ISO 4335, Geneva, 1980.
10. *Code Independent Error Control Systems*, CCITT Rec. V.41, Fascicle VIII.1, IXth Plenary Assembly, Melbourne, 1988.
11. R. L. Freeman, *Reference Manual for Telecommunications Engineering*, 3rd ed., John Wiley & Sons, New York, 2002.
12. *Interface between Data Terminal Equipment (DTE) and Data-Circuit-Terminating Equipment (DCE) for Terminals Operating in the Packet Mode and Connected to the Public Data Networks by Dedicated Circuit*, ITU-T Rec. X.25, ITU-T Organization, Geneva, October 1996.
13. *Interface Between Data Terminal Equipment and Data Circuit-Terminating Equipment (DCE) for Synchronous Operation on Public Data Networks*, CCITT Rec. X.21, Geneva, September 1992.
14. *Use on Public Data Networks of Data Terminal Equipment (DTE) Which Is Designed for Interfacing Synchronous V-Series Modems*, CCITT Rec. X.21 bis, Fascicle VIII.2, IXth Plenary Assembly, Melbourne, 1988.
15. *Packet-Switched Signaling System Between Public Networks Providing Data Transmission Services*, ITU-T Rec. X.75, ITU-T Organization, Geneva, October 1996.
16. *Internet Protocol*, RFC 791, DDN Network Information Center, SRI International, Menlo Park, CA, September 1981.

17. *Assigned Numbers*, RFC 1060, DDN Network Information Center, SRI International, Menlo Park, CA, March 1990.

18. *Internet Protocol Transition Workbook*, SRI International, Menlo Park, CA, March 1982.

19. *A Standard for the Transmission of IP Datagrams over IEEE 802 Networks*, RFC 1042, DDN Network Information Center, SRI International, Menlo Park, CA, February 1988.

20. *An Ethernet Address Resolution Protocol*, RFC 826, DDN Information Center, SRI International, Menlo Park, CA, June 1984.

21. *A Reverse Address Resolution Protocol*, RFC 903, DDN Network Information Center, SRI International, Menlo Park, CA, June 1984.

22. *Internet Control Message Protocol*, RFC 792, DDN Network Information Center, SRI International, Menlo Park, CA, September 1981.

23. *Transmission Control Protocol*, RFC 793, DDN Network Information Center, SRI International, Menlo Park, CA, September 1981.

24. Military Standard, *Transmission Control Protocol*, MIL-STD-1778, U.S. Department of Defense, Washington, DC, August 1983.

25. *Internet Protocol, Version 6 (IPv6), Specification*, RFC-1883, December 1995, Ohio State University (from the web).

26. J. M. Rosenberg, *Dictionary of Computers, Data Processing and Telecommunications*, John Wiley & Sons, New York, 1984.

27. *Multiprotocol Label Switching Architecture* (MPLS), RFC-3031, Network Working Group, Jan. 2001. From the web: www.ietf.org/rfc/rfc3031.txt.

28. Rick Gallaher, "An Introduction to MPLS," *Converge! Network Digest*, September 10, 2001, from the web: www.convergedigest.com/Bandwidth/archive/010910TUTO-RIAL-rgallaher.htm.

29. "Leveraging MPLS to Enhance Network Capabilities in Service Provider and Enterprise Environments," Extreme Networks (white paper), Version 1.0 (July 2001).

30. "Interior Gateway Routing Protocol," Jupiterresearch, from the web: http://networking. webopedia.com/TERM/I/Interior Gateway Routing Protocol.html.

31. "Virtual Private Networks," from the web, www.iec.org, The International Engineering Consortium, Web ProForum Tutorials, 3/21/03.

12

VOICE-OVER IP

1 DATA TRANSMISSION VERSUS CONVENTIONAL TELEPHONY

Conventional voice telephony is transported in a full duplex mode on PSTN circuits optimized for voice. By the *full duplex mode* we mean that there are actually two circuits, one for "send" and one for "receive," to support a normal telephone conversation between two parties. Today, once we depart the local area, all of these circuits are digital. The descriptive word *digital* may seem ambiguous to some.

Let's say that 10 years ago we looked ahead to our present time. All the circuits would consist of 8-bit words, which represent voltage samples of analog voice conversations in a PCM format. This is often characterized in the literature as G.711 service (i.e., ITU-T Rec. G.711). Data are also commonly transported in 8-bit sets called bytes, but more properly called *octets*. It is comparatively simple to replace 8-bit voltage samples of voice with 8-bit octets of data.

However, there remained essential philosophical differences between voice in 8-bit octets and data transmission. A voice circuit is established when a subscriber desires to converse by telephone with some other telephone subscriber. The circuit between the two is set up by a signaling routine. The distant subscriber has a telephone address represented by a distinct telephone number consisting of 7 to 12 digits. The digit sequence of the number sets up a circuit route and connectivity for conversation. The circuit is maintained in place for the duration of the conversation, and it is terminated and taken down when one or the other party hangs up ("goes on hook"). The address sequence of dialed digits is sent just once, at the initiation of the connectivity. This whole process of setting up a circuit, holding the connectivity in place, and then taking down the circuit is called *signaling*.

Signaling on data circuits is approached quite differently. There is the *permanent virtual circuit* (PVC), which has all the trappings similar to a voice

Telecommunication System Engineering, by Roger L. Freeman
ISBN 0-471-45133-9   Copyright © 2004 Roger L. Freeman

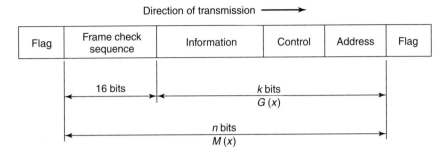

Figure 12.1. A typical data frame.

circuit. The similarities stop here. Data transmission consists of frames or packets of data. A frame (or data packet) is made up of a header and payload. In some cases a portion of the "header" may be appended at the end or on the tail of the data frame (or packet). But every data frame (or packet) has a header consisting of a destination address (or addresses) and the originator's address. It nearly always will also contain some control information. This may be a word (or byte/octet) count of the payload, a CRC sequence for error detection and/or correction, message priority, or some other type of control sequence or sequences.

Digital circuits on the PSTN have either 24-octet frames in the case of T1 (DS1) or 32-octet frames in the case of E1. Each 8-bit octet represents a voice circuit. Such a circuit may be set up using an Initial Address Message of CCITT Signaling System No. 7 or a sequence of DTMF tones where each frequency pair represents a digit in the range of 0 through 9. Once a circuit is set up, no more address messages or DTMF tones are required until the circuit is taken down.

This is not the case on a data circuit. Such a circuit also uses frames, but each frame has a standard header. A typical data frame is illustrated in Figure 12.1. The frame structure and how the various octets of the header (and tail) are utilized are governed by a *protocol*. Various data protocols were discussed in Chapter 11.

We can clearly see that there are two differing philosophies here, one for data communication and the other for digital voice. Digital voice is sometimes called "circuit-switched voice." A majority in the telecommunication community saw how advantageous it would be if we could marry the two and make them one. That is one singular approach for both voice and data. Meanwhile, data hobbyists were trying to transmit voice using data packets. The internet protocol (IP) became the data protocol of choice, but there were many drawbacks.

2 DRAWBACKS AND CHALLENGES FOR TRANSMITTING VOICE ON DATA PACKETS

We have come to measure "quality" of packetized voice service by the equivalent service offered by the switched digital network, sometimes called G.711 [1] service. The user expects a quality of service (QoS) as good as he/she would get on a PSTN dial-up connection.

To achieve this goal, voice-over IP (VoIP) designers were faced with the following degradations:

- Mouth-to-ear delay
- Impact of errored frames (packets)
- Lost frames (packets)
- Variation of packet arrival time, jitter buffering
- Prioritizing VoIP traffic over regular internet and data services
- Talker echo
- Distortion
- Sufficient bit rate capacity on interconnecting transmission media
- Voice coding algorithm standardization
- Optimized standard packet payload size
- Packet overhead
- Silence suppression

3 VOIP, INTRODUCTORY TECHNICAL DESCRIPTION

Figure 12.2 shows a simplified block diagram of VoIP operation from an analog signal deriving from a standard telephone, which is digitized and transmitted over the internet via a conversion device. Then, at the distant end, it is converted back to analog telephony using a similar device suitable for input to a standard telephone. The *gateway* is placed between the voice codec and the digital data transport circuit. An identical device will also be found at the far end of the link. This equipment carries out the signaling role on a telephone call among other functions.

Moving from left to right in Figure 12.2, we have the spurty analog signal developed by the standard telephone. The signal is then converted to a digital counterpart using one of the seven or so codecs [coder–decoder(s)] that the VoIP system designer has to select from. Some of the more popular codecs for this application listed in Table 12.1. The binary output of the codec is then applied

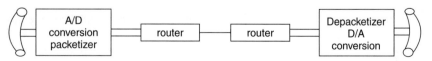

Figure 12.2. Elements of basic operation of VoIP where the input signal derives from a conventional analog telephone.

TABLE 12.1   Characteristics of Speech Codecs Used on Packet Networks

Coding Algorithm	Voice Bit Rate (kbits/s)	Voice Frame Size (bytes)	Header (bytes)	Packets per Second	Packet Bit Rate (kbits/s)
G.711 8-bit PCM [1]	64	80	40	100	96
G.723.1 MPMLQ[a] [4]	6.3	30	40	26	14.6
G.723.1 ACELP[b] [5]	5.3	30	40	22	12.3
G.726 ADPCM[c] [6]	32	40	40	100	64
G.728 LD-CELP[d] [7]	16	20	40	100	48
G.729a CS-ACELP[e] [8]	8	10	40	100	40

[a] MPMLQ, multipulse maximum likelihood quantization.
[b] ACELP, algebraic code-excited linear prediction.
[c] ADPCM, adaptive differential PCM.
[d] LD-CELP, low delay code-excited linear prediction.
[e] CS-ACELP, conjugate structure algebraic code-excited linear prediction.

to a conversion device (i.e., a "packetizer") that loads these binary 1s and 0s into an IP payload of from 20 to 40 octets in length.

The output of this converter consists of IP packets[*] which are transmitted on the web or other data circuit for delivery to the distant end.

At the far end the IP packets or frames are input to a converter (i.e., a "depacketizer") which strips off the IP header, stores the payload, and then releases it in a constant bit stream to a codec. Of course this codec must be compatible with its near-end counterpart. The codec converts the digital bit stream back to an analog signal which is input to a standard telephone.

The insightful reader will comment that many steps of translation and interface have been left out. Most of these considerations will be covered in Section 3.1 in our discussion of the *gateway*.

3.1   VoIP Gateway

Gateways are defined in different ways by different people. A gateway is a server; it may also be called a *media gateway* Figure 12.3 illustrates a typical gateway. It sits on the edge of the network and carries out a switching function of a local, tandem, or toll-connecting PSTN switch described in Chapters 3 and 9. Media gateways are part of the physical transport layer. They are controlled by a call control function housed in a media gateway controller. A media gateway with its associated gateway controller is at the heart of the network transformation to packetized voice. Several of the media gateway functions are listed below:

- Carries out A/D conversion of the analog voice channel (called compression in many texts).
- Converts a DS0 or E0 to a binary signal compatible with IP or ATM.

---

[*] The output may be ATM cells (see Chapter 15) if the intervening network is an ATM network.

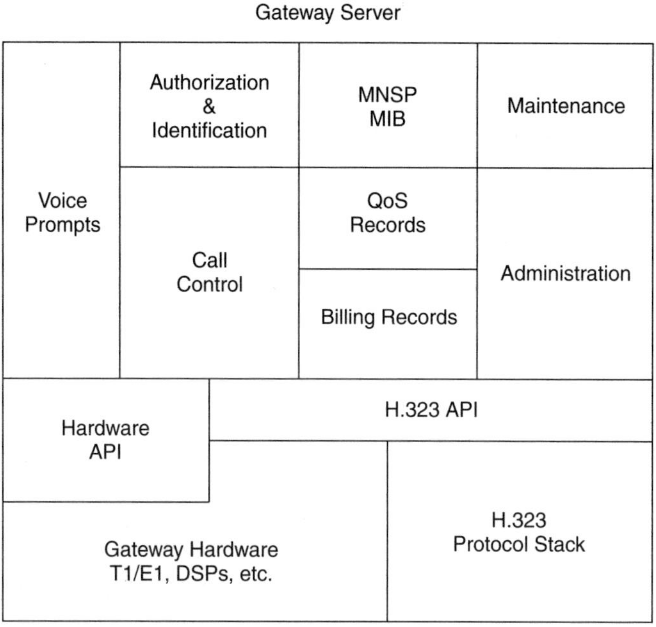

Gateway Server

Voice Prompts	Authorization & Identification	MNSP MIB	Maintenance
	Call Control	QoS Records	Administration
		Billing Records	

| Hardware API | H.323 API | |
| Gateway Hardware T1/E1, DSPs, etc. | | H.323 Protocol Stack |

Figure 12.3. A media gateway, from one perspective. API, applications program interface. From IEC on-line. www.iec.org/online/tutorials/int_tele/topic03.html (January 2003).

- Supports several types of access networks including media such as copper (including various DSL regimes), fiber, radio (wireless), and CATV cable. It is also able to support various formats found in PDH and SDH hierarchies.
- Competitive availability (99.999%).
- Capable of handling several voice and data interface protocols.
- Multivendor interoperability.
- It must provide interface between media gateway control device and the media gateway. This involves one of four protocols: SIP [2], H.323 [3], MGCP and Megaco (H.248).
- Can handle switching and media processing based on standard network PCM, ATM, and traditional IP.
- Transport of voice. There are four transmission categories involved:
  1. Standard PCM (E0/E1 or DS0/DS1)
  2. ATM over AAL1/AAL2
  3. IP-based RTP/RTCP
  4. Frame relay

The most powerful gateway supports the public network or PSTN requiring a high reliability device to meet the PSTN availability requirements. It will be required to process many thousands of digital circuits. As shown in

Figure 12.3, it has a network management capability most often based on SNMP (see Chapter 19).

A somewhat less formidable gateway is employed to provide VoIP for small and medium-sized business. Some texts call this type of gateway an *integrated access device* (IAD) if it can handle data and video products as well. An IAD will probably be remotely configurable.

The least powerful and most economic gateways are residential. They can be deployed in at least five settings:

- POTS (telephony)
- Set-top box (CATV), which provides telephony as well
- PC/modem
- XDSL termination
- Broadband last mile connectivity (to the digital network)

Figure 12.4 shows gateway interface functions via a block diagram. On the left are time slots of a PCM bit stream (T1, in this case). The various signal functions are shown to develop a stream of data packets carrying voice or data. The output on the right consists of IP packets.

The first functional block of the gateway analyzes the content on a timeslot basis. The timeslot may contain an 8-bit data sequence where we must be hands-off regarding the content. A gateway senses the presence of data by the presence of a 2100-Hz tone in the timeslot. The next signal type in the timeslot it looks for is DTMF signaling tones (see Chapter 4). If there is no modem tone nor DTMF tones in the timeslot, then the gateway assumes the timeslot contains human speech. Three actions now have to be accomplished. "Silence" is removed; the standard PCM compression algorithm is applied; an echo canceler is switched in. There are three digital formats used for voice-over packet:

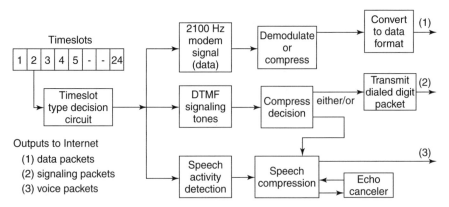

Figure 12.4. A simplified functional block diagram of a gateway providing an interface between a PCM bit stream deriving from the PSTN on the left and an IP network.

1. IP (internet protocol, Chapter 11, Section 7)
2. Frame relay (Chapter 15, Sections 1–3)
3. ATM (asynchronous transfer mode, Chapter 16)

3.2 An IP Packet as Used for VoIP

Assume for argument that we use either a G.711 or G.726 IP packet. The packet consists of a header and a payload. Figure 12.5 shows a typical IP packet. Of interest, as one may imagine, is its payload.

In the case of G.711 (standard PSTN PCM), there may be a transmission rate of 100 packets per second with 80 bytes in the payload of each packet. Of course our arithmetic comes out just right and we get 8000 samples per second, the Nyquist sampling rate for a 4-kHz analog voice channel. Another transmission rate for G.711 is 50 packets per second where each packet will have 160 bytes, again achieving 8000 samples per second per voice channel.

The total raw bytes (octets) per channel come out as follows: 40 bytes for layers 3 and 4 overhead (IP), plus 8 bytes for layer 2 (link layer) overhead. So we add 48 to 80 or 160 bytes (from the previous paragraph) and we get 128 or 208 bytes for a raw packet. The efficiency is nothing to write home about. Keep in mind that the primary concern of the VoIP designer is delay.

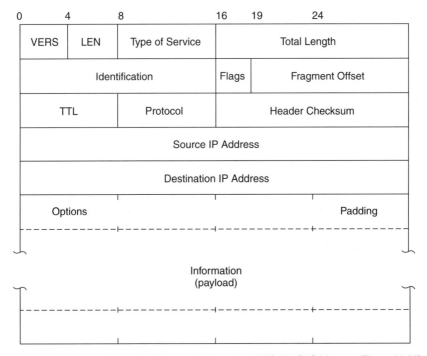

Figure 12.5. A typical IP packet (datagram). Based on RFC 791 [22] (also see Figure 11.28).

3.3 The Delay Trade-off

Human beings are intolerant of delay on a full-duplex circuit, typical of standard PSTN telephony. ITU-T Rec. G.114 [10] recommends the total delay (one-way) in a voice connectivity as follows:

- 0–150 ms acceptable.
- 150–400 ms acceptable but not desirable. Connectivity through a geostationary satellite falls into this category.
- Above 400 ms, unacceptable.

The delay objective (one-way) for a VoIP voice connectivity is < 100 ms. With bridging for conference calls, that value doubles due to the very nature of bridging.

One-way components of delay are as follows:

- Packetization or encapsulation delay based on G.711 or other compression algorithm. In the case of G.711, we must build from 80 PCM samples at 125 μs per sample, so we have consumed 80 × 125 μs or 10,000 μs or 10 ms plus time for the header or 48 × 125 μs or 6 ms, for a total of 16 ms. If we use 160 PCM samples in the payload, then allow 20 ms plus 6 ms for the header or 26 ms. This is a fixed delay.
- Buffer delay is variable. As a minimum, there must be buffering of one frame or packet period. Routers have buffers, by definition. Buffer delay varies with the number of routers in tandem. For G.711 the packet buffer size is 16 or 26 ms.
- Look-ahead delay. This is used by the coder to help in compression. "Look-ahead" is a period of time where the coder looks at packet $N + 1$ for patterns on which it can compress while coding packet N. With G.711 the look-ahead is 0.
- Dejitterizer is a buffer installed at the destination. It injects at least 1 frame (packet) duration (1–20 ms) in the total delay to smooth out the apparent arrival times of packets (frames).
- Queueing delay. Time spent in queue because it is a shared network. One method to reduce this delay is to prioritize voice packets (vis à vis data). Objective <50 ms.
- Propagation delay. Variable. Major contributor to total delay. Geostationary satellite relay of circuits is a special problem. The trip to the satellite and back is budgeted at 250 ms.

One way to speed things up is to increase the bit rate per voice data stream. To do this, the aggregate bit rate may have to be increased. Or the number of voice streams may be reduced on the aggregate bit rate so that each stream can be transmitted at a faster rate.

3.4 Lost Packet Rate

A second concern of the VoIP designer is *lost packet rate*. There are several ways a packet can be "lost."

For example, Section 3.3 described a dejitterizing buffer. It has a finite size. Once the time is exceeded by a late packet, the packet in question is lost. In the case of G.711, this would be the time equivalent 16 or 26 ms (duration of a packet including its header). Another reason for a packet to be lost may be excessive error rate on a packet whereby it is deleted. When the lost (discarded) packet rate begins to exceed 10%, quality of voice starts to deteriorate. If high-compression algorithms are employed such as G.723 or G.729, it is desirable to maintain the packet loss rate below 1%. Router buffer overflow is another source of packet loss.

IP through TCP has excellent retransmission capabilities for errored frames or packets. However, they are not practical for voice-over IP because of the additional delay involved. When there is a packet or frame in error, the receive end of the link transmits a request (RQ) to the transmit end for a packet retransmission and its incumbent propagation delay. This must be added to the transmission delay (and some processing delay) to send the offending packet back to the receiver again.

3.4.1 Concealment of Lost Packets. A lost packet causes a gap in the reception stream. For a single packet we are looking at a 20- to 40-ms gap. The simplest measure to take for lost packets and the resulting gaps is to disregard. The absolute silence of a gap may disturb a listener. In this case, often artificial noise is inserted.

There are packet loss concealment (PLC) procedures which can camouflage gaps in the output voice signal. The simplest techniques require a little extra processing power, and the most sophisticated techniques can restore speech to a level approximating the quality of the original signal. Concealment techniques are most effective for about 40 to 60 ms of missing speech. Gaps longer than 80 ms usually have to be muted.

One of the most elementary PLCs simply smoothes the edges of gaps to eliminate audible clicks. A more advanced algorithm replays the previous packet in place of the lost one, but this can cause harmonic artifacts such as tones or beeps. Good concealment methods use variation in the synthesized replacement speech to make the output more like natural speech. There are better PLCs to preserve the spectral characteristics of the talker's voice and to maintain a smooth transition between estimated signal and surrounding original. The most sophisticated PLCs use CELP (codebook-excited linear predictive) or similar technique to determine the content of the missing packet by examining the previous one [11].

Lost packets can be detected by packet sequence numbering.

3.5 Echo and Echo Control

Echo is commonly removed by the use of echo cancelers and are incorporated on the same DSP chips the perform the voice coding. A good source for information and design of echo cancelers is ITU-T Rec. G.168 [12]. However, most vendors of VoIP equipment have their own proprietary designs. A common design approach is to have the echo canceler store the outgoing speech in a buffer. It then monitors the stored speech after a delay to see whether it contains a component that matches up against the stored speech after a delay. If it does, that component of the incoming speech is canceled out instead of being passed back to the user since it is an echo of what the user originally said. Echo cancelers can be tuned or can tune themselves to the echo delay on any particular connection. Each echo canceler design has a limit as to the maximum delay of echo that it can identify. Echo cancelers are bypassed if a fax signal or modem data is on the line.

4 MEDIA GATEWAY CONTROLLER AND ITS PROTOCOLS

The gateway controller or media gateway controller (MGC) carries out the signaling function on VoIP circuits. Some texts call an MGC a *soft switch* even though they are not truly switches but are servers that control gateways. This function is illustrated in Figure 12.6.

An MGC can control numerous gateways. However, to improve reliability and availability several MGCs may be employed in separate locations with function duplication on the gateways they control. Thus if one MGC fails others can take over its functions. We must keep in mind that the basic topic of Section 4 of this chapter is signaling. That is the establishing (setting up) of telephone connectivities, maintaining that connectivity and the taking down of the circuit when the users are finished with conversation. There is a basic discussion of signaling in Chapter 4 of this text.

There are four possible signaling protocol options between an MGC and gateways. These are:

- ITU-T Rec. H.323. This is employed where all network elements (NEs) have software intelligence.

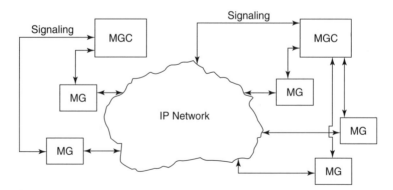

Figure 12.6. A media gateway controller (MGC) provides a signaling interface for media gateways (MG) and thence to the IP network.

- SIP (session initiation protocol [2]) is used when the end devices have software intelligence and the network itself is without such intelligence.
- MGCP (media gateway control protocol) is another gateway control protocol.
- Megaco (ITU-T Rec. H.248 [13]) is a gateway control protocol applicable when end devices are without software intelligence and the network has software intelligence.

4.1 Overview of the ITU-T Rec. H.323 Standard

In May 1996 the ITU ratified the H.323 specification, which defines how voice, video, and data traffic should be transported over IP-based LANs. It also incorporates the ITU-T Rec. T.120 [21] data-conferencing standard. The H.323 recommendation is based on RTP/RTCP (real-time protocol/real-time control protocol) for managing audio and video signals.

What sets H.323 apart is that it addresses core-internet applications by defining how delay-sensitive traffic such as voice and video get priority transport to ensure real-time communication service over the internet. A related protocol is ITU-T Rec. H.324 [14] specification, which defines the transport of voice, data, and video over regular telephone networks. Another related protocol is ITU-T Rec. H.320 [15], which covers the transport of voice, video, and data over the integrated services digital network (ISDN).

H.323 deals with three basic functional elements of VoIP. These are:

- Media gateway
- Media gateway controller (MGC) (in some settings this is called the *gatekeeper*)
- Signaling gateway

H.323 is an umbrella protocol covering:

- H.225 [16], which covers the setup of multimedia channels and
- H.245 [17], which deals with the setup of single channel medium

The standard H.323 [3] prefers the use of the term gatekeeper (versus media gateway controller). Some of the more important responsibilities of a gatekeeper are:

- Security. It authenticates users of the H.323 network.
- It performs address translation between internet addresses and ITU-T Rec. E.164 [18] addresses.
- It polices the capacity of the network in question, whether that network can accept another call.
- H.323 determines call routing, to route through a gateway or be sent directly to the destination.
- It keeps track of the network's bit rate capacity.

H.323 assumes that the transmission medium is a LAN that does not provide guaranteed delivery of packets. In the ITU H.323 standard we will find the term *entity*. An entity carries out a function. For example, a terminal is an endpoint on a LAN that can support real-time communications with another entity on that LAN. It has a capability provided by a voice or audio codec such as a G.711 or G.728 codec. It will also provide a signaling function for VoIP circuit setup, maintain, and take-down. A VoIP terminal optionally can support video and data streams including compression and decompression of these streams. Media streams are carried on RTP (real-time protocol) or RTCP (real-time control protocol). RTP deals with media content while RTCP works with the signaling functions of status and control. This protocol information is embedded in UDP, which is reliably transported by TCP.

Other VoIP entities are gateways and there is a gatekeeper that is optional.

The leading issue in VoIP implementation is guaranteed quality of service (QoS). H.323 is based on RTP (real-time protocol) which is comparatively new. RTP-compliant equipment includes control mechanisms for synchronizing different traffic streams. On the other side of the coin, RTP has no mechanisms for ensuring on-time delivery of traffic signals or for recovering lost packets. It does not address the QoS issue related to guaranteed bit rate availability for specific applications. The IEC [19] reports that there is a draft signaling proposal to strengthen the internet's ability to handle real-time traffic reliably. This would dedicate end-to-end transport paths for specific sessions much like the circuit-switched PSTN does. This is the resource reservation protocol (RSVP). It will be implemented in routers to establish and maintain requested transmission paths and QoS levels.

4.2 Session Initiation Protocol (SIP)

SIP is based on RFC 2543 [2] and is an application layer signaling protocol. It deals with interactive multimedia communication sessions between end-users, called *user agents*. It defines their initiation, modification, and termination. SIP calls may be terminal-to-terminal, or they may require a server to intercede. If a server is to be involved, it is only required to locate the called party. For inter-working with non-IP networks, Megaco and H.323 are required. Often vendors of VoIP equipment integrate all three protocols on a single platform.

SIP is closely related to IP. SIP borrows most of its syntax and semantics from the familiar HTTP (hypertext transfer protocol). An SIP message looks very much like an HTTP message, especially with message formatting, header, and multipurpose internet mail extension support. It uses addresses that are very similar to URLs (uniform resource locators) and to email. For example, a call may be made to so-in-so@such-and-such. SIP messages are text-based rather than binary. This makes writing easier and the debugging of software more straightforward.

There are two modes with which a caller can setup a call with SIP. These are called *redirect* and *proxy*, and servers are designed to handle these modes.

Both modes issue an "invite" message for another user to participate in a call. The redirect server is used to supply the address (URL) of an unknown called addressee. In this case the "invite" message is sent to the redirect server, which consults the location server for address information. Once this address information is sent to the calling user, a second "invite" message is issued, now with the correct address.

One specific type of SIP is called SIP-T (T for telephone). This is a function that allows calls from CCITT Signaling System 7 (SS7) to interface with telephone in an IP-based network. The particular user part of SS7 for this application is ISUP (see Chapter 16).

4.3   Media Gateway Control Protocol (MGCP)

This protocol was the predecessor to Megaco (see Section 4.4 of this chapter) and still holds sway with a number of carriers and other VoIP users. MGCP [20] assumes a call control architecture where the call control "intelligence" is outside the gateways (i.e., at the network edge) and handled by external call control elements. Thus, the MGCP assumes that these call control elements, or "call agents," will synchronize with each other to send coherent commands to the gateways under their command. There is no mechanism defined in MGCP for synchronizing "call agents." It is, in essence, a master/slave protocol where the gateways are expected to execute commands sent by the "call agents."

In the MGCP protocol an assumption is made that the connection model consists of constructs that are basic endpoints and connections. Endpoints are sources or sinks of data and could be physical or virtual. The following are two examples of endpoints:

1. An interface on a gateway that terminates a trunk connected to a PSTN switch (e.g., local or toll-connecting, etc.). A gateway that terminates trunks is called a *trunk gateway*.
2. An interface on a gateway that terminates an analog POTS (plain old telephone service) connection to a telephone, a key system, PABX, and so on. A gateway that terminates residential POTS lines (to telephones) is called a *residential gateway*.

An example of a virtual endpoint is an audio source in an audio-content server. Creation of physical endpoints requires a hardware installation, while creation of virtual endpoints can be done in software [20].

4.4   Megaco or ITU-T Rec. H.248 [13]

Megaco is a call-control protocol that communicates between a gateway controller and a gateway. It evolved from and replaces SGCP (simple gateway control protocol) and MGCP (media gateway control protocol). Megaco addresses the relationship between a media gateway (MG) and a media gateway controller (MGC). An MGC is sometimes called a *softswitch* or *call agent.*

Both Megaco and MGCP are relatively low-level devices that instruct MGs to connect streams coming from outside the cell or packet data network onto a packet or cell stream governed by RTP (real-time transport protocol).

A Megaco (H.248) connection model is illustrated in Figure 12.7. There are two principal abstractions relating to the model: *terminations* and *contexts.* A termination acts as sources and/or sinks for one or more data streams. In a

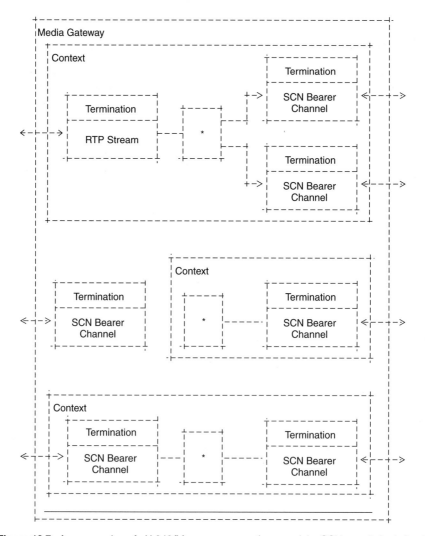

Figure 12.7. An example of H.248/Megaco connection model. SCN = switched circuit network. The asterisk in each box in each of the contexts represents the logical association of terminations implied by the context. Based on Figure 1, RFC 3015 [9].

multimedia conference a termination can be multimedia, and it sources and sinks multiple media streams. The media stream parameters, as well as modem, and bearer parameters are encapsulated within the termination.

A context is an association between a collection of terminations. There is a special type of context called the *null context*, which contain all terminations that are not associated with any other termination. For example, in a decomposed access gateway, all idle lines are represented by terminations in the null context.

Let's look at three context possibilities. (1) A context with just one termination is call waiting. The caller does not hear anyone else. (2) A context with two terminations is a regular phone call. Of course each person is expected to hear the other. (3) An example of more than two terminations is a conference call. Each party hears each and every other one.

The maximum number of terminations in a context is a media gateway (MG) property. MGs that offer only point-to-point connectivity might allow at most two terminations per context. MGs that support multipoint conferences might allow three or more terminations per context.

The attributes of contexts are:

- Context ID.
- The topology (who hears/sees whom).
  The topology of a context describes the flow of media between the terminations within a context. In contrast, the mode of a termination (send/receive___) describes the flow of the media at the ingress/egress of the media gateway.
- The priority is used for a context in order to provide the MG with information about a certain precedence handling for a context. The MGC can also use the priority to control autonomously the traffic precedence in the MG in a smooth way in certain situations (e.g., restart), when a lot of context must be handled simultaneously.
- An indicator for an emergency call is also provided to allow a preference handling in the MG.

Megaco uses a series of commands to manipulate terminations, contexts, events, and signals. For example, the *add command* adds a termination to a context and may be used to create a new context at the same time. Of course we would expect the *subtract command* to remove a termination from a context and may result in the context being released if no terminations remain.

There is also the *modify command* used to modify the description of a termination (e.g., the type of voice compression in use). *Notify* is used to inform the gateway controller if an event occurs on a termination such as a telephone in an off-hook condition, or digits being dialed. There is also a *service change command*.

Terminations are referenced by a TerminationID, which is an arbitrary schema selected by the MG. TerminationIDs of physical terminations are provisioned by the media gateway. The TerminationIDs may be chosen to have structure. For example, a TerminationID may consist of a trunk group and a trunk within the group [9].

REVIEW QUESTIONS

1. How many bits are in a conventional PCM voice sample?

2. How is a telephone call on the PSTN usually terminated?

3. Give the two basic components of a data message?

4. Every data frame or packet carries two components (at least) in the header. What are they?

5. In question 4, how does this differ from a PCM frame?

6. What is a CRC sequence used for?

7. Give at least two different ways the PSTN may setup a voice circuit.

8. List at list five drawbacks or challenges designers faced in making VoIP a reality.

9. List at least four tasks the gateway may do.

10. List at least four transmission media and formats that a media gateway should be able to support.

11. A media gateway interfaces a PSTN PCM bit stream. Name three types of traffic that may be transported in that bit stream, each requiring some sort of special handling.

12. Given an IP packet used for voice transport. Assume G.711 or G.726 coders. How many bytes (octets) can we expect in the payload?

13. Give some data you have learned about one-way delay. What is the one-way delay (in ms) that should not be exceeded as a design goal?

14. What one-way delay (in ms) should never be exceeded?

15. What would you say is the biggest contributor to one-way delay?

16. What is "look-ahead" used for?

17. There is buffer delay. What is its cause and it varies with the number of xxxxx in tandem?

18. What is one very obvious way to reduce delay?

19. On G.711 circuits, the lost packet rate objective is?

20. Name at least two ways to handle the presence of a lost packet.

21. What is the basic function of a media gateway controller?

22. Name at least three of the four signaling protocol options between an MGC and a media gateway.

23. Give a primary function of an MGC regarding delay using the H.323 protocol on a VoIP circuit.

24. Differentiate RTP and RTCP.

25. In the SIP protocol, what are end-users called.

26. What are the two modes with which a caller can set up a call with SIP.

27. In MGCP gateways are expected to execute commands sent by _____.

28. Megaco (H.248) is a call-control protocol that communicates between a _____and a _____?

29. What are the two basic abstractions relating to the Megaco model?

30. A point-to-point connectivity must have at least _____terminations.

31. How are terminations referenced in Megaco?

REFERENCES

1. *Pulse Code Modulation of Voice Frequencies*, ITU-T Rec. G.711, ITU, Geneva, November 1988.
2. G. Malkin, *Routing Information Protocol*xs, RIP Version 2, RFC 2543.
3. *Packet-based Multimedia Communication System*, ITU-T Rec. H.323, ITU Geneva, November 2000.
4. *Speech Coders, Dual Rate for Speech, 5.3 and 6.3 kbps*, ITU-G.723.1, Multipulse Maximum Likelihood Quantization (MPMLQ), ITU, Geneva, 1996.
5. *Speech Coders, Algebraic Code-Excited Linear Prediction Coder*, ITU-G.723.1, ITU, Geneva, March 1996.
6. *40, 32, 24 and 16 kbps Adaptive Differential Pulse Code Modulation*, ITU-T G.726, ITU, Geneva, December 1990.
7. *Coding of Speech at 16 kbps Low-Delay Code-Excited Linear Prediction (LD-CELP)*, ITU-T G.728, ITU, Geneva, September 1992.
8. *Reduced Complexity 8 kbps CS-ACELP Speech Coder*, ITU-T Rec. G.729A, ITU, Geneva, March 1996.
9. *Megaco Protocol*, Version 1.0, RFC 3015, IETF, November 2000, www.RFC.editor.org.
10. *One-Way Transmission Time*, ITU-T Rec. G.114, ITU, Geneva, May 2000.
11. *Nortel Paper on VoIP Performance*, Nortel, Ottawa, Ontario, Canada, November 2000.
12. *Digital Network Echo Cancellers*, ITU-T Rec. G.168, ITU, Geneva, June 2002.
13. *Gateway Control Protocol*, ITU-T Rec. H.248, ITU, Geneva, July 2000.
14. *Terminal for Low Bit-Rate Multimedia Communications*, ITU-T Rec. H.324, ITU, Geneva, March 2002.

15. *Narrowband Audio/Visual Communication Systems and Terminal*, ITU-T Rec. H.320, ITU, Geneva, May 1999.

16. *Call Signaling Protocols and Media Stream Packetization for Packet-Based Communication Systems*, ITU-T Rec. H.225, ITU, Geneva, July 2001.

17. *Control Protocol for Multimedia Communication*, ITU-T Rec. H.245, ITU Geneva 1998.

18. *The International Public Telecommunication Numbering Plan*, ITU-T Rec. E.164, ITU, Geneva, May 1997.

19. IEC Reports On-Line. www.iec.org/online/tutorials, VoIP, January 2003.

20. *Media Gateway Control Protocol (MGCP)*, Version 1.0, RFC 3445, January 2003.

21. *Data Protocols for Multimedia Conferencing*, ITU-T Rec. T.120, ITU, Geneva, June 1996.

22. *Internet Protocol*, RFC 791, DDDN Network Information Center, SRI International, Menlo Park, CA, September 1981.

13

LOCAL AREA NETWORKS

1 DEFINITION AND APPLICATIONS

Local area networks (LANs) use a common transmission medium to interconnect workstations, servers, computers, and/or other related assets over a limited geographical area. Several LAN standards specify capability to serve up to a thousand or more devices on a single LAN. The geographical extension or "local area" may extend no more than several hundred feet (<100 m) to over 6 miles (>10 km) or more in other cases. The transmission media providing this connectivity may be wire pair, coaxial cable, or fiber-optic cable. Local area radio (wireless) schemes based on IEEE 802.11 standards are gaining wide acceptance. Certain LAN schemes accommodate other devices as well such as digital telephones, facsimile, and video equipment. A basic rule on LANs was that only one user at a time may have access to the medium. Switching hubs and LAN routers have now changed this rule.

Data rates on current LANs vary from 1 Mbit/s to 10,000 Mbits/s. LAN data rates, the number of devices connected to a LAN, the spacing of those devices, and the network extension depend on:

- The transmission medium employed
- Transmission technique (i.e., baseband or broadband)
- Network access protocol

Many LANs operate without error correction with bit error rates (BERs) specified in the range of 1×10^{-8} to 1×10^{-12} or better.

The most common application of a LAN is to interconnect data terminals (workstations) with processing resources, where all the devices reside in a single building or complex of buildings, and usually these resources have a common owner. Cost containment is a driving force toward the implementation of LANs.

Telecommunication System Engineering, by Roger L. Freeman
ISBN 0-471-45133-9   Copyright © 2004 Roger L. Freeman

A LAN permits effective cost sharing of high-value data processing equipment, such as mass storage, mainframe or minicomputers, and high-speed printers. There are other benefits as well. One, of course, is resource sharing. Another is e-mail and similar messaging services leading to a "paperless" environment. A LAN may also be considered an aggregator of data for eventual transport of this data over a WAN.

LANs can be extended, up to a certain point, with repeaters or bridges. They can be segmented by means of switches, switching hubs, smart bridges, or routers. Segmenting of a LANs can notably improve performance, especially on a LAN with many users.

The interconnection of LANs in the local area with a high-speed backbone is current practice. LAN interconnection with the outside world such as with distant LANs via a wide area network is becoming prevalent. This is frame relay's principal application. The interface is carried out with a router or gateway.

There are two generic transmission techniques utilized by LANs: baseband and broadband. Baseband transmission can be defined as the direct application of the baseband signal to the transmission medium. Broadband transmission, in this context, is where the baseband signal from the data device is translated in frequency to a particular frequency slot in the RF spectrum. Broadband transmission requires a modem to carry out the translation. Baseband transmission may require some sort of signal conditioning device. With broadband LAN transmission, we usually think of simultaneous multiple RF carriers that are separated in the frequency domain. Present broadband technology comes from the cable television (CATV) industry.

Use of the asynchronous transfer mode (ATM) is beginning to find favor in the LAN community and could radically change many of the concepts introduced in this chapter.

2 LAN TOPOLOGIES

There are three types of LAN topology: bus, ring, and star. These are shown in Figure 13.1 along with the *tree network*, which is a subset of the conventional bus topology. All three topology configurations have been shown previously in Figure 11.4. They are repeated here for convenience of the reader.

A bus is a stretch of transmission medium from which users tap into as shown in Figure 13.1A. Originally, the medium was coaxial cable. Today it can also be UTP or STP (unshielded twisted pair or shielded twisted pair). For high data rate (e.g., ≥ 100 Mbits/s) LANs, fiber-optic cable is now widely used.

A ring is simply a bus that is folded back onto itself. A ring topology is shown in Figure 13.1B. User traffic flows in one direction around the ring. In one approach, which we discuss in this chapter, a second ring is added where the traffic flow is in the opposite direction. Such a dual counterrotating ring concept improves reliability in case of a failed station or a cut of the ring.

A star network is shown in Figure 13.1C. At the center of the star is a switching device. This could be a switching hub. Users can be paired, two at a time, three

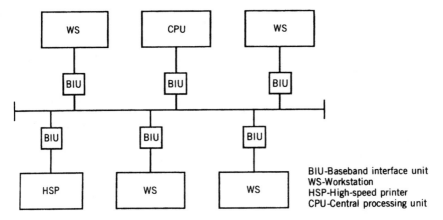

BIU-Baseband interface unit
WS-Workstation
HSP-High-speed printer
CPU-Central processing unit

Figure 13.1A. A bus network.

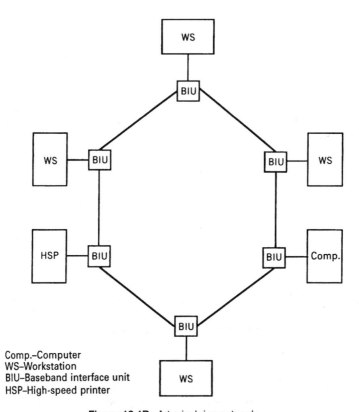

Comp.–Computer
WS–Workstation
BIU–Baseband interface unit
HSP–High-speed printer

Figure 13.1B. A typical ring network.

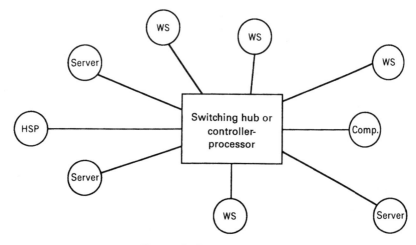

Figure 13.1C. A star network.

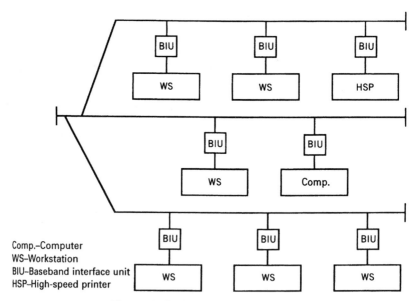

Comp.–Computer
WS–Workstation
BIU–Baseband interface unit
HSP–High-speed printer

Figure 13.1D. A typical tree configuration.

at a time, or all at a time, segmented into temporary families of users depending on the configuration of the switch at that moment in time. Such a concept lends itself particularly well to ATM. In this case each user is connected to the switch on a point-to-point basis for the period of connectivity.

A tree network is illustrated in Figure 13.1D.

3   THE TWO BROAD CATEGORIES OF LAN TRANSMISSION TECHNIQUES

Baseband and broadband are the two basic transmission techniques employed by LANS, which were introduced above.

With both these types of LANs (i.e., baseband and broadband), we are dealing with multipoint operation. Two transmission problems arise as a result. The first deals with signal level and signal-to-noise ratio, and the second deals with standing waves. Each access on a common medium must have sufficient signal level and S/N such that copied signals have a BER in the range of 1×10^{-8} to 1×10^{-12}. If the medium is fairly long in extension and there are many accesses, the signal level must be sufficiently high for a transmitting access to reach its most distant destination. The medium is lossy, particularly affecting the higher bit rates, and each access tap has an insertion loss. This leads to very high signal levels which may be rich in harmonics and spurious emissions, degrading bit error rate. On the other hand, with insufficient signal, the S/N degrades, which will degrade error performance. A good level balance must be achieved for all users. Each and every multipoint connectivity must be examined. The number of multipoint connectivities can be expressed by $n(n - 1)$, where n is the number of accesses. If, on a particular LAN, 100 accesses are planned, there are 9900 possible connectivities to be analyzed to carry out signal level balance. One way to simplify the job is to segment the network, placing a regenerative repeater or bridge at each boundary. This reduces the signal balance job to reasonable proportions and ensures that a clean signal of proper level is available at reach access tap. For baseband LANs, 50-Ω coaxial cable is favored over the more common 75-Ω cable. The lower impedance cable is less prone to signal reflections from access taps and provides better protection against low-frequency interference.

The effects of standing waves can be reduced by controlling the spacing between access taps. For example, the Ethernet technical summary [1] recommends spacing no less than 2.5 m, for the 10-Mbit/s standard Ethernet LAN. The technical summary says that by following this placement rule, the chance the objectionable standing waves will result is reduced to a very low (but not zero) probability. Again for Ethernet, up to 100 devices may be placed on a cable segment and the maximum segment length is 500 m. The segments can be connected through regenerative repeaters, with a maximum total end-to-end length of 2.5 km [1, 4, 21].

The baseband technique incorporates single signal transmission of a digital waveform on a transmission medium. Broadband transmission has the capability of transmitting multiple signals simultaneously on a medium, typically coaxial cable. Each signal is assigned a frequency slot in a frequency division multiplex plan. Broadband technology derives from the CATV (cable television) industry. Baseband lends itself to bus and ring topologies, and broadband to bus and

TABLE 13.1 Comparison of Baseband and Broadband LAN Transmission

Item	Baseband	Broadband
Waveform	Digital: NRZ, RZ, Manchester	RF/FDM
Baseband	Bidirectional	Unidirectional
Topology	Bus or ring	Bus or tree
Access to medium	Tap	Modem or tap
Media	Wire pair, coaxial cable	Coaxial cable
LAN extension	Up to 2 km	Tens of km
Information type	Data only	Data, voice, facsimile, video
Utilization of bandwidth	Single signal occupies entire bandwidth	Multiple simultaneous signals in FDM structure

tree topologies. Broadband systems require a modem at each access; baseband systems do not.* Baseband and broadband transmission techniques are compared in Table 13.1.

One extremely important consideration for baseband transmission is that only a single thread (transmission line) exists. It can accommodate only one user at a time; otherwise there is a high probability of data message collision. Collision is where the electrical signals bearing the traffic of two or more users interfere one with the other, corrupting the traffic of each. To help avoid collisions, segmenting has become an excellent alternative using bridges, hubs, or switches. These devices are discussed at the end of this chapter.

3.1 Broadband Transmission Considerations

Broadband transmission permits multiple users to access the medium without collision. Broadband means that we take advantage of the medium's wide bandwidth. This wide bandwidth is broken down into smaller bandwidth segments in an analogous fashion to FDM. Each of these segments is assigned to a family of users. The statement regarding collision is correct if there are no more than two accesses per frequency segment connected on a point-to-point basis, where one access receives while the other transmits. If, in this case, we assume contention as the access protocol, then as the family increases in number, the chances of collision start to increase. With a little imagination, we can see that, with the proper switching scheme implemented, a user can join any family by simply switching to the proper frequency band of that family. All that is required is a change in modem frequency and possibly modulation waveform. There may also be certain protocol considerations as well.

---

* The semantics of local area network technology are rather loose. Many baseband systems are not truly baseband. One "baseband" system we discuss uses light; another uses RF. Certainly, if we are truly talking about broad bandwidth, the fiber-optic "baseband" system we describe has an available bandwidth that is extremely broad.

Unlike their baseband counterparts, broadband systems can be designed to accommodate digital or analog voice, data from kilobit to multimegabit rates, video, and facsimile. Thus broadband systems are versatile. They are also much more expensive than their baseband counterparts and require a higher level of design engineering effort.

As mentioned, much of our present broadband technology derives from cable television technology. Total system bandwidths are on the order of 300–500 MHz. Each access requires a modem to modulate and demodulate the data or other user signal and to translate the modulated frequency to the assigned frequency slot on the cable.

This, then, is RF transmission and, by its very nature, must be one way or unidirectional. Thus a user can only access another user "downstream" from it. If we assume a single medium, usually a 75-Ω coaxial cable, then how does one access another user "upstream"? This is done using a similar approach to that of a two-way or interactive CATV (cable television) system, where two paths are provided on a single coaxial cable. This is accomplished by splitting the cable spectrum into two frequency segments, one segment for one direction and the other for the opposite direction. At a cable terminating point, which some even call a *head end* from CATV terminology, a frequency translator converts and amplifies signals from one direction (frequency segment) into signals for transmission in the opposite direction (frequency segment). Another term for "head end" for broadband LANs is *central retransmission facility* (CRF).

There are two choices of topology for broadband LANs: bus and tree. The head end or CRF is located at some termination point on the bus, and in the case of tree topology the CRF is located at its root, so to speak.

Another approach to achieve dual path operation is to use two cables; one provides the "go" path and the other the "return" path. For single-cable split-band operation, a rather large guard band is left in the center of the cable spectrum to ensure isolation between the two paths. Then we can see that with the provision of two-cable operation the usable bandwidth can be more than doubled. Modem operation is also simpler because, to access a particular net, only one frequency operation is required (i.e., the send and receive frequencies can be the same). With single-cable split-band operation, send and receive frequencies must necessarily be different [2].

Broadband services make up about 5% of the total LAN market. Baseband predominates.

Table 13.2 gives broadband channel allocations for CSMA/CD services and Table 13.3 provides channel allocations for a broadband token bus configuration. Recommended channelization for both single and dual cable operation is given.

3.2 Fiber-Optic LANs

Fiber-optic LANs may be considered broadband* in that a class of modem is required to place the digital signal on the fiber. The modem, of course, consists of

* We will see a contradiction in terminology here when consulting Section 5.6 dealing with FDDI.

TABLE 13.2 Broadband Channel Allocations for CSMA/CD Services

	Transmit MHz	Receive MHz		
		156.25 Offset		192.25 Offset
Transmit and receive channel options for IEEE 802.3b services (single cable plant)	(1) 35.75–53.75	192.00–210.00	or	228.00–246.00
	(2) 41.75–59.75	198.00–216.00	or	234.00–252.00
	(3) 47.75–65.75	204.00–222.00	or	240.00–258.00
	(4) 53.75–71.75	210.00–228.00	or	246.00–264.00 (preferred)
	(5) 59.75–77.75	216.00–234.00	or	252.00–270.00
	(6) 65.75–83.75	222.00–240.00	or	258.00–276.00
		Transmit and Receive MHz		
Transmit and receive channel options for IEEE 802.3b services (dual cable plant)		(1) 36.00–54.00		
		(2) 42.00–60.00		
		(3) 48.00–66.00		
		(4) 54.00–72.00		
		(5) 60.00–78.00		
		(6) 66.00–84.00 (preferred)		
		(7) 228.00–246.00		
		(8) 234.00–252.00		
		(9) 240.00–258.00		
		(10) 246.00–264.00 (preferred)		
		(11) 252.00–270.00		
		(12) 258.00–276.00		

Source: Table 8, IEEE Std. 802.7 [5].

TABLE 13.3 Broadband Channel Allocations for Token Ring Services

	Transmit MHz	Receive MHz
Transmit and receive channel options for IEEE 802.4 services (single cable plant)	(1) 59.75–65.75	252.00–258.00
	(2) 65.75–71.75	258.00–264.00
	(3) 71.75–77.75	264.00–270.00
	(4) 77.75–83.75	270.00–276.00
	(5) 83.75–89.75	276.00–282.00
	(6) 89.75–95.75	282.00–288.00
	Transmit and Receive MHz	
Transmit and receive channel options for IEEE 802.4 services (dual cable plant)	(1) 59.75–65.75	
	(2) 65.75–72.75	
	(3) 72.75–77.75	
	(4) 77.75–83.75	
	(5) 83.75–89.75	
	(6) 89.75–95.75	
	(7) 252.00–258.00	
	(8) 258.00–264.00	
	(9) 264.00–270.00	
	(10) 270.00–276.00	
	(11) 276.00–282.00	
	(12) 282.00–288.00	

Note: For 10-Mbit/s transmission, channels are paired as follows: 1–2, 3–4, 5–6.
Source: Table 9, IEEE Std. 802.7 [5].

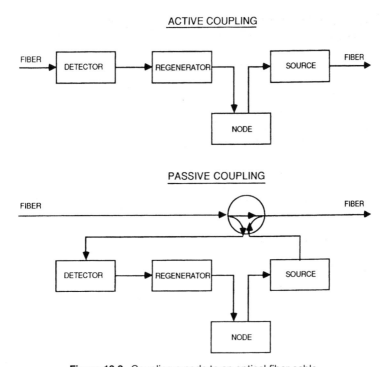

Figure 13.2. Coupling a node to an optical fiber cable.

a light source, detector, and the necessary driver and signal conditioning circuitry. With wavelength-division multiplexing (WDM) we have a true broadband system. At this time, single-wavelength operation prevails.

The type of fiber-optic cable selected for a LAN is a cost trade-off. The extension of a LAN is generally short such that multimode fiber can be used, and the mature short-wavelength technology permits other cost savings. Even plastic fiber may be considered. The losses of the fiber itself will generally be low due to the short-distance operation even at 820-nm operation. A LAN that is 1 km long might display a fiber loss from 2 dB to 5 dB. The major contributor to loss is the taps if a fully passive network is to be implemented.

Fiber lends itself to all three LAN topologies: bus, ring, and star. The use of passive couplers leads to a more reliable system, but the insertion loss of passive couplers is a consideration. With active coupling, the loss of one source or detector can cause the entire network to crash unless bypass switches are used. Figure 13.2 shows passive and active coupler implementations.

4 OVERVIEW OF IEEE/ANSI LAN PROTOCOLS

4.1 General

Many of the widely used LAN protocols have been developed in North America through the offices of the Institute of Electrical and Electronic Engineers (IEEE).

The American National Standards Institute (ANSI) has subsequently accepted and incorporated these standards, and they now bear the ANSI imprimatur.

The IEEE develops LAN standards in the IEEE 802 committee, which is currently organized into the following subcommittees:

802.1*	Bridging & Management
802.2*	Logical Link Control
802.3*	CSMA/CD Access Method
802.4*	Token Passing Bus Access Method
802.5*	Token Ring Networks
802.6*	Metropolitan Area Networks (MANs)
802.7*	Broadband Technical Advisory Group
802.10*	LAN Security
802.11*	Wireless LAN
802.12*	Demand Priority Access Working Group
802.15*	Wireless Personal Area Networks
802.16*	Broadband Wireless Metropolitan Area Networks
802.17	Resilient Packet Ring
802.18	The Radio Regulatory TAG (Technical Advisory Group)
802.19	Coexistence Advisory Group
802.20	Mobile Broadband Wireless Access (Working Group)

The fiber distributed data interface (FDDI) standard, which is discussed in Section 5.6, has been developed directly by ANSI.

4.2 How LAN Protocols Relate to OSI

LAN protocols utilize only OSI layers 1 and 2, the physical and data-link layers, respectively. The data-link layer is split into two sublayers: medium access control (MAC) and logical link control. These relationships are shown in Figure 13.3.

Stallings [3] presents an interesting and rational argument on the reasoning for limiting the layering to the first two OSI layers. There is no question that the functions of OSI layers 1 and 2 must be incorporated in a LAN architecture. We now ask, Why not layer 3? Layer 3, the network layer, is concerned with routing. There is no routing involved with LANs. There is a direct link involved between any two points. The other functions carried out by OSI layer 3—addressing, sequencing, and flow control—are carried out by layer 2 in LANs. The difference is that layer 2 performs these functions across a single link. OSI layer 3 carries out these functions across a sequence of links required to traverse a network. Of course, there is only one link required to traverse a LAN.

It would seem that layer 3 is required when viewed through an attached device. The reason is that the device sees itself attached to a network connecting multiple devices. One would think that ensuring delivery of a message to one or more accesses would be a layer 3 function. It was decided that, although the network

---

* Indicates that one or more published standards or draft standards are available.

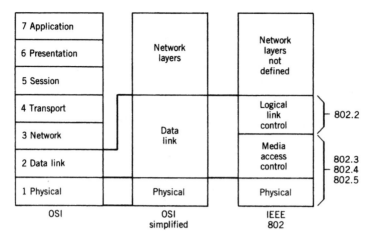

Figure 13.3. LAN 802 architecture related to OSI.

provides services through layer 3, the characteristics of the network allow these functions to be performed in the first two layers.

As shown in Figure 13.3, the OSI data-link layer is divided into two sublayers: logical link control (LLC) and medium access control (MAC). These sublayers carry out four functions:

1. Provide one or more service access points (SAPs). A SAP is a logical interface between two adjacent layers.

2. Before transmission, assemble data into a frame with address and error-detection fields.

3. On reception, disassemble the frame and perform address recognition and error detection.

4. Manage communications over the link.

The first function and those related to it are performed by the LLC sublayer. The last three functions are handled by the MAC sublayer.

In the following subsections we will describe four common IEEE and ANSI standardized protocols. Logical link control (LLC) is common to all four. They differ in the medium access control (MAC) protocol.

A station on a LAN may have multiple users; oftentimes these are just processes, such as processes on a host computer. These processes may wish to pass traffic to another LAN station which may have more than one "user" in residence. We will find that LLC produces a PDU with its own source and destination address. The source address, in this case, is the address of the originating user. The destination address is the address of a user in residence at a LAN station. Such a user is connected through a service access point (SAP) at the upper boundary of the LLC layer. The resulting LLC PDU is then embedded in the information field of a MAC frame. This is shown in Figure 13.4.

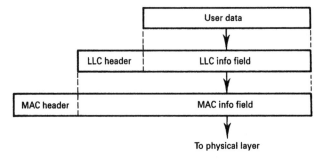

Figure 13.4. A user passes traffic to an LLC where an encapsulation takes place forming an LLC PDU. This traffic is then embedded in a MAC frame. The MAC frame is passed to the physical layer which transmits the traffic on the LAN.

The MAC frame also has a source and destination address. These direct the traffic to a particular LAN station or stations.

4.3   Logical Link Control (LLC)

The LLC provides services to the upper layers at a LAN station. The upper layers are user defined. The LLC provides two forms of services for its users:

1. Unacknowledged connectionless service
2. Connection mode service

Some brief comments are required to clarify the functions and limitations of each service. With unacknowledged connectionless service a single service access initiates the transmission of a data unit to the LLC, the service provider. From the viewpoint of the LLC, previous and subsequent data units are unrelated to the present unit. There is no guarantee by the service provider of the delivery of the data unit to its intended user, nor is the sender informed if the delivery attempt fails. Furthermore, there is no guarantee of ordered delivery. This type of service supports point-to-point, multipoint, and broadcast modes of operation.

As we might imagine with connection mode service, a logical connection is established between two LLC users. During the data transfer phase of the connection, the service provider at each end of the connection keeps track of the data units transmitted and received. The LLC guarantees that all data will be delivered and that the delivery to the intended user will be ordered (e.g., in the sequence as presented to the source LLC for transmission). When there is a failure to deliver, it is reported to the sender.

IEEE [6] defines the LLC as that part of a data station that supports the LLC functions of one or more logical links. The LLC generates command PDUs (protocol data units) and response PDUs for transmission and interprets received command PDUs and response PDUs. Specific responsibilities assigned to the LLC include:

1. Initiation of control signal interchange
2. Interpretation of received command PDUs and generation of appropriate response PDUs
3. Organization of data flow
4. Actions regarding error-control and error-recovery functions in the LLC sublayer

LLC is another derivative of HDLC, which was covered in Chapter 11. It is based on the balanced mode of that link-layer protocol with similar formats and functions. This is especially true when operating in the connection mode.

4.3.1 LLC Generic Primitives. The IEEE [7] defines a (service) primitive as *an abstract, implementation-independent interaction between a service user and service provider.* In general, the services of a layer or sublayer are the capabilities which it offers to a user in the next higher layer or sublayer. In order to provide its service, a layer or sublayer builds its functions on the services it requires from the next lower layer or sublayer.

Services are specified by describing the service primitives and parameters that characterize each service. A service may have one or more related primitives that constitute the activity that is related to a particular service. Each service primitive may have zero or more parameters that convey the information required to provide the service.

The generic primitives used in LLC are identical to the generic primitives employed in HDLC (Chapter 11). These are:

REQUEST   The request primitive is passed from the N-user to the N-layer (or sublayer) to request that a service be initiated.

INDICATION The indication primitive is passed from the N-layer (or sublayer) to the N-user to indicate an internal N-layer (or sublayer) event which is significant to the N-user. This event may be logically related to a remote service request, or may be caused by an event internal to the N-layer (or sublayer).

RESPONSE   The response primitive is passed from the N-user to the N-layer (or sublayer) to complete a procedure previously invoked by an indication primitive.

CONFIRM   The confirm primitive is passed from the N-layer (or sublayer) to the N-user to convey the results of one or more associated previous service request(s).

Examples of primitives with parameters is a sequence used for a successful connection: DL_CONNECT request, DL_CONNECT indication, DL_CONNECT response and DL_CONNECT confirm, in that order [6].

4.3.2 LLC PDU Structure. As shown in Figure 13.4, user data are passed down to the LLC, which appends a header (i.e., encapsulates it). The LLC PDU

DSAP Address	SSAP Address	Control	Information
8 bits	8 bits	8 or 16 bits	M * 8 bits

DSAP Address = destination service access point address field
SSAP Address = source service access point address field
Control = control field (16 bits for formats that include sequence numbering,
         and 8 bits for formats that do not
Information = information field
* = multiplication
M = an integer value equal to or greater than 0. (Upper bound of M is a func-
    tion of the medium access control methodology used.)

Figure 13.5. LLC PDU format [6].

frame format is shown in Figure 13.5. The header consists of address and control information; the information field contains the user data. The control field is identical to the HDLC control field, shown in Figure 11.15 in Section 6.2.3 of Chapter 11. However, the LLC control field is 2 octets long, and there is no provision to extend it to 3 or 4 octets in length as there is in HDLC.

As we mentioned previously, LLC destination address is the user address at an SAP inside the LAN station. It is called the destination service access point (DSAP). The SSAP is the source service access point, and it indicates the message originator inside a particular LAN station. Each has a field of 8 bits as shown in Figure 13.6. However, only the last 7 of those bits are used for actual address. The first bit in the destination address field indicates whether the address is an individual address or a group address (i.e., addressed to more than one SAP). The first bit in the SSAP is the C/R bit which indicates whether a frame is a command frame or a response frame. The control field is briefly described in Section 4.3.4.

4.3.3 Types and Classes of LLC Operation.
Two types of operation are defined in the referenced standard [6]:

(a) TYPE I operation is where PDUs are exchanged between peer LLCs without the need for the establishment of a data-link connection. In the LLC sublayer these PDUs are not acknowledged, nor is there any error recovery or flow control.

(b) TYPE 2 operation requires a data-link connection to be established between the two peer LLCs prior to the exchange of any information-bearing PDUs. The normal cycle of communication between two TYPE 2 LLCs on a data-link connection consists of the transfer of PDUs containing information from the source LLC to the destination LLC, acknowledged by PDUs in the opposite direction.

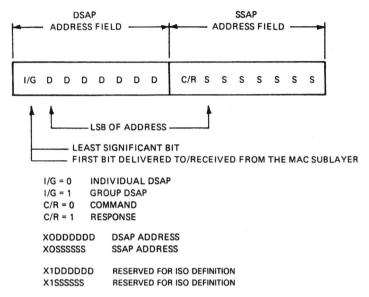

Figure 13.6. DSAP and SSAP address field formats [6].

With TYPE 2 operation, the control of traffic between a source LLC and destination LLC is by means of a numbering scheme which is cyclic within a modulus of 128 and measured in terms of PDUs. This is the same type of numbering used with HDLC with its modulus-128 option. See Section 6.2.3.1 in Chapter 11 for a description of its operation.

There are two classes of LLC operation:

(a) Class I LLCs support TYPE 1 operation only. Class I service is applicable to individual, group, global, and null DSAP addressing, as well as to applications requiring no data-link layer acknowledgment or flow control procedures.

(b) Class II LLCs support both TYPE 1 and TYPE 2 operation. In a Class II station, the operation of TYPE 1 procedures and TYPE 2 procedures are completely independent. A Class II LLC is capable of going back and forth between TYPE 1 and TYPE 2 operation on a PDU-to-PDU basis in the same SAP, if necessary.

4.3.4 LLC Control Field and Its Function. The LLC control field is illustrated in Figure 13.7. It is 16 bits long for formats that include sequence numbering, and it is 8 bits long for formats that do not. The sequence numbering and windows are identical to those used with HDLC and described in Chapter 11, Section 6.2.3.1.

The three formats defined for the control field are used to perform numbered information transfer, numbered supervisory transfer, unnumbered control, and

LLC PDU CONTROL FIELD BITS

	1	2	3	4	5	6	7	8	9	10 - 16

	1	2	3	4	5	6 7 8	9	10 - 16
INFORMATION TRANSFER COMMAND/RESPONSE (I-FORMAT PDU)	0		N(S)				P/F	N(R)
SUPERVISORY COMMANDS/RESPONSES (S-FORMAT PDUs)	1 0	S	S	X	X X X		P/F	N(R)
UNNUMBERED COMMANDS/RESPONSE (U-FORMAT PDUs)	1 1	M	M	P/F	M M M			

N(S)	=	Transmitter send sequence number (Bit 2 = low-order bit)
N(R	=	Transmitter receive sequence number (Bit 10 = low-order bit)
S	=	Supervisory function bit
M	=	Modifier function bit
X	=	Reserved and set to zero
P/F	=	Poll bit—command LLC PDU transmissions
		Final bit—response LLC PDU transmissions
		(1 = Poll/Final)

Figure 13.7. LLC PDU control field formats [6].

unnumbered information transfer functions. The numbered information transfer and supervisory transfer functions apply only to TYPE 2 operation. The unnumbered control and unnumbered information transfer functions apply either to TYPE 1 or TYPE 2 operation (but not both) depending on the specific function selected.

As shown in Figure 13.7, there are three types of frames as there are in HDLC. These are: I-frames, S-frames, and U-frames, which we describe below.

The I-Frame. The information transfer format (I-format) is used to perform a numbered information transfer in TYPE 2 operation. Except where otherwise specified (e.g., command/response frames UI, TEST, FRMR, and XID*), it is the only LLC PDU that contains an information field. The functions of the sequence numbers N(S) and N(R) and the P/F are independent; that is, each I-format PDU has an N(S) sequence number, an N(R) sequence number which shall or shall not acknowledge additional I-format PDUs at the receiving LLC, and a P/F (poll/final) bit that is set to 1 or 0.

The S-Frame. The supervisory format (S-format) is used to perform datalink supervisory control functions in TYPE 2 operation, such as acknowledging I-format PDUs or requesting retransmission of I-format PDUs and requesting a temporary suspension of transmission of I-format PDUs. The functions of N(R) and P/F are independent; that is, each S-format PDU has an N(R)

---

* These are described in Chapter 11, Section 6.2.3. Remember that the C/R bit in the SSAP defines whether a frame is a command or response.

sequence number that does acknowledge or does not acknowledge additional I-format PDUs at the receiving LLC and a P/F bit that is set to 1 or 0.

The U-Frame. The unnumbered format (U-format) PDUs are used in either TYPE 1 or TYPE 2 operation, depending upon the specific function utilized, to provide additional data-link control functions and to provide unsequenced information transfer. The U-format PDUs contain no sequence numbers, but include a P/F bit which is set to 1 or 0. The P/F bit operates in a manner similar to that in HDLC.

5 LAN ACCESS PROTOCOLS

5.1 Introduction

In this context a protocol includes a means of permitting all users to access a LAN fairly and equitably. Access can be random or controlled. The random access schemes to be discussed include CSMA (carrier sense multiple access) and CSMA/CD, where CD stands for collision detection. The controlled access schemes that are described are token bus and token ring. It should be kept in mind that users accessing the network are unpredictable and the transmission capacity of the LAN should be allocated in a dynamic fashion in response to those needs.

5.2 Background: Contention and Polling

Contention and polling were briefly introduced in Chapter 6. Contention is where there is no discipline at all. On any data communication system where more than two users share a common medium, the risk of collision is always present. Collision occurs where two or more accesses transmit simultaneously on a common medium. The result is that the traffic of both contenders is corrupted. Contention is a form of random access. CSMA and CSMA/CD, which is described in the following subsections, are other forms of random access.

Polling is a form of controlled access. It is commonly used on multipoint configurations. Such a configuration is shown in Figure 11.4B. One station, usually the CPU, is assigned the responsibility of master station. The master station polls each remote station periodically, often sequentially, to determine if there is traffic to be transmitted. When a reply from the remote is in the affirmative, the traffic is then transmitted. Each station on the network has a unique address (e.g., bit sequence). This address is incorporated into the poll. Thus only one indicated remote station will respond to that address. In a similar manner, traffic from the master station to a remote station incorporates the address of that remote station in the header of the message. In such a way that remote station and only that remote station will copy the traffic. This type of polling is called *roll-call polling* by some and *broadcast polling* by others. Because each remote station must respond to the poll (either negative or positive) with that additional overhead, roll-call polling is considered inefficient.

Loop polling is another form of polling. In this case the master station sends a poll request to the first remote station in the loop. If that station has traffic, it is sent to the master station. Once the traffic transmission terminates and all the traffic is sent and acknowledged, that remote station forwards the poll request to the next station in the loop. If the second station has traffic, it is transmitted; if not, the poll request is sent to the third station, and so on, until all remote stations have been polled. The process then starts all over again. The advantages of loop polling are that message transmission and polling operations overlap and no negative responses are transmitted.

Token-passing LAN access methods are a direct outgrowth of polling. Such access schemes can also be called a form of time-division multiple access (TDMA).

5.3 CSMA and CSMA/CD Access Techniques

Carrier sense multiple access (CSMA) is a LAN access technique that some simplistically call "listen before transmit." This "listen before transmit" idea gives insight into the control mechanism. If user 2 is transmitting, user 1 and all others hear that the medium is occupied and refrain from using it. In actuality, when an access with traffic senses that the medium is busy, it backs off for a period of time and tries again. How does one control that period of time? There are three methods of control commonly used. These methods are called *persistence algorithms* and are outlined briefly below:

- *Nonpersistent.* The accessing station backs off a random period of time and then reattempts access.
- *1-persistent.* The station continues to sense the medium until it is idle and then proceeds to send its traffic.
- *p-persistent.* The accessing station continues to sense the medium until it is idle, then transmits with some preassigned probability p. Otherwise it backs off a fixed amount of time, then transmits with a probability p or continues to back off with a probability of $(1 - p)$.

The algorithm selected depends on the desired efficiency of the medium usage and the complexity of the algorithm and resulting impact on firmware and software. With the nonpersistent algorithm collisions are effectively avoided because the two stations attempting to access the medium will back off, most probably with different time intervals. The result is wasted idle time following each transmission. The 1-persistent algorithm is more efficient by allowing one station to transmit immediately after another transmission. However, if more than two stations are competing for access, collision is virtually assured. The p-persistent algorithm lies between the other two and is a compromise, attempting to minimize collisions and idle time.

It should also be noted that with CSMA, after a station transmits a message, it must wait for an acknowledgment from the destination. Here we must take into account the round-trip delay (2 × propagation time) and the fact that the acknowledging station must also contend for medium access. Another important point is that collisions can occur only when more than one user begins transmitting within the period of propagation time. Thus CSMA is an effective access protocol for packet transmission systems where the packet transmission time is much longer than the propagation time.

The inefficiency of CSMA arises from the fact that collisions are not detected until the transmissions from the two offenders have been completed. With CSMA/CD, which has collision detection, a collision can be recognized early in the transmission period and the transmissions can be aborted. As a result, channel time is saved and overall available channel utilization capability is increased.

CSMA/CD is sometimes called "listen while transmitting." It must be remembered that collisions can occur at any period during channel occupancy, and this includes the total propagation time from source to destination. Even at multimegabit data rates, propagation time is not instantaneous; it remains constant for a particular medium, no matter what the bit rate is. With CSMA the entire channel is wasted. With CSMA/CD one offending station stops transmitting as soon as it detects the second offending station's signal. It can do this because all accesses listen *while* transmitting.

5.3.1 CSMA/CD Description.
Carrier sense multiple access with collision detection is defined by ISO/IEC 8802-3 international standard and also by the reference ANSI/IEEE Std 802.3 [4, 5]. It is based on the Ethernet approach initiated by Xerox Corporation, Digital Equipment Corporation, and Intel. The IEEE 802.3 version closely resembles Ethernet with changes in packet structure and an expanded set of physical layer options. Figure 13.8 relates CSMA/CD protocol layers to the conventional OSI reference model (see Chapter 11, Section 6.2.1) and identifies acronyms that we use in this description. In the late 1980s, early 1990s the bit rates generally encompassed in CSMA/CD are between 1 and 20 Mbits/s. In 2004, there are over 22 standards in the IEEE 802.3 series covering all the way to 10 Gbits/s. The higher bit rates are more practical on optical fiber; midrange rates may be carried on specialized wire-pair (e.g., CAT 5 or above). The model used for the discussion below covers the traditional 10-Mbits/s rate.

The medium is coaxial cable. A user connects to the cable by means of a medium access unit (MAU). This connects through an attachment unit interface (AUI) to the data terminal equipment (DTE). As shown in Figure 12.8, the DTE consists of the physical signaling sublayer (PLS), the medium access control (MAC), and the logical link control (LLC). The PLS is responsible for transferring bits between the MAC and the cable. It uses differential Manchester encoding for the data transfer. With such coding the datum 0 has a transition from high to low at midcell, while the datum 1 has the opposite transition. The line idle is a steady high condition with no transitions.

CSMA/CD LAN systems (or Ethernet) are probably the most widely used type of LANs worldwide. We would say that this is due to their relatively low cost

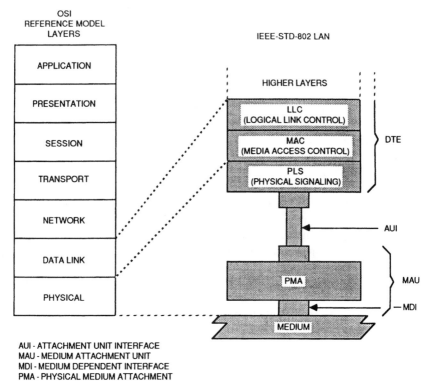

AUI - ATTACHMENT UNIT INTERFACE
MAU - MEDIUM ATTACHMENT UNIT
MDI - MEDIUM DEPENDENT INTERFACE
PMA - PHYSICAL MEDIUM ATTACHMENT

Figure 13.8. CSMA/CD LAN relationship to the OSI model. From IEEE/ANSI Std. 802.3 [4]. Courtesy of IEEE New York.

to implement and maintain and to their simplicity. The down side is that their efficiency starts to drop off radically as the number of users increases as well as increased user activity. Thus the frequency of collisions and backoffs increases to a point that throughput can drop to zero. Some users argue that efficiency starts to drop off at around 30% capacity, while others argue that that point is nearer 50%. We will discuss ways to mitigate this problem in our coverage of bridges.

5.3.1.1 System Operation

TRANSMISSION WITHOUT CONTENTION. A MAC frame is generated from data from the LLC sublayer. This frame is handed to the transmit media access management component of the MAC sublayer for transmission. To avoid contention with other traffic on the medium, the transmit medium access management monitors the carrier sense signal provided by the physical layer signaling (PLS) component. When the medium is clear, frame transmission is initiated through the PLS interface. When the transmission has been completed without contention (a collision event), the MAC sublayer informs the LLC and awaits the next request for frame transmission.

RECEPTION WITHOUT CONTENTION. At each receiving station, the arrival of a frame is first detected by the PLS, which responds by synchronizing with the incoming preamble and by turning on the carrier sense signal. The PLS passes the received bits up to the MAC sublayer where the leading bits are discarded, up to and including the end of the preamble and start frame delimiter (SFD). In this period the receive media access management component of the MAC sublayer has detected the carrier sense and is waiting for the incoming bits to be delivered. As long as the carrier sense is on, the receive media access management collects bits from the PLS. Once the carrier sense signal has been removed, the frame is truncated at an octet boundary, if required, and then passed to the receive data decapsulation for processing.

It is in receive data decapsulation where the destination address is checked to determine if this frame is destined for this particular LAN station. If it is, the destination address (DA) and source address (SA) and the LLC data unit are passed to the LLC sublayer. It also passes along the appropriate status code indicating that reception is complete or reception too long. It also checks for invalid MAC frames by inspecting the frame check sequence to detect any damage to the frame enroute, as well as by checking for proper octet-boundary alignment of the end of frame.

COLLISION HANDLING. A collision is caused by multiple stations attempting to transmit at the same time, in spite of their attempts to avoid this by deferring. A given station can experience a collision during the initial part of its transmission (the collision window) before its transmitted signal has had time to propagate to all stations on the CSMA/CD medium. Once the collision window has passed, a transmitting station is said to have acquired the medium. Once all stations have noticed that there is a signal on the medium (by way of carrier sense), they defer to it by not transmitting, avoiding any chance of subsequent collision. The time to acquire the medium is thus based on the round-trip propagation time of the physical layer whose elements include the PLS, the physical medium attachment (PMA), and the physical medium itself.

In the event of collision, the transmitting station's physical layer notices a notable increase in standing waves on the medium* and turns on the collision detect (CD) signal. The collision-handling process now starts. First, the transmit media access management enforces the collision by transmitting a bit sequence called jam. This "jam," specified as 32 bits long, ensures that all stations involved in the collision are aware that a collision has occurred. After the jam has been sent, the transmit media access management component terminates the transmission and schedules another transmission attempt after a randomly selected time interval. Retransmission is attempted again in the face of repeated collisions. If, on this second attempt, another collision occurs, the transmit media access management attempts to reduce the medium's load by backing off, meaning it voluntarily delays its own retransmissions to reduce the load on the medium. This

---

* This is called "interference" in the ISO/IEC reference standard.

is accomplished by expanding the interval from which the random retransmission time is selected on each successive transmission attempt. Eventually, either the transmission succeeds or the attempt is abandoned on the assumption that the medium has failed or has become overloaded.

5.3.1.2 MAC Frame Structure. The MAC frame is shown in Figure 13.9. There are eight fields in a frame: preamble, start frame delimiter (SFD), the addresses of the frame's source and destination(s), a length field to indicate the length of the following field containing the LLC data to be transmitted, a field that contains padding if required, and the frame check sequence (FCS) field containing a cyclic redundancy check value to detect errors in received frames. All eight fields are of fixed size except the LLC data and PAD fields, which may contain any integer number of octets between the minimum and maximum values determined by a specific implementation that may be selected.

The minimum and maximum frame size limits refer to that portion of the frame from the destination address field through the frame check sequence field, inclusive. The default maximum frame size is 1518 octets; the minimum size is 64 octets.

The preamble field is 7 octets in length and is used so that the receive PLS can synchronize to the transmitted symbol stream. The SFD is the binary sequence 10101011. It follows the preamble and delimits the start of frame.

There are two address fields: the source address and the destination address. The address field length is an implementation decision. It may be 16 or 48 bits long. In either field length, the first bit specifies whether the address is an individual address (bit set to 0) or group address (bit set to 1). In the 48-bit address field, the second bit specifies whether the address is globally administered (bit set to 0) or locally administered (bit set to 1). For broadcast address, the bit is set to 1.

The length field is 2 octets long and indicates the number of LLC data octets in the data field. If the value is less than the minimum required for proper operation of the protocol,* a PAD field (sequence of octets) is appended at the end of the data field and prior to the FCS field. The length field is transmitted and received with the high-order octet first.

PREAMBLE (7 OCTETS)	SFD (1 OCTET)	DESTINATION ADDRESS (2 OR 6 OCTETS)	SOURCE ADDRESS (2 OR 6 OCTETS)	LENGTH (2 OCTETS)	LLC DATA	PAD	FCS (4 OCTETS)

LENGTH - GIVES NUMBER OF OCTETS IN DATA FIELD

LLC - LOGICAL LINK CONTROL (OSI LAYER 3 AND ABOVE)

FCS - FRAME CHECK SEQUENCE - A 32-BIT CRC

PAD - ADDS OCTETS TO ACHIEVE MINIMUM FRAME LENGTH WHERE NECESSARY

SFD - START FRAME DELIMITER

Figure 13.9. MAC frame format.

* Minimum frame length is 64 octets.

The data (LLC data) field contains a sequence of octets that is fully transparent in that any arbitrary sequence of octet values may appear in the data field up to the maximum number specified by the implementation of this standard that is used. The maximum size of the data field supplied by the LLC is determined by the maximum frame size and address size parameters of a particular implementation.

The frame check sequence (FCS) field contains four octets (32 bits) CRC value. This value is computed as a function of the contents of the source address, destination address, length, LLC data, and pad—that is, all fields except the preamble, SFD, and FCS. The encoding is defined by the following generating polynomial:

$$G(x) = x^{32} + x^{26} + x^{23} + x^{22} + x^{16} + x^{12} + x^{11} + x^{10} + x^8 + x^7 + x^5$$
$$+ x^4 + x^2 + x + 1$$

An invalid MAC frame meets at least one of the following conditions:

1. The frame length is inconsistent with the length field.
2. It is not an integral number of octets in length.
3. The bits of the received frame (exclusive of the FCS itself) do not generate a CRC value identical to the one received. An invalid MAC frame is not passed to the LLC.

The minimum frame size is 512 bits for the 10-Mbit/s data rate [4]. This requires a data field of either 46 or 54 octets, depending on the size of the address field used. The minimum frame size is based on the *slot time*, which for the 10-Mbit/s data rate is 512 bit times. Slot time is the major parameter controlling the dynamics of collision handling and it is:

- An upper bound on the acquisition time of the medium.
- An upper bound on the length of a frame fragment generated by a collision.
- The scheduling quantum for retransmission.

To fulfill all three functions, the slot time must be larger than the sum of the physical round-trip propagation time and the MAC sublayer jam time. The propagation time for a 500-m segment of 50-Ω coaxial cable is 2165 ns, assuming that the velocity of propagation of this medium is $0.77 \times 300 \times 10^6$ m/s [4].

5.3.1.3 Transmission Requirements

SYSTEM MODEL. Propagation time is critical for the CSMA/CD access method. The major contributor to propagation time is the coaxial cable and its length. The characteristic impedance of the coaxial cable is 50 Ω ± 2 Ω. The attenuation of a 500-m (1640-ft) segment of the cable should not exceed 8.5 dB (17 db/km) measured with a 10-MHz sine wave. The velocity of propagation is $0.77c$.* The

---

* c = velocity of light in a vacuum.

referenced maximum propagation times were derived from the physical configuration model described here. The maximum configuration is as follows:

1. A trunk coaxial cable, terminated in its characteristic impedance at each end, constitutes a coax segment. A coax segment may contain a maximum of 500 m of coaxial cable and a maximum of 100 MAUs. The propagation velocity of the coaxial cable is assumed to be $0.77c$ minimum ($c = 300,000$ km/s). The maximum end-to-end propagation delay for a coax segment is 2165 ns.

2. A point-to-point link constitutes a link segment. A link segment may contain a maximum end-to-end propagation delay of 2570 ns and shall terminate in a repeater set at each end. It is not permitted to connect stations to a link segment.

3. Repeater sets are required for segment interconnection. Repeater sets occupy MAU positions on coax segments and count toward the maximum number of MAUs on a coax segment. Repeater sets may be located in any MAU position on a coax segment but shall only be located at the ends of a link segment.

4. The maximum length, between driver and receivers, of an AUI cable is 50 m. The propagation velocity of the AUI cable is assumed to be $0.65c$ minimum. The maximum allowable end-to-end delay for the AUI cable is 257 ns.

5. The maximum transmission path permitted between any two stations is five segments, four repeater sets (including optional AUIs), two MAUs, and two AUIs. Of the five segments, a maximum of three may be coax segments; the remainder are link segments.

The maximum length transmission path consists of 5 segments, 4 repeater sets (with AUIs), 2 MAUs, and 2 AUIs as shown in Figure 13.10. If there are two link segments on the transmission path, there may be a maximum of three coaxial cable segments on that path. If there are no link segments on a transmission path, there may be three coaxial cable segments on that path given current repeater technology. Figure 13.11 shows a large system with maximum transmission paths. It also shows the application of link segments versus coaxial

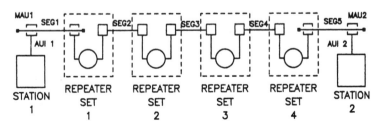

Figure 13.10. Maximum transmission path.

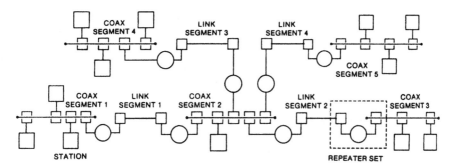

Figure 13.11. An example of a large system with maximum transmission paths.

cable segments. The bitter ends of coaxial cable segments should be terminated in the coaxial cable characteristic impedance. The coaxial cable segments are marked at 2.5-m intervals. The MAUs should be attached at these 2.5-m interval points. This assures nonalignment at fractional wavelength boundaries.

NOTES ON CSMA/CD SYSTEM PARAMETERS

Scheduling of Backoff Attempts. The dynamics of collision handling are largely determined by a single parameter called the *slot time.* This single parameter describes three important aspects of collision handling:

1. It is an upper bound on the acquisition time of the medium.
2. It is an upper bound on the length of a frame fragment generated by a collision.
3. It is the scheduling quantum for retransmission.

To fulfill all three functions, the slot time must be larger than the sum of the physical layer round-trip time and the media access layer maximum jam time. The slot time is determined by the parameters of implementation. Two examples are given further below.

The scheduling of reattempts is determined by a controlled randomization procedure called *truncated binary exponential backoff.* This defines the delay time before reattempt. The delay is an integer multiple of the slot time. The number of slot times to delay before the nth retransmission attempt is chosen as a uniformly distributed integer r in the range of $0 \leq r < 2^k$, where $k = \min(n, 10)$. Algorithms used to generate the integer r should be designed to minimize correlation between the numbers generated by any two stations at a given time.

Other System Parameters. Two sets of CSMA/CD implementation parameters are given below. The first is for 10BASE5, the 10-Mbit/s system described herein.

Parameters	Values
slotTime	512 bit times
interFrameGap	9.6 μs
attemptLimit	16
backoffLimit	10
jamSize	32 bits
maxFrameSize	1518 octets
minFrameSize	512 bits (64 octets)
addressSize	48 bits

The second group of parameter values is for 1BASE5, similar to the above but with a 1-Mbit/s transmission rate.

Parameters	Values
slotTime	512 bit times
interFrameGap	96 μs
attemptLimit	16
backoffLimit	10
jamSize	32 bits
maxFrameSize	1518 octets
minFrameSize	512 bits (64 octets)
addressSize	48 bits

The baseband signal transmitted to the medium by the AUI uses Manchester coding. With Manchester coding there is always a transition in the middle of a bit interval. The first half represents the bit value, and the second half of the bit is the complement of the first half. Thus Manchester coding is rich in transitions for timing extraction and is compatible with the coaxial cable transmission medium [4].

There are a number of different transmission techniques either recommended by Standard 802.3 or derived therefrom. Some of these are shown below:

      10BASE5   10BASE2   10BASET
      10BROAD36   1BASE5

The convention used for identifying these techniques uses the first number to indicate the data rate on the medium in megabits per second (Mbits/s), BASE or BROAD indicates whether it is baseband or broadband transmission, and the final number gives the actual link or segment length in hundreds of meters. 10BASET is a 10-Mbit/s system on twisted pair.

5.4 Token Bus

A token bus LAN in its simplest version is a length of 75-Ω coaxial cable terminated at each end in its characteristic impedance. Users tap the cable with a coupler and connect to the coupler with a 37.5-Ω stub no longer than 350 mm. Three different transmission regimes are described in IEEE Std 802.4 [8]: phase continuous FSK, phase coherent FSK, and multilevel duo-binary AM/PSK. The coaxial cable can be extended by use of regenerative repeaters. In a similar manner tree topologies can be developed. The discussion below covers data rates of 1 Mbit/s, 5 Mbits/s, and 10 Mbits/s.

The 1990 standard [8] also has provision for fiber-optic medium using 62.5-μm/125-μm fiber with data rate options of 5, 10, and 20 Mbits/s.

With the token bus access technique a short control packet known as the token regulates the right of access to the bus. A station holding the token has the exclusive right to use the network for a specified time period. During this period the station may poll other stations, receive responses, and pass data traffic. When the token holding station has completed operations or when its time period is up, it passes the token to the next station, its successor, in logical sequence.

Figure 13.12 illustrates a typical token bus, showing logical connectivity (dashed line) and that the access method is always sequential in a logical sense. The right to access the medium passes from user to user. It should be noted that physical connectivity has little impact on the order of the logical ring. In fact, stations can respond to a query from a token holder without being part of the *logical* ring. For example, stations H and F can receive data frames but cannot initiate a transmission because they cannot receive the token.

Token passing ensures equitable access to the network. Such a control also ensures against collision because only the user holding the token can access the medium.

The user access control equipment is shown functionally in Figure 13.13. "Station management," an important part of the control function, is shown on the right. The figure also shows the layer relationship with the OSI model. Also, the similarity with CSMA/CD is apparent. Key, again, is the medium access control.

Specific responsibilities of the MAC include ordered access to the medium, providing a means of admission and deletion of stations on the LAN and the handling of fault recovery. Among the faults handled are:

- Multiple tokens
- Lost tokens

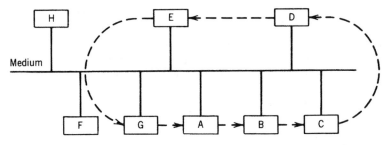

Figure 13.12. Token bus showing logical connectivity (dashed line).

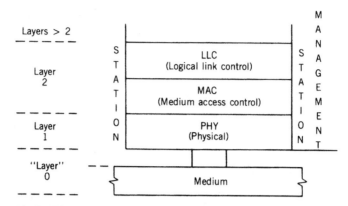

Figure 13.13. IEEE 802.4 model: functional relationship of access control equipment.

- Token pass failure
- Nonresponsive stations (i.e., a station with an inoperative receiver)
- Duplicate station addresses

It should be noted that no station takes on an exclusive monitoring and control function.

As with the CSMA/CD protocol, slot time is also an important parameter with the token-passing bus protocol. Slot time in this case is defined as the maximum time any station need wait for an immediate answer from another station. It is measured in octet times and is an integer. Slot time is twice the sum of the propagation delay, station delay, and safety margin. The *response window* equals the slot time. If a station waiting for a response hears a transmission start during the response window, that station does not transmit again until at least the received transmission terminates.

Holding the token gives the right to transmit. The token is passed from station to station in descending numerical order of station address. When a station hears a token frame addressed to itself, it "has the token" and may transmit data frames. After a station has completed transmitting data frames and has completed maintenance functions where necessary, the station passes the token to its successor by sending a *token MAC frame*. It then goes through a procedure to ensure that its successor has the token. For instance, after sending the token MAC frame, the station listens for evidence that the successor has heard the token frame and is active. If the sender hears a valid frame following the token, it can assume that its successor has the token and is transmitting. If the token-sending station does not, it attempts to assess the state of the network.

If the token-sending station hears a noise burst or a frame with an incorrect FCS, it cannot be sure from the source address which station sent the frame. If a noise burst is heard, the token-passing station sets an internal indicator and continues to listen in the *check token pass state* up to four more slot times.

If nothing is heard during the four-slot time delay, the station assumes that its successor has the token.

If the token holder does not hear a valid frame after sending the first time, it repeats the token pass procedure once more. If the successor does not transmit after the second attempt, it assumes that the successor has failed. The sender then sends a *who follows frame* with the successor's address in the data field of the frame. All active stations on the LAN then compare the value of the data field of a "who follows frame" with the address of their own predecessor, which is the station that normally would send them the token. The station whose predecessor is the successor of the sending station responds to the "who follows frame" by sending its address in a *set successor frame*. The station holding the token thus establishes a new successor, bridging the failed station out of the logical ring.

Figure 13.14 shows the MAC frame format. The number of octets between the start delimiter (SD) and the end delimiter (ED) should be no greater than 8191. The abort sequence consists of an SD and an ED, each of which is 1 octet long [6].

We define three acronyms dealing with a particular LAN station in question, its predecessor, and its successor in the logical ring of token passing. Again, that logical ring is shown in the example given in Figure 13.12.

TS—this station's address

NS—next station's address

PS—previous station's address

A token has seven fields. These are standard preamble, start delimiter (SD), the binary sequence 00001000 for the FC, a DA equal to the value of the station's NS, its SA, an FCS, and an ending delimiter (ED).

An LLC data frame has eight fields: preamble, SD, FC, DA and SA, then the LLC_data_unit followed by the FCS and an ED.

Four different transmission/modulation schemes are presented in the reference specification [6]. Three of these use 75-Ω coaxial cable as the medium and one

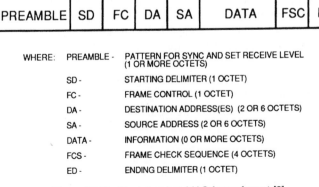

WHERE: PREAMBLE - PATTERN FOR SYNC AND SET RECEIVE LEVEL
(1 OR MORE OCTETS)

SD - STARTING DELIMITER (1 OCTET)

FC - FRAME CONTROL (1 OCTET)

DA - DESTINATION ADDRESS(ES) (2 OR 6 OCTETS)

SA - SOURCE ADDRESS (2 OR 6 OCTETS)

DATA - INFORMATION (0 OR MORE OCTETS)

FCS - FRAME CHECK SEQUENCE (4 OCTETS)

ED - ENDING DELIMITER (1 OCTET)

Figure 13.14. The token bus MAC frame format [8].

uses fiber optics. The bit error rate at the receive physical interface is 1×10^{-8}, with a mean undetected bit error rate of 1×10^{-9}. Extension of the network is accomplished by means of regenerative repeaters. Branched topologies may be implemented by impedance-matched nondirectional splitters, which are three-port passive networks that divide the signal incident at one port into two equal parts that are transmitted to the other two ports. Drop cables generally should not exceed 50 m (160 ft).

Phase continuous FSK is one of the several transmission methods given in IEEE Std 802.4 [8]. Successive symbols presented to the physical layer at the MAC interface are encoded, producing a three-PHY-symbol code: H, L, and OFF. These symbols feed a two-tone FSK modulator where the H symbol is the higher frequency tone, the L symbol is the lower frequency tone, and the OFF symbol is no tone. The output signal is then ac-coupled to the coaxial cable.

IEEE Std 802.4 standardizes the line data rate at 1 Mbit/s with a tolerance of ±0.01% for an originating station and ±0.015% for a repeater station. There are five possible symbols for the FSK implementation: 0, 1, nondata, pad-idle, and silence. Each of these MAC symbols is encoded into a pair of PHY symbols from a different three-symbol H, L, OFF code as follows:

1. Silence (OFF OFF).
2. Pad-idle. Pad-idle symbols are always octets. Each pair of pad-idle symbols is encoded as a sequence of LH, HL.
3. 0 (HL).
4. 1 (LH).
5. Nondata. The MAC layer transmits nondata symbols in pairs, which are encoded as the sequence LL HH.

The start frame delimiter subsequence is *nondata nondata 0 nondata nondata 0*, which is encoded as the sequence LL HH HL LL HH HL. The end-frame delimiter subsequence is *nondata nondata 1 nondata nondata 1*, which is encoded as the sequence LL HH LH HH LH.

The line signal corresponds to an FSK signal with its carrier frequency at 5.00 MHz, varying smoothly between two signaling frequencies of 3.75 MHz ± 80 kHz and 6.25 MHz ± 80 kHz. When transitioning between two signaling frequencies, the FSK modulator changes its frequency in a continuous and monotonic manner within 100 ns. The output signal level of the modulated carrier frequency is between +54 dB and +60 dB relative to 1 mV across 37.5 Ω. An S/N ratio of 20 dB at the receiver produces a BER of 1×10^{-8} or better. This is based on an in-band noise floor of +4 dB or less relative to 1 mV across 37.5 Ω.

Phase coherent FSK is another transmission method specified in IEEE Std 802.4. The waveform in this case is somewhat different from that of continuous FSK. Again successive MAC symbols presented to the physical layer entity at its MAC interface are applied to an encoder that produces a three-PHY-symbol code: H, L, and OFF. The output is then applied to a two-tone FSK modulator that represents each H symbol as one full cycle of a tone whose period is exactly one-half of the MAC symbol period, each L as one-half cycle of a tone whose full-cycle period is exactly the MAC symbol period, and each OFF as no tone for the same half MAC symbol period. This modulated signal is then ac-coupled to the coaxial cable medium.

The standard data signaling rates for phase coherent FSK systems are 5 Mbits/s and 10 Mbits/s. The permitted tolerance for each signaling rate is ±0.01% for an originating station and ±0.015% for a repeater station. The coding is somewhat different from its continuous FSK counterpart.

For the 5-Mbit/s data rate, the lower-tone FSK frequency is 5.0 MHz and the higher-tone FSK frequency is 10.0 MHz. For the 10-Mbit/s data rate the lower-tone FSK frequency is 10.0 MHz and the higher-tone FSK frequency is 20.0 MHz. The output level of the transmitted signal into a 75-Ω resistive load is between +63 dBmV and +66 dBmV, inclusive. The BER at the receiver of 1×10^{-8} will be achieved when a received signal is in the range of +10 dBmV to +66 dBmV and the maximum noise floor is −10 dBmV.

Loss budgets are important when engineering any LAN to ensure that each and every LAN station will be provided the desired performance. Based on phase coherent FSK, a loss budget might be worked out as follows.[*] Both overall length of cable and number of taps should be explicitly combined in the system signal loss budget. Each tap makes a known contribution to the network loss budget, as does each length of the trunk cable. In addition, each tap includes a definite tap loss between the trunk and the drop, which should also be included in the loss budget.

The referenced standard [8] specifies the tap trunk to drop loss value of 20 ± 0.5 dB. Typical values for commercially available two-drop taps conforming to the referenced specification are 0.3-dB insertion loss over a frequency range of 1–30 MHz. An example calculation of the signal loss budget for a simple, unbranched network, including 20 such two-drop taps, might be as follows:

Drop loss at first tap:	20.5 dB
Insertion loss of 18 intervening taps:	5.4 dB
Drop loss at 20th tap:	20.5 dB
Total lumped losses:	46.6 dB

The minimum transmit level is +63 dBmV and the minimum receive level is +10 dBmV. Thus, the cable loss in this example should not exceed 63 − 10 − 46.4 = 6.6 dB. Of course, when we specify attenuation for coaxial cable, it always should be specified for the highest frequency of interest. For a 5-Mbit/s data rate, an RG-11-type cable with an attenuation of 1.5 dB/100 m at 10 MHz limits the cable length to 440 m (1430 ft). A longer cable length would be possible if semirigid cable were used.

If fewer taps were used, the cable could be longer. However, the designer should be advised that phase delay distortion may also limit cable length.

---

[*] Based on Chapter 13 Appendix of ANSI/IEEE Std 802.4 specification of 1990 [8].

The IEEE 802.4 standard also specifies a broadband alternative for token-passing bus using a CATV-like 75-Ω single-cable mid-split configuration. In this configuration, the unidirectional forward and reverse channels are paired to provide bidirectional channels. A frequency offset of 192.25 MHz between forward and reverse channels is used in both the mid-split and high-split configuration. A dual-cable configuration may also be considered.

The following are the three specified data rates, their respective bandwidths, transmitter output power, and receiver noise floors for broadband token bus operation:

Data Rate	Bandwidth	Transmit Power	Receive Noise Floor
1.0 Mbit/s	1.5 MHz	+41 dBmV	−40 dBmV
5.0 Mbits/s	6.0 MHz	+47 dBmV	−34 dBmV
10 Mbits/s	12.0 MHz	+50 dBmV	−31 dBmV

The modulation employed is a 3-state duo-binary AM/PSK system. A 30-dB S/N is required to comply with BER performance requirements of 1×10^{-8} error rate.

5.5 Token Ring

A typical token passing ring LAN is shown in Figure 13.15. The token ring operation, as specified in IEEE Std 802.5 [9], has the capability of 4-Mbit/s or 6-Mbit/s date rate. A ring is formed by physically folding the medium back onto itself. Each LAN station regenerates and repeats each bit and serves as a means of attaching one or more data terminals (e.g., workstations, computers) to the ring for the purpose of communicating with other devices on the network. As a traffic frame passes around the ring, all stations, in turn, copy the traffic. Only those stations included in the destination address field pass that traffic on to the appropriate users that are attached to the station. The traffic frame continues onward back to the originator, who then removes the traffic from the ring. The

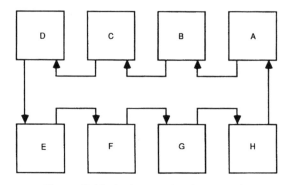

Figure 13.15. A token passing ring network.

pass-back to the originator acts as a form of acknowledgment that the traffic had at least passed by the destination(s).

The sequential connection of stations removes the need to form a logical ring, as in token bus operation. A reservation scheme is used to accommodate priority traffic. Also, one station acts as a ring monitor to ensure correct network operation. A monitor devolvement scheme to other stations is provided in case a monitor fails or drops off the ring (i.e., shuts down). A station on the ring can become inactive (i.e., close down), and a physical bypass is provided for this purpose.

A station gains the right to transmit frames onto the medium when it detects a token passing on the medium. Any station with traffic to transmit, on detection of the appropriate token, may capture the token by modifying it to a start-of-frame sequence and append the proper fields to transmit the first frame. At the completion of its information transfer and after appropriate checking for proper operation, the station initiates a new token, which provides other stations with the opportunity to gain access to the ring. Each station has a token-holding timer that controls the maximum period of time a station may occupy the medium before passing the token on.

Figure 13.16 shows the frame format specified in IEEE Std 802.5 [9] and Figure 13.17 shows the token format. In these figures, the leftmost bit is transmitted first. The frame format in Figure 13.16 is used for transmitting both MAC and LLC messages to destination station(s). It may or may not contain an INFO field.

The starting delimiter (SD) consists of the symbol sequence JK0JK000, where J and K are nondata symbols and are described below. Both frames and tokens start with the SD sequence.

The access control (AC) is 1 octet long and contains 8 bits that are formatted PPPTMRRR. The first three bits, PPP, are the priority bits. These are used to indicate the priority of a token and therefore which stations are allowed to use the token. In a system designed for multiple priority, there are eight levels of priority

	<- SFS ->	<-	FCS COVERAGE	->	<- EFS ->	
SD	AC	FC	DA	SA	INFO	FCS	ED	FS

WHERE:
SFS – START-OF-FRAME SEQUENCE
SD – STARTING DELIMITER (1 OCTET)
AC – ACCESS CONTROL (1 OCTET)
FC – FRAME CONTROL (1 OCTET)
DA – DESTINATION ADDRESS (2 OR 6 OCTETS)
SA – SOURCE ADDRESS (2 OR 6 OCTETS)

INFO – INFORMATION (0 OR MORE OCTETS)*
FCS – FRAME CHECK SEQUENCE (4 OCTETS)
EFS – END-OF-FRAME SEQUENCE
ED – ENDING DELIMITER (1 OCTET)
FS – FRAME STATUS (1 OCTET)

Figure 13.16. Frame format for token ring based on IEEE Std 802.5 [9].

SD	AC	ED

SD = Starting Delimiter (1 octet)
AC = Access Control (1 octet)
ED = Ending Delimiter (1 octet)

Figure 13.17. Token format for token ring operation [9].

available, where the lowest priority is PPP = 000 and the highest is PPP = 111. The AC field contains the token bit T and the monitor bit M. If T = 0, then the frame is a token (Figure 13.17). The T bit is a 1 on all other frames. The M bit is set by the monitor as part of the procedures for recovering from malfunctions. The bit is transmitted as a 0 in all frames and tokens. The active monitor inspects and modifies this bit. All other stations repeat this bit as received. The three R bits are reservation bits. These bits allow stations with high-priority protocol data units (PDUs) to request that the next token issued be at the requested priority.

The next field in Figure 13.16 is the frame control (FC) field. It is one octet long and defines the type of frame and certain MAC and information frame functions. The first two bits in the FC are designated FF bits and the last six bits, the ZZZZZZ bits. If FF is 00, the frame contains a MAC PDU, if FF is 01, the frame contains an LLC PDU.

If the frame-type bits indicate a MAC frame, all stations on the frame interpret and act upon the ZZZZZZ control bits. If the frame-type bits indicate and LLC frame, the ZZZZZZ bits are designated rrrYYY. The rrr bits are reserved and are transmitted as 0s in all the transmitted frames and ignored upon reception. The YYY bits may be used to carry the priority (Pm) of the PDU from the source LLC entity to the target LLC entity or entities.

Each frame (not token) contains a destination address (DA) field and a source address (SA) field. Depending on the token ring LAN system, the address fields may be 16 or 48 bits in length. In either case, the first bit indicates whether the frame is directed to an individual station (0) or to a group of stations (1).

The INFO field carries zero, one, or more octets of user data intended for the MAC, NMT (network management), or LLC. Although there is no maximum length specified for the INFO field, the time required to transmit a frame may be no greater than the token holding period that has been established for the station. For LLC frames, the format of the information field is not specified in the referenced standard [7]. However, in order to promote interworking among stations, all stations should be capable of receiving frames whose information field is up to and including 133 octets in length.

The frame check sequence (FCS) is a 32-bit sequence based on the standard generator polynomial of degree 32 given in Section 5.3.1. It encompasses the FC, DA, SA, and INFO fields. Its transmission commences with the coefficient of the highest term.

The end delimiter (ED) is one octet in length and is transmitted as the sequence JK1JK1IE. The transmitting station transmits the delimiter as shown. Receiving stations consider the ending delimiter (ED) valid if the first six symbols JK1JK1 are received correctly. The I is the intermediate frame bit and is used to indicate whether a frame transmitted is a singular frame or whether it is a multiple frame transmission. The I bit is set at 0 for the singular frame case. The E bit is the error-detected bit. The E bit is transmitted as 0 by the station that originates the token, abort sequence, or frame. All stations on the ring check tokens and frames for errors such as FCS errors and nondata symbols. The E bit of tokens and frames that are repeated is set to 1 when a frame with an error is detected; otherwise the E bit is repeated as received.

The last field in the frame is the frame status (FS) field. It consists of one octet of the sequence ACrrACrr. The r bits are reserved for future standardization and are transmitted as 0s, and their value is ignored by the receiver. The A bit is the address-recognized bit, and the C bit is the frame-copied bit. These two bits are transmitted as 0 by the frame originator. The A bit is changed to 1 if another station recognizes the destination address as its own or relevant group address. If it copies the frame into its buffer, it then sets the C bit to 1. When the frame reaches the originator again, it may differentiate among three conditions:

1. Station nonexistent or nonactive on the ring
2. Station exists but frame was not copied
3. Frame copied

Fill is used when a token holder is transmitting preceding or following frames, tokens, or abort sequences to avoid what would otherwise be an inactive or indeterminate transmitter state. Fill can be either 1s or 0s or any combination thereof and can be any number of bits in length within the constraints of the token-holding timer.

IEEE Std 802.5 describes a true baseband-transmitting waveform using differential Manchester coding. It is characterized by the transmission of two line signal elements per symbol. An example of this coding is shown in Figure 13.18.

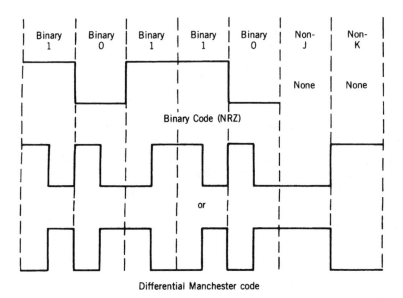

Differential Manchester code

Figure 13.18. Differential Manchester coding format for symbols 1, 0, nondata J, and nondata k. From IEEE Std 802.5 [9].

The figure shows only the data symbols 1 and 0 where a signal element of one polarity is transmitted for one-half the duration of the symbol to be transmitted, followed by the contiguous transmission of a signal element of the opposite polarity for the remainder of the symbol duration. The following advantages accrue from using this type of coding:

- The transmitted signal has no dc component and can be inductively or capacitively coupled.
- The forced midsymbol transition provides inherent timing information on the channel.

The nondata symbols J and K depart from the rule in that a signal element of the same polarity is transmitted for both signal elements of the symbol and therefore there is no midsymbol transition. A J symbol has the same polarity as the preceding symbol. The transmission of nondata symbols occurs in pairs (i.e., JK) to avoid accumulating a dc component.

All stations on the LAN ring are slaved to the active monitor station. They extract timing from the received data by means of a phase-locked loop. *Latency* is the time, expressed in number of bits transmitted, for a signal element to proceed around the entire ring. In order for the token to circulate continuously around the ring when all stations are in the repeat mode, the ring must have a latency of at least the number of bits in the token sequence—that is, 24. Since the latency of the ring varies from one system to another and no a priori knowledge is available, a delay of at least 24 bits should be provided by the active monitor.

IEEE Std 802.5 [9] specifies that the connection of a station to the ring trunk cable be via shielded cable containing two balanced 150-Ω twisted pairs. The magnitude of the transmitted signal measured at the medium interface cable when terminated in 150-Ω resistive is specified between 3.0 and 4.5 V peak-to-peak. The amplitude of the positive and negative transmitted signal should be symmetrical within 5%. The trunk cable itself may be twisted-pair, coaxial cable, or fiber-optic cable, at the user's discretion.

A LAN station provides an output with an error rate of less than or equal to 1×10^{-9} when the signal-to-noise ratio at the output of the equalizer, specified in paragraph 7.5.2 of the reference document [7], is 22 dB.

As mentioned above, under normal operation there is one station on the ring that is the active monitor. All other stations on the ring are frequency- and phase-locked to this station. They extract timing from the received data by means of a phased-locked loop. The phase-locked loop design is based on the requirement to accommodate a combined total of at least 250 stations and repeaters on the ring [9, 10].

5.6 Fiber Distributed Data Interface

5.6.1 Overview. A fiber distributed data interface (FDDI) network consists of a set of nodes (e.g., LAN stations) connected by an optical transmission medium

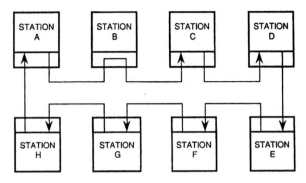

All stations are active except B (illustrated in bypass mode)

Figure 13.19. FDDI token ring, example of logical configuration [12].

(or other medium) into one or more logical rings. A logical ring consists of a set of stations connected as an alternating series of nodes and transmission medium to form a closed loop. This is shown in Figure 13.19. Information is transmitted as a stream of suitably encoded symbols from one active node to the next. Each active node regenerates and repeats each symbol and serves as a means for attaching one more device to the ring for the purpose of communicating with other devices on the ring.

FDDI provides equivalent bandwidth to support a peak data rate of 100 Mbits/s and a sustained data transfer rate of at least 80 Mbits/s. It provides connectivity for many nodes over distances of many kilometers in extent. Certain default parameter values for FDDI (such as timer settings) are calculated on the basis of up to 1000 transmission links or up to 200-km total fiber path length (typically corresponding to 500 nodes and 100 km of dual fiber cable, respectively). However, the FDDI protocols can support much larger networks by increasing these parameter values.

Two kinds of data service can be provided in a logical ring: packet service and circuit service. With packet service, a given station that holds the token transmits information on the ring as a series of data packets, where each packet circulates from one station to the next. The stations that are addressed copy the packets as they pass. Finally, the station that transmitted the packets effectively removes them from the ring.

In the case of circuit service, some of the logical ring bandwidth is allocated to independent channels. Two or more stations can simultaneously communicate via each channel. The structure of the information stream within each channel is determined by the stations sharing the channel.

Conventional FDDI provides packet service via a token ring. A station gains the right to transmit its information on to the medium when it detects a token passing on the medium. The token is a control signal comprised of a unique symbol sequence that circulates on the medium following each series of transmitted packets. Any active station, upon detection of a token, may capture the token by removing it from the ring. The station may then transmit one or more data packets. After transmitting its packets, the station issues a new token, which provides other stations the opportunity to gain access to the ring.

Each station has a token holding timer (or equivalent means) incorporated which limits the length of time a station may occupy the medium before passing the token onwards.

FDDI provides multiple levels of priority for independent and dynamic assignment depending on the relative class of service required. The classes of service may be synchronous, which may be typically used for applications such as real-time packet voice; asynchronous which is typically used for interactive applications; or immediate which is used for such extraordinary applications such as ring recovery. The allocation of ring bandwidth occurs by mutual agreement among users of the ring.

Error detection and recovery mechanisms are provided to restore ring operation in the event that transmission errors or medium transients (e.g., those resulting from station insertion or deletion) cause the access method to deviate from normal operation. Detection and recovery of these cases utilize a recovery function that is distributed among the stations attached to the ring.

One of the more common topologies of FDDI is the counter-rotating ring. Here two classes of station may be defined: dual (attachment) and single (attachment). FDDI trunk rings may be composed only of dual attachment stations which have two PMD (physical layer medium dependent) entities (and associated PHY entities) to accommodate the dual ring. Concentrators provide additional PMD entities beyond those required for their own attachment to the FDDI network, for the attachment of single attachment stations which have only one PMD and thus cannot directly attach to the FDDI trunk network. A dual attachment station, or one-half of it, may be substituted for a single attachment station in attaching to a concentrator. The FDDI network consists of all attached stations.

The example of Figure 13.20 shows the concept of multiple physical connections used to create logical rings. As shown in the figure, the logical sequence of MAC connections is stations 1, 3, 5, 8, 9, 10, and 11. Stations 2, 3, 4, and 6 form an FDDI trunk ring. Stations 1, 5, 7, 10, and 11 are attached to the ring by lobes branching out from stations that form it. Stations 8 and 9 are in turn attached by lobes branching out from station 7. Stations 2, 4, 6, and 7 are concentrators, serving as the means for attaching multiple stations to the FDDI ring. Concentrators may or may not have MAC entities and station functionality. The concentrator examples of Figure 13.20 do not show any MACs, although their presence is implied by the designation of these concentrators as stations.

5.6.2   FDDI Facilities. In this section we will discuss symbol sets, protocol data units (PDUs) utilized in FDDI and coding.

5.6.2.1   Symbol Set. A symbol is the smallest signaling element used by the FDDI MAC. Symbols can be used to convey three types of information:

1. Line states such as halt (H), quiet (Q), and idle (I).

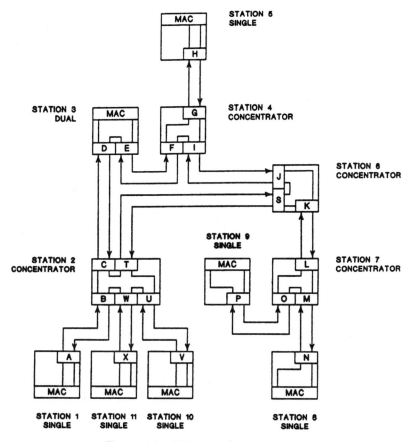

Figure 13.20. FDDI topology example [12].

2. Control sequences such as the starting delimiter (SD), ending delimiter (ED) symbol (T), initial SD symbol (J), and final symbol (K).

3. Data quartets, each representing a group of four ordered data bits.

Peer MAC entities on the ring communicate via a set of fixed-length symbols. These symbols are passed across the MAC-to-PHY interface via defined primitives. The MAC generates PDUs as matched pairs of symbols in accordance with the referenced FDDI standards.

LINE STATE SYMBOLS. These three symbols are reserved for use on the medium between MAC PDUs. The only line state symbol generated by a MAC is Idle. However, the MAC can receive other line state symbols. Detection of any of these symbols within a MAC PDU constitutes an error in the PDU.

Quiet (Q). The quiet symbols indicate the absence of any transitions in the code group.

Halt (H). The halt symbol indicates CMT (Connection Management function) signaling sequences (in the form of line states). It is also used for filtering line state or code violation symbols from the repeated symbol stream while minimizing the dc component of the NRZI signal on the transmission medium.

Idle (I). The idle symbol indicates the normal condition of the medium between MAC PDUs. It provides a continuous fill pattern to establish and maintain clock synchronism.

CONTROL SYMBOLS

Starting Delimiter (SD). A starting delimiter (SD) is used to show the starting boundary of a data transmission sequence (e.g., a MAC PDU). The PDU is normally preceded by a preamble of idle symbols, although it may succeed or preempt a previous PDU. The starting delimiter may also succeed or preempt a previous transmission. The SD sequence consists of the uniquely recognizable symbol sequence (JK) that can be recognized independently of previously established symbol boundaries.

Ending Delimiter (ED) Symbol (T). An ending delimiter symbol (T) terminates all MAC PDUs. The T symbol is not necessarily the last symbol in a transmission sequence, since the ED may be followed by one or more control indicator symbols. A sequence of ending delimiter and control indicator symbols are generated by the data-link layer (DLL) as a balanced sequence of symbol pairs (i.e., an even number of R, S, and T symbols). When no control indicators are generated, this sequence is generated as a pair of T symbols.

CONTROL INDICATORS. Control indicators specify logical conditions with a data transmission sequence. They may be independently altered by repeating nodes without altering the normal data in the transmission sequence. A sequence of ending delimiter and control indicator symbols is generated by the MAC as a balanced sequence of symbol pairs (i.e., an even number of R, S, and T symbols). A single ending delimiter symbol followed by an odd number of control indicator symbols is a balanced symbol pair sequence. However, an ending delimiter symbol followed by an even number of control indicator symbols is balanced by adding a final ending delimiter symbol.

Reset (R). The reset symbol indicates a logical "off" or "false" condition.

Set (S). The set symbol indicates a logical "on" or "true" condition.

DATA QUARTETS (0-F). A data quartet symbol conveys one quartet of binary data within a data transmission sequence. The 16 data quartet symbols are denoted by the hexadecimal digital values (0-F), and a generic member of the set is denoted by the character "n." A sequence of data quartet symbols is generated as a balanced sequence of symbol pairs (i.e., an even number of n symbols).

5.6.2.2 Coding

CODE BIT. Peer PHY entities communicate via fixed-length code bits. A code bit is the smallest signaling element used by PHY. In the NRZI code, a code bit is represented as a transition (one), or absence of a transition (zero), in the polarity of the signal on the medium.

CODE GROUP. A code group is a consecutive sequence of five code bits. It is used to represent a symbol on the medium. Implicit in the definition of code group is the establishment of code group boundaries. The process of establishing a code group boundary is known as "framing," and the established boundary is known as the "framing boundary." Table 13.4 defines the mapping of symbols to code groups.

5.6.3 FDDI Protocol Data Units.
Two types of PDUs are used by a MAC: tokens and frames. In the figures that follow, formats are depicted of PDUs in the order of transmission on the medium, with the leftmost symbol transmitted first.

5.6.3.1 Token.
Figure 13.21 shows the token format. The token is the means by which the right to transmit MAC SDUs (service data units) (as opposed to the normal process of repeating) is passed from one MAC to another.

5.6.3.2 The FDDI Frame.
Figure 13.22 shows the FDDI frame format. The frame format is used for transmitting both MAC recovery information and MAC SDUs between peer MAC entities. A frame may or may not have an information field.

PA = Preamble (4 or more symbols)
SD = Starting Delimiter (2 symbols)
FC = Frame Control (2 symbols)
ED = Ending Delimiter (2 symbols)

Figure 13.21. The FDDI token format [13].

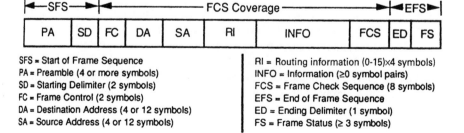

SFS = Start of Frame Sequence
PA = Preamble (4 or more symbols)
SD = Starting Delimiter (2 symbols)
FC = Frame Control (2 symbols)
DA = Destination Address (4 or 12 symbols)
SA = Source Address (4 or 12 symbols)

RI = Routing information (0-15)×4 symbols)
INFO = Information (≥0 symbol pairs)
FCS = Frame Check Sequence (8 symbols)
EFS = End of Frame Sequence
ED = Ending Delimiter (1 symbol)
FS = Frame Status (≥ 3 symbols)

Figure 13.22. The FDDI frame format.

TABLE 13.4 FDDI Symbol Coding

Code Group		Symbol		
Decimal	Binary	Name		Assignment
Line State Symbols				
00	00000	Q	Quiet	
04	00100	H	Halt	
31	11111	I	Idle	
Starting Delimiter				
24	11000	J	First symbol of JK pair	
17	10001	K	Second symbol of JK pair	
Embedded Delimiter				
05	00101	L	Second symbol of IL pair	
Data Quartets			**Hexadecimal**	**Binary**
30	11110	0	0	0000
09	01001	1	1	0001
20	10100	2	2	0010
21	10101	3	3	0011
10	01010	4	4	0100
11	01011	5	5	0101
14	01110	6	6	0110
15	01111	7	7	0111
18	10010	8	8	1000
19	10011	9	9	1001
22	10110	A	A	1010
23	10111	B	B	1011
26	11010	C	C	1100
27	11011	D	D	1101
28	11100	E	E	1110
29	11101	F	F	1111
Ending Delimiter				
13	01101	T	Terminate	
Control Indicators				
07	00111	R	Reset(logical Zero or Off)	
25	11001	S	Set(logical One or On)	

Derived and revised from page 19, ANSI X3.231-1994 [12].

FRAME LENGTH. In basic mode the physical layer of FDDI requires limiting the maximum frame length to 9000 symbols, including the four symbols of the preamble; 9000 symbols (quartets) equals 4500 octets.

Preamble (PA). The preamble is transmitted by a PDU originator with a minimum of 4 symbols of the idle pattern when in the Hybrid mode,* or 16 symbols of idle pattern in the basic mode.

---

* The hybrid mode is another mode of FDDI ring operation involving multiplexing of packet and circuit data and the provision of access to circuit switched channels.

C = Class Bit
L = Address Length Bit
FF = Format Bits
WXYZ = Control Bits

Figure 13.23. Frame control field (FC) [13].

Starting Delimiter (SD). The starting delimiter was described in Section 5.6.2. A basic mode frame or token starts with a JK symbol pair. No basic mode frame or token is considered valid unless it starts with this explicit sequence.

Frame Control (FC). The frame control field defines the type of frame and associated control functions. This field is shown in Figure 13.23.

FRAME CLASS BIT. The frame class bit indicates the class of service as follows:

$C = 0$ indicates an asynchronous frame
$C = 1$ indicates a synchronous frame

FRAME ADDRESS BIT LENGTH. The frame address length indicates the length of both MAC addresses (DA and SA) as follows:

$L = 0$ indicates 16-bit addresses
$L = 1$ indicates 48-bit addresses

FRAME FORMAT BITS. The FF bit in conjunction with the CL bits and the WXYZ control bits indicate the frame type as follows:

CLFF	WXYZ	to	WXYZ	
0L00	0000			Void frame
1000	0000			Nonrestricted token
1100	0000			Restricted token
0L00	0001	to	1111	Station management frame
1L00	0001	to	1111	MAC frame
CL01	0000	to	1111	LLC frame
CL10	0000	to	1111	Reserved for implementer
CL11	0000	to	1111	Reserved for future standardization

where the W bit is reserved for future standardization in all frames except void frames and tokens. It is transmitted as 0, with the exception of SMT next station addressing frames. It is ignored when received, but is included in the FCS computation for all frames. The referenced specification [11–13] states that the XYZ bits are reserved for future assignment. These reserved bits are transmitted as 0 and are ignored when received, but are included in the FCS computation.

$FF = 00$. Frames with $FF = 00$ are used for management of the FDDI network. They are transmitted with the SA field set to individual_MAC_address of the transmitting MAC, and the RI set to 0. FF = 00 frames are not to be forwarded by bridges.

Control Bits. The control bits are used with associated CLFF bits (see Figure 13.23). All other values (not shown below) and those designated by r, are reserved for future assignment.

> **MAC Beacon Frame** *(1L00 r010).* This frame is transmitted to indicate a serious ring fault and assists in locating persistent faults.
>
> **MAC Claim Frame** *(1LL00 r011).* This frame is transmitted during error recovery in the basic mode to determine which MAC creates the token.
>
> **MAC Purge Frame** *(1L00 r100).* This frame is used in the Hybrid mode during error recovery.
>
> **SMT Next Station Addressing Frame** *(0L00 r111).* This frame is transmitted by SMT (station management) to address the next downstream station that is a member of an addressed group. When the broadcast address is used, this is the next downstream MAC.
>
> **Asynchronous LLC Frame** *(0L01 WPPP).* This frame contains an LLC PDU using the asynchronous transmission service. The last three control bits (PPP) indicate the SDU (user) priority, with PPP = 111 being the highest priority and PPP = 000 being the lowest priority.
>
> **Synchronous LLC Frame** *(1L01 rrrr).* This frame contains an LLC PDU using the synchronous transmission service.
>
> **Implementer Frame** *(CL10 rUUU).* This frame contains an implementer defined MAC SDU. The control bits, UUU, are implementer defined.

DESTINATION AND SOURCE ADDRESSES. The approach here is nearly identical to the token ring approach in Section 5.5 of this chapter.

Addresses may be either 16 or 48 bits in length. However, all MACs implement 48-bit address capability. Both 16-bit and 48-bit addresses may be disabled for specific protocol operations. If an address is disabled or not implemented, it is equivalent to the null address. A null address is an address field of all 0s. The group and broadcast address approach is identical to token ring (Section 5.5). However, specific group addresses have been assigned for SMT protocols.

The source address rules used in FDDI have some variances with token ring. There are both 16-bit and 48-bit source addresses. However, bit 1 of the FDDI source address is the routing indicator (RI). If that bit is 0, there is no routing information; if it is 1, there is routing information. This routing information, if present, is contained in a field after the source address (SA) field. The second bit of the 48-bit source address field is the U/L bit and indicates with the address is locally or universally administered.

ROUTING INFORMATION (RI) FIELD. The RI field is used when bridging is available. When the RI bit of a 48-bit address field is set to 1, this indicates that the routing information field is included in the frame. The RI field contains 2 to 30 octets (symbol pairs) whose format and meaning is specified by ISO/IEC 10038 standard on bridging. The length of the RI field in octets is contained in the first octet of the RI field.

INFORMATION (INFO) FIELD. The INFO field contains zero, one, or more data symbol pairs whose meaning is determined by the FC field and whose interpretation is made by the destination entity (e.g., MAC, LLC, or SMT). The length of the INFO field is variable, but must conform to both (a) the maximum frame length criteria given above and (b) the frame validity criteria defined in the referenced standard.

FRAME CHECK SEQUENCE. The frame check sequence generating polynomial is the same as used for CSMA/CD, token bus, and token ring, Sections 5.3, 5.4, and 5.5.

ENDING DELIMITER (ED). The symbol T is the ending delimiter of tokens and frames. Ending delimiters and optional control indicators form a balanced symbol sequence (i.e., be transmitted in pairs so as to maintain octet boundaries). This is accomplished by adding a trailing T symbol as required.

FRAME STATUS (FS). The frame status (FS) field consists of an arbitrary length sequence of control indicator symbols (R or S) that follows an ending delimiter of a frame. It ends if any symbol other than R or S is received. If an expected indicator is not received as R or S, it is reported as not received. A trailing T symbol, if present, is repeated as part of the FS field. The first three control indicators of the frame status field are mandatory, indicating error detected (E), address recognized (A), and frame copied (C). The use of additional control indicators in the frame status field after the C indicator is optional and is implementer-defined. Although the use of optional control indicators in the frame status field is undefined the ISO/IEC 9314 standard, all conforming FDDI MACs must repeat the entire frame status field.

Error-Detected Indicator (E). The error-detected indicator (E) is transmitted as R by the MAC that originates the frame. All MACs on the ring inspect repeated frames for errors. If an error is detected and the received E indicator is not S, then an error is counted. The E indicator is transmitted as S by a repeating MAC when either an error to be counted is detected, the received E indicator is S, or the E indicator was not received.

Address Recognized Indicator (A). The address recognized indicator is transmitted as R by the MAC that originates the frame. If another MAC recognizes the destination address as its own individual or group address, it sets the A indicator to S. A MAC does not set the A indicator for a partially filtered group address (i.e., it only sets the A indicator for a precise match). A bridge performing source routing which recognizes a received frame as one it intends to forward by means of the information in the routing information (RI) field sets the A indicator of the repeated frame to S. In all other cases, a repeating MAC transmits this indicator as received.

Frame-Copied Indicator (C). The frame-copied indicator is transmitted as R by the MAC that originates the frame. If another MAC has set the A indicator and copied the frame into its receive buffer, it sets the C indicator to S. If a transparent bridge, including optional capability for setting the C indicator, recognizes the destination address as one to be forwarded, has not set the A indicator, and has copied the frame into its receive buffer for forwarding, it sets the C indicator to S if the A indicator is received as R. If a MAC, including optional capability for clearing the C indicator, has set the A indicator and has not copied the frame into its receive buffer, it transmits the C indicator as R if the received A indicator is R.

5.6.4   FDDI Timers.
Each MAC maintains three timers to regulate the operation of the ring. These timers are: token holding timer (THT), valid-transmission timer (TVX), and token-rotation timer (TRT).

The token-holding timer controls how long the MAC may transmit asynchronous frames. The valid-transmission timer is used to recover from transient ring error situations. The token-rotation timer is used to control ring scheduling during normal operation and to detect and recover from serious ring error situations.

5.6.5   FDDI Operation.
Access to the physical medium (the ring) is controlled by passing a token around the ring. The token gives the downstream MAC (receiving relative to the MAC passing the token) the opportunity to transmit a frame or a sequence of frames. If a MAC wants to transmit, it strips the token from the ring before the frame control field of the token is repeated. After the token is completely received, the MAC begins transmitting its eligible queued frames. After transmission, the MAC issues a new token for use by a downstream MAC.

MACs that are not transmitting repeat the incoming symbol stream. While repeating the incoming symbol stream, the MAC determines whether frames are intended for this MAC. This is done by matching the DA to its own address or a relevant group address. If a match occurs, the frame is processed by the MAC or sent to SMT or LLC.

FRAME TRANSMISSION. Upon request for service data unit (SDU) transmission, the MAC constructs the PDU or frame from the SDU by placing the SDU in the INFO field of the frame. The SDU remains queued by the requesting entity awaiting the receipt of a token that may be used to transmit it. Upon reception and capture of an appropriate token, the MAC begins transmitting its queued frame(s) in accordance with the rules of token holding. During transmission, the FCS for each frame is generated and appended to the end of the PDU.

After transmission of the frame(s) is completed, the MAC immediately transmits a new token.

FRAME STRIPPING. Each transmitting station is responsible for stripping from the ring the frames that it originated. A MAC strips each frame that it transmits beginning not later than the seventh symbol after the end of the SA field. Normally, this is accomplished by stripping the remainder of each frame whose source address matches the MAC's address from the ring and replacing it with idle symbols.

The process of stripping leaves remnants of frames, consisting at most of PA, SD, FC, DA, and SA and six symbols after the SA field, followed by idle symbols. These remnants exist because the decision to strip a frame is normally based upon recognition of the MAC's address in the SA field, which cannot occur until after the initial part of the frame has already been repeated. These remnants are not recognized as frames because they lack an ending delimiter (ED). The limit of remnant length also prevents remnants from satisfying the minimum frame length criteria. To the level of accuracy required for statistical purposes, they can be distinguished from error or lost frames because they are always followed by the idle symbol. Remnants are removed from the ring when they encounter a transmitting MAC. Remnants may also be removed by the smoothing function of PHY.

RING SCHEDULING. Transmission of normal PDUs (i.e., PDUs formed from SDUs) on the ring is controlled by a timed token rotation protocol. This protocol supports two major classes of service:

(a) *Synchronous*. Guaranteed bandwidth and response time.

(b) *Asynchronous*. Dynamic bandwidth sharing.

The synchronous class of service is used for those applications whose bandwidth and response limits are predictable in advance, permitting them to be preallocated (via the SMT). The asynchronous class of service is used for those applications whose bandwidth requirements are less critical (e.g., bursty or potentially unlimited) or whose response time requirements are less critical. Asynchronous bandwidth in FDDI is instantaneously allocated from the pool of remaining bandwidth that is unallocated, unused, or both.

5.6.6 Clocking. A local clock is used to synchronize both the internal operation of PHY and its interface to the data-link layer. This clock is derived from a fixed frequency reference. This reference may be created internally within the PHY implementation or supplied to PHY. (A crystal oscillator may be used for this purpose.)

Characteristics of the local clock are as follows:

(a) Nominal symbol time (UI) = 40 ns($1/\text{UI} = 25$ MHz)

(b) Nominal code bit cell time (UI) = 8 ns($1/\text{UI} = 125$ MHz)

(c) Frequency accuracy $< \pm 0.005\%(\pm 50$ ppm)

(d) Harmonic content (above 125.02 MHz) < -20 dB

(e) Phase jitter (above 20 kHz) $< \pm 8°$ (0.044 UI pp)

(f) Phase jitter (below 20 kHz in hybrid mode) $< \pm 270°$ (1.5 UI pp)

The receive function derives a clock by recovering the timing information from the incoming serial bit stream. This clock is locked in frequency and phase to the transmit clock of the upstream node. The maximum difference between the received bit frequency and the local bit frequency is 0.01% of the nominal frequency. The received frequency can be either slower or faster than the local frequency, resulting in an excess or a deficiency of bits unless some compensation is included. The elasticity buffer function provides this compensation by adding or dropping idle bits in the preamble between DLL PDUs.

The operation of the elasticity buffer function produces variations in the lengths of the preambles between DLL PDUs as they circulate around the logical ring. The cumulative effect on preamble size of PDU propagation through many elasticity buffers can result in excessive preamble erosion and, in hybrid mode, excessive cycle clock jitter. The smoothing function serves to filter out these undesirable effects.

Source: Section 5.6 is based on ANSI X3.231-1994 [12] and ANSI X3.239-1994 [13]. The figures and tables used in Section 5.6 derive directly from these ANSI referenced standards.

5.7   LAN Performance

For this discussion, LANs can be broken down into two categories:

1. *CSMA/CD Approach with Its Random Access.* Unpredictable performance.
2. *Token Passing.* Predictable performance.

CSMA/CD. There are far more CSMA/CD (Ethernet type) LANs in operation than any other type. There are several reasons why this is so. First, CSMA/CD is very economic. Second, it is simple to implement. There no token monitor or token monitor requirement at each station.

Operational Advantages

A. Once a station is sending packets or frames, it can continue to send them until there are no others to send. There is no time limit as to usage.
B. There is no waiting time once the bus is free. A station can transmit frames or packets as soon as it has an opportunity.
C. Performance for any station is independent from the total number of stations on the bus. Only those stations attempting to transmit at the same time or nearly the same time have any impact on performance.
D. No protocol interchanges are needed. Stations do not need to exchange messages with each other in order to manage access to the bus. This reduces overhead and simplifies station design.

Operational Disadvantages

A. The most serious drawback to CSMA/CD type LANs is the activity factor versus collision problem. Performance starts to degrade as the load grows.

As collisions increase, throughput degrades. It is argued at what point this performance degradation really starts to take hold. One school of thought suggests a point around 25%, whereas another suggests that between 50% and 60%, loading is where trouble starts [14].

B. There is an inability to incorporate priority traffic for voice or real-time needs. There is no guaranteed access to the bus.

C. With a maximum extension (with repeaters) of 2500, thus the maximum number of stations that can be accommodated is about 1000. We doubt that 1000 users could be accommodated operationally with any small usage factor at all.

Note: One way this situation can be eased is by segmenting the LAN into user families, where each family has a high community of interest amongst its members. The separation is done physically/operationally using bridges or possibly routers.

TOKEN PASSING. Token passing schemes have a controlled access. We can readily calculate the performance of a token passing type LAN and access can be guaranteed under any load condition.

Operational Advantages

A. Under heavy traffic loading, each station will still receive a fixed level of service. Throughput stabilizes.

B. There is a means of handling traffic with varying priorities.

Operational Disadvantages

A. There is a limitation on the traffic a station can transmit before it must pass the token and wait for its next turn.

B. Token-passing protocols require an exchange of messages—at least the token. This adds to overhead and time consumed in non-revenue-bearing traffic.

C. All stations must be able to act as the token monitor requiring a replication of function and cost.

D. Any station's performance is a function of the total number of active stations regardless of activity. The more stations, the more each station has to wait for the token.

E. Once a station passes the token, it must wait a minimum time before it can transmit more traffic.

5.8 LAN Internetworking via Spanning Devices

In this section we will briefly describe repeaters, bridges, routers, switches, hubs, and switching hubs.

5.8.1 Repeaters. A repeater is nothing more than a regenerative repeater. It extends a LAN. It does not provide any kind of segmentation of a LAN, except the physical regeneration of the signal. Multiple LANs (with common protocols) can be interconnected with repeaters, in effect making just one large segment. A network using repeaters must avoid multiple paths, as any kind of loop would cause data traffic to circulate indefinitely and could ultimately make the network crash. This concept is shown in Figure 13.24.

The following example shows how a loop can be formed. Suppose two repeaters connect CSMA/CD LAN segments as shown in Figure 13.24. Station 1 initiates an interchange with station 3, both on the same segment (upper in the figure). As data packets or frames are transmitted on the upper segment, each repeater will transmit them unnecessarily to the lower segment. Each repeater will receive the repeated packet on the lower segment and retransmit it once again on to the upper segment. As one can see, any traffic introduced into this network will circulate indefinitely around the loop created by the two repeaters. On larger networks the effects can be devastating, although perhaps less apparent.

5.8.2 Hubs. A hub is a multiport device that allows centralization. A hub is usually mounted in a wiring closet or other central location. Signal leads are brought in from workstations/PCs and other data devices, one for each hub port. Physical rings or buses are formed by internally configuring the hub ports. A typical hub may have 8 or 16 ports. Suppose we wished to incorporate 24 devices on our LAN using the hub. We can stack two hubs, one on top of the other (stackables), using one of the hub ports on each for interconnection. In this case we would have a hub with 30-port capacity $(32 - 2)$. Hubs may also have a certain amount of intelligence such as incorporation of a network management capability. Each hub can incorporate a repeater. However, for CSMA/CD (Ethernet)

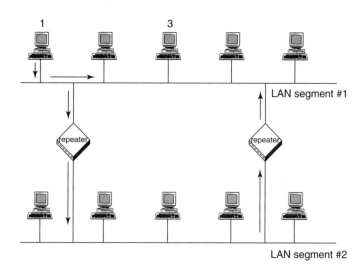

Figure 13.24. Repeaters in multiple paths. Courtesy of Hewlett-Packard Company [15].

there is a 4-repeater limit (i.e., a maximum of five 500-m segments). With FDDI and token ring the repeater limit rule does not apply.

5.8.3 LAN Bridges.
Whereas repeaters have no intelligence, bridges do. Bridges can connect two LANs at the data link or MAC protocol level. There are several varieties of bridges depending on the intelligence incorporated.

There is the transparent bridge. It builds a list of nodes it sees transmitting on either side. It isolates traffic. It will not forward traffic that it knows is destined to another station on the same side of the bridge as the sending station. It is able to isolate traffic according to the MAC source and destination address(es) of each individual data frame. MAC-level broadcasts, however, are propagated through the network by the bridges. A bridge can be used for segmenting and extending LAN coverage. Thus it lowers traffic volume for each segment.

There are four varieties of bridges. We covered the transparent bridge. It does not modify any part of a message that it forwards.

The second bridge is the translation bridge. It is used to connect two dissimilar LANs, such as a token ring to CSMA/CD. In order to do this, it must modify the MAC level header and FCS of each frame it forwards in order to make it compatible with the receiving LAN segment. The MAC addresses and the rest of the data frame are unchanged. Translation bridges are far less common than transparent bridges.

The third type of bridge, as shown in Figure 13.25, is the encapsulation bridge. It is also used to connect LANs of dissimilar protocols. But rather than translate MAC header and FCS fields, it simply appends a second MAC layer protocol around the original frame for transport over the intermediate LAN with a different protocol. There is a destination bridge which strips off this additional layer and extracts the original frame for delivery to the destination network segment.

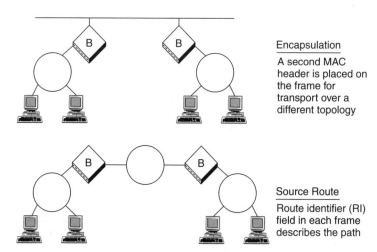

Encapsulation

A second MAC header is placed on the frame for transport over a different topology

Source Route

Route identifier (RI) field in each frame describes the path

Figure 13.25. The concept of bridging. Upper drawing: encapsulation bridge, lower drawing: source routing bridge. Courtesy of Hewlett-Packard Company [15].

The fourth type of bridge is a source routing bridge. It is commonly used in token-ring networks. With source routing bridges, each frame carries within it a route identifier (RI) field which specifies the path that that frame is to take through the network. This concept is also shown in Figure 13.25.

Up to this point we have been discussing local bridges. A local bridge spans LANs in the same geographic location. A remote bridge spans LANs in different geographic locations. In this case, an intervening WAN is required. The remote bridge consists of two separate devices that are connected by a wide area network affording transport of data frames between the two. This concept is shown in Figure 13.26.

As shown in Figure 13.26, the LAN data packet/frame is encapsulated by the remote bridge adding the appropriate WAN header and trailer. The WAN transports the data packet/frame to the distant end remote bridge which strips the WAN header and trailer, and it delivers the data packet/frame to the far-end LAN.

Remote bridges typically use proprietary protocols such that, in most cases, remote bridges from different vendors do not interoperate.

Bridges are good devices to segment LANs, particularly CSMA/CD LANs. Segmenting breaks a LAN up into user families. It is expected that there is a high community of interest amongst members of a family, and a low community of interest amongst different families. There will be large traffic volumes intra-segment and low traffic volumes inter-segment. It should be pointed out that routers are more efficient at segmenting than bridges.

A major limitation of bridges is the inability to balance traffic across two or more redundant routes in a network. The existence of multiple paths in a bridged network can prove to be a bad problem. In such a case, we are again faced with the endless route situation as we were with repeaters. One way to avoid the problem is to use the *spanning tree algorithm*. This algorithm is implemented by having bridges communicate with each other to establish a subset of the actual

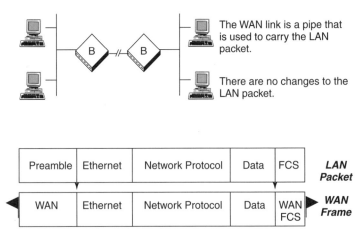

Figure 13.26. The concept of remote bridging. The LAN frame/packet is encapsulated in a WAN frame. Courtesy of Hewlett-Packard Company [15].

network topology that is loop-free (often called a "tree"). The idea, of course, is to eliminate duplicate paths connecting one LAN to another, or one segment to another. If there is only one path from one LAN to another, there can be no loop formed. (Also consult Refs. 16 and 17.)

5.8.4   Routers. Routers carry more intelligence than bridges. Like a bridge, it forwards data packets/frames. Routers make forwarding decisions based on the destination network layer address. Whereas a bridge worked on the data-link layer, a router operates at the network layer level. Routers commonly connect disparate LANs such as CSMA/CD to token ring and FDDI to CSMA/CD.

Routers are addressable nodes in a network. They carry their own MAC address(es) as well as a network address for each protocol handled. Because routers are addressable, a station desiring the facility of a router must direct their packets/frames to the router in question so that the traffic can be forwarded to the appropriate network. As one would expect, networking software at each station is more complex with a network using router than one using bridges.

Routers handle only traffic addressed to them. They make decisions about forwarding data packets/frames based on one or several criteria. The decisions may be based on the cost of the link, the number of hops on each path, and the time-to-live.

Routers change packets/frames that pass through them such as MAC source and destination address; they may also modify the network protocol header of each frame (typically decrementing the time-to-live in the case of IP, and other protocol fields).

Because routers have more intelligence than bridges, routers will typically have better network management agents installed. This enables them to be remotely configured, to be programmed to pass or not to pass data for security purposes, and to be monitored for performance, particularly error performance. Due to the additional processing performed at routers, they tend to be slower than bridges. Reference 18 suggests that some protocols do not lend themselves to routing such as DEC LAT, SNA, and NetBIOS, among others.

5.8.5   LAN Switching with CSMA/CD (Ethernet) Applications. Switching in a CSMA/CD network can reduce the incidence of collisions and notably improve network performance. This is done primarily by segmenting a network. However, the use of switching does not cure all the problems involved with the use of Ethernet.

Unlike hubs, switches map LAN addresses of the nodes residing in each network segment and then allow only the necessary traffic to pass through the switch. When a packet is received by a switch, the switch examines the source and destination hardware addresses in the MAC frame. It compares the addresses to a resident table of network segments and addresses. If the segments are the same, the packet is dropped ("filtered"). If the segments are different, then the packet is forwarded to the proper segment. Switches may also be programmed to prevent bad or misaligned packets from spreading by not forwarding them.

CSMA/CD is a contention-based protocol. It is a shared network, prone to collisions as we discussed in Section 5.3 of this chapter. The greater the utilization factor, the higher the probability of collision. This is because all users on a shared network are competitors for the Ethernet bus. Generally, an Ethernet-based network can sustain a utilization factor no greater than about 35% and for a 10-Mbit/s CSMA/CD network, a throughput of no greater than 2.5 Mbits/s can be achieved.

Minimizing collisions is crucial in the design and operation of CSMA/CD networks. Increased collisions are often the result of too many users or too much traffic, or both, on the network. As a consequence, there is a lot of contention for the network's bit rate capacity. On a well-run Ethernet, collision rates should be under 10%. That is less than 10% of packets sent end up in a collision. Segmenting a network can reduce collision rate.

Two Basic Switch Architectures. A LAN switch comes in one of two varieties: store-and-forward and cut-through. Cut-through switches only examine the destination address before forwarding the packet on to its destination segment. On the other hand, a store-and-forward switch accepts and analyzes the entire packet before forwarding it to its destination. It takes more time to analyze an entire packet, but by this detailed examination the switch is able to catch certain errors and collisions and keep them from propagating bad packets through the network. Modern processing has speeded up to the point where the speed of store-and-forward switches has virtually caught up with cut-through switches, where the difference between the two is minimal. Also, there are a number of switches with hybrid architectures available off-the-shelf today.

Switches with Blocking. Calculate a switch's theoretical maximum speed and capacity adding up all the ports. We then have the theoretical sum totals of a switch's throughput. If the switching bus or switching components cannot handle the theoretical totals of all ports, the switch is considered to be blocking. Ideally, we would like a nonblocking switch, but it may end up quite costly. It has been found that a switching with blocking can work well if it has a reasonable throughput for the application. Lantronix in Ref. 18 cites the following example. Consider an eight-port 10/100 (Mbits/s) switch. It can handle 200-Mbit/s full-duplex. Theoretically, with this switch there is a requirement of 1600 Mbits/s or 1.6 Gbits/s. But probably in the real world each port utilization will not exceed 50%. An 800-Mbit/s switch would end up being adequate. When dimensioning a switch, one must consider actual loads expected rather than desirable theoretical loads.

Switch Buffer Capacities. One of the principal elements of a switch is buffers. Packets are held in buffers for their processing turn. A switch allows segmentation of a CSMA/CD network. If a destination segment is congested, a switch will hold the packet in a buffer as it waits for bit rate capacity to become available on the overburdened segment. Buffers can fill to capacity and then overflow, a very undesirable condition. A network designer should analyze switch buffer sizes and strategies for handling overflows. If the network design is done right, there will not be crowded network segments.

If crowded segments do occur, there are two strategies to be considered for the full buffer problem. Reference 18 calls the first "back pressure flow control." This sends packets back upstream to the source nodes of packets that then find a full buffer, compounding the problem. The second approach is to just drop packets relying on error correction techniques for packet retransmission, increasing network load. The final solution involves oversized buffers and advice to users to avoid overcrowded segments in network design.

Routers and Switches. Routers work in a manner similar to that of switches and bridges in that they filter network traffic. A router does this by specific protocol rather than by packet addresses. Routers divide a network logically rather than physically. A router in an IP network divides that network into subnets so that only traffic destined for particular IP addresses can pass between segments. Expect a router to recalculate the FCS (frame check sequence) and rewrite the MAC header of each packet. This causes an increase in latency of the device, slowing the network down. Much of this delay is due to the filter in a router. Such filtering takes more time than that required in a switch or bridge, which only looks at the CSMA/CD address. An additional benefit of routers is their automatic filtering of broadcasts, but overall they are more complicated to set up.

5.9 Switching Hubs

There are "intelligent" hubs. These are typically modular, multiprotocol, multimedia, multichannel, fault-tolerant, manageable devices where one can concentrate all the LAN connections into a wiring closet or data center. Since these type hubs are modular (i.e., it has various numbers of slots to install LAN interface boards), it can support CSMA/CD, token ring, FDDI, or ATM (ATM is covered in Chapter 16) simultaneously as well as various transmission media such as twisted pair, fiber cable, and others.

Switching hubs are high-speed interconnecting devices with still more intelligence than the garden variety hub or the intelligent hub. They typically interconnect entire LAN segments and nodes. Full LAN data rate is provided at each port of a switching hub. They are commonly used on CSMA/CD LANs providing a node with the entire 10-Mbit/s data rate. Because of a hub's low latency, high data rates and throughputs are achieved.

Source: Ref. 18.

6 WIRELESS LANS (WLANs)

A wireless LAN is a local area network that uses radio frequency as the transmission medium rather than wire-pair, coaxial cable or fiber-optic cable. It is very attractive when there are frequent office rearrangements.

Most wireless LANs are governed by IEEE standard 802.11 [20]. There are, nevertheless, a goodly number of versions of that standard. The different versions are identified by letters. Originally, there were only two versions: IEEE 802.11a and 802.11b. They now have extended all the way through 802.11i. We will very briefly review the highlights of each version through the letter i, below. First we cover some basic IEEE 802.11 WLAN concepts.

A WLAN can extend an existing wired network or replace the wired network in its entirety. There are two basic 802.11 configurations:

1. Peer-to-peer, or ad hoc WLAN
2. Infrastructure networks

The more common application of 802.11 technology is the second item, infrastructure networks.

An infrastructure WLAN consists of two basic components: *access points* and *PC (or PCI) cards*. Access points in the infrastructure network are connected to the wire CSMA/CD network, usually using RJ-45 cable. Access points can be simply software, but are more commonly identifiable hardware. Once the access point is connected to the network, it functions henceforward as a wireless hub, passing data back and forth between the wired network and wireless clients. An access point translates digital data from the network into radio signals that wireless clients can understand.

A typical wireless client is either a laptop or desktop computer equipped with PC or PCI card, respectively. The PC or PCI card serves as the air interface receiving radio signals from an access point with which it is communicating and translates these signals into a digital data bit stream that the PC can operate with.

Let's use IEEE 802.11b as our example standard. It requires operation in the 2.4-GHz band in that portion where "no license is necessary." When we refer to a "channel," it is a specific frequency within a given frequency band. For example, channel 2 in the 2.4-GHz band runs specifically on 2.402 GHz and channel 3 on 2.403 GHz, and so on, for 80 channels, depending in what country you are in. For example, the United States and Canada use channels 1 through 11.

In this example, a WLAN channel is set for each access point, not for each wireless client. Wireless clients search for and automatically set their channels to the channel associated with the access point that has the highest level signal, principally based on the client's location.

For good coverage, a system designer sets up multiple access points. Each access point provides radio connectivity to a wireless client within the access point's range. Experience and analysis tells us that an 802.11b system typically can cover a range from 100 to 300 ft, but has been extended to 500 ft. We may call this coverage set a *cell*, but the 802.11b document calls it a *basic service set* (BSS). Overlapping cells are desirable to ensure continuous connectivity. This is shown in Figure 13.27.

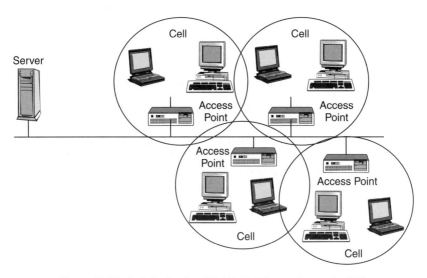

Figure 13.27. An infrastructure WLAN. Note how cells overlap [19].

To avoid co-channel interference, different cells must be set to different frequencies. The system planner must select the modulation. All 802.11a and 802.11b systems use spread spectrum modulation, of which there are two garden varieties: frequency-hop spread spectrum (FHSS) and direct sequence spread spectrum (DSSS). Some 802.11a systems use FHSS, but all 802.11b systems use DSSS.

FHSS system transmitter changes frequency very rapidly in a pattern of changing known to both the transmitter and the intended receiver. The transmitter and the companion receiver are synchronized such that it appears they maintain a single, continuous channel. To unintended receivers, the frequency hop spread spectrum signal appears to be just a short noise burst. FHSS systems are less efficient than DSSS systems. Whereas a FHSS system can maintain a data throughput about 1 Mbit/s, its DSSS companion can achieve 11 Mbits/s.

Direct sequence spread spectrum creates a redundant bit pattern called a "chip" or chipping code of each transmitted bit. Again, the transmitter and its intended receiver are synchronized and they both know the chipping code. They thus are able to filter out signals that do not use the same chipping code.

The chipping code spreads signals across the entire usable and assigned band. The longer the chip, the more bandwidth consumed. The redundancy in these long chips allows a more robust transmission where even badly corrupted data can be recovered without the need for retransmission.

It should be well kept in mind when employing these higher frequencies (e.g., 2.4 and 5 GHz) that line-of-sight conditions should exist between the access point antenna and the antenna(s) of wireless clients. Oftentimes, when one access point displays poor performance, another access point can be selected that may show excellent reception. The indoors propagation problem is further covered in Chapter 18.

6.1   The Different 802.11 Standards Issued as of March 2002

There are two 802.11 standards governing the physical layer: 802.11a and 802.11b. The 802.11a standard is specified for operation in the 5-GHz frequency band. The 802.11b, as we know, specifies operation in the 2.4-GHz band.

Other 802.11 standards are being developed that extend the physical layer options, improve security, add QoS (quality of service) features, or provide better interoperability.

IEEE 802.11a. The 802.11a standard authorizes RF eight channels in the 5-GHz band, and some countries allow up to 12 channels. Each channel can support up to 54 Mbits/s per channel. Because the channel is shared by other 802.11 users, only about half the 54-Mbit/s data rate value can be achieved. As performance degrades due to increase in distance to the radio access point, data rate also decreases

IEEE 802.11b. The IEEE 802.11b only authorizes three RF channels in its assigned 2.4-GHz band. The maximum data rate on a link is 11 Mbits/s, but maximum user data rate will be about half this value because the channel may be shared by two or considerably more 802.11b users. Again, the data rate will decrease as the distance increases to the radio access point.

IEEE 802.11c. Bridge operations procedures.

IEEE 802.11d. 802.11d is supplementary to the media access control (MAC) layer of 802.11 to promote greater worldwide use of WLANs. It allows access points to communicate information on the permissible radio channels with acceptable power levels for user devices. In some countries, operation of 802.11 WLAN devices is prohibited. The purpose of the 802.11d variant is to add features and restrictions to allow WLANs to operate within the rules and regulations of these countries.

IEEE 802.11e. 802.11e is supplementary to the 802.11 MAC layers providing QoS support for LAN applications. It applies to the physical standards of 802.11a, 802.11b, and 802.11g. The purpose is to provide classes of service with managed levels of QoS for data, voice, and video.

IEEE 802.11f. This variant of IEEE 802.11 objective is to achieve better interoperability with radio access points in a multivendor environment. The standard defines the registration of 802.11 access points within a network, and it also defines a method of handoff from one access point to another.

IEEE 802.11g. This is another physical layer standard in the IEEE 802.11 series for both the 2.4-GHz and 5-GHz frequency bands. It specifies three radio channels and its maximum link bit rate is 54 Mbits/s per channel, compared to only 11 Mbits/s for the 802.11b standard. IEEE 802.11g uses orthogonal frequency-division multiplexing as the modulation type. It also supports complementary code keying (CCK) modulation, and, as an option, it allows packet binary convolutional coding (PBCC) modulation.

Potential users beware. There are three modulation types, and the one selected should be chosen with care. Different vendors offer one, another, or the third, so interoperability may be a problem.

IEEE 802.11h. This is another standard that is supplementary to the MAC layer to comply with European regulations for 5-GHz WLAN operation. The Europeans require that 5-GHz LAN products have transmission power control (TPC) and dynamic frequency selection (DFS). TPC limits transmit power to just that required to reach the further LAN user. DFS selects the frequency channel which is least prone to interference, usually some form of radar interference.

IEEE 802.11i. This is another "supplementary to the MAC layer" for 802.11 to improve security. It applies to 802.11a, 802.11b, and 802.11g physical layer standards. Here we have an alternative to wired equivalent privacy (WEP) with new encryption methods and authentication procedures.

Security remains a headache for WLANs. The "i" version incorporates temporal key integrity protocol (TKIP). This will be followed by some new chips with AES (an iterated block cipher) which will be TKIP backwards compatible.

REVIEW QUESTIONS

1. Define a local area network. Contrast it with a wide area network.

2. Discuss how LANs can make multivendor processing equipment compatible.

3. What are the two basic underlying transmission techniques used for LANs? Compare these using a minimum of eight characteristics.

4. Name the three basic LAN topologies (i.e., network types). Name a fourth type that is a subset of one of the basic topologies.

5. What are the general ranges of bit error rates that can be expected from a well-designed LAN? How is the BER achieved (e.g., by ARQ, channel coding, S/N ratio)?

6. What are the two basic transmission problems that must be faced regarding the medium when designing a LAN?

7. If a LAN has 50 accesses, how many transmission connectivities must be analyzed?

8. Consider conventional broadband LANs. Discuss one- and two-cable operation. Include bandwidth utilization.

9. What is the function of a CRF or head end on a broadband LAN?

10. There are two methods of connecting user facilities to optical fiber LANs. What are they? Discuss the pros and cons of each.

11. What is a LAN access protocol?

12. Compare contention and polling. Describe how each leads to a particular LAN access protocol, such as token passing and CSMA/CD.

13. Why is loop polling more efficient than roll-call polling?

14. Relate the IEEE 802 LAN standards to the seven-layer OSI model. Describe the sublayering involved.

15. What is a SAP, and what is its function?

16. LLC derives from what familiar link-layer protocol?

17. Describe the three basic communication services offered by the LLC. In the description show similarities and differences among the three.

18. Name at least three responsibilities of the LLC.

19. Describe the four fields of the LLC PDU and the function of each field.

20. How are collisions detected on CSMA/CD?

21. When a collision occurs on a CSMA/CD LAN, what happens? This should lead to a discussion of persistence algorithms and their relative efficiencies.

22. Given a 500-m length of coaxial cable with access at each extreme, in what time period can a collision occur after station 1 transmits? Assume a velocity of propagation of 2×10^8 m/s.

23. What is the function of a frame check sequence (FCS)?

24. What is the purpose of the "jam" signal on CSMA/CD?

25. Give three reasons why a CSMA/CD MAC frame may be invalid.

26. In CSMA/CD operation, what is the function of the PAD field? Why is it necessary to have a minimum frame length?

27. Describe differential Manchester coding and its application to baseband LAN transmission. Compare it to conventional NRZ.

28. Define *slot time* regarding CSMA/CD. Why is it an important parameter?

29. How are collisions avoided using a token-passing scheme?

30. The MAC handles fault recovery with the token bus access scheme. Name at least four faults that can be handled by the MAC in this case.

31. Define *slot time* for token bus.

32. How does a LAN user know that a traffic frame is directed to it?

33. Describe how a token holder handles a situation when its successor does not answer a token passed to it.

34. Differentiate between logical connectivity and physical connectivity. On which of the four LAN types (discussed in this chapter) can logical connectivity be used effectively?

35. In the token bus frame format we see the data field. For a "data unit frame," what would be inserted in this field? Be specific.

36. Following the IEEE standards, is token bus truly a baseband system as we have defined it?

37. How does token ring differ from token bus? Name at least three differences.

38. Aside from the token, what are the three types of frames used on token ring?

39. Is token ring, as described, really a baseband system?

40. Define *latency* with respect to the token ring LAN.

41. How is the "I" used in the token ring ending delimiter? It is just one bit.

42. What is the baud rate of FDDI? The bit rate?

43. For both token ring operation and FDDI, name one simple way a transmitting station can assume that its frame has been received by the intended destination station.

44. What is *frame stripping*?

45. Regarding timing, what is the function of the elasticity buffer in FDDI?

46. Compare the performance of token-passing schemes versus CSMA/CD schemes under light traffic loading and under heavy traffic loading conditions.

47. What is the function of LAN repeaters?

48. Four types of bridges were discussed in the text. Name three of them, and with one sentence, describe each of the three.

49. Discuss routers in light of bridges.

50. What is the function of a hub? An "intelligent" hub? A switching hub?

51. Differentiate "store-and-forward" from "cut-through" switches.

52. What is the principal advantage of segmenting a CSMA/CD network?

53. On IP what might a router do on the "time-to-live" (TTL) field?

54. Why might an oversized buffer be recommended in a LAN switch?

55. Differentiate LAN routers with LAN switches.

56. Give one specific environment where WLANs are attractive.

57. What are the two basic components of an infrastructure WLAN?

58. Based on IEEE 802.11 standard, what would the principal advantage of a DSSS over a FHSS WLAN system?

59. What are the two principal frequency bands of WLANs?

60. What would be the principal thing we are looking for on a propagation survey for a WLAN system?

REFERENCES

1. J. F. Schoch, Y. K. Delai, D. D. Redell, and R. C. Crane, "Evolution of the Ethernet Local Computer Network," *Computer* (August 1982).
2. R. N. Dunbar, "Design Considerations for Broadband Coaxial Cable Systems," *IEEE Commun. Mag.* **24** (6) (June 1986).
3. W. Stallings, *Handbook of Computer-Communications Standards*, Vol. 2, Local Area Networks, Macmillan, New York, 1987.
4. *Carrier Sense Multiple Access with Collision Detection (CSMA/CD) Access Method and Physical Layer Specifications*, ANSI/IEEE Std. 802.3, 3rd ed, IEEE, New York, March 1992 (ISO/IEC 8802-3).
5. *Broadband Local Area Networks*, IEEE Std. 802.7, IEEE, New York, 1989.
6. *Logical Link Control*, ANSI/IEEE Std 802.2 ISO 8802-2, 1st ed., IEEE, New York, December 1989.
7. *The New IEEE Standard Dictionary of Electrical and Electronics Terms*, 6th ed., IEEE Std 100-1996, IEEE, New York, 1996.
8. *Token-Passing Bus Access Method and Physical Layer Specifications*, ANSI/IEEE Std 802.4, ISO/IEC 8802-4, 1st ed., IEEE, New York, August 1990.
9. *Token Ring Access Method*, IEEE Std 802.5-1989, IEEE, New York, October 1991.
10. *Recommended Practice for Dual Ring Operation with Wrapback Reconfiguration*, IEEE Std. 802-5c-1991, IEEE, New York, March 1991.
11. *Fiber Data Distributed Interface (FDDI)—Token Ring Physical Layer Medium Dependent (PMD)*, ANSI X3.166-1990, ANSI, New York, 1990.
12. *Fiber Distributed Data Interface (FDDI)—Physical Layer Protocol (PHY-2)*, ANSI X.231-1994, ANSI, New York, 1994.
13. *Fiber Distributed Data Interface (FDDI)—Token Ring Media Access Control—2 (MAC-2)*, ANSI X.239-1994, ANSI, New York, 1994.
14. J. McDonnel, *Internetworking and Advanced Protocols*, a seminar, Network Technologies Group, Inc., Boulder, CO, 1985.
15. *Internetwork Troubleshooting Seminar Presentation*, Hewlett-Packard Company, Tempe, AZ, January 1995.
16. ChipCom promotional material 1995, ChipCom, Southborough, MA, February 1995.

17. *High Speed Networking—Options and Implications*, ChipCom, Southborough, MA, 1995.

18. *Network Switching*, Lantronix Networking Tutorials, from the web, www.lantronix.com/learning/tutorials/switching 4/8/03.

19. *What is a WLAN?* Published March 2001 from the web: www.ncmag.com/2001 07/wireless71/whatis.html 4/12/03.

20. *IEEE 802.11 Standards*, IEEE, New York, 2001.

21. *IEEE 802.3 CSMA/CD Standard*, 1998 edition, IEEE, New York, 1998. (*Note*: There are 22 variants of this standard contained in the volume.)

14

INTEGRATED SERVICES DIGITAL NETWORKS

1 BACKGROUND AND GOALS OF INTEGRATED SERVICES DIGITAL NETWORK (ISDN)

The analog public switched telecommunications network (PSTN) was based on the 4-kHz voice (VF) channel. It served well in providing speech telephony since the 1880s. In the 19th century the only other service was telegraph, which predated the telephone by some 35 years. The two services evolved separately and distinctly. In some respects they were competitive. Before World War II there was some melding where telegraph and telex were carried as subcarriers on PSTN telephone channels leased by telephone companies or administrations. This might be viewed as the first move toward integrated services. However, it was probably done more for convenience and economy than for any forward thinking regarding integration.

Looking backward over that period, telephony became ubiquitous, with a telephone in every office and in nearly every home. On the other hand, telegraphy evolved into telex (a switched telegraph service), but still took a back seat to telephony. Historically, facsimile was the next service that was integrated rapidly into the telephone network. Facsimile (fax) required a modem to make it compatible with the analog telephone channel. In the office environment, facsimile has almost completely replaced telex for the transmission of record traffic. Then, in the 1950s computer-related data began to emerge, requiring some method of point-to-point relay. This relay facility was carried out again by the ubiquitous telephone network. Once more, a modem was required to integrate the service into the analog telephone network.

By this time the worldwide telephone network was in place and pervasive. Using that network turned out to be a cost-effective method to communicate

Telecommunication System Engineering, by Roger L. Freeman
ISBN 0-471-45133-9   Copyright © 2004 Roger L. Freeman

other information (i.e., other than speech telephony) from point X to point Y. Dial-up telephone connections provided one way of achieving switched service to transport that "other" information on a point-to-point basis, given the transmission limits of the analog VF channel traversing the PSTN.

Digital telephony began to take hold after the discovery, and later development, of the transistor in 1948.* Solid-state circuitry, particularly LSI, made pulse-code modulation (PCM) transmission and later PCM switching cost-effective. The first application of PCM was in the expansion of the trunk cable plant. In the United States and Canada the entire long-distance plant is digital. We expect that the local switching and trunk plant will be all-digital before the year 2000, and the remainder of the world perhaps around 2010.

The present digital network is based on PCM as described in Chapter 8. PCM standards developed along two, or some could argue three, distinct routes. North America and Japan use a basic 24-channel system with in-channel signaling and added-bit framing. Europe and much of the remainder of the world use a 30-channel format with two extra channels for separate channel signaling and synchronization/framing. Japan has a distinct higher-level PCM hierarchy.

PCM was designed to serve speech telephony. Even today more than 80% of the traffic on the PSTN is voice telephony. Voice traffic optimization of PCM design is even more evident in North America. In-channel signaling and framing corrupted the basic 64-kbit/s channel, thus integrating other services such as computer data, required a drop-back to 56 kbits/s for North American PCM.† Typical of such a drop-back is "switched 56" and AT&T DDS (digital data system).

ISDN has been developed to ease integration of all services, except full motion video. It is based on the 64-kbit/s channel, variously called DS0 or E0, depending on the standard followed (i.e., European or North American). Whereas 4 kHz was the basic building block of analog telephony, 64 kbits/s is the basic building block of the digital network and ISDN. The ISDN basic building block is designed to serve, among other services [1]:

- Digital voice
- 64-kbit/s data, both circuit- and packet-switched
- Telex/teletext
- Facsimile
- Slow-scan video

The goal of ISDN was to provide an integrated facility to incorporate each of the facilities listed above on a common 64-kbit/s channel. The concept was born in the 1960s. A lot has happened in telecommunications development since then. Certainly the most mighty event was the internet, which is quickly becoming ubiquitous. Our concept of a network has become quite different. Because of this, ISDN has had to take a back seat.

---

* The basis of modern digital (PCM) telephony was laid out in 1937. It was not until the development of the IC (integrated circuit) that PCM became feasible.
† 64-kbit/s clear-channel service is available in North America with special conditioning.

Our original intent of this chapter was to provide a solid technical overview of ISDN. Our thinking has changed. We decided against removing ISDN from the book because so much has derived from it. For example, two basic protocols remain an important part of the scene: LAPB (link access protocol, B-channel) and LAPD (link access protocol, D-channel). Both are ISDN derivatives. A number of data-link protocols are related to LAPB, and frame relay is a direct descendant of LAPD. ISDN itself is still around, particularly in Europe. However, it has taken on a secondary role nearly everywhere. It should be noted that ISDN has evolved into B-ISDN and the asynchronous transfer mode (ATM). These are described in Chapter 16 of this text.

In this chapter, we give an early view of "integration," and the transmission and switching involved in ISDN. The reader should note that what we describe here is now often called "narrow-band ISDN" to distinguish it from B-ISDN or Broadband ISDN as covered in Chapter 16.

As the reader proceeds through this chapter, he/she should be aware that we are dealing with user interfaces into an *existing* digital network (Chapter 10) in which CCITT Signaling System No. 7 (Chapter 15) is operational. In this chapter we are not dealing with the digital network itself, only how it affects the user and how the ISDN user affects the network. We will also discuss the two flavors of ISDN: CCITT and North American.

It is suggested that the reader consult Reference 2 covering the ITU-T I. recommendations that govern overall ISDN.

2   ISDN STRUCTURES

2.1   ISDN User Channels

Here we look from the user into the network. We consider two user classes: residential and commercial. The following are the standard bit rates for user access links:

- B-channel: 64 kbits/s
- D-channel: 16 or 64 kbits/s
- H-channels (discussed below)

The B-channel is the basic user channel. It is transparent to bit sequences. It serves all the traffic types listed in Section 1.

In one configuration, called the *basic rate*, the D-channel has a 16-kbit/s data rate; in another, called the *primary rate*, it is 64 kbits/s. Its primary use is for signaling. The 16-kbit/s version, besides signaling, may serve as transport for low-speed data applications, particularly those using X.25 packet data (see Chapter 11).

There are a number of H-channels:

- H_0 channel: 384 kbits/s
- H_1 channels: 1536 kbits/s (H_{11}) and 1920 kbits/s (H_{12})

The H-channel is intended to carry a variety of user information streams. A distinguishing characteristic is that an H-channel does not carry signaling information for circuit switching by the ISDN. User information streams may be carried on a dedicated, alternate (within one call or separate calls), or simultaneous basis, consistent with the H-channel bit rates. The following are examples of user information streams:

- Fast facsimile
- Video, such as video conferencing
- High-speed data
- High-quality audio or sound program channel
- Information streams, each at rates lower than the respective H-channel bit rate (e.g., 64-kbit/s voice), which have been rate-adapted or multiplexed together
- Packet-switched information

Source: Section 2.1 is based on Ref. 1.

2.2   Basic and Primary User Interfaces

The *basic* rate interface structure is composed of two B-channels and a D-channel referred to as "2B + D." The D-channel at this interface is 16 kbits/s. The B-channels may be used independently (i.e., two different simultaneous connections). Industry and much of the literature call the basic rate interface the *BRI*.

Appendix I to CCITT Rec. I.412 [1] states that alternatively the basic access may be just one B-channel and a D-channel or just a D-channel.

The *primary* rate interface (PRI) structures are composed of n B-channels and one D-channel, where the D-channel in this case is 64 kbits/s. There are two primary data rates:

- 1.544 Mbits/s = 23B + D (from the North American T1 configuration)
- 2.048 Mbits/s = 30B + D (from the European E1 configuration)

For the user–network access arrangement containing multiple interfaces, it is possible for the D-channel in one structure not only to serve the signaling requirements of its own structure but also to serve another primary rate structure without an activated D-channel. When a D-channel is not activated, the designated time slot may or may not be used to provide an additional B-channel, depending on the situation, such as 24B with 1.544 Mbits/s.

The primary rate interface H_0-channel structures are composed of H_0 channels with or without a D-channel. When present in the same interface structure the bit rate of the D-channel is 64 kbits/s.

At the 1544-kbit/s primary rate interface, the H_0-channel structures are $4H_0$ and $3H_0 + D$. When the D-channel is not provided, signaling for the H_0-channels is provided by the D-channel in another interface.

At the 2048-kbit/s primary rate interface, the H_0 structure is $5H_0 + D$. In the case of a user–network access arrangement containing multiple interfaces, it is possible for the D-channel in one structure to carry the signaling for H_0-channels in another primary rate interface without a D-channel in use.

The 1536-kbit/s H_{11}-channel structure is composed of one 1536-kbit/s H_{11}-channel. Signaling for the H_{11}-channel, if required, is carried on the D-channel of another interface structure within the same user–network access arrangement.

The 1920-kbit/s H_{12} structure is composed of one 1920-kbit/s H_{12}-channel and a D-channel. The bit rate of the D-channel is 64 kbits/s. Signaling for the H_{12}-channel, if required, is carried in this D-channel or the D-channel of another interface structure within the same user–network access arrangement.

3   USER ACCESS AND INTERFACE

3.1   General

The objective of the ISDN designers was to provide a telecommunications service which would be ubiquitous and universal. Whether this ambitious goal is being met can be argued. ISDN transmission rates were to be accommodated on copper wire pairs. Within the next 20 years most of the copper plant will be replaced by fiber optics. Some futurists argue 10 years and others 30 years. Nevertheless, enterprise networks now demand transmission rates in excess of what ISDN has been designed for. The demand has been essentially to provide LAN connectivity over WAN distances. Frame relay (Chapter 15) has removed some of this demand pressure.

Also it is suggested that the user consult Refs. 7, 8, and 10.

Figure 14.1 shows generic ISDN user connectivity to the network. We can select either the basic or primary rate service (e.g., $2B + D$, $23B + D$, or $30B + D$) to connect to the ISDN network. The objectives of any digital interface design, and specifically of ISDN access and interface, are as follows:

1. Electrical and mechanical specification
2. Channel structure and access capabilities

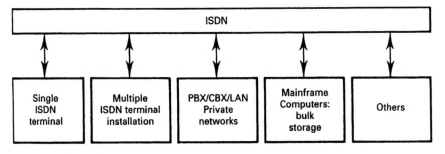

Figure 14.1. ISDN generic users.

3. User–network protocols
4. Maintenance and operation
5. Performance
6. Services

ISDN specifications as set out by the ITU-T I recommendations and relevant Telcordia(Bellcore)/ANSI specifications covers the six items listed above.

Figure 14.2 shows the conventional ISDN reference model. It delineates interface points for the user. In the figure NT1, or network termination 1, provides the physical layer interface; it is essentially equivalent to OSI layer 1. The functions of the physical layer include:

- Transmission facility termination
- Layer 1 maintenance functions and performance monitoring
- Timing
- Power transfer
- Layer 1 multiplexing
- Interface termination, including multidrop termination employing layer 1 contention resolution

Network termination 2 (NT2) can be broadly associated with OSI layers 1, 2, and 3. Among the examples of equipment that provide NT2 functions are user controllers, servers, LANs, and PABXs. Among the NT2 functions are:

- Layers 1, 2, and 3 protocol processing
- Multiplexing (layers 2 and 3)
- Switching
- Concentration
- Interface termination and other layer 1 functions
- Maintenance functions

A distinction must be drawn here between North American and European practice. Historically, telecommunication administrations in Europe have been, in general, national monopolies that are government-controlled. In North America (i.e.,

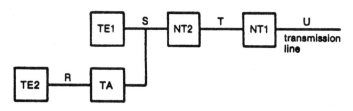

Figure 14.2. ISDN reference model.

United States and Canada) they are private enterprises, often very competitive. Thus in Europe, NT1 is considered as part of the digital network and belongs to the telecommunications administration. The customer ISDN equipment starts at the T interface. In North America, both NT1 and NT2 belong to the ISDN user, and the U interface defines the network entry point.

It should be noted that there is an overall trend outside of North America to privatize telecommunications such as has happened in the United Kingdom and is scheduled to take place in other countries such as Germany, Mexico, and Venezuela.

TE1 in Figure 14.2 represents the terminal equipment and has an interface that complies with ISDN terminal–network interface specifications at the S interface. We will call this equipment *ISDN compatible*. TE1 covers functions broadly belonging to OSI layer 1 and higher OSI layers. Among the equipment items are digital telephones, computer workstations, and other devices in the user end-equipment category that are ISDN compatible.

TE2 in Figure 14.2 refers to equipment that does *not* meet ISDN terminal–network interface at point S. TE2 adapts the equipment to meet the ISDN terminal–network interface. This process is assisted by TA, the terminal adapter.

Reference points T, S, and R are used to identify the interface available at those points. T and S are identical electrically, mechanically, and from the point of view of protocol. Point R relates to the TA interface or, in essence, it is the interface of that nonstandard (i.e., non-ISDN) device. The U-interface is peculiar to the North American version of ISDN.

We will return to user–network interfaces once the stage is set for ISDN protocols looking into the network from the user.

4   ISDN PROTOCOLS AND PROTOCOL ISSUES

ISDN was originally designed for both circuit-switching and packet-switching. In most countries packet-switching did not mature. Circuit-switching is available in a number of telephone companies and administrations, particularly in Europe. In all cases, the B-channel is fully transparent to the network, permitting the user to utilize any protocol or bit sequence so there is end-to-end agreement on the protocol utilized. Of course, the protocol itself, in the body of the info field, should be transparent to bit sequences. We assume 8-bit sequences.

It is the D-channel that carries the circuit-switching control function for its related B-channels. Whether it is the 16-kbit/s D-channel associated with BRI or the 64-kbit/s D-channel associated with PRI, it is that channel which transports the signaling information from the user's ISDN terminal from NT to the first serving telephone exchange of the telephone company or administration. Here the D-channel signaling information is converted over to CCITT No. 7 signaling data employing ISUP (ISDN User Part) of SS No. 7. Thus it is the D-channel's responsibility for call establishment (setup), supervision, termination (takedown), and all other functions dealing with network access and signaling control.

Figure 14.3 is a simplified conceptual diagram of ISDN circuit switching. It shows the B-channel riding on the public digital network and the D-channel, which is used for signaling. Of course, the D-channel is a separate channel. It is converted to CCITT Signaling System No. 7 signaling structure and in this form may traverse several signal transfer points (STPs) and may be quasi-associated or disassociated from its companion B-channel(s). Figure 14.4 is a more detailed diagram of the same ISDN circuit-switching concept. The reader should note the following in the figure. (1) Only users at each end have a peer-to-peer relationship available for all seven OSI layers of the B-channel. As the call is routed through

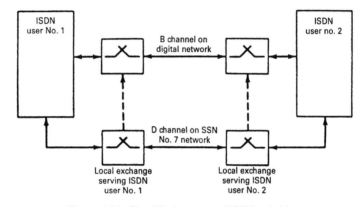

Figure 14.3. Simplified concept of ISDN switching.

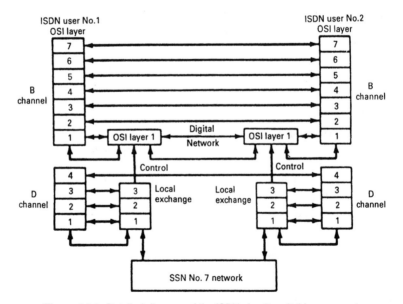

Figure 14.4. Detailed diagram of the ISDN circuit-switching concept.

the system, there is only layer 1 (physical layer) interaction at each switching node along the call route. (2) The D-channel requires the first three OSI layers for call setup to the local switching center at each end of the circuit. (3) The D-channel signaling data are turned over to CCITT Signaling System No. 7 (SS No. 7) at the near- and far-end local switching centers. (4) SS No. 7 also utilizes the first three OSI layers for circuit establishment, which requires the transfer of control information. In SS No. 7 terminology, this is called the *message transfer part*. There is a fourth layer of SS No. 7 called the *user part*. Today, among the three user parts, the most common is the ISDN user part (ISUP).

5   ISDN NETWORKS

In this context, ISDN networking is seen as a group of access attributes connecting an ISDN user at either end of the connection to the local serving exchange (i.e., the local switch). This is shown in Figure 14.5, the basic architectural model of ISDN.

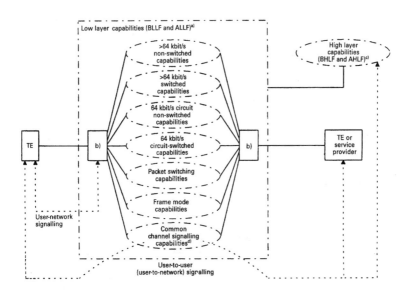

TE   Terminal equipment
BLLF   Basic low layer functions
ALLF   Additional low layer functions
BHLF   Basic high layer functions
AHLF   Additional high layer functions

a) In certain national situations, ALLF may also be implemented outside the ISDN, in special nodes or in certain categories of terminals.
b) The ISDN local functional capabilities correspond to functions provided by a local exchange and possibly including other equipment, such as electronic cross connect equipment, muldexes, etc.
c) These functions may either be implemented within ISDN or be provided by separate networks. Possible applications for basic high layer functions and for additional high layer functions are contained in Recommendation 1.210.
d) For signalling between international ISDNs, CCITT Signalling System No. 7 shall be used.

Figure 14.5. The basic architectural model of ISDN. From Figure I/I.324, page 3, ITU-T Rec. I.324, 1991 [3].

Figure 14.6 shows the ISDN reference configuration of public connection type. Figure 14.7 illustrates the access connection element model. Figure 14.8 shows the national tandem/transit connection element model. Figure 14.9 illustrates the private ISDN access connection element.

The national transit network (ITU-T terminology) is the public switched digital network. The network, whether DS1-DS4-based or E1-E5-based, provides the two necessary attributes for ISDN compatibility:

1. 64-kbit/s channelization
2. Separate channel signaling based on CCITT Signaling System No. 7

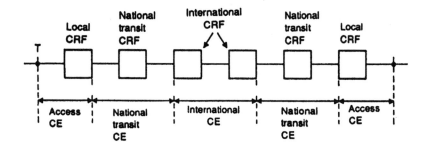

IRP   Internal reference point
CRF   Connection related functions
CE   Connection element

Figure 14.6. Reference configuration of public ISDN connection type. From Figure 3/I.324, page 8, ITU-T Rec. I.324, 1991 [3].

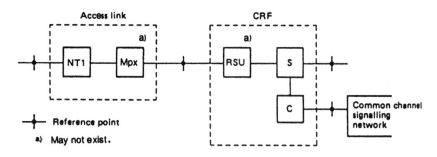

NT1   Network termination 1
S   64 kbit/s circuit switch
C   Signalling handling and exchange control functions
Mpx   (Remote) multiplexer
RSU   Remote switching unit and/or concentrator
CRF   Connection related function

Figure 14.7. Access connection element model. From Figure 4/I.324, page 9, ITU-T Rec. I.324, 1991 [3].

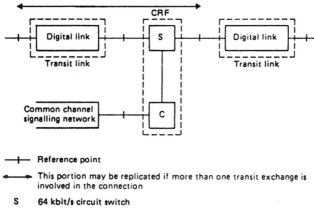

Figure 14.8. National tandem/transit element model. From Figure 5/I.324, page 10, ITU-T Rec. I.324, 1991 [3].

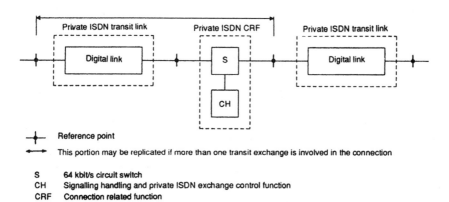

Figure 14.9. Private ISDN access connection element. From Figure 8/I.324, page 11, ITU-T Rec. I.324, 1991 [3].

Connections from the user at the local connecting exchange interface include:

- Basic service (BRI) $2B + D = 192$ kbits/s (ITU-T specified)
  $$160 \text{ kbits/s (North American ISDN)}$$
- Primary service (PRI) $23B + D/30B + D = 1.544$ Mbits/s/2.048 Mbits/s

6   ISDN PROTOCOL STRUCTURES

6.1   ISDN and OSI

Figure 14.10 shows the ISDN relationship with OSI. OSI was discussed in Chapter 11. As seen in the figure, ISDN concerns itself with only the first three

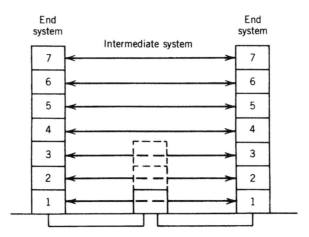

Figure 14.10. A generic communication context showing ISDN's relationship with the seven-layer OSI model. Note that the end system protocol blocks may reside in the subscriber's TE or network exchanges or other ISDN equipment.

OSI layers. OSI layers 4 to 7 are peer-to-peer connections and the end-user's responsibility. Remember that the B-channel is concerned with OSI layer 1 only. We showed the one exception; that is, when the B-channel is used for packet service, it has to interface with the first three OSI layers.

The D-channel with its signaling and control functions is the exception to the above statement. The D-channel interfaces with CCITT Signaling System No. 7 at the first serving exchange. D-channels handle three types of information: signaling (s), interactive data (p), and telemetry (t).

The layering of the D-channel has followed the intent of the OSI reference model. The handling of the p and t data can be adapted to the OSI model; the s data, by its very nature, cannot.* Figure 14.11 shows the correspondence between the D-channel signaling protocols, SS No. 7 levels, and the OSI 7-layer model.

6.2 Layer 1 Interface, Basic Rate

The S/T interface of the reference model, Figure 14.2 (or layer 1 physical layer), requires a balanced metallic transmission medium (i.e., copper pair) for each direction of transmission (four-wire) capable of supporting 192 kbits/s. Again, this is the NT interface of the ISDN reference model.

---

* For CCITT No. 7 signaling system, like any other signaling system, the primary quality-of-service measure is "post dial delay." This is principally the delay in call setup. To reduce the delay time as much as possible, it is incumbent upon system engineers to reduce processing time as much as possible. Thus SS No. 7 truncates OSI to 4 layers, because each additional layer implies more processing time.

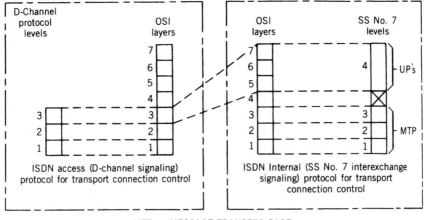

MTP = MESSAGE TRANSFER PART
UP = USER PART

Figure 14.11. Correspondence among the ISDN D-channel, CCITT Signaling System No. 7 and the OSI model. From Ref. 4, copyright IEEE, 1985.

Layer 1 provides the following services to layer 2 for ISDN operation.

- The transmission capability by means of appropriately encoded bit streams of the B- and D-channels and also any timing and synchronization functions that may be required.
- The signaling capability and the necessary procedure to enable customer terminals and/or network terminating equipment to be deactivated when required and reactivated when required.
- The signaling capability and necessary procedures to allow terminals to gain access to the common resource of the D-channel in an orderly fashion while meeting the performance requirements of the D-channel signaling system.
- The signaling capability and procedures and necessary functions at layer 1 to enable maintenance functions to be performed.
- An indication to higher layers of the status of layer 1.

6.2.1 *Primitives Between Layer 1 and Other Entities.* A *primitive* is data relating to the development or use of software that is employed in developing measures or quantitative descriptions of software. Primitives are directly measurable or countable, or may be given a constant value or condition for a specific measure (IEEE definition [5]).

In this discussion, primitives represent in an abstract way the logical exchange of information and control between layer 1 and other entities or interfaces.

TABLE 14.1 Primitives Associated with Layer 1

Generic	Specific Name		Parameter		Message Unit Content
	Request	Indication	Priority Indicator	Message Unit	
L1 ↔ L2					
PH-Data	X[a]	X	X[b]	X	Layer 2 peer-to-peer message
PH-Activate	X	X	—	—	
PH-Deactivate	—	X	—	—	
M ↔ L1					
MPH-Error	—	X	—	X	Type of error or recovery from a previously reported error
MPH-Activate	—	X	—	—	
MPH-Deactivate	X	X	—	—	
MPH-Information	—	X	—	X	Connected/disconnected

[a] PH-Data request implies underlying negotiation between layer 1 and layer 2 for the acceptance of the data.
[b] Priority indication applies only to the request type.
Source: Table 1/I.430, page 2, ITU-T Rec. I.430, November 1995 [6].

The primitives to be passed across the boundary between layers 1 and 2 or to the management entity are defined and summarized in Table 14.1. The parameter values associated with these primitives are also summarized in the table. For further information the reader may consult ITU-T Rec. I.211 [19] which describes the syntax and use of these primitives.

6.2.2 Interface Functions. The S and T functions for the BRI consist of three bit streams that are time-division multiplexed: two 64-kbit/s B-channels and one 16-kbit/s D-channel for an aggregate bit rate of 192 kbits/s. Of this 192 kbits/s, the 2B + D configuration accounts for only 144 kbits/s. The remaining 48 kbits/s are overhead bits whose function will be briefly described below.

The functions covered at the interface include bit timing at 192 kbits/s to enable the TE and NT to recover information from the aggregate bit stream. This octet timing provides 8-kHz octet timing for the NT and TE to recover the time-division multiplexed channels (i.e., 2B + D multiplexed). Other functions include D-channel access control, power feeding, deactivation, and activation.

Interchange circuits are required of which there is one in either direction of transmission (i.e., to and from the NT); they are used to transfer digital signals across the interface. All of the functions described above, except for power feeding, are carried out by means of a digitally multiplexed signal described in the next section.

6.2.3 Frame Structure. In both directions the bits are grouped into frames of 48 bits each. The frame structure is identical for all configurations whether point-to-point or point-to-multipoint. However, the frame structure is different for

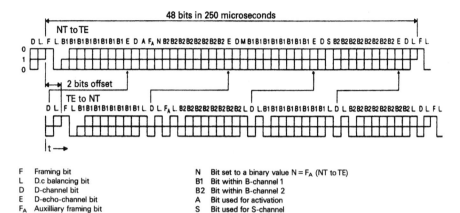

Figure 14.12. Frame structures at reference points S and T. From Figure 3/I.430, page 7, ITU-T Rec. I.430, November 1995 [6].

TABLE 14.2 Explanatory Notes to Frame Structure, Direction TE to NT

Bit Position	Group
1 and 2	Framing signal with balance bit
3 to 11	B1-channel (first octet) with balance bit
12 and 13	D-channel bit with balance bit
14 and 15	F_A auxiliary framing bit for Q bit with balance bit
16 to 24	B2-channel (first octet) with balance bit
25 and 26	D-channel bit with balance bit
27 to 35	B1-channel (second octet) with balance bit
36 and 37	D-channel bit with balance bit
38 to 46	B2-channel (second octet) with balance bit
47 and 48	D-channel bit with balance bit

Source: Table 2/I.430, page 7, ITU-T Rec. I.430, November 1995 [6].

each direction of transmission. These structures are shown in Figure 14.12, with explanatory notes given in Table 14.2 for direction TE to NT and in Table 14.3 for direction NT to TE.

6.2.4 Line Code (ITU-T). For both directions of transmission, pseudoternary coding is used with 100% pulse width as shown in Figure 14.13. Coding is performed such that a binary 1 is represented by no line signal, whereas a binary 0 is represented by a positive or negative pulse. The first binary signal following

TABLE 14.3 Explanatory Notes to Frame Structure, Direction NT to TE

Bit Position	Group
1 and 2	Framing signal with balance bit
3 to 10	B1-channel (first octet)
11	E-, D-echo channel bit
12	D-channel bit
13	Bit A used for activation
14	F_A auxiliary framing bit
15	N bit (coded as defined in Section 6.3 of the reference document)
16 to 23	B2-channel (first octet)
24	E-, D-echo channel bit
25	D-channel bit
26	M, multiframing bit
27 to 34	B1-channel (second octet)
35	E-, D-echo channel bit
36	D-channel bit
37	S
38 to 45	B2-channel (second octet)
46	E-, D-echo channel bit
47	D-channel bit
48	Frame balance bit

Note: The use of the S bit is optional, and when not used it is set to binary ZERO.

Source: Table 3/I.430, page 8, ITU-T Rec. I.430, November 1995 [6].

Figure 14.13. Pseudoternary line code, example of application.

the framing balance bit is the same polarity as the balance bit. Subsequent binary 0s alternate in polarity. A balance bit is a 0 if the number of 0s following the previous balance bit is odd. A balance bit is a binary 1 if the number of 0s following the previous balance bit is even. The balance bits tend to limit the buildup of a DC component on the line.

6.2.5 Timing Considerations.
The NT derives its timing from the network clock. A TE synchronizes its bit, octet, and frame timing from the NT, which has derived its timing from the ISDN bit stream being received from the network. The NT uses this derived timing to synchronize its transmitter clock.

6.2.6 BRI Differences in the United States.
Bellcore and ANSI prepared ISDN BRI standards fairly well modified from the CCITT I Recommendation counterparts. The various PSTN administrations in the United States are

at variance with most other countries. Furthermore, it was Bellcore's* intention to produce equipment that was cost effective and marketable and that would easily interface with existing North American telephone plant.

One point, of course, is where the telephone company responsibilities end and customer responsibilities begin. This is called the "U" interface (see Figure 14.2). The tendency toward a two-wire interface rather than a four-wire interface is another difference. The line waveform is another. Rather than a pseudoternary line waveform, the United States uses a waveform called 2B1Q with an example shown in Figure 14.14. The line bit rate is 160 kbits/s rather than that recommended by the ITU-T Organization, namely 192 kbits/s. The 2B + D + overhead frames differ significantly [7].

It is convenient to express the 2B1Q waveform as $+3$, $+1$, -1, -3 because this indicates symmetry about zero, equal spacing between states and convenient integer magnitudes. The block synchronization word (SW) contains nine quaternary elements repeated every 1.5 ms shown as follows:

$$+3, +3, -3, -3, +3, -3, +3, +3$$

6.3 Layer 1 Interface, Primary Rate

This interface is applicable for the 1.544- or 2.048-Mbit/s data rates.

6.3.1 Interface at 1.544 Mbits/s

6.3.1.1 Bit Rate and Synchronization

NETWORK CONNECTION CHARACTERISTICS. The network delivers (except as noted below) a signal synchronized from a clock having a minimum accuracy of

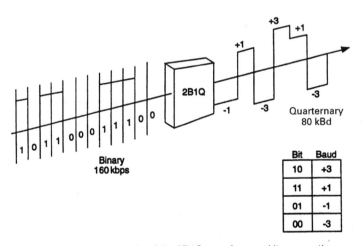

Bit	Baud
10	+3
11	+1
01	-1
00	-3

Figure 14.14. An example of the 2B1Q waveform and its generation.

* Bellcore is now called Telcordia.

1×10^{-11} (stratum 1; see Chapter 9 for stratum definition). When synchronization by a stratum 1 clock has been interrupted, the signal delivered by the network to the interface will have a minimum accuracy of 4.6×10^{-6} (stratum 3). [8]

While in normal operation, the TE1/TA/NT2 transmits a 1.544-Mbits/s signal having an accuracy equal to that of the received signal by locking the frequency of its transmitter signal to the long-term average of the incoming 1.544-Mbits/s signal, or by providing equal signal frequency accuracy from another source. ITU-T Rec. I.431 [9] advises against this latter alternative.

RECEIVER BIT STREAM SYNCHRONIZED TO A NETWORK CLOCK

(a) *Receiver Requirements.* Receivers of signals across interface I_a operate with an average transmission rate in the range of 1.544 Mbits/s $\pm$ 4.6 ppm. However, operation with a received signal transmission rate in the range of 1.544 Mbits/s $\pm$ 32 ppm is required in any maintenance state controlled by signals/messages passed over the m-bits and by AIS (alarm indication signal). In normal operation the bit stream is synchronized to stratum 1.

(b) *Transmitter Requirements.* The average transmission rate of signals transmitted across interface I_a by the associated equipment is the same as the average transmission rate of the received bit stream.

Note: The I_a and I_b interfaces are located at the input/output port of TE or NT.

TE1/TA OPERATING BEHIND AN NT2 THAT IS NOT SYNCHRONIZED TO A NETWORK CLOCK.

(a) *Receiver Requirements.* Receivers of signals across interface I_a operate with a transmission rate in the range of 1.544 Mbits/s $\pm$ 32 ppm.

(b) *Transmitter Requirements.* The transmitted signal across interface I_a is synchronized to the received bit stream.

SPECIFICATION OF OUTPUT PORTS. The signal specification for output ports is summarized in Table 14.4.

6.3.1.2 Frame Structure. The frame structure is shown in Figure 14.15.

Each time slot consists of consecutive bits, numbered 1 through 8. Each frame is 193 bits long and consists of an F-bit (framing bit) followed by 24 consecutive

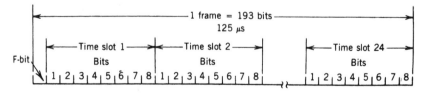

Figure 14.15. Frame structure at 1.544-Mbit/s interface.

TABLE 14.4   Digital Interface, Primary Rate, at 1.544 Mbits/s

Bit rate:		1544 kbits/s
Pair(s) in each direction of transmission:		One symmetrical pair
Code:		B8ZS[a]
Test load impedance:		100-Ω resistive
Nominal pulse shape:		See pulse mask[b]
Signal level[b,c]:	Power at 772 kHz	+12 dBm to +19 dBm
	Power at 1544 kHz	At least 25 dB below the power at 772 kHz

[a] B8ZS is modified AMI code in which eight consecutive binary ZEROs are replaced with $000 + -0 - +$ if the preceding pulse was positive $(+)$ and with $000 - +0 + -$ if the preceding pulse was negative $(-)$.

[b] The pulse mask and power level requirements apply at the end of a pair having a loss at 772 kHz of 0 to 1.5 dB.

[c] The signal level is the power level measured in a 3-kHz bandwidth at the output port for an all binary ONEs pattern transmitted.

Source: Table 4/I.431, page 13, ITU-T Rec. I.431, March 1993 [9].

time slots. The frame repetition rate is 8000 frames per second. This is described in Chapter 8, Sections 4 and 9.

Table 14.5 shows the multiframe structure (called *extended superframe* in the United States), which is 24 frames long. It takes advantage of the more advanced search algorithms for frame alignment. There are 8000 F-bits (frame alignment bits) per second (i.e., 8000 frames per second, 1 F-bit per frame). In 24 frames, with these newer strategies, only 6 bits are required for frame alignment as shown in Table 14.5. The remaining 18 bits are used as follows. There are 12 m-bits used for control and maintenance. The 6 e-bits are used for CRC6 error checking.

6.3.1.3   Time-Slot Assignment. Time slot 24 is assigned to the D-channel, when this channel is present.

A channel occupies an integer number of time slots and in the same time-slot positions in every frame. A B-channel may be assigned any time slot in the frame, an H_0-channel may be assigned any six slots in a frame in numerical order (not necessarily consecutive), and an H_{11}-channel may be assigned slots 1 to 24. The assignments may vary on a call-by-call basis.

CODES FOR IDLE CHANNELS, IDLE SLOTS, AND INTERFRAME TIME FILL.   A pattern including at least 3 binary 1s in an octet is transmitted in every time slot that is not assigned a channel (e.g., time slots awaiting channel assignment on a per-call basis, residual slots on an interface that is not fully provisioned, etc.) and in every time slot of a channel that is not allocated to a call in both directions. Interframe (layer 2) time fill consists of contiguous HDLC flags transmitted on the D-channel when its layer 2 has no frames to send.

6.3.2   Interface at 2.048 Mbits/s

6.3.2.1   Frame Structure. There are 8 bits per time slot and 32 time slots per frame, numbered 0 through 31. The number of bits per frame is 256 (i.e., 32×8) and the frame repetition rate is 8000 frames per second. Time slot 0

TABLE 14.5   Multiframe Structure (Extended Superframe)

Multiframe Frame Number	Multiframe Bit Number	F-bits Assignments		
		FAS	m-bits	CRC
1	1	—	m	—
2	194	—	—	e_1
3	387	—	m	—
4	580	0	—	—
5	773	—	m	—
6	966	—	—	e_2
7	1159	—	m	—
8	1352	0	—	—
9	1545	—	m	—
10	1738	—	—	e_3
11	1931	—	m	—
12	2124	1	—	—
13	2317	—	m	—
14	2510	—	—	e_4
15	2703	—	m	—
16	2896	0	—	—
17	3089	—	m	—
18	3282	—	—	e_5
19	3475	—	m	—
20	3668	1	—	—
21	3861	—	m	—
22	4054	—	—	e_6
23	4247	—	m	—
24	4440	1	—	—

Source: Table 5/I.431, page 15, ITU-T Rec. I.431, March 1993 [9].

provides frame alignment and time slot 16 is assigned to the D-channel when that channel is present. A channel occupies an integer number of time slots and the same time-slot position in every frame. A B-channel may be assigned any time slot in the frame and an H_0-channel may be assigned any six time slots, in numerical order, not necessarily consecutive. The assignment of time slots may vary on a call-by-call basis. An H_{12}-channel is assigned time slots 1 to 15 and 17 to 31 in a frame. Time slots 1 to 31 provide bit-sequence-independent transmission. See Chapter 8 for more complete discussion of DS1 and E1.

6.3.2.2   Timing Considerations. The NT derives its timing from the network clock. The TE synchronizes its timing (bit, octet, framing) from the signal received from the NT and synchronizes accordingly the transmitted signal. In an unsynchronized condition—that is, when the access that normally provides network timing is unavailable—the frequency deviation of the free-running clock shall not exceed ±50 ppm. A TE shall be able to detect and to interpret the input signal within a frequency range of ±50 ppm.

Any TE which provides more than one interface is declared to be a multiple access TE and is capable of taking the synchronizing clock frequency from its internal clock generator from one or more than one access (or all access links) and synchronize the transmitted signal at each interface accordingly.

6.3.2.3   Codes for Idle Channels and Idle Time Slots, Interframe Fill. A pattern including at least three binary 1s in an octet is transmitted in every time slot that is not assigned to a channel (e.g., time slots awaiting channel assignment on a per-call basis, residual slots on an interface that is not fully provisioned, etc.), and in every time slot of a channel that is not allocated to a call in both directions.

Interframe (layer 2) time fill consists of contiguous HDLC flags (01111110) which are transmitted on the D-channel when its layer 2 has no frames to send.

Frame alignment and CRC procedures can be found in ITU-T Rec. G.706, paragraph 4. Also see Refs. 10 and 18.

7   LAYER 2 INTERFACE: LINK ACCESS PROCEDURE FOR THE D-CHANNEL

The link access procedure (LAP) for the D-channel (LAPD) is used to convey information between layer 3 entities across the ISDN user–network interface using the D-channel.

A *service access point* (SAP) is a point at which the data-link layer provides services to its next higher OSI layer or layer 3. Associated with each data-link layer (OSI layer 2) is one or more data-link connection endpoints (see Figure 14.16). A data-link connection endpoint is identified by a data-link connection endpoint identifier, as seen from layer 3, and by a data-link connection identifier (DLCI), as seen from the data-link layer.

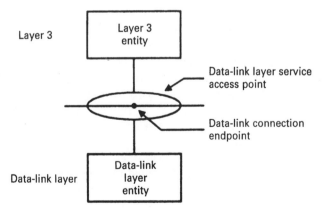

Figure 14.16. Entities, service access points (SAPs), and endpoints. From Figure 2/Q.920, page 2, ITU-T Rec. Q.920, March 1993 [11].

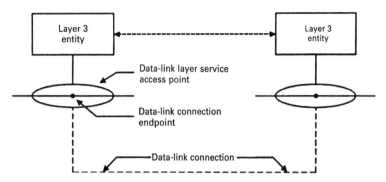

Figure 14.17. Peer-to-peer relationship. From Figure 3/Q.920, page 3, ITU-T Rec. Q.920, March 1993 [11].

Cooperation between data-link entities is governed by a specific protocol to the applicable layer. In order for information to be exchanged between two or more layer 3 entities, an association must be established between layer 3 entities in the data-link layer using a data-link layer protocol. This association is provided by the data-link layer between two or more SAPs, as shown in Figure 14.17. Data-link message units are conveyed between data-link layer entities by means of a physical connection. Layer 3 uses *service primitives* to request service from the data-link layers. A similar interaction takes place between layer 2 and layer 1.

Between the data-link layer and its adjacent layers there are four types of service primitives:

1. Request
2. Indication
3. Response
4. Confirm

These functions are shown diagrammatically in Figure 14.18.

The REQUEST primitive is used where a higher layer is requesting service from the next lower layer. The INDICATION primitive is used by a layer providing service to notify the next higher layer of activities related to the REQUEST primitive. The RESPONSE primitive is used by a layer to acknowledge receipt from a lower layer of the INDICATION primitive. The CONFIRM primitive is used by the layer providing the requested service to confirm that the requested activity has been completed.*

Figure 14.19 shows the data-link layer reference model. The data-link layer messages are transmitted in frames delimited by flags, where a flag is a unique sequence bit pattern. The frame structure, as defined in ITU-T Rec. I.441 (Q.921 [12]), is briefly described in this section.

---

* Remember that LAPD is a direct derivative of HDLC described in Chapter 11.

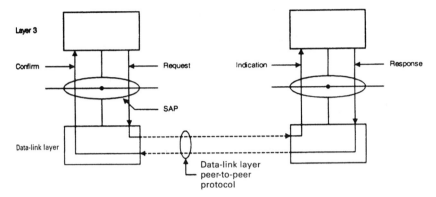

Figure 14.18. Functions of service primitives action sequence. *Note*: The same principle applies for data-link layer–physical layer interactions. From Figure 4/Q.920, page 4, ITU-T Rec. Q.920, March 1993 [11].

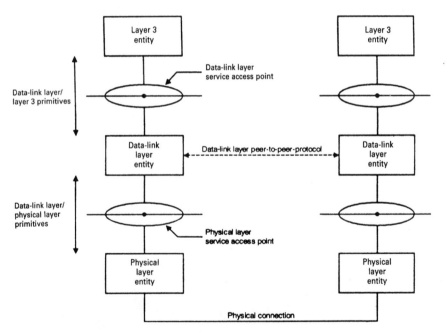

Figure 14.19. Data-link layer reference model. From Figure 5/Q.920, page 5, ITU-T Rec. Q.920, March 1993 [11].

The LAPD includes functions for:

1. The provision of one or more data-link connections on a D-channel. Discrimination between the data-link connections is by means of a data-link connection identifier (DLCI) contained in each frame.

2. Frame delimiting, alignment, and transparency, allowing recognition of a sequence of bits transmitted over a D-channel as a frame.

3. Sequence control, which maintains the sequential order of frames across a data-link connection.

4. Detection of transmission, format, and operational errors on a data link.

5. Recovery from detected transmission, format, and operational errors. Notification to the management entity of unrecovered errors.

6. Flow control.

There is unacknowledged and acknowledged operation. With unacknowledged operation, information is transmitted in unnumbered information (UI) frames. At the data-link layer the UI frames are unacknowledged. Transmission and format errors may be detected, but no recovery mechanism is defined. Flow control mechanisms are also not defined. With acknowledged operation, layer 3 information is transmitted in frames that are acknowledged at the data-link layer. Error recovery procedures based on retransmission of unacknowledged frames are specified. For errors that cannot be corrected by the data-link layer, a report to the management entity is made. Flow control procedures are also defined.

Unacknowledged operation is applicable for point-to-point and broadcast information transfer. However, acknowledged operation is applicable only for point-to-point information transfer.

There are two forms of acknowledged information that are defined:

- Single-frame operation
- Multiframe operation

For single-frame operation, layer 3 information is sent in sequenced information 0 (SI0) and sequenced information 1 (SI1) frames. No new frame is sent until an acknowledgment has been received for a previously sent frame. This means that only one unacknowledged frame may be outstanding at a time. With multiple-frame operation, layer 3 information is sent in numbered information (I) frames. A number of I frames may be outstanding at the same time. Multiple-frame operation is initiated by a multiple-frame establishment procedure using set asynchronous balanced mode/set asynchronous balanced mode extended (SABM/SABME) command.

7.1 Layer 2 Frame Structure for Peer-to-Peer Communication

There are two frame formats used on layer 2 frames:

1. Format A, for frames where there is no information field.
2. Format B, for frames containing an information field.

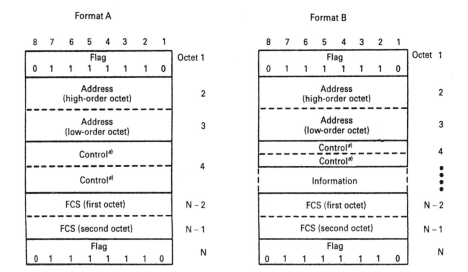

Figure 14.20. Frame formats or layer 2 frames. From Figure 1/Q.921, page 20, ITU-T Rec. Q.921, September 1997 [12].

These two frame formats are illustrated in Figure 14.20. The following discussion briefly describes the frame content (sequences and fields) for the LAPD (layer 2) frame.

Flag Sequence. All frames start and end with a flag sequence consisting of one 0 bit followed by six contiguous 1 bits and one 0 bit. These flags are called the *opening* and *closing* flags. If sequential frames are transmitted, the closing flag of one frame is the opening flag of the next frame.

Address Field. As illustrated in Figure 14.21, the address field consists of two octets and identifies the intended receiver of a command frame and the transmitter of a response frame. A single octet address field is reserved for LAPB operation to allow a single LAPB data-link connection to be multiplexed along with the LAPD data-link connections.

Control Field. The control field consists of one or two octets. It identifies the type of frame, either command or response. It contains sequence numbers where applicable. Three types of control field formats are specified:

1. Numbered information transfer (I format)

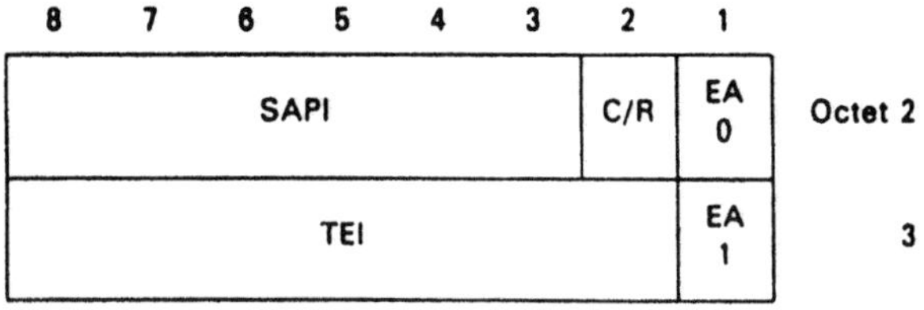

8	7	6	5	4	3	2	1	
SAPI						C/R	EA 0	Octet 2
TEI							EA 1	3

EA   — Address field extension bit
C/R — Command/response field bit
SAPI — Service access point identifier
TEI — Terminal endpoint identifier

Figure 14.21. LAPD address field format. From Figure 5/Q.921, page 23, ITU-T Rec. Q.921, September 1997 [12].

2. Supervisory functions (S format)
3. Unnumbered information transfers and control functions (U format)

Information Field. The information field of a frame, when present, follows the control field and precedes the frame check sequence (FCS). The information field contains an integer number of octets:

- For a SAP supporting signaling, the default value is 128 octets.
- For SAPs supporting packet information, the default value is 260 octets.

Frame Check Sequence (FCS) Field. The FCS field is a 16-bit sequence and is the 1's complement of the modulo-2 sum of:

1. The remainder of X raised to the k power:

$$X^{15} + X^{14} \ldots X^{1} + 1$$

divided by the generating polynomial

$$X^{16} + X^{12} + X^{5} + 1$$

where k is the number of bits in the frame existing between but not including the final bit of the opening flag and the first bit of the FCS, excluding bits inserted for transparency, and

2. The remainder by modulo-2 division by the generating polynomial given above of the product of X^{16} by the content of the frame defined above.

Transparency, mentioned above, ensures that a flag or abort sequence is not initiated within a frame. On the transmit side the data-link layer examines the frame content between the opening and closing flag sequences and inserts a 0 bit after all sequences with five contiguous 1 bits (including the last five bits of the FCS). On the receive side the data-link layer examines the frame contents between the opening and closing flag sequences and discards any 0 bit that directly follows five contiguous 1 bits. [See Ref. 10]

ADDRESS FIELD FORMAT. The address field is illustrated in Figure 14.21. It contains address field extension bits (EA), a command-response indication bit (C/R), a data-link layer service access point identifier (SAPI) subfield, and a terminal endpoint identifier (TEI) subfield.

Address Field Extension Bit (EA). The address field range is extended by reserving the first transmitted bit of the address field to indicate the final octet of the address field. The presence of a 1 in the first bit position of an address field octet signals that it is the final octet of the address field. The double octet address field for LAPD operation has bit 1 of the first octet set to a 0 and bit 1 of the second octet set to a 1.

Command/Response Field Bit (C/R). The C/R bit identifies a frame as either a command or a response. The user side sends commands with the C/R bit set to 0, and it sends responses with the C/R bit set to 1. The network side does the opposite; that is, commands are sent with the C/R bit set to 1, and responses are sent with the C/R bit set to 0.

In keeping with HDLC rules,* commands use the address of the peer data-link entity while responses use the address of their own data-link layer entity. In accordance with these rules, both peer entities on a point-to-point data-link connection use the same data-link connection identifier (DLCI) composed of an SAPI-TEI where SAPI and TEI conform to the definitions given below.

Service Access Point Identifier (SAPI). The SAPI identifies a point at which data-link services are provided by a data-link layer entity to a layer 3 or management entity. Consequently, the SAPI specifies a data-link layer entity that should process a data-link layer frame. The SAPI allows 64 service access points (SAPs) to be specified, where bit 3 of the address field octet containing the SAPI is the least significant binary digit and bit 8 is the most significant. The SAPI values are allocated as shown in Table 14.6.

Terminal Endpoint Identifier (TEI). The TEI for a point-to-point link connection may be associated with a single terminal equipment (TE). A TE may contain one or more TEIs used for point-to-point data transfer. The TEI for a broadcast data-link connection is associated with all user side data-link layer entities containing the same SAPI. The TEI subfield allows 128 values where bit 2 of the address field octet containing the TEI is the least significant bit

---

* LAPD, as we know, is a derivative of HDLC.

TABLE 14.6   Allocation of SAPI Values

SAPI Value	Related Layer 3 or Management Entity
0	Call control procedures
1–15	Reserved for future standardization
16	Packet communication conforming to X.25 level 3 procedures
17–31	Reserved for future standardization
63	Layer 2 management procedures
All others	Not available for Q.921 procedures

Source: ITU-T Rec. Q.921, Table 2/Q.921, page 7, September 1997 [12].

and bit 8 is the most significant bit. The following conventions apply for assignment of these values:

1. TEI for broadcast data-link connection. The TEI subfield bit pattern 111 1111(= 127) is defined as the group TEI. The group TEI is assigned to the broadcast data-link connection associated with the addressed SAP.
2. TEI for point-to-point data-link connection. The remaining TEI values are used for point-to-point data-link connections associated with the addressed SAP. The range of TEI values is as follows:

Values 0–63:   Nonautomatic TE1 assignment user equipment
Values 64–126:   Automatic TEI assignment user equipment

CONTROL FIELD FORMATS. Control field formats are shown in Table 14.7. The control field identifies the type of frame, either a command or a response. The control field contains sequence numbers, where applicable. Three types of control field formats are specified: numbered information transfer (I-format), supervisory functions (S-format) and unnumbered information transfers and control functions (U-format).

TABLE 14.7   Control Field Formats

Control Field Bits (Modulo 128)	8	7	6	5	4	3	2	1	
I format				N(S)				0	Octet 4
				N(R)				P	5
S format	X	X	X	X	S	S	0	1	Octet 4
				N(R)				P/F	5
U format	M	M	M	P/F	M	M	1	1	Octet 4

N(S)	Transmitter send sequence number	M	Modifier function bit
N(R)	Transmitter receive sequence number	P/F	Poll bit when issued as a command, final bit when issued as a response
S	Supervisory function bit	X	Reserved and set to 0

Source: Table 4/Q.921, page 28, CCITT Rec. Q.921, September 1997 [12].

7.2 LAPD Primitives

The following comments clarify the semantics and usage of primitives. Primitives consist of commands and their respective responses associated with the services requested of a lower layer. The general syntax of a primitive is

XX-generic name-type: parameters

where XX designates the layer providing the service. For the data-link layer, XX is DL, PH for the physical layer, or MDL for the management entity to the data-link layer interface, Table 14.8 gives the primitives associated with the data-link layer.

8 OVERVIEW OF LAYER 3

The layer 3 protocol, of course, deals with the D-channel and its signaling capabilities. It provides the means to establish, maintain, and terminate network connections across an ISDN between communicating application entities. A more detailed description of the layer 3 protocol may be found in ITU-T Rec. Q.931 [14].

Layer 3 utilizes functions and services provided by its data-link layer, as described in Section 7 under LAPD functions. These necessary layer 2 support functions are listed and briefly described below:

- Establishment of the data-link connection
- Error-protected transmission of data
- Notification of unrecoverable data-link errors
- Release of data-link connections
- Notification of data-link layer failures
- Recovery from certain error conditions
- Indication of data-link layer status

Layer 3 performs two basic categories of functions and services in the establishment of network connections. The first category directly controls the connection establishment. The second category includes those functions relating to the transport of messages in addition to the functions provided by the data-link layer. Among these additional functions are the provision of rerouting of signaling messages on an alternative D-channel (where provided) in the event of D-channel failure. Other possible functions include multiplexing and message segmenting and blocking. The D-channel layer 3 protocol is designed to carry out establishment and control of circuit-switched and packet-switched connections. Also, services involving the use of connections of different types, according to user specifications, may be provided through "multimedia" call control procedures. Functions performed by layer 3 include:

TABLE 14.8 Primitives Associated with the Data-Link Layer

Generic name	Type				Parameters		Message Unit Contents
	Request	Indication	Response	Confirm	Priority Indicator	Message Unit	
L3 ↔ L2							(Note 1)
DL-Establish	X	X	—	X	—	—	
DL-Release	X	X	—	X	—	—	
DL-Data	X	X	—	—	—	X	Layer 3 (peer-to-peer message)
DL-Unit Data	X	X	—	—	—	X	Layer 3 (peer-to-peer message)
M ↔ L2							
MDL-Assign	X	X	—	—	—	X	TEI value, CES (Note 2)
MDL-Remove	X	—	—	—	—	X	TEI value, CES
MDL-Error	—	X	X	—	—	X	Reason for error message
MDL-Unit Data	X	X	—	—	—	X	Management function peer-to-peer message
MDL-XID	X	X	X	X	—	X	Connection management PDU (peer-to-peer XIO frame)
L2 ↔ L1							
PH-Data	X	X	—	—	X	X	Data-link layer peer-to-peer frame
PH-Activate	X	X	—	—	—	—	
PH-Deactivate	—	X	—	—	—	—	
M ↔ L1							
MPH-Activate	—	X	—	—	—	—	
MPH-Deactivate	X	X	—	—	—	—	
MPH-Information	—	X	—	—	—	X	Connected/ disconnected

L3 ↔ L2: Layer 3/data-link layer boundary   X Exists
L2 ↔ L1: Data-link layer/physical layer boundary   — Does not exist
M ↔ L2: Management entity/data-link layer boundary
M ↔ L1: Management entity/physical layer boundary
Note 1: Although not shown below, the CES is implicitly associated with each *L3 ↔ L2* primitive, indicating the applicable connection endpoint.
Note 2: TEI value is included only in the MDL-ASSIGN request.

Source: Table 6/Q.921, page 15, CCITT Rec. Q.921, September 1997 [12].

1. The processing of primitives for communicating with the data-link layer.
2. Generation and interpretation of layer 3 messages for peer level communications.
3. Administration of timers and logical entities (e.g., call references) used in call control procedures.
4. Administration of access resources, including B-channels and packet-layer logical channels (e.g., ITU-T X.25).
5. Checking to ensure that services provided are consistent with user requirements, such as compatibility, address, and service indicators.

The following functions may also be performed by layer 3:

1. *Routing and Relaying.* Network connections exist either between users and ISDN exchanges or between users. Network connections may involve intermediate systems which provide relays to other interconnecting sub-networks and which facilitate interworking with other networks. Routing functions determine an appropriate route between layer 3 addressees.
2. *Network Connection.* This function includes mechanisms for providing network connections making use of data-link connections provided by the data-link layer.
3. *Conveying User Information.* This function may be carried out with or without the establishment of a circuit-switched connection.
4. *Network Connection Multiplexing.* Layer 3 provides multiplexing of call control information for multiple calls onto a single data-link connection.
5. *Segmenting and Reassembly (SAR).* Layer 3 may segment and reassemble layer 3 messages to facilitate their transfer across user–network interface.
6. *Error Detection.* Error detection functions are used to detect procedural errors in the layer 3 protocol. Error detection in layer 3 uses, among other information, error notification from the data-link layer.
7. *Error Recovery.* This includes mechanisms for recovering from detected errors.
8. *Sequencing.* This includes mechanisms for providing sequenced delivery of layer 3 information over a given network connection when requested. Under normal conditions, layer 3 ensures the delivery of information in the sequence it is submitted by the user.
9. *Congestion Control and User Data Flow Control.* Layer 3 may indicate rejection or unsuccessful indication for connection establish requests to control congestion within a network. Typical is the congestion control message to indicate the establishment or termination of flow control on the transmission of User Information messages.
10. *Restart.* This function is used to return channels and interfaces to an idle condition to recover from certain abnormal conditions.

8.1 Layer 3 Specification

The layer 3 specification is contained in ITU-T Rec. Q.930/931 [13, 14]. It includes both circuit-switched and packet-switched operation. There are 23 message types for circuit-mode connection control. These are shown in Table 14.9, and the content elements of each is given in over 50 tables in the reference specification. One typical table from the group is presented in Table 14.10, Setup Message Content. Additional tables are provided in the recommendation for packet switching.

The following are several explanatory notes for the tables found in ITU-T Rec. Q.931 and are valid for Table 14.10 as well. The letters "M" and "O" mean *mandatory* and *optional*, respectively. The letter "n" and "u" refer to *network* and *user*, respectively, and give the direction of traffic such as n → u (network to user) and u → n (user to network). An asterisk (*) in the table means undefined length.

TABLE 14.9   Messages for Circuit-Mode Connection Control

Call establishment messages:
Alerting
Call Proceeding
Connect
Connect Acknowledge
Progress
Setup
Setup Acknowledge

Call information phase messages:
Resume
Resume Acknowledge
Resume Reject
Suspend
Suspend Acknowledge
Suspend Reject
User Information

Call clearing messages:
Disconnect
Release
Release Complete

Miscellaneous messages:
Congestion Control
Facility
Information
Notify
Status
Status Enquiry

Source: Table 3-1/Q.931, page 10, CCITT Rec. Q.931, May 1998 [14].

TABLE 14.10   Setup Message Content

Information Element	Direction	Type	Length
Protocol discriminator	Both	M	1
Call reference	Both	M	2–*
Message type	Both	M	1
Sending complete	Both	O (Note 1)	1
Repeat indicator	Both	O (Note 2)	1
Bearer capability	Both	M (Note 3)	4–13
Channel identification	Both	M (Note 4)	2–*
Facility	Both	O (Note 5)	2–*
Progress indicator	Both	O (Note 6)	2–4
Network-specific facilities	Both	O (Note 7)	2–*
Display	n → u	O (Note 8)	Note 9
Keypad facility	u → n	O (Notes 10, 12)	2–34
Signal	n → u	O (Note 11)	2–3
Switchhook	u → n	O (Note 12)	2–3
Feature activation	u → n	O (Note 12)	2–4
Feature indication	n → u	O (Note 12)	2–5
Calling party number	Both	O (Note 13)	2–*
Calling party subaddress	Both	O (Note 14)	2–23
Called party number	Both	O (Note 15)	2–*
Called party subaddress	Both	O (Note 16)	2–23
Transit network selection	u → n	O (Note 17)	2–*
Low layer compatibility	Both	O (Note 18)	2–16
High layer compatibility	Both	O (Note 19)	2–4
User–user	Both	O (Note 20)	Note 21

(continued overleaf)

TABLE 14.10 *(continued)*

Source: Table 3-16/Q.931, page 30, CCITT Rec. Q.931, May 1998 [14].

Explanatory Notes to Table 14.10

Note 1: Included if the user or the network optionally indicates that all information necessary for call establishment is included in the Setup message.

Note 2: The Repeat indicator information element is included immediately before the first Bearer capability information element when either the in-call modification procedure or the bearer capability negotiation procedure is used.

Note 3: May be repeated if the bearer capability negotiation procedure is used. For bearer capability negotiation, either two or three Bearer capability information elements may be included in descending order of priority — that is, highest priority first.

Note 4: Mandatory in the network-to-user direction. Included in the user-to-network direction when the user wants to indicate a channel. If not included, its absence is interpreted as "any channel acceptable."

Note 5: May be included for functional operation of supplementary services.

Note 6: Included in the event of interworking or in connection with the provision or in-band information/patterns.

Note 7: Included by the calling user or the network to indicate network-specific facilities information.

Note 8: Included if the network provides information that can be presented to the user.

Note 9: The minimum length is 2 octets; the maximum length is network dependent and is either 34 or 82 octets.

Note 10: Either the Called party number or the Keypad facility information element is included by the user to convey called party number information to the network. The Keypad facility information element may also be included by the user to convey other call establishment information to the network.

Note 11: Included if the network optionally provides additional information describing tones.

Note 12: As a network option, may be used for stimulus operation of supplementary services.

Note 13: May be included by the calling user or the network to identify the calling user.

Note 14: Included in the user-to-network direction when the calling user wants to indicate the calling party subaddress. Included in the network-to-user direction if the calling user included a Calling party subaddress information element in the Setup message.

Note 15: Either the Called party number or the Keypad facility information element is included by the user to convey called party number information to the network. The Called party number information element is included by the network when called party number information is conveyed to the user.

Note 16: Included in the user-to-network direction when the calling user wants to indicate the called party subaddress. Included in the network-to-user direction if the calling user included a Called party subaddress information element in the Setup message.

Note 17: Included by the calling user to select a particular transit network.

Note 18: Included in the user-to-network direction when the calling user wants to pass Low layer compatibility information to the called user. Included in the network-to-user direction if the calling user included a Low layer compatibility information element in the Setup message.

Note 19: Included in the user-to-network direction when the calling user wants to pass High layer compatibility information to the called user. Included in the network-to-user direction if the calling user included a High layer compatibility information element in the Setup message.

Note 20: Included in the user-to-network direction when the calling user wants to pass user information to the called user. Included in the network-to-user direction if the calling user included a user–user information element in the Setup message.

Note 21: The minimum length is 2 octets; the standard default maximum length is 131 octets.

8.1.1 General Message Format and Information Elements Coding.

Within this protocol, every message consists of the following parts:

(a) Protocol discriminator
(b) Call reference
(c) Message type
(d) Other information elements, as required

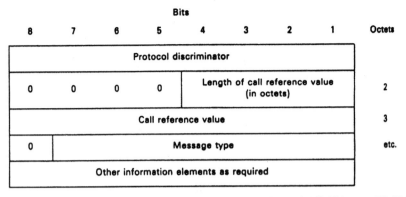

Figure 14.22. General message organization example. From Figure 4-1/Q.931, page 53, ITU-T Rec. Q.931, May 1998 [14].

Information elements (a), (b), and (c) are common to all messages and are always present, while information element (d) is specific to each message type. This organization is illustrated in the example shown in Figure 14.22.

The term "default" implies that the value defined should be used in the absence of any assignment or in the negotiation of alternative values. When a field, such as the call reference value, extends over more than one octet, the order of bit values progressively decreases as the octet number increases. The least significant bit of the field is represented by the lowest-numbered bit of the highest-numbered octet field.

PROTOCOL DISCRIMINATOR. The purpose of the protocol discriminator is to distinguish messages for user–network call control from other messages (to be defined) with ITU-T Rec. Q.931 from those OSI network layer protocol units which are coded to other ITU-T Recommendations and other standards.

CALL REFERENCE. The purpose of the call reference is to identify the call or facility registration/cancellation request at the local user–network interface to which the particular message applies. The call reference does not have end-to-end significance across ISDNs.

As a minimum, all networks and users must be able to support (1) a call reference value of one octet for a basic user–network interface and (2) a call reference value of two octets for a primary rate interface. The call reference information element includes the call reference value and the call reference flag.

Call reference values are assigned by the originating side of the interface for a call. These values are unique to the originating side within a particular D-channel layer 2 logical link connection. The call reference value is assigned at the beginning of a call and remains fixed for the lifetime of the call (except in the case of call suspension). After a call ends or after a successful suspension, the associated call reference value may be reassigned to a later call. Two identical call reference values on the same D-channel layer 2 logical link connection may be used when each value pertains to a call originated at opposite ends of the link.

The call reference flag can take the values of 0 or 1. The call reference flag is used to identify which end of the layer 2 logical link originated a call reference. The origination side always sets the call reference flag to "0." The destination always sets the call reference flag to "1."

Hence the call reference flag identifies who allocated the call reference value for this call, and the only purpose of the call reference flag is to resolve simultaneous attempts to allocate the same call reference value.

MESSAGE TYPE. The purpose of the message type is to identify the function of the message being sent. For instance, 00000101 indicates a call-setup message.

OTHER INFORMATION ELEMENTS. Forty "other information elements" are listed, such as "sending complete," "congestion level," "call identity," "date/time," and "calling party number." For numbering and numbering plan the reader should consult Refs. 15, 16, and 17.

REVIEW QUESTIONS

1. Name at least three services other than speech telephony that already had been integrated into the PSTN prior to the advent of ISDN.

2. Identify at least two shortcomings of PCM as implemented today regarding its suitability for ISDN. (*Hint*: Think standards.)

3. PCM, as designed, is optimized for speech telephony. Approximately what percentage of traffic on the PSTN is speech telephony?

4. Name at least five communication services that ISDN will support.

5. Distinguish *primary rate* and *basic rate*.

6. Define the B-channel, the D-channel (two answers here), and the three basic variants of the H-channel.

7. Give at least three applications of the H-channel (different from question 4).

8. 30B + D is the primary rate for E1 configurations. What ever happened to E1's second separate channel?

9. What is the aggregate bit rate of 2B + D? First for CCITT, then for North American ISDN.

10. Distinguish TE1 and TE2 in the standard ISDN network interface. European ISDN would have an NT12 interface. The equipment at this interface would belong to whom? In North America, this interface is split: NT1 and NT2. Discuss the difference.

11. Which OSI layers are involved with the B-channel and D-channel for a voice connection?

12. What is a TA and what purpose does it serve?

13. For ISDN to be a reality, what signaling system has to be implemented in the intervening digital PSTN?

14. Define *primitive* in the context of ISDN protocols.

15. How does an ISDN user (e.g., a TE) derive its timing?

16. Describe *pseudoternary coding* as used as a line signal for ISDN.

17. What is the function of the balancing bit in an ISDN BRI frame?

18. Why must the simulation of a frame alignment signal be prevented?

19. In the United States, a 2B1Q line signal is used where the bit rate is 160 kbits/s. What is the modulation rate (baud rate) of this signal?

20. North American ISDN practice for the BRI is two-wire for full-duplex operation. What unique feature at the U-interface allows this operation without mutual interference from the outgoing and incoming bit streams on the same wire pair?

21. ISDN PRI operates at 1.544 Mbits/s $\pm$ 4.6 ppm at the receiver. How can it maintain such an excellent stability?

22. How many B-channels can carry traffic in the normal and conventional North American PRI configuration?

23. The 1920-kbit/s H_{12}-channel occupies which time slots in an E1 configuration?

24. What are the four LAPD service primitives? Where have we seen them before?

25. Name and describe at least three of the functions carried out by ISDN layer 3.

26. What are the two types of operation which can be carried out by LAPD?

27. What are the three types of LAPD control field formats?

28. What does an SAPI specify?

REFERENCES

1. *ISDN User–Network Interfaces—Interface Structures and Access Capabilities*, CCITT Rec. I.412, Fascicle III.8, IXth Plenary Assembly, Melbourne, 1988.
2. *Guidance to the I. Series Recommendations*, CCITT Rec. I.200, Fascicle III.8, IXth Plenary Assembly, Melbourne, 1988.
3. *ISDN Network Architecture*, ITU-T Rec. I.324, ITU Telecommunications Standardization Sector, Geneva, 1991.

4. W. Stallings, ed., *Tutorial: Integrated Services Digital Networks (ISDN)*, IEEE Computer Society Press, Washington, DC, 1985.

5. *The IEEE Standard Dictionary of Electrical and Electronic Terms*, 6th ed., IEEE Press, New York, 1996.

6. *Basic User–Network Interface Layer 1 Specification*, ITU-T Rec. I.430, ITU Telecommunications Standardization Sector, Geneva, November 1995.

7. *ISDN Basic Access Transport System Requirements*, Technical Reference TR-TSY-000397, Issue 1, October 1988, Bellcore, Morristown, NJ, 1988.

8. R. L. Freeman, *Telecommunication Transmission Handbook*, 4th ed., John Wiley & Sons, New York, 1998.

9. *Primary Rate User–Network Interface—Layer 1 Specification*, ITU-T Rec. I.431, ITU Telecommunications Standardization Sector, Geneva, March 1993.

10. *Frame Alignment and Cyclic Redundancy Procedures Relating to Basic Frame Structures Defined in Rec. G.704*, CCITT Rec. G.706, Geneva, 1991.

11. *Digital Subscriber Signalling System No. 1 (DSS1): ISDN User–Network Interface Data Link Layer—General Aspects*, ITU-T Rec. Q.920, ITU Telecommunications Standardization Sector, Geneva, March 1993.

12. *ISDN User–Network Interface—Data Link Layer Specification*, ITU-T Rec. Q.921, ITU, Geneva, September 1997.

13. *ISDN User–Network Interface: Layer 3—General Aspects*, CCITT Rec. Q.930, Fascicle VI.11, IXth Plenary Assembly, Melbourne, 1988.

14. *ISDN User–Network Interface: Layer 3—For Basic Call Control*, ITU-T Rec. Q.931, ITU, Geneva, May, 1998.

15. *Numbering Plan for the ISDN Era*, ITU-T Rec. E.164, ITU, Geneva, March, 1998.

16. *BOC Notes on the LEC Networks—1994*, Issue 2, SR-TSV-002275, Bellcore, Piscataway, NJ, April 1994.

17. *International Numbering Plan for Public Data Networks*, ITU-T Rec. X.121, ITU, Geneva, October 2000.

18. R. L. Freeman, *Reference Manual for Telecommunication Engineering*, 3rd ed., John Wiley & Sons, New York, 2002.

19. *B-ISDN Service Aspects*, ITU-T Rec. I.211, ITU, Geneva, March 1993.

20. *ISDN—Network Functional Principles*, ITU-T Rec. I.310, ITU, Geneva, 1993.

15

SPEEDING THINGS UP WITH FRAME RELAY

1 INTRODUCTION

In this chapter we discuss the one high-throughput digital transmission and switching technique that is currently in wide use and will remain so into the foreseeable future. This is frame relay. Its basic application is to provide a means of LAN interconnects over longer distances [i.e., from several miles (km) to virtually around the world]. It is a higher efficiency system than X.25 packet service, with much less latency. We must admit that it is less robust having no means of inherent error correction.

This chapter is also a lead-in to Chapter 16, where we discuss the Asynchronous Transfer Mode (ATM). Many of the revolutionary concepts of ATM/B-ISDN had prior application to frame relay. These will be enumerated as we move forward in our discussion.

2 HOW CAN THE NETWORK BE SPEEDED UP?

2.1 Background and Rationale

It would seem from the terminology that somehow we were making bits travel faster down the pipe. By some means we'd broken the speed barrier by dramatically increasing the velocity of propagation. Of course this is eminently not true.

If bandwidth permits,* of course, the bit rate can be increased. That certainly will speed things up. One way to get out of the bandwidth bind is to work around the analog voice channel; use some other means. ISDN was a good step in that direction. Note that 64 kbits/s gets four or five times as many bits per second than analog voice channel data rates.

---

* There may be other constraints such as group delay, bandwidth coherence, and so forth.

Telecommunication System Engineering, by Roger L. Freeman
ISBN 0-471-45133-9   Copyright © 2004 Roger L. Freeman

Probably the greatest pressure to speed up the network came from LAN users that wished to extend LAN traffic to distant destinations. Ostensibly this traffic, as described in Chapter 13, has local transmission rates from 1 to 100 Mbits/s. X.25 WAN connectivity was one possible answer. Its packet circuits are robust, but slow and tedious. There must be a better way.

What slows down X.25 service? X.25 is feasible at 64 kbits/s and even at T1/E1 transmission rates. It was the intensive processing at every node (see Table 15.1) and continual message exchange as to the progress of packets, from node to originator and node to destination. On many X.25 connectivities, multiple nodes are involved slowing this down still further. One clue is that X.25 was designed for circuits with poor transmission performance degrading error rates, typically BERs in the range of 1×10^{-4}. Meanwhile, the underlying digital networks in North America have error performance in the range of 5×10^{-10} or better. This begs the question of removing the responsibility of error recovery from the service provider. If errors statistically appear in about 1 in 200 million bits, there is a strong argument for removing error recovery.

TABLE 15.1 Functional Comparison of X.25 and Frame Relay

Function	X.25 in ISDN (X.31)	Frame Relay
Flag recognition/generation	X	X
Transparency	X	X
FCS checking/generation	X	X
Recognize invalid frames	X	X
Discard incorrect frames	X	X
Address translation	X	X
Fill inter-frame time	X	X
Manage V(S) state variable	X	
Manage V(R) state variable	X	
Buffer packets awaiting acknowledgment	X	
Manager timer T-1	X	
Acknowledge received I-frames	X	
Check received N(S) against V(R)	X	
Generation of rejection message	X	
Respond to poll/final bit	X	
Keep track of number of retransmissions	X	
Act upon reception of rejection message	X	
Respond to receiver not ready (RNR)	X	
Respond to receiver ready (RR)	X	
Multiplexing of logical channels	X	
Management of D bit	X	
Management of M bit	X	
Management of W bit	X	
Management of P(S) packets sent	X	
Management of P(R) packets received	X	
Detection of out-of-sequence packets	X	
Management of network layer RR	X	
Management of network layer RNR	X	

In fact with frame relay the following salient points emerge:

- There is no process for error recovery by the frame relay service provider.
- The service provider does not guarantee delivery, nor are there any sort of acknowledgments provided.
- It only uses the first two OSI layers (physical and data link layer), thus removing layer 3 and its intensive processing requirements.
- Frame overhead is kept to a minimum to minimize processing time and to increase useful throughput.
- There is no control field, no sequence numbering.
- Discarding of frames: Frames are discarded without notifying originator for such reasons as congestion and having encountered an error.
- Operates on a statmultiplex concept.

In sum, the service that the network provides can be speeded up by increasing data rate, eliminating error recovery procedures and reducing processing time. One source states that a frame relay frame takes some 20 ms to reach the distant end (statistically), where an X.25 packet of similar size takes in excess of 200 ms on terrestrial circuits inside CONUS.

Another advantage of frame relay over a conventional static TDM connection is that it uses virtual connections. Data traffic is often bursty and normally would require much greater bit rate capacity to support the short data messages, and much of the time that bit rate capacity would remain idle. Virtual connections of frame relay only use the required bit rate capacity for the period of the burst or usage. This is one reason frame relay is used so widely to interconnect LANs over a WAN (wide area network). Figure 15.1 shows a typical frame relay network.

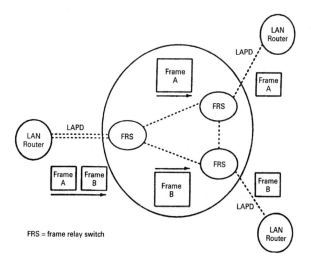

Figure 15.1. A typical frame relay network.

2.2 The Genesis of Frame Relay

Frame relay derives from the ISDN D-channel LAPD, or D-channel layer two. LAPD was discussed in Chapter 14. Its importance has taken on such a magnitude that the ITU-T organization has formulated I.122, *Framework for Mode Bearer Services* "[1] and I.233, *Frame Mode Bearer Services*" [2]. Even the term LAPD, although modified in many cases, continues to be used for many frame relay applications. However, in some implementations, LAPF has gained favor to differentiate it from its ISDN counterpart.

Frame relay has become an ANSI initiative. There is also the Frame Relay Forum (FRF), consisting of manufacturers of frame relay equipment, that many feel is leading the pack for application. So when we discuss frame relay, we must consider what specifications a certain system is designed around:

- ANSI, based on ANSI specifications and their publication dates
- Frame Relay Forum (FRF) with publication dates
- ITU-T organization and its most current recommendations

There are also equivalent ANSI specifications directly derived from ITU-T recommendations such as ANSI T1.617-1991 [3]. We will see the term *core aspects* of ISDN LAPD or *DL-CORE*. This refers to a reduced subset of LAPD found in Annex A of ITU-T Rec. Q.922 [4]. The basic body of Q.922 presents CCITT/ITU-T specification for frame relay. This derivative is called LAPF, referred to above, rather than LAPD. The material found in ANSI T1.618 [5] is identical for all intents and purposes with Annex A of Q.922.

To properly describe frame relay from our perspective, we will briefly give an overview of the ANSI T1.618-1991 and T1.606-1990 [6]. This will be followed by some fairly well identified variants.

Also see Ref. 13.

2.3 Introduction to Frame Relay

Frame relay may be considered a cost-effective outgrowth of ISDN, meeting high data rate (e.g., 2 Mbits/s) and low delay data communications requirements. Frame relay encapsulates data files. These may be considered "packets," although they are called frames. Thus frame relay is compared to CCITT Rec. X.25 packet service.

Frame relay was designed for current transmission capabilities of the network with its relatively wider bandwidths* and excellent error performance (e.g., BER better than 1×10^{-7}).

The incisive reader will note the use of the term *bandwidth*. It is used synonymously with bit rate. If we were to admit at first approximation 1 bit per hertz of bandwidth, such use is acceptable. We are mapping frame relay bits into bearer channel bits probably on a one-for-one basis. The bearer channel may be a DS0/E0 64-kbit/s channel, a 56-kbit/s channel of a DS1 configuration, or multiple

---

* We'd rather use the term *greater bit rate capacity*.

DS0/E0 channels in increments of 64 kbits/s up to 1.544/2.048 Mbits/s. We may also map the frame relay bits into a SONET or SDH configuration (Chapter 7). The final bearer channel may require more or less bandwidth than that indicated by the bit rate. This is particularly true for such bearer channels riding on radio systems and, to a lesser extent, on a fiber-optic medium or other transmission media. The reader should be aware of certain carelessness of language used in industry publications.

Frame relay works well in the data rate range from 56 kbits/s up to 45 Mbits/s. It is being considered for SONET STS-3, with a 155-Mbit/s rate for still additional *speed.*

ITU-T's use of the ISDN D-channel for frame relay favors X.25-like switched virtual circuits (SVCs). However, ANSI recognized that the principal application of frame relay was interconnection of LANs, and not to replace X.25. Because of the high data rate of LANs (megabit to gigabit range), dedicated connections are favored. ANSI thus focused on permanent virtual connections (PVCs). With PVCs, circuits are set up by the network,* not by the end points. This notably simplified the signaling protocol. Also, ANSI frame relay does not support video.

As mentioned above, the ANSI frame relay derives from ISDN LAPD core functions. The core functions of the LAPD protocol that are used in frame relay (as defined here) are as follows:

- Frame delimiting, alignment, and transparency provided by the use of HDLC flags and 0-bit insertion/extraction.[†]
- Frame multiplexing/demultiplexing using the address field.
- Inspection of the frame to ensure that it consists of an integer number of octets prior to 0-bit insertion or following 0-bit extraction.
- Inspection of the frame to ensure that it is not too long or too short.
- Detection of (but *not* recovery from) transmission errors.
- Congestion control functions.

In other words, ANSI has selected certain features from the LAPD structure/protocol, rejected others, and added some new features. For instance, the control field was removed, but certain control functions have been incorporated as single bits in the address field. These are the C/R bit (command/response), DE (discard eligibility), FECN bit (forward explicit congestion notification), and BECN bit (backward explicit congestion notification).

2.4 The Frame Structure

User traffic passed to a FRAD (frame relay access device) is segmented into frames with a maximum length information field or with a default length of 262 octets and a recommended length (ANSI) of at least 1600 octets when the

---

* Meaning set up by the switching nodes in the network.
[†] LAPD is a derivative of HDLC.

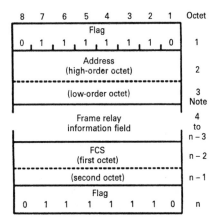

Figure 15.2. Frame relay ANSI frame format with a two-octet address [5].

application is LAN interconnectivity. The minimum information field length is one octet.

Figure 15.2 shows the frame relay frame structure. As mentioned before, it uses HDLC flags (01111110) as opening and closing flags. The closing flag may also serve as the opening flag of the next frame; however, receivers must be able to accommodate reception of one or more consecutive flags on a bearer channel.

ADDRESS FIELD. This consists of at least two octets, but may be extended to three or four octets, as shown in Figure 15.3. There is no control field as in HDLC, LAPD, or LAPB. In its most reduced version, there are just 10 bits allocated to the address field in two octets (the remainder of the bits serve as control functions) supporting up to 1024 logical connections.

It should be noted that the number of addressable logical connections is multiplied because they can be reused at each nodal (switch) interface. That is, an address in the form of a data-link connection identifier (DLCI) has meaning only on one trunk between adjacent nodes. The switch (node) that receives a frame is free to change the DLCI before sending the frame onwards over the next link. Thus, the limit of 1024 DLCIs applies to the link, not the network.

INFORMATION FIELD. This follows the address field and precedes the frame check sequence (FCS). The maximum size of the information field is an implementation parameter, and the default maximum is 262 octets. ANSI chose this default

8 7 6 5 4	3	2	1	Octet
DLCI (high-order)		C/R 0/1	EA 0	1
DLCI (low-order)	FECN BECN	DE	EA 1	2

Figure 15.3. Address field ANSI default format [5].

maximum to be compatible with LAPD on the ISDN D channel which has a two-octet control field and a 260-octet maximum information field. All other maximum values are negotiated between users and networks and between networks. The minimum information field size is one octet. The field must contain an integer number of octets; partial octets are not allowed. A maximum of 1600 octets is encouraged for applications such as LAN interconnects to minimize the need for segmentation and reassembly by user equipment.

TRANSPARENCY. As with HDLC, X.25 (LAPB), and LAPD, the transmitting data-link layer must examine the frame content between opening and closing flags and inserts a 0 bit after all sequences of five contiguous 1s (including the last five bits of the FCS) to ensure a flag or an abort sequence is not simulated within the frame. At the other side of the link, a receiving data-link layer must examine the frame contents between the opening and closing flags and must discard any 0 bit that directly follows five contiguous 1s.

FRAME CHECK SEQUENCE (FCS) is based on the generator polynomial $X^{16} + X^{12} + X^5 + 1$. The CRC processing includes the content of the frame existing between, but not including, the final bit of the opening flag and the first bit of the FCS, excluding the bits inserted for transparency. The FCS, of course, is a 16-bit sequence. If there are no transmission errors (detected), the FCS at the receiver will have the sequence 00011101 00001111.

ORDER OF BIT TRANSMISSION. The order of bit transmission is that the octets are transmitted in ascending numerical order and inside an octet bit 1 is the first bit transmitted.

FIELD MAPPING CONVENTION. When a field is contained within a single octet, the lowest bit number of the field represents the lowest-order value. When a field spans more than one octet, the order of bit values progressively decreases as the octet number increases within each octet. The lowest bit number associated with the field represents the lowest-order value.

An exception to the preceding field mapping convention is the data-link layer FCS field, which spans two octets. In this particular case, bit 1 of the first octet is the high-order bit while bit 8 of the second octet is the low-order bit.

INVALID FRAMES. An invalid frame is a frame that:

- is not properly bounded by two flags (e.g., a frame abort);
- has fewer than three octets between the address field and the closing flag;
- does not consist of an integral number of octets prior to 0-bit insertion or following 0-bit extraction;
- contains a frame check sequence error;
- contains a single-octet address field;

- contains a data-link connection identifier (DLCI) that is not supported by the receiver.

Invalid frames are discarded without notification to the sender, with no further action.

FRAME ABORT. This consists of seven or more contiguous 1 bits. Upon receipt of an ABORT, the data-link layer ignores the frame currently being received.

2.4.1 Address Field Discussion.
Figure 15.3 shows the ANSI-defined address field formats. Included in the field are: the address field extension bits; a reserved bit to support a command/response (C/R) indication bit; forward and backward explicit congestion indicator (FECN and BECN) bits, discard eligibility indicator (DE); a data-link connection identification (DLCI) field; and, finally, a bit to indicate whether the final octet of a three- or four-octet address field is the low-order part of the DLCI or DL-CORE control information. The minimum and default length of the address field is two octets. However, the address field length may be extended to three or four octets. To support a larger DLCI address range, the three- or four-octet address fields may be supported at the user–network interface or network–network interface based on bilateral agreement.

2.4.2 Address Field Variables
ADDRESS FIELD EXTENSION BIT (EA). The address field range is extended by reserving the first transmitted bit of the address field octets to indicate the final octet of the address field. If there is a 0 in this bit position, it indicates that another octet of the address field follows this one. If there is a 1 in the first bit position, it indicates that this octet is the final octet of the address field. As an example, for a two-octet address field, bit 1 of the first octet is set to 0 and bit 1 of the second octet is set to 1. It should be understood that a two-octet address field is specified by ANSI. It is a user's option whether a three- or four-octet field is desired.

COMMAND/RESPONSE BIT (C/R). The C/R bit is not used by the DL-CORE protocol, and the bit is conveyed transparently.

FORWARD EXPLICIT CONGESTION NOTIFICATION (FECN) BIT. This bit may be set by a congested network to notify the user that congestion avoidance procedures should be initiated, where applicable, for traffic in the direction of the frame carrying the FECN indication. This bit is set to 1 to indicate to the receiving end-system that the frames it receives have encountered congested resources. The bit may be used to adjust the rate of destination controlled transmitter. While setting this bit by the network or user is optional, no network shall ever clear this bit (i.e., set to 0). Networks that do not provide FECN shall pass this bit unchanged.

BACKWARD EXPLICIT CONGESTION NOTIFICATION (BECN). This bit may be set by a congested network to notify the user that congestion avoidance procedures should be initiated, where applicable, for traffic in the opposite direction of the frame carrying the BECN indicator. This bit is set to 1 to indicate to the receiving end-system that the frames it transmits may encounter congested resources. The bit may be used to adjust the rate of source-controlled transmitters.

While setting this bit by the network or user is optional according to the ANSI specification, no network shall ever clear (i.e., set to 0) this bit. Networks that do not provide BECN shall pass this bit unchanged.

DISCARD ELIGIBILITY INDICATOR (DE) BIT. This bit, if used, is set to 1 to indicate a request that a frame should be discarded in preference to other frames in a congestion situation. Setting this bit by the network or user is optional. No network shall ever clear (i.e., set to 0) this bit. Networks that do not provide DE capability shall pass this bit unchanged. Networks are not constrained to only discard frames with DE equal to 1 in the presence of congestion.

DATA-LINK CONNECTION IDENTIFIER (DLCI). This is used to identify the logical connection, multiplexed within the physical channel, with which a frame is associated. All frames carried within a particular physical channel and having the same DLCI value are associated with the same logical connection.

The DLCI is an unstructured field. For two-octet addresses, bit 5 of the second octet is the least significant bit. For three- and four-octet addresses, bit 3 of the last octet is the least significant bit. In all cases, bit 8 of the first octet is the most significant bit.

The structure of the DLCI field may be established by the network at the user–network interface subject to bilateral agreements.

In order to allow for compatibility of call control and layer management between B/H and D channels, the following ranges of DLCIs are reserved and preassigned. See Table 15.2. The DLCIs have local significance only.

TABLE 15.2   DLCI Values for B-Channel and H-Channel Applications

DLCI Values	Function
0	In-channel signaling
1–15	Reserved
16–991	Assigned to connections
992–1007	Layer 2 management of frame relay bearer service
1008–1022	Reserved
1023	In channel layer management

TABLE 15.3 DLCI Values for D-Channel (Two-Octet Address Format)

DLCI Values	Function
512–991	Assigned using frame relay connection procedures

Source: Ref. 5.

DLCI ON THE D-CHANNEL. The six most significant bits (bits 8 to 3 of first octet) correspond to the service access point identifier (SAPI) field in ANSI (standard) T1.602 [9]. The DLCI subfield (bits 8 to 3 of the first octet) values that apply on a D-channel are reserved for specific functions to ensure compatibility with operation of the D-channel that may also use ANSI T1.602 protocols. A two-octet address format for the DLCI is assumed when used on the D-channel. *Note*: For frame relay in the D-channel, only DLCI values in the range 512–991 (SAPI = 31–61) will be assigned. Table 15.3 gives DLCI values for the D-channel.

DLCI OR DL-CORE CONTROL INDICATOR (D/C). The D/C indicates whether the remaining six usable bits of that octet are to be interpreted as the lower DLCI bits or as DL-CORE control bits. This bit is set to 0 to indicate that the octet contains DLCI information. When this bit is set to 1, it indicates that the octet contains DL-CORE control information. The D/C is limited to use in the last octet of the three- or four-octet type address field. The use of this indication for DL-CORE control is reserved as there have not been any additional control functions defined that need to be carried in the address field. Thus this indicator has been added to provide possible future expansion of the protocol.

Currently, standards support both PVC and SVC. However, most frame relay services support PVC only. For an SVC, the DLCI is assigned during call setup process. For PVCs, DLCIs are assigned by the network providers at the subscription time.

Source: Ref. 5.

2.5 DL-CORE Parameters (As Defined by ANSI)

DLCI VALUE parameter conveys the DLCI agreed to be used between core entities in support of DL-CORE connection. Its syntax and usage are described above.

DL-CORE connection endpoint identifier (CEI) uniquely defines the DL-CORE connection.

Physical connection endpoint identifier (PH-CEI) uniquely identifies a physical connection to be used in support of a DL-CORE connection.

2.6 Procedures

For permanent frame relay bearer connections, information related to the operation of the DL-CORE protocol in support of DL-CORE connection is maintained by DL-CORE management. For demand frame relay bearer connections, layer 3 establishes and releases DL-CORE connections on behalf of the DL-CORE sublayer. Therefore, information related to the operation of the DL-CORE protocol is maintained by coordination of layer 3 management and DL-CORE sublayer management through the operation of the local system environment.

CONNECTION ESTABLISHMENT. When it is necessary to notify the DL-CORE sublayer entity (either because of establishment of a demand frame relay call, because of notification of reestablishment of a permanent frame relay bearer connection, or because of system initialization) that a DL-CORE connection is to be established, the DL-CORE layer management entity signals an MC-ASSIGN request primitive to the DL-CORE sublayer entity. The DL-CORE sublayer entity establishes the necessary mapping between supporting ph-connection, the CORE-CEI, and the DLCI. In addition, if it has not already done so, it begins to transmit flags on the physical connection except on the D-channel.

CONNECTION RELEASE. When it is necessary to notify the DL-CORE sublayer entity (either because of release of a demand frame relay call or because of notification of failure of a permanent frame relay bearer connection) that a DL-CORE connection is to be released, the DL-CORE layer management entity signals the MC-REMOVE request primitive to the DL-CORE sublayer entity.

2.7 Traffic and Billing on Frame Relay

Figure 15.4 shows a typical traffic profile on a conventional public telephone network, whereas Figure 15.5 shows a typical profile of bursty traffic over a frame relay network. Such bursty traffic is typical for a LAN. The primary employment of frame relay is to interconnect LANs at a distance.

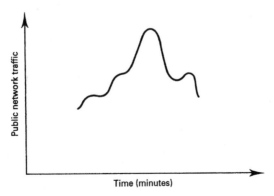

Figure 15.4. Typical traffic profile of a public switched telephone network. Courtesy of Hewlett-Packard Company [7].

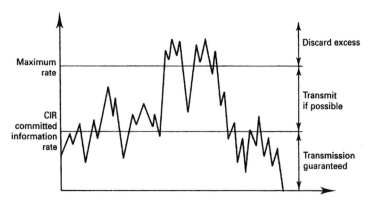

Figure 15.5. Typical bursty traffic of frame relay. Note the traffic levels indicated. Courtesy of Hewlett-Packard Company [7].

Turning to Figure 15.5, the traffic is bursty. With conventional leased data circuits, we have to pay for the bandwidth* whether it is used or not. On the other hand, with frame relay, we only pay for the "time" used. Billing can be handled in one of three ways:

1. CIR (committed information rate) is a data rate subscribed to by a user. This rate may be exceeded for short bursts during the peak period as shown in Figure 15.5.
2. We can pay a flat rate.
3. We can pay per packet (i.e., frame).

Note on the right-hand side of Figure 15.5 there is the guaranteed transmission bit rate equivalent to the CIR. Depending on the traffic load and congestion, during short periods a user may exceed the CIR. However, there is a point where the network cannot sustain further increases in traffic without severe congestion resulting. Traffic above such levels is arbitrarily discarded by the network without informing the originator.

2.8 Congestion Control

Congestion in the user plane occurs when traffic arriving at a resource exceeds the network's capacity. It can also occur for other reasons such as equipment failure. Network congestion affects the throughput, delay, and frame loss experienced by the end-user.

End-users should reduce their offered load in the face of network congestion. Reduction of offered load by an end-user may well result in an increase in the effective throughput available to the end-user during congestion. Congestion avoidance procedures, including optional explicit congestion notification, are

---

* I prefer the use of the expression *bit rate capacity*.

used at the onset of congestion to minimize its negative effects on the network and its users.

Explicit notification is a procedure used for congestion avoidance and is part of the data transfer phase. Users should react to explicit congestion notification (i.e., optional but highly desirable). Users that are not able to act on explicit congestion notification shall have the capability to receive and ignore explicit notification generated by the networks.

Congestion recovery, along with the associated implicit congestion indication due to frame discard, is used to prevent network collapse in the face of severe congestion. Implicit congestion detection involves certain events available to the protocols operating above the core function to detect frame loss (e.g., receipt of a REJECT frame, timer recovery). Upon detection of congestion, the user reduces the offered load to the network. Use of such reduction by users is optional.

2.8.1 Network Response to Congestion. Explicit congestion signals are sent in both forward (toward frame destination) and backwards (toward frame source) directions. Forward explicit congestion notification is provided by using the FECN bit in the address field. Backward explicit congestion notification is provided by one of two methods. When timely reverse traffic is available, the BECN bit in an appropriate address field may be used. Otherwise, a single consolidated link layer management message may be generated by the network. The consolidated link layer management (CLLM) message travels on the U-plane physical path. The generation and transport of CLLM by the network is optional. All networks transport the FECN and BECN bits without resetting.

2.8.2 User Response to Congestion. Reaction by the end user to the receipt of explicit congestion notification is rate-based. Annex A to ANSI T1.618-1991 [5] gives an example of user reaction to FECN and BECN.

END-USER EQUIPMENT EMPLOYING DESTINATION-CONTROLLED TRANSMITTERS. End-user reaction to implicit congestion detection or explicit congestion notification (FECN indications), when supported, is based on the values of FECN indications that are received over a period of time. The method is consistent with commonly used destination-controlled protocol suites, such as OSI class 4 Transport protocol operated over the OSI connectionless service.

END-USER EQUIPMENT EMPLOYING SOURCE-CONTROLLED TRANSMITTERS. End-user reaction to implicit congestion notification (BECN indication), when supported, is immediate when a BECN indication or a CLLM is received. This method is consistent with implementation as a function of data-link layer elements of procedure commonly used in source-controlled protocols such as CCITT Rec. Q.922 elements of procedure.

See also Ref. 14.

2.8.3 Consolidated Link Layer Management (CLLM) Message. The CLLM uses XID frames for the transport of functional information. We may remember in the discussion of HDLC (Chapter 11) that an XID frame was an *exchange identification* frame. It was for reporting its station identification. In frame relay, CLLM with its XID frames is used for network management as an alternative for congestion control. CLLM messages originate at network nodes to the frame relay interface usually housed in a router near or incorporated with the user.

As we mentioned, BECN/FECN bits in frames must pass congested nodes in the forward or backward direction. Suppose that for a given user no frames pass in either direction and that the user has no knowledge of network congestion because at that moment the user is not transmitting or receiving frames. Frame relay standards do not permit a network to generate frames with the DLCI of the congested circuit. CLLM covers this contingency. It has DLCI = 1023 reserved.

The use of CLLM is optional. If it is used, it may or may not operate in conjunction with BECN/FECN. The CLLM frame format has one octet for the cause of congestion such as excessive traffic, equipment or facility failure, preemption, or maintenance action.

This same octet indicates whether the cause is expected to be short or long term. Short term is on the order of seconds or minutes, and anything greater is long term. There is also a bit sequence in this octet indicating an unknown cause of congestion and whether short or long term.

CLLM octets 19 and above give the DLCI values that identify logical links that have encountered congestion. This field must accommodate DLCI length such as two-octet, three-octet, and four-octet DLCI fields.

2.8.4 Action at a Congested Node.

When a node is congested, it has several alternatives it may use to mitigate or eliminate the problem. It may set the FECN and BECN bits to "1" in the address field and/or use the CLLM message. Of course the purpose of explicit congestion notification is:

- to inform the "edge" node at the network ingress of congestion so that that edge node can take the appropriate action to reduce the congestion or
- to notify the source that the negotiated throughput has been exceeded or
- to do both.

One of the strengths of the CLLM is that it contains a list of DLCIs that correspond to the congested frame relay bearer connections. These DLCIs indicate not only the sources currently active causing the congestion but also those sources that are not active. The reason for the latter is to prevent those sources that are not active from becoming active and thus causing still further congestion. It may be necessary to send more than one CLLM message if all the affected DLCIs cannot fit into a single frame.

2.9 Policing a Frame Relay Network

2.9.1 Introduction. Frame relay switches may carry out a policing function on accessing users. The result of such action is discarded frames. Similar policing actions show up in ATM (asynchronous transfer mode) discussed in Chapter 16.

2.9.2 Definitions

Access Rate (AR). The data rate, expressed in bits per second, of the user access channel (D, B, or H channel). The rate at which users can offer data to the network is bounded by the access rate.

Excess Burst Size (B_e). The maximum amount of uncommitted data (in bits) that the network will attempt to deliver over measurement interval (T). This data may or may not be contiguous (i.e., may appear in one frame or in several frames, possibly with interframe idle flags). B_e is negotiated at call establishment (for demand establishment of communication) or at service subscription time (for permanent establishment of communication). Excess burst data may be marked for discard eligibility (with the DE bit) by the network.

Measurement Interval (T). The time interval over which rates and burst sizes are measured. In general, the duration of T is proportional to the *burstiness* of the traffic. Except as noted below, T is computed as $T = B_c/\text{CIR}$ or $T = B_e/\text{AR}$.

Committed Information Rate (CIR). The rate, expressed in bits per second, at which the network agrees to transfer information under normal conditions. This rate is measured over the measurement interval T. CIR is negotiated at call establishment or service subscription time. Data market *discard eligible* (*DE*) is not accounted for in CIR.

Committed Burst Size (B_c). The maximum amount of data (in bits) that a network agrees to transfer under normal conditions over a measurement interval T. These data may or may not be contiguous (i.e., it may appear in one frame or in several frames, possibly with interframe idle flags). B_c is negotiated at call establishment (for demand establishment of communication) or service subscription time (for permanent establishment of communication).

Fairness. An attempt by the network to maintain the negotiated quality-of-service for all users operating under normal conditions (i.e., within their CIR and B_c, without discard due to congestion). For example, the network may discard frames offered in excess of CIR and may reject any call attempts that would cause the network resources to be overcommitted.

Offered Load. The bits offered to the network by an end-user, to be delivered to the selected destination. The information rate and burst length offered to the network could exceed the negotiated class of service parameters. The offered load consists of the user data portion of frames and therefore excludes flags, FCS, and inserted zeros.

2.9.3 Relationship Among Parameters. Figure 15.6 illustrates the relationship among access rate, excess burst, committed burst, committed information rate, discard eligibility indicator, and measurement interval parameters. The CIR, B_c, and B_e parameters are negotiated at call establishment time for demand establishment of communication or established by subscription for permanent establishment of communication. Access rate is established by subscription for permanent access connections or during demand access connection establishment. Each end-user and the network participate in the negotiation of these parameters to agreed-upon values. These negotiated values are then used to determine the measurement interval parameter, T, and when the discard eligibility indicator (if used) is set. These parameters are also used to determine the maximum allowable end-user input levels. The relationship among parameters can be used at any instant of time, T_0, to measure the offered load over the interval $(T_0, T_c + T)$. Similarly, the offered load over any interval $(t - T, t)$ may be measured at any instant of time t, as long as the measurement function retains memory of user activity over the previous interval T. One way of doing this is by use of a "leaky bucket" algorithm described below.

The measurement interval is determined as shown in Table 15.4. The network and the end-users may control the operation of the discard eligibility indicator (DE) and the rate enforcer functions by adjusting at call setup the CIR, B_c, and B_e parameters in relation to the access rate. If both of the CIR and B_c

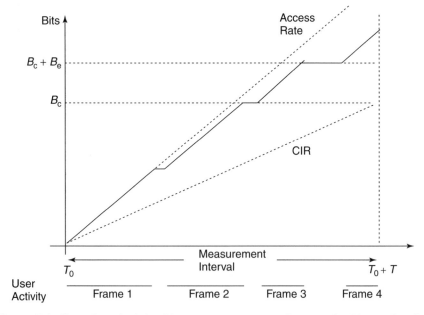

Figure 15.6. Illustration of relationships among parameters. B_c, committed burst size; B_e, excess burst size; CIR, committed information rate. From figure found on FR-21, EE6364 DCW from web www.-ee.edu/online/wang/FR.PDF, 5/7/03 [23].

TABLE 15.4 Congestion Parameter State

CIR	B_c	B_e	Measurement Interval (T)
>0	>0	>0	$T = B_c/\text{CIR}$
>0	>0	=0	$T = B_c/\text{CIR}$
=0	=0	>0	T is a network-dependent value

Source: Ref. 8.

parameters are not equal to zero, then $T = B_c/\text{CIR}$. In addition, there are two special conditions:

(a) When CIR = access rate, $B_c = 0$, and $B_e = 0$, both access rates must be equal (i.e., ingress = egress).
(b) When CIR = 0 (B_c must = 0) and $B_e > 0$, then T is a *network-dependent value*.

Notes

1. The ingress and egress access rate do not have to be equal; however, when the ingress access rate is substantially higher than the egress access rate, continuous input of B_e frames at the ingress interface may lead to persistent congestion of the network buffers at the egress interface, and a substantial amount of the input B_e data will be discarded.
2. When a frame, entering the ingress node, consumes the remaining capacity of B_e or B_c, reducing it to zero, the action taken on that frame is network-dependent.

Figure 15.6 is a static illustration of the relationship among time, cumulative bits of user data, and rate. In this example, the user sends four frames during the measurement interval ($T_0 + T$). The *slope* of the line marked "CIR" is B_c/T. Bits are received at the access rate (by the ingress node) of the access channel. Since the sum of the number of bits contained in frames 1 and 2 is not greater than B_c, the network does not mark these frames with the discard eligibility indicator (DE). The sum of the number of bits in frames 1, 2, and 3 is greater than B_c, but not greater than $B_c + B_e$; therefore frame 3 is marked discard eligible. Since the sum of the number of bits received by the network in frames 1, 2, 3, and 4 exceed $B_c + B_e$, frame 4 is discarded at the ingress node. This figure does not address the case in which the end-user sets the DE bit.

Source: Based on ANSI T1.606a-1992 paragraph 10.2 [8, 15 and 16]

2.10 Quality of Service Parameters

The quality that frame-relaying service provides is characterized by the values of the following parameters. ANSI adds in Ref. 6 that the specific list of negotiable parameters is for further study.

1. Throughput
2. Transit delay
3. Information integrity
4. Residual error rate
5. Delivered error(ed) frames
6. Delivered duplicated frames
7. Delivered out-of-sequence frames
8. Lost frames
9. Misdelivered frames
10. Switched virtual call establishment delay
11. Switched virtual call clearing delay
12. Switched virtual call establishment failure
13. Premature disconnect
14. Switched virtual call clearing failure

2.10.1 Network Responsibilities. The frames are routed through the network on the basis of an attached label (i.e., the DLCI value in the frame). This label is a logical identifier with local significance. In the virtual call case, the value of the logical identifier and other associated parameters such as layer 1 channel delay, and so on, may be requested and negotiated during call setup. Depending on the value of the parameters, the network may accept or reject the call. In the case of the permanent virtual circuit, the logical identifier and other associated parameters are defined by means of administrative procedures (e.g., at the time of subscription).

The user–network interface structure allows for the establishment of multiple virtual calls or permanent virtual circuits, or both, to many destinations over a single access channel.

Specifically, for each connection, the bearer service:

1. Provides bidirectional transfer of frames
2. Preserves their order as given at one user–network interface if and when they are delivered at the other end. (*Note*: No sequence numbers are kept by the network. Networks are implemented in such a way that frame order is preserved.)
3. Detects transmission, format, and operational errors such as frames with an unknown label.
4. Transports the user data contents of a frame transparently. Only the frame's address and FCS fields may be modified by network nodes.
5. Does not acknowledge frames.

At the user–network interface, the FRAD (frame access device), as a minimum, has the following responsibilities:

1. Frame delimiting, alignment, and transparency provided by the use of HDLC flags and 0-bit insertion.
2. Virtual circuit multiplexing/demultiplexing using the address field of the frame.
3. Inspection of the frame to ensure that it consists of an integer number of octets prior to 0-bit insertion or following 0-bit extraction.
4. Inspection of the frame to ensure that it is not too short or too long.
5. Detection of transmission, format, and operational errors.

A frame received by a frame handler may be discarded if the frame:

1. Does not consist of an integer number of octets prior to 0-bit insertion or following 0-bit extraction.
2. Is too long or too short.
3. Has a frame check sequence (FCS) that is in error.

The network will discard a frame if it:

1. Has a DLCI value that is unknown.
2. Cannot be routed further due to internal network conditions.

A frame can be discarded for other reasons such as exceeding negotiated throughput.

Source: Sections 2.3 through 2.8 were extracted from ANSI T1.618-1991 [5]. Sections 2.9 and 2.10 were extracted from ANSI T1.606-1990 [6] and T1.606a-1992 [8]. Also see Refs. 10, 11 and 17.

3 FRAME RELAY STANDARDS

3.1 ANSI T1.618

This is the model standard covered up to this point in our discussion of frame relay. T1.618 also includes the consolidated link layer mechanism (CLLM). Its use is optional and when in operation, DLCI = 1023 and applies when sending link layer control messages.

T1.618 is a subset of ANSI T1.602 and is called the "core aspects" of the standard and is used for both PVCs and SVCs.

3.2 ANSI T1.617

When using a switched virtual circuit connection, frame relay users establish a dialog with the network using the signaling specification in ANSI T1.617. This procedure results in the issuance of a DLCI. Once this dialog is established, T1.618 procedures apply.

To establish a PVC, a setup protocol is used which is identical to the ISDN D-channel protocol defined in T1.617. If the user is ISDN-compatible, the D-channel is employed. For noncompatible ISDN users, there is no D-channel, so that the dialog between the user and the network is separated from the ordinary data transfer procedures. In T1.617, DLCI = 0 is reserved. T1.617 also has specification on how frame relay parameters are to be negotiated.

3.3   ANSI LMI

Annex D of ANSI T1.617 defines the LMI for a PVC management system. The ANSI LMI is almost identical to the manufacturers' LMI, without the optional extensions. The ANSI LMI uses DLCI = 0.

3.4   Manufacturers' LMI

The manufacturers' LMI is a frame relay specification with extension-document number 001-208966, dated September 18, 1990. This specification defines a generic frame relay services based on PVCs for interconnecting DTE devices with frame relay network equipment. In addition to the ANSI standard, the manufacturers' LMI includes extensions and LMI functions and procedures. DLCI = 1023 is used.

3.5   Frame Relay NNI PVC

Frame Relay Forum 2 (FRF.2) describes the network-to-network interface (NNI) PVC frame relay implementation. The NNI deals with the transfer of C-plane and U-plane information between two network nodes belonging to two different frame relay networks.

3.6   FRF.3

FRF.3 provides multiprotocol encapsulation over frame relay using an ANSI T1.618 frame. The FRF.3 frame structure is shown in Figure 15.7. In the figure, octet 6 is the network level protocol ID (NLPID) field that designates encapsulation or the protocol that follows. For example, the value of $0 \times CC$ indicates an encapsulated IP frame based on Q.922 control.

3.7   FRF.4 UNI SVC

FRF.4 applies to non-ISDN frame relay equipment or to an ISDN network which only uses Case A. The following are valid SVC message types:

- Status
- Release
- Disconnect

8	7	6	5	4	3	2	1	Octet
Flag (7E hex)								1
T1.618 Address (including 10-bit DLCI)								2 3
Q.922 Control (UI or I frame)								4
Optional Pad (0x00)								5
NLPID								6
Data								7
Frame Check Sequence								n-2 n-1
Flag (7E hex)								n

Figure 15.7. Frame structure, FRF.3 [12].

- Status enquiry
- Release complete
- Connect acknowledge
- Setup
- Progress
- Connect
- Call proceeding

3.8   FRF.10 NNI SVC

This is the network-to-network interface (NNI) switched virtual connection which is covered in FRF.10. It applies to SVCs over frame relay NNIs and to switched PVCs. It covers public networks, private networks, and a mix of each.

3.9   FRF.11

FRF.11 covers a minimal set of switching functions to forward variable sized data payloads through a frame relay network. It augments the basic Frame Relay Forum UNI and NNI implementation agreements by detailing techniques for structuring application data over the basic information field. The techniques described in FRF.11 enable support for data applications such as LAN bridging, IP routing and IBM SNA. It also extends support for transport of digital voice payloads. Specifically, FRF.11 addresses the following requirements:

- Support of data subchannels on a multiplexed frame relay DLCI
- Effective use of low-bit rate frame relay connections

- Multiplexing of up to 255 subchannels on a single frame relay DLCI
- Supports a large set of voice compression algorithms
- Transport of compressed voice with the payload of a frame relay frame
- Support of multiple voice payloads on the same or different subchannel within a single frame [23]

3.10   Frame Relay Fragmentation Implementation Agreement, FRF.12

Large data packets can be the cause of delay and delay variation of frame relay circuits. Fragmenting long packets and reassembling them at the far end can mitigate the delay problem. Delay and delay variation can be particularly troublesome where there is a mix of both packet voice and data on frame relay permanent virtual circuits (PVCs).

FRF.12 supports three fragmentation applications:

1. End-to-end between two frame relay DTEs interconnected by one or more frame relay networks.
2. Locally across a frame relay UNI interface between DTE/DTE peers.
3. Locally across a frame relay NNI interface between DCE peers.

3.11   Timeplex (BRE2)

The bridge relay encapsulation protocol is proprietary to Ascom Timeplex. It extends bridging across WAN links by means of encapsulation. BRE2 is an improved form of the bridging standard in that it sits directly on the link layer protocol, requires less configuration, and provides its own routing. Figure 15.8 shows the BRE2 frame format.

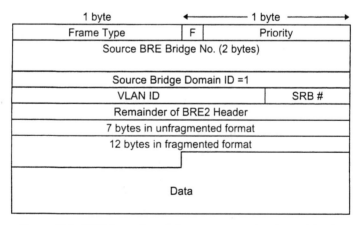

Figure 15.8. BRE2 frame format. *Source*: www.protocols.com/pbook.

In the figure, if $F = 0$, the frame is unfragmented and the header is 17 octets long. If $F = 1$, the frame is fragmented and the BRE2 header is 22 octets long. SRB# = source route bridge number (4 bits).

3.12 Cascade

"Cascade" describes a family of frame relay switches installed by Regional Bell Operating Companies in multiple LATAs. These switches are interconnected to provide service to customers across the LATAs, as well as the capability to manage the switches in multiple LATAs from a single network management station.

The trunk header format for the cascade STDX switch family conforms to ANSI T1.618-1991, ISDN core aspects of the frame (relay) protocol. The header format is illustrated in Figure 15.9. The following is the legend for the header format shown in Figure 15.9:

R. Reserved.

C/R. Command/response field bit.

Version. Header version. Defines the version of the trunk header format for the Cascade STDX family of switches. Current value for this field is 0.

ODE. Set to 1 if the ingress rate is greater than the excess burst size.

DE. Discard eligibility indicator as specified in ANSI T1.618.

FECN. Forward explicit congestion notification as specified in ANSI T1.618.

BECN. Backward explicit congestion notification as specified in ANSI T1.618.

VC Priority. Virtual circuit priority. This is used to differentiate sensitive traffic from traffic not so sensitive to delays, such as file transfers or batch traffic. The priority may be 1, 2, or 3 where 1 is the highest priority.

For management data, the fifth octet contains information dealing with the type of PDU information to follow. Values are as follows:

0	call request PDU
1	confirmation PDU
2	rejected PDU
3	clear PDU

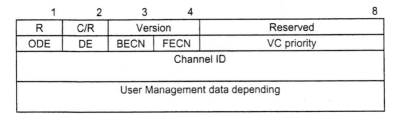

Figure 15.9. Cascade trunk header format. *Source*: www.protocols.com/pbook/pdf/frame_relay.pdf [12].

	8	7	6	5	4	3	2	1
Default address field format		Upper DLCI					C/R	EA 0
(2 octets)		Lower DLCI			FECN (Note)	BECN (Note)	DE (Note)	EA 1

Figure 15.10. LAPF header address format. EA, address field extension bit. Expands address space in 1-octet increments. C/R, command response bit; FECN, forward explicit congestion notification; DE, discard eligibility; BECN, backward explicit congestion notification; DLCI, data-link connection identifier. *Source*: www.protocols.com/pbook.

Control field bits (Modulo 128)	8	7	6	5	4	3	2	1	
I Format			N(S)					0	Octet 4 (Note)
			N(R)					P/F	Octet 5
S Format	X	X	X	X	Su	Su	0	1	Octet 4
			N(R)					P/F	Octet 5
U Format	M	M	M	P/F	M	M	1	1	Octet 4

Figure 15.11. The three LAPF frame formats. N(S), transmitter send sequence number; N(R), transmitter receive sequence number; P/F, poll bit when used as a command, final bit when used as response bit; X, reserved; Su, supervisory function bit; M, modifier function bit. *Source*: www.protocols.com/pbook/frame.

4	disrupt PDU
5	hello PDU
6	hello acknowledgment PDU
7	defined path hello PDU
8	defined path hello acknowledgment PDU

3.13 LAPF

Link access procedure frame (relay) (LAPF) is so named to distinguish it from LAPD. It is used to convey data-link service data units between DL-service users in the U-plane for frame mode bearer services across the ISDN user–network interface on B-, D-, or H-channels. Frame mode bearer connections are established using either procedures established in ITU-T Rec. Q.933, or for permanent virtual circuits by subscription. LAPF uses a physical layer service, and it allows for statistical multiplexing of one or more frame mode bearer connections over a single ISDN B-, D- or H-channel by use of LAPF and compatible HDLC procedures. The header format of LAPF is illustrated in Figure 15.10.

Figure 15.11 shows the three LAPF frame formats.

3.14 Multiprotocol over Frame Relay (Based on RFC 1490 and RFC 2427)

This standard allows the user to encapsulate several LAN protocols over frame relay. The ITU-T Rec. Q.922 Annex A type of frame which we show in

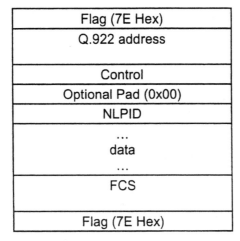

Flag (7E Hex)
Q.922 address
Control
Optional Pad (0x00)
NLPID
...
data
...
FCS
Flag (7E Hex)

Figure 15.12. Frame format for multiprotocol over a frame relay frame. Q.922 address = two-octet address field containing the 10-bit DLCI field. Under a certain situation, the address field may be optionally 3 or 4 octets long. Control = Q.922 control field. The UI value 0 × 03 is used unless negotiated otherwise. The use of XID (0 × AF or 0 × BF) is permitted. Pad = used to align the remainder of the frame on a two-octet boundary. There may be 0 or 1 pad octet within the pad field. The value is always 0. NLPID = network level protocol ID, administered by ISO and ITU-T. Identifies encapsulated protocol. FCS = frame check sequence. *Source*: www.protocols.com/pbook/frame/pdf.

Figure 15.12. To allow the far-end receiver to correctly process the frame, the Annex A frame must contain information within the PDU to identify the LAN protocol encapsulated.

A frame relay network supports two types of packets: routed packets and bridged packets. These two packet types have distinct formats. Therefore they must contain an indicator that the destination may employ to correctly interpret a frame's contents. The indicator is embedded within the NLPID and SNAP (subnetwork access protocol) header information.

For those protocols that do not have NLPID capability, it will be necessary to provide a mechanism to allow easy protocol identification. There is an NLPID value defined to indicate that there is a SNAP header present. The format of the SNAP header is shown in Figure 15.13.

All stations on the frame relay network must be able to accept and properly interpret both the NLPID encapsulation and the SNAP header encapsulation for a routed packet.

In Figure 15.13 the first field is the organizationally unique identifier (OUI) and is 3 octets long. This field identifies the organization which administers the

Organizationally Unique Identifier (OUI)	Protocol Identifier (PID)
3 bytes	2 bytes

Figure 15.13. SNAP header.

meaning of the protocol identifier (PID) field that follows. Taken together, they identify a distinct protocol. For example, if OUI is $0 \times$ 00-00-00, it specifies that the following PID is an Ethertype.

Not all protocols have specific NLPID values assigned to them. When packets involving such protocols are transmitted over frame relay networks, they are sent using NLPID 0×80 followed by SNAP. If the protocol has an Ethertype assigned, the OUI is $0 \times$ 00-00-00, which indicates that an Ethertype follows, and PID is the Ethertype of the protocol in use. Here one pad octet is used to align the protocol data on a two-octet boundary.

The second type of frame relay traffic is bridged packets. Such packets are encapsulated using the NLPID value of 0×80 indicating SNAP. As with other SNAP encapsulated protocols, there is one pad octet to align the data portion of the encapsulated FRAME. The SNAP header which follows the NLPID identifying the format of the bridged packet. The OUI value used for this encapsulation is the 802.1 organization code $0 \times$ 00-80-C2. The PID portion of the SNAP header (the two octets immediately following the OUI) specifies the form of the MAC header, which immediately follows the SNAP header. Additionally, the PID indicates whether the original FCS is preserved with the bridged frame.

Source: Ref. 23.
   Also see Refs. 18–22.

REVIEW QUESTIONS

1. What was the driving force to speed up data network connectivity over WAN distances? From what has been learned in another chapter, give a second, even more cogent reason, to speed things up.

2. X.25 connectivity was one alternative to interconnect LANs over long distances. What are the drawbacks of X.25 for LAN interconnectivity?

3. Frame relay operation relies on the excellent performance of the intervening digital network. What performance parameter is so important here? Quantify it for minimum performance.

4. Three important factors are mentioned which are necessary to speed up the network applying to frame relay. Name these three factors. Can you think of a fourth?

5. Frame relay relates to which OSI layers?

6. There are two possible reasons why a frame may be discarded in frame relay. What are they?

7. How does an end-user know that a frame or frames have been lost or discarded? How do these end-users recoup lost or discarded frames?

8. Name at least three of the standardization agencies involved with frame relay. Name a fourth organization issuing frame relay standards.

9. Frame relay operation derives from what predecessor?

10. How is a frame relay address field extended (based on the ANSI standard)? Explain how it works.

11. Name four of the six causes of declaring a frame relay frame invalid.

12. How are the FECN and BECN bits used? Distinguish between them. (*Hint*: Use direction of transmission).

13. Discuss the use of the DE bit. Under what circumstances will the network change a DE bit setting?

14. What are the three possible ways of handling billing on frame relay?

15. Discuss the use of CLLM as an alternative for congestion control?

16. Give at least two advantages of CLLM over FECN/BECN for congestion control.

17. What are the three basic parameters that are negotiated at call establishment or at subscription for frame relay?

18. If sequence numbers are not used in frame relay, how is sequence maintained at the destination end of a connection?

19. Name four of the five reasons a frame may be discarded by the FRAD and/or frame relay switch.

20. When dealing with frame relay, what sort of organization would use CASCADE?

21. What function does SNAP carry out for a bridged packet in frame relay?

22. What sort of signaling function does ANSI T1.617 carry out for frame relay?

REFERENCES

1. *Framework for Frame Mode Bearer Services*, ITU-T Rec. I.122, ITU Telecommunications Standardization Sector, Geneva, March 1993.
2. *Frame Mode Bearer Services*, ITU-T Rec. I.233, ITU Telecommunications Standardization Sector, Geneva, 1992.
3. *ISDN Signaling Specification for Frame Relay Bearer Service for Digital Subscriber Signaling System Number 1 (DSS1)*, ANSI T1.617-1991, ANSI, New York, 1991.
4. *ISDN Data Link Layer Service for Frame Mode Bearer Services*, ITU-T Rec. Q.922, ITU Telecommunications Standardization Sector, Geneva, 1992.
5. *Integrated Services Digital Network (ISDN)—Core Aspects of Frame Protocol for Use with Frame Relay Bearer Service*, ANSI T1.618-1991, ANSI, New York, 1991.
6. *ISDN—Architectural Framework and Service Description for Frame Relaying Bearer Service*, ANSI T1.606-1990, ANSI, New York, 1990.

7. *Frame Relay & SMDS*, a seminar, Hewlett-Packard, Burlington, MA, October 1993.

8. *Integrated Services Digital Network (ISDN)—Architectural Framework and Service Description for Frame-Relaying Bearer Service (Congestion Management and Frame Size)*, ANSI T1.606a-1992, ANSI, New York, 1992.

9. *Telecommunications: Integrated Services Digital Network (ISDN)—Data Link Layer Signaling Specification for Application at the User–Network Interface*, ANSI T1.602-1989, ANSI New York, 1989.

10. *Distributed Queue Dial Bus (DQDB) Subnetwork of a Metropolitan Area Network (MAN)*, IEEE Std 802.6-1990, IEEE Computer Society, IEEE, New York, 1991.

11. W. A. Flanagan, *Frames, Packets and Cells in Broadband Networking*, Telcom Library, New York 1991.

12. From the web: www.protocols.com/pbook/pdf/frame_relay.pdf, April 2003.

13. *Frame Mode Bearer Services. Frame Relay Multicast*, CCITT Rec. I.233.1-2, ITU Geneva, October 1991.

14. *Congestion Management for the ISDN Frame Relaying Bearer Services*, CCITT Rec. I.370, ITU, Geneva, October 1991.

15. *Frame Relay Bearer Services, Network–Network Interface Requirements*, CCITT Rec. I.372, ITU, Geneva, March 1993.

16. *Frame Relaying Bearer Service Interworking*, ITU-T Rec. I.555, ITU, Geneva, September 1997.

17. *Frame Relaying Operation and Maintenance Principles and Functions*, ITU-T Rec. I.620, ITU, Geneva, October 1996.

18. *Digital Signaling System No. 1 (DSS1)—Signalling Specifications for Frame Mode Switched Virtual Circuits and Permanent Virtual Circuits Control and Monitoring*, ITU-T Rec. Q.933 (in particular, Annex A), ITU, Geneva, October 1995.

19. *Abstract Test Suite—Signaling Specification for Frame Mode Basic Call Control Conformance Testing for Permanent Virtual Circuits*, ITU-T Rec. Q.933 bis, ITU, Geneva, October 1995.

20. J. Weinberger, *Frame Relay Encoding, Encapsulation, Framing and Management Methods*, Lucent Technologies Network Care, Bell Labs 2000, from the web.

21. *Multiprotocol Interconnect over Frame Relay*, RFC 1490, July 1993, from www.cis.ohio-state.edu/cgi-bin/rfc/rfc1490.html, 5/6/03.

22. From the web: www.angelfire.com/ca/framerelay, April 30, 2003.

23. From the web EE6364 DCW FR11: www.-ee.edu/online/wang/FR.PDF.

16

THE ASYNCHRONOUS TRANSFER MODE (ATM) AND BROADBAND ISDN

1 WHERE ARE WE GOING?

Frame relay (Chapter 15) was the beginning of the march toward an optimized* format for multimedia transmission (voice, data, video, facsimile). There were new concepts in frame relay. There was a trend toward simplicity where the header was notably shortened. The header was pure overhead, so it was cut back as much as practically possible. The header also implied processing. By reducing the processing, delivery time could be speeded up.

In the effort to speed up delivery, operation was unacknowledged (at least at the frame relay level); there was no operational error correction scheme. It was unnecessary because it was assumed that the underlying transport system had excellent error performance (better than 1×10^{-7}). There was error detection for each frame, and a frame found in error was thrown away. Now that is something that we never did for those of us steeped in old time data communication. It is assumed that the higher OSI layers would request repeats of the few frames missing (i.e., thrown away).

Frame relay also moved into the flow control arena with the BECN and FECN bits and the CLLM. The method of handling flow control has a lot to do with its effectiveness in this case. It also uses a DE or discard eligibility bit which set a type of priority to a frame. If the DE bit was set, the frame would be among the first to be discarded in a time of congestion.

2 INTRODUCTION TO ATM

ATM is an outgrowth of the several data transmission format systems discussed above, although some may argue this point. Whereas the formats described above

---

* *Compromise* might be a more appropriate word.

Telecommunication System Engineering, by Roger L. Freeman
ISBN 0-471-45133-9   Copyright © 2004 Roger L. Freeman

ostensibly were to satisfy the needs of the data world, ATM (according to some) provides an optimum format or protocol family for data, voice, and image communications, where cells of each can be intermixed as shown in Figure 16.1. It would really seem to be more of a compromise. Typically, these ATM cells can be transported on SONET, SDH, E1/T1, and other popular digital formats. Cells can also be transported contiguously without an underlying digital network format.

Philosophically, voice and data are worlds apart regarding time sensitivity. Voice cannot wait for long processing and ARQ delays. Most types of data can. So ATM must distinguish the type of service such as constant bit rate (CBR) and variable bit rate (VBR) services. Voice service is typical of constant bit rate or CBR service.

Signaling is another area of major philosophical difference. In data communications, "signaling" is carried out within the header of a data frame (or packet). As a minimum the signaling will have the destination address, and quite often the source address as well. And this signaling information will be repeated over and over again on a long data file that is heavily segmented. On a voice circuit, a connectivity is set up and the destination address, and possibly the source address, is sent just once during call set up. There is also some form of circuit supervision to keep the circuit operational throughout the duration of a telephone call. ATM is a compromise, stealing a little from each of these separate worlds.

Like voice telephony, ATM is fundamentally a connection-oriented telecommunication system. Here we mean that a connection must be established between two stations before data can be transferred between them. An ATM connection specifies the transmission path, allowing ATM cells to self-route through an ATM network. Being connection-oriented also allows ATM to specify a guaranteed quality of service (QoS) for each connection.

By contrast, most LAN protocols are connectionless. This means that LAN nodes simply transmit traffic when they need to, without first establishing a specific connection or route with the destination node. In that ATM is a connection-oriented protocol, bandwidth (bit rate capacity) is allocated only when the originating end-user requests a connection. This allows ATM to efficiently support a network's aggregate demand by allocating bit rate capacity on demand based

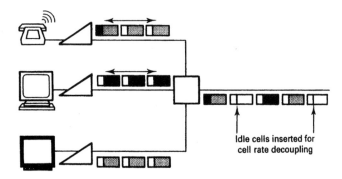

Figure 16.1. ATM links simultaneously carry a mix of voice, data, and image information.

on immediate user need. Indeed it is this concept which lies in the heart of the word *asynchronous*. An analogy would help. New York City is connected to Washington, DC, with a pair of railroad tracks for passenger trains headed south and another pair of tracks for passenger trains headed north. On those two pair of tracks we'd like to accommodate everybody we can when they'd like to ride. The optimum for reaching this goal is to have a continuous train of coupled passenger cars. As the train enters Union Station, it disgorges its passengers and connects around directly for the northward run to Pennsylvania Station. Passenger cars are identical in size, and each has the same number of identical seats.

Of course at 2 A.M. the train will have very few passengers and many empty seats. Probably from 7 to 9 A.M. the train will be full, no standees allowed, so we'll have to hold potential riders in the waiting room. They'll ride later; those few who try to be standees will be bumped. Others might seek alternate transportation to Washington, DC.

Here we see that the railroad tracks are the transmission medium. Each passenger car is a SONET/SDH frame. The seats in each car are our ATM cells. Each seat can handle a person no bigger than 53 units. Because of critical weight distribution, if a person is not 53 units in size/weight, we'll stick some bricks in the seat to bring the size/weight to 53 units exactly. Those bricks are removed at the destination. All kinds of people ride the train because America is culturally diverse, analogous to the fact that ATM handles all forms of traffic. The empty seats represent idle or unassigned cells. The header information is analogous to the passengers' tickets. Keep in mind that the train can only fill to its maximum capacity of seats. The SONET/SDH frame we can imagine as full of cells in the payload, some cells busy and some idle/unassigned. At the peak traffic period, all cells will be busy, and some traffic (passengers) may have to be turned away.

We can go even further with this analogy. Both Washington, DC, and New York City attract large groups of tourists, and other groups travel to business meetings or conventions. A tour group has a chief tour guide in the lead seat (cell) and an assistant guide in the last seat (cell). There may be so many in the group that they extend into a second car or may just intermingle with other passengers on the train. The tour guide and assistant tour guide keep an exact count of people on the tour. The lead guide wears a badge that says BOM, all tour members wear badges that say COM and the assistant tour guide wears a badge that says EOM. Each group has a unique MID (message ID). We also see that service is connection-oriented (Washington, DC, to New York City).

Asynchronous means that we can keep filling the seats on the train until we reach its maximum capacity. If we look up the word, it means *nonperiodic*, whereas the familiar E1/T1 are periodic (i.e., synchronous). One point that seems to get lost in the literature is that the train has a maximum capacity. Thus the concept "bandwidth on demand" is that we can use the "bandwidth" until we fill to rated capacity. Again the unfortunate use of the word *bandwidth*, because our capacity will be measured in octets, not hertz. For example, SONET's STS-1 has a payload capacity of 87×9 octets (see Chapter 7), not 87×9 Hz.

3 USER–NETWORK INTERFACE (UNI) CONFIGURATION AND ARCHITECTURE

ATM is the underlying packet technology of Broadband ISDN (B-ISDN). At times in this section, we will use the terms ATM and B-ISDN interchangeably. Figures 16.2 and 16.3 interrelate the two. Figure 16.2 relates the B-ISDN access reference configuration with ATM user–network interface (UNI). Note the similarities of this figure with Figure 14.2. The only difference is that the block nomenclature has a "B" placed in front to indicate *broadband*. Figure 16.3 is the traditional ITU-T Rec. I.121 [1] B-ISDN protocol reference model showing the extra layer necessary for the several services.

Returning to Figure 16.2, we see that there is an upper part and lower part. The lower part shows the UNI boundaries. The upper part is the B-ISDN reference configuration with four interface points. The interfaces at reference points U_B, T_B, and S_B are standardized. These interfaces support all B-ISDN services.

There is only one interface per B-NT1 at the U_B and one at the T_B reference points. The physical media is point-to-point (in each case) in the sense that there is only one receiver in front of one transmitter.

One or more interfaces per NT2 are present at the S_B reference points. The interface at the S_B reference point is point-to-point at the physical layer in the sense that there is only one receiver in front of one transmitter and may be point-to-point at other layers.

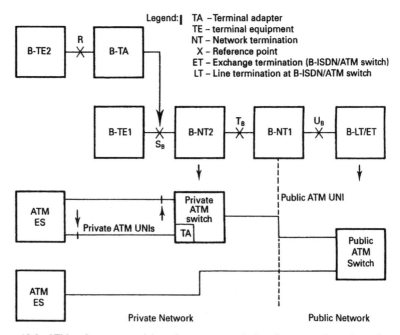

Figure 16.2. ATM reference model and user–network interfaces configuration. *Sources*: Refs. 2–5.

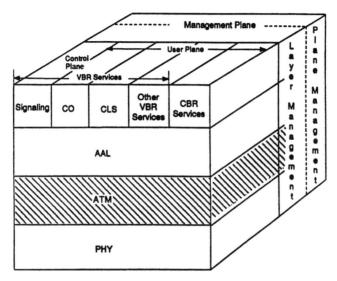

Figure 16.3. B-ISDN protocol reference model. *Source*: Refs. 6, 19 & 20.

Consider now the functional groupings in Figure 16.2. B-NT1 includes functions broadly equivalent to OSI layer 1, the physical layer. These functions include:

- Line transmission termination
- Interface handling at T_B and U_B
- OAM functions

The B-NT2 functional group includes functions broadly equivalent to OSI layer 1 and higher OSI layers. The B-NT2 may be concentrated or distributed. In a particular access arrangement, the B-NT2 functions may consists of physical connections. Examples of B-NT2 functions are:

- Adaptation functions for different media and topologies
- Cell delineation
- Concentration; buffering
- Multiplexing and demultiplexing
- OAM functions
- Resource allocation
- Signaling protocol handling

The functional group B-TE (TE = terminal equipment) also includes roles of OSI layer 1 and higher OSI layers. Some of these roles are:

- User/user and user/machine dialog and protocol
- Protocol handling for signaling

- Connection handling to other equipment
- Interface termination
- OAM functions

B-TE1 has an interface that complies with the B-ISDN interface. B-TE2, however, has a noncompliant B-ISDN interface. Compliance refers to ITU-T Recs. I.413 and I.432 as well as ANSI T1.624 [2–4].

The terminal adapter (B-TA) converts the B-TE2 interface into a compliant B-ISDN user–network interface.

Four bit rates are specified at the UB, TB, and SB interfaces based on Ref. 4 (ANSI T1.624). These are:

51.840 Mbits/s (SONET STS-1)

155.520 Mbits/s (SONET STS-3 and SDH STM-1)

622.080 Mbits/s (SONET STS-12 and SDH STM-3)

44.736 Mbits/s (DS3)

These interfaces are discussed subsequently in this chapter.

The following definitions refer to Figure 16.3.

User Plane (in other literature called the U-plane). The user plane provides for the transfer of user application information. It contains physical layer, ATM layer, and multiple ATM adaptation layers required for different service users such as constant bit rate service (CBR) and variable bit rate service (VBR).

Control Plane (in other literature called the C-plane). The control plane protocols deal with call establishment and call release and other connection control functions necessary for providing switched services. The C-plane structure shares the physical and ATM layers with the U-plane as shown in Figure 16.3. It also includes ATM adaptation layer (AAL) procedures and higher layer signaling protocols.

Management Plane (called the M-plane in other literature). The management plane provides management functions and the capability to exchange information between the U-plane and the C-plane. The M-plane contains two sections: layer management and plane management. The layer management performs layer-specific management functions, while the plane management performs management and coordination functions related to the complete system.

We return to Figure 16.3 and B-ISDN/ATM layering and layer descriptions in Section 6 of this chapter.

4   THE ATM CELL — KEY TO OPERATION

4.1   ATM Cell Structure

The ATM cell consists of 53 octets, 5 of which makeup the header and 48 octets are in the payload or "info" portion of the cell. Figure 16.4 shows an ATM

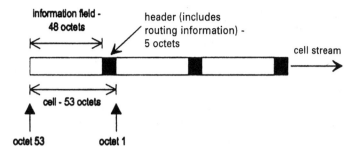

Figure 16.4. An ATM cell stream illustrating the basic makeup of a cell.

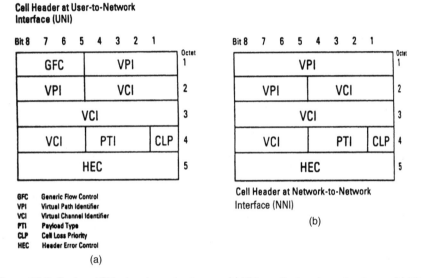

cell stream delineating the 5-octet header and 48-octet information field of each cell. Figure 16.5 shows the detailed structure of the cell headers at the user–network interface (UNI) (Figure 16.5a) and at the network–node interface (NNI)* (Figure 16.5b).

We digress a moment to discuss why a 53-octet cell was standardized. The cell header contains only 5 octets. It was shortened as much as possible containing the minimum address and control functions for a working system. It is also non-revenue-bearing overhead. It is the information field that contains the revenue-bearing payload. For efficiency, we'd like the payload to be as long as possible. Yet the ATM designer team was driven to shorten the payload as much as possible.

---

* NNI is variously called network–node interface or network–network interface. It is the interface between two network nodes or switches.

The issue in this case was what is called *packetization delay*. This is the amount of time required to fill a cell at a rate of 64 kbits/s—that is, the rate to fill the cell with digitized voice samples. According to Ref. 8, the design team was torn between efficiency and packetization delay. One school of thought fought for a 64-octet cell, and another argued for a 32-octet cell size. Thus, the ITU-T opted for a fixed-length 53-octet compromise.

Now let's return to the discussion of the ATM cell and its headers. The left-hand side of Figure 16.5 shows the structure of a UNI header, whereas the right-hand side illustrates the NNI header. The only difference is the presence of the GFC field in the UNI header. The following paragraphs define each header field. By removing the GFC field, the NNI has four additional bits for addressing.

GFC—GENERIC FLOW CONTROL. The GFC field contains 4 bits. When the GFC function is not used, the value of this field is 0000. This field has local significance only and can be used to provide standardized local flow control functions on the customer side. In fact the value encoded in the GFC is not carried end-to-end and will be overwritten by ATM switches (i.e., the NNI interface).

Two modes of operation have been defined for operation of the GFC field. These are *uncontrolled access* and *controlled access*. The "uncontrolled access" mode of operation is used in the early ATM environment. This mode has no impact on the traffic which a host generates. Each host transmits the GFC field set to all zeros (0000). In order to avoid unwanted interactions between this mode and the "controlled access" mode where hosts are expected to modify their transmissions according to the activity of the GFC field, it is required that all CPE (customer premise equipment) and public network equipment monitor the GFC field to ensure attached equipment is operating in "uncontrolled mode." A count of the number of nonzero GFC fields should be measured for nonoverlapping intervals of $30,000 \pm 10,000$ cell times. If 10 or more nonzero values are received within this interval, an error is indicated to layer management [2].

ROUTING FIELD (VPI/VCI). Twenty-four bits are available for routing a cell. There are 8 bits for virtual path identifier (VPI) and 16 bits for virtual channel identifier (VCI). Preassigned combinations of VPI and VCI values are given in Table 16.1. Other preassigned values of VPI and VCI are for further study according to the ITU-T organization. The VCI value of zero is not available for user virtual channel identification. The bits within the VPI and VCI fields used for routing are allocated using the following rules:

- The allocated bits of the VPI field are contiguous.
- The allocated bits of the VPI field are the least significant bits of the VPI field, beginning at bit 5 of octet 2.
- The allocated bits of the VCI field are contiguous.
- The allocated bits of the VCI field are the least significant bits of the VCI field, beginning at bit 5 of octet 4.

TABLE 16.1   Combinations of Preassigned VPI, VCI, and CLP Values at the UNI

Use	VPI	VCI	PT	CLP
Meta-signaling	XXXXXXXX	00000000 00000001	0A0	C
(refer to Rec. I.311)	(Note 1)	(Note 5)		
General broadcast signaling	XXXXXXXX	00000000 00000010	0AA	C
(refer to Rec. I.311)	(Note 1)	(Note 5)		
Point-to-point signaling	XXXXXXXX	00000000 00000101	0AA	C
(refer to Rec. I.311)	(Note 1)	(Note 5)		
Segment OAM F4 flow cell	YYYYYYYY	00000000 00000011	0A0	A
(refer to Rec. I.610)	(Note 2)	(Note 4)		
End-to-end OAM F4 flow cell	YYYYYYYY	00000000 00000100	0A0	A
(refer to Rec. I.610)	(Note 2)	(Note 4)		
Segment OAM F5 flow cell	YYYYYYYY	ZZZZZZZZ ZZZZZZZZ	100	A
(refer to Rec. I.610)	(Note 2)	(Note 3)		
End-to-end OAM F5 flow cell	YYYYYYYY	ZZZZZZZZ ZZZZZZZZ	101	A
(refer to Rec. I.610)	(Note 2)	(Note 3)		
Resource management cell	YYYYYYYY	ZZZZZZZZ ZZZZZZZZ	110	A
(refer to Rec. I.371)	(Note 2)	(Note 3)		
Unassigned cell	00000000	00000000 00000000	BBB	0

The GFC field is available for use with all of these combinations.

A Indicates that the bit may be 0 or 1 and is available for use by the appropriate ATM layer function.

B Indicates the bit is a "don't care" bit.

C Indicates the originating signaling entity shall set the CLP bit to 0. The value may be changed by the network.

Notes

1 XXXXXXXX: Any VPI value. For VPI value equal to 0, the specific VCI value specified is reserved for user signaling with the local exchange. For VPI values other than 0, the specified VCI value is reserved for signaling with other signaling entities (e.g., other users or remote networks).

2 YYYYYYYY: Any VPI value.

3 ZZZZZZZZ ZZZZZZZZ: Any VCI value other than 0.

4 Transparency is not guaranteed for the OAM F4 flows in a user-to-user VP.

5 The VCI values are preassigned in every VPC at the UNI. The usage of these values depends on the actual signaling configurations. (See ITU-T Rec. I.311.)

Source: Table 2/I.361, ITU-T Rec. I.361, page 3, March 1993 [7].

PAYLOAD-TYPE (PT) FIELD. Three bits are available for PT identification. Table 16.2 gives the payload-type identifier coding. The main purpose of the PTI is to discriminate between user cells (i.e., cells carrying user information) from nonuser cells. The first four code groups (000–011) are used to indicate user cells. Within these four, 2 and 3 (010 and 011) are used to indicate congestion has been experienced. The 5th and 6th code groups (100 and 101) are used for VCC level management functions.

HEADER ERROR CONTROL (HEC) FIELD. The HEC is an 8-bit field and it covers the entire cell header. The code used for this function is capable of either single-bit error correction or multiple-bit error detection. Briefly, the transmitting side computes the HEC field value. The receiver has two modes of operation as

TABLE 16.2   PTI Coding

	PTI Coding			Interpretation
Bits	4	3	2	
	0	0	0	User data cell, congestion not experienced. ATM-user-to-ATM-user indication = 0
	0	0	1	User data cell, congestion not experienced. ATM-user-to-ATM-user indication = 1
	0	1	0	User data cell, congestion experienced. ATM-user-to-ATM-user indication = 0
	0	1	1	User data cell, congestion experienced. ATM-user-to-ATM-user indication = 1
	1	0	0	OAM F5 segment associated cell
	1	0	1	OAM F5 end-to-end associated cell
	1	1	0	Resource management cell
	1	1	1	Reserved for future functions

Source: ITU-T Rec. I.361, page 4, March 1993 [7].

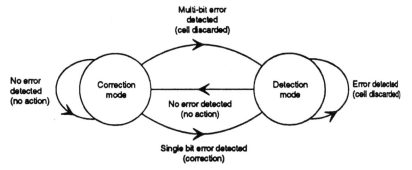

Figure 16.6. HEC: receiver modes of operation. Based on ITU-T Rec. I.432, March 1993 [3].

shown in Figure 16.6. In the default mode there is the capability of single-bit error correction. Each cell header is examined and, if an error is detected, one of two actions takes place. The action taken depends on the state of the receiver. In the *correction mode*, only single-bit errors can be corrected and the receiver switches to the *detection mode*. In the "detection mode," all cells with detected header errors are discarded. When a header is examined and found not to be in error, the receiver switches to the "correction mode." The term *no action* in Figure 16.6 means no correction is performed and no cell is discarded.

Figure 16.7 is a flowchart showing the consequence of errors in the ATM cell header. The error protection function provided by the HEC provides for both recovery from single-bit errors and a low probability of delivery of cells with errored headers under bursty error conditions. ITU-T Rec. I.432 [3] states that error characteristics of fiber-optic transmission systems appear to be a mix of single-bit errors and relatively large burst errors. Thus, for some transmission systems the error correction capability might not be invoked.

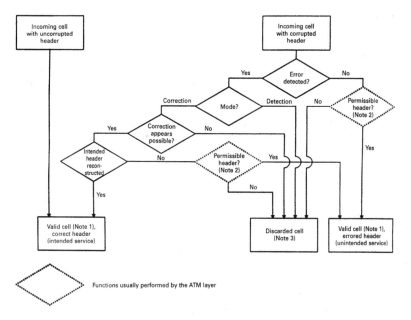

Functions usually performed by the ATM layer

Notes
1 Definition of "valid cell": A cell where the header is declared by the header error control process to be free of errors (ITU-T Rec. I.113).

2 An example of an impermissible header is a header whose VPI/VCI is neither allocated to a connection nor preassigned to a particular function (idle cell, OAM cell, etc.). In many instances, the ATM layer will decide if the cell header is permissible.

3 A cell is discarded if its header is declared to be invalid; or if the header is declared to be valid and the resulting header is impermissible.

Figure 16.7. Consequences of errors in an ATM cell header. From Figure 12/I.432, page 18, ITU-T Rec. I.432, March 1993 [3].

Any congested network element, upon receiving a user data cell, may modify the PTI as follows. Cells received with PTI = 000 or PTI = 010 are transmitted with PTI = 010. Cells received with PTI = 001 or PTI = 011 are transmitted with PTI = 011. Noncongested network elements should not change the PTI.

CELL LOSS PRIORITY (CLP) FIELD. Depending on network conditions, cells where the CLP is set (i.e., CLP value is 1) are subject to discard prior to cells where the CLP is not set (i.e., CLP value is 0). The concept here is identical to that of frame relay and the DE (discard eligibility) bit. ATM switches may tag CLP = 0 cells detected by the UPC (usage parameter control) to be in violation of the traffic contract by changing the CLP bit from 0 to 1.

4.2 Idle Cells

Idle cells cause no action at a receiving node except for cell delineation including HEC verification. They are inserted and extracted by the physical layer in order to adapt the cell flow rate at the boundary between the ATM layer and the physical layer to the available payload capacity of the transmission media. This

TABLE 16.3   Header Pattern for Idle Cell Identification

	Octet 1	Octet 2	Octet 3	Octet 4	Octet 5
Header pattern	00000000	00000000	00000000	00000001	HEC = Valid code 01010010

Source: Table 4/I.432, page 19, ITU Rec. I.432, March 1993 [3].

is called *cell rate decoupling*. Idle cells are identified by the standardized pattern for the cell header as shown in Table 16.3. The content of the information field is 01101010 repeated 48 times for an idle cell.

There is some variance in this area between the ITU-T Organization documentation and that of the ATM Forum. McDysan and Spohn [8] point out the following. ITU-T Rec. I.321 places this function in the TC (transmission convergence) sublayer of the PHY (physical layer) and uses idle cells, whereas the ATM Forum places it in the ATM layer and uses unassigned cells. This presents a potential low-level incompatibility if different systems use different cell types for cell rate decoupling.

5   CELL DELINEATION AND SCRAMBLING

5.1   Delineation and Scrambling Objectives

Cell delineation allows identification of the cell boundaries. The cell header error control (HEC) field achieves cell delineation. Keep in mind that the ATM signal must be self-supporting in that it has to be transparently transported on every network interface without any constraints from the transmission systems used.

Scrambling is used to improve security and robustness of the HEC cell delineation mechanism discussed below. In addition, it helps the randomizing of data in the information field for possible improvement in transmission performance. Any scrambler specification must not alter the ATM header structure, header error control, and cell delineation algorithm.

5.2   Cell Delineation Algorithm

Cell delineation is performed by using the correlation between the header bits to be protected (32 bits or 4 octets) and the HEC octet which are the relevant control bits (8 bits) introduced in the header using a shortened cyclic code with the generating polynomial $X^8 + X^2 + X + 1$. Figure 16.8 shows the state diagram of the HEC cell delineation method. A discussion of the figure is given below.

1. In the HUNT state, the delineation process is performed by checking bit by bit for the correct HEC (i.e., syndrome equals zero) for the assumed header field. For the cell-based* physical layer, prior to scrambler synchronization,

---

* Only cell-based and SDH-based interfaces are covered by current ITU-T recommendations. Besides these, we will cover cells riding on other transport means at the end of this chapter.

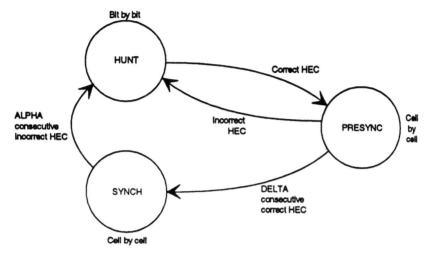

Figure 16.8. Cell delineation state diagram. From Figure 13/I.432, page 20, ITU-T Rec. I.432, March 1993 [3].

only the last six bits of the HEC are used for cell delineation checking. For the SDH-based* interface, all 8 bits are used for acquiring cell delineation. Once such an agreement is found, it is assumed that one header has been found, and the method enters the PRESYNCH state. When octet boundaries are available within the receiving physical layer prior to cell delineation as with the SDH-based interface, the cell delineation process may be performed octet by octet.

2. In the PRESYNCH state, the delineation process is performed by checking cell by cell for the correct HEC. The process repeats until the correct HEC has been confirmed *Delta* times consecutively. If an incorrect HEC is found, the process returns to the HUNT state.

3. In the SYNCH state the cell delineation will be assumed to be lost if an incorrect HEC is obtained *Alpha* times consecutively.

The parameters *Alpha* and *Delta* are chosen to make the cell delineation process as robust and secure as possible while satisfying QoS (quality of service) requirements. Robustness depends on *Alpha* when it is against false misalignments due to bit errors. And robustness depends on *Delta* when it is against false delineation in the resynchronization process.

For the SDH-based physical layer, values of *Alpha* = 7 and *Delta* = 6 are suggested by the ITU-T organization (Rec. I.432 [3]); and for cell-based physical layer, values of *Alpha* = 7 and *Delta* = 8. Figures 16.9 and 16.10 give performance information of the cell delineation algorithm in the presence of random bit errors, for various values of *Alpha* and *Delta*.

* See footnote on previous page

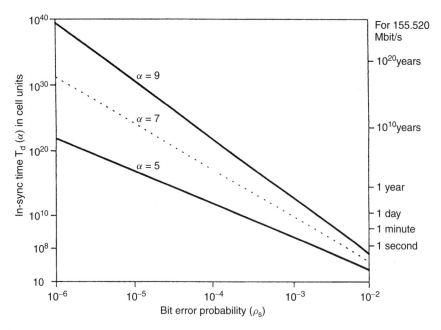

Figure 16.9. In-sync time versus bit error probability. From Figure B.1/I.432, page 33, ITU-T Rec. I.432, March 1993 [3].

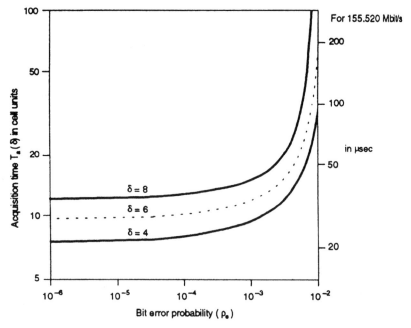

Figure 16.10. Acquisition time versus bit error probability. From Figure B.2/I.432, page 34, ITU-T Rec. I.432, March 1993 [3].

6   ATM LAYERING AND B-ISDN

The B-ISDN reference model is given in Figure 16.3, and its several planes are described. This section provides brief descriptions of the ATM layers and sublayers.

6.1   Functions of Individual ATM/B-ISDN Layers

Figure 16.11 illustrates B-ISDN/ATM layering and sublayering of the protocol reference model. It identifies the functions of the physical layer, the ATM layer and the AAL, and related sublayers.

6.1.1   Physical Layer. The physical layer consists of two sublayers. The physical medium (PM) sublayer includes only physical medium-dependent functions. The transmission convergence (TC) sublayer performs all functions required to transform a flow of cells into a flow of data units (i.e., bits) which can be transmitted and received over a physical medium. The service data unit (SDU) crossing the boundary between the ATM layer and the physical layer is a flow

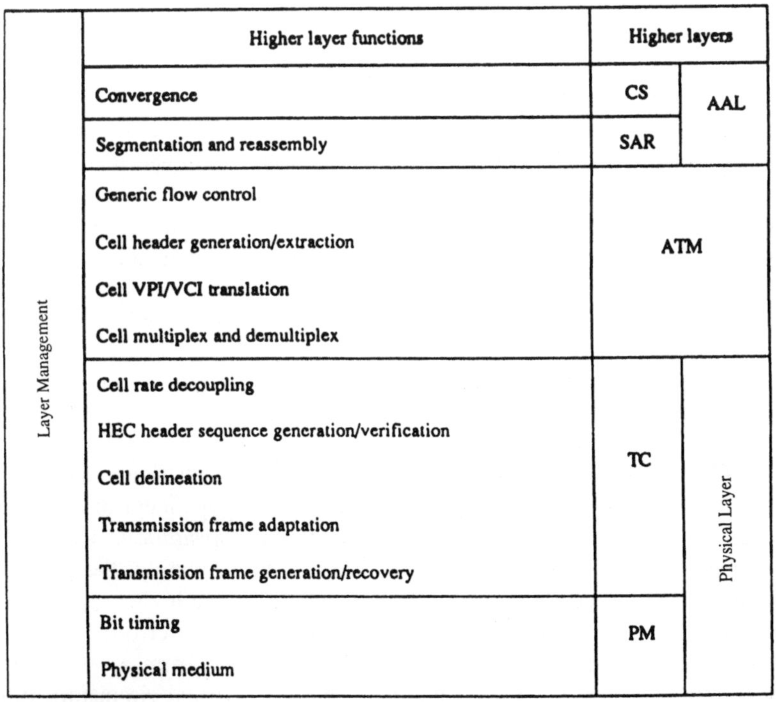

	Higher layer functions	Higher layers	
	Convergence	CS	AAL
	Segmentation and reassembly	SAR	
	Generic flow control		
	Cell header generation/extraction	ATM	
	Cell VPI/VCI translation		
	Cell multiplex and demultiplex		
	Cell rate decoupling		
	HEC header sequence generation/verification	TC	
	Cell delineation		
	Transmission frame adaptation		
	Transmission frame generation/recovery		
	Bit timing	PM	
	Physical medium		

Left column label: Layer Management. Right column label: Physical Layer (spanning TC and PM rows).

CS Convergence sublayer
PM Physical medium
SAR Segmentation and reassembly sublayer
TC Transmission convergence

Figure 16.11. B-ISDN/ATM functional layering.

of valid cells. The ATM layer is unique (meaning independent of the underlying physical layer). The data flow inserted in the transmission system payload is physical medium-independent and self-supported. The physical layer merges the ATM cell flow with the appropriate information for cell delineation, according to the cell delineation mechanism described above and carries the operations and maintenance (OAM) information relating to this cell flow.

The physical medium sublayer provides bit transmission capability including bit transfer and bit alignment as well as line coding and electrical-optical transformation. Of course, the principal function is the generation and reception of waveforms suitable for the medium, the insertion and extraction of bit timing information, and line coding where required. The primitives identified at the border between the PM and TC sublayers are a continuous flow of logical bits or symbols with this associated timing information.

TRANSMISSION CONVERGENCE SUBLAYER FUNCTIONS. Among the important functions of this sublayer is the generation and recovery of transmission frame. Another function is transmission frame adaptation which includes the actions necessary to structure the cell flow according to the payload structure of the transmission frame (transmit direction) and to extract this cell flow out of the transmission frame (receive direction). The transmission frame may be a cell equivalent (i.e., no external envelope is added to the cell flow), an SDH/SONET envelope, an E1/T1 envelope, and so on. In the transmit direction, the HEC sequence is calculated and inserted in the header. In the receive direction, we include cell header verification. Here cell headers are checked for errors and, if possible, header errors are corrected. Cells are discarded where it is determined that headers are errored and are not correctable.

Another transmission convergence function is cell rate decoupling. This involves the insertion and removal of idle cells in order to adapt the rate of valid ATM cells to the payload capacity of the transmission system. In other words, cells must be generated to exactly fill the payload of SDH/SONET, as an example, whether the cells are idle or busy.

Section 12 of this chapter gives several examples of transporting cells using the convergence sublayer.

6.1.2 The ATM Layer.

Table 16.4 shows the ATM layer functions supported at the UNI (U-plane). The ATM layer is completely independent of the physical medium. One important function of this layer is *encapsulation*. This includes cell header generation and extraction. In the transmit direction, the cell header generation function receives a cell information field from a higher layer and generates an appropriate ATM cell header except for the header error control (HEC) sequence. This function can also include the translation from a service access point (SAP) identifier to a VP (virtual path) and VC (virtual circuit) identifier.

In the receive direction, the cell header extraction function removes the ATM cell header and passes the cell information field to a higher layer. As in the transmit direction, this function can also include a translation of a VP and VC identifier into an SAP identifier.

TABLE 16.4   ATM Layer Functions Supported at the UNI

Functions	Parameters
Multiplexing among different ATM connections	VPI/VCI
Cell rate decoupling (unassigned cells)	Preassigned header field values
Cell discrimination based on predefined header field values	Preassigned header field values
Payload type discrimination	PT field
Loss priority indication and selective cell discarding	CLP field, network congestion state
Traffic shaping	Traffic descriptor

Source: Based on Refs. 2 and 7.

In the case of the NNI (network–node interface) the GFC (generic flow control) is applied at the ATM layer. The flow control information is carried in assigned and unassigned cells. Cells carrying this information are generated in the ATM layer.

In a switch the ATM layer determines where the incoming cells should be forwarded to, resets the corresponding connection identifiers for the next link and forwards the cell. The ATM layer also handles traffic management functions and buffers incoming and outgoing cells. It indicates to the next higher layer (the AAL) whether or not there is congestion during transmission. The ATM layer monitors both transmission rates and conformance to the service contract—called traffic shaping and traffic policing.

CELL DISCRIMINATION BASED ON PREDEFINED HEADER FIELD VALUES. The predefined header field values defined at the UNI and NNI are given in ANSI T1.627-1993 [6]. Of interest are the VPI and VCI values reserved as shown in Ref. 6 tables for six of the seven rows (less invalid pattern).

Meta-signaling cells are used by the meta-signaling protocol for establishing and releasing signaling virtual channel connections (VCCs). For virtual channels allocated permanently (PVC), meta-signaling is not used.

General broadcast signaling cells are used by the ATM network to broadcast signaling information independent of service profiles.

The virtual path connection (VPC) operation flow (F4 flow) is carried via specially designated OAM cells. F4 flow OAM cells have the same VPI value as the user-data cells transported by the VPC but are identified by two unique preassigned virtual channels within this VPC. At the UNI, the virtual channel identified by a VCI value = 3 is used for VP level management functions between ATM nodes on both sides of the UNI (i.e., single VP link segment) while the virtual channel identified by a VCI value = 4 can be used for VP level end-to-end (user ↔ user) management functions.

What are flows such as "F4 flows?" OAM (operations, administration, and management) flows deal with cells dedicated to fault and performance management of the total system. Consider ATM as a hierarchy of levels, particularly in

SDH/SONET, which are the principal bearer formats for ATM. The lowest level where we have F1 flows is the regenerator section (called the section level in SONET). This is followed by F2 flows at the digital section level (called the line level in SONET). There are the F3 flows for the transmission path (called the path level in SONET). ATM adds F4 flows for virtual paths (VPs) and F5 flows for virtual channels (VCs), where multiple VCs are completely contained within a single VP. We discuss VPs and VCs later.

6.1.2.1 ATM Layer Management (M-Plane). Management functions at the UNI require some level of cooperation between customer premises equipment and network equipment. To minimize the coupling required between equipment on both sides of the UNI, the functional requirements have been reduced to a minimal set. The ATM layer management functions supported at the UNI are grouped into the categories under the general heading of fault management.

Fault management contains alarm surveillance and connectivity verification functions. OAM cells are used for exchanging related operation information.

6.1.3   The ATM Adaptation Layer (AAL). The basic purpose of the AAL is to isolate the higher layers from the specific characteristics of the ATM layer by mapping the higher layer protocol data units (PDUs) into the information field of the ATM cell and vice versa.

6.1.3.1 Sublayering of the AAL. To support services above the AAL, some independent functions are required of the AAL. These functions are organized in two logical sublayers: the convergence sublayer (CS) and the segmentation and reassembly sublayer (SAR). The prime functions of these sublayers are:

- SAR. The segmentation of higher layer information into a size suitable for the information field of an ATM cell.
- Reassembly of the contents of ATM cell information fields into higher layer information.
- CS. Here the prime function is to provide the AAL service at the AAL-SAP (SAP = service access point). This sublayer is service-dependent.

6.1.3.2 Service Classification for the AAL. Service classification is based on the following parameters:

- Timing relation between source and destination (this refers to urgency of traffic): required or not required
- Bit rate: constant or variable
- Connection mode: connection-oriented or connectionless

When we combine these parameters, four service classes emerge as shown in Figure 16.12. Examples of services in the classes shown in Figure 16.12 are as follows:

Service Parameters	Class A	Class B	Class C	Class D
Timing Compensation	Required		Not Required	
Bit Rate	Constant		Variable	
Connection Mode	Connection-oriented			Connectionless
AAL Types	AAL1	AAL2	AAL3/4 or AAL5	AAL3/4 or AAL5
Examples	DS1, E1, n × 64-kbps emulation	Packet video, audio	Frame relay X.25	IP, SMDS

Figure 16.12. Service classification for AAL. Based on Refs. 8–10 and 15.

- Class A: constant bit rate such as uncompressed voice or video
- Class B: variable bit rate video and audio, connection-oriented synchronous traffic
- Class C: connection oriented data transfer, variable bit rate, asynchronous traffic
- Class D: connectionless data transfer, asynchronous traffic such as SMDS

6.1.3.3 AAL Categories or Types. There are five different AAL types or categories. The simplest of these is AAL-0. It just transmits cells down a pipe. That pipe is commonly a fiber-optic link. Ideally we would like the bit rate to be some multiple of 53×8 or 424 bits. For example, 424 Mbits/s would handle 1 million cells per second.

6.1.3.3.1 AAL-1. AAL-1 is used to provide transport for synchronous bit streams. Its primary application is to adapt ATM cell transmission to typically E1/T1 and SDH/SONET circuits. Typically, AAL-1 is for voice communications (POTS—plain old telephone service). AAL-1 robs one octet from the payload and adds it to the header, leaving only a 47-octet payload. This octet includes two major fields: sequence number (SN) and sequence number protection (SNP). The principal purpose of these two fields is to check that mis-sequencing of information does not occur by verifying a 3-bit sequence counter. It also allows for regeneration of the original clock timing of the data received at the far end of the link. The SAR-PDU format of AAL-1 is shown in Figure 16.13. The 4-bit sequence number (SN) is broken down into a 1-bit CSI (convergence sublayer indicator) and sequence count. The SNP (sequence number protection) contains a 3-bit CRC and a parity bit. End-to-end synchronization is an important function for the type of traffic carried on AAL-1. With one mode of operation, clock recovery is via a synchronous residual time stamp (SRTS) and common network clock by means of a 4-bit residual time stamp extracted from CSI of cells with

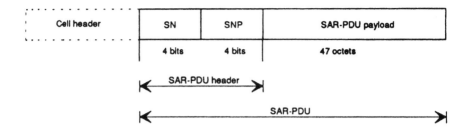

SN Sequence number (4 bits); to detect lost or misinserted cells. A specific value of the sequence number may indicate a special purpose, e.g. the existence of convergence sublayer functions. The exact counting scheme is for further study.

SNP Sequence number protection (4 bits). The SNP field may provide error detection and correction capabilities. The polynomial to be used is for further study.

Figure 16.13. SAR-PDU format for AAL-1. From Figure 1/I.363, page 3, CCITT Rec. I.363, 1991 [9].

odd sequence numbers. The residual time stamp is transmitted over 8 cells. It supports DS1 and DS3 and E1 digital streams. Another mode of operation is structured data transfer (SDT). SDT supports an octet-structured nXDS0 service.

Alarm indication in this adaptation layer is via a check of the one's density. When the one's density of the received cell stream becomes significantly different than the density used for the particular PCM line coding scheme in use, it is determined that the system has lost signal and alarm notifications are given.

6.1.3.3.2 AAL-2. AAL-2 handles the variable bit rate (VBR) scenario such as MPEG* video. Functions in AAL type 2 include:

(a) Segmentation and reassembly (SAR) of user information
(b) Handling of cell delay variation
(c) Handling of lost and misinserted cells
(d) Source clock frequency recovery at the receiver
(e) Monitoring and handling of AAL-PCI bit errors (PCI = protocol control information)
(f) Monitoring of user information field for bit errors and possible corrective action

AAL-2 is still in ITU-T definitive stages. However, an example of a SAR-PDU format for AAL type 2 is given in Figure 16.14.

6.1.3.3.3 AAL-3/4. Initially, in ITU-T Rec. I.363 1991, there were two separate AALs, one for connection-oriented variable bit rate data services (AAL-3) and

---

* MPEG is a set of video compression schemes. MPEG stands for Motion Picture Experts Group.

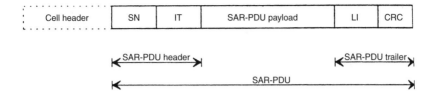

The need of each of the above fields, the position of those fields and their size are for further study.

SN   Sequence number, to detect lost or misinserted cells. A specific value of the sequence number may indicate a specified purpose

IT   Information type, used to indicate beginning of message (BOM), continuation of message (COM), end of message (EOM), timing information and also component of the video or audio signal

LI   Length indicator, check to indicate that the number of octets of the CS-PDU are included in the SAR-PDU payload field

CRC   Cyclic redundancy check code, to correct up to two correlated bit errors.

Figure 16.14. Example of an SAR-PDU format for AAL-2. From Figure 2/I.363, page 5, ITU-T Rec. I.363, 1991 [9].

one for connectionless service (AAL-4). As the specifications evolved, the same procedures turned out to be necessary for both of these services, and the specifications were merged to become the AAL-3/4 standard. AAL-3/4 is used for ATM shipping of SMDS, CBDS (connectionless broadband data services, an ETSI initiative), IP (internet protocol), and frame relay.

AAL-3/4 has been designed to take variable length frames/packets and segment them into cells. The segmentation is done in a way that protects the transmitted data from corruption if cells are lost or mis-sequenced.

Variable length packets (up to 64 kbytes) from SMDS/CLNAP (CLNAP stands for connectionless network access protocol) or frame relay frames are padded to an integral word length and encapsulated with a header and a trailer to form what is called the convergence sublayer PDU (CS_PDU) and then is segmented into cells. The passing is done to make sure that fields align themselves to 32-bit boundaries, allowing the efficient implementation of the operations in hardware at lower layers. The added header and trailer contain a tag to match the end and the packet length so that the receiving end may allocate the buffer for this packet upon reception of the first fragment. In practice this Buffer Allocation size (BAsize) is mostly used as a length check for integrity verification, because the most efficient algorithm is to allocate fixed size maximum length buffers. This AAL convergence sublayer protocol data unit (CS_PDU) is shown in Figure 16.15.

Figure 16.16 shows the SAR-PDU format for AAL-3/4 from ITU-T Rec. I.363, and Figure 16.17 shows this same format when used for transmitting SMDS frames.

Turning to Figure 16.17, the segmented portions are 44 octets long (except possibly for the last segment). These portions are then encapsulated with another header (2 octets) and trailer (2 octets) to become a segmentation and reassembly PDU (SAR_PDU) which is inserted into cell payloads. The header at this level with the segment-type field identifies what kind of a cell it is [i.e., BOM (beginning of message), COM (continuation of message), or EOM (end of message)]

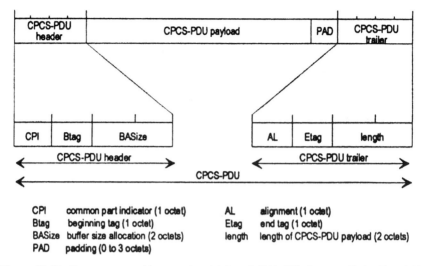

CPI   common part indicator (1 octet)   AL   alignment (1 octet)
Btag   beginning tag (1 octet)   Etag   end tag (1 octet)
BASize buffer size allocation (2 octets)   length length of CPCS-PDU payload (2 octets)
PAD   padding (0 to 3 octets)

Figure 16.15. Convergence sublayer protocol data unit (AAL-3/4). *Sources:* Refs. 16 and 17.

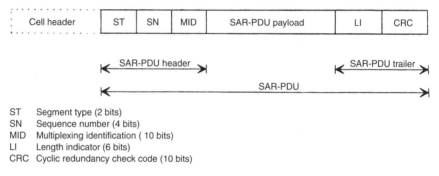

ST   Segment type (2 bits)
SN   Sequence number (4 bits)
MID Multiplexing identification (10 bits)
LI   Length indicator (6 bits)
CRC Cyclic redundancy check code (10 bits)

Figure 16.16. SAR-PDU format for AAL-3/4. From Figure 6/I.363, page 13, ITU-T Rec. I.363, 1991 [9].

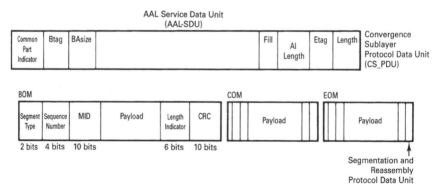

Figure 16.17. AAL-3/4 SAR-PDU as applied to the transmission of SMDS frames. Courtesy of Hewlett-Packard Company [10].

so that the individual CS_PDUs can be delineated. The header also includes a sequence number for protection against misordered delivery. It also includes the MID (message identification in SMDS, multiplexing identifier for ATM). The SAR_PDU trailer contains a length indicator to identify how much of the payload is filled. It also has a CRC-10 error check to protect against cell corruption. A complete message contains a BOM cell, zero or more COM cells, and an EOM cell. If the entire message can fit into one cell, it is called a single segment message (SSM), where the CS_PDU is less than 44 octets long.

AAL-3/4 has several measures to ensure the integrity of the data which has been segmented and transmitted as cells. The contents of the cell are protected by the CRC-10; sequence numbers protect against misordering. Still another measure to ensure against corrupted PDUs being delivered is EOM/BOM protection. If the EOM of one CPCS_PDU and the BOM of the next are dropped for some reason, the resulting cell stream could be interpreted as a valid PDU. To protect against these kinds of errors, the BEtag numeric values in the CPCS_PDU headers and trailers are compared to ensure that they match (CPCS = common part convergence sublayer). Two modes of service are defined for AAL-3/4:

- *Message Mode Service.* This provides for the transport of one or more fixed size AAL service data units in one or more convergence sublayer protocol data units (CS-PDUs).
- *Streaming Mode Service.* Here the AAL service data unit is passed across the AAL interface in one or more AAL interface data units (IDUs). The transfer of these AAL-IDUs across the AAL interface may occur separated in time, and this service provides the transport of variable length AAL-SDUs. The streaming mode service includes an abort service by which the discarding of an AAL-SDU partially transferred across the AAL interface can be requested. In other words, in the streaming mode, a single packet is passed to the AAL layer and transmitted in multiple CPCS-PDUs, as and when pieces of the packet are received. Streaming mode may be used in intermediate switches or ATM-to-SMDS routers so they can begin retransmitting a packet being received before the entire packet has arrived. This reduces the latency experienced by the entire packet.

6.1.3.3.4   AAL-5.   This type of AAL was designed specifically to carry data traffic typically found in today's LANs. AAL-5 evolved after AAL-3/4, which was found to be too complex and inefficient for LAN traffic. Thus AAL-5 got the name "SEAL" for simple and efficient AAL layer. Only a small amount of overhead is added to the CPCS-PDU and no extra overhead is added when the AAL-5 segments them into SAR-PDUs. There is no AAL level cell multiplexing. In AAL-5 all cells belonging to an AAL-5 CPCS-PDU are sent sequentially.

As shown in Figure 16.18, the CPCS-PDU has only a payload and a trailer. The trailer contains padding, a length field, and a CRC-32 field for error detection. The CPCS-PDUs are padded to become integral multiples of 48, ensuring that there will never be a need to send partially filled cells after segmentation. A bit

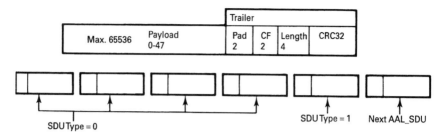

Figure 16.18. AAL Type 5. Courtesy of Hewlett-Packard Company [10].

in the PTI field in the cell headers is used to indicate when the last cell of a PDU is transmitted, so that one PDU can be distinguished from the one that follows.

7   SERVICES: CONNECTION-ORIENTED AND CONNECTIONLESS

The issues such as routing decisions and architectures have a major impact on connection-oriented services, where B-ISDN/ATM end nodes have to maintain or get access to look-up tables which translate destination addresses into circuit paths. These circuit path look-up tables which differ at every node must be maintained in a quasi-real-time fashion. This will have to be done by some kind of routing protocol.

One way to resolve this problem is to make it an internal network problem and use a connectionless service as described in ITU-T Rec. I.364 [11]. We must keep in mind that ATM is basically a connection-oriented service. Here we are going to adapt it to provide a connectionless service.

7.1   Functional Architecture

The provision of connectionless data service in the B-ISDN is carried out by means of ATM switches and connectionless service functions (CLSF). ATM switches support the transport of connectionless data units in the B-ISDN between specific functional groups where the CLSF handles the connectionless protocol and provides for the adaptation of the connectionless data units into ATM cells to be transferred in a connection-oriented environment. As shown in Figure 16.19, CLSF functional groups may be located outside the B-ISDN, in a private connectionless network or in a specialized service provider, or inside the B-ISDN.

The ATM switching is performed by the ATM nodes (ATM switch/cross-connect) which are a functional part of the ATM transport network. The CLSF functional group terminates the B-ISDN connectionless protocol and includes functions for the adaptation of the connectionless protocol to the intrinsically connection-oriented ATM layer protocol. These latter functions are performed by the ATM adaptation layer type 3/4 (AAL-3/4), while the CLSF group terminations are carried out by the services layer above the AAL called the CLNAP (connectionless network access protocol). The CL protocol includes functions such as

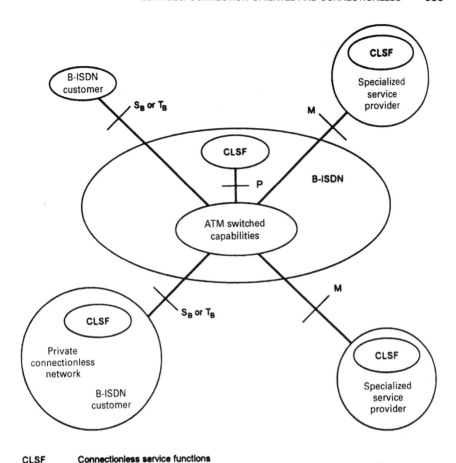

CLSF   **Connectionless service functions**
P, M, S, T   **Reference point**

Figure 16.19. Reference configuration for the provision of the CL (connectionless) data service in the B-ISDN. From Figure 1/I.364, page 2, ITU-T Rec. I.364, March 1993 [11].

routing, addressing, and QoS (quality of service) selection. In order to perform the routing of CL data units, the CLSF has to interact with the control/management planes of the underlying ATM network.

The general protocol structure for the provision of connectionless (CL) data service is shown in Figure 16.20. Figure 16.21 shows the protocol architecture for supporting connectionless layer service. The CLNAP (connectionless network access protocol) layer uses the type 3/4 AAL unassured service and includes the necessary functionality to provide the connectionless layer service.

The connectionless service layer provides for transparent transfer of variable size data units from source to one or more destinations in a manner such that lost or corrupted data units are not retransmitted. This transfer is performed using a connectionless technique, including embedding destination and source addresses into each data unit.

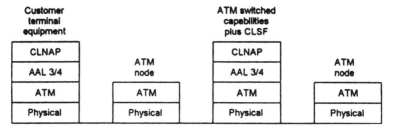

Figure 16.20. General protocol structure for provision of CL data service in B-ISDN.

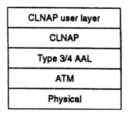

Figure 16.21. Protocol architecture for supporting connectionless service. CLNAP, connectionless network access protocol.

7.2 CLNAP Protocol Data Unit (PDU) and Encoding

Figure 16.22 shows the detailed structure of the CLNAP-PDU which contains the following fields:

Destination Address and Source Address. These 8-octet fields each contain a 4-bit address-type subfield, followed by the 60-bit *address* subfield. The "*address-type*" subfield indicates whether the "address" subfield contains a publicly administered 60-bit individual address or a publicly administered 60-bit group address. The "address" subfield indicates to which CLNAP-entity(ies) the CLNAP-PDU is destined; and in the case of the source address, it indicates the CLNAP-entity that sourced the CLNAP-PDU. The encoding of the "address-type" and "address" are shown in Figures 16.23A and 16.23B. The address is structured in accordance with ITU-T Rec. E.164 [12].

Higher-Layer-Protocol-Identifier (HLPI). This 6-bit field is used to identify the CLNAP user layer entity which the CLNAP-SDU is to be passed to at the destination node. It is transparently carried end-to-end by the network.

PAD Length. This 2-bit field gives the length of the PAD field (0–3 octets). The number of PAD octets is such that the total length of the user-information field and the PAD field together is an integral multiple of four octets (32 bits).

QoS (Quality of Service). 4 bits (assignment under study).

CRC Indication Bit (CIB). This 1-bit field indicates the presence (if CIB = 1) or absence (if CIB = 0) of a 32-bit CRC field.

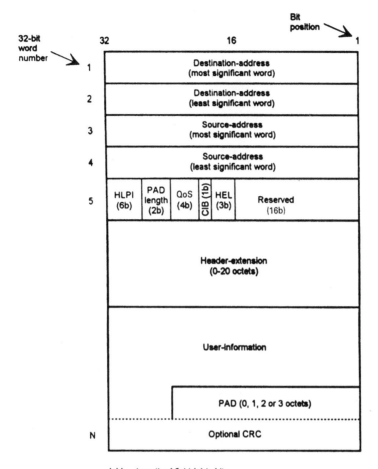

(nb)   Length of field (n) in bits

Figure 16.22. Structure of the CLNAP-PDU. From Figure 5/I.364, page 7, ITU-T Rec. I.364, March 1993 [11].

Address type	Meaning
1100	60-bit publicly administered individual address
1110	60-bit publicly administered group address

Figure 16.23A. Destination address field.

Address type	Meaning
1100	60-bit publicly administered individual address

Figure 16.23B. Source address field.

Header Extension Length (HEL). This 3-bit field can take on any value from 0 to 5 and indicates the number of 32-bit words in the header extension field.

Reserved. This 16-bit field is reserved for future use. Its default value is 0.

Header Extension. This variable-length field can range from 0 to 20 octets. Its length is indicated by the value of the header extension length field (see above). In the case where the header extension length (HEL) is not equal to zero, all unused octets in the header extension are set to zero. The information carried in the header extension is structured into information entities. An information entity (element) consists of (in this order) of element length, element type, and element payload.

> *Element Length*: This is a 1-octet field and contains the combined lengths of the element length, element type, and element payload in octets.
>
> *Element Type*: This is also a 1-octet field and contains a binary coded value which indicates the type of information found in the element payload field.
>
> *Element Payload*: This is a variable length field and contains the information indicated by the element-type field.

TABLE 16.5   ATM Service Classes

Service Class	Quality of Service Parameter
Constant bit rate (CBR)	This class is used for emulating circuit switching. The cell rate is constant with time. CBR application are quite sensitive to cell delay variation. Examples of applications that can use CBR are telephone traffic (i.e., $n \times 64$ kbits/s), videoconferencing and television.
Variable bit rate — nonreal time (VBR-NRT)	This class allows users to send traffic at a rate that varies with time, depending on the availability of user information. Statistical multiplexing is provided to make optimum use of network resources. Multimedia e-mail is an example of VBR-NRT.
Variable bit rate — real time (VBR-RT)	This class is similar to VBR-NRT but is designed for applications that are sensitive to cell-delay variation. Examples for real-time VBR are voice with speech activity detection (SAD) and interactive compressed video.
Available bit rate (ABR)	This class of ATM services provides rate-based flow control and is aimed at data traffic such as file transfer and e-mail. Although the standard does not require the cell transfer delay and cell-loss ratio to be guaranteed or minimized, it is desirable for switches to minimize delay and loss as much as possible. Depending upon the state of congestion in the network, the source is required to control its rate. The users are allowed to declare a minimum cell rate, which is guaranteed to the connection by the network.
Unspecified bit rate (UBR)	This class is the catch-all. It is widely used today for TCP/IP.

User Information. This field is variable length up to 9188 octets and is used to carry the CLNAP-SDU.

PAD. This field is 0, 1, 2, or 3 octets in length and is coded as all zeros. Within each CLNAP-PDU the length of this field is selected such that the length of the resulting CLNAP-PDU is aligned on a 32-bit boundary.

CRC. This optional 32-bit field may be present or absent as indicated by the CIB field. The field contains the result of a standard CRC-32 calculation performed over the CLNAP-PDU with the "Reserved" field always treated as if it were coded as all zeros.

7.3 ATM Classes of Service

ATM is connection oriented and allows the user to specify the resources required on a per-connection basis (per SVC) dynamically. There are five classes of service defined for ATM (as per ATM Forum 4.0 specification [22]). The QoS parameters for these services are defined in Table 16.5. Also see Figure 16.12.

8 ASPECTS OF A B-ISDN/ATM NETWORK

8.1 ATM Routing and Switching

An ATM transmission path supports virtual paths (VPs), and inside virtual paths are virtual channels (VCs) as shown in Figure 16.24.

As we discussed in Section 4.1, each ATM cell contains a label in its header to explicitly identify the VC to which the cell belongs. This label consists of two parts: a virtual channel identifier (VCI) and a virtual path identifier (VPI).

8.1.1 The Virtual Channel Level. A virtual channel (VC) is a generic term used to describe a unidirectional communication capability for the transport of ATM cells. A VCI identifies a particular VC link for a given virtual path connection (VPC). A specific value of VCI is assigned each time a VC is switched in the network. A VC link is a unidirectional capability for the transport of

Figure 16.24. Relationship between the VC, the VP and the transmission path.

ATM cells between two consecutive ATM entities where the VCI value is translated. A VC link is originated or terminated by the assignment or removal of the VCI value.

Routing functions of virtual channels are done at a VC switch/cross-connect.* The routing involves translation of the VCI values of the incoming VC links into the VCI values of the outgoing VC links.

Virtual channel links are concatenated to form a virtual channel connection (VCC). A VCC extends between two VCC endpoints or, in the case of point-to-multipoint arrangements, more than two VCC endpoints. A VCC endpoint is the point where the cell information field is exchanged between the ATM layer and the user of the ATM layer service. At the VC level, VCCs are provided for the purpose of user–user, user–network, or network–network information transfer. Cell sequence integrity is preserved by the ATM layer for cells belonging to the same VCC.

8.1.2   Virtual Path Level. A virtual path (VP) is a generic term for a bundle of virtual channel links; all the links in a bundle have the same endpoints.

A VPI identifies a group of VC links, at a given reference point, that share the same VPC. A specific value of VPI is assigned each time a VP is switched in the network. A VP link is a unidirectional capability for the transport of ATM cells between two consecutive ATM entities where the VPI value is translated. A VP link is originated or terminated by the assignment or removal of the VPI value.

Routing functions for VPs are performed at a VP switch/cross-connect. This routing involves translation of the VPI values of the incoming VP links into the VPI values of the outgoing VP links.

VP links are concatenated to form a VPC. A VPC extends between two VPC endpoints or, in the case of point-to-multipoint arrangements, there are more than two VPC endpoints. A VPC endpoint is the point where the VCIs are originated, translated, or terminated. At the VP level, VPCs are provided for the purpose of user–user, user–network, and network–network information transfer.

When VPCs are switched, the VPC supporting the incoming VC links are terminated first and a new outgoing VPC is then created. Cell sequence integrity is preserved by the ATM layer for cells belonging to the same VPC. Thus cell sequence integrity is preserved for each VC link within a VPC.

Figure 16.25 is a representation of the VP and VC switching hierarchy where the physical layer is the lowest layer composed of, from bottom up, a regenerator section level, a digital section level, and a transmission path level. The ATM layer resides just above the physical layer and is composed of a VP level, and just above that is the VC level.

Source: Section 8 is based on ITU-T I.311 [13].

---

* A VC cross-connect is a network element which connects VC links. It terminates VPCs and translates VCI values and is directed by management plane functions, not by control plane functions.

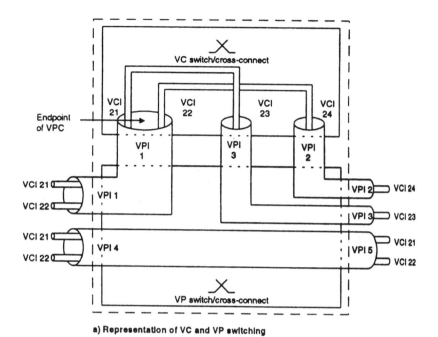

a) Representation of VC and VP switching

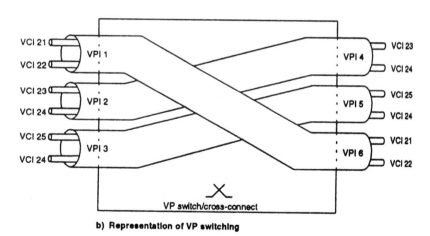

b) Representation of VP switching

Figure 16.25. Representation of the VP and VC switching hierarchy. From Figure 4/I.311, page 5, ITU-Rec. I.311, March 1993 [13].

9 SIGNALING REQUIREMENTS

9.1 Setup and Release of VCCs

The setup and release of VCCs at the user–network interface (UNI) can be performed in various ways:

- Without using signaling procedures. Circuits are set up at subscription with permanent or semipermanent connections.
- By meta-signaling procedures where a special VCC is used to establish or release a VCC used for signaling. Meta-signaling is a simple protocol used to establish and remove signaling channels. All information interchanges in meta-signaling are carried out via single cell messages.
- User-to-network signaling procedures such as a signaling VCC to establish or release a VCC used for end-to-end connectivity.
- User-to-user signaling procedures such as a signaling VCC to establish or release a VCC within a preestablished VPC between two UNIs.

9.2 Signaling Virtual Channels

9.2.1 Requirements for Signaling Virtual Channels.
For a point-to-point signaling configuration, the requirements for signaling virtual channels are as follows:

- One virtual channel connection in each direction is allocated to each signaling entity. The same VPI/VCI value is used in both directions. A standardized VCI value is used for point-to-point signaling virtual channel (SVC).
- In general, a signaling entity can control, by means of associated point-to-point SVCs (signaling virtual channels), user-VCs belonging to any of the virtual paths (VPs) terminated in the same network element.
- As a network option, the user-VCs controlled by a signaling entity can be constrained such that each controlled user-VC is in either upstream or downstream VPs containing the point-to-point SVCs of the signaling entity.

For point-to-multipoint signaling configurations, the requirements for signaling virtual channels are as follows:

(a) *Point-to-Point Signaling Virtual Channels.* For point-to-point signaling, one virtual channel connection in each direction is allocated to each signaling entity. The same VPI/VCI value is used in both directions.

(b) *General Broadcast Signaling Virtual Channel.* The general broadcast signaling virtual channel (GBSVC) may be used for call offering in all cases. In cases where the "point" does not implement service profiles or where "the multipoints" do not support service profile identification, the GBSVC is used for call offering. The specific VCI value for general broadcast signaling is reserved per VP at the UNI. Only when meta-signaling is used in a VP is the GBSVC activated in the VP.

(c) *Selective Broadcast Signaling Virtual Channels.* Instead of the GBSVC, a virtual channel connection for selective broadcast signaling (SBS) can be used for call offering, in cases where a specific service profile is used. No other uses for SBSVCs are foreseen.

9.3 Meta-Signaling

9.3.1 Meta-Signaling Requirements. A meta-signaling channel manages signaling virtual channels only within its own VP pair. In VPI = 0, the meta-signaling virtual channel is always present and has a standardized VCI value.

Meta-signaling VC is activated at VP establishment. The signaling virtual channel (SVC) is assigned and removed when necessary.

A specific VCI value for meta-signaling is reserved per VP at the UNI. For a VP with point-to-multipoint signaling configuration, meta-signaling is required and the meta-signaling VC within this VP is activated.

The user negotiates the SVC bandwidth (bit rate capacity) parameter value. The meta-signaling virtual channel (MSVC) bandwidth has a default value. The bandwidth can be changed by mutual agreement between a network operator and user.

9.3.2 Meta-Signaling Functions at the User Access. In order to establish, check, and release the point-to-point and selective broadcast signaling virtual channel connections, meta-signaling procedures are provided. For each direction, meta-signaling is carried out in a permanent virtual channel connection having a standardized VCI value. The channel is called the meta-signaling virtual channel. The meta-signaling protocol is terminated in the ATM layer management entity.

The meta-signaling function is required to:

- Manage the allocation of capacity to signaling channels
- Establish, release, and check the status of signaling channels
- Provide a means to associate a signaling endpoint with a service profile if service profiles are supported
- Provide a means to distinguish between simultaneous requests

Meta-signaling should be able to be supported on any VP; however, meta-signaling can only control signaling VCs within its VP.

Source: Section 9 is based on Ref. 11.

10 QUALITY OF SERVICE (QOS)

10.1 ATM Service Quality Review

A basic performance measure for any digital data communication system is bit error rate (BER). Well-designed fiber-optic links will predominate now and into the foreseeable future. We may expect BERs from such links on the order of 1×10^{-10} and with end-to-end performance better than 1×10^{-9} [10].* Thus other performance issues may dominate the scene. These may be called ATM unique QoS items, namely:

---

* These seem like ambitious goals if end-to-end. They should be considered in light of ITU-T Rec. G.821 and Rec. G.826.

- Cell transfer delay
- Cell delay variation
- Cell loss ratio
- Mean cell transfer delay
- Cell error ratio
- Severely errored cell block ratio
- Cell misinsertion rate

10.2 QoS Parameter Descriptions

10.2.1 Cell Transfer Delay. In addition to the normal delay through network elements and lines, extra delay is added to an ATM network at an ATM switch. The cause of the delay at this point is the statistical asynchronous multiplexing. Using this method, two cells can be directed toward the same output of an ATM switch or cross-connect resulting in output contention.

The result is that one cell or more is held in a buffer until the next available opportunity to continue transmission. We can see that the second cell will suffer additional delay. The delay of a cell will depend upon the amount of traffic within a switch and thus the probability of contention.

The asynchronous path of each ATM cell also contributes to cell delay. Cells can be delayed one or many cell periods, depending on traffic intensity, switch sizing, and the transmission path taken through the network.

10.2.2 Cell Delay Variation (CDV). ATM traffic by definition is asynchronous magnifying transmission delay. Delay is also inconsistent across the network. It can be a function of time (i.e., a moment in time), network design/switch design (such as buffer size), and traffic characteristics at that moment in time. The result is cell delay variation (CDV).

CDV can have several deleterious effects. The dispersion effect, or spreading out, of cell inter-arrival times can impact signaling functions or the reassembly of cell user data. Another effect is called *clumping*. This occurs when the inter-arrival times between transmitted cells shorten. One can imagine how this could affect the instantaneous network capacity and how it can impact other services using the network.

There are two performance parameters associated with cell delay variation: 1-point delay variation (1-point CDV) and 2-point cell delay variation (2-point CDV).

The 1-point CDV describes variability in the pattern of cell arrival events observed at a single boundary with reference to the negotiated peak rate $1/T$ as defined in ITU-T Rec. I.371 [14]. The 2-point CDV describes variability in the pattern of cell arrival events as observed at the output of a connection portion (MP$_2$) with reference to the pattern of the corresponding events observed at the input to the connection portion (MP$_1$).

10.2.3 Cell Loss Ratio. Cell loss may not be uncommon in an ATM network. There are two basic causes of cell loss: error in cell header or network congestion.

Cells with header errors are automatically discarded. This prevents misrouting of errored cells, as well as the possibility of privacy and security breaches.

Switch buffer overflow can also cause cell loss. It is in these buffers that cells are held in prioritized queues. If there is congestion, cells in a queue may be discarded selectively in accordance with their level of priority. Here enters the CLP (cell loss priority) bit discussed in Section 15.3. Cells with this bit set to 1 are discarded in preference to other, more critical cells. In this way, buffer fill can be reduced to prevent overflow [10].

Cell loss ratio is defined for an ATM connection as

Lost cells/Total transmitted cells

Lost and transmitted cells counted in severely errored cell blocks should be excluded from the cell population in computing cell loss ratio [2].

10.2.4 Mean Cell Transfer Delay. Mean cell transfer delay is defined as the arithmetic average of a specified number of cell transfer delays.

10.2.5 Cell Error Ratio. Cell error ratio is defined as follows for an ATM connection:

Errored cells/(Successfully transferred cells + Errored cells)

Successfully transferred cells and errored cells contained in cell blocks counted as severely errored cell blocks should be excluded from the population used in calculating cell error ratio.

10.2.6 Severely Errored Cell Block Ratio. The severely errored Cell Block Ratio for an ATM connection is defined as

Total severely errored cell blocks/Total transmitted cell blocks

A cell block is a sequence of N cells transmitted consecutively on a given connection. A severely errored cell block outcome occurs when more than M errored cells, lost cells, or misinserted cell outcomes are observed in a received cell block.

For practical measurement purposes, a cell block will normally correspond to the number of user information cells transmitted between successive OAM cells. The size of a cell block is to be specified.

10.2.7 Cell Misinsertion Rate. The cell misinsertion rate for an ATM connection is defined as

Misinserted cells/Time interval

This rate may be expressed equivalent as the number of misinserted user information cells per virtual connection.

A misinserted cell is a received cell that has no corresponding transmitted cell on that connection. Cell misinsertion on a particular connection can be caused by an undetected or miscorrected error in the header of a cell originated on a different connection or by an incorrectly programmed translation of VPI or VCI values for cells originated on a different connection.

Source: Refs. 2, 10, and 18.

11 TRAFFIC CONTROL AND CONGESTION CONTROL

11.1 Generic Functions

The following functions form a framework for managing and controlling traffic and congestion in ATM networks and are to be used in appropriate combinations from the point of view of ITU-T Rec. I.371 [14]:

1. *Network Resource Management (NRM)*. Provision is used to allocate network resources in order to separate traffic flows in accordance with service characteristics.

2. *Connection Admission Control (CAC)*. This is defined as a set of actions taken by the network during the call setup phase or during the call renegotiation phase in order to establish whether a virtual channel (VC) or virtual path (VP) connection request can be accepted or rejected, or whether a request for re-allocation can be accommodated. Routing is part of connection admission control actions.

3. *Feedback Controls*. These are a set of actions taken by the network and by users to regulate the traffic submitted on ATM connections according to the state of network elements.

4. *Usage/Network Parameter Control (UPC/NPC)*. This is a set of actions taken by the network to monitor and control traffic, in terms of traffic offered and validity of the ATM connection, at the user access and network access, respectively. Their main purpose is to protect network resources from malicious as well as unintentional misbehavior which can affect the QoS of other already established connections by detecting violations of negotiated parameters and taking appropriate actions.

5. *Priority Control*. The user may generate different priority traffic flows by using the CLP. A congested network element may selectively discard cells with low priority, if necessary, to protect as far as possible the network performance for cells with higher priority.

Figure 16.26 is a reference configuration for traffic and congestion control.

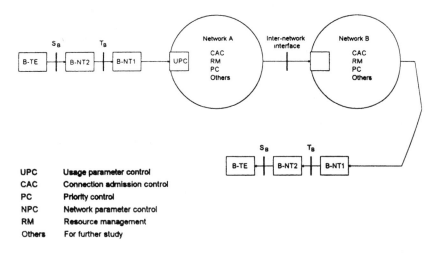

UPC	Usage parameter control
CAC	Connection admission control
PC	Priority control
NPC	Network parameter control
RM	Resource management
Others	For further study

NOTES

1 NPC may apply as well at some intra-network NNIs.

2 The arrows are indicating the direction of the cell flow.

Figure 16.26. Reference configuration for traffic control and congestion control. From Figure 1/I.371, page 3, ITU-T Rec. I.371, March 1993 [14].

11.2 Events, Actions, Time Scales, and Response

Figure 16.27 shows the time scales over which various traffic control and congestion control functions can operate. The response time defines how quickly the controls react. For example, call discarding can react on the order of the insertion time of a cell. Similarly, feedback controls can react on the time scale of round-trip propagation times. Since traffic control and resource management functions are needed at different time scales, no single function is likely to be sufficient.

11.3 Quality of Service, Network Performance, and Cell Loss Priority

QoS at the ATM layer is defined by a set of parameters such as cell delay, cell delay variation sensitivity, cell loss ratio, and so forth.

A user requests a specific ATM layer QoS from the QoS classes which a network provides. This is part of the *traffic contract* at connection establishment. It is a commitment for the network to meet the requested QoS as long as the user complies with the traffic contract. If the user violates the traffic contract, the network need not respect the agreed-upon QoS.

A user may request at most two QoS classes for a single ATM connection, which differ with respect to the cell loss ratio objectives. The cell loss priority (CLP) bit in the ATM header allows for two cell loss ratio objectives for a given ATM connection.

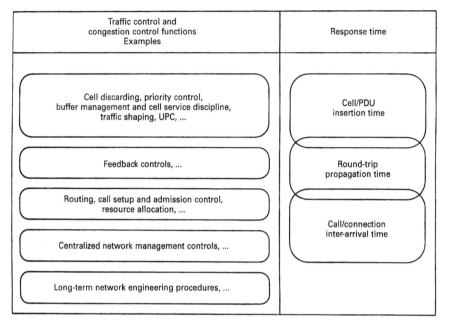

Figure 16.27. Control response times. From Figure 2/I.371, page 4, ITU-T Rec. I.371, March 1993 [14].

Network performance objectives at the ATM SAP (service access point) are intended to capture the network ability to meet the requested ATM layer QoS. It is the role of upper layers, including the AAL, to translate this ATM layer QoS to any specific application requested QoS.

11.4   Traffic Descriptors and Parameters

Traffic parameters describe traffic characteristics of an ATM connection. Traffic parameters are grouped into source traffic descriptors for exchanging information between the user and the network. Connection admission control procedures use source traffic descriptors to allocate resources and derive parameters for the operation of the UPC/NPC (user parameter control/network parameter control).

We now define the terms *traffic parameters* and *traffic descriptors* [14]:

Traffic Parameters. A traffic parameter is a specification of a particular traffic aspect. It may be qualitative or quantitative. Traffic parameters, for example, may describe peak cell rate, average cell rate, burstiness, peak duration, and source type (such as telephone, videophone). Some of these traffic parameters are interdependent such as burstiness with average and peak cell rate.

Traffic Descriptors. The ATM traffic descriptor is the generic list of traffic parameters which can be used to capture the intrinsic traffic characteristics of an ATM connection. A *source traffic descriptor* is the set of traffic parameters belonging to the ATM traffic descriptor used during the connection setup to capture the intrinsic traffic characteristics of the connection requested by the source.

Connection Traffic Descriptor. This specifies the traffic characteristics of the ATM connection at the public or private UNI. The connection traffic descriptor is the set of traffic parameters in the source traffic descriptor, cell delay variation (CDV) tolerance, and the conformance definition that is used to unambiguously specify the conforming cells of an ATM connection. Connection admission control (CAC) procedures will use the connection traffic descriptor to allocate resources and to derive parameter values for the operation of the UPC. The connection traffic descriptor contains the necessary information for conformance testing of cells of the ATM connection at the UNI.

Any traffic parameter and the CDV tolerance in a connection traffic descriptor should fulfill the following requirements:

- They should be understandable by the user or terminal equipment and conformance testing should be possible as stated in the traffic contract.
- They should be useful in resource allocation schemes meeting Network Performance requirements as described in the traffic contract.
- They should be enforceable by the UPC.

Source: Ref. 14.

11.5   User–Network Traffic Contract

11.5.1   Operable Conditions. CAC and UPC/NPC procedures require the knowledge of certain parameters to operate efficiently. For example, they should take into account the source traffic descriptor, the requested QoS, and the CDV tolerance (defined below) in order to decide whether the requested connection can be accepted.

The source traffic descriptor, the requested QoS for any given ATM connection, and the maximum CDV tolerance allocated to the CEQ (customer equipment) define the traffic contract at the T_B reference point (see Figure 16.26). Source traffic descriptors and QoS are declared by the user at connection setup by means of signaling or subscription. Whether the maximum allowable cell delay variation (CDV) tolerance is also negotiated on a subscription or on a per connection basis is for further study by the ITU-T organization.

The connection admission control (CAC) and usage/network parameter control (UPC/NPC) procedures are operator-specific. Once the connection has been accepted, the value of the CAC and UPC/NPC parameters are set by the network on the basis of the network operator's policy. ITU-T Rec. I.371 [14] notes that all ATM connections handled by network connection related functions (CRF) have to be declared and enforced by the UPC/NPC. ATM layer QoS can only be assured for compliant ATM connections. As an example, individual VCCs inside user end-to-end VPC are neither declared nor enforced at the UPC and hence no ATM layer QoS can be assured for them.

11.5.2 Source Traffic Descriptor, Quality of Service, and Cell Loss Priority.
If a user requests two levels of priority for an ATM connection, as indicated by the CLP bit value, the intrinsic traffic characteristics of both cell flow components have to be characterized in the source traffic descriptor. This is by means of a set of traffic parameters associated with the $CLP = 0$ component and a set of traffic parameters associated with the $CLP = 0 + 1$ component.

As discussed above, the network provides an ATM layer QoS for each of the components ($CLP = 0$ and $CLP = 0 + 1$) of an ATM connection. The traffic contract specifies the particular QoS choice (from those offered by the network operator) for each of the ATM connection components. There may be a limited offering of QoS specifications for the $CLP = 1$ component, according to Ref. 14.

11.5.3 Impact of Cell Delay Variation on UPC/NPC and Resource Allocation.
ATM layer functions such as cell multiplexing may alter the traffic characteristics of ATM connections by introducing cell delay variation (CDV) as shown in Figure 16.28. When cells from two or more ATM connections are multiplexed, cells of a given ATM connection may be delayed while cells of another ATM connection are being inserted at the output of the multiplexer. Similarly, some cells may be delayed while physical layer overhead or OAM cells are inserted. Therefore, some randomness affects the time interval between reception of ATM cell data requests at the endpoint of an ATM connection to the time that an ATM cell data-indication is received at the UPC/NPC. Besides, AAL multiplexing may cause CDV.

The UPC/NPC mechanism will not discard or tag cells in an ATM connection if the source conforms to the source traffic descriptor negotiated at connection establishment. However, if the CDV is not bounded at a point where the UPC/NPC function is performed, it is not possible to design a suitable UPC/NPC mechanism and to allocate resources properly. Therefore, ATM specifications (e.g., I.371) require that a maximum allowable value of CDV be standardized edge-to-edge (e.g., between the ATM connection endpoint and T_B, between T_B and an inter-network interface, and between inter-network interfaces). (See Figure 16.28.)

UPC/NPC should accommodate the effect of the maximum CDV allowed on ATM connections within the limit resulting from the accumulated CDV allocated to upstream subnetworks including customer equipment (CEQ). Traffic shaping partially compensates for the effects of CDV on the peak cell rate of the ATM connection. Examples of traffic shaping mechanisms are re-spacing cells of individual ATM connections according to their peak cell rate or suitable queue service schemes. Values of the CDV are network performance issues.

It has been found that the definition of a source traffic descriptor and the standardization of a maximum allowable CDV may not be sufficient for a network to allocate resources properly. When allocating resources, the network should take into account the worst-case traffic passing through the UPC/NPC in order to

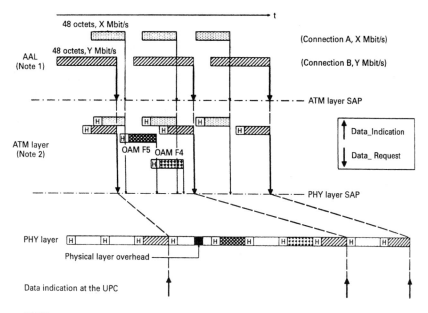

Figure 16.28. Origins of cell delay variation. From Figure 3/I.371, page 7, ITU-T Rec. I.371, March 1993 [14].

avoid impairments to other ATM connections. The worst-case traffic depends on the specific implementation of the UPC/NPC. The trade-offs between UPC/NPC complexity, worst-case traffic, and optimization of network resources are made at the discretion of network operators. The quantity of available network resources and the network performance to be provided for meeting QoS requirements can influence these trade-offs.

11.5.4 Traffic Contract Parameter Specification.

Peak cell rate for CLP $= 0 + 1$ is a mandatory traffic parameter to be explicitly or implicitly declared in any source traffic descriptor. In addition to the peak cell rate of an ATM connection, it is mandatory for the user to declare either explicitly or implicitly the cell delay variation tolerance τ within the relevant traffic contract [2].

PEAK CELL RATE (PCR). The following definition applies to ATM connections supporting both CBR and VBR services [14]:

> The peak cell rate in the source traffic descriptor specifies an upper bound on the traffic that can be submitted on an ATM connection. Enforcement of this bound by the UPC/NPC allows the network operator to allocate sufficient resources to ensure that the performance objectives (e.g., for cell loss ratio) can be achieved.

For switched ATM connections, the peak cell rate for $CLP = 0 + 1$ and the QoS class must be explicitly specified for each direction in the connection-establishment SETUP message. The cell delay variation tolerance must be either explicitly specified at subscription time or implicitly specified.

The sustainable cell rate (SCR) and burst tolerance is an optional traffic parameter set in the source traffic descriptor. If either SCR or burst tolerance is specified, then the other must be specified within the relevant traffic contract.

12 TRANSPORTING ATM CELLS

12.1 In the DS3 Frame

One of the most popular higher speed digital transmission systems in North America is DS3 operating at a nominal transmission rate of 45 Mbits/s. It is also being widely implemented for transport of SMDS. The system used to map ATM cells into the DS3 format is the same as used for SMDS.

DS3 uses the physical layer convergence protocol (PLCP) to map ATM cells into its bit stream. A DS3 PLCP frame is shown in Figure 16.29.

There are 12 cells in a frame. Each cell is preceded by a 2-octet framing pattern (A1, A2), to enable the receiver to synchronize to cells. After the framing pattern there is an indicator consisting of one of 12 fixed bit patterns used to identify the cell location within the frame (POI). This is followed by an octet of overhead information used for path management. The entire frame is then padded with

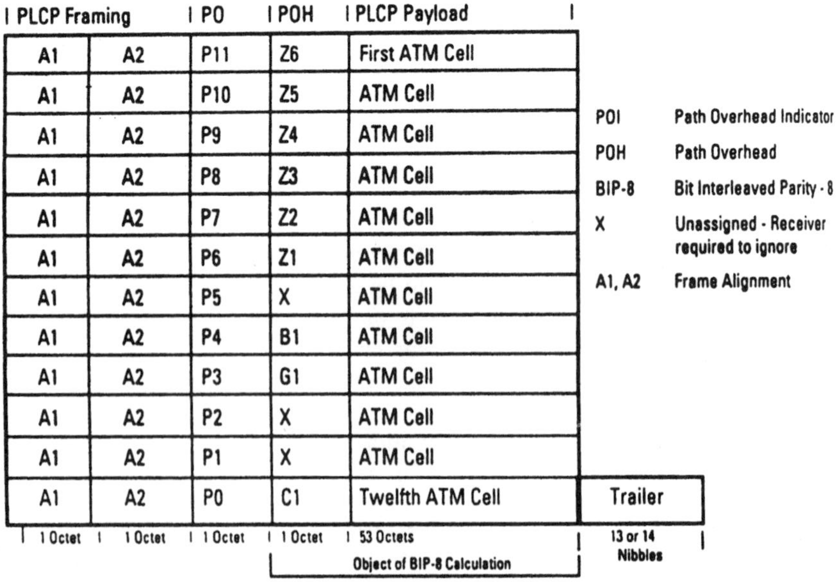

PLCP Framing		PO	POH	PLCP Payload	
A1	A2	P11	Z6	First ATM Cell	
A1	A2	P10	Z5	ATM Cell	
A1	A2	P9	Z4	ATM Cell	
A1	A2	P8	Z3	ATM Cell	
A1	A2	P7	Z2	ATM Cell	
A1	A2	P6	Z1	ATM Cell	
A1	A2	P5	X	ATM Cell	
A1	A2	P4	B1	ATM Cell	
A1	A2	P3	G1	ATM Cell	
A1	A2	P2	X	ATM Cell	
A1	A2	P1	X	ATM Cell	
A1	A2	P0	C1	Twelfth ATM Cell	Trailer
1 Octet	1 Octet	1 Octet	1 Octet	53 Octets / Object of BIP-8 Calculation	13 or 14 Nibbles

POI — Path Overhead Indicator
POH — Path Overhead
BIP-8 — Bit Interleaved Parity - 8
X — Unassigned - Receiver required to ignore
A1, A2 — Frame Alignment

Figure 16.29. Format of DS3 PLCP frame. Courtesy of Hewlett-Packard Company [10].

either 13 or 14 nibbles (a nibble = 4 bits) of trailer to bring the transmission rate up to the exact DS3 bit rate. The DS3 frame has 125-µs duration.

DS3 has to contend with network slips (added/dropped frames to accommodate synchronization alignment). Thus PLCP is padded with a variable number of stuff (justification) bits to accommodate possible timing slips. The C1 overhead octet indicates the length of padding. The BIP (bit interleaved parity) checks the payload and overhead functions for errors and performance degradation. This performance information is transmitted in the overhead.

12.2 DS1 Mapping

One approach to mapping ATM cells into a DS1 frame is to use a similar procedure as used on DS3 with PLCP. In this case, only 10 cells are bundled into a frame, and two of the Z overheads are removed. The padding in the frame is set at 6 octets. The entire frame takes 3 ms to transmit and spans many DS1 ESF (extended superframe) frames. This mapping is shown in Figure 16.30. Note the reference to L2_PDU, taken directly from SMDS (see Ref. 21). One must also consider the arithmetic. Each DS1 time slot is 8 bits long, 1 octet. There are then 24 octets in a DS1 frame. This, of course, can lead to the second method of carrying ATM cells in DS1, by directly mapping in ATM cells octet for octet (time slot). This is done in groups of 53 octets (1 cell) and would, by necessity, cross DS1 frame boundaries to accommodate an ATM cell.

12.3 E1 Mapping

E1 PCM has a 2048-Mbit/s transmission rate. An E1 frame has 256 bits representing 32 channels or time slots, 30 of which carry traffic. Time slots (TS) 0 and 16 are reserved. TS0 is used for synchronization and TS16 for signaling.

1	1	1	1	←——53 Octets——→
A1	A2	P9	Z4	L2_PDU
A1	A2	P8	Z3	L2_PDU
A1	A2	P7	Z2	L2_PDU
A1	A2	P6	Z1	L2_PDU
A1	A2	P5	F1	L2_PDU
A1	A2	P4	B1	L2_PDU
A1	A2	P3	G1	L2_PDU
A1	A2	P2	M2	L2_PDU
A1	A2	P1	M1	L2_PDU
A1	A2	P0	C1	L2_PDU

OH Byte	Function
A1, A2	Framing Bytes
P9-P0	Path Overhead Identifier Bytes
PLCP Path Overhead Bytes	
Z4-Z1	Growth Bytes
F1	PLCP Path User Channel
B1	BIP-8
G1	PLCP Status
M2-M1	SMDS Control Information
C1	Cycle/Stuff Counter Byte

Trailer = 6 Octets

3 msec

Figure 16.30. DS1 mapping with PLCP. Courtesy of Hewlett-Packard Company [10].

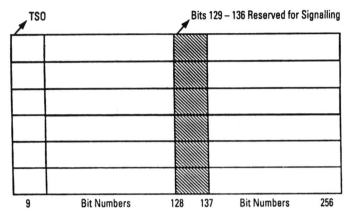

Figure 16.31. Mapping ATM cells directly into E1. Courtesy of Hewlett-Packard Company [10].

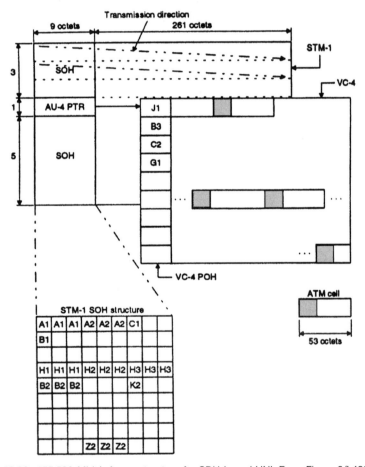

Figure 16.32. 155.520-Mbit/s frame structure for SDH-based UNI. From Figure 8/I.432, page 13, ITU-T Rec. I.432, March 1993 [3].

The E1 frame is shown in Figure 16.31. Bits 9 to 128 and bits 137 to 256 may be used for ATM cell mapping.

ATM cells can also be directly mapped into special E3 and E4 frames. The first has 530 octets available for cells (i.e., 10 cells) and the second has 2160 octets (not evenly divisible).

12.4 Mapping ATM Cells into SDH

12.4.1 At STM-1 (155.520 Mbits/s). SDH is described in Chapter 7. Figure 16.32 shows the mapping procedure. The ATM cell stream is first mapped into the C-4 and then mapped into the VC-4 container along with the VC-4 path overhead. The ATM cell boundaries are aligned with the STM-1 octet boundaries. Since the C-4 capacity (2340 octets) is not an integer multiple of the cell length (53 octets), a cell may cross a C-4 boundary. The AU-4 pointer (octets H1 and H2 in the SOH) is used for finding the first octet of the VC-4.

12.4.2 At STM-4 (622.080 Mbits/s). As shown in Figure 16.33, the ATM cell stream is first mapped into C-4-4c and then packed into the VC-4-4c container

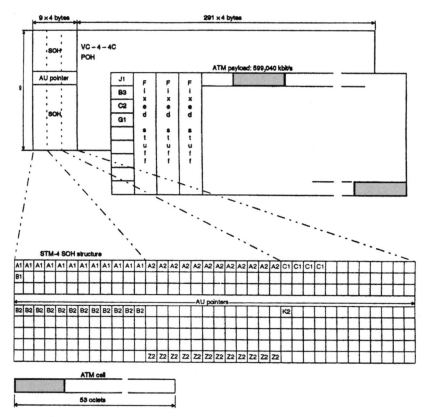

Figure 16.33. The 622.080-Mbit/s frame structure for SDH-based UNI. From Figure 10/I.432, page 15, ITU-T Rec. I.432, March 1993 [3].

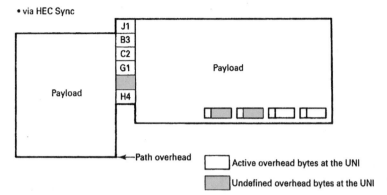

Figure 16.34. Mapping ATM cells into a SONET STS-1 frame. Courtesy of Hewlett-Packard Company [10].

along with the VC-4-4c path overhead. The ATM cell boundaries are aligned with STM-4 octet boundaries. Since the C-4-4c capacity (9360 octets) is not an integer multiple of the cell length (53 octets), a cell may cross a C-4-4c boundary. The AU pointers are used for finding the first octet of the VC-4-4c.

12.5 Mapping ATM Cells into SONET

ATM cells are mapped directly into the SONET payload (49.54 Mbits/s). As with SDH, the payload in octets is not an integer multiple of the cell length, thus a cell may cross an STS frame boundary. This mapping concept is shown in Figure 16.34. The H4 pointer can indicate where the cells begin inside an STS frame. Another approach is to identify cell headers and, thus, the first cell in the frame.

REVIEW QUESTIONS

1. During the design of ATM, one issue was the length of the cell header. Name and discuss at least two advantages in reducing the size of a frame/block/packet/cell header. Discuss at least one disadvantage.

2. ATM is an "optimum" format for which types of media which are to be carried and switched on the telecommunications network?

3. Relate the media to be transmitted over ATM with the services offered by ATM, namely VBR and CBR.

4. Compare signaling philosophy with standard POTS service and data network service.

5. Discuss *asynchronous* in general, and what it means regarding ATM.

6. Leaving aside Telcordia (Bellcore), there are three ATM standardizing operations bodies. What are they?

7. Give five of the NT-2 functions in B-ISDN/ATM.

8. Give the size, in octets, of the ATM cell. How much is header, how much is payload? Why must we be careful when we specify this?

9. Describe at least two of the functions of the HEC in the ATM header.

10. Define *packetization delay*. How did packetization delay force ATM designers to shorten cell payload as much as possible?

11. What is probably the most cogent reason for setting the CLP bit?

12. Discuss *cell rate decoupling*.

13. Describe how the cell delineation algorithm works.

14. Layer 6 of ATM is the services layer. Name and briefly describe the functions of the other five layers/sublayers of ATM.

15. What is the purpose of meta-signaling in ATM?

16. Explain what F1, F2–F4 flows are in ATM.

17. What is the principal purpose of the AAL layer?

18. The service classifications of AAL are based on three parameters. What are the three parameters?

19. What are the five AAL categories? Tell how they fit into classes A, B, C, and D.

20. AAL0 is the simplest of the AALs. What does it do?

21. AAL1 robs one octet of payload and places that octet in the header. This octet has two fields. Name and discuss the two fields.

22. What would be an application of AAL2? Remember what kind of service AAL2 offers, or is restricted to in this case.

23. Why would we want fields to be aligned to 32-bit boundaries?

24. What is the principal application of AAL5? Why is AAL5 called "SEAL?"

25. Regarding ATM routing and switching, what happens at a VCC endpoint?

26. What does a VPI identify?

27. The setup and release of VCCs can be performed in four ways. Describe each of the four.

28. Meta-signaling channels are provided for what?

29. There are seven ATN-unique QoS items, Name six of them.

30. Describe/define CDV (cell delay variation) and its deleterious effects.

31. What happens to cells with error(s) found in the header?

32. Define CAC (connection admission control).

33. What is the purpose of user parameter control/network parameter control?

34. Define *traffic parameters* and *traffic descriptors*.

35. There are three items basic to the *traffic contract*. What are they?

36. The *peak cell rate* in the traffic descriptor describes what?

37. Name at least five common digital transmission formats into which ATM cells can be mapped.

38. Name a method of transmitting ATM cells, which we could call a non-format.

39. Consider mapping 53-octet ATM cells into DS1. Describe the two ways of doing this as suggested in the text.

REFERENCES

1. *Broadband Aspects of ISDN*, CCITT Rec. I.121, CCITT, Geneva, 1991.

2. *ATM User–Network Interface Specification*, Version 3.0, The ATM Forum. PTR Prentice-Hall, Englewood Cliffs, NJ, 1993.

3. *B-ISDN User–Network Interface—Physical Layer Specification*, ITU-T Rec. I.432, ITU Telecommunication Standardization Sector, Geneva, March 1993.

4. *Broadband ISDN User–Network Interfaces—Rates and Formats Specifications*, ANSI T1.624-1993, American National Standards Institute, New York, 1993.

5. *B-ISDN User–Network Interface*, CCITT Rec. I.413, Geneva, 1991.

6. *Broadband ISDN—ATM Layer Functionality and Specification*, ANSI T1.627–1993, American National Standards Institute, New York, 1993.

7. *B-ISDN ATM Layer Specification*, ITU-T Rec. I.361, ITU, Geneva, March 1993.

8. D. E. McDysan and D. L. Spohn, *ATM Theory and Application*, McGraw-Hill, New York, 1995.

9. *B-ISDN ATM Adaptation Layer (AAL) Specification*, CCITT Rec. I.363, CCITT, Geneva, 1991.

10. *Broadband Testing Technologies*, An H-P seminar, Hewlett-Packard, Burlington, MA, October 1993.

11. *Support of Broadband Connectionless Data Service on B-ISDN*, ITU-T Rec. I.364, ITU Telecommunication Standardization Sector, Geneva, March 1993.

12. *Numbering Plan for the ISDN Era*, CCITT Rec. E.164, CCITT, Geneva, 1991.

13. *B-ISDN General Network Aspects*, ITU-T Rec. I.311, ITU Telecommunication Standardization Sector, Geneva, March 1993.

14. *Traffic Control and Congestion Control in B-ISDN*, ITU-T Rec. I.371, ITU Telecommunication Standardization Sector, Geneva, March 1993.

15. *B-ISDN ATM Adaptation Layer (AAL) Functional Description*, ITU-T Rec. I.362, ITU, Geneva, March 1993.

16. L. C. Cuthbert and J.-C. Sapanel, *ATM The Broadband Telecommunications Solution*, IEE Telecommunication Series 29, London, 1993.

17. *Broadband ISDN ATM Adaptation Layer 3/4 Common Part Functions and Specification*, ANSI T1-629-1993, American National Standards Institute, New York, 1993.

18. *B-ISDN ATM Layer Cell Transfer—Performance Parameters*, ANSI T1.511-1994, American National Standards Institute, New York, 1994.

19. *B-ISDN Protocol Reference Model and Its Application*, CCITT Rec. I.321, ITU, Geneva, 1991.

20. *ATM Switching*, from the Web, Cisco Systems, www.cisco.com/universal/cc/td/doc/CISintwk/ITO_doc/atm.htm, Cisco, 01/05/03.

21. *Telcordia Notes on the Network*, Telcordia Special Report SR 2275, Issue 4, Piscataway, NJ, October 2000.

22. *ATM User–Network Interface (UNI) Signaling Specification Version 4.1*, af-sig-0061.002, April 2002, The Presidio, San Francisco.

17

CCITT SIGNALING SYSTEM NO. 7

1 INTRODUCTION

CCITT Signaling System No. 7 (SS No. 7) was developed to meet the advanced signaling requirements of the all-digital network based on the 64 kbits/s channel. It operates in a quite different manner than the signaling discussed in Chapter 4. Nevertheless, it must provide supervision of circuits, address signaling, carry call progress signals, and alerting notification to be eventually passed to the called subscriber. These requirements are no different from those of Chapter 4. The difference is in how it is done. CCITT No. 7 is a data network entirely dedicated to interswitch signaling.*

Simply put, CCITT SS No. 7 is described as an international standardized general-purpose common-channel signaling system that:

- Is optimized for operation with digital networks where switches use stored-program control (SPC), such as the DMS-100 and AT&T 4ESS and 5ESS switches discussed in Chapter 9.

- Can meet present and future requirements of information transfer for inter-processor transactions with digital communications networks for call control, remote control, network data base access and management, and maintenance signaling.

- Provides a reliable means of information transfer in correct sequence without loss or duplication.

Source: CCITT 1980 Yellow Book [2].

---

* This would be called *interoffice signaling* in North America.

Telecommunication System Engineering, by Roger L. Freeman
ISBN 0-471-45133-9   Copyright © 2004 Roger L. Freeman

CCITT SS No. 7, in the years since 1980, has become known as the signaling system for ISDN. This it is. Without the infrastructure of SS No. 7 embedded in the digital network, there will be no ISDN with ubiquitous access. One important point is to be made. CCITT SS No. 7, in itself, is the choice for signaling in the digital PSTN without ISDN. It can and does stand on its own in this capacity.

As we mentioned above, SS No. 7 is a data communication system designed for only one purpose: signaling. It is *not* a general-purpose system. We then must look at CCITT SS No. 7 as (1) a specialized data network and (2) a signaling system.

2   OVERVIEW OF SS NO. 7 ARCHITECTURE

The SS No. 7 network model consists of network nodes, termed *signaling points* (SPs), interconnected by point-to-point signaling links, with all the links between two SPs called a link set. When the model is applied to a physical network, most commonly there is a one-to-one correspondence between physical nodes and logical entities. But when there is a need (for example, a physical gateway node needs to be a member of more than one network), a physical network node may be logically divided into more than one SP, or a logical SP may be distributed over more than one physical node. These artifices require careful administration to ensure that management procedures within the protocol work correctly.

Messages between two SPs may be routed over a link set directly connecting the two points. This is referred to as the *associated mode* of signaling. Messages may also be routed via one or more intermediate SPs that relay messages at the network layer. This is called *nonassociated mode* of signaling. SS No. 7 supports only a special case of this routing, called *quasi-associated mode*, in which routing is static except for relatively infrequent changes in response to events such as link failures or addition of new SPs. SS No. 7 does not include sufficient procedures to maintain in-sequence delivery of information if routing were to change completely on a packet-by-packet basis.

The function of relaying messages at the network layer* is called the *signaling transfer point* (STP) function. Although this practice results in some confusion, the logical and physical network nodes at which this function is performed are frequently called STPs, even though they may provide other functions as well. An important part in designing an SS No. 7 network is including sufficient equipment redundancy and physical-route diversity so that the stringent availability objectives of the system are met. The design is largely a matter of locating signaling links and SPs with the STP function, so that performance objectives can be met for the projected traffic loads at minimum cost.

Many telephone companies and administrations deploying SS No. 7 have opted to implement a physical network structure similar to the model illustrated in

---

* Bellcore (now Telcordia) [3] reports that "purists restrict this further to MTP relaying." (MTP stands for message transfer part.)

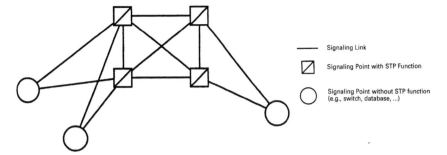

Figure 17.1. Signaling System No. 7 network structure model.

Figure 17.1 due to the cost of links and STPs. The STP function is concentrated in a relatively small number of nodes that are essentially dedicated to that function. The STPs are paired or *mated*, and pairs of STPs are interconnected with the *quad* configuration of signaling links as shown in the figure. This has proved to be an extremely reliable backbone network. Other nodes, such as switching systems and service control points (SCPs), are typically homed on one of the mated pairs of STPs with one or more links to each of the mates, depending on traffic volumes.

Source: Ref. 4.

3   SS NO. 7 RELATIONSHIP TO OSI

CCITT SS No. 7 relates to OSI (Chapter 11, Section 6.2.2) up to a certain point. One group believes that SS No. 7 should be fully compatible with the seven layers of OSI. However, the CCITT working groups responsible for the SS No. 7 concept and design were concerned with delay, whether for the data or telephone user of the digital PSN or ISDN. Recall from Chapter 4 that postdial delay is one of the principal measures of performance of a signaling system. To minimize delay, the seven layers of OSI were truncated at layer 4. In fact, CCITT Rec. Q.709 specifies no more than 2.2 s of postdial delay for 95% of calls. To accomplish this, a limit is placed on the number of relay points, called STPs, that can be traversed by a signaling message and by the inherent design of SS No. 7 as a four-layer system. Figure 17.2 relates SS No. 7 protocol layers to OSI.

We should note that SS No. 7 layer 3 signaling network functions include signaling message-handling functions and network management functions. Figure 17.3 shows the general structure of SS No. 7 signaling system.

Schlanger [5] makes the following pertinent observations:

- "Signaling is typically performed to create a communications subnetwork for a 'network end user.' As such, some argue that the entire reference model of SS No. 7, as a protocol within the communications subnetwork, should only exist at OSI layer 3 (the network layer) and below."

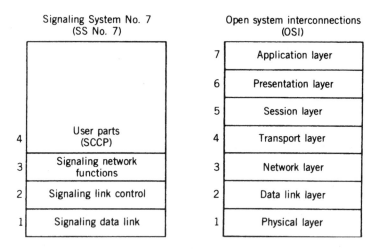

Figure 17.2. How SS No. 7 relates to OSI.

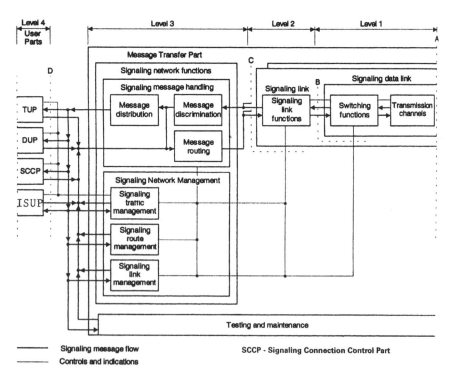

Figure 17.3. General structure of signaling system functions. From Figure 6/Q.701, page 8, ITU-T Rec. Q.701, March 1993 [6].

- "The applications processes within a communications network invoke protocol functionality to communicate with one another in much the same way as 'end users.' Thus, the same seven-layer reference model is felt to apply in this application context."

- "The signaling system protocol is felt to encompass operations, administration, and maintenance (OA & M) activities related to telecommunications. Because craftspeople can be involved in such activities (truly end users) as well as OA & M application processes, the distinction between network layer entities and end users becomes 'fuzzy.'"

There seem to be various efforts to force-fit SS No. 7 into OSI from layer 4 upwards. These efforts have resulted in the sublayering of layer 4 into user parts and SCCP (signaling connection control part).

In Section 4 we briefly describe the basic functions of the four SS No. 7 layers, which are covered in more detail in Sections 5 through 7 [12].

4   SIGNALING SYSTEM STRUCTURE

Figure 17.3, which illustrates the basic structure of SS No. 7, shows two parts of the system: the message transfer part (MTP) and user parts. There are three user parts: telephone user (TUP), data user (DUP), and ISDN user part (ISUP). Figures 17.2 and 17.3 show OSI layers 1, 2, and 3, which make up the MTP. The next paragraphs describe the functions of each of these layers from a system viewpoint.

Layer 1 defines the physical, electrical, and functional characteristics of the signaling data link and the means to access it. In the digital network environment the 64-kbit/s digital path is the normal basic connectivity. The signaling link may be accessed by means of a switching function that provides the capability of automatic reconfiguration of signaling links.

Layer 2 carries out the signaling link function. It defines the functions and procedures for the transfer of signaling messages over one individual signaling data link. A signaling message is transferred over the signaling link in variable-length signal units. A signal unit consists of transfer control information in addition to the information content of the signaling message. The signaling link functions include:

- Delimitation of a signal unit by means of flags.
- Flag imitation prevention by bit stuffing.
- Error detection by means of check bits included in each signal unit.
- Error control by retransmission and signal unit sequence control by means of explicit sequence numbers in each signal unit and explicit continuous acknowledgments.

- Signaling link failure detection by means of signal unit error monitoring and signaling link recovery by means of special procedures.

Layer 3, signaling network functions, in principle, defines such transport functions and procedures that are common to and independent of individual signaling links. There are two categories of functions in layer 3:

1. *Signaling Message-Handling Functions.* During message transfer, these functions direct the message to the proper signaling link or user part.
2. *Signaling Network Management Functions.* These control real-time routing, control, and network reconfiguration, if required.

Layer 4 is the user part. Each user part defines the functions and procedures peculiar to the particular user, whether telephone, data, or ISDN user part.

The *signal message* is defined by CCITT Rec. Q.701 as an assembly of information, defined at layer 3 or 4, pertaining to a call, management transaction, and so on, which is then transferred as an entity by the message transfer function. Each message contains "service information," including a service indicator identifying the source user part and possibly whether the message relates to international or national application of the user part.

The *signaling information* of the message contains user information, such as data or call control signals, management and maintenance information, and type and format of message. It also includes a "label." The label enables the message to be routed by layer 3 through the signaling network to its destination and directs the message to the desired user part or circuit.

On the signaling link such signaling information is contained in the *message signal units* (MSUs), which also include transfer control functions related to layer 2 functions on the link.

There are a number of terms used in SS No. 7 literature that should be understood before we proceed further.

Signaling Points. Nodes in the network that utilize common-channel signaling.

Signaling Relation (similar to traffic relation). Any two signaling points for which the possibility of communication between their corresponding user parts exist are said to have a signaling relation.

Signaling Links. Signaling links convey signaling messages between two signaling points.

Originating and Destination Points. The originating and destination points are the locations of the source user part function and location of the receiving user part function, respectively.

Signaling Transfer Point (STP). An STP is a point where a message received on one signaling link is transferred to another link.

Message Label. Each message contains a label. In the standard label, the portion that is used for routing is called the *routing label*. The routing label includes:

- Destination and originating points of the message.
- A code used for load sharing, which may be the least significant part of a label component that identifies a user transaction at layer 4.

The standard label assumes that each signaling point in a signaling network is assigned an identification code according to a code plan established for the purpose of labeling.

Message Routing. Message routing is the process of selecting the signaling link to be used for each signaling message. Message routing is based on analysis of the routing label of the message in combination with predetermined routing data at a particular signaling point.

Message Distribution. Message distribution is the process that determines to which user part a message is to be delivered. The choice is made by analysis of the service indicator.

Message Discrimination. Message discrimination is the process that determines, on receipt of a message at a signaling point, whether or not the point is the destination point of that message. This decision is based on analysis of the destination code of the routing label in the message. If the signaling point is the destination, the message is delivered to the message destination function. If not, the message is delivered to the routing function for further transfer on a signaling link [13].

4.1 Signaling Network Management

4.1.1 Scope. The three signaling network management functional blocks are shown in Figure 17.3. These are signaling traffic management, signaling route management, and signaling link management.

4.1.2 Signaling Traffic Management. The signaling traffic management functions are:

1. To control message routing. This includes modification of message routing to preserve, when required, accessibility of all destination points concerned or to restore normal routing.
2. In conjunction with modifications of message routing, to control the resulting transfer of signaling traffic in a manner that avoids irregularities in message flow.
3. Flow control.

Control of message routing is based on analysis of predetermined information about all allowed potential routing possibilities in combination with information, supplied by the signaling link management and signaling route management functions, about the status of the signaling network (i.e., current availability of signaling links and routes).

Changes in the status of the signaling network typically result in modification of current message routing and thus in the transfer of certain portions of the signaling traffic from one link to another. The transfer of signaling traffic is performed in accordance with specific procedures. These procedures are: *changeover*, *changeback*, *forced rerouting*, and *controlled rerouting*. The procedures are designed to avoid, as far as circumstances permit, such irregularities in message transfer as loss, mis-sequencing, or multiple delivery of messages.

The changeover and changeback procedures involve communication with other signaling point(s). For example, in the case of changeover from a failing signaling link, the two ends of the failing link exchange information (via an alternative path) that normally enables retrieval of messages that otherwise would have been lost on the failing link.

A signaling network has to have a signaling traffic capacity that is higher than the normal traffic offered. However, in overload conditions (e.g., due to network failures or extremely high traffic peaks) the signaling traffic management function takes flow control actions to minimize the problem. An example is the provision of an indication to the local user functions concerned that the MTP is unable to transport messages to a particular destination in the case of total breakdown of all signaling routes to that destination point. If such a situation occurs at a signaling transfer point (STP), a corresponding indication is given to the signaling route management function for further dissemination to other signaling points in the network.

4.1.3 Signaling Link Management. Signaling link management controls the locally connected signaling link sets. In the event of changes in the availability of a local link set, it initiates and controls actions with the objective of restoring the normal availability of that link set.

The signaling link management interacts with the signaling link function at level 2 by receipt of indications of the status of signaling links. It also initiates actions, also at level 2, such as initial alignment of an out-of-service link.

The signaling system can be applied in the method of provision of signaling links. Consider that a signaling link probably will consist of a terminal device and data link. It is also possible to employ an arrangement in which any switched connection to the far end may be used in combination with any local signaling terminal device. Here the signaling link management initiates and controls reconfigurations of terminal devices and signaling data links to the extent such reconfigurations are automatic. This implies some sort of switching function at level 1.

4.1.4 Signaling Route Management. Signaling route management only relates to the quasi-associated mode of signaling. It transfers information about changes in availability of signaling routes in the signaling network to enable remote signaling points to take appropriate signaling traffic actions. For example, a signaling transfer point may send messages indicating inaccessibility of a particular signaling point via that signaling transfer point, thus enabling other signaling points to stop routing message to an inoperative route.

5   THE SIGNALING DATA LINK (LAYER 1)

A signaling data link is a bidirectional transmission path for signaling, comprising two data channels operating together in opposite directions at the same data rate. It constitutes the lowest level (level 1) in the SS No. 7 functionality hierarchy.

A digital signaling data link is made up of digital transmission channels and digital switches or their terminating equipment providing an interface to SS No. 7 signaling terminals. The digital transmission channels may be derived from a digital multiplex signal at 1.544, 2.048, or 8.448 Mbits/s having a frame structure as defined in CCITT Rec. G.704 (see Chapter 8) or from digital multiplex bit streams having a frame structure specified for data circuits in CCITT Recs. X.50, X.51, X.50 bis, and X.51 bis.

The operational signaling data link is exclusively dedicated to the use of SS No. 7 signaling between two signaling points. No other information may be carried by the same channels together with the signaling information.

Equipment such as echo suppressors, digital pads, or A/μ-law converters attached to the transmission link must be disabled in order to ensure full duplex operation and bit count integrity of the transmitted data stream. Here 64-kbit/s digital signaling channels are used which are switchable as semipermanent channels in the exchange.

The standard bit rate on a digital bearer is 64 kbits/s. The minimum signaling bit rate for telephone call control applications is 4.8 kbits/s. For other applications such as network management, bit rates lower than 4.8 kbits/s may also be used.

The following is applicable for a digital signaling data link derived from a 2.048-Mbit/s digital path (i.e., E1). At the input–output interface, the digital multiplex equipment or digital switch block will comply with CCITT Rec. G.703 for electrical characteristics and Rec. G.704 for the functional characteristics—in particular the frame structure. The signaling bit rate is 64 kbits/s. The standard time slot for signaling is time slot 16. When time slot 16 is not available, any time slot available for 64-kbit/s user transmission rate may be used. No bit inversion is performed.

For a signaling data link derived from 8.448-Mbit/s (E2) digital link, the following applies. At the multiplex input–output interface, there should be compliance with CCITT Rec. 703 for electrical characteristics and Rec. G.704 for functional characteristics—in particular, frame structure. The signaling bit rate is 64 kbit/s. The standard time slots for use of a signaling data link are time slots 67–70 in descending order of priority. When these time slots are not available, any channel time slot available for 64-kbits/s user transmission rate may be used. No bit inversion is performed.

Source: Ref. 13.

For North American applications of SS No. 7, *Telcordia Notes on the Networks—2000* [3] states that data rates from 56 to 64 kbits/s may be used with up to 16 such circuits in parallel.

6 THE SIGNALING LINK (LEVEL 2)

This section deals with the transfer of signaling messages over one signaling link directly connecting two signaling points. Signaling messages delivered by superior hierarchical levels are transferred over the signaling link in variable-length signal units. The signal units include transfer control information for proper operation of the signaling link in addition to the signaling information. The signaling link (level 2) functions include:

(a) Signaling unit delimitation
(b) Signal unit alignment
(c) Error detection
(d) Error correction
(e) Initial alignment
(f) Signal link error monitoring
(g) Flow control

All these functions are coordinated by the link state control as shown in Figure 17.4.

Signal Unit Delimitation and Alignment. The beginning and end of a signal unit are indicated by a unique 8-bit pattern, called the *flag*. Measures are taken to ensure that the pattern cannot be imitated elsewhere in the unit. Loss of alignment occurs when a bit pattern disallowed by the delimitation procedure (i.e., more than six consecutive 1s is received, or when a certain maximum length of signal unit is exceeded. Loss of alignment will cause a change in the mode of operation of the signal unit error rate monitor.

Error Detection. The error detection function is performed by the standard CRC (cyclic redundancy check) with the 16 bits appended to the end of each SS No. 7 signal unit (frame). If the presence of errors is indicated, the signal unit is discarded.

Error Correction. Two forms of error correction are provided: the *Basic Method* and the *Preventative Cyclic Retransmission Method.* The basic method applies to (a) signaling links using terrestrial transmission means and (b) intercontinental signaling links where one-way propagation is less than 15 ms. Preventive cyclic retransmission method applies to (a) intercontinental signaling links where the one-way delay is equal to or greater than 15 ms and (b) signaling links established via satellite.

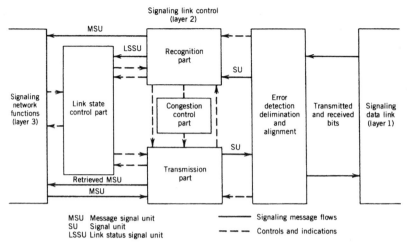

Signaling link control
(layer 2)

MSU   LSSU   Recognition part   SU

Signaling network functions (layer 3)   Link state control part   Congestion control part   Error detection delimination and alignment   Transmitted and received bits   Signaling data link (layer 1)   SU

Transmission part   Retrieved MSU   SU

MSU

MSU   Message signal unit
SU   Signal unit
LSSU Link status signal unit

———— Signaling message flows

— — — Controls and indications

Figure 17.4. Interactions of functional specification blocks for signaling link control. *Note*: The MSUs, LSSU, and SUs do not include error-control information. From Figure 1/Q.703, page 2, ITU-T Rec. Q.703, March 1993 [7].

The basic method is a noncompelled, positive/negative acknowledgment, retransmission error correction system. A signal unit which has been transmitted is retained at the transmitting signaling link terminal until a positive acknowledgment for that signal unit is received. If a negative acknowledgment is received, then the transmission of new signal units is interrupted and those signal units which have been transmitted but not yet positively acknowledged (starting with that indicated by the negative acknowledgment) will be transmitted once, in the order in which they were first transmitted.

The preventive cyclic retransmission method is a noncompelled, positive acknowledgment, cyclic retransmission forward error correction system. A signal unit which has been transmitted is retained at the transmitting signaling unit terminal until a positive acknowledgment for that signaling unit is received. During the period when there are no new signal units to be transmitted, all signal units which have not been positively acknowledged are retransmitted cyclically.

The forced retransmission procedure is defined to ensure that forward error correction occurs in adverse conditions (e.g., degraded BER and/or high-traffic loading). When a predetermined number of retained, unacknowledged signal units exist, the transmission of new signal units is retransmitted cyclically until the number of unacknowledged signal units is reduced.

Initial Alignment. The initial alignment procedure is used for both first-time initialization (e.g., after "switch-on") and alignment in association with restoration after link failure. The procedure is based on (a) the compelled exchange of status information between the two signaling points concerned and (b) the provision of a proving period. No other signaling link is involved in the initial alignment of any particular link, and the exchange occurs only on the link to be aligned.

Signaling Link Error Monitoring. Two signaling link error rate monitoring functions are provided: one which is employed while a signaling link is in service and which provides one of the criteria for taking the link out of service, and one which is employed while a link is in the proving-in state of the initial alignment procedure. They are called the *signal unit error rate monitor* and the *alignment error rate monitor*, respectively. The characteristics of the signal unit error rate monitor are based on a signal unit error count, incremented and decremented using the "leaky bucket" principle, whereas the alignment error rate monitor is a linear count of signal unit errors. During loss of alignment, the signal unit error rate monitor is incremented in proportion to the period of the loss of alignment.

Link State Control Functions. Link state control is a function of the signaling link which provides directives to the other signaling link functions.

Flow Control. Flow control is initiated when congestion is detected at the receiving end of the signaling link. The congested receiving end of the link notifies the remote transmitting end of the condition by means of an appropriate link status signal and it withholds acknowledgments of all incoming message signal units. When congestion abates, acknowledgments of all incoming signal units are resumed. When congestion exists, the remote transmitting end is periodically notified of this condition. The remote transmitting end will indicate that the link has failed if the congestion continues too long.

6.1   Basic Signal Unit Format

Signaling and other information originating from a user part is transferred over the signaling link by means of signal units. There are three types of signal units used in SS No. 7:

1. Message signal unit (MSU)
2. Link status signal unit (LSSU)
3. Fill-in signal unit (FISU)

These units are differentiated by means of the *length indicator*. MSUs are retransmitted in case of error; LSSUs and FISUs are not. The MSU carries signaling information; the LSSU provides link status information, and the FISU is used during the link idle state—they fill in.

The signaling information field is variable in length and carries the signaling information generated by the user part. All other fields are of fixed length. Figure 17.5 shows the basic formats of the three types of signal units. As shown in the figure, the message transfer control information encompasses eight fixed-length fields in the signal unit that contains information required for error control and message alignment. These eight fields are described below. In the figure we start from right to left, which is the direction of transmission.

The opening *flag* indicates the start of a signal unit. The opening flag of one signal unit is normally the closing flag of the previous signal unit. The flag

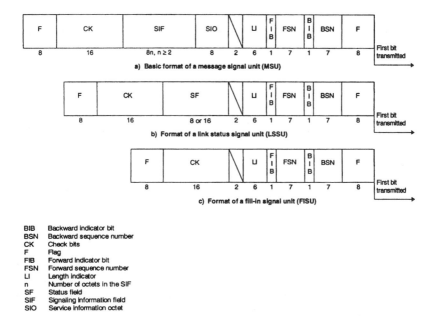

Figure 17.5. Signal unit formats. From Figure 3/Q.703, page 5, ITU-T Rec. Q.703, March 1993 [7].

bit pattern is 01111110. The *forward sequence number* (FSN) is the sequence number of the signal unit in which it is carried. The *backward sequence number* (BSN) is the sequence number of a signal unit being acknowledged. The value of the FSN is obtained by incrementing (modulo 128) the last assigned value by 1. The FSN value uniquely identifies a message signal unit until its delivery is accepted without errors and in correct sequence by the receiving terminal. The FSN of a signal unit other than an MSU assumes the value of the FSN of the last transmitted MSU. The maximum capacity of sequence numbers is 127 message units before reset (modulo 128).

Positive acknowledgment is accomplished when a receiving terminal acknowledges the acceptance of one or more MSUs by assigning an FSN value of the latest accepted MSU to the BSN of the next signal unit sent in the opposite direction. The BSNs of subsequent signal units retain this value until a further MSU is acknowledged, which will cause a change in the BSN sent. The acknowledgment to an accepted MSU also represents an acknowledgment to all, if any, previously accepted, though not yet acknowledged, MSUs.

Negative acknowledgment is accomplished by inverting the backward indicator bit (BIB) value of the signal unit transmitted. The BIB value is maintained in subsequently sent signal units until a new negative acknowledgment is to be sent. The BSN assumes the value of the FSN of the last accepted signal unit.

As we can now discern, the forward indicator bit (FIB) and the backward indicator bit together with the FSN and BSN are used in the basic error-control method to perform signal unit sequence control and acknowledgment functions.

The *length indicator* (LI) is used to indicate the number of octets following the length indicator octet and preceding the check bits and is a binary number in the range of 0–63. The length indicator differentiates between three types of signal units as follows:

Length indicator $= 0$ Fill-in signal unit
Length indicator $= 1$ or 2   Link status signal unit
Length indicator ≥ 2 Message signal unit

The *service information octet* (SIO) is divided into a *service indicator* and a *subservice field*. The service indicator is used to associate signaling information for a particular user part and is present only in MSUs. Each is 4 bits long. For example, a service indicator with a value 0100 relates to the telephone user part, and 0101 relates to the ISDN user part. The subservice field portion of the SIO contains two network indicator bits and two spare bits. The network indicator discriminates between international and national signaling messages. It can also be used to discriminate between two national signaling networks, each having a different routing label structure. This is accomplished when the network indicator is set to 10 or 11.

The *signaling information field* (SIF) consists of an integral number of octets greater than or equal to 2 and less than or equal to 62. In national signaling networks it may consist of up to 272 octets. Of these 272 octets, information blocks of up to 256 octets in length may be accommodated, accompanied by a label and other possible housekeeping information that may, for example, be used by layer 4 to link such information blocks together.

The *link status signal unit* (LSSU) provides link status information between signaling points. The status field can be made up of one or two octets. CCITT Rec. Q.703 shows application of the one-octet field in which the first three bits (from right to left) are used (bits A, B, and C) and the remaining five bits are spare. These first 3-bit values are given in Table 17.1.

Source: Ref. 7.

TABLE 17.1   Three-Bit Status Indications

Bits				
C	B	A	Status Indication	Meaning
0	0	0	0	Out of alignment
0	0	1	N	Normal alignment
0	1	0	E	Emergency alignment
0	1	1	OS	Out of service
1	0	0	PO	Processor outage
1	0	1	B	Busy

Source: paragraph 11.1.3, ITU-T Rec. Q.703 [7].

7   SIGNALING NETWORK FUNCTIONS AND MESSAGES (LAYER 3)

7.1   Introduction

In this section we describe the functions and procedures relating to the transfer of messages between signaling points (i.e., signaling network nodes). These nodes are connected by signaling links involving layers 1 and 2 described in Sections 5 and 6. Another important function of layer 3 is to inform the appropriate entities of a fault and, as a consequence, carry out a rerouting of messages through the network. The signaling network functions are broken down into two basic categories:

- Signaling message handling
- Signaling network management

7.2   Signaling Message-Handling Functions

The signaling message-handling function ensures that a signaling message originated by a particular user part at an originating signaling point is delivered to the same user part at the destination point as indicated by the sending user part. Depending on the particular circumstances, the delivery may be made through a signaling link directly interconnecting the originating and destination points or via one or more intermediate signaling transfer points (STPs).

The signaling message-handling functions are based on the label contained in the messages which explicitly identifies the destination and origination points. The label part used for signaling message handling by the MTP is called the *routing label*.

As shown in Figure 17.6, the signaling message handling is divided into the following:

- The *message routing* function, used at each signaling point to determine the outgoing signaling link on which a message is to be sent toward its destination point.
- The *message discrimination* function, used at a signaling point to determine whether or not a received message is destined to that point itself. When the signaling point has the transfer capability, and a message is not destined for it, that message is transferred to the message routing function.
- The *message distribution* function, used at each signaling point to deliver the received messages (destined to the point itself) to the appropriate user part.

7.2.1   Routing Label. The label contained in a signaling message and used by the relevant user part to identify the particular task to which the message refers (e.g., a telephone circuit) is also used by the message transfer part to route the message toward its destination point.

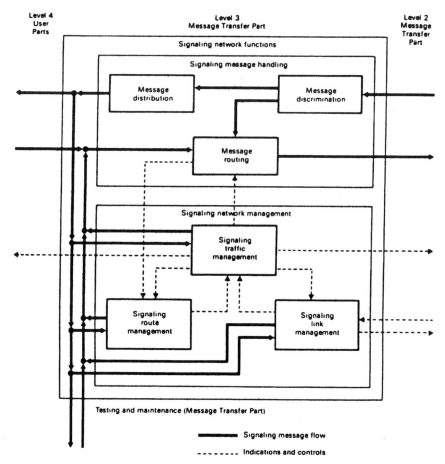

Figure 17.6. Signaling network functions. From Figure 1/Q.704, page 125, ITU-T Rec. Q.704, July 1986 [14].

The part of the message that is used for routing is called the *routing label*, and it contains the information necessary to deliver the message to its destination point. Normally the routing label is common to all services and applications in a given signaling network, national or international. (However, if this is not the case, the particular routing label of a message is determined by means of the service indicator.) The standard routing label should be used in the international signaling network and is applicable in national applications.

The standard routing label field is 32 bits long and is placed at the beginning of the signaling information field (SIF). Its structure is shown in Figure 17.7.

The *destination point code* (DPC) indicates the destination of the message. The *originating point code* (OPC) indicates the originating point of the message. The coding of these codes is pure binary. Within each field, the least significant bit occupies the first position and is transmitted first.

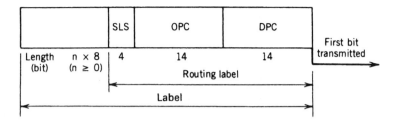

DPC Destination point code
OPC Originating print code
SLS Signaling link selection

Figure 17.7. Routing label structure. Based on ITU-T Rec. Q.704, July 1996 [14].

A unique numbering scheme for the coding of the fields is used for the signaling points of the international network irrespective of the user parts connected to each signaling point. The *signaling link selection* (SLS) field is used, where appropriate, in performing load sharing. This field exists in all types of messages and always in the same position. The only exception to this rule is some message transfer part level 3 messages (e.g., changeover order) for which the message routing function in the signaling point of origin of the message is not dependent on the field. In this particular case the field does not exist as such, but is replaced by other information (e.g., in the case of the changeover order, the identity of the faulty link).

In the case of circuit-related messages of the TUP, the field contains the least significant bits of the circuit identification code [or the bearer identification code in the case of the data user part (DUP)], and these bits are not repeated elsewhere. In the case of all other user parts, the SLS is an independent field. In these cases it follows that the signaling link selection of messages generated by any user part will be used in the load-sharing mechanism. As a consequence, in the case of the user parts which are not specified (e.g., transfer of charging information) but for which there is a requirement to maintain order of transmission of messages, the field is coded with the same value for all messages belonging to the same transaction, sent in a given direction.

In the case of message transfer part level 3 messages, the signaling link selection field exactly corresponds to the signaling link code (SLC) which indicates the signaling link between destination point and originating point to which the message refers.

7.2.2 Message Routing Function. Each signaling point will have routing information that allows it to determine the signaling link over which a message has to be sent on the basis of the destination point code and the signaling link selection field, and, in some cases, of the network indicator. Typically, the destination point code is associated with more than one signaling link that may be used to carry the message; the selection of the particular signaling link is made by means of the signaling link selection field, thus effecting load sharing.

Two basic cases of load sharing are defined:

1. Load sharing between links belonging to the same link set
2. Load sharing between links not belonging to the same link set

A load-sharing collection of one or more link sets is called a *combined link set*. The capability to operate in load sharing according to both cases is mandatory for any signaling point in the international network.

In case 1, the traffic flow carried by a link is shared (on the basis of the signaling link selection field) between different signaling links belonging to the link set. An example of such a case is given by a link set directly interconnecting the originating and destination points in the associated mode of operation. This is shown in Figure 17.8.

In case 2, traffic relating to a given destination is shared (on the basis of the signaling link selection field) between different signaling links not belonging to the same link set. This situation is shown in Figure 17.9. The load sharing rule used for a particular signaling relation may or may not apply to all the signaling relations which use one of the signaling links involved. In the example, traffic destined to B is shared between signaling links DE and DF with a given signaling link selection field assignment, whereas that destined to C is sent only on link DF, due to the failure of link EC.

As a result of the message routing function, in normal conditions all messages having the same routing label (e.g., call setup message related to a given circuit) are routed via the same signaling links and STPs.

Source: Refs. 13 & 14.

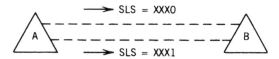

Figure 17.8. Example of load sharing within a link set.

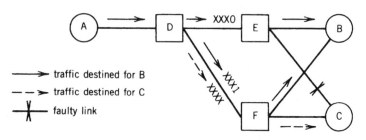

Figure 17.9. Example of load sharing between link sets. Based on ITU-T Rec. Q.704, July 1996 [14].

7.3 Signaling Network Management

This basic function deals with maintaining signaling service in view of three possible fault or degraded operation conditions:

- Loss of signaling link
- Loss of signaling point
- Degraded operation due to congestion

As mentioned previously, there are three subsidiary functions used to carry out signaling network management: traffic management, link management, and route management. The signaling traffic management function is used to direct signaling traffic from a link or route to one or more different links or routes or to deload temporarily a link or route in the case of congestion at a signaling point. The following procedures are involved with this function:

- Changeover
- Change back
- Forced rerouting
- Controlled rerouting
- Signaling traffic flow control

The signaling link management function is used to restore failed links, activate idle links (not yet aligned), and deactivate aligned signal links. This function involves the following procedures:

- Signaling link activation
- Link set activation
- Automatic allocation of signaling terminals and signaling data links

The signaling route management function distributes information about signaling network status in order to block or unblock signaling routes. Six procedures for route management are described in ITU-T Rec. Q.704 [14].

8   SIGNALING NETWORK STRUCTURE

8.1   Introduction

In this section several aspects in the design of signaling networks are treated. These networks may be national or international networks. The national and international networks are considered to be structurally independent and, although a particular signaling point may belong to both networks, signaling points are allocated *signaling point codes* according to the rules of each network.

Signaling links are basic components in a signaling network connecting signaling points. The signaling links encompass *level 2* functions which provide for message error control. In addition, provision for maintaining the correct message sequence is provided.

Signaling links connect signaling points at which signaling network functions such as message routing are provided at *level 3* and at which the user functions may be provided at *level 4* if it is also an originating or destination point. A signaling point that *only* transfers messages from one signaling link to another at level 3 serves as a signaling transfer point (STP). The signaling links, STPs and signaling (originating or destination) points may be combined in many different ways to form a *signaling network*.

8.2 International and National Signaling Networks

The worldwide signaling network is structured into two functionally independent levels: international and national levels. This is shown in Figure 17.10. Such a structure allows a clear division of responsibility for signaling network management and permits numbering plans of signaling points of the international network and the different national networks to be independent of one another.

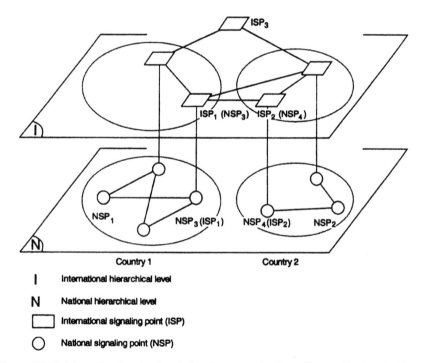

Figure 17.10. International and national signaling networks. From Figure 1/Q.705, ITU-T Rec. Q.705, page 2, March 1993 [8].

A signaling point (SP), including a signaling transfer point (STP), may be assigned to one of three categories:

- National signaling point (NSP) (an STP) which belongs to the national signaling network (e.g., NSP_1) and is identified by a signaling point code (OPC or DPC) according to the national numbering plan for signaling points.
- International signaling point (ISP) (an STP) which belongs to the international signaling network (e.g., ISP_3) and is identified by a signaling point code (OPC or DPC) according to the international numbering plan for signaling points.
- A node that functions both as an international signaling point (STP) and a national signaling point (STP) and therefore belongs to both the international signaling network and a national signaling network and accordingly is identified by a specific signaling point code (OPC or DPC)* in each of the signaling networks.

If discrimination between international and national signaling point codes is necessary at a signaling point, the network indicator is used [3, 12, 13].

9 SIGNALING PERFORMANCE – MESSAGE TRANSFER PART

9.1 Basic Performance Parameters

ITU-T Rec. Q.706 [9] breaks down SS No. 7 performance into three parameter groups:

- Message delay
- Signaling traffic load
- Error rate

Consider the following parameters and values:

Availability. The unavailability of a signaling route set should not exceed 10 min per year.

Undetected Errors. Not more than 1 in 10^{10} of all signal unit errors will go undetected in the message transfer part.

Lost Messages. Not more than 1 in 10^7 messages will be lost due to failure of the message transfer part.

Messages Out of Sequence. Not more than 1 in 10^{10} messages will be delivered out of sequence to the user part due to failure in the message transfer part. This includes message duplication.

* OPC and DPC are discussed in Section 10.

9.2 Traffic Characteristics

Labeling Potential. There are 16,384 identifiable signaling points.

Loading Potential. Loading potential is restricted by the following factors:

- Queuing delay
- Security requirements (redundancy with changeover)
- Capacity of sequence numbering (127 unacknowledged signal units)
- Signaling channels using bit rates under 64 kbits/s

9.3 Transmission Parameters

The message transfer part operates satisfactorily with the following error performance:

- Long-term error rate on the signaling data links of less than 1×10^{-6}
- Medium-term error rate of less than 1×10^{-4}

9.4 Signaling Link Delays over Terrestrial and Satellite Links

Data channel propagation time depends on data rate, the distance between nodes, repeater spacing, and the delays in the repeaters. Data rate (in bits/s) and repeater delays depend on the type of medium used to transmit the messages. The velocity of propagation of the medium is a most important parameter. Three types of medium are considered in Table 17.2, which gives transmission delays for various call distances. These are wire, fiber-optic cable, and terrestrial radio.

Delay is also a function of the following: processing delays in SPs, number of SPs, number of STPs, number of signaling links, propagation delays of each signaling link, signaling link loading, and message length mix.

Source: Ref. 9.

TABLE 17.2 Calculated Terrestrial Transmission Delays for Various Call Distances

	Delay Terrestrial (ms)		
Arc Length (km)	Wire	Fiber	Radio
500	2.4	2.50	1.7
1,000	4.8	5.0	3.3
2,000	9.6	10.0	16.6
5,000	24.0	25.0	16.5
10,000	48.0	50.0	33.0
15,000	72.0	75.0	49.5
17,737	85.1	88.7	58.5
20,000	96.0	100.0	66.0
25,000	120.0	125.0	82.5

Source: Table 1/Q.706, page 3, ITU-T Rec. Q.706 March 1993 [9].

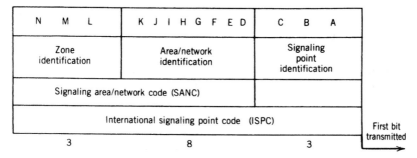

Figure 17.11. Format for international signaling point code (ISPC). From Figure 1/Q.708, page 1, ITU-T Rec. Q.708, March 1993 [10].

10   NUMBERING PLAN FOR INTERNATIONAL SIGNALING POINT CODES

The numbering plan described in ITU-T Rec. Q.708 [10] has no direct relationship with telephone, data, or ISDN numbering. A 14-bit binary code is used for identification of signaling points. An international signaling point code (ISPC) is assigned to each signaling point in the international signaling network. The breakdown of these 14 bits into fields is shown in Figure 17.11.

The assignment of signaling network codes is administered by the ITU Telecommunication Standardization Section (previously CCITT).

All international signaling point codes (ISPCs) consist of three identification subfields as shown in Figure 17.11. The world geographical zone is identified by the NML field consisting of 3 bits. A geographical area or network in a specific zone is identified by the 8-bit field K–D. The 3-bit subfield CBA identifies a signaling point in a specific geographical area or network. The combination of the first and second subfields is called a signaling area/network code (SANC).

Each country (or geographical area) is assigned at least one SANC. Two of the zone identifications, namely 1 and 0 codes, are reserved for future allocation.

The ISPC system provides for $6 \times 256 \times 8(12288)$ ISPCs.

If a country or geographical area should require more than eight international signaling points, one or more additional signaling area/network code(s) would be assigned to it by the ITU-T Organization.

A list of SANCs and their corresponding countries can be found in Annex A to ITU-T Rec. Q.708 [10]. The first number of the code identifies the zone. For example, zone 2 is Europe and zone 3 is North America and its environs.

11   HYPOTHETICAL SIGNALING REFERENCE CONNECTIONS

The hypothetical signaling reference connection (HSRC) is composed of signaling points and STPs that are connected in series by signaling data links to produce a signaling connection. The number of signaling points and STPs depends on the size of the network. There are two important parameters we derive from the HSRC: (1) the number of signaling points and STPs in a signaling connection and (2) the signaling message transfer delay, which will directly affect postdial delay.

Definition of Average-size Country. When the maximum distance between an international switching center and a subscriber who can be reached from it does not exceed 1000 km or, exceptionally, 1500 km, and when the country has less than $n \times 10^7$ subscribers, the country is considered to be average size. A country with a larger distance between an international switching center and a subscriber, or with more than $n \times 10^7$ subscribers, is considered to be of large size. The ITU-T Organization states in Ref. 11 that the value of n is for further study.

The number of signaling points in the HSRC (hypothetical signaling reference connection) has been determined by the ITU-T Organization by considering the maximum number of links allowed by the Telephone Routing Plan of ITU-T Rec. E.171 (see Chapter 6). Limiting the number of signaling points required reduces signaling delay, because signaling point delay forms the largest component of signaling delay.

The number of STPs in a HRSC is a function of the number of signaling points and the signaling network topology used to connect these signaling points. For an international signaling relationship, no more than 2 STPs should be used in a signaling relation.

The total unavailability for any signaling route set should not exceed 10 minutes per year. The overall maximum signaling delays for link-by-link signaling for 50% and 95% of connections are given in Table 17.3 for various combinations of large-size and average-size countries. Average signal point and STP delays at normal loading are assumed. The values given in the table are mean values. These values (in Table 17.3) must be increased by transmission propagation delays.

TABLE 17.3   Maximum Overall Signaling Delays[a]

| | | Delay (ms); Message Type | |
Country Size	Percent of Connections	Simple (e.g., Answer)	Processing Intensive (e.g., IAM)
Large-size to Large-size	50	1170	1800
	95	1450	2220
Large-size to Average-size	50	1170	1800
	95	1450	2220
Average-size to Average-size	50	1170	1800
	95	1470	2240

[a] The values given in the table are mean values.

Source: Table 5/Q.709, page 5, ITU-T Rec. Q.709, March 1993 [11].

TABLE 17.4   Maximum Overall Delay at Signaling Nodes[a,b]

Country Size	Percent of Connections	Delay (ms); Message Type	
		Processing Simple	Processing Intensive
Large-size	50	900	1320
to			
Large-size	95	1270	1900
Large-size	50	900	1320
to			
Average-size	95	1180	1740
Average-size	Mean	900	1320
to			
Average-size	95	1200	1760

[a] The maximum signaling nodes delay is the sum of all cross-office involved.
[b] All values are provisional.
Source: Table 10/Q.709, page 9, ITU-T Rec. Q.709, March 1993 [11].

For end-to-end signaling, 50% and 95% delays are given in Table 17.4 for various combinations of large-size and average-size countries. Average signaling delays at normal loading are assumed.

12   SIGNALING CONNECTION CONTROL PART (SCCP)

12.1   Introduction

The signaling connection control part (SCCP) provides additional functions to the message transfer part (MTP) for both connectionless and connection-oriented network services to transfer circuit-related and non-circuit-related signaling information between switches and specialized centers in telecommunication networks (such as for management and maintenance purposes) via a Signaling System No. 7 network.

Turn now to Figure 17.3 to see where the SCCP appears in a functional block diagram of an SS No. 7 terminal. It is situated above the MTP in level 4 with the user parts. The MTP is transparent and remains unchanged when SCCP services are incorporated in an SS No. terminal. From an OSI perspective, the SCCP carries out the network layer function.

The overall objectives of the SCCP are to provide the means for:

(a) Logical signaling connections within the Signal System No. 7 network.
(b) A transfer capability for network service signaling data units (NSSDUs) with or without the use of logical signaling connections.

Functions of the SCCP are also used for the transfer of circuit-related and call-related signaling information of the ISDN user part (ISUP) with or without setup of end-to-end logical signaling connections.

12.2   Services Provided by the SCCP

The overall set of services is grouped into:

- Connection-oriented services
- Connectionless services

Four classes of service are provided by the SCCP protocol, two for connectionless services and two for connection-oriented services. The four classes are:

0 Basic connectionless class
1 Sequenced connectionless class
2 Basic connection-oriented class
3 Flow control connection-oriented class

For connection-oriented services, a distinction has to be made between temporary signaling connections and permanent signaling connections.

Temporary signaling connection establishment is initiated and controlled by the SCCP user. Temporary signaling connections are comparable with dialed telephone connections.

Permanent signaling connections are established and controlled by the local or remote O&M function or by the management function of the node and they are provided for the SCCP user on a semipermanent basis. They can be compared with leased telephone lines.

12.3   Peer-to-Peer Communication

The SCCP protocol facilitates the exchange of information between two peers of the SCCP. The protocol provides the means for:

- Setup of logical signaling connection
- Release of logical signaling connections
- Transfer of data with or without logical signaling connections

A signaling connection is modeled in the abstract by a pair of queues. The protocol elements are objects on the queue added by the origination service user. Each queue represents a flow control function. Figure 17.12 illustrates the SCCP model for connection-oriented service.

12.4   Primitives and Parameters

Primitives consists of commands and their respective responses associated with the services requested, in this case for the SCCP. Such primitives and parameters are an attempt to conform to OSI, as expressed in CCITT Rec. X.200 series, especially CCITT Rec. X.213. Table 17.5 gives an overview of the primitives to the upper layers and the corresponding parameters for the (temporary) connection-oriented network service. Figure 17.13 shows an overview state transition diagram for the sequence of primitives at a connection endpoint. Here one should

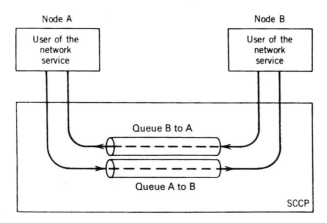

Figure 17.12. Model for internode communication with the SCCP, connection-oriented services. From Figure 4/Q.711, page 4, ITU-T Rec. Q.711, March 1993 [15].

TABLE 17.5   Network Service Primitives for Connection-Oriented Services

| Generic Name | Primitives | | Parameters |
	Specific Name	
N-CONNECT	Request	Called address
	Indication	Calling address
	Response	Responding address
	Confirmation	Receipt confirmation election
		Expedited data selection
		Quality of service parameter set
		User data
		Connection identification[a]
N-DATA	Request	Confirmation request
	Indication	User data
		Connection identification[a]
N-EXPEDITED DATA	Request	User data
	Indication	Connection identification[a]
N-DATA ACKNOWLEDGE	Request	Connection identification[a]
(for further study)	Indication	
N-DISCONNECT	Request	Originator
	Indication	Reason
		User data
		Responding address
		Connection identification[a]
N-RESET	Request	Originator
	Indication	Reason
	Response	Connection identification[a]
	Confirmation	

[a] In 5.3/X.213, this parameter is implicit. This parameter is for further study.

Source: Table 1/Q.711, page 7, ITU-T Rec. Q.711, March 1993 [15].

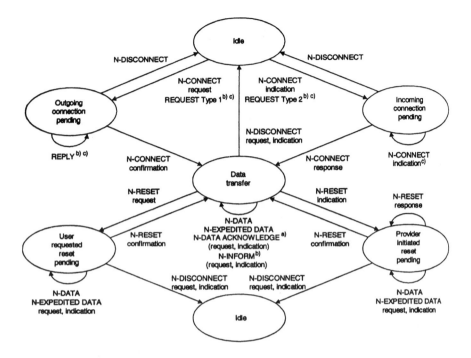

^{a)} The need for this primitive is for further study.
^{b)} This primitive is not in Recommendation X.213.
^{c)} For user part type A only.

Figure 17.13. State transition diagram for sequence of primitives at a connection endpoint (basic functions). From Figure 7/Q.711, page 8, ITU-T Rec. Q.711, March 1993 [15].

refer to CCITT Rec. X.213, *Network Layer Service Definition of OSI*, for CCITT application.

12.5 Connection-Oriented Functions: Temporary Signaling Connections

12.5.1 Connection Establishment. A connection setup uses service primitives described in Section 12.4 above. The following are the principal functions used in the connection establishment phase by the SCCP to set up a signaling connection:

- Setup of a signaling connection
- Establishment of the optimum size of NPDUs (network protocol data units)
- Mapping network address onto signaling relations
- Selecting operational functions during data transfer phase (for instance, layer service selection)

- Providing means to distinguish network connections
- Transporting user data (within the request)

12.5.2 *Data Transfer Phase.* The data transfer phase functions provide the means of a two-way simultaneous transport of messages between two endpoints of a signaling connection. The principal data transport phase functions are listed below. These are used or not used in accordance with the result of the selection function performed in the connection establishment phase. The item with an asterisk requires further study by the ITU-T Organization.

- Segmenting/reassembling
- Flow control
- Connection identification
- NSDU delimiting (M-bit)
- Expedited data
- Mis-sequence detection
- Reset
- Receipt confirmation*
- Others

12.5.3 *Connection Release Functions.* Release functions disconnect the signaling connection regardless of the current phase of the connection. The release may be performed by an upper layer stimulus or by maintenance of the SCCP itself. The release can start at each end of the connection (symmetric procedure). Of course the principal function of this phase is disconnection.

12.6 SCCP Formats and Codes

The SCCP messages are carried on the signaling data link by means of signal units which was described in Sections 6 and 7. The service indicator is coded 0011 for the SCCP. The Signaling Information Field carries the SCCP message. A message consists of:

- The routing label
- The message type code
- The mandatory fixed part
- The mandatory variable part
- The optional part, which may contain fixed length or variable length fields

The routing label is discussed in Section 7.2.1.

The message type code consists of a one-octet field and is mandatory for all messages. The message type code uniquely defined the function and format of each SCCP message. Message type codes can be found in Ref. 17, Table 1. A general SCCP message format is shown in Figure 17.14.

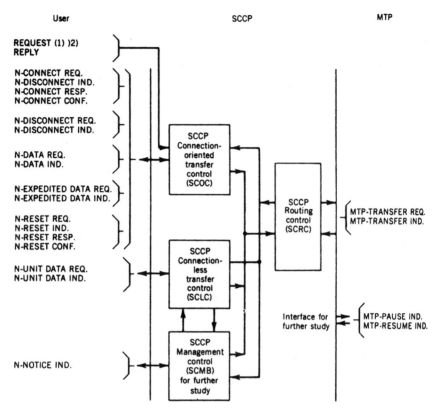

Figure 17.14. General SCCP message format. From Figure 2/Q.713, page 3, ITU-T Rec. Q.713, ITU Geneva, March 2001 [17].

13   USER PARTS

13.1   Introduction

SS No. 7 user parts, along with the routing label, carry out the basic signaling functions. Turn again to Figure 17.5. There are two fields in the figure we should now discuss: the SIO (service information octet) and the SIF (signaling information field). In the paragraphs that follow, we briefly cover the telephone user part (TUP) and the ISDN user part (ISUP). As shown in Figure 17.15, the user part, OSI layer 4, is contained in the signaling information field to the left of the routing label. ITU-T Rec. Q.723 [20] deals with the sequence of three sectors (fields and subfields of the standard basic message signal unit shown in Figure 17.5).

Turning now to Figure 17.15, we have from right to left, the service information octet (SIO), the routing label, and user information subfields (after the routing label in the SIF). The SIO is an octet in length made up of two subfields: the service indicator (4 bits) and the subservice field (4 bits). The service

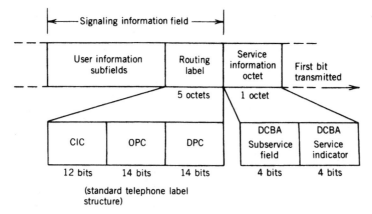

Figure 17.15. Signaling information field (SIF) preceded by the service information octet (SIO). The sequence runs from right to left with the least significant bit transmitted first. DPC, destination point code; OPC, originating point code; CIC, circuit identification code.

indicator, being 4 bits long, has 16 bit combinations with the following meanings (read from right to left):

Bits DCBA	Meaning
0000	Signaling network management message
0001	Signaling network testing and maintenance
0010	Spare
0011	SCCP
0100	Telephone user part
0101	ISDN user part
0110	Data user part (call- and circuit-related message)
0111	Data user part (facility registration and cancellation)
Remainder (8 sequences)	Spare

It is evident that the SIO directs the signaling message to the proper layer 4 entity, whether SCCP or user part. This is called *message distribution*.

The subservice indicator contains the network bits C and D and two spare bits, A and B. The network indicator is used by signaling message-handling functions determining the relevant version of the user part. If the network indicator is set at 00 or 01, the two spare bits, coded 00, are available for possible future needs. If these two bits are coded 10 or 11, the two spare bits are for national use, such as message priority as an optional flow procedure. The network indicator provides discrimination between international and national usage (bits D and C).

The routing label forms part of every signaling message:

- To select the proper signaling route.
- To identify the particular transaction by the user part (the call) to which the message pertains.

The label format is shown in Figure 17.15. The DPC is the destination point code (14 bits) which indicates the signaling point for which the message is intended. The OPC (originating point code) indicates the source signaling point. The CIC (circuit identification code) indicates the one circuit (speech circuit in the TUP case) among those directly interconnecting the destination and originating points.

For the OPC and DPC, unambiguous identification of signaling points is carried out by means of an allocated code. Separate code plans are used for the international and national networks. The CIC, as shown in the figure, is applicable only to the TUP. CCITT Rec. Q.704 shows a signaling link selection (SLS) field following (to the left) the OPC. The SLS is 4 bits long and is used for load sharing. The ISDN user part address structure is capable of handling E.164 addresses in the calling and called number and is also capable of redirecting address information elements [16].

13.2   Telephone User Part (TUP)

The core of the signaling information is carried in the SIF (see Figure 17.15). The TUP label was described briefly in Section 7.2.1. Several signal message formats and codes are described in what follows. These follow the label.

One typical message of the TUP is the initial address message (IAM); its format is shown in Figure 17.16. A brief description is given of each subfield, providing further insight of how SS No. 7 operates.

Common to all signaling messages are the subfields H0 and H1. These are the heading codes, each consisting of 4 bits, giving 16 code possibilities in pure binary coding. H0 identifies the specific message group to follow. "Message group" means the type of message. Some samples of message groups are:

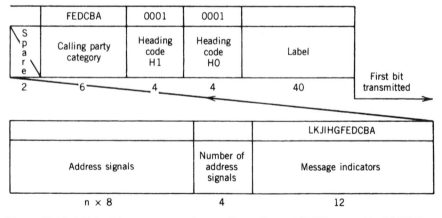

Figure 17.16. Initial address message format. From Figure 3/Q.723, page 23, CCITT Rec. Q.723, 1988 [21].

Message Group Type	H0 Code
Forward address messages	0001
Forward setup messages	0010
Backward setup messages	0100
Unsuccessful backward setup messages	0101
Call supervision messages	0110
Node-to-node messages	1001

H1 contains a signal code or identifies the format of more complex messages. For instance, there are four types of address message identified by H0 = 0001, and H1 identifies the type of message, such as:

Address Message Type	H0	H1
Initial address message	0001	0001
IAM with additional information	0001	0010
Subsequent address message	0001	0011
Subsequent address message with signal unit	0001	0100

Moving from right to left in Figure 17.16, after H1 we have the calling party subfield consisting of 6 bits. It identifies the language of the operator (Spanish, English, Russian, etc.). An English-speaking operator is coded 000010. It also differentiates the calling subscriber from one with priority, a data call, or a test call. A data call is coded 001100 and a test call is coded 001101. Fifty of the 64 possible code groups are spare.

Continuing to the left in the figure, 2 bits are spare for international allocation. Then there is the message indicator, where the first 2 bits, B and A, give the nature of the address. This is information given in the forward direction indicating whether the associated address or line identity is an international, national (significant), or subscriber number. A subscriber number is coded 00, an international number is coded 11, and a national (significant) number is coded 10.

Bits D and C are the circuit indicator. In this location 00 indicates that there is no satellite circuit in the connection. Remember that the number of space satellite relays in a speech telephone connection is limited to one relay link through a satellite because of propagation delay.

Bits F and E are significant for common-channel signaling systems such as CCIS, CCS No. 6, and SS No. 7. The associated voice channel operates on a separate circuit. Does this selected circuit for the call have continuity? The bit sequence FE is coded:

Bits F and E	Meaning
00	Continuity check not required
01	Continuity check required on this circuit
10	Continuity check performed on previous circuit
11	Spare

Bit G gives echo suppressor information. When coded 0 it indicates that the outgoing half-echo suppressor is not included, and when coded 1 it indicates that the outgoing half-echo suppressor is included. Bit I is the redirected call indicator. Bit J is the all-digital path required indicator. Bit K tells whether any path may be used or whether only SS No. 7 controlled paths may be used. Bit L is spare.

The next subfield has 4 bits and gives the number of address signals contained in the initial address message. The last subfield contains address signals where each digit is coded by a 4-bit group as follows:

Code	Digit	Code	Digit
0000	0	1000	8
0001	1	1001	9
0010	2	1010	Spare
0011	3	1011	Code 11
0100	4	1100	Code 12
0101	5	1101	Spare
0110	6	1110	Spare
0111	7	1111	ST

The most significant address signal is sent first. Subsequent address signals are sent in successive 4-bit fields. As shown in Figure 17.16, the subfield contains n octets. A filler code of 0000 is sent to fill out the last octet, if needed. Recall in Chapter 4 that the ST signal is the "end of pulsing" signal and is often used on semi-automatic circuits.

Besides the initial address message, there is the subsequent address message used when all address digits are not contained in the IAM. The subsequent address message is an abbreviated version of the IAM. There is a third type of address message, the initial address message with additional information. This is an extended IAM providing such additional information as network capability, user facility data, additional routing information, called and calling address, and closed under group (CUG). There is also the forward setup message, which is sent after the address messages and contains further information for call setup.

CCITI SS No. 7 is rich with backward information messages. In this group are backward setup request; successful backward setup information message group, which includes charging information; unsuccessful backward setup information message group, which contains information on unsuccessful call setup; call supervision message group; circuit supervision message group; and the node-to-node message group (CCITT Recs. Q.722 and Q.723 [20, 21]).

Label capacity for the telephone user part is given in CCITT Rec. Q.725 [19] as 16,384 signaling points and up to 4096 speech circuits for each signaling point.

13.3   ISDN User Part (ISUP)

13.3.1   Introduction. CCITT Signaling System No. 7 is an integral segment of ISDN. Without SS No. 7 implemented in the PSTN, ISDN does not work. The

ISUP encompasses the signaling functions required to provide switched services and user facilities for voice and nonvoice applications in ISDN. We must not forget that ISDN handles voice as well as data and other digital services. The ISUP is also suited for application to dedicated telephone and circuit-switched data networks and in analog and mixed analog and digital networks. In particular the ISUP meets the requirements defined by CCITT for worldwide international semiautomatic and automatic telephone and circuit switched data traffic.

The ISUP is furthermore suitable for national applications. Most signaling procedures, information elements and message types specified for international use are also required in typical national applications. Moreover, coding space has been reserved in order to allow PSTNs to introduce specific signaling messages and elements of information within the internationally standardized protocol structure.

Numbering requirements are described in ITU-T Rec. E.164 [27].

In the sections that follow we provide a brief description of several key elements of ISUP.

13.3.2 Basic ISUP Signaling Procedures

ADDRESS SIGNALING. In general, the call setup procedure described is standard for both speech and nonspeech connections using en-bloc address signaling for calls between ISDN terminals. Overlap address signaling is also specified.

BASIC PROCEDURES. The basic call control is divided into three phases: call set-up, the data/conversation phase, and call cleardown (takedown). Messages on the signaling link are used to establish and terminate different phases of a call. Standard in-band supervisory tones and/or recorded announcements are returned to the caller on appropriate connection types to provide information on call progress. Calls originating from ISDN terminals may be supplied with more detailed call progress information by means of additional messages in the access protocol supported by a range of messages in the network.

Two signaling methods are used with ISUP:

- link-by-link
- end-to-end*

The link-by-link method is primarily used for messages that need to be examined at each intervening exchange. The end-to-end methods are used for messages of endpoint significance. The link-by-link method may also be used for messages of endpoint significance.

ISUP INTERWORKING. In call control interworking between two (ISUP) protocols, the call control provides interworking logic. Peer-to-peer interworking takes place between two exchanges that support different implementations of the same protocol. Interworking is carried out following interpretation of the protocol information received by either exchange.

---

* See Chapter 4 for conceptual definitions of link-by-link and end-to-end signaling.

For this the 1992 version of ISUP (ISUP '92), only one ISUP protocol implementation may be present in an exchange since ISUP '92 is backwards compatible with previous versions of ISUP as a result of:

- The basic call procedures and supplementary service procedures of ISUP '92 ensure backwards compatibility with the ISUP procedures conforming to the 1988 version (IXth Plenary Assembly, Melbourne, 1988—Blue Books) and those conforming to CCITT Rec. Q.767. No knowledge is required to be stored in the exchange to this effect.
- From ISUP '92 onwards, the forward compatibility is ensured by the guidelines for future protocol enhancements and compatibility procedures as outlined in Section 6 of ITU-T Rec. 761 [22].

13.3.3   Formats and Codes of the ISUP. The format of and codes used in the service information octet (SIO) are described in Section 13.1. The service indicator for the ISUP is coded 0101. The signaling information field of each message signal unit containing and ISUP message consists of an integral number of octets and includes the parts as shown in Figure 17.17.

ROUTING LABEL. The format and codes of the routing label are discussed in Section 7.2.1 and 13.1 and in ITU-T Rec. Q.704 (paragraph 2.2). For each individual circuit connection, the same routing label must be used for each message that is transmitted for that connection.

CIRCUIT IDENTIFICATION CODE (CIC). The CIC is an octet long. Bits 1 through 4 are the CIC bits, and bits 5 through 8 are spare. The allocation of circuit identification codes to individual circuits is determined by bilateral agreement and/or in accordance with applicable predetermined rules.

For international applications, the four spare bits of the circuit identification field are reserved for CIC expansion, provided that bilateral agreement is obtained before any increase in size is performed. For national applications, the four spare bits can be used as required.

A typical allocation for the 2.048-Mbit/s digital path is as follows. For circuits derived from a 2.048-Mbit/s digital path, the circuit identification code contains

Routing label
Circuit identification code
Message type code
Mandatory fixed part
Mandatory variable part
Optional part

Figure 17.17. ISUP message parts.

in the five least significant bits a binary representation of the actual number of the time slot which is assigned to the communication path. The remaining bits in the CIC are used, where necessary, to identify these circuits uniquely among all other circuits of other systems interconnecting an originating and destination point.

MESSAGE TYPE CODE. The message type code consists of a one-octet field and is mandatory for all messages. The message type code uniquely defines the function and format of each ISDN user part message. The allocation is summarized in Table 17.6.

FORMATTING PRINCIPLES. Each message consists of a number of PARAMETERS described in ITU-T Rec. Q.763 [24], Section 2. Each parameter has a NAME which is coded as a single octet. There are 59 parameter names listed in Table 5 in Q.763, plus 10 reserved parameter items. For instance, the first parameter listed is "access delivery information" and is coded 00101110. Between parameters there should be no unused (dummy) octets.

A general format diagram is shown in Figure 17.18.

MANDATORY FIXED PART. Those parameters that are mandatory and of fixed length for a particular message type are contained in the mandatory fixed part. The position, length, and order of the parameters are uniquely defined by the message type; thus the names of the parameters and the length indicators are not included in the message.

MANDATORY VARIABLE PART. Mandatory parameters of variable length are included in the mandatory variable part. Pointers are used to indicate the beginning of each parameter. Each pointer is encoded as a single octet.

The pointer value (in binary) gives the number of octets between the pointer itself (included) and the first octet (not included) of the parameter associated with that pointer. The pointer value of all zeros is used to indicate that, in the case of optional parameters, no optional parameter is present. Parameter names are, therefore, not included in the message. The number of parameters, and thus the number of pointers, is uniquely defined by the message type.

A pointer is also included to indicate the beginning of the optional part. If the message type indicates that no optional part is allowed, then this pointer is not present. If the message type indicates that an optional part is possible, but there is no optional part included in this particular message, then a pointer field containing all zeros is used. The ITU-T Organization recommends that all future message types with a mandatory variable part indicate that an optional part is allowed.

All pointers are sent consecutively at the beginning of the mandatory variable part. Each parameter contains the parameter length indicator followed by the contents of the parameters. If there are no mandatory variable parameters, but optional parameters are possible, the start of optional parameters pointer (coded all zeros if no optional parameter is present and coded 00000001 if any optional parameter is present) is included.

TABLE 17.6 Message-Type Code Allocation

Message Type	Reference (Table)[a]	Code
Address complete	21	00000110
Answer	22	00001001
Blocking	39	00010011
Blocking acknowledgment	39	00010101
Call progress	23	00101100
Circuit group blocking	40	00011000
Circuit group blocking acknowledgment	40	00011010
Circuit group query @	41	00101010
Circuit group query response @	24	00101011
Circuit group reset	41	00010111
Circuit group reset acknowledgment	25	00101001
Circuit group unblocking	40	00011001
Circuit group unblocking acknowledgment	40	00011011
Charge information @	b	00110001
Confusion	26	00101111
Connect	27	00000111
Continuity	28	00000101
Continuity check request	39	00010001
Facility @	45	00110011
Facility accepted	42	00100000
Facility reject	29	00100001
Facility request	42	00011111
Forward transfer	37	00001000
Identification request	47	00110110
Identification response	48	00110111
Information @	30	00000100
Information request @	31	00000011
Initial address	32	00000001
Loop back acknowledgment @	39	00100100
Network resource management	46	00110010
Overload @	39	00110000
Pass-along @	43	00101000
Release	33	00001100
Release complete	34	00010000
Reset circuit	39	00010010
Resume	38	00001110
Segmentation	49	00111000
Subsequent address	35	00000010
Suspend	38	00001101
Unblocking	39	00010100
Unblocking acknowledgment	39	00010110
Unequipped CIC @	39	00101110
User part available	44	00110101
User part test	44	00110100
User-to-user information	36	00101101
Reserved (used in 1984 version)		00001010
		00001011
		00001111
		00100010
		00100011

TABLE 17.6 *(continued)*

Message Type	Reference (Table)[a]	Code
Reserved (used in 1988 version)		00100101
		00100110
		00011101
		00011100
		00011110
		00100111

[a] Reference to table number in ITU-T Rec. Q.763.
[b] The format of this message is a national matter.
Source: Table 4/Q.763, page 2, ITU-T Rec. Q.763 [24].

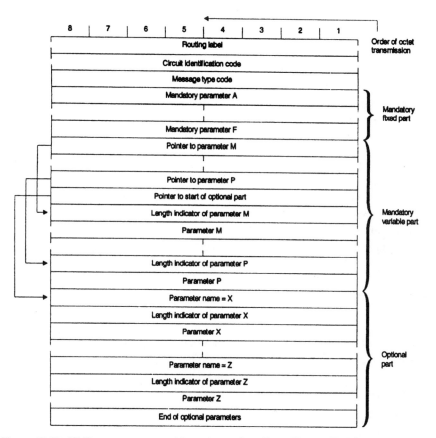

Figure 17.18. ISUP message, general format overview. From Figure 3/Q.763, page 5, ITU-T Rec. Q.763, December 1999 [24].

OPTIONAL PART. The optional part consists of parameters that may or may not occur in any particular message type. Both fixed-length and variable-length parameters may be included. Unless it is explicitly stated to the contrary within the ISUP 1992 Recommendations, an optional parameter cannot occur multiple times within one message. Optional parameters may be transmitted in any order. Each optional parameter includes the parameter name (one octet) and the length indicator (one octet) followed by the parameter contents.

If optional parameters are present and after all optional parameters have been sent, an "end of optional parameters" octet containing all zeros is transmitted. If no optional parameter is present, an "end of optional parameter" octet is not transmitted.

13.3.4 *Example of ISUP Signaling Procedure.* An example of basic call control and signaling procedures is given below for *successful call setup*, forward address signaling, en-bloc operation.

Actions Required at the Originating Exchange

(A) CIRCUIT SELECTION. When the originating gateway exchange has received the complete selection information from the calling party, and has determined that the call is to be routed to another exchange, selection of a suitable, free, interexchange circuit takes place and an initial address message is sent to the succeeding exchange.

Appropriate routing information is either stored at the originating exchange or at a remote database to which a request may be made.

The selection of the route will depend on the called party number, connection type required, and the network signaling capability required. This selection process may be performed at the exchange or with the assistance of the remote database.

In addition, in the case of a subscriber with digital access, the setup message contains bearer capability information which is analyzed by the originating exchange to determine the correct connection type and network signaling capability. The bearer capability information will be mapped into the user service information parameter of the initial address message. The high layer capability information will be mapped into the user teleservice information parameter of the initial address message. The information received from the access interface is used to set the value of the transmission medium requirement parameter.

The connection types allowed are:

- Speech
- 3.1-kHz audio
- 64 kbits/s unrestricted
- 64 kbits/s unrestricted preferred
- 2×64 kbits/s unrestricted; multirate connection type
- 384 kbits/s unrestricted; multirate connection type
- 1536 kbits/s unrestricted; multirate connection type
- 1920 kbits/s unrestricted; multirate connection type

The information used to determine the routing of the call by the originating exchange will be included in the initial address message (as transmission medium requirement and forward call indicators) to enable correct routing at intermediate exchanges. The initial address message conveys implicitly the meaning that the indicated circuit has been seized.

(B)   ADDRESS INFORMATION SENDING SEQUENCE. The sending sequence of address information on international calls will be the country code followed by the national (significant) number. On national connections, the address information may be the subscriber number or the national (significant) number as required by the administration concerned. For calls to international operator positions (Code 11 and Code 12) refer to Rec. Q.107.

The end-of-pulsing (ST) signal will be used whenever the originating exchange is in a position to know by digit analysis that the final digit has been sent.

(C)   INITIAL ADDRESS MESSAGE. The initial address message in principle contains all the information that is required to route the call to the destination exchange and connect the call to the called party.

If the initial address message (IAM) exceeds the 272-octet limit for MTP transfer, it is segmented by the use of the segmentation message (segmentation is discussed in Section 13.3.4.1).

All IAMs include a protocol control indicator and (in the forward call indicator parameter) a transmission medium requirement parameter.

The originating exchange will set the parameters in the protocol control indicator and in the ISDN-user part preference indicator to indicate:

(i) The type of end-to-end method that can be accommodated (see Rec. Q.730)

(ii) The availability of Signaling System No. 7 signaling

(iii) The use of the ISDN-user part

(iv) Network signaling capability required—for example, ISDN-user part required all the way

The ISDN-user part preference indicator is set according to the bearer service, teleservice, and supplementary service(s) requested. The exact setting depends on the service demand conditions and may be different depending on individual cases. In principle, if the service demand requires ISDN-user part to be essential, then the indicator is set to "required," and if the service required is optional but preferred it is set to "preferred." Otherwise it is set to "not required." The indicator is set to either "required" or "preferred," or "not required," according to the most stringent condition required by one or more of the parameters in the initial address message.

The nature of connection indicators is set appropriately based on the characteristics of the selected outgoing circuit.

The transmission medium requirement parameter contains the connection type required information—for example, 3.1-kHz audio.

The propagation delay counter is included according to paragraph 2.6 of Ref. 25.

(D) COMPLETION OF TRANSMISSION PATH. Through-connection of the transmission path will be completed in the backward direction (the transmission path is completed in the forward direction on receipt of a connect or answer message) at the originating exchange immediately after the sending of the initial address message, except in those cases where conditions on the outgoing circuit prevent it.

It is also acceptable that on speech or 3.1-kHz audio calls, through-connection of the transmission path will be completed in both directions immediately after the initial address message has been sent, except in those cases where conditions on the outgoing circuit prevent it.

(E) NETWORK PROTECTION TIMER. When the originating exchange has sent the initial address message, the awaiting address complete timer (T7) is started. If timer (T7) expires, the connection is released and an indication is returned to the calling subscriber.

Actions Required at an Intermediate National Exchange

(A) CIRCUIT SELECTION. An intermediate national exchange, on receipt of an initial address message, will analyze the called party number and the other routing information to determine the routing of the call. If the intermediate national exchange can route the call using the connection type specified in the transmission medium requirement parameter, a free interexchange circuit is seized and an initial address message is sent to the succeeding exchange. Within a network if the intermediate national exchange does not route the call using just the connection type specified in the transmission medium requirement parameter, the exchange may also examine the user service information containing the bearer capability information and/or the user teleservice information containing the high layer capability information, if available, to determine if a suitable route can be selected. In this case if a new connection type is provided, the transmission medium requirement parameter is modified to the new connection type.

(B) PARAMETERS IN THE INITIAL ADDRESS MESSAGE. An intermediate national exchange may modify signaling information received from the preceding exchange according to the capabilities used on the outgoing route. Signaling information that may be changed is the nature of connection indicator and propagation delay counter. Other signaling information is passed on transparently—for example, the access transport parameter, user service information, and so on.

The satellite indicator in the nature of connection parameter should be incremented if the selected outgoing circuit is a satellite circuit. Otherwise, the indicator is passed on unchanged.

(C) COMPLETION OF TRANSMISSION PATH. Through-connection of the transmission path in both directions will be completed at an intermediate national exchange immediately after the initial address message has been sent, except in those cases where conditions on the outgoing circuit prevent it.

Actions Required at an Outgoing International Exchange

(A) CIRCUIT SELECTION. On receipt of an initial address message, an outgoing international exchange will analyze the called party number and the other routing information to determine the routing of the call. If the outgoing international exchange can route the call using the connection type specified in the transmission medium requirement parameter, a free interexchange circuit is seized and an initial address message is sent to the succeeding exchange.

If the outgoing international exchange cannot trust that the transmission medium requirement value received from the national network reflects the minimum value of the information transfer susceptance, then the transmission medium requirement value may be modified according to the contents of the information transfer capability and information transfer rate fields of the user service information parameter (if available).

The outgoing international exchange must ensure that the transmission medium requirement parameter is set according to the service requested by the customer (see CCITT Rec. E.172). More specifically this parameter is carried unchanged within the international network.

(B) PARAMETERS IN THE INITIAL ADDRESS MESSAGE. An outgoing international exchange may modify signaling information received from the preceding exchange according to the capabilities used on the outgoing route. Signaling information that may be changed is the nature of connection indicator and propagation delay counter; the most significant digits in the called party number may be amended or omitted (country code is removed at the last exchange before the incoming international exchange). Other signaling information is passed on transparently—for example, the access transport parameter, user service information, and so on.

If the outgoing international exchange belongs to a country using μ-law PCM encoding nationally and the transmission medium requirement indicates speech or 3.1-kHz audio, then the user information layer 1 protocol identification field of the user service information parameter must be checked. If it indicates "Rec. G.711 μ-law" this must be changed to "Rec. G.711 A-law" and a μ-law to A-law convertor must be enabled.

The satellite indicator in the nature of connection parameter should be incremented if the selected outgoing circuit is a satellite circuit. Otherwise, the indicator is passed on unchanged.

The outgoing international gateway exchange should include the originating ISC point code parameter in the initial address message. This information is used for statistical purposes—for example, accumulation of the number of incoming calls on an originating international switching center basis.

If a location number parameter is received, the nature of address indicator is checked, and if the nature of address indicator is set to "international number," then the parameter is passed unchanged; otherwise the number is modified to the international number format and the nature of address is set to "international number" before being passed.

The end-of-pulsing (ST) signal will be used whenever the outgoing exchange is in a position to know by digit analysis that the final digit has been sent.

(C) COMPLETION OF TRANSMISSION PATH. Through-connection of the transmission path in both directions will be completed at an outgoing international exchange immediately after the initial address message has been sent, except in those cases where conditions on the outgoing circuit prevent it.

(D) NETWORK PROTECTION TIMER. When an outgoing international exchange has sent the initial address message, the awaiting address complete timer (T7) is started. If timer (T7) expires, the connection is released and an indication is returned to the calling subscriber.

Actions Required at an Intermediate International Exchange

(A) CIRCUIT SELECTION. On receipt of an initial address message, an intermediate international exchange will analyze the called party number and the other routing information to determine the routing of the call. If the intermediate international exchange can route the call using the connection type specified in the transmission medium requirement parameter, a free interexchange circuit is seized and an initial address message is sent to the succeeding exchange.

(B) PARAMETERS IN THE INITIAL ADDRESS MESSAGE. An intermediate international exchange may modify signaling information received from the preceding exchange according to the capabilities used on the outgoing route. Signaling information that may be changed is the nature of connection indicator and propagation delay counter; the most significant digits in the called party number may be amended or omitted (country code is removed at the last exchange before the incoming international exchange). Other signaling information is passed on transparently—for example, the access transport parameter, user service information, and so on.

The satellite indicator in the nature of connection parameter should be incremented if the selected outgoing circuit is a satellite circuit. Otherwise, the indicator is passed on unchanged.

(C) COMPLETION OF TRANSMISSION PATH. Through-connection of the transmission path in both directions will be completed at an intermediate international exchange immediately after the initial address message has been sent, except in those cases where conditions on the outgoing circuit prevent it.

When an intermediate international exchange has sent the initial address message, the awaiting address complete timer (T7) is started. If timer (T7) expires, the connection is released and an indication is returned to the calling subscriber.

Actions Required at an Incoming International Exchange

(A) CIRCUIT SELECTION. An incoming international exchange, on receipt of an initial address message, will analyze the called party number and the other routing information to determine the routing of the call. If the incoming international exchange can route the call using the connection type specified in the transmission medium requirement parameter, a free interexchange circuit is seized and an initial address message is sent to the succeeding exchange.

(B) PARAMETERS IN THE INITIAL ADDRESS MESSAGE. An incoming international exchange may modify signaling information received from the preceding exchange according to the capabilities used on the outgoing route. Signaling information that may be changed are nature of connection indicator and propagation delay counter. Other signaling information is passed on transparently—for example, the access transport parameter, user service information, and so on.

The satellite indicator in the nature of connection parameter should be incremented if the selected outgoing circuit is a satellite circuit. Otherwise, the indicator is passed on unchanged.

If the incoming international exchange belongs to a country using μ-law PCM encoding nationally and the transmission medium requirement indicates speech or 3.1-kHz audio, then the user information layer 1 protocol identification field of the user service information parameter must be checked. If it indicates "Rec. G.711 A-law," this must be changed to "Rec. G.711 μ-law" and a μ-law to A-law convertor must be enabled.

The incoming international gateway exchange should delete the originating ISC point code parameter from the initial address message and set up a connection to the national network. This information is used for statistical purposes—for example, accumulation of the number of incoming calls on an originating international switching center basis.

(C) COMPLETION OF TRANSMISSION PATH. Through-connection of the transmission path in both directions will be completed at an incoming international exchange immediately after the initial address message has been sent, except in those cases where conditions on the outgoing circuit prevent it.

(D) NETWORK PROTECTION TIMER. When an incoming international exchange has sent the initial address message, the awaiting address complete timer (T7) is started. If timer (T7) expires, the connection is released and an indication is returned to the calling subscriber.

Actions Required at the Destination Exchange

(A) SELECTION OF CALLED PARTY. Upon receipt of an initial address message, the destination exchange will analyze the called party number to determine to which party the call should be connected. It will also check the called party's line condition and perform various checks to verify whether or not the connection is allowed. These checks will include correspondence of compatibility checks—for example, checks associated with supplementary services.

In this case where the connection is allowed, the destination exchange will set up a connection to the called party. If a continuity check has to be performed on one or more of the circuits involved in a connection, setting up of the connection to the called party must be prevented until the continuity of such circuits has been verified.

(B)   SEGMENTED INITIAL ADDRESS MESSAGE. If the initial address message had been segmented by the use of the segmentation message, the remainder of the call setup information is awaited.

13.3.4.1   Segmentation. The simple segmentation procedure uses the segmentation message to convey an additional segment of an overlength message. Any message containing either the optional forward or backward call indicators can be segmented using this method. This procedure provides a mechanism for the transfer of certain messages whose contents are longer than 272 octets but not longer than 544 octets.

The procedure is as follows:

(a) The sending exchange, on detecting that the message to be sent exceeds the 272-octet limit of the message transfer part, can reduce the message length by sending some parameters in a segmentation message sent immediately following the message containing the first segment.

(b) The parameters that may be sent in the second segment using the segmentation message are the user-to-user information, generic digit, generic notification, generic number, and access transport parameters. If the user-to-user information and access transport parameters cannot be carried in the original message and the two together do not fit in the segmentation message, the user-to-user information parameter is discarded.

(c) The sending exchange sets the simple segmentation indicator in the optional forward or backward call indicators to indicate that additional information is available.

(d) When a message is received, at a local exchange, with the simple segmentation indicator set to indicate additional information is available, the exchange starts timer T34 to await the segmentation message. This action may also take place at incoming or outgoing international exchanges if policing of information is required.

(e) When the segmentation message is received, timer T34 is stopped and the call continues.

(f) In case any other message except the ones listed below is received before the segmentation message containing the second segment, the exchange should react as if the second segment is lost; that is, timer T34 is stopped and the call continues. The messages are:

- Continuity
- Blocking
- Blocking acknowledgment
- Circuit group blocking
- Circuit group blocking acknowledgment
- Unblocking
- Unblocking acknowledgment
- Circuit group unblocking
- Circuit group unblocking acknowledgment
- Circuit group query
- Circuit group query response

(g) After expiration of timer T34, the call proceeds and a received segmentation message containing the second segment of a segmented message is discarded.

(h) At an incoming or outgoing international exchange, when following the simple segmentation procedure, it is possible that the exchange has to reassemble an incoming message and subsequently resegment it for onward transmission. In this case it has to be ensured that any unrecognized parameters received in the first, or second, segment are transmitted in the first, or second, segment, respectively, when the passing of the parameter is required by the compatibility procedure.

13.3.5 Continuity Check. Because the signaling information in SS No. 7 does not pass over the speech/data channel (i.e., the B-channel), facilities are provided for making a continuity check of the traffic circuit in the circumstances described below.

The application of the continuity check depends on the type of transmission system used for the circuit. For transmission systems having some inherent fault indication features giving an indication to the switching system in case of fault, a continuity check is not required. However, a per-call continuity check may be needed on fully digital circuits when circuits or bundles of circuits in primary multiplex groups are dropped and inserted en route between switches and alarm indications carried on bits of the primary multiplex frame structure are lost in passing through an intermediate transmission facility that does not relay them transparently. Typical, per-call continuity checks may be needed when the transmission link between switches contains a TDMA satellite system, a digital circuit multiplication system, or a digital access and cross-connection system, where fault indications are lost (see CCITT Rec. Q.33).

When an initial address message is received with a request for a continuity check, a continuity-check loop is connected.

Means should be provided in Signaling System No. 7 to detect circuit identification code misunderstandings between Signaling System No. 7 exchanges.

For exchanges having both analog and digital circuits served by Signaling System No. 7, the continuity check initiated by a continuity-check request message could be used to test for proper alignment of circuit code identities. On those exchanges, reception of a continuity-check request message should always cause a loop to be attached to the circuit.

Alternative methods for detection of circuit identity misunderstandings in exchanges with all-digital circuits may be employed.

The continuity check is not intended to eliminate the need for routine testing of the transmission path.

The continuity check of the circuit will be done, link-by-link, on a per-call basis or by a statistical method prior to the commencement of conversation. Procedures and requirements are specified in 7/Q.724 [20].

The actions to be taken when pilot supervision is used are described in 9/Q.724 [20].

When an initial address message is received with a request for continuity check (either on this circuit or on a previous circuit), timer T8 is started. On receipt of a successful indication of continuity check in a continuity message, timer T8 is stopped. However, if timer T8 expires, the connection is cleared.

If an indication of continuity-check failure is received in a continuity message, timer T27 is started awaiting a continuity recheck request. Also, the connection to the succeeding exchange is cleared. Timer T27 is stopped when the continuity-check request message is received and timer T36 is started awaiting a continuity or release message.

If either timer T27 or timer T36 expires, a reset circuit message is sent to the preceding exchange. On reception of the release complete message, the circuit is set to idle.

Where circumstances require per-call continuity checking for multirate connection type calls, the continuity of the single 64-kbit/s circuit whose circuit identification code is contained in the initial address message shall be checked.

Source: Sections 13.3.4 and 13.3.5 have been extracted from ITU-T Rec. Q.764 [25].

14   SS7 SIGNALING DATA CONNECTIVITY OVER THE INTERNET

14.1   New IP Transport Protocol

The internet has broad application. It can now be used to transport signaling information. For this chapter, that interest lies in CCITT Signaling System No. 7 and the transport of its messages over IP. To define ways to carry out this function, the IETF set up the working group Sigtran (signaling transport). The goal of Sigtran is to develop methods of transporting PSTN signaling over IP network. Sigtran is a protocol defined by the IETF under RFC 2719. (*Note*: The reader should consult Chapter 11, Section 7, for an overview of TCP/IP protocol family. SCTP is a part of this family.)

There are three components of Sigtran: standard IP network, a common channel signaling transport called SCTP, and an adaptation sublayer that supports certain specific primitives. Several adaptation sublayer protocols have been developed by the IETF for this purpose, including M2PA, M2UA, M3UA, SUA, and IUA.

ACRONYM DEFINITIONS

IETF	Internet engineering taskforce
M2PA	MTP2 peer-to-peer adaptation
M2UA	MTP2 user adaptation
M3UA	MTP3 user adaptation
IUA	IP user adaptation
SUA	SCCP user adaptation
MTP	Message transfer part (SS7 term)

The remainder of this section deals with SCTP (stream control transport protocol) and defined by IETF RFC 2960 and RFC 3309 [32, 33], under the umbrella of the IETF Sigtran Working Group.

14.2 Stream Control Transport Protocol (SCTP)

SCTP is a third transport protocol in the same IP stack level as the familiar TCP and UDP. In the IP protocol stack we can view SCTP as taking the place of TCP when IP is to carry out certain functions, especially signaling. SCTP is connection oriented (vis à vis connectionless).

Like TCP, SCTP provides a reliable transport service, ensuring that data are transported across the network without error and in sequence. Also like TCP, SCTP is a session-oriented mechanism, meaning that a relationship is created between endpoints of an SCTP association prior to data being transmitted, and this relationship is maintained until all data transmission has been successfully completed.

Unlike TCP, SCTP provides a number of functions that are critical for telephony signaling transport (e.g., SS7 messages), and at the same time can potentially benefit other applications needing transport with additional performance and reliability.

SCTP is a unicast protocol, and supports data exchange between exactly two endpoints, although these may be represented by multiple IP addresses. SCTP transmission is full-duplex.

SCTP is message-oriented and supports framing of individual message boundaries. In comparison, TCP is byte oriented and does not preserve any implicit structure within a transmitted byte stream without enhancement.

SCTP affords reliable transmission. It detects when data are discarded, reordered, duplicated, or corrupted. It then retransmits damaged and missing data as necessary.

Similar to TCP, SCTP is rate adaptive and scales back data throughput to meet prevailing conditions of the network. It acts cooperatively with TCP sessions attempting to use the same bit rate.

14.2.1 Multistreaming, a Feature of SCTP. Multistream is a characteristic of SCTP which allows data to be partitioned into multiple streams that have the property of independently sequenced delivery. As a result, message loss in any one stream will only initially affect delivery within that stream, and not delivery in other streams.

TCP presents a different situation. It assumes a single stream of data and ensures that delivery of that stream takes place with byte sequence preservation. While this is desirable for the transport of a file or record which are not so time-sensitive, it causes additional delay when message loss or sequence error occurs within the network. When such occurs, TCP delays delivery of data until the correct sequencing is restored, either by receipt of an out-of-sequence message or by retransmission of a lost message. *Note*: This adds considerably to *postdial delay*, the principal performance parameter of a telephony signaling system.

For a number of applications, the characteristic of strict message (frame) sequence preservation is not mandatory. In telephony signaling, such as SS7, it is only necessary to maintain sequencing of messages that affect the same resource (e.g., the same call or the same channel). Other messages (frames) are only loosely correlated and can be delivered without having to maintain overall sequence integrity.

SCTP achieves multistreaming by creating independence between data transmission and data delivery. In particular, each payload DATA *chunk* in the protocol uses two sets of sequence numbers, a transmission sequence number that governs the transmission of messages and the detection of message loss, and the stream ID/stream sequence number pair, which is used to determine the sequence of delivery of received data.

Such independence of mechanisms allows the receiver to determine immediately when a gap in the transmission sequence occurs (e.g., such as message loss), and also whether or not messages received following the gap are within the affected stream. If a message is received within the affected stream, there will be a corresponding gap in the stream sequence number, while messages from other steams will not show a gap. This allows the receiver to continue delivering messages (frames) to the unaffected streams while buffering messages in the affected stream until retransmission occurs.

Source: Sections 14.1 and 14.2 have been based on IETF RFCs 3286 and 2960 [30, 32]. These RFCs have copyright notices by the IETF.

14.3 Message Format of SCTP

Figure 17.19 shows the overall message format of SCTP. Figure 17.20 shows the frame structure of an SCTP header. After the checksum in the SCTP header, the message payload, called "CHUNK," follows. "*Chunk*" is illustrated, by fields, in

MAC header
IP header
SCTP header
Data – – – –

Figure 17.19. Overall general format of an SCTP message (frame).

Source port	Destination port
Verification tag	
Check sum	
Chunk 1 – – – –	
– – – – Chunk n	

Figure 17.20. SCTP header.

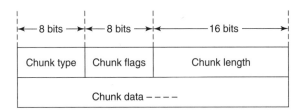

Figure 17.21. Chunk field structure.

Figure 7.21. It should be noted that the *source port* and *destination port* fields are each 16 bits wide. The remainder of the fields are all 32 bits wide.

14.3.1 SCTP Summary. SCTP is designed to transport PSTN signaling messages (such as SS7) over IP networks, but is capable of broader applications. SCTP is a reliable transport protocol operating on top of a connectionless packet network such as IP. It offers the following services to users:

- Acknowledged error-free nonduplicated transfer of user data
- Data fragmentation to conform to discovered MTU (maximum transfer unit) size.
- Sequenced delivery of user messages within multiple streams, with an option for order-of-arrival delivery of individual user messages
- Optional bundling of multiple user messages into a single SCTP packet
- Network-level fault tolerance through supporting of multihoming at either or both ends of an association

The design of SCTP includes appropriate congestion avoidance behavior and resistance to flooding and masquerade attacks.

Source: Section 14.3.1 is based on Ref. 34.

REVIEW QUESTIONS

1. What is the principal rationale for developing and implementing CCITT Signaling System No. 7?

2. Describe SS No. 7 relationship with OSI. Why does it truncate at OSI layer 4?

3. Give the two primary "parts" of SS No. 7. Briefly describe the function of each generic part.

4. With CCITT Signaling System No. 7, OSI layer 4 is subdivided into two sublayers. Name them.

5. Name the three user parts of SS No. 7.

6. What is the normal basic connectivity of SS No. 7? (*Hint*: bit rate).

7. Layer 2 of SS No. 7 functions as the signaling link. Name five of the seven functions of layer 2. In one sentence describe what layer 2 does.

8. What are the two basic categories of functions of layer 3 of SS No. 7? What is the basic purpose of layer 3? (*Hint*: Think OSI).

9. What is a signaling relation?

10. How are signaling points defined (identified)?

11. Interrelate message routing, message discrimination, and message distribution.

12. What does a routing label do?

13. What is the purpose of signaling unit delimitation?

14. What is the *basic* method of error correction in SS No. 7?

15. How is flow control initiated when congestion takes place at a signaling point(s)?

16. Give the three signaling network management functional blocks.

17. There are three types of signal units used in SS No. 7. What are they? Define the basic function of each.

18. Differentiate forward and backward sequence numbers. When does reset occur for FSN? In essence, what do FSN and BSN accomplish?

19. What is the function of the service indicator field?

20. The routing label is analogous to what in our present telephone system? Name the three basic pieces of information that the routing label provides.

21. Give three of the fault conditions handled by the signaling network management.

22. Define labeling potential.

23. What is the function of an STP? Differentiate between an STP and a signaling point.

24. Why do we wish to limit the number of STPs in a signaling connection on a particular relation?

25. What are the three basic measures of performance of SS No. 7?

26. Give the long-term and medium-term error rates for SS No. 7.

27. What is the biggest contributor to signaling delay in the SS No. 7 network?

28. At what OSI layer does the SCCP reside? What is the basic rationale of implementing SCCP?

29. How can we use a hypothetical signaling reference connection in practice?

30. Regarding user parts, what is the function of the SIO? Of the network indicator?

31. What does the heading code (H0, H1) tell us?

32. What is the purpose of circuit continuity? Explain.

33. Address signals (such as telephone number digits) are sent digit by digit embedded in the last subfield in the SIF. How are they represented? (*Hint*: per digit).

34. Give two examples of backward information in SS No. 7.

35. What facilitates end-to-end signaling typically for ISUP (i.e., what "part" of SS No. 7)?

36. Give at least three functions of the protocol control indicator, as used, for example, in the IAM.

37. Give the actions taken at an intermediate exchange during call setup using SS No. 7.

38. Give the actions taken by a destination exchange during call setup using SS No. 7.

39. How is an IAM handled that has contents greater than 272 octets?

40. Through what means does SS No. 7 handle A-law/μ-law conversion (i.e., in what message type)?

41. Where would it really be advantageous to use the internet for SS No. 7 connectivity? (*Hint*: another chapter in this text).

42. What are the principal objectives of the work by Sigtran?

43. What is the principal application of SCTP (not that it doesn't have other applications)?

44. How does SCTP achieve multistreaming?

45. SCTP is not designed for multicasting but it can handle multiple ___.

REFERENCES

1. W. Stallings, *Tutorial, Integrated Services Digital Network (ISDN)*, IEEE Computer Society, Washington, DC, 1985.

2. *Specifications of Signaling System No. 7* (Q.700 series), Fascicle VI.6, CCITT Recommendations (Yellow Books), VIIth Plenary Assembly, Geneva, 1980.

3. *Telcordia Notes on the Networks—2000*, Telcordia SR-2275, Issue 4, October 2000, Telcordia, Piscataway, NJ, 1994.

4. *Introduction to CCITT Signaling System No. 7*, ITU-T Rec. Q.700, Helsinki, Mar. 1993.

5. G. G. Schlanger, "An Overview of Signaling System No. 7," *IEEE J. on Selected Areas in Commun.*, **7** (3) (May 1986).

6. *Functional Description of the Message Transfer Part (MTP) of Signaling System No. 7*, ITU-T Rec. Q.701, Helsinki, March 1993.

7. *Signaling System No. 7—Signaling Link*, ITU-T Rec. Q.703, ITU, Geneva, July 1996.

8. *Signaling System No. 7—Signaling Network Structure*, ITU-T Rec. Q.705, Helsinki, March 1993.

9. *Signaling System No. 7—Message Transfer Part Signaling Performance*, ITU-T Rec. Q.706, Helsinki, March 1993.

10. *Signaling System No. 7—Numbering of International Signaling Point Codes*, ITU-T Rec. Q.708, ITU, Geneva, March 1999.

11. *Signaling System No. 7—Hypothetical Signaling Reference Connection*, ITU-T Rec. Q.709, Helsinki, March 1993.

12. W. C. Roehr, Jr., "Signaling System No. 7," in *Tutorial: Integrated Services Digital Network (ISDN)*, W. Stallings, ed., IEEE Computer Society, Washington, DC, 1985.

13. R. L. Freeman, *Reference Manual for Telecommunication Engineering*, 3rd ed., John Wiley & Sons, New York, 2001.

14. *Signaling Network, Functions and Messages*, ITU-T Rec. Q.704, ITU, Geneva, July 1996.

15. *Signalling System No. 7—Functional Description of the Signalling Connection Control Part*, ITU-T Rec. Q.711, ITU, Geneva, March 2001.

16. *Signalling System No. 7—Definition and Function of the Signalling Connection Control Part (SCCP) Messages*, ITU-T Rec. Q.712, ITU, Geneva, July 1996.

17. *Signaling System No. 7—Signaling Connection Control Part (SCCP) Formats and Codes*, ITU-T Rec. Q.713, ITU, Geneva, March 2001.

18. *Signaling System No. 7—Signaling Connection Control Part (SCCP) Performance*, ITU-T Rec. Q.716, ITU, Helsinki, March 1993.

19. *Signaling System No. 7—Signaling Performance in the Telephone Application*, ITU-T Rec. Q.725, ITU, Helsinki, March 1993.

20. CCITT Recommendations, Volume VI, Fascicle VI.8, *Specifications of Signalling System No. 7*, Recs. Q.721-Q.766, IXth Plenary Assembly, Melbourne, 1988.

21. *Formats and Codes (Telephone User Part)*, CCITT Rec. Q.723, Fascicle VI.8, IXth Plenary Assembly, Melbourne, 1988.

22. *Functional Description of the ISDN User Part of Signalling System No. 7*, ITU-T Rec. Q.761, ITU, Geneva, December 1999.

23. *General Function of Messages and Signals of the ISDN User Part of Signalling System No. 7*, ITU-T Rec. Q.762, ITU, Geneva, December 1993 and Addendum, June 2000.

24. Formats and Codes of the ISDN User Part of Signalling System No. 7, ITU-T Rec. Q.763, ITU, Geneva, December 1999.

25. *Signaling System No. 7 ISDN User Part Signaling Procedures*, ITU-T Rec. Q.764, ITU, Geneva, December 1999 and Amendment 1, July 2001.

26. *Signaling System No. 7—Signaling Connection Control Part (SCCP) Procedures*, ITU-T Rec. Q.714, ITU, Geneva, June 2001.

27. *The International Public Telecommunication Numbering Plan*, ITU-T Rec. E.164, ITU, Geneva, May 1997.

28. *Signaling Connection Control Part (SCCP) User Guide*, ITU-T Rec. Q.715, ITU, Geneva, April 2002.

29. *SS7 over IP Signaling Transport & SCTP*, IEC Tutorials, from the web: www.iec.org, January 2002.

30. *An Introduction to the Stream Transmission Control Protocol*, RFC 3286, IETF, Ann Arbor, MI, May 2002.

31. *Framework Architecture for Signaling Transport*, RFC 2719, IETF, Ann Arbor, MI, October 1999 (Sigtran).

32. *Stream Control Transmission Protocol (SCTP)*, RFC 2960, IETF, Ann Arbor, MI, October 2000.

33. *Stream Control Transmission Protocol Checksum Change*, RFC 3309, IETF, Ann Arbor, MI, September 2002.

34. From the web: http:/www.networksorcery.com/enp/protocol/sctp.htm. Copied 9/22/03.

18

WIRELESS AND CELLULAR/MOBILE RADIO

1 INTRODUCTION

1.1 Background

Mobile radio communications dates back to Marconi. For a time it was the principal application of radio. This was ship-to-shore and ship-to-ship communications. The pioneer in this was the Marconi Company of the United Kingdom. It spread to land vehicles and aircraft in the 1920s.

Since 1980, mobile radio communication has taken on a more personal flavor. Cellular radio systems have extended the telephone network to the car, to the pedestrian, and even into the home and office. A new and widely used term in our vocabulary is *personal communications*. It is becoming the universal tether. No matter where we go, on land, at sea, and in the air, we can have near instantaneous two-way communications by voice, data, and facsimile. At some time it will encompass video.

Personal radio terminals are becoming smaller. There is the potential of becoming wristwatch size. However, the human interface requires input–output devices that have optimum usefulness. A wristwatch-size keyboard or keypad is rather difficult to operate. A hard-copy printer requires some minimum practical dimensions, and so forth.

A new name has entered our vocabulary, *wireless*. The British have been using the term since Marconi. It is relatively new in North America, and with a different flavor in meaning. I think we can define wireless as a telecommunication method that does not require wires to communicate. From our perspective, wireless and radio are synonymous.

Reference 28 states that there were 1 billion (1×10^9) by the end of 2002; and Reference 27 expects there to be 3 billion (3×10^9) traditional telephone

Telecommunication System Engineering, by Roger L. Freeman
ISBN 0-471-45133-9   Copyright © 2004 Roger L. Freeman

and wireless users in the world by the year 2010. It is our opinion that this latter estimate may be on the low side.

1.2 Scope and Objective

This chapter presents an overview of "personal communications" or what many call *wireless*. Much of the discussion deals with cellular radio and wireless LANs (WLANs), and it extends this thinking inside of buildings. The coverage most necessarily includes propagation for the several environments, propagation impairments, and methods to mitigate these impairments, access techniques, bandwidth limitations, and ways around this problem. It will cover several mobile radio standards and compare a number of existing and planned systems. The chapter objective is to provide an appreciation of mobile/personal communications. Space limitations force us to confine our discussion to what might be loosely called "land mobile systems."

2 BASIC CONCEPTS OF CELLULAR RADIO

Cellular radio systems connect a mobile terminal to another user, usually through the PSTN. The "other user" most commonly is a telephone subscriber of the PSTN. However, the other user may be another mobile terminal. Most of the connectivity is extending POTS to mobile users. Data and facsimile services are in various stages of implementation (see Chapter 13, Section 6). Some of the terms used in this section have a strictly North American flavor.

Figure 18.1 shows a conceptual layout of a cellular system. The heart of the system for a specific serving area is the mobile telephone switching office (MTSO). The MTSO is connected by a trunk group to a nearby telephone exchange providing an interface to, and connectivity with, the PSTN.

The area to be served by a *cellular geographic serving area* (CGSA) is divided into small geographic cells which ideally are hexagonal. Cells are initially laid out with centers spaced about 4 to 8 miles (6.4 to 12.8 km) apart. The basic system components are the cell sites, the MTSO, and mobile units. These mobile units may be hand-held or vehicle-mounted terminals.

Each cell has a radio facility housed in a building or shelter. The facility's radio equipment can connect and control any mobile unit within the cell's responsible geographic area. Radio transmitters located at the cell site have a maximum effective radiated power (ERP*) of 100 watts. Combiners are used to connect multiple transmitters to a common antenna on a radio tower, usually between 50 and 300 ft (15 and 92 m) high. Companion receivers use a separate antenna system mounted on the same tower. The receive antennas are often arranged in a space-diversity configuration.

---

* Care must be taken with terminology. In this instance, ERP and EIRP are not the same. The reference antenna in this case is the dipole which has a 2.16-dBi gain.

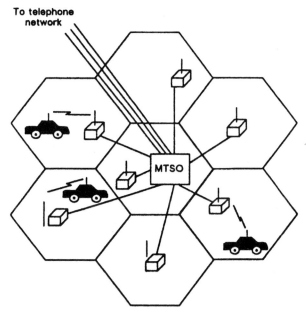

To telephone network

MTSO

Figure 18.1. Conceptual layout of a cellular system.

The MTSO provides switching and control functions for a group of cell sites. A method of connectivity is required between the MTSO and the cell site facilities. The MTSO is an electronic switch and carries out a fairly complex group of processing functions to control communications to and from mobile units as they move between cells as well as to make connections with the PSTN. Besides making connectivity with the public network, the MTSO controls cell site activities and mobile actions through command and control data channels. The connectivity between cell sites and the MTSO is often via DS1 on wire pairs or on microwave facilities, the latter being the most common.

A typical cellular mobile unit consists of a control unit, a radio transceiver, and an antenna. The control unit has a telephone handset, a pushbutton keypad to enter commands into the cellular/telephone network, and audio and visual indications for customer alerting and call progress. The transceiver permits full duplex transmission and reception between a mobile and cell sites. Its ERP is nominally 6 watts. The unit is usually vehicle-mounted. Hand-held terminals combine all functions into one small package that can be easily held in one hand. The ERP of a hand-held is a nominal 0.6 watts. It seems that this package is being made smaller and smaller.

In North America, cellular communication is assigned a 25-MHz band between 824 and 849 MHz for mobile unit-to-base transmission and a similar band between 869 and 894 MHz for transmission from base to mobile. The original North American cellular radio systems was called AMPS (advanced mobile telephone system). The original system description was contained in an entire issue of the

Bell System Technical Journal (BSTJ) of January 1979. The present AMPS is based on 30-kHz channel spacing using frequency modulation. The peak deviation is 12 kHz. The cellular bands are each split into two to permit competition. Thus only 12.5 MHz is allocated to one cellular operator for each direction of transmission. With 30-kHz spacing, this yields 416 channels. However, nominally 21 channels are used for control purposes, with the remaining 395 channel available for cellular end-users.

Common practice with AMPS is to assign 10 to 50 channel frequencies to each cell for mobile traffic. Of course, the number of frequencies used depends on the expected traffic load and the blocking probability. Radiated power from a cell site is kept at a relatively low level with just enough antenna height to cover the cell area. This permits frequency reuse of these same channels in nonadjacent cells in the same CGSA with little or no co-channel interference. A well-coordinated frequency reuse plan enables tens of thousands of simultaneous calls over a CGSA.

Figure 18.2 illustrates a frequency reuse method. Here four channel frequency groups are assigned in a way that avoids the same frequency set used in adjacent cells. If there were uniform terrain contours, this plan could be applied directly. However, real terrain conditions dictate further geographic separation of cells that use the same frequency set. Reuse plans with 7 or 12 sets of channel frequencies provide more physical separation and are often used depending on the shape of the antenna pattern employed.

With user growth in a particular CGSA, cells may become overloaded. This means that grade of service objectives are not being met due to higher than planned traffic levels during the busy hour (BH). In these cases, congested cells can be subdivided into small cells, each with its own base station, as shown in Figure 18.3. These smaller cells use lower transmitter power and antennas with less height, thus permitting greater frequency reuse. These subdivided cells

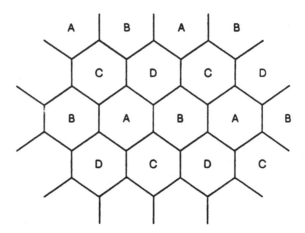

Figure 18.2. Cell separation with four different sets of frequencies.

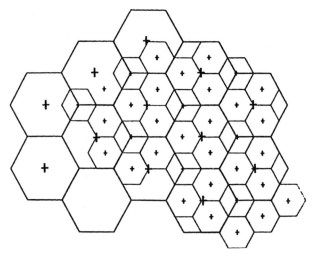

Figure 18.3. Staged growth by cell splitting (subdividing).

can be split still further for still greater frequency reuse. However, there is a practical limit to cell splitting, often with cells with a 1-mile (1.6-km) radius. Under normal, large-cell operation, antennas are usually omnidirectional. When cell splitting is employed, 60° or 120° directional antennas are often used to mitigate interference brought about by increased frequency reuse.

Radio system design for cellular operation differs from that used for LOS microwave operation. For one thing, mobility enters the picture. Path characteristics are constantly changing. Mobile units experience multipath scattering, reflection, or diffraction by obstructions and buildings in the vicinity. There is shadowing, often very severe. The resulting received signal under these conditions varies randomly as the sum of many individual waves with changing amplitude, phase, and direction of arrival. The statistical autocorrelation distance is on the order of one-half wavelength. Space diversity at the base station tends to mitigate these impairments.

Source: Ref. 21.

In Figure 18.1, the MTSO is connected to each of its cell sites by a voice trunk for each of the radio channels at the site. Also two data links (AMPS design) connect the MTSO to each cell site. These data links transmit information for processing calls and for controlling mobile units. In addition to its "traffic" radio equipment, each cell site has installed signaling monitoring equipment and a "setup" radio channel to establish calls.

When a mobile unit becomes operational, it automatically selects the setup channel with the highest signal level. It then monitors that setup channel for incoming calls destined for it. When an incoming call is sensed, the mobile terminal in question again samples signal level of all appropriate setup channels so it can respond through the cell site offering the highest signal level, and then it tunes to that channel for response. The responsible MTSO assigns a vacant voice channel to the cell in question, which relays this information via the setup channel to the mobile terminal. The mobile terminal subscriber is then alerted that there is an incoming call. Outgoing calls from mobile terminals are handled in a similar manner.

While a call is in progress, the serving cell site examines the mobile's signal level every few seconds. If the signal level drops below a prescribed level, the system seeks another cell to handle the call. When a more appropriate cell site is found, the MTSO sends a command, relayed by the old cell site, to change frequency for communication with the new cell site. At the same time, the landline subscriber is connected to the new cell site via the MTSO. The periodic monitoring of operating mobile units is known as *locating*, and the act of changing channels is called *handover*. Of course, the functions of locating and handover are to provide subscribers satisfactory service as a mobile unit traverses from cell to cell. When cells are made smaller, handovers are more frequent.

The management and control functions of a cellular system are quite complex. Handover and locating are managed by signaling and supervision techniques which take place on the setup channel. The setup channel uses a 10-kbit/s data stream which transmits paging, voice channel designation, and overhead messages to mobile units. In turn, the mobile unit returns page responses, origination messages, and order confirmations.

Both digital messages and continuous supervision tones are transmitted on the voice radio channel. The digital messages are sent as discontinuous "blank-and-burst" inband data stream at 10 kbits/s and include order and handover messages. The mobile unit returns confirmation and messages that contain dialed digits. Continuous positive supervision is provided by an out-of-band 6-kHz tone, which is modulated onto the carrier along with the speech transmission.

Roaming is a term used for a mobile unit that travels such distances that the route covers more than one cellular organization or company. The cellular industry is moving toward technical and tariffing standardization so that a cellular unit can operate anywhere in the United States, Canada, and Mexico.

Source: Ref. 21.
Also see Refs. 2–7.

3 PERSONAL COMMUNICATION SYSTEMS

3.1 Defining Personal Communications

A personal communication system (PCS) is wireless. It is an extension of cellular radio operating in a smaller, more confined environment. In most of its applications it covers ranges well under 1 km (625 ft).

The term *wireless* is the popular name for PCS. This simply means that it is radio-based. The user requires no *tether*. The conventional telephone is connected

TABLE 18.1 **Wireless PCS Technologies**

| | High-Power Systems | | | | Low-Power Systems | | | |
| | Digital Cellular (High-Tier PCS) | | | | Low-Tier PCS | | Digital Cordless | |
System	IS-136	IS-95 (DS)	GSM	DCS-1800	WACS/PACS	Handi-Phone	DECT	CT-2
Multiple access	TDMA/FDMA	CDMA/FDMA	TDMA/FDMA	TDMA/FDMA	TDMA/FDMA	TDMA/FDMA	TDMA/FDMA	FDMA
Freq. band (MHz) Uplink (MHz) Downlink (MHz)	869–894 824–849 (USA)	869–894 824–849 (USA)	935–960 890–915 (Europe)	1710–1785 1805–1880 (UK)	Emerg. Tech.[a] (USA)	1895–1907 (Japan)	1880–1900 (Europe)	864–868 (Europe and Asia)
RF ch. spacing Downlink (kHz) Uplink (kHz)	30 30	1250 1250	200 200	200 200	300 300	300	1728	100
Modulation	$\pi/4$ DQPSK	BPSK/QPSK	GMSK	GMSK	$\pi/4$ QPSK	$\pi/4$ DQPSK	GFSK	GFSK
Portable XMTr Power max./avg.	600 mW/ 200 mW	600 mW	1 W/ 125 mW	1 W/ 125 mW	200 mW/ 25 mW	80 mW/ 10 mW	250 mW/ 10 mW	10 mW/ 5 mW
Speech coding	VSELP	QCELP	RPE-LTP	RPE-LTP	ADPCM	ADPCM	ADPCM	ADPCM
Speech rate (kbits/s)	7.95	8 (var.)	13	13	32/16/8	32	32	32
Speech ch./RF ch.	3	—	8	8	8/16/32	4	12	1
Ch. bit rate (kbits/s) Uplink (kbits/s) Downlink (kbits/s)	48.6 48.6		270.833 270.833	270.833 270.833	384 384	384	1152	72
Ch. coding	1/2 rate conv.	1/2 rate fwd 1/3 rate rev.	1/2 rate conv.	1/2 rate conv.	CRC	CRC	CRC (control)	None
Frame (ms)	40	20	4.615	4.615	2.5	5	10	2

[a]Spectrum is 1.85 to 2.2 GHz allocated by the FCC for emerging technologies; DS is direct sequence.
XMTr = transmitter

Source: Table 1, Ref. 26, reprinted with permission of the IEEE.

by a wire pair through to the local serving switch. The wire pair is a tether. We can only walk as far with that telephone handset as the "tether" allows.

The difference between cellular radio and "wireless" is blurred. Both of the systems can be classified as PCS. Cellular radio, particularly with the hand-held terminal, gives the user tetherless telephone communication. Paging systems, discussed in Section 9 of this chapter, provided a mobile/ambulatory user a means of being alerted that someone wishes to talk to that person on the telephone or can leave that person a short message.

The cordless telephone is certainly another example of PCS. It has extremely wide use around the world. By the end of 2004, it is estimated that there will be over 150 million cordless telephones in use in the United States.

New applications of PCS are either on the horizon, going through field tests or reaching maturity. One that seems to offer great promise in the office environment is the wireless PABX. It will almost eliminate the telecommunication manager's responsibilities with office rearrangements. Another is the wireless LAN (WLAN).

Applications are in the early stages of maturity that will provide cellular and PCS (wireless) terminals with facsimile and data, particularly internet access. UMTS (Universal Mobile Telecommunications Systems) will give the PCS/cellular user just such services. This upcoming family of transmission formats and modulation waveforms is described in Section 12 of this chapter. It is an outgrowth of GSM and we expect it to come to fruition before 2010.

Don Cox [26] breaks PCS down into what he calls "high tier" and "low tier." Cellular radio systems are regarded as high-tier PCS, particularly when implemented in the new 1.9-GHz PCS frequency band. Cordless telephones are classified as low-tier.

Table 18.1 summarizes some of the more prevalent wireless technologies. Also see Ref. 8.

4 RADIO PROPAGATION IN THE MOBILE/PCS ENVIRONMENT

4.1 The Propagation Problem

Line-of-sight microwave and satellite communications covered in Chapter 7 dealt with fixed systems. Such systems were and are optimized. They are built up and away from obstacles. Sites are selected for best propagation.

Not so with mobile systems. Motion and a third dimension are additional variables. The end-user terminal often is in motion. Or the user is temporarily fixed, but that point can be anywhere within a serving area of interest. Whereas before we dealt with point-to-point, here we deal with point-to-multipoint.

One goal in line-of-sight microwave design was to stretch the distance as much as possible between repeaters by using high towers. In this chapter there are some overriding circumstances where we try to limit coverage extension by reducing tower heights, what we briefly introduced in Section 2. Even more importantly, coverage is area coverage where shadowing is frequently encountered. Examples are valleys, along city streets with high buildings on either side, in verdure such as trees and inside buildings to name a few situations. Such an environment is rich with multipath scenarios. Paths can be highly dispersive, as much as 10 μs of delay spread [1]. Due to a user's motion, Doppler shift can be expected.

The radio-frequency bands of interest are UHF, especially around 800 and 900 MHz and 1700 to 2100 MHz.

Many of the concepts discussed here derive from Refs. 2, 3, 4, 7, 8, and 9.

4.2 Several Propagation Models

In this section we concentrate on cellular operation. There is a fixed station (FS) and mobile stations (MS) moving through the cell. A cell is the area of responsibility of the fixed station, a cell site. It usually is pictured as a hexagon in shape, although its propagation profile is more like a circle with the fixed station in its center. The cell contour is rather ragged due to obstructions in each path radial. Cell radii vary from 1 km (0.6 mile) to 30 km (19 miles) or somewhat more.

4.2.1 Path Loss.
We recall the free space loss (FSL) formula discussed in Chapter 7. It simply stated that FSL was a function of the square of the distance and the square of the frequency plus a constant. It is a very useful formula if the strict rules of obstacle clearance are obeyed. Unfortunately in the cellular situation, it is impossible to obey these rules. Then to what extent is this free space loss formula modified by atmospheric effects, the presence of the earth bulge, and the effects of trees, buildings, and hills which exist in, or close to, the transmission path?

4.2.1.1 CCIR Formula.
CCIR developed a simple path loss formula (CCIR Rec. 370-5 [12]) for radio and television broadcasting where the frequency term has been factored out. It states that the path loss (L_{dB}):

$$L_{dB} = 40 \log d_m - 20 \log(h_T h_R) \qquad (18.1)$$

Note that the equation is an inverse fourth power law and is fundamental to terrestrial mobile radio. Distance is in meters; h_T is the height of the transmit antenna above plane earth, again in meters; and h_R is the height of the receive antenna above plane earth in meters.

Suppose $d = 1000$ m (1 km) and the product of $h_T \times h_R$ is 10 m^2 (low antennas), then $L = 100$ dB.

Now suppose that $d = 25$ km and $h_T \times h_R = 100$ m^2, where the base station is on high ground.

The loss L is 136 dB.

The CCIR formula only brings in two new, but important, variables: h_T and h_R; it also uses the fourth power rather than the second power, when compared to free space loss discussed in Chapter 7.

4.2.1.2 The Amended CCIR Equation. This amended model takes into account:

(a) Surface roughness
(b) Line-of-sight obstacles
(c) Buildings and trees

The resulting path loss equation is

$$L_{dB} = 40 \log d - 20 \log(h_T h_R) + \beta \qquad (18.2)$$

where β represents the additional losses listed above lumped together.

4.2.1.3 British Urban Path Loss Formula. The following formula was proposed by Allesbrook and Parsons [13]:

$$L_{dB} = 40 \log d_m - 20 \log(h_T h_R) + 20 + f/40 + 0.18L - 0.34H \qquad (18.3)$$

where f is the frequency in MHz; L is the land usage factor, a percentage of the test area covered by buildings of any type, 0–100%; and H is the terrain height difference between T_x and R_x (i.e., T_x terrain height minus R_x terrain height).

Example: Let $d = 2000$ m, $h_T = 30$ m, $h_R = 3.3$ m, $f = 900$ MHz, $L = 50\%$, and $H = 27$ m.

$$h_T \times h_R = 100 \text{ m}^2$$
$$L_{dB} = 132 - 40 + 20 + 22.5 + 5.4 - 9.18$$
$$= 130.76 \text{ dB}$$

4.2.1.4 The Okumura Model. Okumura et al. [14] carried out a detailed analysis for path predictions around Tokyo for mobile terminals. Hata [15] published an empirical formula based on Okumura's results to predict path loss:

$$L_{dB} = 69.55 + 26.16 \log f - 13.82 \log h_t - A(h_r)$$
$$+ (44.9 - 6.55 \log h_t) \log d \qquad (18.4)$$

where f is between 150 and 1500 MHz, h_t is between 30 and 300 m, and d is between 1 and 20 km.

$A(h_r)$ is the correction factor for mobile antenna height and is computed as follows:

For a small or medium-sized city

$$A(h_r) = (1.1 \log f - 0.7)h_r - (1.56 \log f - 0.8) \quad \text{(dB)} \qquad (18.5a)$$

where h_r is between 1 and 10 m.

For a large city

$$A(h_r) = 3.2[\log(11.75h_r)]^2 - 4.97 \text{ dB} \quad \text{(dB)} \quad\quad (18.5b)$$

where $f \geq 400$ MHz

Example: Let

$$f = 900 \text{ MHz}$$
$$h_t = 40 \text{ m}$$
$$h_r = 5 \text{ m}$$
$$d = 10 \text{ km}$$

Calculate $A(h_r)$ for a medium-sized city.

$$A(h_r) = 12.75 - 3.8 = 8.95 \text{ dB}$$
$$L_{dB} = 69.55 + 72.28 - 22.14 - 8.95 + 34.4$$
$$= 145.15 \text{ dB}$$

4.2.1.5 Longley–Rice Model. One of most reliable methods, which dates back to the 1960s, is the Longley–Rice model [31, 32]. The model is applicable from 1- to 2000-km-long paths, for varied path and climatic conditions over the frequency range from 20 MHz to 20 GHz. The Longley–Rice model consists of two modules or algorithms which are employed under different circumstances. One is called the "area prediction" model. It is applied when complete information about the paths in question is unknown. It finds application in the broadcast and mobile communities when only generalized information is available. Its accuracy at short ranges is questionable, often giving higher losses than would typically be measured. At medium and long ranges the algorithm gives good results. The user of this module is asked for the following inputs: terrain roughness factor (ΔH), surface refractivity (N_s), and average terrain heights, depending on the options selected by the user.

The second algorithm of the Irregular Terrain Model (ITM) is the "point-to-point" prediction model. It is designed to be used on a fixed path where certain gross features of the intervening terrain are known. This model finds particularly good application for the evaluation of specific communication links and for solving special interference problems. The model is not recommended for use with microwave line-of-sight links. It should be noted that the treatment of such links is quite different from the ITM approach. Median signal levels are at or near free space values, and variability around the median follows a different set of laws.

4.2.2 Diffraction Modeling. The general case of point-to-point diffraction paths is covered in Ref. 29, Section 5.4. Another source is ITU-R Rec. P.526-5 [30].

P.526-5 gives two general methods:

1. The first method is spherical earth, which gives a numerical approach and a graphical approach, based on the same approximation, namely a three-term equation. The first term is distance-dependent, and the second and third terms represent the antenna height-gain functions. The accuracy of this method depends, to a large extent, on whether the terrain irregularities are small when compared to the first Fresnel zone radius in the middle path.

2. This second method deals with diffraction over irregular terrain. Depending on the type of obstacle(s) in the path, different prediction techniques are provided. The simplest and most well-known is a path with a single, knife-edge obstacle. This idealized model is relevant only in cases where the radius of the curvature of the obstacle can be neglected.

The problem of multiple diffraction is much more complicated. In the case of two knife edges, ITU-R Rec. P.526 recommends the use of one of the following two approaches: (1) Epstein and Peterson (Ref. 30) which is then corrected by Millington (30). (2) However, if one obstacle is predominant, the Deygout method is recommended (30). Either method may be applied to rounded obstacles.

4.2.3 Fresnel Zone Breakpoint Defined. Prior to our discussions dealing with indoor and indoor–outdoor transmission (propagation) loss, we must define a phenomenon called the *Fresnel zone breakpoint*, which is illustrated in Figure 18.4.

Let us allow that signal level varies with distance R as A/R^n. For distances greater than 1 km, n is typically between 3.5 and 4. The parameter A describes the effects of environmental features in a highly averaged manner [11].

Typical PCS radio paths can be of an LOS nature, particularly near the fixed transmitter where $n = 2$. Such paths may be down the street from the transmitter. The other types of paths are shadowed paths. One type of shadowed path is found in highly urbanized settings where the signal may be reflected off of high-rise buildings. Another is found in more suburban areas where buildings are often just two stories high.

Propagation between floors of a modern office building can be very complex. If the floors are constructed of reinforced concrete or prefabricated concrete, transmission loss can be 10 dB or more. Floors constructed of concrete poured over steel panels show much greater loss. In this case [11], signals may propagate over other paths involving diffraction rather than transmission through the floors. For instance, signals can exit the building through windows and reenter on higher floors by diffraction mechanisms along the face of the building.

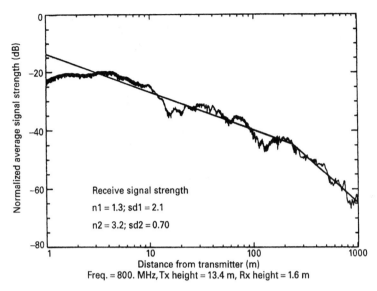

Figure 18.4. Signal variation on a line-of-sight path in a rural environment. From Figure 3, Ref. 11.

When a signal at 800 MHz is plotted versus R on a logarithmic scale as in Figure 18.5, there are distinctly different slopes before and after the Fresnel break-point. We call this the break distance (from the transmit antenna) R_B. This is the point for which the Fresnel ellipse about the direct ray just touches the ground. This model is illustrated in Figure 18.5. The distance R_B is approximated by

$$R_B = 4h_1h_2/\lambda \qquad (18.6a)$$

For $R < R_B$, N is less than 2, and for $R > R_B$, n approaches 4.

It was found that on non-LOS paths in an urban environment with low base station antennas and with users at street level, propagation takes place down streets and around corners rather than over buildings. For these non-LOS paths

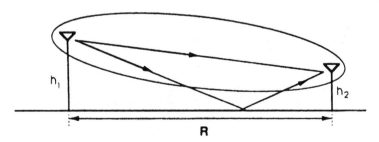

Figure 18.5. Direct and ground reflected rays, and showing the Fresnel ellipse about the direct ray. From Figure 18, Ref. 11.

the signal must turn corners by multiple reflections and diffraction at vertical edges of buildings. Field tests reveal that signal level decreases by about 20 dB when turning a corner.

In the case of propagation inside buildings where the transmitter and receiver are on the same floor, the key factor is the clearance height between the average tops of furniture and the ceiling. Bertoni et al. [11] call this clearance W. Here building construction consists of drop ceilings of acoustical material supported by metal frames. That space between the drop ceiling and the floor above contains light fixtures, ventilation ducts, pipes, support beams, and so on. Because the acoustical material has a low dielectric constant, the rays incident on the ceiling penetrate the material and are strongly scattered by the irregular structure, rather than undergo specular reflection. Floor-mounted building furnishings such as desks, cubicle partitions, filing cabinets, and work benches scatter the rays and prevent them reaching the floor, except in hallways. Thus it is concluded that propagation takes place in the clear space, W.

Figure 18.6 shows a model of a typical floor layout. When both the transmitter and receiver are located in the clear space on the same floor, path loss can be related to the Fresnel ellipse. If the Fresnel ellipse associated with the path lies entirely in the clear space, the path loss has LOS properties $(1/L^2)$. Now as the separation between the transmitter and receiver increases, the Fresnel ellipse grows in size so that scattering lie within it. This is shown in Figure 18.6. Now the path loss is greater than free space.

Bertoni et al. reported on one measurement program where the scatters have been simulated using absorbing screens. It was recognized that path loss will be highly dependent on nearby scattering objects. Figure 18.7 was developed from this program. It was found that the path loss in excess of free space calculated at 900 and 1800 MHz for $W = 1.5$ is plotted in Figure 18.7 as a function of path length L. The figure shows that the excess path loss (over LOS) is small at each frequency out to distances of about 20 and 40 m, respectively, where it increases dramatically.

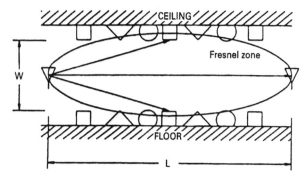

Figure 18.6. Fresnel zone for propagation between transmitter and receiver in clear space between building furnishings and ceiling fixtures. From Figure 35, Ref. 11.

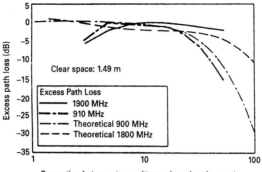

Figure 18.7. Measured and calculated excess path loss at 900 and 1800 MHz for a large office building having head-high cubicle partitions, but no floor-to-ceiling partitions. From Ref. 11, Figure 36.

4.2.4 Indoor Propagation Loss. ITU-R Rec. P.1238 [34] provides a site-general model of the indoor propagation loss, $L_{\text{Total(dB)}}$:

$$L_{\text{Total(dB)}} = 20 \log f + N \log d + L_f(n) - 28 \qquad (18.6b)$$

where f is the frequency (MHz), N is the path loss coefficient, d is the separation distance between the base station and the portable (m) ($d > 1$ m), L_f is the floor penetration loss (dB), and N is the number of building floors between the base station and the portable ($n \geq 1$).

The reference handbook [33] adds the following comments: (1) The path loss coefficient, N, and the floor penetration loss, L_f, are functions of the frequency and building type. (2) As one might expect, the floor penetration loss, L_f, is neglected when the base station and the portable are on the same building floor. Typical values of N and L_f are given in Tables 18.2 and 18.3.

TABLE 18.2 Power Loss Coefficients, N, for Indoor Transmission Loss Calculations

Frequency	Residential	Office	Commercial
900 MHz	—	33	20
1.2–1.3 GHz	—	32	22
1.8–2.0 GHz	28	30	22
4 GHz	—	28	22
5.2 GHz	—	31	—
60 GHz[a]	—	22	17

[a] 60-GHz values assume propagation within a single room or space, and they do not include any allowance for transmission through walls. Gaseous absorption around 60 GHz is also significant for distances greater than about 100 m which may influence frequency reuse distances. (See ITU-R Rec. P.676.)

Source: Table 6.1, page 60, *ITU-R Handbook* [33].

TABLE 18.3 Floor Penetration Loss Factors, L_f, with n Being the Number of Floors Penetrated

Frequency	Residential	Office	Commercial
900 MHz	—	9 (1 floor)	
		19 (2 floors)	—
		24 (3 floors)	
1.8–2.0 GHz	$4n$	$15 + 4(n-1)$	$6 + 3(n-1)$
5.2 GHz	—	16 (1 floor)	—

Source: Table 6.2, page 61, *ITU-R Handbook* [33].

The ITU-R organization in Ref. 33 offers the following guidance, taken originally from ITU-R Rec. P.1238 [34]. These comments are particularly applicable to the frequency band 900 MHz to 2 GHz:

1. Paths dominated by a pure line-of-sight (LOS) component have a path loss coefficient (see equation 18.6) of approximately 20. However, it is also necessary that the building walls, ceiling, and floor be suitably distant for this condition to occur.

2. For long unobstructed paths, the first Fresnel zone breakpoint may occur, particularly at the lower frequencies giving rise to sub-path diffraction. In these cases, the path loss coefficient will increase rapidly from a value of 20 to a value of approximately 40, in the vicinity of the breakpoint.

3. For paths in long narrow corridors or hallways on a single floor, the path loss coefficient may be less than 20. This is a result of the fact that the corridor behaves roughly as a resonant cavity. This is also true for paths in single rooms with moderately reflective walls and suitable dimensions.

4. For those paths in buildings in which the rooms are separated by full floor-to-ceiling walls (e.g., closed plan office buildings), the path loss coefficient for room-to-room paths is typically found to be near the value of 40.

5. For paths traversing multiple floors, we can expect that the floor penetration loss will be limited by paths through multifloor atria, stairwells, or other lower loss mechanisms than direct transmission through the building's floors.

Source: Ref. 33.

4.2.5 Combined Indoor to Outdoor and Outdoor-to-Indoor Propagation.

In situations involving propagation over combined paths, either indoor-to-outdoor or vice versa, it has been found to be appropriate to concentrate on the points of entry or exit that allow for the transmission of electromagnetic radiation between the two environments. Points of entries can be apertures such as windows, skylights and doors, ventilation intake and exhaust penetrations, other structural openings, and, if the electromagnetic shielding provided by the building materials is low, even the exterior skin of the building itself. If the distributions of the electromagnetic fields across the points of entry are known or, alternatively, can be inferred, then these points of entry can replace the actual sources of radiation (either indoor or outdoor) as equivalent radiation sources (i.e., either outwardly or inwardly, respectively). The total field at the desired observation point will then be the result of superposition of the fields at the observation point due to each of the equivalent sources, taken individually. If it happens that a single point of entry is expected, on physical grounds, to dominate the response at the observation point, then it is often sufficient to concentrate on obtaining the desired electromagnetic fields' distributions for this single point of entry, excluding the others.

Source: Ref. 33.

An alternative, empirically based approach is to use empirical data for building entry loss (see ITU-R Rec. P.1411 [35]).

Another source [16] provided the following information. At 1650 MHz the floor loss factor was 14 dB, while the wall losses were 3–4 dB for double plasterboard and 7–9 dB for breeze block or brick. The parameter $L(v)$, clutter-loss, was 29 dB. When the propagation frequency was 900 MHz, the first floor factor was 12 dB and $L(v)$ was 23 dB. The higher value for $L(v)$ at 1650 MHz was attributed to a reduced antenna aperture at this frequency compared to 900 MHz. For a 100-dB path loss, the base station and mobile terminal distance exceeded 70 m on the same floor, was 30 m for the floor above, and 20 m for the floor above that, when the propagation frequency was 1650 MHz. The corresponding distances at 900 MHz were 70 m, 55 m, and 30 m. Results will vary from building to building, depending on the type of construction of the building, the furniture and equipment it houses, and the number and deployment of the people who populate it.

4.3 Microcell Prediction Model According to Lee

For this section a microcell is defined as a cell with a 1-km or less radius. Such cells are used in heavily urbanized areas where demand for service is high and where large cell coverage would be spotty at best. With this model, line-of-sight conditions are seldom encountered; shadowing is the general rule, as shown in Figure 18.8. The major contribution to loss in such situations is due to the dimensions of intervening buildings.

Lee's model [6] also includes an antenna height-gain function. Reference 6 reports a 9 dB/oct or 30 dB/dec for an antenna height change. This would mean that if we doubled the height of an antenna, 9-dB transmission loss improvement would be achieved.

The Lee model for a microcell breaks the prediction process down into a received signal level (dBm) for the line-of-sight component (P_{LOS}) and then the attenuation due to the building blockage component.

Figure 18.9 provides a microcell scenario which we will use to understand Lee's model. The receive signal level P_r is equal to the receive signal level

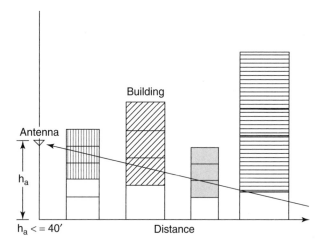

Figure 18.8. A propagation model typical of an urban microcell. From Figure 2.24, Ref. 6.

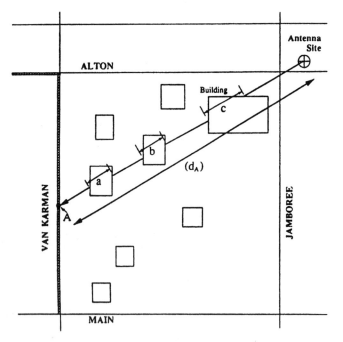

Figure 18.9. Path loss model for a typical microcell in an urban area. Mobile terminal is at location A. With antenna site so situated, there is blockage by buildings a, b, and c. Thus $B = a + b + c$. After Lee [6], Figure 2.26, page 90.

for LOS conditions (P_{LOS}) minus the blockage loss due to buildings, α_B. B is the blockage distance in feet. From Figure 18.5, $B = a + b + c$, the sum of the distances (in feet) through each building.

$$P_{LOS} = P_t - 77 \text{ dB} - 21.5 \log(d/100') + 30 \log(h_1/20') \tag{18.7a}$$
$$\text{for } 100' \ d < 200'$$
$$= P_t - 83.5 \text{ dB} - 14 \log(d/200') + 30 \log(h_1/20') \tag{18.7b}$$
$$\text{for } 200' \ d < 1000'$$
$$= P_t - 93.3 \text{ dB} - 36.5 \log(d/1000') + 30 \log(h_1/20') \tag{18.7c}$$
$$\text{for } 1000' \ d < 5000'$$

$$\alpha_B = 0 \qquad\qquad 1' \le B \tag{18.8a}$$
$$\alpha_B = 1 + 0.5 \log(B/10') \qquad 1' \le B < 25' \tag{18.8b}$$
$$\alpha_B = 1.2 + 12.5 \log(B/25') \qquad 25' \le B < 600' \tag{18.8c}$$
$$\alpha_B = 17.95 + 3 \log(B/600') \qquad 600' \le B < 3000' \tag{18.8d}$$
$$\alpha_B = 20 \text{ dB} \qquad\qquad 3000' \le B$$

where P_t is the ERP* in dBm, d is the total distance in feet, and h is the antenna height in feet.

Example 1: Mobile terminal is 500 ft from the cell site antenna, which is 30 ft high. There are three buildings in line between the mobile terminal and the cell site antenna with cross section (in line with the ray beam) distance of 50, 100, and 150 ft, respectively. Thus $B = 300$ ft. The ERP $= +30$ dBm (1 watt).
    First use equation 18.7b:

$$P_{LOS} = +30 \text{ dBm} - 83.5 \text{ dB} - 14 \log(500'/200') + 30 \log(30'/20')$$
$$= +30 \text{ dBm} - 83.5 \text{ dB} - 5.57 \text{ dB} + 5.28 \text{ dB}$$
$$= -53.79 \text{ dBm}$$

Use equation 18.8c:

$$\alpha_B = 1.2 + 12.5 \log(300'/25')$$
$$= 1.2 \text{ dB} + 13.5 \text{ dB}$$
$$= 14.7 \text{ dB}$$

---

* ERP stands for effective radiated power (over a dipole).

The receive signal level, P_r, is

$$P_r = -53.79 \text{ dBm} - 14.7 \text{ dB}$$

$$= -68.47 \text{ dBm}$$

Of course, we must assume that the sum of the gain of the receive antenna and the transmission line loss equals 0 dB so that P_r is the same as isotropic receive level.

Example 2: There are 4000 ft separating the cell site transmit antenna from the receive terminal. The transmit antenna is 40 ft high. There are four buildings causing blockage of 150 ft, 200 ft, 140 ft, and 280 ft. These distances are measured along the ray beam line. Thus $B = 770$ ft. The ERP is +20 dBm. What is the receive signal level (P_r) assuming no gain or loss in the receive antenna system?

Use equation 18.7c for P_{LOS}:

$$P_{LOS} = +20 \text{ dBm} - 93.3 \text{ dB} - 36.5 \log(4000/1000) + 30 \log(40/20)$$

$$= -73.3 \text{ dBm} - 21.4 \text{ dB} + 9 \text{ dB}$$

$$= -85.67 \text{ dBm}$$

Now use equation 18.8d to calculate α_B:

$$\alpha_B = 17.95 + 3 \log(770/600)$$

$$= 18.28 \text{ dB}$$

$$P_r = -85.67 \text{ dBm} - 18.28 \text{ dB}$$

$$= -103.94 \text{ dBm}$$

5   IMPAIRMENTS — FADING IN THE MOBILE ENVIRONMENT

5.1   Introduction

Fading in the mobile situation is quite different from in the static line-of-sight (LOS) microwave setting discussed in Chapter 7. Radio paths are not optimized as in the LOS environment. The mobile terminal may be fixed throughout a telephone call, but it is more apt to be in motion. Even the hand-held terminal may well have micromotion. When a terminal is in motion, the path characteristics are constantly changing.

Multipath propagation is the rule. Consider the simplified multipath pictorial model in Figure 18.10. Commonly, multiple rays reach the receive antenna, each with its own delay. The constructive and destructive fading can become quite complex. We must deal with both reflection and diffraction. Energy will arrive

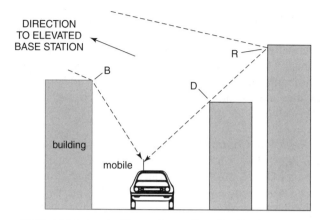

Figure 18.10. Mobile terminal in an urban setting. R, reflection; D, diffraction.

at the receive antenna reflected off sides of buildings, streets, lakes, and so on. Energy will also arrive diffracted from knife edges (e.g., building corners) and rounded obstacles (e.g., water tanks, hilltops).

Because the same signal arrives over several paths, each with a different electrical length, the phases of each path will be different resulting in constructive and destructive amplitude fading. Fades of 20 dB are common, and even 30 dB fades can be expected.

On digital systems, the deleterious effects of multipath fading can be even more severe. Consider a digital bit stream to a mobile terminal with a transmission rate of 1000 bits/s. Assuming NRZ coding, the bit period would be 1 ms (bit period = 1/bit rate). We find the typical multipath delay spread may be on the order of 10 μs. Thus delayed energy will spill into a subsequent bit (or symbol) for the first 10 μs of the bit period and will have no negative effect on the bit decision. If the bit stream is 64,000 bits/s, then the bit period is 1/64,000 or 15 μs. Destructive energy from the previous bit (symbol) will spill into the first two-thirds of the bit period, well beyond the mid-bit sampling point. This will cause typical intersymbol interference (ISI), and in this case there is a high probability that there will be a bit error. The reader should appreciate that the destructive potential of ISI increases as the bit rate increases (i.e., as the bit period decreases).

5.2 Classification of Fading

We consider three types of channels to place bounds on radio system performance. These are:

- Gaussian channel
- Rayleigh channel
- Rician channel

5.2.1 *The Gaussian Channel.* The Gaussian channel can be considered the ideal channel, and it is only impaired by "additive white Gaussian noise" (AWGN) developed internally by the receiver. We hope to achieve a BER typical of a Gaussian channel when we have done everything we can to mitigate fading and its results. These measures we take could be diversity, equalization, FEC coding with interleaving, and so forth. The ideal Gaussian channel is very difficult to achieve in the mobile radio environment.

5.2.2 *The Rayleigh Channel.* The Rayleigh channel is at the other end of the line, often referred to as a *worst-case channel*. Remember, in Chapter 7, we treated fading on LOS microwave as Rayleigh fading and gave us the very worst case fading scenario. Figure 18.11 shows a channel where the signal approaches Rayleigh fading characteristics. Of course, we are dealing with multipath here.

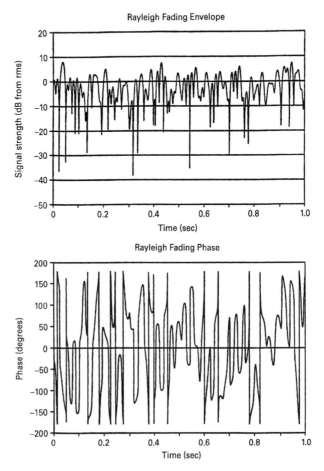

Figure 18.11. Typical Rayleigh fading envelope and phase in a mobile scenario. Vehicle speed about 30 mph; frequency 900 MHz. From Steele, [1], Figure 1.1.

We showed that in the mobile radio scenario, multipath reception commonly had many components. Thus if each multipath component is independent, the PDF (probability density function) of its envelope is Rayleigh.

5.2.3 The Rician Channel.
The characteristics of a Rician channel are in between those of a Gaussian channel and those of a Rayleigh channel. The channels can be characterized by a function K (not to be confused with the K-factor in Chapter 7). K is defined as follows:

$$K = \text{power in the dominant path/power in the scattered paths} \qquad (18.9)$$

As cells get smaller, the LOS component becomes more and more dominant. There are many cases; in fact, in nearly all cases where there is no full shadowing, there is an LOS component and scattered components. This is a typical multipath scenario. Turning now to equation 18.9, when $K = 0$, the channel is Rayleigh (i.e., the numerator is 0 and all the received energy derives from scattered paths). When K is infinity, the channel is Gaussian and the denominator is zero. Figure 18.12 gives BER values for some typical values of K. It shows that those intermediate values of K provide a superior BER than for the Rayleigh channel where $K = 0$. For a microcell mobile scenario, values of K vary from 5 to 30 [1]. Larger cells tend more toward low values of K.

There is also an advantage for Rician fading with higher values of K regarding co-channel interference performance for a desired BER. The smaller the cell, the more fading becomes Rician approaching the higher values of K.

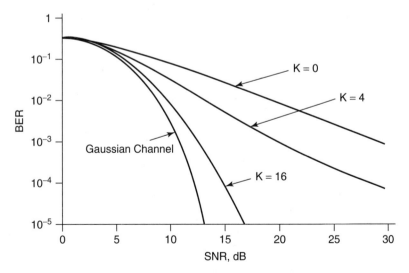

Figure 18.12. BER versus channel SNR for various values of K; noncoherent FSK. From Steele [1], Figure 1.7.

5.3   Diversity — A Technique to Mitigate the Effects of Fading and Dispersion

5.3.1   Scope. We discuss diversity to reduce the effects of fading and to mitigate dispersion. Diversity was briefly covered in Chapter 7 where we dealt with LOS microwave radio system. In that chapter we discussed frequency and space diversity. There is a third diversity scheme called time diversity, which can be applied to digital cellular radio systems.

In principle, such techniques can be employed at the base station and/or at the mobile unit, although different problems have to be solved for each. The basic concept behind diversity is that when two or more radio paths carrying the same information are relatively uncorrelated, when one path is in a fading condition, often the other path is not undergoing a fade. These separate paths can be developed by having two channels, separated in frequency. The two paths can be separated in space. Also, the two paths can be separated in time.

When the two (or more) paths are separated in frequency, we call this frequency diversity. However, there must be at least 2% or greater frequency separation for the paths to be comparatively uncorrelated. Because, in the cellular situation, we are so short of spectrum, using frequency diversity (i.e., using a separate frequency with redundant information) is essentially out of the question, and it will not be discussed further except for its implicit use in CDMA.

5.3.2   Space Diversity. Space diversity is commonly employed at cell sites, and two separate receive antennas are required, separated in either the horizontal or vertical plane. Separation of the two antennas vertically can be impractical for cellular receiving systems. Horizontal separation, however, is quite practical. The space diversity concept is illustrated in Figure 18.13.

One of the most important factors in space diversity design is antenna separation. There is a set of rules for the cell site, and there is another for the mobile unit.

Space diversity antenna separation, shown as distance D in Figure 18.13, varies not only as a function of the correlation coefficient but also as a function of antenna height, h. The wider the antennas are separated, the lower the correlation coefficient and the more uncorrelated the diversity paths are. Sometimes we find

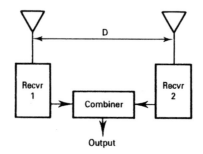

Figure 18.13. The space diversity concept.

that by lowering the antennas as well as adjusting the distance between the antennas we can achieve a very low correlation coefficient. However, we might lose some of the height-gain factor.

Lee [6] proposes a new parameter where

$$\eta = \text{antenna height/antenna separation} = h/d \qquad (18.10)$$

In Figure 18.14 we relate the correlation coefficient (ρ) with $\eta \cdot \alpha$ as the orientation of the antenna regarding the incoming signal from the mobile unit. Lee recommends a value of $\rho = 0.7$. Lower values are unnecessary because of the law of diminishing returns. There is much more fading advantage achieved from $\rho = 1.0$ to $\rho = 0.7$ than from $\rho = 0.7$ to $\rho = 0.1$.

Based on $\rho = 0.7$, $\eta = 11$, from Figure 18.10, we can calculate antenna separation values (for 850-MHz operation). For example, if $h = 50$ ft (15 m), we can calculate d using formula 18.10:

$$d = h/\eta = 50/11 = 4.5 \text{ ft} (1.36 \text{ m})$$

For an antenna 120 ft (36.9 m) high, we find that $d = 120/11 = 10.9$ ft or 3.35 m.

Source: Ref. 6.

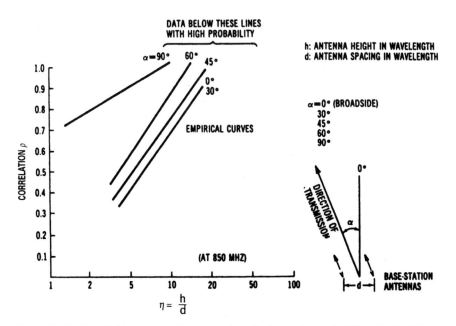

Figure 18.14. Correlation ρ versus the parameter η for two antennas in different orientations. From Lee [6], reprinted with permission.

5.3.2.1   Space Diversity on a Mobile Platform. Lee [6] discusses both vertically separated and horizontally separated antennas on a mobile unit. For the vertical and horizontal separation cases, 1.5λ and 0.5λ, respectively, are recommended. At 850 MHz, $\lambda = 35.29$ cm. Then $1.5\lambda = 1.36$ ft or 52.9 cm. For the 0.5λ, the value is 0.45 ft or 17.64 cm.

5.3.3   Frequency Diversity. We pointed out that conventional frequency diversity was not a practical alternative in cellular systems because of the shortage of available bandwidth. However, with CDMA (direct sequence spread spectrum), depending on the frequency spread, many frequency diversity paths are available, and in most CDMA systems we have what is called *implicit diversity*; and multipath can be resolved with the use of a RAKE filter. This is one of the many advantages of CDMA.

5.3.4   Forward Error Correction — A Form of Time Diversity. Forward error correction (FEC) can be used on digital cellular systems not only to improve bit error rate but to reduce fading. To reduce the effects of fading, an FEC system must incorporate an *interleaver*.

An interleaver pseudorandomly shuffles bits. It first stores a span of bits and shuffles them using a generating polynominal. The span of bits can represent a time period. The rule is that for effective operation against burst errors,* the interleaving span must be much greater than the typical fade duration. The deinterleaver used at the receive end is time-synchronized to the interleaver incorporated at the transmit end of the link. (For FEC, see also Section 5.5 of Chapter 10.)

5.4   Cellular Radio Path Calculations

Consider the path from the fixed cell site to the mobile platform. There are several mobile receiver parameters that must be considered. The first we derive from EIA/TIA-IS-19B [19]. The minimum SINAD (signal + interference + noise and distortion to interference + noise + distortion ratio) is 12 dB. This SINAD equates to a threshold of -116 dBm or 7 μV/m. This assumes a cellular transceiver with an antenna with a net gain of 1 dBd (dB over a dipole). The gross antenna gain is 2.5 dBd with a 1.5-dB transmission line loss. A 1-dBd gain is equivalent to a 3.16 dBi gain (i.e., 0 dBd = 2.16 dBi). Furthermore, this value equates to an isotropic receive level of -119.16 dBm [19].

One design goal for a cellular system is to more or less maintain a cell boundary at the 39-dBμ contour [20]: 39 dB$\mu = -95$ dBm (based on a 50-Ω impedance at 850 MHz). Then at this contour, a mobile terminal would have a 24.16-dB fade margin.

If a cellular transmitter had a 10-watt output per channel and an antenna gain of 12 dBi and 2-dB line loss, the EIRP would be $+20$ dBW or $+50$ dBm. The maximum path loss to the 39-dBμ contour would be $+50$ dBm $- (-119.16$ dBm) or 169 dB.

---

* Burst errors are bunched errors due to fading impulse noise bursts.

6 THE CELLULAR RADIO BANDWIDTH DILEMMA

6.1 Background and Objectives

The present cellular radio bandwidth assignment in the 800- and 900-MHz band cannot support the demand for cellular service, especially in urban areas in the United States and Canada. AMPS, widely used in the United States, Canada, and many other nations of the Western Hemisphere, requires 30 kHz per voice channel. This system can be called an FDMA (frequency division multiple access) system, much like the FDMA/DAMA system described in Chapter 7. We remember that the analog voice channel is a nominal 4 kHz, and 30 kHz is seven times that value.

The trend is to convert to digital. As we discussed in Chapters 8 and 9, digital is notorious for being wasteful of bandwidth, when compared to the 4-kHz analog channel. We showed that PCM has a 16-times multiplier of the 4-kHz analog channel.

One goal of system designers, therefore, is to reduce the required bandwidth of the digital voice channel without sacrificing too much voice quality. As we will show, they have been quite successful.

The real objective is to increase the ratio of users to unit bandwidth. We will describe two distinctly different methods, each claiming to be more bandwidth conservative than the other. The first method is TDMA (time division multiple access) and the second is CDMA (code division multiple access). The former was described in Chapter 7 and the latter was mentioned.

6.2 Bit Rate Reduction of the Digital Voice Channel

6.2.1 ADPCM. Adaptive differential PCM. (ADPCM) provides nearly equal quality voice but at only 32 kbits/s. Key to the operation of ADPCM is the predictor, as shown in Figure 18.15. Predictions are based on the decoded sequence (identical to the sequence at the output of the receive decoder in the absence of transmission errors) that consists of the input speech sequence contaminated by quantization noise. With ADPCM the predictors have adaptive coefficients.

6.2.2 Vocoders. Vocoder operation is based on algorithms which attempt to describe the speech production mechanism in terms of a few independent parameters serving as the information-bearing signals. The concept behind vocoder design considers that speech is produced from a *source-filter* arrangement. It is modeled after the human generation of speech. Voiced-speech* is the result of exciting the vocal tract (a *filter*) with a series of quasi-periodic glottal pulses generated by the vocal cords which is considered the *source.*

As its name implies, a vocoder codes speech. It uses an analysis process based on a speech production model and extracts a set of source-filter parameters

---

* Unvoiced speech is generated by exciting a filter patterned after the vocal tract with random white noise.

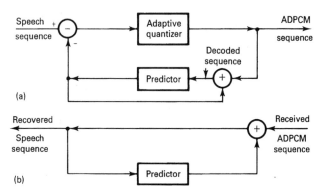

Figure 18.15. An ADPCM codec: (a) coder; (b) decoder.

which are encoded and transmitted. At the distant-end decoder, the parameters are decoded and are used to control a speech synthesizer based on the speech production model at the transmit end. The synthesized signal at the receiver resembles the original speech signal.

There are two types of vocoders that operate in the frequency domain: the channel and the formant vocoders. The well-known LPC (linear predictive coder) vocoder is based on a time-domain algorithm. The U.S. Department of Defense uses 2400-bit/s LPC10 vocoders. They have good intelligibility, but voice quality is considered poor.

6.2.3 Sub-Band Coding (SBC). In sub-band coding, the speech signal is filtered into a relatively small number of sub-bands (2 to 16), and each sub-band signal is translated to zero frequency and sampled at the Nyquist rate and adaptively encoded. The number of bits used in the encoding process differs for each sub-band signal, with bits assigned as quantizers in accordance with perceptual criteria. By encoding each sub-band individually, the quantization noise* is confined within its sub-band. Sub-band encoders produce near toll quality speech at 16 kbits/s. At 9.6 kbits/s the speech quality is reduced as the "effective" bandwidth of the recovered signal is decreased.

Communication quality speech can be achieved at the bit rates between 4.8 and 8 kbits/s by using sophisticated time-domain coding for the sub-band signals and high-frequency regeneration of "inactive" frequency bands.

6.2.4 RELP, a Hybrid Coder. Many of the residual excited hybrid coding schemes use linear predictive modeling of the synthesis filter. One of the more well-known schemes is the residual excited linear predictive (RELP) type coder. These coders operate at bit rates between 4.8 and 16 kbits/s.

RELP coders use short-term (and sometimes long-term) linear prediction to derive a difference signal, called "residual" in a feed-forward manner. Macario [5]

---

* Quantization noise or quantization distortion is described in Chapter 8.

reports that earlier RELP systems used "baseband" coding and transmitted a low-pass version of the signal. The decoder recovered an approximation of the full band residual signal by employing *high-frequency regeneration* (HFR), which was then used to synthesize speech.

Simple HFR techniques (such as full-wave rectification, spectral folding, and spectral translation) generate considerable distortion in the restored speech signal.

The quality of the recovered speech can be considerably improved for rates below 9.6 kbits/s if, instead of the "feed-forward" RELP approach, an *analysis by synthesis* (AbS) optimization technique is used to define the excitation signal. This approach leads to the *AbS predictive coder*. In these systems both the "filter" and the "excitation" are defined on a short-term basis using a "closed-loop" optimization process which minimizes a perceptually weighted error measure formed between the input and the decoded output speech signals [5].

6.2.5 CELP Techniques.

One of the most popular voice coding techniques is the codebook excitation linear predictive or CELP coder. These types of coders provide surprisingly good voice quality at low bit rates. Reference 1 states that the 4.8-kbit/s CELP coder gives "communication quality" speech, whereas the the 8-kbit/s coder provided near-toll-quality speech.

Schroeder and Atal [17] describe CELP as coders that employ a vocal tract linear predictive (LP)-based model, a codebook-based excitation model, and an error criterion which serves to select an appropriate excitation sequence using an AbS optimization process. As a result, a CELP system selects that excitation sequence which minimizes a "perceptually" weighted mean square error formed between the input and the "locally" decoded signal.

Reference 5 reports that vocal tract model utilizes both a short-term filter, which models the spectral envelope of speech, and a long-term filter, which accounts for pitch periodicity in voiced speech.

Much work has been done on developing codebooks for CELP to reduce the excessive computational load needed to search for the optimum code sequence. These are structured codebooks where the codebook structure enables fast search procedures. There are sparse, ternary, overlapping, and algebraic codebooks.

Figure 18.16 is a schematic diagram of the CELP synthesis model, and some basic notions on how CELP operates are given below.

After short-term prediction and long-term prediction of the speech signal, redundancies in the speech signal are removed and the residual signal has very little correlation. A Gaussian process with slowly varying power spectrum can be used to represent the residual signal, and the speech waveform is generated by filtering white Gaussian innovation sequences through the time-varying linear long-term and short-term synthesis filters. The optimum innovation sequence is selected from a codebook of random white Gaussian sequence by minimizing the subjectively weighted error between the original and the synthesized speech. An adaptive overlapping codebook carries out the pitch correlation filter function. The address selected from the adaptive codebook and the corresponding gain (i.e., the pitch delay and gain), along with the address selected from the stochastic

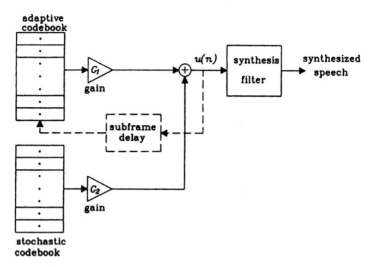

Figure 18.16. Schematic diagram of the CELP synthesis model. From Ref. 1, Figure 3.54. Reprinted with permission.

codebook and the corresponding scaling gain, are transmitted to the decoder. The decoder uses the same codebooks to determine the excitation signal at the input of the LPC synthesis filter (assuming no channel errors) to produce the synthesized speech.

The excitation codebook contains L codewords (stochastic vectors) of length N samples. Typically, $L = 1024$ and $N = 40$ corresponding to a 5-ms excitation frame. The excitation signal of a speech frame of length N is selected by the exhaustive search of the codebook after scaling the Gaussian vectors by a gain factor.

Source: Ref. 1.

7 NETWORK ACCESS METHODS

7.1 Introduction

The objective of a cellular radio operation is to provide a service where mobile subscribers can communicate with any subscriber in the PSTN, where any subscriber in the PSTN can communicate with any mobile subscriber, and where mobile subscribers can communicate amongst themselves via the cellular radio system. In all cases the service is full duplex.

A cellular service company is allotted a radio bandwidth segment to provide this service. Ideally, for full duplex service, a portion of the bandwidth is assigned for transmission from a cell site to mobile subscriber, and another portion is assigned for mobile user to cell site direction. Our goal here is to select an "access" method to provide this service given our bandwidth constraints.

We will discuss three generic methods of access: frequency division multiple access (FDMA), time division multiple access (TDMA), and code division multiple access (CDMA). It might be useful to the reader to review our discussion of satellite access in Chapter 7 where we describe FDMA and TDMA. However, in this section, the concepts are the same, but some of our constraints and operating parameters will be different.

7.2   Frequency Division Multiple Access (FDMA)

With FDMA, our band of frequencies is divided into segments and each segment is available for one user access at a time. Half of contiguous segments are assigned cell site outbound to mobile users and the other half inbound. A guard band is usually provided between outbound and inbound contiguous channels. This concept is shown in Figure 18.17.

Because of our concern to optimize the number of users per unit bandwidth, the key question is the actual width of one user segment. The North American AMPS system was described in Section 2 where each segment width was 30 kHz. The bandwidth of a user segment is greatly determined by the information bandwidth and modulation type. With AMPS the information bandwidth was a single voice channel with a nominal bandwidth of 4 kHz. The modulation was FM and the bandwidth was then determined by Carson's rule (Chapter 7). As we pointed out, AMPS is not exactly spectrum-conservative. On the other hand, it has a lot of the redeeming features that FM provides, such as the noise and interference advantage (FM capture).

Another approach to FDMA would be to convert the voice channel to its digital equivalent using CELP, for example (Section 4.2.5) with a transmission rate of 4.8 kbits/s. The modulation could be BPSK using a raised cosine filter where the bandwidth could be 1.25% of the bit rate of 6 kHz per voice channel. This alone would increase voice channel capacity five times over AMPS with its 30 kHz per channel. It should be noted that a radio carrier is required for each frequency slot.

7.3   Time Division Multiple Access (TDMA)

With TDMA we work in the time domain rather than the frequency domain of FDMA. Each user is assigned a time slot rather than a frequency segment; and

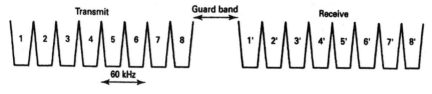

Figure 18.17. A conceptual drawing of FDMA.

during the user's turn, the full frequency bandwidth is available for the duration of the user's assigned time slot.

Let's say that there are n users and so there are n time slots. In the case of FDMA, we had n frequency segments and n radio carriers, one for each segment. For the TDMA case, only one carrier is required. Each user gains access to the carrier for $1/n$ of the time and there is generally an ordered sequence of time slot turns. A TDMA frame can be defined as cycling through n users' turns just once.

A typical TDMA frame structure is shown in Figure 18.18. One must realize that TDMA is only practical with a digital system such as PCM or any of those discussed in Section 4. TDMA is a store and burst system. Incoming user traffic is stored in memory; and when that user's turn comes up, that accumulated traffic is transmitted in a digital burst.

Suppose there are 10 users. Let each user's bit rate be R, then a user's burst must be at least $10R$. Of course, the burst will be greater than $10R$ to accommodate a certain amount of overhead bits as shown in Figure 18.18.

We define *downlink* as outbound where traffic goes from the base station to mobile station(s), and we define *uplink* where traffic goes from mobile station to base station. Typical frame periods are:

North American IS-54*	40 ms for six time slots
European GSM:	4.615 ms for eight time slots

One problem with TDMA, often not appreciated by the novice is *delay*. In particular, this is the delay on the uplink. Consider Figure 18.19, where we set up a scenario. A base station receives mobile time slots in a circular pattern and the radius of the circle of responsibility of that base station is 10 km. Let the velocity of a radio wave be 3×10^8 m/s. The time to traverse 1 km is $1000/3 \times 10^8$ or 3.333 μs. Making up an uplink frame is a mobile right on top of the base station with essentially no delay and another mobile right at 10 km with 10×3.33-μs or 33.3-μs delay. A GSM time slot is about 576 μs in duration. The terminal at the 10-km range will have its time slot arriving 33.3 μs late compared to the terminal with no delay. A GSM bit period is about 3.69 μs so that the late arrival eats up roughly 10 bits, and unless something is done, the last bit of the burst will overlap the next burst.

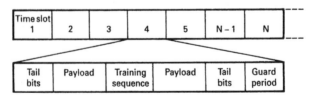

Figure 18.18. A simplified TDMA frame with blow-up of typical time slot. Tail bits assist in data recovery and act as time-slot marker. Training sequence assists in synchronization and bit timing recovery. Note similarity to the GSM frame.

* Also TIA-136.

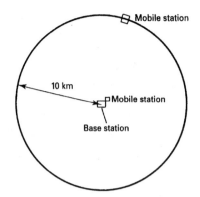

Figure 18.19. A TDMA delay scenario.

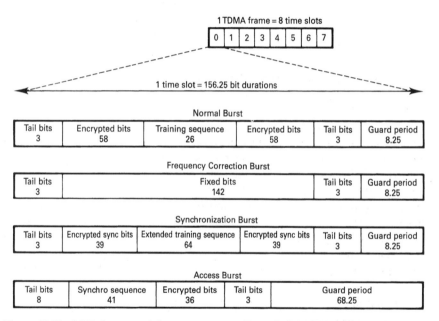

Figure 18.20. GSM frame and burst structures. From Ref. 1, Figure 8.7, reprinted with permission.

Refer now to Figure 18.20, which illustrates GSM burst structures. Note that the access burst has a guard period of 68.25-bit durations or a "slop" of 3.69×68.25 μs which will well accommodate the late arrival of the 10-km mobile terminal of only 33.3 μs.

To provide the same long guard period in the other bursts is a waste of valuable "spectrum."* The GSM system overcomes this problem by using adaptive frame

---

* We are equating bit rate, or bit durations to bandwidth. One could assume 1 bit per hertz as a first-order estimate.

alignment. When the base station detects a 41-bit random access synchronization sequence with a long guard period, it measures the received signal delay relative to the expected signal from a mobile station with zero range. This delay, called the timing advance, is transmitted to the mobile station using a 6-bit number. As a result, the mobile station advances its time base over the range of 0 to 63 bits (i.e., in units of 3.69 µs). By this process the TDMA bursts arrive at the base station in their correct time slots and do not overlap with adjacent ones. As a result, the guard period in all other bursts (Figure 18.16) is reduced to 8.25×3.69 µs or approximate 30.46 µs, the equivalent of 8.25 bits only. Under normal operations, the base station continuously monitors the signal delay from the mobile station and thus instructs the mobile station to update its time advance parameter. In very large traffic cells there is an option to actively utilize every second time slot only to cope with the larger propagation delays. This is spectrally inefficient, but in large, low-traffic rural cells it is admissible.

Source: Ref. 1.

For GSM see Ref. 10.

7.3.1 Comments on TDMA Efficiency.

Multichannel FDMA can operate with a power amplifier for every channel, or with a common wideband power amplifier for all channels. With the latter, we are setting up a typical generator of intermodulation products as these carriers mix in a comparatively nonlinear common power amplifier. To reduce the level of IM products, just like in satellite communications discussed in Chapter 7, back-off of power amplifier drive is required. This back-off can be on the order of 3–6 dB.

With TDMA (downlink), only one carrier is present on the power amplifier at any moment in time, thus removing most of the causes of IM noise generation. Thus with TDMA, the power amplifier can be operated to full saturation, a distinct advantage. FDMA required some guardbands between frequency segments; there are no guard bands with TDMA. However, as we saw above, a guard time between uplink time slots is required to accommodate the following situations:

- Timing inaccuracies due to clock instabilities
- Delay spread due to propagation*
- Transmission time delay due to propagation distance (see Section 5.3 above)
- Tails of pulsed signals due to transient response

The longer guard times are extended, the more inefficient a TDMA system is.

7.3.2 Advantages of TDMA.

The introduction of TDMA results in a much improved system signaling operation and cost. Assuming a 25-MHz bandwidth, up to 23.6 times capacity can be achieved with North American TDMA compared to FDMA (AMPS) (see Table II, Ref. 24).

---

* Lee [6] reports that a typical urban delay spread is 3 µs.

A mobile station can exchange system control signals with the base station without interruption of speech (or data) transmission. This facilitates the introduction of new network and user services. The mobile station can also check the signal level from nearby cells by momentarily switching to a new time slot and radio channel. This enables the mobile station to assist with handover operations and thereby improve the continuity of service in response to motion or signal fading conditions. The availability of signal strength information at both the base and mobile stations, together with suitable algorithms in the station controllers, allows further spectrum efficiency through the use of dynamic channel assignment and power control.

The cost of base stations using TDMA can be reduced as radio equipment shared by several traffic channels. A reduced number of transceivers leads to a reduction of multiplexer complexity. Outside the major metropolitan areas, the required traffic capacity for a base station may, in many cases, be served by one or two transceivers. The saving in the number of transceivers results in a significantly reduced overall cost.

A further advantage of TDMA is increased system flexibility. Different voice and nonvoice services may be assigned a number of time slots appropriate to the service. For example, as more efficient speech CODECs are perfected, increased capacity may be achieved by the assignment of a reduced number of time slots for voice traffic. TDMA also facilitates the introduction of digital data and signaling services as well as the possible later introduction of such further capacity improvements as digital speech interpolation (DSI).

Table 18.4 compares three operational digital TDMA systems.

7.3.3   Notes on GSM. By the end of 2002 the total worldwide GSM terminal population exceeded 700 million subscribers, over two-thirds of the total cellular radio count globally. It has deep penetration in the United States [28].

Its access method is TDMA. The modulation type employed is GMSK with a transmission rate per RF carrier of 270.833 kbits/s. Its maximum range (i.e., maximum cell radius) is 52 miles. This limit has nothing to do with the link budget. It is brought about by the maximum time delay due to propagation which can be compensated for (see Figure 18.19). GSM standards are published by ETSI in their 300, 500, and 700 series.

Additional information on GSM performance is given in Ref 10.

7.4   Code Division Multiple Access (CDMA)

CDMA means spread-spectrum multiple access. There are two types of spread spectrum: frequency hop and direct sequence (sometimes called pseudo-noise). In the cellular environment, CDMA means direct sequence spread spectrum [1]. However, the GSM system uses frequency hop, but not as an access technique.

Using spread-spectrum techniques accomplishes just the opposite of what we were trying to accomplish in Chapter 7. Bit packing is used to conserve bandwidth by packing as many bits as possible in 1 Hz of bandwidth. With spread

TABLE 18.4 Three TDMA Systems Compared

Feature	GSM	North America (TIA-136)	Japan
Class of emission			
Traffic channels	271KF7W	40K0G7WDT	tbd[a]
Control channels	271KF7W	40K0G1D	tbd
Transmit frequency bands (MHz)			
Base stations	935–960	869–894	810–830 (1.5 GHz tbd)
Mobile stations	890–915	824–849	940–960 (1.5 GHz tbd)
Duplex separation (MHz)	45	45	130
			48 (1.5 GHz)
RF carrier spacing (kHz)	200	30	25 interleaved
			50
Total number of RF duplex channels	124	832	tbd
Maximum base station erp (W)			
Peak RF carrier	300	300	tbd
Traffic channel average	37.5	100	tbd
Nominal mobile station transmit power (W): peak and average	20 and 2.5 8 and 1.0 5 and 0.625 2 and 0.25	9 and 3 4.8 and 1.6 1.8 and 0.6 tbd and tbd	tbd
Cell radius (km)			
Min	0.5	0.5	0.5
Max	35	20	20
	(up to 120)		
Access method	TDMA	TDMA	TDMA
Traffic channels/RF carrier			
Initial	8	3	3
Design capability	16	6	6
Channel coding	Rate one-half convolutional code with interleaving plus error detection	Rate one-half convolutional code	tbd
Control channel structure			
Common control channel	Yes	Shared with AMPS	Yes
Associated control channel	Fast and slow	Fast and slow	Fast and slow
Broadcast control channel	Yes	Yes	Yes

TABLE 18.4 *(continued)*

Feature	GSM	North America (TIA-136)	Japan
Delay spread equalization capability (μs)	20	60	tbd
Modulation	GMSK[b] (BT = 0.3)	$\pi/4$ diff.[c] encoded QPSK (roll-off = 0.25)	$\pi/4$ diff. encoded QPSK (roll-off = 0.5)
Transmission rate (kbits/s)	270.833	48.6	37–42
Traffic channel structure			
Full-rate speech codec			
Bit rate (kbits/s)	13.0	8	6.5–9.6
Error protection	9.8-kbit/s FEC + speech processing	5 kbits/s FEC	~3 kbits/s FEC
Coding algorithm	RPE-LTP	CELP	tbd
Half-rate speech codec			
Initial	tbd	tbd	tbd
Future	Yes	Yes	Yes
Data			
Initial net rate (kbits/s)	Up to 9.6	2.4, 4.8, 9.6	1.2, 2.4, 4.8
Other rates (kbits/s)	Up to 12	tbd	8 and higher
Handover			
Mobile assisted	Yes	Yes	Yes
Intersystem capability with existing analog system	No	Between digital and AMPS	No
International roaming capability	Yes > 16 countries	Yes	Yes
Design capability for multiple system operators in same area	Yes	Yes	Yes

[a] tbd, to be defined.
[b] GMSK, Gaussian minimum shift keying.
[c] diff., differentially.

Source: CCIR Rep. 1156, Table 1, pages 120–123 [24].

spectrum we do the reverse by spreading the information signal over a very wide bandwidth.

Conventional AM requires about twice the bandwidth of the audio information signal with its two sidebands of information (i.e., ± 4 kHz). On the other hand, depending on its modulation index, frequency modulation could be considered a type of spread spectrum in that it produces a much wider bandwidth than its transmitted information. As with all other spread-spectrum systems, a signal-to-noise advantage is gained with FM, depending on its modulation index. For example, with AMPS, a typical FM system, 30 kHz is required to transmit the nominal 4-kHz voice channel.

If we are spreading a voice channel over a very wide frequency band, it would seem that we are defeating the purpose of frequency conservation. With spread spectrum, with its powerful anti-jam properties, multiple users can transmit on the same frequency with only some minimal interference one to another. This assumes that each user is employing a different key variable (i.e., in essence, using a different code). At the receiver, the CDMA signals are separated using a correlator that accepts only signal energy from the selected key variable binary sequence (code) used at the transmitter and then despreads its spectrum. CDMA signals with unmatching codes are not despread and only contribute to the random noise.

CDMA reportedly provides increases in capacity 15 times that of its analog FM counterpart. It can handle any digital format at the specified input bit rate such as facsimile, data, and paging. In addition, the amount of transmitter power required to overcome interference is comparatively low when utilizing CDMA. This translates into savings on infrastructure (cell site) equipment, as well as longer battery life for hand-held terminals. CDMA also provides so-called soft hand-offs from cell site to cell site that make the transition virtually inaudible to the user.

Source: Ref. 25.

Dixon [18] develops from Claude Shannon's classical relationship an interesting formula to calculate the spread bandwidth given the information rate, signal power, and the noise power:

$$C = W \log_2(1 + S/N) \qquad (18.11)$$

where C is the capacity of a channel in bits per second, W is the bandwidth in hertz, N is the noise power, and S is the signal power. This equation shows the relationship between the ability of a channel to transfer error-free information, compared with the signal-to-noise ratio existing in the channel, and the bandwidth used to transmit the information.

If we let C be the desired system information rate and changing the logarithm base to the natural base (e), the result is

$$C/W = 1.44 \log_e(1 + S/N) \qquad (18.12)$$

and, for an S/N that is very small (e.g., 0.1) which would be used in an anti-jam system,* we can say

$$C/W = 1.44[S/N] \qquad (18.13)$$

From this equation we find that

$$N/S = 1.44W/C \approx W/C \qquad (18.14)$$

---

* If we think about it, a cellular scenario where one user is transmitting right on the top of others on the same frequency plus adjacent channel interference, we are indeed in an anti-jam situation.

and

$$W \approx NC/S \qquad (18.15)$$

This exercise shows that for any given S/N we can have a low information-error rate by increasing the bandwidth used to transfer that information.

Suppose we had a cellular system using a data rate of 4.8 kbits/s and an S/N of 20 dB (numeric of 100), then the bandwidth for this 4.8-kbits/s channel would be

$$W = 100 \times 4.8 \times 10^3 / 1.44$$

$$= 333.333 \text{ kHz}$$

There are two common ways that information can be embedded in the spread-spectrum signal. One way is to add the information to the spectrum-spreading code before the spreading modulation stage. It is assumed that the information to be transmitted is binary because modulo-2 addition is involved in this process. The second method is to modulate the RF carrier with the desired information before spreading the carrier. The modulation is usually PSK or FSK or other angle modulation scheme [18].

Dixon [18] lists some advantages of spread spectrum:

1. Selective addressing capability
2. Code division multiplexing is possible for multiple access
3. Low-density power spectra for signal hiding
4. Message security
5. Interference rejection.

And of most importance for the cellular user [18, page 10]:

> When codes are properly chosen for low cross correlation, minimum interference occurs between users, and receivers set to use different codes are reached only by transmitters sending the correct code. Thus more than one signal can be unambiguously transmitted at the same frequency and at the same time; selective addressing and code-division multiplexing are implemented by the coded modulation format.

Figure 18.21 shows a direct sequence (pseudo-noise) spread-spectrum system with waveforms.

Processing gain is probably the most commonly used parameter to describe the performance of a spread-spectrum system. It quantifies the signal-to-noise ratio improvement when a spread signal is passed through a "processor." For instance, a certain processor had an input S/N of 12 dB and an output S/N of 20 dB; thus the processing gain is 8 dB.

Processing gain is expressed by the following:

$$G_p = \text{Spread bandwidth in Hz/information bit rate} \qquad (18.16)$$

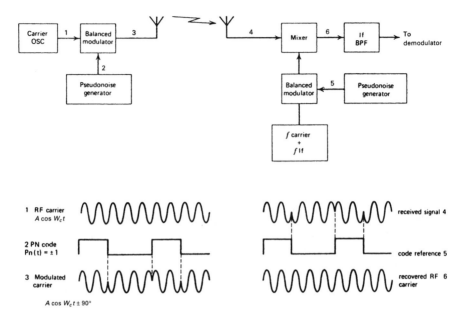

Figure 18.21. A direct sequence spread-spectrum system showing waveforms. From Ref. 18, Figure 2.3, reprinted with permission.

More commonly, processing gain is given in a dB value; then

$$G_{p(dB)} = 10 \log [\text{spread bandwidth in Hz/information bit rate}] \quad (18.17)$$

Example: A certain cellular system voice channel information rate is 9.6 kbits/s and the RF spread bandwidth is 9.6 MHz. What is the processing gain?

$$G_{p(dB)} = 10 \log 9.6 \times 10^6 - 10 \log 9600$$
$$= 69.8 - 39.8 \text{ (dB)}$$
$$= 30 \text{ dB}$$

It has been pointed out in Ref. 1 that the power control problem held back the implementation of CDMA for cellular application. If the standard deviation of the received power from each mobile at the base station is not controlled to an accuracy of approximately ±1 dB relative to the target receive power, the number of users supported by the system can be significantly reduced. Other problems to be overcome were synchronization and sufficient codes available for a large number of mobile users [1].

Qualcomm, a North American company, has a CDMA design that overcomes these problems and has fielded a cellular system based on CDMA. It operates at the top of the AMPS band using 1.23 MHz for each uplink and downlink. This is the equivalent of 41 AMPS channels (i.e., 30 kHz × 41 = 1.23 MHz) deriving up to 62 CDMA channels (plus one pilot channel and one synchronization channel) or some 50% capacity increase. The Qualcomm system also operates in the 1.7- to 1.8-GHz band [1].

Consult Ref. 25 for further reading on CDMA.

8 FREQUENCY REUSE

Because of the limited bandwidth allocated in the 800-MHz band for cellular radio communications, frequency reuse is crucial for its successful operation. A certain level of interference has to be tolerated. A major source of interference is co-channel interference from a "nearby" cell using the same frequency group as the cell of interest. For the 30-kHz bandwidth AMPS system, Ref. 5 suggests that C/I be at least 18 dB. The primary isolation derives from the distance between the two cells with the same frequency group. In Figure 18.2 there is only one cell diameter for protection.

Refer to Figure 18.22 for the definition of the parameters R and D, where D is the distance between cell centers of repeating frequency groups, and R is the "radius" of a cell. We let

$$a = D/R$$

The D/R ratio is a basic frequency reuse planning parameter. If we keep the D/R ratio large enough, co-channel interference can be kept to an acceptable level.

Lee [6] calls a the co-channel reduction factor and relates path loss from the interference source to R^{-4}.

A typical cell in question has six co-channel interferers, one on each side of the hexagon. So there are six equidistant co-channel interference sources. The goal is C/I 18 dB or a numeric of 63.1. So

$$C/I = C/I = C/6I = R^{-4}/6D^{-4} = a^4/6 = 63.1, \quad \text{then} \quad a = 4.4$$

This means that D must be 4.4 times the value of R. If R is 6 miles (9.6 km), then $D = 4.4 \times 6 = 26.4$ miles (42.25 km).

Lee [6] reports that co-channel interference can be reduced by other means such as directional antennas, tilted beam antennas, lowered antenna height, and an appropriately selected site.

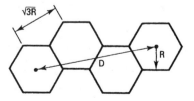

Figure 18.22. Definition of R and D.

If we consider a 26.4-mile path, what is the height of earth curvature at mid-path? From Chapter 7, $h = 0.667(d/2)^2/1.33 = 87.3$ ft (26.9 m). Provided that the cellular base station antennas are kept under 87 ft, the 40-dB/decade rule of Lee holds. Of course, we are trying to keep below line-of-sight conditions.

The total available (one-way) bandwidth is split up into N sets of channel groups. The channels are then allocated to cells, one channel set per cell on a regular pattern, which repeats to fill the number of cells required. As N increases, the distance between channel sets (D) increases, reducing the level of interference. As the number of channel sets (N) increases, the number of channels per cell decreases, reducing the system capacity. Selecting the optimum number of channel sets is a compromise between capacity and quality. Note that only certain values of N lead to regular repeat patterns without gaps. These are $N = 3, 4, 7, 9$, and 12, and then multiples thereof. Figure 18.23 shows a repeating 7-pattern for frequency reuse. This means that $N = 7$ or there are 7 different frequency sets for cell assignment.

Cell splitting can take place especially in urban areas in some point in time because the present cell structure cannot support the busy hour traffic load. Cell splitting, in effect, provides more frequency slots for a given area. Macario in [5] reports that cells can be split as far down as a 1-km radius.

Co-channel interference tends to increase with cell splitting. Cell sectorization can cut down the interference level. Figure 18.24 shows a three- and six-sector plan. Sectorization breaks a cell into three or six parts each with a directional antenna. With a standard cell, co-channel interference enters from six directions.

A six-sector plan can essentially reduce the interference to just one direction. A separate channel set is allocated to each sector.

The three-sector plan is often used with a seven-cell repeating pattern resulting in an overall requirement for 21 channel sets. The six-sector plan with

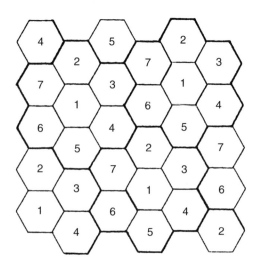

Figure 18.23. A cell layout based on $N = 7$.

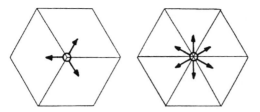

Figure 18.24. Breaking a cell up into three sectors (*left*) and six sectors (*right*).

its improved co-channel performance and the rejection of secondary interferers allows a four-cell repeat plan (Figure 18.2) to be employed. This results in an overall 24-channel set requirement. Sectorization requires a larger number of channel sets and fewer channels per sector. Outwardly it appears that there is less capacity with this approach; however, the ability to use much smaller cells results in actually a much higher capacity operation [5].

9 PAGING SYSTEMS

9.1 What Are Paging Systems?

Paging is a one-way radio alerting system. The direction of transmission is from a fixed paging transmitter to an individual. It is a simple extension of the PSTN. Certainly, paging can be classified as one of the first PCS (personal communication system) operations. The paging receiver is a small box, usually carried on a person's belt. As a minimum, a pager alerts the user that someone wishes to reach him/her by telephone. The person so alerted goes to the nearest telephone and calls a prescribed number. Some pagers have a digital readout which provides the calling number, whereas others give the number and a short message.

Most paging systems now operate in the VHF and UHF bands with a 3-kHz bandwidth. Transmitters have 1 to 5 watts output, and paging receivers have sensitivities in the range of 10 μV/m to 100 μV/m.

Source: Ref. 5.

9.2 Radio-Frequency Bands for Pagers

All three ITU regions have some or all of the following frequency bands allocated to mobile services:

26.1–50 MHz	68–88 MHz
146–174 MHz	450–470 MHz
806–960 MHz	

TABLE 18.5   Propagation Loss Suffered by Signals in Penetrating Buildings

Frequency:	150 MH	250 MHz	400 MHz	800 MHz
Building penetration loss[a]:	22 dB	18 dB	18 dB	17 dB[b]

[a]The loss is given as the ratio between the median value of the field strengths measured over the lower floors of buildings and the median value of the field strengths measured on the street outside. Similar measurements made in other countries confirm the general trend, but the values of building penetration loss vary about those shown. For instance, measurements made in the United Kingdom indicate that building penetration loss at 160 MHz is about 14 dB and about 12 dB at 460 MHz.
[b]Somewhat less accurate than the other results.
Source: CCIR Rep. 499-5, Table 1, page 59 [23].

9.3   Radio Propagation into Buildings

Measurement results submitted to the CCIR [21] have indicated that frequencies in the range of 80–460 MHz are suitable for personal radio paging in urban areas with high building densities. It is possible that frequencies in the bands allocated around 900 MHz may also be suitable but that higher frequencies are less suitable. From measurements made in Japan, the following median values of propagation loss suffered by signals in the penetration of buildings (building penetration loss) have been derived. These results are summarized in Table 18.5

9.4   Techniques Available for Multiple Transmitter Zones

To cover a service are a effectively, it is often necessary to use a number of radio-paging transmitters. When the required coverage area is small, a single RF channel should be used so as to avoid the need for multichannel receivers. In these circumstances, the separate transmitters may operate sequentially or simultaneously. In the latter case, the technique of offsetting carrier frequencies, by an amount appropriate to the coding system employed, is often used. It is also necessary to compensate for the differences in the delay to the modulating signals arising from the characteristics of the individual landlines to the paging transmitters. One way to do this is to carry out synchronization of the code bits via the radio-paging channel. Of course, information will be required about the bit rates which this synchronization method would permit.

It is preferable that the frequency offset of the transmitter carrier frequencies in a binary digital radio-paging system be at least twice the signal fundamental frequency. It is also preferable that delay differences between modulation of the transmitters in a binary digital paging system should be less than a quarter of the duration of a bit if direct FSK, NRZ modulation is employed. For subcarrier systems, the corresponding limit should be less than one-eighth of a cycle of the subcarrier frequency.

9.5   Paging Receivers

Built-in antennas can be designed for 150-MHz operation with reasonable efficiency. A typical radio-paging receiver antenna using a small ferrite rod exhibits a loss factor of about 16 dB relative to a half-wave dipole.

The majority of wide-area paging systems use some form of angle modulation. Repeated transmission of calls can be used to improve the paging success rate of tone alert pagers. If p is the probability of receiving a single call, then $1 - (1 - p)^n$ is the probability of receiving a call transmitted n times, provided that the calls are uncorrelated. Correlations under Rayleigh fading conditions can be largely removed by spacing the calls more than 1 s apart. Longer delays between subsequent transmission (about 20 s) are required to improve the success rate under shadowing conditions.

Receivers with numeric or alphanumeric message displays can only take advantage of call repetitions if the supplementary messages are used to detect and correct errors.

9.6 System Capacity

The capacity of any paging system is affected by the following:

- Number and characteristics of the radio channels used
- The number of times each channel is reused within the system
- The actual paging location requirements of the individual users
- The peak information (address and message) requirement in a location(s)
- Tolerable paging delay
- Data transmission rate
- Code efficiency
- Method of using the total code capacity throughout the system (this may also affect the system's capabilities for roaming)
- Any inefficiency introduced by battery-saving provisions
- Possible telephone system input restrictions

9.7 Codes and Formats for Paging Systems

The U.S. paging system is broadly employed across North America. It is a binary digital code and has a format with a signal address capacity of up to 400,000 with noncoded battery saving and 4,000,000 with coded preamble. The code is a Golay (23:12) cyclic code with two code words representing an address. Messages are encoded using a BCH (15:7) code. The code and format provide queueing and numeric alphanumeric message flexibility and the ability to operate in a mixed mode of transmission with other formats.

Japan uses a BCH (31:16) codeword with a Hamming distance of 7. The format gives approximately 65,000 addresses, 15 groups for battery economy, and a total cycle length of 4185 bits. Each group contains 8 address code words headed by a 31-bit synchronizing and group-indicating signal.

The UK paging system employs a BCH (31:21) code plus even-parity code word with a Hamming distance of 6. The code format can handle over 8 million addresses and can be expanded. It can also handle any type of data message such as hexadecimal and CCITT alphabet No. 5. It is designed to share a channel with other codes and to permit mixed simultaneous and sequential multitransmitter operation at the normal 512-bit/s transmission rate. This code is sometimes referred to as POCSAG, and it has been adopted as CCIR Radio-Paging Code No. 1 (RPC1).

9.8 Considerations for Selecting Codes and Formats

- Number of subscribers to be served
- Number of addresses assigned to each subscriber
- The calling rate expected including that from any included message facility
- Zoning arrangement
- The data transmission rates possible over the linking network and radio channel(s), taking into account the propagation factors of the radio frequencies to be used
- The type of service: vehicular or personal, urban or rural

Once the data are provided from the above listing of topics, codes may be compared by their characteristics with respect to:

- Code address capacity
- Number of bits per address
- Code word Hamming distance
- Code efficiency, such as number of information bits compared to the total number of bits per code word
- Error-detecting capability; error-correcting capability
- Message capability and length
- Battery-saving capability
- Ability to share a channel with other codes
- The capability of meeting the needs of paging systems which vary with respect to size and transmission mode (e.g., simultaneous versus sequential)

Source: Sections 9.2 through 9.8 are based on CCIR Reps. 900-2 and 499-5 [22, 23].

10 MOBILE SATELLITE COMMUNICATIONS

10.1 Background and Scope

In our earlier discussions on cellular mobile radio and PCS, there seemed to be no clear demarcation where one ended and the other began. Cellular hand terminals certainly are used inside of all types of buildings with some fair success. Granted some of the connections are marginal. On the other hand, PCS/wireless terminal can have mobile connectivities a mile or more from a fixed facility.

Often we see PCS in the bigger picture of cellular mobile radio. Even CCIR (ITU-R Organization) describes IMT-2000* as an integrated system where there is no dividing line between PCS and cellular. There is a third element in the PCS/mobile communication picture. This is mobile satellite service (MSS), using ITU-R terminology.

Mobile satellite systems essentially service three markets: maritime mobile, aeronautical mobile and land mobile. MSS provides communications to and from mobile platforms for voice telephony, slower speed data (up to 64 kbits/s), internet/email, and, in some cases, slow-speed video. Some connectivities are constrained to voice only or a distress message because of small size and other mobile terminal characteristics.

In this section we briefly describe how mobile satellite system work and then give an overview of a number of existing mobile satellite services that provide PCS and cellular mobile radio on a worldwide basis. Early MSS systems gave service only to ships. This then expanded to aircraft, and service included a safety branch for mobile stations in distress. Soon services were expanded to land mobile and fixed users.

10.2 How MSS Operates

Conventional cellular/PCS requires two over-the-air links: (1) a downlink from the cell site or master station to the mobile platform and (2) an uplink from the mobile platform to the cell site or master station. This is aptly described in Section 2 of this chapter.

The mobile satellite service (MSS) requires four links for connectivity. If a call is initiated in the terrestrial network (e.g., a PSTN telephone), it is connected through a mobile switching center to the appropriate earth station which provides a feeder link to the satellite. The satellite then connects the call through a down-link to the mobile platform. In the other direction, the mobile platform transmits on an uplink to the satellite, and the call is run through the satellite transponder. The satellite forwards the connectivity on a feeder link to the earth station which connects through to the PSTN. Thus we have full-duplex communication. The operation is illustrated graphically in Figure 18.25.

Feeder links commonly uplink in the familiar 6-GHz band and downlink in the 4-GHz band. *Service links*, those links that connect the mobile platform to the satellite, often use the 1.5/1.6-GHz band or the 1.9/2.1-GHz bands.

Feeder links, of which there are few, are usually sophisticated and overbuilt to compensate for the modest mobile platform terminals of which there are many. Inmarsat, for example, had 260,000 subscribers as of 2002. These terminals are small, cost-effective, and very modest in size. Many service link mobile accesses are just hand-held terminals. However, in the case of Inmarsat and some other

---

* IMT-2000. A third-generation (3G) integrated cellular–PCS mobile communication system. It includes a satellite component for both PCS and cellular service. It is the forerunner of UMTS discussed in Section 11.

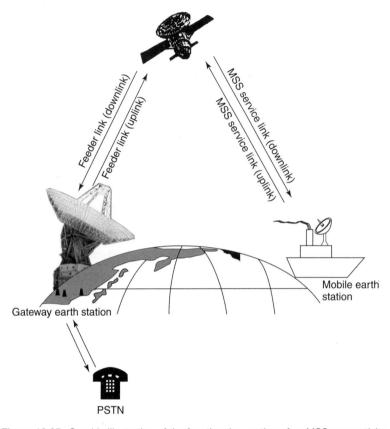

Figure 18.25. Graphic illustration of the functional operation of an MSS connectivity.

mobile service providers, the mobile platform terminals can be fairly large and sophisticated depending largely on the type of service offered.

With a mobile platform the antenna subsystem is the principal limiting factor to performance and bit rate supported. Typical antenna gains used in MSS vary between 0 dBi (a hand-held terminal, for example) to +21 dBi (85-cm reflector antenna). Receiver system gain-noise temperature ratios (G/T) typically range between −26 dB-K and −4 dB-K. Of course, terminals with larger antenna apertures will require tracking of the satellite. Nearly all MSS terminals provide full duplex operation, with safety systems being a prime exception (e.g., EPIRB).

Until about 1990, MSS systems utilized satellites in a geostationary orbit (GSO). Now systems can be found in other constellation/orbital configurations such as LEO (low earth orbit), MEO (medium earth orbit), and elliptical orbits. The Van Allen belts are the principal gating factor of orbital place. The Van Allen belts are belts of intense radiation that circle the earth between 3200- and 7600-km altitude. The higher the orbital placement, the less satellites that are needed for complete earth coverage. *Iridium*, for example, has a mean altitude of 780 km. It has a constellation of 66 satellites in six orbital places, whereas *Globalstar* has been placed at a 1400-km altitude with a constellation of 48 satellites in eight orbital planes. A typical MEO may be placed at a 10,000-km altitude requiring 10 satellites in two orbital planes for complete earth coverage.

10.3 Safety Systems Associated with Mobile Platforms

10.3.1 Satellite EPIRBs

EPIRBs (emergency position indicating radio beacons) have been in use since World War II. Satellite EPIRBs are comparatively recent, and the concept has evolved considerably. Their principal purpose is to notify authorities of ship or aircraft distress. The latest position of the vessel or aircraft is taken from GPS and relayed along with the distress message.

There are two satellite-based EPIRB systems that are operational. The first is operated by COSPAS-SARSAT. This system supports a range of beacons for distress alerting and position location including that for GMDSS (Global Maritime Distress and Safety System). The other is operated by Inmarsat (Section 10.4). This system concentrates on providing GMDSS capability for use on board vessels only. Both services are operated free of charge by the operators.

10.3.2 COSPAS-SARSAT. The COSPAS-SARSAT is a satellite system dedicated to search and rescue (SAR) which uses satellite-based sensors and ground processing facilities to detect and locate emergency distress beacons. COSPAS-SARSAT comprises:

- SAR payloads on board LEO satellites (LEOPSAR) and GEO satellites (GEOSAR)
- Distress beacons on mobile platforms which transmit at 121.5, 243, and 406 MHz (*Note*: These are standard international VHF/UHF distress frequencies)
- Satellite ground receiving stations located throughout the world
- An extensive ground command and control network for distributing distress alerting information and data required to operate the system

10.3.3 INMARSAT-E. For those mobile platforms operating with the INMARSAT system (see Section 10.4.1), there is the INMARSAT-E emergency service. It operates with EPIRBs designed to uplink in a reserved portion of the 1.6-GHz band. The EPIRBs incorporate an integral radio navigation receiver which utilizes the Global Positioning System (GPS), which determines the geographic position of an EPIRB within 200 m. When the EPIRB is activated, the distress information is transmitted via one or more INMARSAT satellites to two INMARSAT land earth stations where an alarm is triggered to alert search and rescue authorities.

10.4 Operational or Near-Term Planned MSS Systems

Section Note: Reference 36, from which this section has been excerpted, gives descriptions of ten MSS systems. Due to space limitations, our discussion covers only three of these systems.

10.4.1 INMARSAT. INMARSAT [International Maritime Satellite (consortium)] has been providing worldwide full-duplex voice, data, and record traffic service with ships since 1979. It extended its service to a land-mobile market and to aircraft. By 2001 there were over 260,000 INMARSAT terminals. About 30% of these are land-transportable. INMARSAT satellites are in geostationary earth orbit (GEO). The present series of INMARSAT-3 satellites support up to 64 kbits/s of data traffic; INMARSAT-4, due to be operational in 2005, will provide up to 432 kbits/s data connectivity. INMARSAT-3 satellites are placed over four strategic locations above the equator to provide full earth coverage in GSO from 80°S to 80°N latitude. The subsatellite points of the INMARSAT-3 series are: 178° E, 54° W, 15.5° W and 64° E longitude.

There is also a fifth INMARSAT-3 satellite on station as an operational spare. There are four INMARSAT-2 satellites that are operational. These are stationed at other orbital locals and are used as operational spares or to carry additional traffic to supplement the traffic load of the INMARSAT-3 satellites. There is an INMARSAT-4 satellite, two of which will be launched initially. These are similar to the INMARSAT-3 space vehicles but have additional spot beam antennas for coverage of major land masses and principal ship and aircraft routes. Table 18.6 lists the various terminal types offered by INMARSAT, their respective antenna gains, and their bit rate capabilities. Table 18.7 summarizes the performance of INMARSAT-4 satellite series.

10.4.2 The Globalstar Mobile Satellite System. Globalstar provides mobile cellular for voice and data throughout the world, from 70°S latitude to 70°N latitude. The space segment consists of (a) 48 operational satellites at 1414-km low earth orbits (LEO) distributed in 8 orbital planes and (b) 4 on-station spare satellites. There are 6 satellites per orbital plane. A satellite makes one complete rotation around an orbital plane every 114 minutes.

Globalstar satellites carry subscriber traffic between Globalstar telephone and data modems and Globalstar gateways. The gateways then interface with the terrestrial PSTN and cellular networks via the gateway's mobile switching center (MSC) in the host country.

From the following information we can derive available Globalstar operational bandwidth (in Hz):

- The feeder frequencies are as follows. Uplink: 5091 − 5250 MHz (159 MHz); downlink: 6875–7055 MHz (180 MHz).
- The satellite downlink to mobile platform is: 2483.5 − 2500 MHz (16.5 MHz); uplink from a mobile platform is 1610 − 1621.35 MHz (11.35 MHz).

TABLE 18.6   INMARSAT Terminal Types

Terminal	Typical Antenna Gain (dBi)	Voice	Data Rate	Beam of Operation
Inm-A maritime	22	√	9.6–64 kbits/s	Global
Inm-A transportable	22	√	9.6–64 kbits/s	Global
Inm-M portable	14	√	4 kbits/s	Global/spot
Inm-M maritime	16	√	4 kbits/s	Global/spot
Inm-M vehicular	12	√	4 kbits/s	Global/spot
Inm-M fixed	29	√	4 kbits/s	Global/spot
Inm-B transportable	22	√	64 kbits/s (ISDN)	Global/spot
Inm-B maritime	22	√	64 kbits/s (ISDN)	Global/spot
Inm-B fixed	29	√	64 kbits/s (ISDN)	Global/spot
Inm-mini M portable	10–12	√	2.4 kbits/s	Spot
Inm-mini M vehicular	7–10	√	2.4 kbits/s	Spot
Inm-mini M maritime	7–10	√	2.4 kbits/s	Spot
Inm-mini M rural	18	√	2.4 kbits/s	Spot
Inm-mini M aero	6	√	2.4 kbits/s	Spot
Inm-C maritime	Omni	x	600 bits/s	Global
Inm-C land mobile	Omni	x	600 bits/s	Global
Inm-D/D+	Omni	x	—	Global/spot
Inm-E maritime	Omni	x	—	Global
Aero-H	12	√	G3 fax at 4.8-kbits/s packet data at 10.5 kbits/s	Global
Aero-H+	12	√	G3 fax at 2.4-kbits/s packet data at 10.5 kbits/s	Global/spot
Aero-I	6	√	600–4800 bits/s	Global/spot

TABLE 18.6 *(continued)*

Terminal	Typical Antenna Gain (dBi)	Voice	Data Rate	Beam of Operation
Aero-L	Omni	x	600–1200 bits/s compatible with X.25 networks	Global
Aero-C	Omni	x	600 bits/s	Global
Swift64	12	√	64-kbits/s circuit mode (ISDN) and packet data	Spot
Mobile satcom unit (GAN)	18	√	64-kbits/s circuit mode (ISDN) and packet data	Spot
PMC (B-GAN)	7.5–16.5	√	72–432 kbits/s	Inm-4 small spot

Source: Table 29, pages 123–124, *ITU-R MSS Handbook*, 2002 [36]. Reprinted with permission of the ITU.

Reference 36 indicates that Globalstar, assuming bent pipe operation, can interface with the following cellular formats: CDMA (assume IS-95), GSM 900, AMPS 800, and "Globalstar modes."

For further information on mss see Ref. 9.

10.5 Advantages and Disadvantages of a Low Earth Orbit

The low earth orbit (LEO) offers a number of advantages over the geostationary orbit, and at least one serious disadvantage.

Delay. One-way delay to a GEO satellite is budgeted at 125 ms; one-way up and down is double this value of 250 ms. Round-trip delay is about 0.5 s. Delay to a typical LEO is 2.67 ms and round-trip delay is 4 × 2.67 ms or about 10.66 ms. Calls to/from mobile users of such systems may be relayed still again by conventional satellite services. Data services do not have to be so restricted on the use of "handshakes" and stop-and-wait ARQ as with similar services via a GEO system.

Higher Elevation Angles and "Full Earth Coverage." The GEO orbit provides no coverage above about 80° latitude, and it gives low angle coverage of many of the world's great population centers because of its comparatively high latitude. Typically, cities in Europe and Canada face this dilemma. LEO satellites, depending on orbital plane spacing, can all provide elevation angles >40°. This is particularly attractive in urban areas with tall buildings. Coverage would only be available on the south side of such buildings in the Northern Hemisphere with a clear shot to the horizon. Properly designed

TABLE 18.7 Performance Summary of INMARSAT-4 Satellite Communication Subsystem

Repeater	Transparent bent-pipe repeater utilizing a digital channelizer, providing 630 × 200-kHz channels on each of the forward and return links, with a digital L-band beamformer allowing the generation of different types of beams. The basic L-band coverage relies on around 200 narrow spot beams (1.2°, 3-dB beamwidth), 19 wide spot beams (4.5°, 3-dB beamwidth), and one global beam. All power amplifiers are solid state and utilize linearizers.	
Service link options	C-L, L-C, C-C, L-L and navigation	
Antennas	Separate transmit and receive for C-band and single antenna for L-band. Single L-band center-fed navigation antenna provides global coverage. Focal plane array, offset-fed, deployed unfurlable reflector for L-band transmit/receive, producing the global and spot beam coverages. C-band horns provide global coverage. Single navigation antenna	
G/T	L-band = 10 dB/K for narrow spot beams = 0 dB/K for wide spot beams = −10 dB/K for global beam C-band = −11 dB/K	
e.i.r.p.	L-band	67-dBW aggregate e.i.r.p. for narrow spot beams. Can be continuously shared with wide spot beams and global beam.
	C-band	31-dBW/polarization
	Navigation	28.5 dBW

Source: Table 28, page 123, *Handbook Mobile Satellite Service (MSS)*, 2002 [36]. Reprinted with permission of the ITU.

LEO systems will not have such drawbacks. Coverage will be available at any orientation.

*Tracking, a Disadvantage of LEOs, MEOs.** At L-band quasi-omnidirectional antennas for the mobile user are fairly easy to design and produce. Although such antennas display only modest gain of several decibels, links to LEO satellites can be easily closed with hand-held terminals. However, large feeder, fixed earth terminals will require a good tracking capability as LEOs pass overhead. Hand-off is also required as an LEO disappears over the horizon to another satellite just as it appears over the opposite horizon. The hand-off should be seamless.

* *MEO stands for medium earth orbit (i.e., 5000- to 13,000-km orbits).*

The quasi-omnidirectional user terminal antennas will not require tracking, and the hand-off should not be noticeable to the mobile user.

11   1G, 2G, 2-1/2G, AND 3G, THAT IS THE QUESTION

The G-designation stands for *generation*. Thus we have 1G for first generation, 2G for second generation, and so on. In this case we are referring to mobile communications. The first-generation mobile or cellular radio was AMPS in North America (Section 2 of this chapter), Nordic Mobile Telephone (NMT), and Total Access Communication System (TACS). There were 20 million 1G subscribers by 1990.

11.1   Second Generation (2G)

The second-generation (2G) development came about by the requirement for better transmission quality, system capacity, and coverage. Here was the transformation to digital cellular with a number of different systems fielded such as GSM, Digital AMPS (D-AMPS), CDMA (e.g., IS-95), and Personal Digital Communication (PDC).

2G has become extremely prevalent throughout the world, yet some analog systems continue to hang on (e.g., AMPS). These standards serve different purposes and requirements such as wireless local loop (WLL), paging, cordless telephones, private mobile radio, and mobile satellite systems besides conventional 2G cellular radio. Many standards are used in only one country or region and most are incompatible one with another.

Certainly, the most successful and widespread of these standards is the GSM family including GSM900, GSM-railway (GSM-R), GSM1800, GSM1900, and GSM400. GSM can be found in over 140 countries and has widespread use in the United States.

11.2   Evolution from 2G to 3G

In 1996 ETSI* decided to further enhance GSM in Phase 2+ releases that incorporate 3G capabilities. These 3G features include intelligent network (IN) services with customized application for mobile enhanced logic (CAMEL), enhanced speech compression–decompression (CODEC), enhanced full rate (EFR) and adaptive multirate, high data-rate services and new transmission principles with high-speed circuit switched data (HSCSD), general packet radio service (GPRS), and enhanced data rates for GSM evolution (EDGE). UMTS is a 3G GSM successor standard that is downward-compatible with GSM, using GSM Phase 2+ enhanced core network.

* ETSI - European Telecommunication Standardization Institute.

11.2.1   3G and IMT2000. IMT2000 is the collective name for 3G systems consisting of a family of compatible standards. The following lists the principal characteristics of these standards:

- Worldwide use
- Employed for all mobile applications
- Supports both packet-switched and circuit-switched data
- Offers data rates up to 2 Mbits/s
- Displays excellent spectral efficiency

IMT2000 means "International Mobile Telecommunications," and 2000 carries two meanings. First, the year 2000 was the initial year for IMT2000 (3G) trials, and then 2000 stands for the 2000-MHz frequency range.

The most important IMT2000 development is that UMTS (W-CDMA) is to be the successor to GSM. Meanwhile, CDMA2000 is to be the interim IS-95 successor. Then time division-synchronous CDMA (TD-SCDMA) universal wireless communication-136 ([UWC-136]/EDGE) provides TDMA enhancements to D-AMPS/GSM, all leading toward the ultimate goal of IMT-2000.

12   UNIVERSAL MOBILE TELECOMMUNICATIONS SYSTEM (UMTS)

12.1   Introduction

UMTS is being developed by the Third-Generation Partnership Project (3GPP), a joint venture of a number of European and North American entities such as ETSI, ANSI, and Association of Radio Industries and Business/Telecommunication Technology Committee, among others. The 3GPP releases UMTS standards in phases and annual releases.

One of the most significant changes was brought about in Release 1999 for the new UMTS terrestrial radio access (UTRA), a W-CDMA radio interface for land-based communications. UTRA supports time division duplex (TDD) and frequency division duplex (FDD). The first mode, TDD, is designed for the micro-cell and pico-cell operations and for cordless telephones. The FDD mode is optimized for wide-area coverage (i.e., public micro- and macro-cells). Both modes offer data rates up to 2 Mbits/s. There is now a newly defined UTRA mode that is called *multi-carrier* (MC) which provides compatibility between UMTS and CDMA2000.

12.2   Architecture of a UMTS Network

As we mentioned above, Release 1999 incorporates enhanced GSM phase 2+ Core Networks with GPRS and CAMEL. This provides UMTS with a new radio

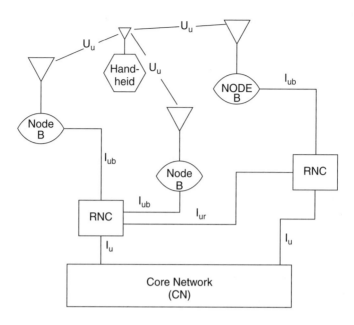

Figure 18.26. Functional block diagram of UMTS illustrating several of its interfaces. RNC, radio network controller; I, interface.

access network (RAN), and it is connected via the I_u to the GSM Phase 2+ core network (CN). The I_u is the UTRAN interface between RNC and the packet-switched domain of the CN (I_u-PS) and is used for PS data, and the UTRAN interface between RNC and the circuit-switched domain of the CN (I_u-CS) is used for CS data. Figure 18.26 is a functional block diagram of UMTS illustrating several of these interfaces.

There is also the problem of the "GSM-only" mobile facilities. Such facilities will be connected to the network via the GSM air (radio) interface (Um). UMTS/GSM dual mode user equipment (UE) will be connected to the network via UMTS air–radio interface (U_u) at data rates up to 2 Mbits/s. Outside the UMTS service area, UMTS/GSM UE will be connected to the network at data rates something less than 2 Mbits/s. Outside of the UMTS service area, UMTS dual mode user equipment (Ue) will be connected to the network at reduced data rates via the Um.

The following are the maximum data rates that can be expected in the UMTS environment:

- Circuit-switched (CS) data by HSCSD (high-speed circuit-switched data): 115 kbits/s
- Packet-switched data by the GPRS (general packet radio services): 171 kbits/s

- Enhanced data rates for GSM evolution (EDGE): 553.6 kbits/s
- UMTS terrestrial radio access (UTRA): 1920 kbits/s

In UMTS Release 1999, there are three major categories of network elements described for the public land mobile network (PLMN):

- GSM Phase $\frac{1}{2}$ core network elements which include: mobile services switching center (MSC), visitor location register (VLR), home location register (HLR), authentication center (AC), and equipment identity register (EIR).
- GSM Phase 2+ enhancements: GGPRS (serving GPRS support node (SGSN) and gateway GPRS support node (GGGSN) and CAMEL (CAMEL service environment [CSE]).
- UMTS specific modifications and enhancements, particularly UTRAN.

Table 8.8 lists expected QoS parameter values for UMTs when it is fully implemented.

Source: Table 6.2, page 449, *Mobile Radio Networks*, 2nd ed., 2002 [40]. Also see Ref. 39.

NETWORK ELEMENTS FROM GSM PHASE $\frac{1}{2}$. GSM Phase $\frac{1}{2}$ consists of three subsystems:

- Base station subsystem (BSS)

TABLE 18.8   Expected QoS Parameters of UMTS

Service	Call Duration	Data Rate (kbits/s)	Residual Bit Error Ratio	Delay (ms)
Telephony				
Voice	2 min	8–32	10^{-4}	40
Teleconferencing	1 h	32–128	10^{-4}	40
Video telephony	2 min	64–384	10^{-7}	40–90
Video conferencing	1 h	384–768	10^{-7}	90
Message services				
SMS and paging	cl	1.2–9.6 (1.2–2.4 type)	10^{-6}	100
Voice mail	2 min	8–32	10^{-4}	90
Facsimile mail	1 min	32–64	10^{-6}	90
Video mail	tbd	64	10^{-7}	90
e-mail	cl	1.2–64	10^{-6}	100
Distribution services	tbd	1.2–9.6 (2.4 type)	10^{-6}	100
Database use	tbd	2.4–768	10^{-6}	200+
Teleshopping	tbd	2.4–768	10^{-6}	90
Electronic mail	tbd	2.4–2000	10^{-6}	200
Message distribution	cl	2.4–2000	10^{-6}	300
Tele-action services	tbd	1.2–64	10^{-6}	100–200

tbd, to be defined; cl, connectionless.

- Network and switching subsystem (NSS)
- Operations support system (OSS)

The BSS consists of: base station controller (BSC), base transceiver station (BTS), and the transcoder and rate adapter unit (TRAU). The NSS consists of the MSC, VLR, HLR, EIR, and AC. The heart of the OSS are the operation and maintenance centers (OMCs) which carry out centralized and remote OA&M tasks.

GSM Phase 2+ Network Elements. The introduction of GPRS (general packet radio service) and CAMEL (customized application for mobile enhanced logic) brought GSM one step closer to the desired UMTS goal.

GPRS brings packet switching (PS) into the GSM core network (CN). It allows direct access to packet data networks and brings transmission speeds well beyond the ISDN 64 kbits/s up to the desired UMTS transmission rate of 2 Mbits/s. GPRS is seen as the necessary prerequisite to the introduction of UMTS.

Unique to UMTS is the *virtual home environment* (VHE) concept. CAMEL enables the incorporation of VHE through access to telco-specific intelligent network (IN) applications. Among these applications we would expect to find prepaid, call screening, and supervision. VHE may be seen as a collection of service creation tools that enable a telco to modify or enhance existing services or define new services. In addition, VHE permits worldwide access to these telco-specific services available in every GSM and UMTS public land mobile network (PLMN) as well as the introduction of location-based services through the interaction of the GSM/UMTS mobility management function. To successfully bring about these changes, the CN will require a new common control signaling system No. 7 CAMEL application part (CAP) and a CSE (CAMEL service environment).

12.3 Changes and Requirements for UMTS Phase 1

The major difference between GSM Phase 2+ and UMTS lies in the air interface. The access and modulation method used with UMTS is wideband CDMA (W-CDMA) instead of the TDMA employed by GSM. This involves a new radio access network (RAN) called UTRAN or UMTS terrestrial radio access network. One modification required is the allocation of the transcoder (TC) function for speech compression to the CN.

12.4 UMTS Network Elements

12.4.1 UTRAN (UMTS Terrestrial Access Network). UTRAN consists of a number of *Radio Network Subsystems* (RNSs), each of which is connected over the I_u interface (see Figure 18.26) to the transport network CN. An RNS consists of an RNC and one or more Node Bs. *Node B* is a logical unit responsible for radio transmission in one or more cells and communicates with the RNC over

reference interface point I_{ub}. A Node B on the user side only contains the physical layer of the U_u protocol stack.

UTRAN carries out numerous functions in the UMTS. Some of these functions are described below.

Admission Control. This prevents overloading of the radio network. Based on available interference and load measurements, it decides which radio resources should be reserved in the system. These include such things as access control for new connections, the reconfiguration of existing connections and the reservation of resources for macrodiversity and handover. The admission control function is embedded in the SRNS. The SRNS is described in Section 12.4.2 dealing with the RNC.

Congestion Control. This handles network overloads and is responsible for returning the system top a stable state in a way that is largely unnoticeable to the users.

Radio Channel Encryption. This is carried out by UTRAN to protect against the undesired decoding of signals on the radio interface. The encryption algorithm used is with a session-specific key. Air interface encryption is a function of UTRAN as well as UE (user equipment).

Handover. This is a function that implements mobility management at the air interface. The handover function is also used to ensure compliance with the QoS requested by the CN. The function can be controlled either by the network or from the UE. Thus, it can be embedded in both the SRNS and the UE. The handover function in an UTRAN network also allows connections to be switched to other networks or from other networks to the UTRAN. An example of this would be GSM/UMTS handover.

SRNS Relocation. This may be required when the role of a respective RNC might change during the course of the connection due to the mobility of the mobile station. Since end-to-end connections between user equipment (UE) and core network (CN) always only exist over the I_u reference point on the SRNS, the UTRAN must always change the I_u reference interface point if there is a change in the SRNS. The SNRS initiates the change of SRNS through functions in the RNC and the CN.

Radio Channel Measurements. For evaluating radio channel quality, the following parameters are measured by the UTRAN;

- The receive signal level (RSL) of serving and neighboring cells
- The measured or estimated BERs on connections to serving or neighboring cells
- The estimated distance to the serving BS
- Doppler shift

- Propagation conditions such as slow and fast fading
- Current level of synchronization
- Received interference power
- The total power received by each UE on the downlink

Because CDMA is used at the radio interface, interference signals can be differentiated from wanted signals. (Note that this cannot be done with GSM.) In CDMA systems the different physical channels are separated by codes and each code corresponds to a certain power. The measurement function is embedded both in the UE and in the UTRAN.

Macro-diversity is a function that allows data streams to be duplicated and sent simultaneously over various physical channels in different cells to a terminal. In the reverse, macro-diversity also enables the data stream sent by a mobile station to more than one base station to be received and merged again. The data stream can be retried both in the *Serving Radio Network Controller* (SRNC) and in the *Drift Radio Network Controller* (DRNC), or even in Node B. The macro-diversity function is used only in the frequency division duplex mode (FDD).

Radio Resource Management. The allocation and release of radio resources are functions of the RNC. This function is necessary when resources for macro-diversity or for improvement of QoS of bearer services are needed. In the time division duplex (TDD) mode, the switching point between the uplink and the downlink can be varied to enable the transport of asymmetrical traffic loads. This can cause interference between all base stations and mobile stations in a system. Such is only preventable through dynamic channel allocation. Interference in a system can also be controlled through what is called *fast dynamic channel allocation*, which controls time-slot capacity and therefore the number of simultaneously activated codes. As can readily be seen, this function is also related to the function of access control. Dynamic channel allocation is not mandatory. TDD networks with fixed channel allocation (FCA) will have to use the same switching point in all cells.

Data Transmission over the Radio Interface. A UTRAN network transmits user and control data over the radio interface. This requires the mapping of the bearer services onto that radio interface. This functionality includes segmentation and reassembly of messages as well as acknowledged and unacknowledged transmission.

Power Control of a Transmitter. This reduces interference and maintains the prescribed QoS on connection quality. There are two types of power control in a UTRAN: one with two interleaved control loops and one with an open control loop. Open control loops are used on both the uplink and the downlink to determine transmitter power (for example, for random access). Base stations (BSs) are mobile-controlled, whereas mobile stations either are BS-controlled or use system parameters sent by the BSs to determine the appropriate transmit power. It should be noted that power control with interleaved control loops requires a signaling connection between the UE (user equipment) and the UTRAN.

Random Access. This is a function that is used to detect and handle initial instances of random access by UEs. The random access function has the additional responsibility for resolution of collisions because of the use of a slotted Aloha protocol.

Channel Coding. Systematic redundancy is added to data streams for the protection of data transmission. The type and code rate can vary for the different logical channels and for the different bearer services.

Radio Bearer Control. This is a function used in the reconfiguration of radio bearer services and in the provision or release of radio bearer services for call set up and termination as well as for handover.

12.4.2   Radio Network Controller (RNC). The RNC enables autonomous radio resource management (RRM) by UTRAN. It performs the same functions as the GSM BSC. It provides central control for the RNS (radio network subsystem) such as RBC and Node Bs.

As we can see in Figure 18.26, the RNC handles protocol exchanges between I_u, I_{ur}, and I_{ub} interfaces. One should note that these interfaces use ATM transmission. The RNC is responsible for centralized operation and maintenance (OAM) of the entire RNS with access to the OSS (operation subsystem). Because these interfaces are ATM-based, the RNC switches ATM cells between them. The user's circuit-switched and packet-switched data coming from I_u–CS and I_u–PS interfaces are multiplexed together for multimedia transmission via I_{ur}, I_{ub}, and U_u interfaces to and from the UE.

The RNC employs the I_{ur} interface to autonomously handle all of the RRM, eliminating that burden from the core network (CN). A single serving RNC (SRNC) provides serving control functions such as admission, RRC connection to the UE, congestion, and handover/macro-diversity.

When another RNC is involved in an active connection by means of an inter-RNC soft handover, it is called a drift RNC (DRNC). A DRNC is only responsible for the allocation of code resources. It is also possible for a reallocation of the SRNC functionality to the former DRNC. This is called *serving radio network subsystem (SRNS) relocation.* An RNC that controls logical resources of its UTRAN access points is termed a *controlling RNC (CRNC).*

It should be noted here that UMTS defines four new open interfaces:

- U_u: UE to Node B. This is UTRA, the UMTS W-CDMA air interface
- I_u: RNC to GSM Phase 2+ CN interface (MSC/VLR or SGSN)
  - I_u-CS for circuit-switched data
  - I_u-PS for packet-switched data

- I_{ub}: RNC to Node B interface
- I_{ur}: RNC to RNC interface. It is not comparable to any interface in GSM.

See Figure 18.26 for further clarification.

12.4.3   Node B.

As shown in Figure 18.26, Node B is a physical unit for radio transmission and reception with cells. Depending on the sectoring used, one or more cells may be served by a Node B. Both FDD and TDD modes can be supported by Node B. It can also be co-located with a GSM BTS to reduce implementation costs. Node B is the ATM termination point. Its connection is with the UE via the W-CDMA U_u radio interface and with the RNC via the I_{ub} ATM-based interface.

Node B has a number of tasks. The main one is the conversion of data to and from the U_u radio interface. This includes forward error correction, rate adaptation, W-CDMA spreading/dispreading, and QPSK modulation at the air interface. It measures signal quality and signal strength at the connection, and from that information it determines frame error rate (FER), transmitting this information to the RNC as a measurement report for handover and macro-diversity combining. Node B is also responsible for the FDD soft handover. The micro-diversity combining is carried out independently, eliminating the need for additional transmission capacity in the I_{ub}.

Node B also supports power control. It enables the UE to adjust its power using downlink transmission power control (TPC) commands via the inner-loop power control on the basis of uplink TPC information. There are predefined values of inner-loop power control that are derived from the RNC by means of the outer-loop power control.

Source: Section 12 is based on Refs. 37, 39 and 40.

13   WIRELESS ACCESS PROTOCOL (WAP)

WAP is an open protocol for wireless multimedia messaging. WAP allows the design of advanced, interactive, and real-time mobile services, such as mobile banking or internet-based news and travel services.

Internet protocol is not designed to operate efficiently over mobile networks. Standard HTML web content cannot be displayed fully on the small-size screens of wireless devices, pagers, and mobile telephones. WAP was designed to address these issues.

13.1   Wireless Markup Language (WML) and WAP Proxy

Our present method of accessing the internet is through a server. This is done through interactions using HTTP (hypertext transfer protocol) request and reply messages. This assumes minimal restrictions on storage and processing power.

Conventional wireless devices have distinct limitations in this regard. WAP is based on a technology which has been optimized to address the constraints of wireless links and devices. The services provided through HTML (hypertext markup language) do not work well on hand-held devices with small displays and limited computing capability. As a result, WML (wireless markup language) was developed. An additional advantage of WML is that it can be encoded in a binary format reducing the amount of data required for transmission over a wireless interface.

The actual protocol conversion and data formatting is carried out by a WAP proxy. It acts as an interface between the wireless and wired environments. There are notable differences between the two. One we mentioned is bit rate capacity. Others are BER requirements, storage, and processing capabilities. The wireless device has a WAP interface; a network internet server has an HTTP interface. The WAP proxy carries out an interface conversion between the two. Figure 18.27 graphically illustrates how a WAP proxy serves to provide the intervening interface between a client, a wireless device, and a web server. The WAP proxy resides in the WAP gateway [38].

13.2 Stability Issues

When compared to the conventional wired network, the wireless network may be comparatively unstable. Connectivity may not be available when required. Fading and signal dropouts can be fairly common. In the design of the WAP protocol, stability issues have been taken into account. For instance, we can expect the wireless device in question to have a lower bit rate capacity than the web network and its associated IT equipment.

The data transmission rate for a wireless network is around 9.6 kbits/s, and the resulting download time for one page based on these values is around 17 s. We have not made any allowance here for network congestion nor for latency. The

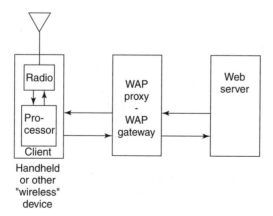

Figure 18.27. The WAP proxy resides in the WAP gateway, which interfaces the web server and the wireless device.

way the WAP handles these situations is by minimizing the data traffic over that interface of interest. One way of doing this is by using WML and WMLScript, which are binary-coded onto a compact form before they are transmitted.

A wired network is much more stable when compared to its wireless counterpart. This is especially true when the wireless device is in motion. Fading and dropouts result. Then there is Doppler shift.

The small display on the wireless device presents another problem. Instead of using the flat document structure that HTML provides, WML structures its document to be displayed in *decks* and *cards*. A "card" is a single unit of interaction with the end-user, such as a selection list, a text screen, an input field, or a combination of those. A card is typically small enough to be displayed even on a very small screen. For the small screen such as found on wireless devices, the user navigates through a series of cards. The series of cards used for making an application are collected in a deck.

Small, hand-held devices that are commonly found in a wireless environment have limited memory capacity and computational power when compared to desktop computers. WAP handles this situation by employing a lightweight protocol stack. The limited set of functionalities provided by WML and WMLScript makes it possible to implement browsers that require only small amounts of computational power and ROM resources. When it comes to RAM, the binary encoding of WML and WMLScript helps to keep the amount of RAM used as small as possible.

WAP is a layered protocol design and is stack-independent of the underlying network, which then could take the form of GSM, CDMA, CDPD, and iDEN. Hence, WAP is essentially an application stack specification; it is not network-centric.

Source: Section 13 is based on Ref. 38.

REVIEW QUESTIONS

1. How does a mobile terminal make connectivity to the PSTN in conventional cellular operations?

2. What are typical radiated power levels for a cell site, mobile unit, and hand-held terminal? Reference here is to the North American AMPS system.

3. In North America, what is the assigned bandwidth for a cellular operator outbound? (Of course, the same bandwidth in another part of the spectrum would be assigned inbound.)

4. Cellular radio service has seen explosive growth. What will a cellular operator do when a group of cell sites are overloaded with traffic?

5. Name at least four ways that cellular radio differs from LOS microwave regarding propagation.

6. Describe and discuss *locating* and *handover*.

7. List and discuss actions that go on over the setup channel.

8. What sort of a dispersion value might be encountered on a cellular link? Admittedly, this value approaches a worst-case value. What type of operation will such dispersion affect?

9. Path loss on LOS microwave was a square (2-power) (i.e., 20 log D) relationship. What is the relationship for cellular?

10. The amended CCIR equation for transmission loss for cellular radio takes into account three additional factors further adding to the loss. What are they?

11. The British urban path loss formula included the 40 log d factor and a factor to compensate for antenna height–gain. What other factors did it take into account, including a constant?

12. Okumura, a famous researcher on urban path loss for cellular systems, developed a method, applied by HATA, to calculate path loss. In this case, path loss (transmission loss) varied with what factors?

13. What were the basic factors involved in the equation for building penetration loss?

14. For microcells in heavily urbanized areas, what is the major contributor to excess loss (above line of sight)?

15. Explain how dispersion can corrupt a digital signal and show that this corruption, showing up as ISI (intersymbol interference), is a function of the bit rate. Use 10 μs as the value for dispersion.

16. Discuss the three fading models given in the text in view of the function K.

17. Why is space diversity favored over frequency diversity, particularly with cellular operations?

18. Discuss space-diversity antenna separation (1) for a cell site and (2) for a mobile platform. The example should be for the 900-MHz band.

19. Space-diversity antenna separation not only varies as a function of the correlation coefficient but also as a function of _____ _____.

20. Where would we use an interleaver and how does it work?

21. Often cells and cell site transmitter–antenna parameters are built around a signal level contour. What is the value of this contour in dBμ? Also give this dBμ value in dBm based on a 50 − Ω impedance at 850 MHz.

22. Convert 1-dBd antenna gain to its equivalent in dBi.

23. Discuss the cellular radio bandwidth dilemma. What is the basic problem anyway?

24. In three sentences, describe how a sub-band coder operates.

25. Describe how a CELP coder works. Down to what bit rate can we expect from a CELP coder without sacrificing too much voice quality/intelligibility?

26. We described three different cellular radio system access techniques. What were they? In two sentences each, describe how they work.

27. TDMA is a leading contender as a digital access technique. Describe how it operates and especially the problem of delay between a nearby user and another user at cell boundary. Describe one way of mitigating that problem.

28. How many time slots does the GSM TDMA have?

29. Based on North American TDMA (IS 54), and comparing it to AMPS, what is the bandwidth utilization improvement with this type of TDMA?

30. Give at least three advantages of using CDMA as a cellular access technique.

31. What is the processing gain of a direct sequence CDMA system with an information rate of 5 kbits/s and a spread bandwidth of 5 MHz?

32. Frequency reuse is now vital in cellular radio operations. In large cellular systems, particularly in urban areas, we have to expect a certain level of interference. What value C/I or better is our goal?

33. In a typical hexagonal cell, from how many directions does interference enter? How can we cut down on these interference levels to really just one direction?

34. When we opt for cell splitting to increase system capacity, what would be the minimum practical cell size?

35. Define a paging system (i.e., What does it do?).

36. Paging system capacity is affected by 10 parameters. List five of them.

37. To improve performance and robustness, almost all paging systems use _____.

38. By approximately how much (dB) does a signal level decrease when turning a corner (urban area)?

39. Discuss the advantages and disadvantage(s) of a LEO system compared to a GEO satellite communication system.

40. UMTS evolved, is an outgrowth of which popular cellular access protocol?

41. At the air interface for UMTS, which type of modulation waveform is used?

42. What is the principal purpose of the WAP (protocol)?

REFERENCES

1. R. Steele, ed., *Mobile Radio Communications*, IEEE Press, New York and Pentech Press, London, 1992.

2. D. Bodson et al., eds, *Land Mobile Communications Engineering*, IEEE Press, New York, 1983.

3. J. D. Parsons and J. G. Gardiner, *Mobile Communication Systems*, Blackie, London and Halsted Press, New York, 1989.

4. J.-P. Linnartz, *Narrowband Land-Mobile Radio Networks*, Artech House, Norwood, MA, 1993.

5. R. C. V. Macario, ed., *Personal and Mobile Radio Systems*, IEE/Peter Peregrinus, London, 1991.

6. W. C. Y. Lee, *Mobile Communications Design Fundamentals*, 2nd ed., John Wiley & Sons, New York, 1993.

7. A. Jagoda and M. de Villepin, *Mobile Communications*, John Wiley & Sons, Chichester, UK, 1991/1992.

8. K. Pahlavan and A. H. Levesque, *Wireless Information Networks*, John Wiley & Sons, New York, 1995.

9. W. W. Wu et al., "*Mobile Satellite Communications*," *Proc. IEEE*, **82** (9), (September 1994).

10. W. F. Fuhrmann and V. Brass, "Performance Aspects of the GSM System," *Proc. IEEE*, **89** (9), (September 1984).

11. H. L. Bertoni et al., "UHF Propagation Prediction for Wireless Personal Communication," *Proc. IEEE*, **89** (9) (September 1994).

12. *VHF and UHF Propagation Curves for the Frequency Range from 30 MHz to 1000 MHz*, CCIR Rec. 370-5, XVIIth Plenary Assembly, Dusseldorf, 1990.

13. K. Allesbrook and J. D. Parsons, "Mobile Radio Propagation in British Cities at Frequencies in the VHF and UHF Bands," *Proc. IEE*, 1977, **124** (2).

14. Y. Okumura et al., "Field Strength and Its Variability in VHF and UHF Land Mobile Service," *Rev. Electrical Commun. Lab. (Tokyo)*, **16** (1968).

15. M. Hata, "Empirical Formula for Propagation Loss in Land-Mobile Radio Services," *IEEE Trans. VT*, **VT-20** (1980).

16. F. C. Owen and C. D. Pudney, *In-Building Propagation at 900 MHz and 1650 MHz for Digital Cordless Telephones*, 6th International Conference on Antennas and Propagation, ICCAP '89, Pt. 2: Propagation, Conference Publication No. 301, 1989.

17. M. R. Schroeder and B. S. Atal, *Code-Excited Linear Prediction, High Quality Speech at Low Bit Rates*, IEEE Proceedings, ICASSP, 1985.

18. R. C. Dixon, *Spread Spectrum Systems with Commercial Applications*, 3rd ed., John Wiley & Sons, New York, 1994.

19. *Recommended Minimum Standards for 800-MHz Cellular Subscriber Units*, EIA Interim Standard EIA/IS-19B.

20. "Cellular Radio Systems," a seminar given at the University of Wisconsin—Madison by Andrew H. Lamothe, Consultant, Leesburg, VA, 1993.

21. Telecommunications Transmission Engineering, 2nd ed, Vol. 2, Bellcore, Piscataway, NJ 1992.

22. *Radio-Paging Systems*, CCIR Rep. 900-2, Vol. VIII-1, XVIIth Plenary Assembly, Dusseldorf, 1990.

23. *Radio-Paging Systems*, CCIR Rep. 499-5, Vol. VIII.1, XVIIth Plenary Assembly, Dusseldorf, 1990.

24. *Digital Cellular Public Land Mobile Telecommunication Systems (DCPLMTS)*, CCIR Rep. 1156, Vol. VIII.1, XVIIth Plenary Assembly, Dusseldorf, 1990.

25. M. Engelson and J. Hebert, "Effective Characterization of CDMA Signals," *Wireless Rep.* (January 1995).

26. Donald C. Cox, "Wireless Personal Communications: What Is It?," *IEEE Personal Communi.*, **2** (2) (April 1995).

27. *Fixed Wireless Access, Handbook on Land Mobile (including Wireless Access)*, Vol. 1, 2nd ed., International Telecommunication Union, Radiocommunication Bureau.

28. A. Dorman, "Are We Better Off without 3G?" *Network Mag.* (June 2003).

29. R. L. Freeman, *Radio System Design for Telecommunications*, 2nd ed., John Wiley, New York, 1997.

30. *Propagation by Diffraction*, ITU-R Rec. P.526-5, ITU-R Volume 1997, P Series, Part 1. (*Note*: We could not find subsequent editions in our 2002 P.Series in CD-ROM.)

31. A. G. Longley, and P. L. Rice, *Prediction of Tropospheric Radio Transmission Loss over Irregular Terrain—A Computer Method*, ESSA Technical Report ERL-79, National Bureau of Standards (now NIST), Washington, DC, 1968.

32. G. A. Hufford, A. G. Longley, and W. A. Kissick, *A Guide to the Use of the ITS Irregular Terrain Model in the Area Prediction Mode*, NTIA Rep. 82–100, NTIA, Washington, DC, April 1982.

33. *Handbook, Terrestrial Land-Mobile Radiowave Propagation in the VHF/UHF Bands*, ITU Radiocommunications Bureau, Geneva, 2002.

34. *Propagation Data and Prediction Models for the Planning of Indoor Radiocommunication Systems and Radio Local Area Networks in the Frequency Range 300 MHz to 100 GHz*, ITU-R Rec. P.1238, Vol. 1997, Part 2, ITU, Geneva, 1997 (does not appear in 2002 CD-ROM).

35. *Propagation Data and Prediction Methods for the Planning of Short Range Outdoor Radiocommunication Systems and Local Area Networks in the Frequency Range of 300 MHz to 100 GHz*, ITU-R Rec. P. 1411, ITU, Geneva, October 15, 1999.

36. *Handbook Mobile Satellite Service (MSS)*, Radiocommunication Bureau, edition 2002, ITU-R, Geneva, 2002.

37. *Tektronix—UMTS Protocol and Protocol Testing*, Web Proforum Tutorials, International Engineering Consortium, www.iec.org.

38. I. Lee, "Wireless Access Protocol (WAP) Architecture," white paper from the web. www.crystal.uta.edu/~kumar/cse6392/termpapers/ihlee_paper.pdf.

39. F. Hillebrand, ed., *GSM and UMTS*, John Wiley & Sons, Chichester, UK, 2002.

40. B. Walke, *Mobile Radio Networks*, 2nd ed, John Wiley & Sons, Chichester, UK, 2002.

19

LAST-MILE BROADBAND CONNECTIVITY AND WIRELESS LOCAL LOOP (WLL)

1 BACKGROUND AND CHAPTER OBJECTIVE

In this chapter we discuss methods of bridging the last mile or first mile, depending on how you look at it. We see this as the principal bottleneck in bit rate transmission of a telecommunication network. In order of increasing transmission rate capacity, we list as follows:

1. Wire pair. As found in the traditional telephone plant
2. Wire pair. In one of several possible DSL configurations
3. Digital loop carrier (DLC)
4. Broadband microwave/millimeter wave, point-to-multipoint downstream (e.g., MMDS, LMDS), point-to-point upstream
5. CATV and CATV modified
6. Fiber optics

Narrow-band radio may also be considered a candidate. It is only cost effective in a rural situation where there is low telephone user density. In this case the loops will usually be very long, possibly over 30 miles. Such loops can only carry voice and comparatively low rate data (e.g., 4800 to 14,400 bits/s).

With fiber-optic cable we can transport great quantities of data anywhere on the earth's surface or even undersea. This can be done in a cost-effective manner so long as the endpoints and/or the tributaries generate sufficient revenue to make such connectivities profitable for the investor. One of the principal

Telecommunication System Engineering, by Roger L. Freeman
ISBN 0-471-45133-9   Copyright © 2004 Roger L. Freeman

advantages of optical fiber is its almost infinite bandwidth. Unfortunately, in many cases it would be difficult to make a cost-effective case for optical fiber because of insufficient demand for service, especially for modest users such as single-family residences.

Even in our own personal case where our home has several high-speed modern PCs connected to the internet and three digital TV sets, we cannot justify a fiber termination in our residence. One way to get around this issue is to have homes and small businesses share or pool the bit rate capacity provided by a fiber terminal. Besides entertainment TV, many CATV operators offer data services, especially internet. This is a very apt example of sharing the large capacity of optical fiber among many users.

2   CONVENTIONAL WIRE PAIR IN THE LAST MILE

This very widely used technique for last-mile transport was described in Chapters 2 and 5 of this text. The context of this discussion was in the support of POTS* where the service was nearly exclusively voice. At some place in the network, beyond the subscriber-pair wire plant a filter would limit voice-channel bandwidth to the region of 300 to 3400 Hz. In the wire pair itself, bridged taps and load coils are also bandwidth limiting. As a result, the conventional wire pair in the subscriber plant can support data transmission up to about 56 kbits/s. This will also encompass typical facsimile (i.e., CCITT Group 3) because it is transmitted to the line as conventional voice-band data.

Wire pair is the most widely deployed transmission medium and may be found worldwide. It is commonly 26 gauge (0.4 mm) in subscriber plant and 24 gauge/22 gauge (0.51 mm/0.64 mm) in the local trunk wire plant that remains. This latter plant is being rapidly replaced by fiber-optic cable. A second important parameter of a wire pair is mutual capacitance which runs between 40 and 50 nf/km. Once load coils and bridged taps are removed from a subscriber loop, it can easily support 1.544 or 2.048-Mbits/s digital transmission but with certain length limitations (i.e., <12,000 ft or <3657 m).

3   WIRE PAIR EQUIPPED WITH DSL MODEMS

A digital subscriber line (DSL) modem allows better use of the available bandwidth of a subscriber line. It is assumed in every case that the wire line so equipped will be transporting a digital configuration of binary 1s and 0s. There are a number of approaches to the design of a DSL modem. Because of space limitations of this text, only four designs will be discussed here:

1. Asymmetric digital subscriber line (ADSL) [1–4].
2. High bit rate digital subscriber line (HDSL) [5].

---

* POTS, "plain old telephone service." A familiar expression used more in North America for conventional telephone service.

3. Rate-adaptive DSL (RADSL) [6].

4. Very high (bit rate) digital subscriber line (VDSL) [8].

Differentiating "Symmetric" and "Asymmetric." Symmetric refers to when the bit rate of a particular data connectivity is the same bit rate in either direction, upstream and downstream. Asymmetric refers to when the data rate is different upstream versus downstream. Subscriber demands on internet connections have essentially brought this about. There is a much greater traffic intensity on an internet connectivity downstream than upstream. We "download" much more than we "upload."

3.1 Asymmetric Digital Subscriber Line (ADSL)

An ADSL circuit provides three services over a single wire pair: a high-speed downstream channel, a medium-speed full duplex channel, and a basic POTS telephone channel. This POTS channel is split off in a FDM fashion from the other two channels by filters (ANSI design). The bit rate of the high-speed channel varies from 1.5 to 6.1 Mbits/s downstream, and the duplex rates are from 16 to 640 kbits/s. Both the latter two are distance-dependent and may be submultiplexed to form lower rate channels.

Different manufacturers offer different variants of the ADSL modem. All are designed around the North American DS1 (T1) and European E1 (2.048 Mbits/s) data rates downstream and at least 16-kbit/s duplex for the second channel as a minimum configuration. There are ADSL modems now off-the-shelf that offer up to 8 Mbits/s or better data rate downstream and up to 640 kbits/s upstream. Some modems can accommodate asynchronous transfer mode (ATM) with variable rates and can compensate for ATM overhead, as well as internet protocols (IPs).

Those all-important downstream data rates vary with loop length, wire gauge, presence of bridged taps, and load coils that have mistakenly not been removed. Just the presence of one load coil will make the loop inoperable for the two data channels. Remember, as gauge number increases (e.g., 24 to 26), loop resistance and loss increase accordingly (see Chapter 3). Table 19.1 gives loop performance based on data rate multiples of DS1 and E1 as well as duplex secondary channels transmission rates. Table 19.2 gives claimed ADSL performance with gauges stated and maximum loop lengths. Figure 19.1 is a simplified functional block diagram of an ADSL terminal at the network end. The remote ADSL terminal at the user side of the loop is a mirror image of this terminal shown in Figure 19.1.

It should be noted that there is a second ADSL configuration that uses an echo control device on the telephone channel, thereby allowing the low-speed upstream and downstream channels to overlap. The overlap region has send and receive separation by means of a local echo cancellation device.

By removing the FDM filter, group delay is reduced, thereby allowing a somewhat greater transmission rate.

The ANSI ADSL modem uses FEC to improve error performance on the loop. The total aggregate downstream data flow is multiplexed together and formed

TABLE 19.1   ADSL Loop Performance as Specified by ANSI

Downstream Bearer Channels $n \times 1.536$ Mbits/s (T1-related)	Data Rate in Mbits/s	Duplex Bearer Channels in kbits/s
	1.536	16
	3.072	64
	4.608	Optional channels 160 bits/s
	6.144	384
$n \times 2.048$ Mbits/s	2.048	544
	4.096	576

Source: Table 19.1 based on Refs. 1 and 2.

TABLE 19.2   Claimed ADSL Performance with Physical Media Limitations Stated

Data Rate (Mbits/s)	Wire Gauge (AWG)	Distance (ft)	Wire Size (mm)	Distance (km)
1.5 or 2	24	18,000	0.5	5.5
1.5 or 2	26	15,000	0.4	4.6
6.1	24	12,000	0.5	3.7
6.1	26	9,000	0.4	2.7

Source: Based on Ref. 1, Table 15-1.

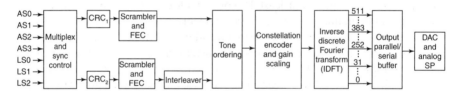

CRC = cyclic redundancy check
SP = signal processing
ASO = Three downstream simplex subchannel
    designators
LSO = Two duplex subchannel designators
DAC = digital-to-analog converter

Figure 19.1. Block diagram of ADSL ANSI modem. Based on Figure 4.20, page 258, Ref. 3. Reprinted with permission.

up into blocks. Subsequently the ADSL modem attaches an error correction code to each block. The far-end receiver then corrects errors that occur during transmission up to the limits dictated by the code and block length. At the user's option, the ADSL unit can also create superblocks by interleaving data within subblocks. This allows the receiver to correct any combination of errors within a specific span of bits.

3.1.1 ADSL Modulation Types. There are two different modulation waveforms that may be employed in an ADSL configuration:

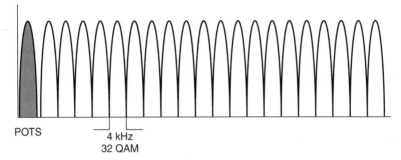

Figure 19.2. Example of discrete multitone (DMT) line code as used in ADSL. POTS = "plain old telephone service" (i.e., telephone channel); QAM = quadrature amplitude modulation.

1. Discrete multi-tone (DMT)
2. Carrierless amplitude/phase (CAP)

DMT separates a DSL signal into 256 individual channels each 4.3125 kHz wide, and each channel has a tone that is M-QAM-modulated. The default value of M is 32. Note that the value "32" allows a theoretical bit packing of 5 bits per hertz ($2^5 = 32$). The DMT uses the fast Fourier transform (FFT) for modulation and demodulation. A sample ADSL DMT waveform is shown in Figure 19.2.

Carrierless amplitude/phase (CAP) is based directly on M-QAM, but here the carrier is suppressed before transmission. (QAM and M-QAM are discussed in Chapter 7, Section 4.4.2.)

CAP Description, Step-by-Step. First the message signal is modulated by a carrier signal and stored in memory. Next, pieces of this modulated signal are reassembled and passed through a band-shaping filter before being transmitted. The band-shaping filter actually imposes a carrier on this assembled signal, converting it into a modulated wave. The advantage of CAP is that it has a lower peak-to-average signal power ratio relative to the competing DMT. As a result, the end-user equipment requires less power than DMT. CAP also tests the quality of the access line before transmitting and implements the most efficient version of QAM so as to optimize performance during transmission.

It seems to boil down to the fact that DMT is more complex when compared to CAP and gives better performance; CAP has a simpler design and is more economic than DMT. DMT is backed by the ANSI/ATIS specification [2], whereas CAP is a design approach of several manufacturers of ADSL equipment.

3.2 High-Bit-Rate Digital Subscriber Line (HDSL)

This is one of the earliest forms of DSL where data rates of 1.544 or 2.048 Mbits/s are transmitted in both directions, meaning that it is symmetrical. HDSL uses two wire pairs, one serving in each direction. The line coding used is 2B1Q (see Section 6.2.6 of Chapter 14).

3.3   Rate-Adaptive DSL (RADSL)

RADSL is a proprietary approach to high-rate DSL developed by Westell. In this design, software determines the data rate at which signals can be transmitted on a given customer wire pair, and that rate is adjustable accordingly. Westell's FlexCap2 uses RADSL to deliver from 640 kbits/s to 2.2 Mbits/s downstream and from 272 kbits/s to 1.088 Mbits/s upstream over an existing wire-pair line.

3.4   Very High Rate DSL (VDSL)

VDSL is a method to transmit multimegabit data over comparatively short distances on wire pair. It is the highest data rate technology over wire pair available today, with a maximum data rate of 52 Mbits/s. It has simpler implementation requirements than ADSL and may be either asymmetric or symmetric. VDSL is supported by the ANSI working group T1E1.4. The trade-off to reach these very high transmission rates (i.e., 52 Mbits/s) is loop length. The VDSL spectrum is specified to range from 200 kHz to 30 MHz. However, Figure 19.3 shows the high-speed downstream channel only going out to 8.0 MHz. Table 19.3 gives downstream line rates for asymmetric VDSL services.

VDSL also provides symmetrical data rate services. It can be utilized, for example, to provide short haul replacements for T1 in an $n \times$ T1 configuration and can provide symmetrical service all the way up to T3 (DS3) rate or 44.736 Mbits/s over a twisted wire pair.

3.5   The DSLAM (Digital Subscriber Line Access Multiplexer)

A DSLAM is a network device, usually housed at the telephone company's local switch, that receives signals from multiple customer DSL connections and inserts the resulting aggregate on a high-speed backbone line using multiplexing techniques. Depending on the product, DSLAM multiplexers connect DSL lines with some combination of asynchronous transfer mode (ATM), frame relay, or IP networks. DSLAM enables a telephone company to offer business or home residence users among some of the fastest phone line technology (DSL) with the fastest backbone network technology available. Only CATV using HFC may offer even faster service.

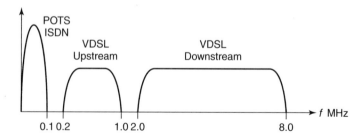

Figure 19.3. An artist's rendition of single-carrier VDSL asymmetric spectral allocation.

TABLE 19.3   Downstream Line Rates for Asymmetric VDSL Services[a]

Typical Service Range	Bit Rate (Mbits/s)	Symbol Rate (Mbaud)	Comments
Short range, 1 kft	51.84	12.96	Baseline
	38.88	12.96	
	29.16	9.72	Optional
	25.92	12.96	
Medium range, 3 kft	25.92	6.48	Baseline
	22.68	5.67	
	19.44	6.48	
	19.44	4.86	Optional
	16.20	4.05	
	14.58	4.86	
	12.96	6.48	
Long range, 4.5 kft	12.96	3.24	Baseline
	9.72	3.24	Optional
	6.48	3.24	

[a]Refer to ANSI T1E1.4 committee [20]. Note that these downstream data rates derive from submultiples of SONET rates.

4   DIGITAL LOOP CARRIER (DLC)

Digital loop carrier (DLC) has evolved from subscriber carrier, which has a long history in the telephone plant. Subscriber carrier was one way we could make two loops out of one, then three or four or more, by stacking FDM carriers one on top of the other where each carrier provided a usable bandwidth of at least 4 kHz [7].

In the 1980s, next-generation digital loop carriers (NGDLC) began to appear. They solved two problems. First, new services could be offered the customer through NGDLC. Second, it provided a cost-effective solution for rural and suburban applications.

The original goal of NGDLC was to deliver ISDN to every customer. This never materialized, but the underlying system was good at delivering high-bit-rate services such as DS1 (T1), E1, DDS (digital data service), and actual ISDN. On the trunk side the system was designed for fiber connectivity to the local serving exchange. On the line side either fiber, copper, or both might be expected. For the case of optical fiber, an ideal concentrator evolved for up to 2000 customers. The resulting equipment also displayed remote software provisioning. One thing that really added to the cost-effectiveness of this technology was the addition of a time-slot interchanger. What held this key advantage back was the lack of a system of customer billing in the local device. Yet the principal advantage was that the NGDLC architecture could be overlaid on the existing plant design with minimum a of modifications and cost outlay.

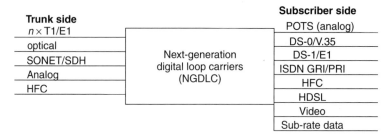

Figure 19.4. Some of the services that can be offered to rural and suburban customers by the incorporation of NGDLC.

Figure 19.4 graphically represents the many services that can be accommodated by an NGDLC device or devices.

5   BROADBAND MICROWAVE/MILLIMETER WAVE LAST-MILE TRANSMISSION

There are two broad groupings of service encompassed here based on U.S. FCC (and later Radio Regulations) nomenclature. These are (1) MMDS (multichannel multipoint distribution system) and (2) LMDS (local multipoint distribution system). The former operates in the 5-GHz band, and the latter operates in the 30- or 40-GHz bands.

5.1   Multichannel Multipoint Distribution Service (MMDS)

This service is covered by CFR 47, FCC "rules and regulations" Part 21, "Domestic Public Fixed Radio Services." The FCC (Part 21.2, Definitions) restricts service to the band 2596 MHz to 2644 MHz and associated 125-kHz channels. On the other hand, Multipoint Distribution Service (MDS), Part 21.901 allows somewhat broader use of the 2.1- to 2.6-GHz spectrum, noncontinuous in segments of 6 MHz. Typically, these bands are 2150–2162 MHz, 2596–2644 MHz, 2650–2656 MHz, 2662–2668 MHz, 2674–2680 MHz, and 2686–2690 MHz, and they are available for assignment to fixed stations in this service.

The intent here is to provide a wireless (radio) alternative to wired CATV. The dearth of spectral space in the lower portions of the RF band (i.e., below 22 GHz) does not truly give much real competition to "wired" CATV. As we will see in Section 5.2, LMDS is a much different story.

MDS/MMDS main stations are allowed a maximum EIRP of +33 dBW (2000 watts). Part 21.904 assumes a bandwidth of 6 MHz, or one analog TV channel. Various types of digital modulation are also permitted. With MPEG2 compression and an average of 3.5 Mbits/s per channel, possibly up to 30 Mbits/s can then be accommodated with 64-QAM, which would allow for eight digital TV programs aggregate in 6 MHz of bandwidth; the total aggregate TV program stream can be quite impressive. Yet it still cannot compete head-to-head with an HFC-wired CATV system, with some transmitting better than 400 simultaneous channels. Rules for MDS response stations are covered in Part 21.909.

5.2 Local Multipoint Distribution System (LMDS)

LMDS is rich in bandwidth, with two license blocks assigned:

- Block A of 1150 MHz
  - 27,500–28,350 MHz
  - 29,100–29,250 MHz
  - 31,075–31,225 MHz
- Block B of 150 MHz
  - 31,000–31,075 MHz
  - 31,225–31,300 MHz

Source: Ref. 21, CFR 47, Part 101, Subpart L.

The original idea for LMDS was to be a radio-based rival to wire-based CATV. It was a one-way, downstream system, point-to-multipoint. The U.S. FCC encouraged the inventor to make it two-way and digital. Unlike MMDS/MDS, excess attenuation due to rainfall becomes a problem as propagation availability requirements increase.

It is a general rule in radio system engineering that we can neglect excess attenuation due to rainfall and atmospheric gaseous absorption when operating below 10 GHz. If we are to operate above 10 GHz, excess attenuation due to rainfall can be a major limiting factor on range given some reasonable availability objective.

The present basic LMDS architecture is point-to-multipoint downstream and point-to-point upstream. It is also cellular to allow frequency reuse. Each cell has a nominal radius of from 1 to 4 km. Much of the required isolation from one cell to another for same-frequency operation derives from polarization isolation, up to some 35 dB.

5.2.1 LMDS Rationale. The high points in arguing for LMDS operation deal with costs and return on investment (ROI). For example, a large part of the radio network's cost is not incurred until the customer premises equipment (CPE) is installed. The network operator can time such capital expenditures to coincide with the signing of new subscribers. LMDS is one of the most cost-effective last mile solutions both for incumbent service providers and for competitive service providers to deliver services directly to end users. Benefits range over the following:

- Ease and speed of deployment.
- Fast realization of revenue stream because of rapid deployment.

- Lower entry and deployment costs (versus wired CATV).
- Demand-based buildout (extension) (services and coverage areas can be easily expanded as customer demand warrants).
- Noticeable cost-shift from fixed plant to variable components (i.e., CPE). CATV must invest in coaxial cable before customers can be taken aboard. Both LMDS and CATV have fixed fiber trunk costs.
- Less stranded capital with subscriber turnover.
- More cost-effective OA&M than its wired CATV counterpart.

Source: Ref. 10.

5.2.2 *LMDS Network Architecture.*

5.2.2 *LMDS Network Architecture.* Many LMDS network operators segment their network in 6-MHz frequency segments. This is more from legacy of the analog NTSC TV channel than anything else. We would continue to argue for the 3.5-Mbit/s single TV program stream using MPEG2 compression. If we were to use QPSK at a practical 1.2 bits per hertz, then nominally 3.5/1.2 (in MHz) could be rounded off to 3.0 MHz per TV channel. At 64-QAM downstream with 6 bits theoretical, 5 bits practical, a TV channel would be 3.5/5- or 700-kHz bandwidth.

Just as its wired CATV counterpart, LMDS can offer two-way voice, video and data to the end-user. However, much of the demand for bit rate capacity will be for downstream internet service. Enterprise VPN structures will also ride the internet both ways. One method to transport these basic services is to convert to ATM cells; another method could involve IP structures.

An LMDS network can be broken out into four major components:

1. Fiber infrastructure.
2. Base or cell station. This is where the fiber terminates and the signals are converted over to 30-GHz radio carriers (and vice versa).
3. Customer premise equipment (CPE).
4. Network operations control center (NOCC).

5.2.2.1 *Fiber Infrastructure.* The most efficient means to transport signals that are to be distributed by LMDS is to use optical fiber. Fiber-optic cable links are discussed in Chapter 7, Section 6, of this text. There are several familiar digital formats to choose from that can be transported by a fiber strand. Embedded in one or several of these formats will be voice, data, and/or video. The digital formats may be SONET/SDH structures, 802.3 frames, IP packets, DSX/E-format structures, ATM cells, or structures based on DOCSIS or DAVIC.* The format will vary with customer traffic-type demands and a most cost-effective

---

* DOCSIS, DAVIC. These are two transmission standards commonly found in the CATV plant. The first favors North America; the second favors Europe and locations under European hegemony. DOCSIS stands for Data over Cable Service Interface Specification; DAVIC stands for Digital Audio Video Council.

solution. Residences, multiple residences, and businesses will have different and varying requirements. Big buildings with numerous large businesses may have far different requirements than equally large buildings that are apartment houses. It may be attractive to the system designer in either case to consider point-to-point microwave where peak traffic density may justify such a move. In some cases the traffic load may be so heavy as to require direct fiber connectivity.

One can envision fiber-optic links connecting a national backbone fiber-optic network to LMDS distribution points. There are two overriding engineering considerations in the design and dimensioning of these fiber-optic links. These are:

1. A realistic and cost-effective growth policy.
2. Availability and survivability within cost constraints (better than 99.9%).

The growth policy should be based on "reasonable" forecasts. The policy can be carried out by one or two ways, or combinations thereof. The first option is to install active fiber and additional dark fiber strands for growth. The second option is a WDM plan where extra wavelengths can be added for growth. It should be kept in mind that it probably will take considerably more complexity than just adding additional optical sources and detectors. If regenerators are required, each optical lightwave will have to be broken down to its electrical equivalent and converted back to a light wave again. The add–drop multiplex plan becomes more complicated than with the dark fiber concept. As complexity increases, the cost of maintainability design increases and reliability decreases.

Implementing improved availability and survivability can become very costly indeed. How much does that additional "9" cost for availability improvement? The designer will turn to one of several ring architectures for that improvement. These ring architectures have been well standardized [6, 7]. The additional cost implication can be between 1.8 and 3 times initial, one fiber cable line expenditure. To achieve credible survivability, the ring cannot be folded back on itself. A new, separate route must be installed, preferably better than 20 miles from the first route. This separation is need in case of flood, forest fire, or earthquake where one life may be affected and the second line is unscathed.

There are two distinctly different architectures for connecting the backbone optical network to LMDS distribution points. These two options are illustrated in Figure 19.5.

In Figure 19.5A at the main-line optical hubs, the traffic is divided up into streams, one stream for each LMDS distribution point. Each stream is delivered by an optical strand. Another strand provides the connectivity in the other direction. If there are 20 LMDS distribution points, 20 + 20 optical strands will be required plus 20 + 20 for growth and breakdown spares.

Figure 19.5B carries out the same concept, but rather than two strands for each LMDS distribution point, we use two strands for the entire extension with an optical add–drop multiplexer at each LMDS distribution point plus growth/spares. One strand is for transmit downstream; the other strand routes traffic upstream. Each LMDS distribution point will have a separate upstream and a downstream wavelength assigned.

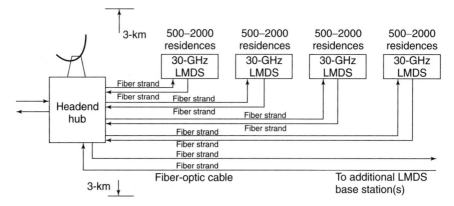

Figure 19.5A. Fiber-optic cable from mainline consists of two strands for each LMDS distribution point plus spares/growth.

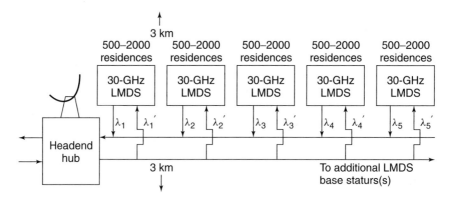

Figure 19.5B. Fiber-optic cable consists of two strands for entire extension plus spare/growth where each LMDS distribution point has a wavelength assigned in each direction.

5.2.2.2 LMDS Base Station (Distribution Point).

Figure 19.6 is a simplified functional block diagram of an LMDS base station. It consists of the fiber-optic facility, radio facility, and optical/electrical test access and patch panel. The two basic design options for the fiber-optic facility have been discussed above. There are a number of architectural alternatives for the design of the radio facility. The selection of an RF design will depend largely on traffic flow in each direction, its burstiness, symmetry, and expected growth.

The network node or base station is where the fiber cable terminates via an optical add–drop multiplex or through an optical patch panel arrangement. These concepts are illustrated in Figure 19.5.

Each LMDS base (cell) station may be fed by a fiber-optic pair from the hub, or the cells may be fed by wire means, coaxial cable, CAT-5 or CAT-6 wire

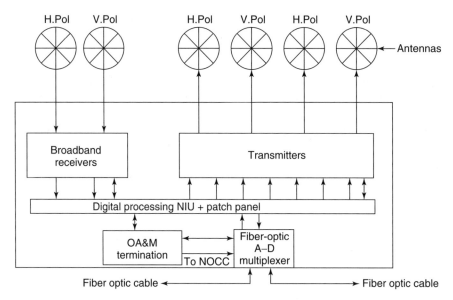

Figure 19.6. A simplified functional block diagram of an LMDS base station.

pair, or possibly some form of LOS microwave. Wide bandwidth, of course, is a major consideration. Because a cell may handle as much as 0.5 Gbits/s in the aggregate transmit and receive, the connecting transmission medium must be carefully selected such that the medium's bandwidth can handle the necessary bit rate capacity. Optical fiber is the easiest choice to deal with in this regard.

Most texts covering LMDS (and MMDS) break equipment out as outdoor and indoor. We have no quarrel with that. However, in the case of LMDS at 30 GHz, and particularly if the 40 GHz band is used, transmission lines at these frequencies are VERY lossy. To minimize these RF losses, certain design rules should be followed. Basically, make RF transmission lines just as short as possible. The receiver front ends should be incorporated in the antenna structure. If distributed receivers are used, an LNA (low noise amplifier) should be placed as close to the feed point as possible. It may even behoove the designer to incorporate a down-converter right after the LNA. By this means, we can nearly completely eliminate all transmission line loss and added thermal noise due to that line and associated elements. If the HPA uses the distributed technique, a similar course may be followed as the receiver.

Another design approach is to use a single broadband HPA (high-power amplifier) for all LMDS transmitted carriers. There may be a serious trade-off between back-off, to maintain linearity as best possible, and using some other approach, such as distributed HPAs. Patch antennas lend themselves to this type of operation and can be very cost-effective. A phase control device on each patch element for the base station allows for a "smart antenna" where beams can be formed and steered. Usually, a base station will need an array of patch antennas to be hemispherically omnidirectional.

When working at these higher frequencies (i.e., 30 GHz rather than 2.5 GHz of MMDS/MDS), our principal concern for a receiver is what sort of noise figure can one expect in off-the-shelf, production-run equipment. Certainly, 5 dB is a good approximation. The designer would then ask himself/herself, Does that 3-dB noise figure* look very good? What will be paid for that extra 2 dB? Is it really worth it in the link budget? It will certainly show up in the link margin. Do we want to "spend" the 2 dB in extending the link length? How much more can we extend it or how much better can we make the link time availability?

A similar exercise can be carried out for the transmitter HPA (high-power amplifier). Off the assembly line, production equipment can be fairly economic. Tailor-made equipment can be costly. If we double the power of an HPA, the link budget will "improve" by 3 dB. Cost will scale exponentially as power output increases linearly at these higher frequencies. The basic reason for this is that the power is generated by combining the output of two smaller amplifiers. The combiners are lossy. We take two steps ahead with power and a step and a half backward because of combiner losses.

5.2.2.3 Typical LMDS CPE. With an LMDS CPE, cost of the equipment is the driving factor in design. In an LMDS cell, one base station may serve 1000 CPE. Once we look at the economic factor in this light, we can see that the multiplier for cost is 1000 in the example, thus we can spend considerably more on the base station to save money on CPE.

As in the base station, the CPE will have outdoor and indoor equipment. The antenna can afford to be less expensive and directional. The equipment designer still has the problem of transmission line loss. Allowing or cost considerations, as much of the RF equipment will be antenna-mounted. Again, this is done to reduce transmission line losses. The indoor equipment will consist of baseband processing which in all probability will include FEC, a modulator and demodulator carrying the signal up to IF and down from IF to baseband. The outdoor–indoor interface will be at IF, often 70 MHz. If costs allow, space diversity may be a consideration. There will be a trade-off whether IF or baseband combining is used with the diversity subsystem. At 30 GHz, we can achieve space diversity in a single antenna structure because the required distance between the normal and diversity antennas is quite short. Another complex issue is placing the antenna. Ideally, we want a line-of-sight condition between the base station antenna and with every CPE antenna. In many instances (i.e., LMDS cells), this may be hard to achieve. For example, a subscriber antenna can be physically hidden behind a large building or behind another structure shadowing the line-of-sight condition. Intervening verdure may be the cause of another shadowing problem. A simplified functional block diagram of an LMDS CPE is illustrated in Figure 19.7.

---

* *Noise figure.* The system designer should be sure to distinguish what is meant by noise figure. Do we mean noise figure of the front-end receiver or noise figure of the entire receiving system including the antenna and transmission line?

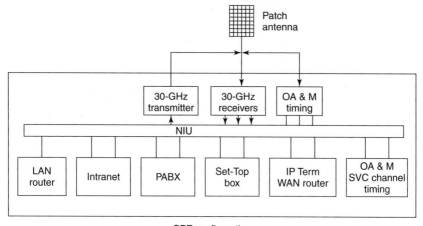

CPE configuration

Figure 19.7. A typical LMDS CPE.

A subscriber CPE can have as many as six baseband subsystems or interfaces:

1. LAN, via a LAN router
2. PABX
3. Intranet. This subsystem will also access the internet via a hub server.
4. Set-top box for video/CATV—conference TV
5. IP termination/WAN router
6. OA&M: orderwire/service channel; timing

5.2.3 Bandwidth (Hz) and System Traffic Load. Under most circumstances the traffic load for LMDS is asymmetrical, with the heavier load being downstream up to an extreme where nearly all the load is internet. Here there is real imbalance between upstream (very light) to downstream (very heavy). Upstream could be covered by the lowest portion of the band and even in that portion TDMA might be the most spectrally conservative. The downstream would service 10BaseT ports on CPEs using the remainder of the upper portion of the 30-GHz band.

An example might be a ten-story office building with 30 employees per floor. We would calculate the POTS service as one telephone per employee with reserve capacity for future growth. We arbitrarily use 50% growth over 5 years for a total capacity of $10 \times 30 + 0.5 \times 10 \times 30 = 450$ POTS stations. These stations would terminate in a PABX. To be conservative, the PABX does 1×1 concentration. It will then require 450 FX ports; with a $2:1$ concentration at the PABX, 225 FX ports would be required. This value is still very conservative. We assume 20 fax terminations, 2 per floor plus 50% growth for 15 fax ports, full duplex CCITT Group 4, backward compatible to Group 3. Eighty percent of the employees have PCs, each with an intranet port. The intranet interfaces with the internet, giving each PC internet access. WAN interconnection is via frame relay using 1.544-Mbit/s trunks. The LAN uses 10BaseT IEEE 802.3 with a router on each floor. The router not only segments the LAN floor by floor, it also provides frame relay interface. Each router has a WAN interface. The WAN may be based on IP or on frame relay. Let's suppose 1.544-Mbit/s circuits. There are then ten 1.544-Mbit/s ports to be accommodated by LMDS with expansion slots to a total of 15. The LMDS hub provides DSLAM-type functions for the data circuits. POTS local circuits are 64 kbits/s terminating in a DS1 (T1) 1.544-Mbit/s circuit via the PABX. Only 12 of the available 24 T1 DS0 circuits are active. The remainder is reserved for growth. This can significantly reduce the number of LMDS ports required.

Consider the following breakdown:

POTS including growth + Fax including growth. Total: 240 circuits with 2 : 1 concentration in PABX @ 64 kbits/s or 20 T1 ports (dividing by 12):

DS1 (T1) ports	20
WAN ports @ 1.544 Mbits/s incl growth	15
Intranet ports, 1.544 Mbits/s, one on each floor + 2 growth	12
Total LMDS ports	**47**

[access to the internet is through the intranet at the fiber hub (base station)].

A port represents an equal transmit/receive capability. In other words, it is symmetrical operation. We decided to make 1.544 Mbits/s the common bit rate, and 64 kbits/s is the common subrate of POTS service. As the reader can appreciate, there are many trade-offs available to the system planner. For one thing, we have assumed FDMA operation. There may be a TDMA trade-off. We have not identified traffic sources that may have bursty traffic. POTS lines are bursty; frame relay is bursty, for example. In fact, if we look at all sources, burstiness can be identified such as at intranet ports. Usage does not have a 100% duty factor, far from it.

How much of the LMDS bandwidth will be required by our building using the present structuring? Remember, it is bandwidth (measured in Hz) which incurs the bottom line. From above we have 47 LMDS ports where each port is handling 1.544 Mbits/s and we use FDMA and this would be downstream. At first blush, assume 1 bit per hertz of spectral occupancy.

Then we would have $47 \times 1.544 \times 106 = 72.568$ MHz. If we were to use TDMA, the value may be cut in 1/3, roughly 24 MHz. Consider now Table 19.4 to see what can be wasted/saved by considering spectral efficiencies versus type of modulation and bit packing.

Up to this point, we have considered all circuits operating with at 100% duty cycle. This is not true for any of the circuits. For a full duplex telephone speech circuit, in theory, in one direction there is talking only half the time, while the other half of the time (in theory) is dedicated to listening. There are also idle spaces between words, sentences, and syllables. The corporate LANs typically have very spurty traffic levels, and, as a result, so do the WAN(s). The fax is

TABLE 19.4 Modulation Type, Spectral Efficiency and Bandwidth Conservation[a]

Modulation Abbreviation	Modulation Type	Theoretical Bandwidth	Practical Bandwidth
BPSK	Binary phase shift keying	1.544 MHz @ 1 bit/Hz	2.1616 MHz @ 0.714 bit/Hz
QPSK	Quaternary phase shift keying	0.772 MHz @ 2 bits/Hz	1.103 MHz @ 1.4 bits/Hz
8-PSK	Octal phase shift keying	0.5146 MHz @ 3 bits/Hz	0.6176 MHz @ 2.5 bits/Hz
16-QAM	Quadrature amplitude modulation, 16 states	0.386 MHz @ 4 bits/Hz	0.463 MHz @ 3.333 bits/Hz
64-QAM	Quadrature amplitude modulation, 64 states	0.257 MHz @ 6 bits/Hz	0.3088 MHz @ 5 bits/Hz
256-QAM	Quadrature amplitude modulation, 256 states	0.193 MHz @ 8 bits/Hz	0.232 MHz @ 6.66 bits/Hz

[a] Bit rate = 1.544 Mbits/s.

also very spurty: a few sheets are sent or received, and then the fax is at rest for minutes or hours. ATM and frame relay can take advantage of these rest periods by sending other parties' traffic. With careful planning, the system designer might get away with 10 MHz or less of the total LMDS bandwidth.

This is just one building. In all probability it will be in just one cell site, physically. There may well be other such buildings sharing the same cell site.

An important consideration is sufficient isolation for frequency reuse. At least 35 dB of isolation can be achieved by orthogonal polarizations. In Figure 19.6 the reader will note the use of sectorized antennas having at least one group with vertical polarization and another, separate group with horizontal polarization. An antenna sector will have a front-to-back ratio providing additional isolation reducing co-channel interference. Free space loss from an offending antenna with a member or the entire frequency family being the same.

Some texts dealing with LMDS recommend placing the base station antenna as high as possible, especially on tops of tall buildings. There are arguments against the position as well. Lowering the antenna, as we learned in cellular frequency reuse, allows us to take more advantage of natural shadowing to improve frequency reuse isolation. If the range of an LMDS facility is limited to 3 km because of rainfall loss, reaching out further than 3 km may be counterproductive. On the other side of the coin, shadowing may make a customer location unreachable from a near antenna and yet reachable on a different radial from a further antenna.

In the LMDS system planner's mind of perfection, an LMDS cell has a perfectly circular radius, but because of shadowing by objects, buildings, and ridges in the signal path, the real cell radius is far from circular. This concept is illustrated in Figures 19.8A and 19.8B.

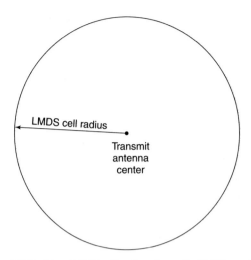

Figure 19.8A. Ideal LMDS cell radius, theoretical LOS conditions.

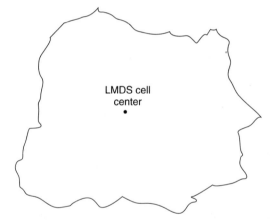

Figure 19.8B. Actual LMDS cell radius under real conditions. Distortions are due to shadowing.

5.2.4   Network Operations Control Center (NOCC). All of an LMDS system OA&M telemetry circuits terminate in the NOCC and all system command (control) circuits originate at the NOCC. LMDS repair teams are dispatched from the NOCC. As shown in Figures 19.6 and 19.7, all CPE facilities and base stations have telemetry circuits routed to the NOCC giving instant updates on system status. Standby equipment can be switched in at these locations by commands issued by the NOCC. In large LMDS systems there usually is a technician on watch 24 hours a day. Figure 19.9 illustrates a typical LMDS NOCC. In many instances the NOCC uses straightforward SCADA* techniques to monitor and control remote LMDS terminals.

---

* SCADA = Supervisory control and data acquisition.

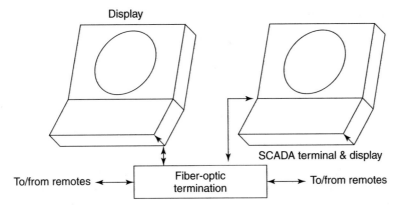

Figure 19.9. A simplified functional block diagram of an LMDS NOCC.

5.2.5 *Excess Attenuation Due to Rainfall.* At 28 or 30 GHz, excess atten-uation due to rainfall limits LMDS link time availability. A link's length may be severely restricted if we place too great an availability performance requirement on the system. Our experience has shown that time availability exceeding 99.99% (e.g., 99.991%) makes an LMDS system cost prohibitive.

Detailed methodology of calculation of excess attenuation due to rainfall may be found in the *Telecommunication Transmission Handbook*, 4th ed. [9]. The procedure used in the reference is taken from ITU-R.

Suppose we use Chicago, Illinois, as an example. Chicago is in rainfall region K. We calculate the excess attenuation for an exceedance of 0.01%. This will give a link in the Chicago region a 99.99% time availability. The rainfall rate in this case is 42 mm/h. Our test frequency is 30 GHz. We now have sufficient information to calculate the excess attenuation due to rainfall for a 1-km link. The excess attenuation formula in dB/km is

$$A \text{ (attenuation, dB)} = aR^b$$

where R is the rain rate in mm per hour and a and b are regression coefficients. Values for a and b are taken from Table 8.2 of Ref. 9. The table distinguishes between horizontal and vertical polarization. Vertical polarization gives a smaller attenuation value of a path than horizontal polarization. We consider the horizontal polarization case first:

$$A = 0.187(42)^{1.021}$$

$$= 8.49 \text{ dB}$$

Vertical polarization:

$$A = 0.167(42)^{1.000}$$

$$= 7.014 \text{ dB}$$

Calculate the path reduction factor, r. First, calculate the parameter L_0:

$$L_0 = 35 \exp(-0.63)$$
$$= 18.64$$

Assume 1-km path,

$$r = 1/(1 + 1/18.64)$$
$$= 0.95$$

To calculate the total excess attenuation due to rainfall for each polarization, we multiply the path attenuation factor A, by the path reduction factor r by the path length in kilometers. Thus, to achieve a 99.99% path time availability (due to rainfall), we need 8.49×0.95 or 8.06-dB link margin for horizontal polarization. For vertical polarization, the excess attenuation value is 7.014×0.95 or 6.66 dB. Table 19.5 provides a summary of excess attenuation values for 99.99% availability for path lengths 1, 2, 3, 4, and 5 km. Operation is at 30 GHz.

Carry out a link budget of a 1-km path at 30 GHz. Assume 1-watt transmitter power output, and also assume that the receiver system noise figure is 5.0 dB. The CPE patch antenna has 35-dBi gain toward the base station antenna; the base station antenna has 18-dBi gain. The base station has 1-watt RF power radiated in a quasi-omnidirectional pattern. Assume that transmission line losses in either direction at CPE are about 1 dB and at hub, 4 dB.

EIRP calculation for base station:

Transmit power:	$+0$ dBW
Line loss:	-4 dB
Antenna gain:	$+18$ dB (Gabriel)
EIRP:	$+14$ dBW

Link budget, 1 downstream carrier, 30 GHz, test distance 1 km, waveform 64-QAM, 1.544 Mbits/s

Free space loss based on km unit distance, 1 km:	-121.94 dB
Atmospheric gas loss:	-0.2 dB
Polar loss:	-0.3 dB

TABLE 19.5 Excess Attenuation Due to Rainfall Values for 1-, 2-, 3-, 4-, and 5-km Paths, 99.99% Time Availability, and 30-GHz Operation

Path Length (km)	Reduction Factor	Attenuation H. Pol (dB)	Attenuation V. Pol (dB)
1	0.95	8.06	6.66
2	0.903	15.33	12.67
3	0.861	21.93	18.12
4	0.82	27.85	23.0
5	0.788	33.45	27.64

Modulation implement loss:	-2.0 dB
IRL* at CPE antenna surface:	-110.44 dBW
Transmitter line loss:	-1.0 dB
CPE antenna gain:	$+35.0$ dBi
RSL at receiver input port:	-76.44 dBW

$$\text{Receiver } N_0 = -204 \text{ dBW} + 5 \text{ dB} = -199 \text{ dBW/Hz}$$

$$E_b = -76.44 \text{ dBW}/10\log(1.544 \times 10^6)$$

$$= -76.44 \text{ dBW} - 61.88 \text{ dB}$$

$$= -138.32 \text{ dBW}$$

$$E_b/N_0 = -138.32 + 199.0 \text{ dBW}$$

$$= 60.68 \text{ dB}$$

Required E_b/N_0 for BER $= 1.6 \times 10^{-8} = 20$ dB without coding gain. With convolutional outer coding and Reed–Solomon inner code, allow 5-dB coding gain thereby reducing required E_b/N_0 by 5 dB to 15 dB. This results in a link margin now 5 dB greater or 60.68 dB $- 15$ dB $= 45.68$ dB. For results of these calculations, see Table 19.6.

Assuming a time availability of 99.99%, we calculate the necessary margin for a 2-, 3-, 4-, and 5-km links. The rain rate of Region K (Chicago, Illinois) is 42 mm/hr. The regression coefficients are the same as the 1-km case. The reduction factor, r, varies with real link length. For the calculation of r, L_0 remains the same as with the 1-km link, calculated above or $L_0 = 18.64$.

When the margin, given in the far right column in Table 19.6, is less in value than the excess attenuation due to rainfall (fourth column from left), the link in question will not meet the availability requirements. This link will not meet the

TABLE 19.6 Results of Link Length Limits to Achieve 99.99% Time Availability, Downstream, 30 GHz

Link Length (km)	FSL (dB)	RSL (dBW)	Excess Attenuation Due to Rainfall (dB)	E_b/N_0 Calculated (dB)	E_b/N_0 Required (dB)	Margin (dB)
1	121.94	-76.44	Vert. pol. 6.66	60.68	15.0	$+45.68$
			Horiz. pol. 8.06	60.68	15.0	$+45.68$
2	128.0	-82.5	Vert. pol. 12.66	54.62	15.0	$+39.62$
			Horiz. pol. 15.33	54.62	15.0	$+39.62$
3	131.48	-85.98	Vert. pol. 18.117	51.14	15.0	$+36.14$
			Horiz. pol. 21.92	51.14	15.0	$+36.14$
4	133.98	-88.48	Vert. pol. 23.0	48.64	15.0	$+33.64$
			Horiz. pol. 27.85	48.64	15.0	$+33.64$
5	135.92	-90.42	Vert. pol. 27.64	46.70	15.0	$+31.7$
			Horiz. pol. 33.45	46.70	15.0	$+31.7$

* IRL stands for isotropic receive level.

99.99% availability with horizontal polarization when the link length of 5 km or greater. From experience, the first place for improvement is in downstream receive antenna aperture. A 50% increase in aperture dimension will provide a 3-dB gain. This will well place the system inside the 5-km operational radius for a time availability (rainfall) of 99.99%. Other measures we can take to increase coverage (i.e., increase coverage area dimensions) are as follows:

1. Increase RF output power of HPA at the other end.
2. Increase antenna gain at main station or hub.
3. Improve subscriber receiver noise performance.
4. Decrease required E_b/N_0.
5. Decrease the bit rate.
6. Reduce bit packing, to 16-QAM or 8-PSK.
7. Accept a reduced time availability number such as 99.95%.

There are numerous trade-offs that should be considered to improve performance, increase service area, and make the system more cost-effective. One such trade-off is bit rate. If the LMDS system that is proposed is to carry mostly video, consider using 3.5 Mbits/s or some multiple of it. This bit rate is developed from the TV MPEG2 compression standard, an average for a typical digital TV channel (NTSC standards). If the underlying system is based predominantly on ATM, some multiple of 53 octets would be in order (e.g., 53 octets per cell). The major selling point on LMDS is that it can carry some really respectable bit rates.

5.2.6 Cell Layout and Frequency Reuse.

Let's first accept that downstream connectivity will be point-to-multipoint (PtM) and that upstream will be point-to-point (PtP). The principal exception to this rule is that when traffic intensity justifies, connectivity in both directions will be point-to-point, possibly using frequency bands other than those given at the beginning of Section 5.2.

The first design step is to layout a traffic matrix. This is like an accounting procedure. From each LMDS port (location), traffic should be broken down in digital voice channels + growth @ 64 kbits/s each, number of LANs broken down into type and bit rate, (e.g., 802.3 at 10 Mbits/s) + growth, ATM @ peak cells/second, and so on. It may behoove the designer to multiplex each traffic source to a common bit rate. Upstream and downstream bit rates must be distinguished.

The layout of an LMDS system is similar to a cellular system layout (see Figure 19.10). A cell has a theoretical hexagonal geometrical shape with the base station in the center. Turning to the beginning of Section 5.2, the LMDS frequency list, one or several transmit and receive frequencies with appropriate bandwidths are assigned to each base station. A number of crucial design decisions have to be made at this juncture, such as the following:

1. From a hub site, determine the traffic level in each direction (i.e., upstream and downstream) associated with each user port. Traffic burstiness.

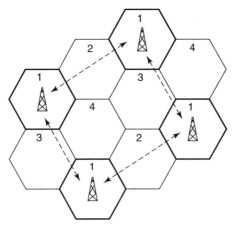

Figure 19.10. Idealized LMDS cell shape, hexagonal. Number in hexagon is a frequency family. For example, in cells with a number 2, all frequencies are identical. See Chapter 18, Figure 18 for definition of D/R ratio. D = distance between cell centers; R = cell radius.

2. Select the downstream operational format, pure FDMA, TDMA, and TDMA characteristics. Modulation type, such as BPSK, QPSK, 8-ary PSK, possibly FSK, M-ary FSK, OFDM. One of several QAM formats.
3. Upstream operational format. TDMA with timing, FDMA.
4. Co-channel interference goals in both directions.
5. Frequency reuse strategy versus antenna types and performance. Co-channel isolation requirements and how achieved. Polarization isolation, antenna height, earth bulge isolation, and so on.
6. Fiber-optic cable termination strategy at each main station, optical or electrical ADM.

We must set out the performance requirements at a master station and typical CPE for each traffic type. We will show how to achieve these requirements. Where the sub-ehf technology stands regarding cost and performance will be a major driver in determining the LMDS system design. What will be most cost effective: distributed HPAs (one for each carrier) or lumped amplification where one broadband HPA serves all downstream data flows from a master station?

One effective method to achieve the necessary isolation is to use a sectorized antenna as shown (for example) in Figure 19.11. Here we combine polarization isolation with frequency isolation (different frequencies) and antenna isolation (different sectors, different antennas using a common frequency) to achieve the desired C/I.

5.2.7 *IEEE 802.16.* One might observe that the IEEE 802.16 standard is the embodiment of LMDS in the IEEE 802 family of standards. Most of our reference sources relate 802.16 to the Metropolitan Area Network (MAN). Another

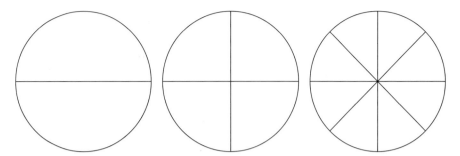

Figure 19.11. Several sectorized antennas. It should be noted that all sectors do not necessarily have to be mounted on a common antenna mount.

name for this standard is the IEEE WirelessMAN air interface providing wireless last-mile broadband access in the MAN. It delivers performance comparable to traditional copper cable, DSL, and T1/E1 offerings. There are also three related 802 series standards: IEEE 802.15 for the Personal Area Network (PAN) having a 10-m range and which is Blue Tooth-related; IEEE 802.11 and its many subsets of wireless LAN offerings; and the proposed IEEE 802.20 for the Wide Area Network (WAN).

The IEEE 802.16 standard operates on assigned frequencies between 10 and 66 GHz and assumes line-of-sight capability. Subsequent to the 802.16 standard, the 802.16a variance was issued covering non-LOS operation from 2 to 11 GHz for last-mile capability where trees and buildings place shadowing conditions in the radio path.

The most common 802.16a configuration consists of a base station mounted on a building or tower that communicates on a point-to-multipoint basis with CPEs located in businesses and homes. Roughly speaking, IEEE 802.16a has an operational range up to 30 miles (50 km) where the more typical cell radius is 4 to 6 miles (6 to 10 km).

With shared data rates up to 75 Mbits/s, a single "sector"* of an 802.16a base station provides sufficient bit rate capacity to simultaneously support more than 60 businesses with DS1 (T1)-level connectivity and hundreds of homes with DSL-rate connectivity using 20 MHz of channel bandwidth. To support a profitable business model, operators and service providers need to sustain a mix of high-revenue business customers and high-volume residential customers. IEEE 802.16a helps meet this requirement by supporting differentiated service levels, which can include guaranteed T1 (E1)-level services for business and best effort DSL-speed service for home consumers. The 802.16 specification also includes robust security features and QoS needed to support services that require low latency, such as voice and video.

Important considerations in the development of the IEEE 802.16 standard were *throughput, scalability, coverage,* and q*uality of service* (QoS). By employing a

---

* Where a sector is defined as a single transmit/receive radio pair at the base station.

TABLE 19.7   IEEE 802.16 Modulation Rates and Channel Sizes for the 10–66 GHz Band

Channel Width (MHz)	Modulation Rate (Msys/s)	QPSK Bit Rate (Mbits/s)	16-QAM Bit Rate (Mbits/s)	64-QAM Bit Rate (Mbits/s)
20	16	32	64	96
25	20	40	80	120
28	22.4	44.8	89.6	134.4

robust modulation scheme, IEEE 802.16 delivers high throughput at long ranges with a high level of spectral efficiency (i.e., bit packing) that is also tolerant of signal impairments. Dynamic adaptive modulation allows the base station to trade off throughput, BER, and range. For example, if a base station cannot establish a robust link to a distant CPE using the highest-order modulation scheme (e.g., 64-QAM), the modulation order is reduced to 16-QAM or even QPSK, reducing throughput while holding BER and increasing effective range [11].

To accommodate easy cell planning in both the licensed and license-exempt spectrum worldwide, IEEE 802.16 supports flexible channel bandwidths. For instance, an operator may be assigned 20 MHz of spectrum which he may wish to divide into two sectors of 10 MHz each or 4 sectors of 5 MHz each. By focusing power on increasingly narrow sectors, the number of users can be increased while maintaining good range and throughput. To scale coverage still further, the operator can reuse the same spectrum in two or more sectors by creating sufficient isolation between base station antennas.

The IEEE 802.16 standard not only provides a robust and dynamic modulation scheme, it also supports technologies that increase coverage, including mesh topology and "smart antenna" techniques. As radio technology improves and costs drop, the ability to increase coverage and throughput by using multiple antennas to create "transmit" and/or "receive" diversity can greatly enhance coverage in extreme environments. Table 19.7 gives modulation rates (baud rates) and channel sizes for the 802.16 standard [11, 13 and 14].

6   CATV AS A BASIC TRANSPORT MEDIUM FOR THE LAST MILE

For most developed countries there will be an existing cable television system at most population centers. In most instances we can depend on the fact that coaxial cable with drops will be in place. The existing system will also probably be an HFC (hybrid fiber coax) system. Whether the existing system has been optimized for broadband* delivery and transmission may be another matter entirely. The architecture and layout of the serving coaxial cable portion of the CATV system is usually the issue. We must keep in mind that a CATV system's original purpose was to deliver multiple TV channels to users. Delivering something to the headend *from* users is a comparatively new idea. Figures 19.12A and 19.12B show some typical CATV coaxial cable spectrum layouts [15].

---

* Broadband. We always like to ask "How *broad* is broad?"

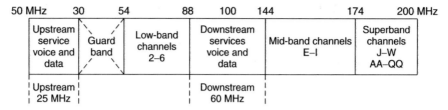

Figure 19.12A. One typical spectrum layout for a two-way CATV system, coaxial cable portion.

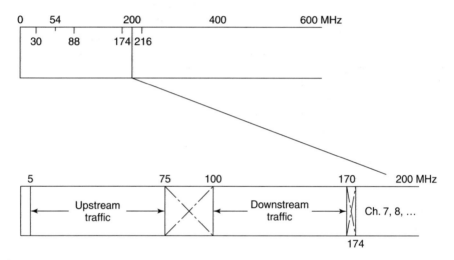

Figure 19.12B. Another possible spectrum layout for a two-way CATV system.

We could say that the "last-mile" connectivity in a CATV system was based on coaxial cable transmission. Early on we found that coaxial cable systems required broadband amplifiers at distance elements that were some "22 dB long." These amplifiers could provide some hefty gains, over 30 dB. However, it was soon realized that excessive signal distortion was introduced when these amplifiers were pushed too hard. The ideal gain was settled at 22 dB. Of course, CATV system designers installed these amplifiers at points along the coaxial cable where the accumulated loss from the previous amplifier point was 22 dB. Then a second fact was driven home: Thermal noise and distortion were cumulative. Too many amplifiers in tandem spoiled the picture, so to speak.

Turning now to that "last mile" (or first mile), the shorter we could make the coaxial cable used to connect to CATV subscribers, the better the picture. Fiber-optic amplifiers add little noise and distortion to the CATV signal; broadband coaxial cable amplifiers add considerable noise. Now enter HFC systems (hybrid coaxial cable). As we add length to our fiber-optic cable, little noise and distortion are added to the video signal. This is particularly so for digital TV transmission where we can take advantage of regeneration (see Chapter 8, Section ·7). Whereas early CATV systems were severely limited in length (really area), HFC systems can cover significantly large areas, many hundreds of square miles.

One key to modern HFC CATV operation is the concept of "passing of residences." A fiber-optic cable is laid along a public street right-of-way. For every 500 to 2000 residences passed by the fiber-optic cable, there are a drop and insert point and conversion to coaxial cable. The drop and insert again has two options: dropping and inserting a fiber strand in each direction or dropping and inserting an optical wavelength in each direction. Figure 19.13 shows this concept. It frees the system up from congestion because there can never be more than 2000 users upstream on a fiber strand or fiber wavelength, depending on the option selected. With proper design, each wavelength (or each strand) can handle 10 Gbits/s. We believe that the value will shortly become 40 Gbits/s. That is plenty of capacity, even for 500 TV channels downstream, allowing as much as 6 Mbits/s per channel (MPEG2 only requires about 3.5 Mbits/s per channel).

Cable Labs, working with the Society of Cable Telecommunication Engineers (SCTE), have developed DOCSIS (Data over Cable Service Interface Specification) which is now for most North American CATV companies the guiding specification for data, voice, and TV transmission. European countries and those under European hegonomy use the DAVIC (Digital Audio Video Council) as the CATV multimedia transmission and operations standard [16–19].

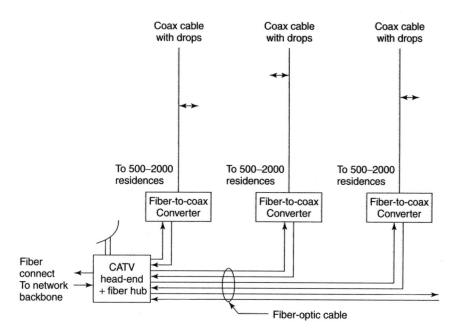

Figure 19.13. Fiber-optic cable along an urban/suburban street. HFC drop and insert point for every 500 to 2000 CATV subscribers. Note similarity with Figures 19.5A and 19.5B.

REVIEW QUESTIONS

1. What is the significance of the "last mile?" Why even mention it?

2. What is the generally accepted bandwidth (Hz) of a wire pair in the telephone network?

3. Differentiate the terms "symmetric" and asymmetric" when discussing local plant.

4. Define "downstream" and "upstream."

5. What is the purpose of FEC on an ANSI DSL signal?

6. Of the two modulation types on ADSL, which is espoused by ANSI? Which costs more yet displays better performance?

7. The maximum bit rate on VDSL is up to 51.84 Mbits/s. To achieve this bit rate, what is the limiting loop parameter?

8. What is the theoretical bit packing on 64-QAM? What would be about the best practical value I could achieve?

9. Differentiate MMDS/MDS with LMDS besides what the initials mean? Which one would get better BLOS (beyond-line-of-sight) performance grades?

10. What is the principal performance impairment, given a time availability value, of LMDS?

11. What is the nominal radius range of an LMDS cell (in km)?

12. How large (in MHz) is a common frequency segment of LMDS? Why that particular bandwidth?

13. What are the four major components of an LMDS network?

14. What would a "reasonable" time availability value should we expect on an LMDS network? It should be measured from a fiber hub to/through the CPE.

15. What are the two different options I have in the fiber-optic distribution network to LMDS?

16. Why would we want to keep RF transmission lines as short as possible in LMDS?

17. Give two methods, usually run concurrently, used to keep RF transmission lines short in LMDS?

18. Over the long run (timewise), what is the duty cycle of the transmit side of a POTS line?

19. About how much isolation (in dB) can be achieved from orthogonal polarizations with LMDS?

20. Why is isolation so important in LMDS?

21. In LMDS, what circuits terminate at the NOCC?

22. What is the EIRP out of an antenna where the antenna gain is 32 dBi, transmission line loss is 4 dB, and the transmit power is 1 watt?

23. If I double the aperture, approximately what increase in antenna gain (dB) can be expected?

24. In HFC CATV, why would we want to have the coaxial portion be as short as possible at the expense of a longer fiber-optic cable portion?

25. Regarding question 24, why not run fiber right to the CPE and eliminate the coax altogether?

26. Compressed TV is more demanding of BER performance than uncompressed. This, then, is one good reason to employ FEC. Can you give another reason.

27. How can an 802.16 circuit improve its range without improving antenna performance or increasing power? This sword cuts two ways. What is the other way it cuts? We lose something in trade.

28. For 802.16 operation, we can apply one of three common modulation waveforms. What are these three?

29. In North America we can expect to find that most CATV systems are ruled by the DOCSIS standard for two-way, voice, video, and data operation. What is its European counterpart? Spell out the acronym.

REFERENCES

1. *Framework Generic Requirements for Asymmetric Digital Subscriber Lines*, FA-NWT-001307, Issue 1, Bellcore (Telcordia), Piscataway, NJ, December 1992.

2. *Asymmetric Digital Subscriber Line Metallic Interface*, ATIS T1.413, ATIS/ANSI, New York, 1998.

3. *Asymmetric Digital Subscriber Line*, ITU-T Rec. G.992.1, ITU, Geneva, July 1999; and ADSL Transceivers, G.992.1 Annex H, Geneva, October 2000.

4. *Splitterless ADSL*, ITU-T Rec. G.992.2, Geneva, July 1999.

5. *High Data Rate Digital Subscriber Line (HDSL) Transceivers*, ITU-T Rec. G.991.1, Geneva, October 1998.

6. *A Business Case for Universal ADSL/VDSL Interfaces*, Ikanos Communications, from the web www.VDSLalliance.com/docs/ April 2003.

7. *Functional Criteria for Digital Loop Carrier Systems*, TR-NWT-000057, Issue 2, Bellcore (Telcordia) Piscataway, NJ, January 1993.

8. K. Maxwell, "Asymmetric Digital Subscriber Line Interim Technology for the Next Forty Years," *IEEE Commun. Mag.* (October 1996).

9. R. L. Freeman, *Telecommunication Transmission Handbook*, 4th ed., John Wiley & Sons, New York, 1998.

10. C. Smith *LMDS*, McGraw-Hill, New York, 2000.

11. *IEEE 802.16 Working Group on Broadband Wireless Access Standardization*, IEEE, New York, 2003.

12. *IEEE Standard for Local and Metropolitan Area Networks*, IEEE 802.16-2001, IEEE, New York, 2001.

13. Documentation on IEEE 802.16, from the web: www.wirelessman.org/published.html/.

14. IEEE 802.16a, Amend 2, *MAC Modifications and Additional Physical Layer Specifications for 2–11 GHz Operation*, IEEE, New York, 2003.

15. W. O. Grant, *Cable Television*, GWG Associates, 3rd ed., Schoharie, New York, 1994.

16. W. Ciciora, J. Farmer, and D. Large, *Modern Cable Television Technology*, Morgan-Kaufmann, San Francisco, 1999.

17. *Data over Cable Service Interface Specification 2.0, DOCSIS* CableLabs, Louisville, CO, 1997.

18. *Digital Audio Video Council (DAVIC)*. DAVIC Specifications 1.0–1.5, other material. On CD-ROM. See www.DAVIC.org, Torino, Italy, 2001.

19. Digital Video Transmission Standard for Cable Television, SCTE DVS 031, Society of Cable Telecommunication Engineers, Exton, PA, 1997.

20. Technical subcommittee T1E1, Working Group T1E1.4 Digital Subscriber Line Transmission, Alliance for Telecommunications Industry Solutions (ATIS), 1200 G St., NW, Suite 500, Washington, DC, 2000.

21. Code of Federal Regulations (CFR), 47, Part 101, Subpart L, "Local Multipoint Distribution Service" (LMDS), Oct. 1, 2000, Office of Federal Register.

20

OPTICAL NETWORKING

1 BACKGROUND AND CHAPTER OBJECTIVE

It is the old adage of supply and demand. The commodity is bit rate capacity, often called bandwidth by IT people. Certainly, the demand is there and probably doubling every one to two years. In the supply chain the bits, now in great quantities, must be transported in some manner. The other requirement is that they be directed to or accepted from a user.

What we are talking about is information represented by bits. The only transport that can handle these great quantities is an optical link. At every node in the optical network the stream of bits must be returned to the electrical domain for switching/routing. The goal is an all-optical network without the laborious return to the electrical domain at switching nodes except at the input–output points.

Optical links are presently carrying 10 Gbits/s per bit stream per wavelength. With dense wavelength division multiplexing (DWDM) a single fiber can carry 8, 16, 32, 40, 80, 160, or 320 wavelengths. Within some years of the publication of this book, we will have 40 Gbits/s per bit stream and a single fiber will support in that same time frame 320 wavelengths or more. If each wavelength carries 40 Gbits/s, then a single fiber will have the capacity to carry 40×160 Gbits/s or 6400 Gbits/s.

As seen from the network provider, a major drawback in fiber networking technology is the costly requirement for repeated conversions from the optical domain to the electrical domain and back again, called OEO, at every regeneration point and for periodic signal monitoring along the line. Amplifiers replacing regenerative repeaters where the add–drop function is not necessary have improved the cost situation somewhat. When optical switching without OEO is employed in the network, the requirement for regenerative repeaters (regens) in the network will be vastly reduced.

Telecommunication System Engineering, by Roger L. Freeman
ISBN 0-471-45133-9   Copyright © 2004 Roger L. Freeman

In the PSTN, fiber networks are primarily based on SONET and SDH infrastructures. In such cases the cost of regenerating optical signals can be very expensive, especially when it requires full SONET or SDH termination equipment at every ADM regeneration point. It has been found that even in these relatively homogeneous all-SONET environments, optical layer management can be a key factor in maintaining system integrity. Present optical networks, unfortunately, require OEO conversion for effective network management.

Even at locations along the fiber line where full conversion is not necessary, partial conversion at key points can be vitally important for monitoring to assure circuit quality. At amplifier locations as in conversion signal points, active signal monitoring capabilities should be included. This will require optical signal splitting and optical-to-electrical conversion of some portion of the light signal.

The move toward direct deployment of gigabit ethernet (GbE) over the MAN/WAN is a factor that may help to temporarily mitigate some of the push for all-optical switching because the cost of interface equipment for GbE fiber transport links is significantly less than for a SONET or SDH link. We see it as highly unlikely that GbE could completely replace SONET/SDH in the foreseeable future for anything other than very limited environments. However, the real-world impact of GbE deployment is likely to be increased traffic heterogeneity that will further drive the need for effective optical-layer management.

The ultimate goal behind DWDM deployment is the provision of more bit rate capacity. As a corollary then, MANs and WANs in the real world require optimization of the use of each wavelength's bit rate capacity. With the PSTN, where long-haul links traverse the network core, this goal of optimized utilization has typically been accomplished by pre-grooming all traffic such that uniform groups of signals can be efficiently transported long distances with a minimum of intervening decision points along the way. However, for traffic traveling nearer the edge of the transport network, new-generation equipment needs to provide a higher level of traffic monitoring and grooming capabilities within the optical domain to achieve a balance of flexibility, performance, and capacity utilization.

We would argue that in most cases it would make little economic and practical sense to invest in DWDM and then map GbE connections across individual wavelength carriers on a one-to-one basis. Therefore, the push for aggregation of multiple connections can very quickly lead to a mix of heterogeneous non-concatenated traffic traveling within a shared wavelength with a multitude of different end-point destinations.

The goal of the industry is to make the network all-optical except at each edge transition point. This point will be on a user's premises. By transition we mean the conversion of light to equivalent electrical information expressed as 1s and 0s.

This chapter's objective is to describe various steps to be taken toward what may now be called the "all-optical" network, its topology, and routing/switching in the optical domain.

2 NEW OPTICAL TECHNOLOGIES REQUIRED

The following is a list of new technology and radical approaches to make an all-optical network a reality:

a. Optical switching
b. Advanced wavelength demultiplexing and multiplexing
c. Tunable filters
d. Stabilized lasers
e. New approaches to modulation
f. Improved optical amplifiers with flat gain characteristics
g. New and larger optical cross-connects (OXCs)
h. Optical add–drop multiplexers (OADMs)
i. Signaling techniques in the light domain

2.1 Derived Technology Applications

The semiconductor optical amplifier (SOA) is one of the most promising technologies for optical networks. By integrating the amplifier functionality into the semiconductor material, the same basic component can perform many different applications. SOAs can perform switching and routing roles in an integrated functionality within the semiconductor material. Other important elements of an all-optical network are space switches, wavelength converters, and wavelength selectors, all of which can be fabricated from SOAs.

3 DISTRIBUTED SWITCHING

The new-generation managed optical network is moving toward a *distributed switching* model in which lambda (λ) switches with intelligent Layer 1 cross-connect capability are distributed at various points along the network border. This concept is illustrated in Figure 20.1. Such an architecture provides seamless and efficient Layer 1 management of heterogeneous traffic types throughout the network, without sacrificing performance or flexibility in either the core or edge environments. This global distributed-switching architecture is equally adaptable to using dedicated wavelengths packed with homogeneous traffic for long-haul point-to-point transport or for flexibly managing heterogeneous traffic on dynamically allocated short-haul wavelengths.

With cross-connects along the edge of the network cloud, there is an emerging need for supporting a managed optical layer within a distributed optical switching environment. This outlines the crux of the matter and presents significant opportunities and challenges for both the semiconductor level and module-level

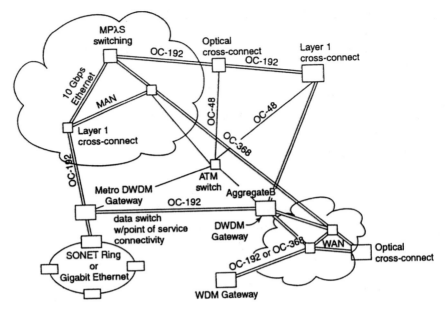

Figure 20.1. Distributed switching architectures. Note the combination of optical switches and Layer 1 cross-connects. Based on Refs. 1, 5, and 6.

developers and manufacturers. To achieve the required performance requirements, next-generation cross-connects need to be closer to the network by providing Layer 1 switching as opposed to the present traditional Layer 2 switching.

There will be two cross-point designs: asynchronous and synchronous. The higher-speed asynchronous cross-points enable heterogeneous MAN implementations to efficiently support different types of native-mode traffic within the same ring. In long-haul networks, there will be innovative uses of synchronous-switching cross-points which will provide the necessary performance requirements. These switches are seen as more of a *time–space–time* switch rather than the more rudimentary *space* switch. These new-generation synchronous cross-points will incorporate Layer 1 grooming capabilities that can selectively switch SONET (or SDH) or any other TDM (time-division multiplex) signals between any combination of inputs and outputs.

It is expected that these optical Layer 1 switching capabilities will use high-speed synchronous ICs. This next-generation synchronous cross-point switch will offer the capability to selectively groom out and switch and STS-1 (STM-1) from within STS-48 (STM-16) or STS-192 (STM-64) bit streams. These devices will provide complete flexibility for provisioning IC-level managed optical cross-connects from any STS-1 input to any STS-1 output. Non-SONET traffic mapped to STS-N-equivalent containers and protocol-independent wrapped traffic can be switched within the same cross-connects.

These high-density, high-speed grooming switches are deployed along the edge of the switching network cloud. They can optimize capacity utilization while efficiently making Layer 1 access decisions to partition out traffic to outlying internet protocol, GbE, ATM, fiber channel, or other Layer 2 switches. Localized Layer 2 functions such as routing and policy management are appropriately handled by the outlying switches while Layer 1 access switches provide high-speed performance IC-level switching/grooming of DWDM wavelengths [1].

4 OVERLAY NETWORKS

Today's data networks typically have four layers:

- IP for carrying applications
- ATM for traffic engineering
- SONET/SDH for transport
- DWDM for capacity

This architecture has been slow to scale, making it ineffective for photonic networks. Multilayering architectures typically suffer from the lowest common denominator effect where any one layer can limit the scalability of the other three layers and the entire network.

4.1 Two-Layer Networks Are Emerging

For the optical network developer, an absolute prerequisite for success is the ability to scale the network and deliver bit rate capacity where a customer needs it. Limitations of the existing network infrastructure are hindering movement to this service-delivery business model. It is the general belief in the industry that a new network foundation is required. This network foundation is seen as one that will easily adapt to support rapid change, growth, and highly responsive service delivery. What is needed is an intelligent, dynamic, photonic transport layer deployed in support of the service layer.

The photonic-network model divides the network into two domains: service and optical transport. The new architecture is seen as combining the benefits of photonic switching with advances in DWDM technology. It delivers a multigigabit bit rate capacity and provides wavelength-level traffic-engineered network interfaces to the service platforms. The service platform includes routers, ATM switches, and SONET/SDH add–drop multiplexers, which are redeployed from the transport layer to the service layer. The service layer is seen as relying completely on the photonic transport layer for the delivery of the necessary bit rate capacity where and when it is needed to peer nodes or to network elements (NEs). The bit rate capacity is provisioned in wavelength granularity rather than in PDH TDM granularities. We expect exponential growth rate of the fiber network; to meet these requirements, rapid provisioning is an integral part of the new architecture. While the first implementations of this model will support error detection, fault isolation, and restoration via SONET, these functions will gradually move to the optical layer.

Expect to see routers, ATM switches, and SONET/SDH ADMs to request bit rate capacity where and when needed via the provisioning capabilities of photonic switching with the traffic engineering capabilities of MPLS.* For the protocol designed for the optical domain, the name $MP\lambda S$ was selected. This protocol is designed to combine recent advances in MPLS traffic engineering control-plane techniques with emerging photonic switching technology to provide a framework for real-time provisioning of optical channels. This will allow the use of uniform semantics for network-management operations control in hybrid networks consisting of photonic switches, label-switched routers (LSRs), ATM switches, and SONET/SDH ADMs. While the proposed approach is particularly advantageous for data-centric optical inter-networking systems, it easily supports basic transmission services. MPλS supports the basic network architectures, overlay, and peer, proposed for designing a dynamically provisionable optical network.

Figure 20.2 illustrates the photonic network model. Here the network is divided into two domains: service and optical transport. The service platform includes routers, ATM switches, and SONET ADMs.

With the overlay model, there are two different control planes. One of these is in the core optical network and the other is the edge interface, variously called the UNI or user–network interface. The interaction between the two planes is virtually minimum. The derived network is very similar to our present IP/ATM networks. It can be dynamically set up through signaling or is statically configured. The internal operation of the network is transparent to the light carriers entering from the edge.

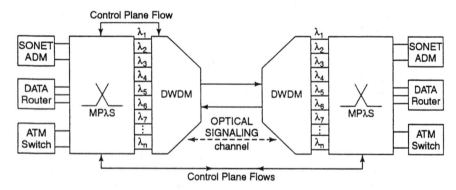

Figure 20.2. The photonic network model illustrating its two platforms: service and optical transport. The service platform at far left and far right shows individual routers, ATM switches, and SONET ADM capabilities. Inside the switch is the photonic transport layer which consists of optical switches and DWDM equipment. There is a standardized control plane used to communicate between the various elements. Based on Refs. 2, 4, and 5.

* MPLS stands for multiprotocol label switching. See Chapter 11, Section 8, for a brief overview of MPLS.

One drawback of an overlay network that has been envisioned is the amount of signaling and control traffic required due to the edge mesh of point-to-point connections. This excessive amount of routing protocol traffic results in limiting the number of edge devices in the network. For example, a single link-state advice flooding event is multiplied, creating a very large number of repetitive messages on the point-to-point mesh.

In the peer-to-peer model, a single action of the control plane spans both the core optical network and the surrounding edge devices as shown in Figure 20.3. Here we see the overlay and peer network models distinguished. In Figure 20.3a the overlay model hides the internal topology of the optical network, thus the optical cloud. In Figure 20.3b the peer model allows edge devices to participate in the routing decisions and eliminates the artificial barriers between the network domains.

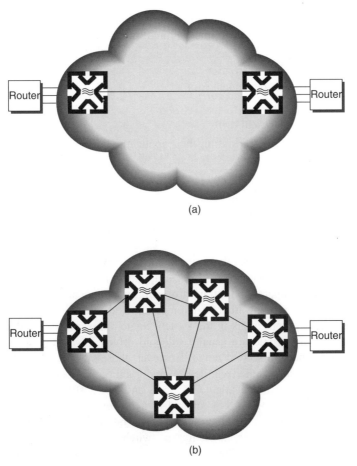

Figure 20.3. (a) Overlay and (b) peer network models. Based on Refs. 2, 4, and 21, courtesy of Calient Networks.

5   OPTICAL SWITCHING

The optical switch is one of the most important fiber-optic components that seamlessly maintains network survivability and is a flexible platform for signal routing. Today's switching in telecommunication systems is done electronically (in the electrical domain); however, as modern photonic networks evolve, routing of fiber-optic signals will be carried out completely in the optical realm.

The most common form of optical switches found off-the-shelf today are either electro-optical or optomechanical models. The electro-optical switch usually consists of optical waveguides with electro-optic crystal materials such as lithium niobate. 1×2 and 2×2 switch configurations are achieved using a Mach–Zehnder interferometer structure with a 3-dB waveguide coupler. The differential phase between the two paths is controlled by a voltage applied to one or both paths. The resulting interference effect routes signals to the desired output as the drive voltage applied to one or both paths of the Mach–Zehnder interferometer structure changes the phase differential between them.

Electro-optical switches have many limitations such as:

- High insertion loss
- High polarization-dependent loss
- High crosstalk
- Sensitivity to electrical drift
- They are nonlatching, limiting their applications in network protection and reconfiguration
- Require a high operating voltage
- Have high manufacturing costs

The principal advantage of this type of switch is switching speed that is in the nanosecond range.

Optomechanical switches rely on mechanically driven moving parts. They are the most widely used switches for optical application relying on mature optical technologies. Their operation is straightforward. Input optical signals are mechanically switched by moving fiber ends or prisms and mirrors to direct or reflect light to different output fibers in the switch. Switch movements must be precise for correct alignment and are usually solenoid driven. Switching speeds unfortunately are in the millisecond range. However, these switches are widely used because of low cost, simplicity in design, and good optical performance. Simple configurations are readily available in 1×2 and 2×2 structures. Also small-scale matrix nonblocking $M \times N$ optical switches are easy to build. By using multistage configurations, switches such as 64×64 can be built which are partially nonblocking. However, larger matrix switches are complex and unwieldy. Table 20.1 gives typical specifications of a 2×2 optomechanical switch.

Switching in this regime will be wavelength switching. A DWDM configuration out of the multiplexer will be in formations of optical wavelengths consisting

TABLE 20.1 Typical Specifications of a 2 x 2 Optome-chanical Switch

Parameter	Unit	Specification
Wavelength range	nm	1260–1600
Insertion loss	dB	0.6
Polarization-dependent loss	dB	0.05
Crosstalk	dB	−60
Switching speed	ms	5
Polarization mode dispersion	ps	0.1
Return loss	dB	55

Source: Courtesy of Yigun Hu, E-TEK Dynamics [3, 23].

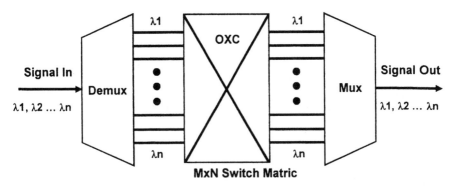

Figure 20.4. MXN switching by wavelength selection using optical cross-connects. Courtesy of Y. Hu, E-TEK Dynamics [3, 23].

of two wavelengths up to 160 or more. Let's think that some wavelengths must be routed to point X, others to point Y, and still others to point Z. This concept is illustrated in Figure 20.4.

Wavelength division multiplexing and λ-switching (wavelength switching) are tightly tied together. In a DWDM aggregate each wavelength must be clearly separable to minimize crosstalk.

5.1 MEMS Switching

MEMS is an acronym for *micro-electromechanical systems*. There are two types of these switches being developed: mechanical and microfluidic. The mechanical variety utilizes an array of micromachined mirrors that can range into the hundreds of thousands on a single chip. The fluidic type switch relies on the movement of liquid in small channels that have been etched into a chip. In the case of a mirror-based switch, an array of micromachined mirrors is fabricated in silicon. The incoming light signal is directed to the desired output port by means of control signals applied to the MEMS chip which fixes the position of each individual mirror.

Mirror-based MEMS switches are classified by mirror movement. There are two-dimensional (2-D) switches and three-dimensional (3-D) switches. In the case of the 2-D switches, a mirror has taken on only one of two positions. This would be typically either up or down or side-by-side. With a 3-D switch, there are a wide variety of positions that a mirror can assume. A mirror can be swiveled in a wide variety of positions and in multiple angles [7, 11, 22].

Reference 7 states that most of the major players in optical switching are following the mirror-based route. This group does not include Agilent Technologies, which is taking advantage of the company's knowledge of microfluidics. They have developed a switch that is based on Hewlett-Packard's inkjet printing technology. This Agilent optical switch is quite unique in that it consists of intersecting silica waveguides, with a trench etched diagonally at each point of intersection. The trenches contain a fluid that, in the default mode, allows the light to travel through the switch. To activate the switch such that it switches light, bubbles are formed and removed hundreds of times per second in the fluid, which then reflect the light to the appropriate output port [7, 22].

5.1.1 Control of Mirrors and Bubbles.

Reference 7 covers three types of actuation mechanisms being used in MEMS switching: electrostatic, electromagnetic, and thermal.

Electrostatic Actuators. This is the most common and well-developed method to actuate MEMS, since IC processes provide a wide selection of conductive and insulating materials. By using conductors as electrodes and insulators for electrical isolation of electrodes, electrostatic forces can be generated by applying a voltage across a pair of electrodes. This type of actuation requires lower power levels than other methods and is the fastest.

Magnetic Actuators. Magnetic actuators usually require relatively higher currents (resulting in higher power), which can limit their use. In addition, these mechanisms use magnetic materials that are not common in IC technology and often require some manual assembly, a distinct drawback. Thus the choice of magnetic material is limited to those that can be easily micromachined. However, electromagnetic microactuators can be actuated much faster and consume less power than their thermal actuator equivalents.

Thermal Actuators. Thermal actuators require heating which is accomplished by passing a current through the device. To its detriment, the heating elements can require high power consumption. Then, of course, the heated material has to cool down to return the actuator to its original position. Then the heat is dissipated in the surrounding structure. All of this takes time limiting the switching speed of the device.

LIGA (lithographic plating and molding) is a very promising fabrication method of MEMS devices. LIGA combines the basic process of IC lithography with electroplating and molding to achieve depth. In LIGA, patterns are created

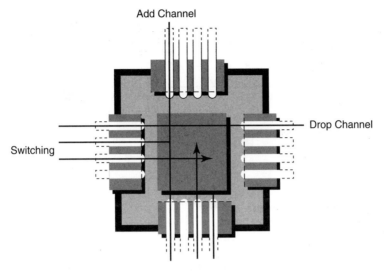

Figure 20.5. A 2-D MEMS optical cross-connect switch with additional third and fourth planes to add/drop functionality. Courtesy of Zeke Kruglic, OMM Inc., San Diego, CA [11, 22].

in a substrate, which are then electroplated to create 3-D molds. The molds can be used to create the final product. However, a variety of materials can also be injected into them to produce a product. Here we have two distinct advantages to this technique: Materials other than silicon can be used (in particular, metal and plastic), and devices with very high aspect ratios can be built.

MEMS switches currently support up to 32 bidirectional ports. The goal in MEMS switching is to develop devices with 1000×1000 arrays. Some optical switching companies are tackling this task head-on. Other companies believe that scaling smaller switches together to make a larger array is the better approach (based on Refs. 7 and 13). Figure 20.5 is an example of a 2-D MEMS optical cross-connect switch with additional third and fourth planes to add/drop functionality.

Note: MEMS cross-connects are described in Section 8.

6 A PRACTICAL OPTICAL ADD–DROP MULTIPLEXER

True optical ADMs will allow provisionable add–drop channels similar to time-slot assignment (TSA) and reassignment of optical channels similar to time-slot interchange (TSI) found in present-day electronic digital switches. Figure 20.6 is a block diagram showing the basic functions of a programmable optical ADM. Since these cross-connects will be provisioned on a wavelength basis, new sites requiring access to the network can be more easily added and the planning burden of network planners can be reduced. Migration to the all optical layer provides new methods of protection for network restoration.

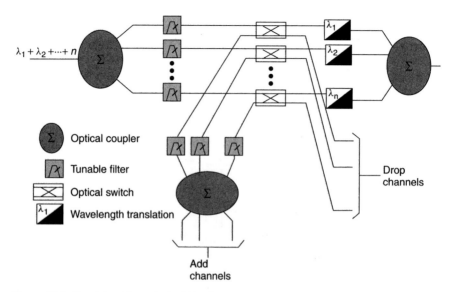

Figure 20.6. Provisionable optical ADMs. Based on Figure 3, Alcatel www web release, July 24, 2000 [4].

The evolution of optical networks will lead to more advanced systems that provide wavelength routing capability. As technology breakthroughs are made in the arena of optical gates and matrices, optical cross-connect systems will begin to appear on the scene. Figure 20.7 is a high-level diagram of an optical cross-connect system (OCCS).

There are two basic types of cross-connect systems: line side and tributary side. The tributary side, or Type 1 OCCS, will provide functions similar to the broadband SONET cross-connects available today. The line side, or Type 2 OCCS, will support high-level network restoration and network reconfiguration on the high-speed transport system.

As new optical network services become available, there will be a dramatic increase in customer base and a similar increase in the demand to transport traffic. Up to now, electronic broadband cross-connects have met the requirements demanded by the network, but the complexity of these systems and their matrix sizes will eventually reach the limit of feasibility. Optical cross-connects can reduce the size and complexity of electronic digital cross-connect (DCSs) with a higher level of traffic loading and routing at the optical wavelength level. Signals can be routed at levels greater than STS-1 can be efficiently handled at the optical layer. An optical matrix is inherently smaller than its electronic counterpart, requires less power, switches at higher speeds, and handles larger configurations of bit rate capacity with less complexity. Because a significant portion of the "bandwidth explosion" is due to customer driven requirements for larger pipes, these connections can be more efficiently managed with an optical matrix rather than its electronic counterpart.

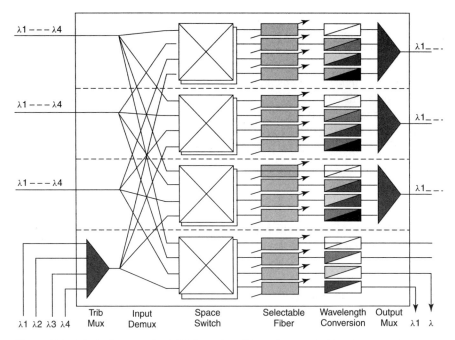

Figure 20.7. Block diagram of OCCS. Based on Figure 4, page 3, Alcatel www web release, July 24, 2000 [4].

Currently, network restoration for fully restorable services is accomplished by two basic methods: mesh protection using digital cross-connects and ring protection by means of SONET/SDH multiplexers. Both methods have weak points such as in restoration time, cost, and management efficiency. Once the optical network is in place, one major efficiency will accrue over its electronic counterpart. Consider SONET (or SDH) rings. All ADMs in the ring must operate at the same data rate. Reference 4 suggests that this will lead to inefficiencies and additional cost in the transport network especially since some routes will be significantly more dense than others. With a ring in the all-optical network, some wavelengths can be operated at one rate, say OC-24, others at OC-192, and still others at OC-48.

6.1 OXCs and OADMs Enhance Availability and Survivability

The major step toward the all-optical network is adding both OADM and OXC* elements. These network elements provide carriers with the ability to reconfigure the network traffic for optimal data transport. It will also have the capability of rapid restoration in the event of link failure, all within the optical layer.

---

* OADM stands for optical add–drop multiplex; OXC stands for optical cross-connect.

OXCs are dynamic switching fabrics with connections between any of M fiber inputs and any of N fiber outputs in a DWDM network. Thus, optical cross-connect switches have nonblocking one-to-many connections in a matrix configuration. OXCs provide good network survivability, lower network management cost, and reconfigurable paths for signal routing in the optical layer. These attributes help eliminate the need for complex and expensive digital switches operating in the electrical domain. Since they operate in the optical realm, optical cross-connects can potentially accommodate terabit data streams due to their wavelength, bit-rate, and protocol transparent characteristics.

Source: Ref. 3.

7 IMPROVEMENTS IN THE MANAGEMENT OF THE NEW NETWORK ARCHITECTURE

As discussed above, the network architecture will be two-layer. Both the IP router community and the optical community have agreed that the way to control both layers is by means of multiprotocol label switching (MPLS). MPLS for this application has been slightly modified; as covered above, it is appropriately called MPλS. Each control plane, the optical and IP router, has two phases in a switching routine. One phase sets up paths, and there is the steady-state traffic phase in which the state information has been set up at each node to define the paths and then forwards packets in a way that provides much of the missing QoS capability.

MPλS will replace the two current protocols operating in the lower layers with their many variants for several reasons. First, these traditional software families are very much vendor-dependent. Second, IP and SONET/SDH are very different from one another; third, they are very slow for the anticipated needs of restoration, provisioning, and protection.

There are two communities, MPLS and the MPλS, and there is only one disagreement. That is, whether the control entity within each set of IP routers forming the IP layer will be topologically aware of just what pattern of OXC traversals will constitute a lightpath across an optical network cloud, or whether the optical layer will set these up autonomously and then tell the IP layer where the endpoints are without saying which sequence of OXCs constitutes a lightpath. Paul Green [5] believes the latter will prevail, at least initially.

Protection switching, discussed in Section 12.4 of Ref. 12, was the first to receive attention in the optical layer integrity processes. To activate protection requires precanned algorithms similar to those used in SONET/SDH. Only a small portion of the network is affected when protection switching is invoked. This is an optical layer function, and the trigger to activate can be the loss of OSNR (optical signal-to-noise ratio).

Similar to present automatic protection switching, there is the restoration phase, which is the replacement of the failed optical path by another. Once repairs have been completed, this failed path, now capable of being operational, becomes the protection path.

Provisioning/reconfiguration becomes an interesting concept. Reference 5 describes the condition of *stranded bit rate capacity* where capacity on a fiber or on a cable lies fallow, unused. This bit rate capacity can now be brokered between service providers who might use this optical facility, setting up a rent-a-wavelength condition. See also Refs. 8, 9, and 10.

8 ALL-OPTICAL CROSS-CONNECTS

During the period of preparation of this book, large all-optical cross-connects arrived on the scene. These are micro-electromechanical systems (MEMS), which now have progressed from 2-D optical switches to 3-D devices. The concept of the 2-D MEMS switch derives from the old analog circuit crossbar switch. It involves N^2 popup mirrors to deflect collimated light from an input port to an output port. This is shown in Figure 20.8a. Figure 20.8b shows a MEMS switching machine consisting of only $2N$ mirrors, with N of them directing the

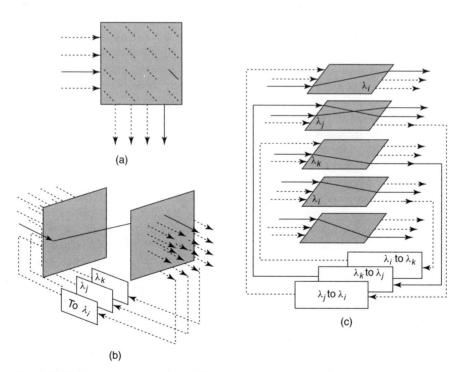

Figure 20.8. Optical cross-connects (OXCs). (a) 2-D using binary (popup) mirrors, (b) 3-D using mirrors with analog control, showing an attached wavelength converter pool; (c) multiplane architecture using multiple 2-D modules, showing an attached wavelength converter pool. From Figure 3, *IEEE Commun. Mag.*, January 2001 [5].

inputs toward distinct outputs and another N directing the outputs to connect to the inputs. Green [3] states that the advantage of 3-D is therefore linear scalability with port count (compared to quadratic for 2-D), but at the expense of analog mirror tilt control versus binary for 2-D.

The 3-D OXC has other advantages. A large port-count device can be used for managing whole fibers as well as many wavelengths carried on a fiber. Its cost is comparatively low and does not suffer the high attenuation suffered due to the interconnections involved to accommodate large nonblocking $N \times N$ structures from small-N 2-D subcomponents. These subcomponents have achieved at most a 32×32 size [5].

Another problem arises when an OXC is to be used just for WDM switching. By WDM switching we mean wavelength shifting all optically that is protocol transparent. Within the OXC we would want a path between wavelength A and any output connected to wavelength B. As the number of wavelengths increases, the number of "from-to" paths goes up radically. The number of lasers in the wavelength converters can be decreased by using tunable lasers.

9   OPTIONS FOR OPTICAL LAYER SIGNALING

We assume that the optical network is *connection oriented*. Thus circuits require setup and release algorithms. These signaling protocols are implemented in software leading to limitations on the call-handling capacity of a switch.

From the signaling and control perspective, two network models have evolved to provide interoperability between the IP and optical layers. There is the peer model which is based on the premise that the optical layer control intelligence can be transferred to the IP layer, which has assumed end-to-end control.

The second model is the client–server model. This model is based on the premise that the optical layer is independently intelligent and serves as an open platform for the dynamic interaction of multiple client layers. This includes IP.

In either case, we assume an optical mesh network. The control plane is IP-compatible based on the MPLS protocol discussed in Chapter 8, Section 11, of this text. The routing protocols are IP and carry out topology discovery. MPLS signaling protocols are used for automatic provisioning. Expect to see the IP-based optical layer control stack to be standardized as the model becomes adopted.

Applications are managed differently. The optical control plane will control *dynamic lambda provisioning* with routers on the edge of the network cloud and linked by optical subnetworks as illustrated in Figure 20.9.

When a router encounters congestion, either the Network Management System or the router itself will request the provisioning of a dynamic lambda, meaning an additional wavelength carrier. This requires that optical switches have the capability to create new or enhanced service channels such as OC-48 or OC-192 capacity to meet the needs of that router. This dynamic lambda provisioning can adapt to traffic flows.

The client–server model handles things differently. It will let each router communicate directly with the optical network using a well-defined UNI

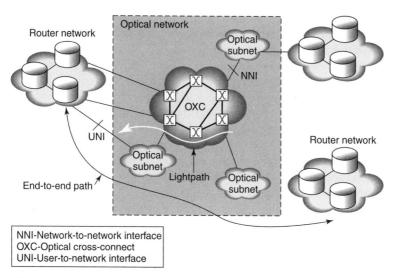

Figure 20.9. This sketch illustrates the client–server model. The optical layer has intelligence which controls lightwave links. The network is made up of subnetworks with well-defined interfaces. Courtesy of Calient Networks, San Jose, CA [2, 21].

(user–network interface). The interconnection between subnetworks would be via an NNI (network-to-network interface). This permits each subnetwork to evolve independently.

In optical networks, as with wire and radio networks, operators wish to take advantage of competition and thereby build multivendor networks. To do this, standardized interoperability is required.

When we compare the two models, the client–server model has a significant advantage over the peer model in that it has a faster path to interoperability. On the other hand, the client–server model is more direct and simplified. To manage end-to-end paths on optical links, additional communication between IP and optical layers is required. These additional communication connectivities will pervade across the entire network cloud.

10   FOUR CLASSES OF OPTICAL NETWORKS

10.1   Generic Networks

Whether optical or electrical, there are three general network types; two are connection-oriented (CO), namely ATM and PSTN, and there is one connectionless (CL) network which is IP. They can also be placed in categories of circuit-switching and packet-switching. The PSTN relies on circuit-switching and ATM and IP can be classified as packet switching networks.

Optical network designers have changed the definitions of circuit-switching and packet switching to meet the special requisites of an optical network. *Circuit switching* is position-based in that bits arriving in a certain position are switched to a different output position. The position is determined by a combination of one or more of three dimensions: space (port number), time, and wavelength. Packet switches are label-based in that they use intelligence in the header, which we call *labels*, to decide how and where to switch the packet. Note how these definitions differ from conventional ones. Still here in the case of data, a circuit may be set up prior to the exchange of packets and thus meets the conventional definition, but this setup is not a necessary attribute. Think of the permanent virtual circuit in X.25 and frame relay.

It is also important to note in the case of optical networks actually whether a circuit is set up prior to data exchange or is a property of whether a circuit is CL (connectionless) or CO (connection-oriented), not whether a circuit is circuit-switched or packet-switched. IP is a typical example of CL packet-switched networks, and ATM is a typical example of CO packet-switched networks to simplify discussion. We recognize that when resource reservation protocol (RSVP) and/or multiprotocol label switching (MPLS) adds a CO mode of operation to IP networks, there may be confusion in semantics. Thus we hold with our ATM and IP examples.

We will briefly review four classes of optical networks based on the types of components used: optical link networks, broadcast-and-select (B&S), wavelength-routed (WR), and photonic packet-switched networks. Table 20.2 organizes these network types based on their optical components.

Optical link networks are defined as a network using electronic switches interconnected by optical links, either single-channel or multichannel. Multichannel links are derived from optical WDM multiplexers/demultiplexers at either end. WDM passive star couplers are used to create broadcast links for shared-medium operation. These two component types are not programmable; as a result, no reconfiguration is possible.

Figure 20.10 illustrates the classes of optical networks. There are (a), (b), and (c) columns, where column (a) lists optical link networks using all-electronic switching. Column (b) is a listing of single-hop B&S networks and photonic packet-switched networks. These network classes give examples using all-optical

TABLE 20.2 Classes of Optical Networks Based on Components Employed

Types of Optical Communications Components	Classes of Optical Networks			
	Optical Link Networks	Broadcast-and-Select Networks	Wavelength-Routed Networks	Photonic Packet-Switched Networks
Nonswitching optical components	√	√	√	√
Tunable transmitters and/or tunable receivers	X	√	May or may not be present	May or may not be present
Optical circuit switches (OADMs and OXCs)	X	X	√	May or may not be present
Optical packet switches	X	X	X	√

Note: The check sign √ indicates nonswitching optical components and the X indicates switching components (author).

Source: Table 1, page 121, *IEEE Commun. Mag.* [6].

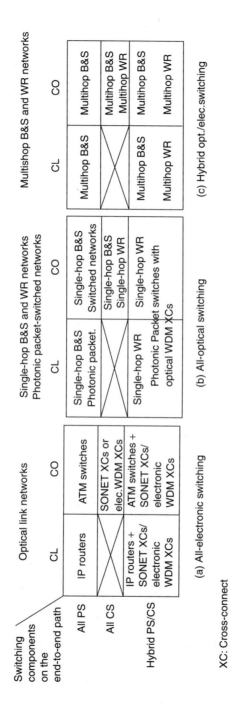

Switching components on the end-to-end path	Optical link networks	
	CL	CO
All PS	IP routers	ATM switches
All CS		SONET XCs or elec. WDM XCs
Hybrid PS/CS	IP routers + SONET XCs/ electronic WDM XCs	ATM switches + SONET XCs/ electronic WDM XCs

(a) All-electronic switching

XC: Cross-connect

Single-hop B&S and WR networks Photonic packet-switched networks	
CL	CO
Single-hop B&S Photonic packet.	Single-hop B&S Switched networks
	Single-hop B&S WR Single-hop WR
Single-hop WR Photonic Packet switches with optical WDM XCs	Single-hop WR

(b) All-optical switching

Multishop B&S and WR networks

CL	CO
Multihop B&S	Multihop B&S
	Multihop B&S Multihop WR
Multihop B&S Multihop WR	Multihop B&S Multihop WR

(c) Hybrid opt./elec.switching

Figure 20.10. Classification of optical networks. B&S, broadcast and select; WR, wavelength routing; CL, connectionless; CO, connection-oriented; CS, circuit-switched; PS, packet-switched. Derived from Figure 3, page 120, *IEEE Commun. Mag.* March 2001 [6].

switching. Column (c) is a listing of multihop and WR networks. These network types use a hybrid switching optoelectronic type.

Single-hop B&S networks have optical transmitters and receivers that can be tuned on a packet-by-packet or call-by-call basis. All three networking techniques are theoretically possible with single-hop B&S networks as shown in Figure 20.10, column (b).

There are also *multihop B&S networks*. With this type of network, data are broadcast on all links. Effectively electronic switches provide wavelength conversion on the desired path between source and destination because not all nodes receive all wavelengths. In that the classification of such networks is B&S, the only optical switching takes place in the tunable transmitters and receivers. The electronic switches can be circuit switches or packet switches because the components can be tuned on a packet-by-packet basis or call-by-call basis. Multihop B&S networks can be operated in all categories of column (c) of Figure 20.10, except for the CS–CL category.

Wavelength-routed (WR) networks include optical circuit switches, which we'll call OADMs and OXCs. On an optional basis, these networks may also have tunable transmitters and receivers. WR networks can be single-hop or multihop. Single-hop networks use only optical switching components, and thus they are listed in column (b).

The final category of optical networks is *Photonic Packet-Switched Networks*. We can consider these networks as having packet switches and, optionally, circuit switches with tunable transmitters and receivers. Look for these networks in column (b).

Of all the networks listed in Figure 20.10, only the optical link network is feasible today and is presently in operation. Of the remaining three, the fiber-optic industry is directing its attention to WR networks. Multihop WR networks with electronic packet switches are the most common [6]. An example is a network of IP routers interconnected by optical circuit switches such as OADMs/OXCs.

11   OPTICAL BIDIRECTIONAL LINE-SWITCHED RINGS

To further improve an optical network's availability and survivability, the familiar bidirectional ring architecture is applied. An optical ring uses the same principles as the fiber ring to protect against equipment and network failures. See Figure 20.11.

There are software devices housed in network elements (NEs) that are capable of sensing a module failure or fiber break. Traffic is then automatically re-routed in the opposite direction around the fiber-optic ring. The optical line-switched rings give the network operator an increased level of confidence against failure. The importance of survivability is so much greater with optical networks because they transport orders of magnitude more traffic, up to possibly hundreds of wavelengths rather than many DS-3 or E-3 configurations. Present switching time is under 50 ms.

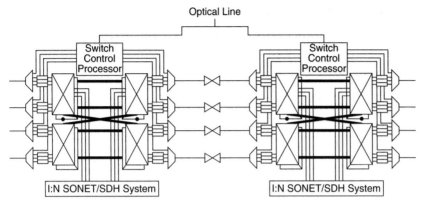

Figure 20.11. Optical ring architecture. Based on Figure 18, paragraph 11, www.iec.org/online/tutorials/opt_net/topic11 [13].

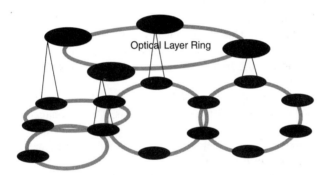

Figure 20.12. Multiple optical rings across an optical network. From Ref. 13, Figure 19.

Optical rings are most effective on large networks, hence switching time is critical. One development underway is *network protection equipment* (NPE), which can significantly reduce the switching time required on large networks. Instead of routing traffic from network elements (NEs) adjacent to the fiber cut, the optical bidirectional line-switched rings (OBLSR) using NPE redirects traffic from the node where it enters the ring. This redirection prevents the traffic from being backhauled across the network, which greatly improves switching time.

Figure 20.12 illustrates multiple optical rings across a fiber cable network.

12 OVERVIEW OF GENERALIZED MULTIPROTOCOL LABEL SWITCHING (GMPLS)

Based on Refs. 24 through 28, from the internet.

12.1 Introduction

An overview of MPLS was given in Section 8 of Chapter 11 of this text. GMPLS extends MPLS to provide the control plane (for signaling and routing) for devices that switch in any of these domains:

- Packet
- Time
- Wavelength
- Fiber

In providing a common control plane, GMPLS resolves issues in operating across dissimilar network types such as packet, plesiochronous network, and optical network. GMPLS is an extension of MPLS protocols to include devices that switch in time, wavelength (e.g., DWDM), and space domains (typically an optical cross-connect). The extension allows GMPLS-based networks to find and provision an optimal path based on user traffic requirements for a flow that potentially starts on an IP network. That flow then is embedded in SONET (or SDH) and then switched through a specific wavelength and physically on a specific fiber. Table 20.3 summarizes the basic framework of GMPLS. Figure 20.13 illustrates a typical optical network architecture.

The challenge for the designer of GMPLS was to devise an all-encompassing control protocol for the establishment, maintenance, and management of traffic-engineered paths to permit that data plane to effectively ship user data from source to destination. A data frame on a GMPLS-controlled network is likely to traverse several network spans which may be SONET/SDH-based through an ATM configuration. There may be edge aggregation along the way of flows from multiple users which may then feed into the long-haul network which uses lambdas to transport these flows into many metro networks.

TABLE 20.3 Review of GMPLS Framework

Switching Domain	Traffic Type	Forwarding Scheme	Example of Device	Nomenclature
Packet, cell	IP, asynchronous transfer mode (ATM)	Label as shim header, virtual channel connection (VCC)	IP router, ATM switch	Packet switch capable (PSC)
Time	TDM/SONET	Time slot in repeating cycle	Digital cross-connect system (DCS), ADM	TDM capable
Wavelength	Transparent	Lambda	DWDM	Lambda switch capable (LSC)
Physical space	Transparent	Fiber, line	OXC	Fiber switch capable (FSC)

Source: Based on Table 1, page 4, Ref. 28.

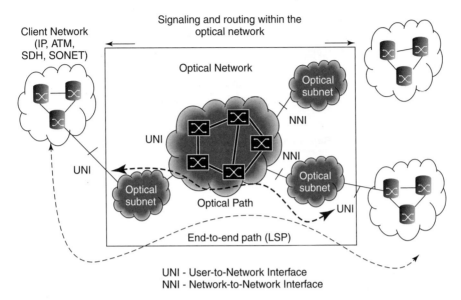

UNI - User-to-Network Interface
NNI - Network-to-Network Interface

Figure 20.13. A typical optical network architecture. Based on Tellium material [28].

12.2   Selected GMPLS Terminology

CR-LDP: Constraint-based routing—Label distribution protocol

FEC: Forwarding equivalence class

IS-IS-TE: Intermediate system-to-intermediate system–traffic engineering

Label: A short, fixed length, physically contiguous identifier which is used to identify an FEC, usually of local significance

Label merging: The replacement of multiple incoming labels for a particular FEC with a single outgoing label

LDP: Label distribution protocol

LMP: Link management protocol

LSP: Label-switch path

LSR: Label-switched router

LER: Label edge router (term is useful, but not found in standards)

OSPF-TE: Open shortest path first—traffic engineering

OXC: Optical cross-connect

RSVP-TE: Resource reservation protocol–traffic engineering

UPSR: Unidirectional path-switched ring

12.3   The GMPLS Protocol Suite

One way GMPLS evolved from MPLS was by extending the signaling protocols (RSVP-TE, CR-LDP) and the routing protocols (OSPF-TE and IS-IS-TE).

These extensions made GMPLS compatible with plesiochronous/synchronous and optical networks. In parallel with these developments a new protocol was introduced, LMP (link-management protocol). It manages and maintains the health of the control and data planes between two neighboring nodes. LMP is IP-based and includes extensions to RSVP-TE and CR–LDP. Table 20.4 recapitulates the protocols and extensions just discussed.

TABLE 20.4   Protocols Related to GMPLS

Protocols		Description
Routing	OSPF-TE, IS-IS-TE	Routing protocols for the auto-discovery of network topology, advertise resource availability (e.g., bandwidth or protection type). The major enhancements are as follows:
		Advertising of link-protection type $(1 + 1, 1:1,$ unprotected, extra traffic)
		Implementing derived links (forwarding adjacency) for improved scalability
		Accepting and advertising links with no IP address — link ID
		Incoming and outgoing interface ID
		Route discovery for back-up that is different from the primary path (shared-risk link group)
Signaling	RSVP-TE, CR-LDP	Signaling protocols for the establishment of traffic-engineered LSPs. The major enhancements are as follows:
		Label exchange to include non-packet networks (generalized labels)
		Establishment of bidirectional LSPs
		Signaling for the establishment of a back-up path (protection Information)
		Expediting label assignment via suggested label
		Waveband switching support — set of contiguous wavelengths switched together
Link Management	LMP	**Control-Channel Management:** Established by negotiating link parameters (e.g., frequency in sending keep-alive messages) and ensuring the health of a link (hello protocol)
		Link-Connectivity Verification: Ensures the physical connectivity of the link between the neighboring nodes using a PING-like test message
		Link-Property Correlation: Identification of the link properties of the adjacent nodes (e.g., protection mechanism)
		Fault Isolation: Isolates a single or multiple faults in the optical domain

Source: Table 2 GMPLS Protocols, page 6/217 [28].

12.4 GMPLS Switching Based on Diverse Formats

GMPLS has introduced new additions to its label format in order to support network devices that switch in different domains. The architects of GMPLS have called this the *generalized label*. It is versatile in that it contains such information to allow a receiving device to program its related switch and forwards the data packet or frame regardless of structure. That structure may be pack, DS1/E1, lambda, and so on. The generalized label may represent a single wavelength, a time slot, or a single fiber strand. IEC, in Ref. 28, states that the information in a generalized label includes the following:

1. LSP encoding type that indicates what type of label is being carried (e.g., packet, lambda, SONET, etc.).
2. Switching type that indicates whether the node is capable of switching packets, time slots, fibers, or wavelengths.
3. There is a general payload identifier to indicate what payload is being carried by the LSP (e.g., virtual tributary [VT], DS-3, ATM, Ethernet, etc.)

GMPLS has a marked similarity with its predecessor, MPLS, where label distribution starts with the upstream LSR requesting a label from the downstream LSR. GMPLS allows the upstream LSR to suggest a label for an LSP that can be overridden by the downstream LSR. The LSP of interest can also suggest its own label. This label selection process is crucial to systems that must configure their own switch fabric, which is time-consuming. An example cited is a DCS that has a high switching granularity, typically found with DS-1 or DS-3 (E1 or E3) involving thousands of ports. We are asked to remember that a label in this case is used to quickly determine the internal path between input and output port. Figure 20.14 shows the GMPLS protocol stack.

Rather than wait to receive a label from the downstream node and carry out hardware configuration, the suggested label is used saving processing delay and

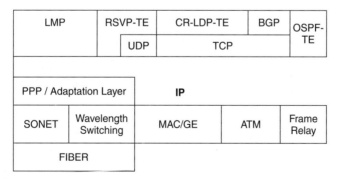

Figure 20.14. GMPLS protocol stack.

setup time. However, there is the eventuality that the downstream device rejects the suggested label and offers its own. In this case the upstream device is required to reconfigure itself with the new label [15].

12.5 Bundling Links

GMPLS has introduced link bundling which allows the mapping of several links into one. This is then advertised in the routing protocol (e.g., IS-IS, OSPF). Reference 25 states that some information is lost caused by the aggregating with the increased level of abstraction. The net result, though, is due to the link-state database size is greatly lowered as well as the number of links that need to be advertised. A bundled link only needs one control channel. The further helps to reduce the number of messages to be exchanged in the related signaling and routing protocols. Both point-to-point links and label-switched paths (LSPs) that have been advertised as links to the OSPF (forward adjacency) may be bundled.

Reference 25 lists the following restrictions for link bundling with GMPLS:

1. All links that comprise a bundled link must have the same link type such as point-to-point or multicast.

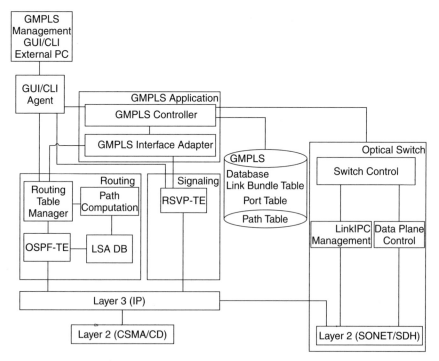

Figure 20.15. An example of GMPLS control plane software architecture. Based on Tellium material [28].

2. All links that comprise a bundled link must begin and end on the same pair of LSRs (label-switched router).

3. All links that comprise a bundled link must have the same traffic metric (e.g., protection type or bit rate).

4. All links that comprise a bundled link must have the same switching capability—PSC, TDMC, LSC, or FSC [16].

Figure 20.15 shows an example of GMPLS control plane software.

13 STANDARDIZATION OF OPTICAL CONTROL PLANE PROTOCOLS

Three international organizations are involved in the standardization of protocols for the optical control plane. These are the ITU with ASON (Automatically Switched Optical Network), the IETF with GMPLS [Generalized Multiprotocol Label (Lambda) Switching], and the OIF (Optical Internetworking Forum) [17].

13.1 GMPLS and ASON Differ

GMPLS. GMPLS switches are seen as operating in a GMPLS-only cloud of peer network elements. However, nodes on the edge of the GMPLS cloud are capable of accepting non-GMPLS traffic and tunneling it across the GMPLS cloud to edge nodes on the other side.

GMPLS implies a trusted environment in that all nodes and links that constitute the GMPLS network share the same IP address space, and this information is shared freely among nodes.

ASON. The key principle with ASON is to build for support of network legacy devices explicitly into the architecture. In the case of ASON, the ITU believes full vendor interoperability to be of low priority and considers it to be unrealistic.

The network, as seen by ASON, is composed of domains which interact with other domains in a standardized manner. However, the internal operation is protocol-independent and not subject to standardization. ASON calls the interfaces between these domains E-NNI or exterior node-to-node interface. E-NNIs can also be classified into "inter-operator" and "intra-operator."

Then, of course, there is the I-NNI (interior NNI), which is a vendor-specific proprietary interface used with a single-vendor domain.

ASON's concept of the network is extended more widely than in GMPLS to allow users to participate in the automated control plane. Specifically, the "user" is an endpoint device that requests the service of the transport network rather than provide it. With ASON, users can request connection services dynamically over a user–network interface (UNI). In GMPLS the closest thing to an ASON user is simply a GMPLS edge node, but this is not an exact mapping of the ASON concept.

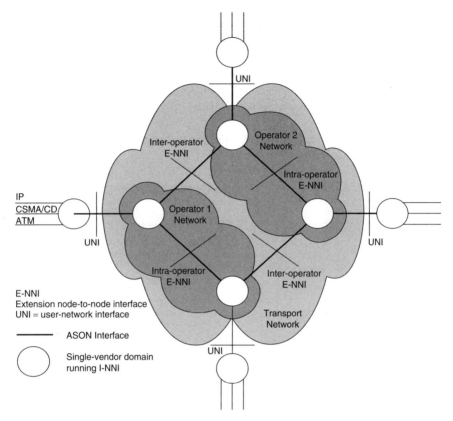

Figure 20.16. Simple ASON network showing UNI and E-NNI reference points. Based on Figure 2, page 13, Ref. 14.

Figure 20.16 illustrates a simplified ASON network showing UNI and E-NNI reference points.

The UNI, E-NNI, and I-NNI (interior-NNI) are known as "reference points," and the UNI and E-NNI indicate locations in the network where standardized protocols will need to be used. Reference points have different requirements on the amount of information kept hidden at each reference location. For example:

- The I-NNI is a trusted reference point. Full routing information can be flooded here.
- The UNI is an untrusted reference point. And it should hide all addressing and routing information pertaining to the interior of the network from the user. The ASON spec makes it clear that users should belong to a different address space from internal network nodes. This means that when GMPLS is mapped onto the ASON UNI reference point, the usual IP address cannot represent a user.

Reference 14 states that the inter-operator E-NNI is a semi-trusted reference point. Some amount of routing information is exchanged to allow routes to be calculated across the network. Network internals, though, are hidden to avoid leakage of confidential information between operators. The reference further states that intra-operator E-NNI is either trusted or semi-trusted, depending on the administrative structure of the particular operator.

CHANGE REQUIREMENTS IN GMPLS. New features are required at the UNI that have not been provided in the core GMPLS. Several of these features are listed below.

There must be complete separation of the use and the network addressing spaces. This is a requirement for the operators who support ASON. It then follows that because there is no routing information allowed to flow across the UNI, the user itself cannot calculate suitable routes. Instead, it must pass its requirements across to its neighbor in the network. Finally, the user has to have an expectation of what requirements the network can actually satisfy in advance, which creates the need for a start-of-day service discovery process.

The OIF has done the initial work to define the UNI profile of GMPLS in their UNI 1.0 specification. This involved creating a profile of the two GMPLS signaling protocols that satisfied the signaling requirements. They also enhanced the LMP (link management protocol) to include service discovery.

ASON makes a distinction between call setup signaling and connection setup. An ASON "call" is an association between two user endpoints. Even the concept of a "call" is a holdover from conventional telephony and is problematic to map onto GMPLS because:

- GMPLS signaling already has an association between endpoints. An ASON call looks like duplication of function.
- There are no "users" in GMPLS as in the ASON sense of the term.

Unmodified GMPLS routing protocols will not support the ASON requirement to flood user address reachability.

LAYERING AND LINKS. "Layering" has a different flavor here than in other chapters of this book. Here layering takes on the meaning of nesting finer granularity, lower bit rate capacity connections over coarser granularity, and higher bit rate capacity connections using a multiplexing function. These are known as hierarchical *label-switched paths* (LSPs) in GMPLS.

In common layered networks, a connection is set up at a lower layer $(n - 1)$ in order to provide a link at a higher layer (n). This is to say that the connection endpoints at layer $n - 1$ become directly adjacent at layer n. This concept is illustrated in Figure 20.17. As a result, connection setup and take-down operations at the $n - 1$ layer are used to modify the network topology at layer n.

The ITU (ASON) and the IETF (ASON) agree that a network consists of a set of nodes connected by a set of links. They start to disagree when we carry

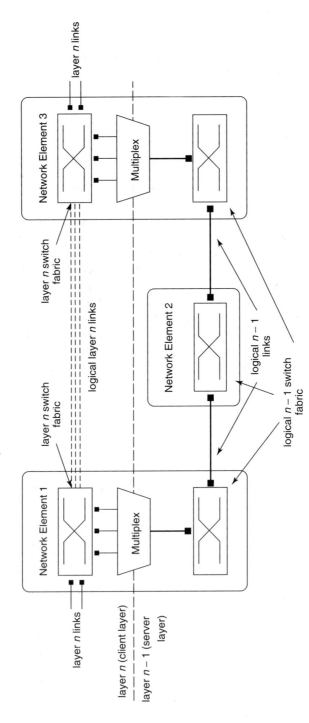

Figure 20.17. An illustration of layering. Based on Figure 3, page 16, Ref. 14.

this further. They do not agree on the function of a link in terms of the types of traffic it can carry.

GMPLS defines a link as being capable of supporting multiple different layers of switched traffic. In the GMPLS scheme of things regarding routing, a node can indicate whether it is any combination of λ-switch capable, TDM capable, or packet-switch capable for a given link. In the language of GMPLS a higher-layer link set up over a lower-layer connection is known as a *virtual link*.

A *link* is defined in ASON as being capable of carrying only a single layer of switched traffic. A link operated over a real physical medium is indistinguishable from one operated over a lower-layer higher bit rate capacity connection from the point of view of signaling, routing, and discovery. Thus with ASON each network layer must be treated separately. The phrase "treated separately" means that for each layer, there is a layer-specific instance of signaling, routing, and discovery protocols running.

This is not so with hierarchical routing where there are actually several instances of routing protocol operating over a single layer. There is one instance for each routing hierarchy level. For discovery, routing controllers maintain and advertise a separate topology for each switching layer in the network.

There seems to be little controversy between the IETF and the ITU when dealing with layered signaling, where both groups view it as intrinsically a single layer. Reference 14 explains that the purpose of signaling is to set up a switched connection, and connections are between endpoints at the same switching layer.

Layered routing is another matter. In GMPLS, a real physical fiber might be represented by OSPF-TE (open shortest path first–terminal endpoint, a protocol) as a single logical link with multiple switching capabilities. ASON, on the other hand, has the multiple logical links supported by the fiber that has to be advertised for discovery at their respective layer in the routing protocol.

The ITU, for their part, believe that this strict requirement on routing layering is crucial for scalable administration of large networks. It allows each layer to operate independently of any other layer. The complexity of route calculations does not increase, nor does information flooding within a particular layer. It only affects the entity that arbitrates between layers at each node.

Many in the IETF take the position that this is an overengineered requirement and is actually unscalable. Each new layer adds additional connectivities with neighbors, with resulting big increase in traffic in the control network. As a result, each node and link at each layer requires its own unique identifier. This results in a need for large address spaces capable of accommodating multiple layers.

13.2 Hierarchical Routing in Optical Networks

Once a network grows beyond a certain size, flooding a complete topological network description becomes unwieldy and impractical. This is primarily due to the frequency of updates and the large number of stations (nodes) requiring those updates. In such circumstances it becomes paramount to limit the amount of information being flooded. The first step is to partition the network into routing areas. Then, the routing databases are populated with more detailed information about the local routing information, and less detail is included about remote routing area. Furthermore, routing areas themselves can be partitioned recursively. This creates a hierarchy of routing information that varies in its level of summarization. A routing protocol instance runs at each level of this hierarchy.

Packet-switched networks already use the two broad approaches to hierarchical routing. These are (a) path vector routing at the top hierarchical level and used in Border Gateway Protocol (BGP) and (b) fully hierarchical link state routing, which is used in PNNI (private network-to-node interface). Now to explain hierarchical routing in these circumstances:

- BGP floods path vector information rather than link state information. It advertises routes to destinations and not network topology. It simplifies. Where multiple destinations are reachable via the same route, they are aggregated so that only one route is advertised. In the case where a single destination is reachable by multiple routes, the least cost route is retained. The others are discarded.
- PNNI establishes a hierarchy of routing controllers, all with a link state view of the network and can be run recursively at each level of the hierarchy. This is unlike BGP which just runs at the top level of the hierarchy. It uses higher-level routing controllers which have a wider view of the network but with more abstract information about nodes and links. There are lower-level routing controllers which give a narrower view of the network but more detailed information regarding nodes and links. It is noted in Ref. 14 that PNNI is not limited to the two or three hierarchical levels used in IP networks. PNNI is mature but has not been widely deployed in multivendor networks.

The ITU has taken a position that path vector information is insufficient to support a large scale, end-to-end optical network. A fully hierarchical link state protocol is recommended in its place. Reference 14 states that it is a lot clearer how constrained path computation works in a fully hierarchical routing scheme. The complexity here lies in the process of abstracting and summarizing a lower-level hierarchy to present a meaningful and useful topology at a higher level (Refs. 17–20).

OIF DDRP (DOMAIN-TO-DOMAIN ROUTING PROTOCOL). The OIF has recommended an enhancement of OSPF to turn it into a hierarchical link state protocol, DDRP. OSPF was selected because of its wide use in the IP world. Two subsets have evolved from DDRP: (1) a protocol-independent description of requirements and architecture and (2) two protocol-specific documents. One is based on OSPF, and the other is based on ISIS (an IETF routing protocol). After completion of its work, the OIF will select which of the two DDRP flavors to adopt for its E-NNI (external-NNI) implementation agreement.

The changes required—whichever flavor is selected—are fairly insignificant. What will be significant is the decision to go with DDRP itself. If this is done, we assume that the organization that adopts DDRP will give up on the use of the IP routing model for optical network provisioning. They will move, instead, to a fully hierarchical model.

Reference 14 expects that the ITU will adopt the OIF's DDRP model as the basis for ASON-compliant routing protocols (as specified in ITU Rec. G.7715). Whether the IETF changes its view that all that is required for optical routing is a three-level hierarchy has yet to be determined as of the end of 2003.

14 SUMMARY

A network consists of switches and transmission links connecting those switches. Switches and transmission links and devices involve hardware. In addition, a network requires a means of access and a method of routing messages. This portion of the network deals with protocols built with software. It would seem that we are describing conventional electronic networks. We are really dealing with optical networks. In this chapter we covered possible hardware and software combinations for feasible optical networks of the present and future. The question was addressed regarding how much of the network will remain electronic and how much can be implemented optically. We believe that the truly all-optical network remains illusive and a future goal. Optical signaling used for circuit setup in a connection-oriented network and in message headers for a connectionless-based network is going to be difficult to achieve with near-term technology.

REVIEW QUESTIONS

1. GbE (gigabit Ethernet) and SONET/SDH are competitors for an all-optical (data) network. Why would you think GbE would be more friendly to our cause than SONET/SDH? (Our "cause" is the development of an all-optical network.)

2. When dealing with an all-optical network, compare layer 1 with layer 2 switching.

3. Name the four layers of today's typical data network as envisioned for optical application.

4. For the next-generation optical network, it is said that two domains are emerging. Name each domain and its application.

5. What are the two basic network architectures supported by MPλS?

6. In a photonic network, name the two basic platforms.

7. Name the principal drawback of an overlay network.

8. What will be the principal change in switching as the telecommunications network evolves from an essentially electrical network to an all-optical network?

9. Given today's technology level, what are the two most common optical switches found that are available off-the-shelf?

10. There are seven basic limitations given for electro-optical switches. Name five of them.

11. One of the most common switches found in optical networks is the optomechanical switch. Give some idea of the switching speed of these switches (nanosecond, microsecond, millisecond, and second range).

12. Why is switching speed so important?

13. DWDM has brought a whole new dimension to optical switching. What will we be routing in one of these switches?

14. We expect MEMS switching to continue to take on greater importance in an optical network. There are two types of MEMS switch, one quite different from the other. What are they?

15. What are the three types of actuation mechanism used in MEMS switches?

16. Presently (circa 2003), MEMS switches can support up to 32 bidirectional ports. What would be an optimum goal for MEMS crosspoint matrix arrays?

17. What are the two basic methods of restoral of an optical network?

18. The future optical network will be two-layer. What will be the basic protocol used? It will evolve to _____ (protocol acronym) when DWDM is used.

19. Similar to present automatic protection switching, there is the restoration phase, which is the replacement of the failed optical path by another. Once the failed path is repaired, what will be its function as operation returns to normal? There is an advantage to the network operator here by doing it the way suggested in the text? What is that advantage?

20. What is *dynamic lambda provisioning*?

21. In the optical mesh network we have assumed that the control plane will be IP-compatible using what specific protocol?

22. Name the four types of optical networks based on the components employed.

23. How are *optical link networks* defined?

24. Wavelength routed (WR) networks include optical circuit switches. There are two types which should be very familiar to the reader. What are they?

25. In a ring network, as with any traffic-bearing fiber, is switching time so important? (*Hint*: Go back to the basics of digital transmission.)

26. We listed four types of optical networks; name three of them.

27. There are four switching domains. Name three of them.

28. In an optical network, traffic starts at the electronic level with ones and zeros. What kind of traffic would we most probably expect? (such as ATM, switched TDM—).

29. An LSP bases its switching on what?

30. In a DCS (digital cross-connect system), what is the purpose of a label (assuming GMPLS)?

31. With GMPLS, what does bundling do?

32. Assume GMPLS—all bundled links must start and end at a what?

33. In the optical network sense, explain the concept of *layering*. It is quite different than our earlier discussions of layering (e.g., OSI, etc.).

34. How does GMPLS define a *virtual link*?

35. Why has the industry turned to hierarchical routing in large/very large optical networks?

36. What would be the first step in turning an optical network into one that uses *hierarchical routing*?

REFERENCES

1. M. Sluyski, AMCC, "The Evolution of Crossconnects within the Emerging Managed Optical Layer," *Lightwave* (June 2000).

2. L. Ceuppens, "Multiprotocol Lambda Switching Comes Together," *Lightwave*, page 80 (August 2000).

3. R. Chua and Y. Hu, "Optical Switches Are Key Components in High-Capacity, Data-Centric Networks," *Lightwave*, page 43 (November 1999).

4. T. Krause, "Migration to All-Optical Networks," Alcatel Raleigh, July 24, 2000, from the web: www.usa.alcatel.com/telecom.

5. P. Green, "Progress in Optical Networking," *IEEE Commun. Mag.* (January 2001).

6. M. Veeraraghavan, R. Karri, et al., "Architectures and Protocols That Enable New Applications on Optical Networks," *IEEE Commun. Mag.* (March 2001).

7. M. Bourne, "MEMS Switching...and Beyond," Cahners In-Stat Group, *Lightwave*, page 204 (March 2001).

8. J. Lawrence, Cisco Systems, "Designing Multiprotocol Label Switching Networks," *IEEE Commun. Mag.*, page 134 (July 2001).

9. E. Rosen, et al., *MPLS Architecture*, RFC 3031, January 2001. From the internet.

10. Internet: www.nanog.org/mtg-9905/ppt/mpls. 10/23/01.

11. M. Fernandez and E. Kruglic, "MEMS Technology Ushers in New Age in Optical Switching," Optical Micromachines Inc., *Lightwave*, page 146 (August 2000).

12. R. L. Freeman, *Fiber-Optic Systems for Telecommunications*, John Wiley & Sons, Hoboken, NJ, 2003.

13. From the web: www.iec.org/online/tutorials/opt_net/topic11 (Alcatel), 8/24/03.

14. N. Larkin, *ASON and GMPLS—The Battle of the Optical Control Plane*, Data Connection Limited, Enfield, UK, 2002.

15. *Generalized Automatic Discovery Techniques*, ITU Rec. G.7714/Y.1705, ITU, Geneva 11-2001 (*Note*: refers to ASON).

16. *Architecture and Specification of Data Communication Network*, ITU Rec. G.7717/Y.1703, ITU, Geneva, November 2001 (*Note*: refers to ASON).

17. *Architecture and Requirements for Routing in the Automatic Switched Optical Network(s)*, ITU, Rec. G.7715/Y.1706, ITU, Geneva, June 2002.

18. *Architecture for the Automatic Switched Optical Networks*, ITU Rec. G.8080/Y.1305, ITU Geneva, Nov. 2001.

19. *Generalized Multi-Protocol Label Switching (GMPLS) Architecture*, draft-ietf-ccamp-gmpls-architecture, IETF, ed., Ann Arbor, MI, September 10, 2001 (see RFC 3160).

20. *Automatic Switched Optical Network (ASON) Architecture and Its Related Protocols*, draft-ietf-ipo-ason, IETF, ed., Ann Arbor, MI, June 2001.
Note to reader: ccamp = common control and measurement protocol (plane).

21. L. Ceuppens, private communication, Calient Networks, San Jose, CA, March 1, 2002.

22. E. Kruglick, private communication, OMM Inc., San Diego, CA, March 8, 2002.

23. Y. Hu, private communication, E-TEK Dynamics, San Jose, CA, March 25, 2002.

24. Multiprotocol Label Switching Architecture, RFC 3031, IETF, January 2001.

25. Generalized Multiprotocol Label Switching (GMPLS), from the web, www.iec.org/online/tutorials/gmpls/topic01/.

26. D. Briera and C. Bacco, *Streamlining Your Network with GMPLS, Parts 1 and 2* (from the web), Network World Fusion, 01/07/02.

27. *GMPLS Technology Primer*, Pioneer Consulting, from the web: www.reasearchand-marketreport/info/ASP.

28. H. Liu et al., *GMPLS-Based Control Plane for Optical Networks, Early Experience*, Tellium, 2 Crescent Place, P.O. Box 901, Oceanport, NJ 07757-0901. From the web: www.tellium.com/documents/spie2000_hang.pdf. hliu@tellium.com.

21

NETWORK MANAGEMENT

1 WHAT IS NETWORK MANAGEMENT?

Effective network management optimizes a telecommunication network's operational capabilities. The key word here is *optimizes*.

These are some of the connotations that can be derived:

- It keeps the network operating at peak performance.
- It informs the operator of impending deterioration.
- It provides easy alternative routing and work-arounds when deterioration and/or failure take place.
- It provides the tools for pinpointing cause/causes of performance deterioration or failure.
- It serves as the front-line command post for network survivability.

There are numerous secondary functions of network management. They are important but, in our opinion, still secondary. Among these items are:

- It informs in quasi-real time regarding network performance.
- It maintains and enforces network security, such as link encryption and issuance and use of passwords.
- It gathers and files data on network usage.
- It performs a configuration management function.
- It also performs an administrative management function.

2 THE BIGGER PICTURE

Many seem to view network management as a manager of data circuits only. There is a much bigger world out there. Numerous enterprise and government

Telecommunication System Engineering, by Roger L. Freeman
ISBN 0-471-45133-9   Copyright © 2004 Roger L. Freeman

networks serve for the switching and transport of *multimedia communications*. The underlying network will direct (switch) and transport voice, data, and image traffic. Each will have a traffic profile notably differing from the other. Nevertheless, they should be managed as an entity. If for no other reason, it is more cost-effective to treat the whole rather than piecemeal by its parts.

There is a tendency in the enterprise scene to separate voice telephony (by calling it *telecommunications*) and another world, that of data communications. Perhaps that is why network management seems to often operate on two separate planes. One is data and very sophisticated, and the other is voice, which may have no management facilities at all. This next section treats network management as a whole consisting of its multimedia parts: voice, image, and data, which includes facsimile, telemetry, and CAD/CAM.

3 TRADITIONAL BREAKOUT BY TASKS

There are five tasks traditionally involved with network management:

- Fault management
- Configuration management
- Performance management
- Security management
- Accounting management

3.1 Fault Management

This is a facility that provides information on the status of the network and subnetworks. The "information on the status" should not only display faults (meaning failures) and their location, but should also provide information on deteriorated performance. One cause of deteriorated performance is congestion. Thus, ideally, we would also like to isolate the cause of the *problem*.

Fault management should include the means to bypass troubled sections of a network, as well as to patch-in new equipment for deteriorated or failed equipment.

The complexity of modern telecommunication networks is such that as many network management tasks as possible should be automated. All displays, read-outs, and hard-copy records should be referenced to a network time base down to 0.1 s.* This helps in correlating events, an important troubleshooting tool.

3.2 Configuration Management

Configuration management establishes an inventory of the resources to be managed. It includes resource provisioning (timely deployment of resources to satisfy

---

* Network time bases are usually much more precise and derive from a network clock, which may be external to the network, such as GPS. For example, Stratum #1 clock has $\pm 1 \times 10^{-11}$ per month stability, which is minimum acceptable for the digital network.

an expected service demand) and service provisioning (assigning services and features to end-users). It identifies, exercises control over, collects data from, and provides data to the network for the purpose of preparing for, initializing, starting, and providing for the operation and termination of services. Configuration management deals with equipment and services, subnetworks, networks, and interfaces. Its functions are closely tied to fault management, as we have defined it above.

3.3 Performance Management

Performance management is responsible for monitoring network performance to assure it is meeting specified performance. Some literature references [1] add growth management. They then state that the objective of performance and growth management is to ensure that sufficient capacity exists to support end-user communication requirements.

Of course, there is a fine line defining *network capacity*. If too much capacity exists, there will probably be few user complaints, but there is excess capacity. Excess capacity implies wasted resources, thus wasted money. Excess capacity, of course, can accommodate short-term growth. Therefore performance/growth management supplies vital information on network utilization. Such data provide the groundwork for future planning.

3.4 Security Management

Security management controls access to and protects both the network and the network management subsystem against intentional or accidental abuse, unauthorized access, and communication loss. It involves link encryption, changes in encryption keys, user authentication, passwords, firewalls, and unauthorized usage of telecommunication resources.

3.5 Accounting Management

Accounting management processes and records service and utilization information. It generates customer billing reports for services rendered. It identifies costs and establishes charges for the use of services and resources in the network. It may also be a repository for plant-in-place investment for telecommunications plant and provides reports to upper management on return on that investment.

4 SURVIVABILITY — WHERE NETWORK MANAGEMENT REALLY PAYS

The network management center is the front-line command post for the battle for network survivability. We can model numerous catastrophic events affecting a telecommunications network, whether the PSTN or a private/enterprise network. Among such events are fires, earthquakes, floods, hurricanes, terrorism, and public disorder. Telecommunications have brought revolutionary efficiencies to the way we do business and are so very necessary even for life itself. Loss of these facilities could destroy a business, even possibly destroy a nation. A properly designed network management system could mitigate losses, even save a network almost in its entirety.

This brings in the first rule toward survivability. A network management center is a place of point failure. Here we mean that if the center is lost, probably nearly all means of reconfiguring the network are lost. To avoid such a situation, a second network management center should be installed. This second center should be geographically separate from the principal center. It is advisable that the second center share the network management load and be planned and sized to be able to take the entire load with the loss of the principal center. There should be a communications orderwire between the two centers. Both centers should be provided with no-break power and backup diesel generators.

One simple expedient for survivability is to backup circuits with an arrangement with the local telephone company. This means that there must be compatibility between the enterprise network and the PSTN as well as one or more points of interface. A major concern is the network clock. One easy solution is to have the enterprise network derive its clock from the digital PSTN.

In the following chapter sections we will describe means to enhance survivability still further.

4.1   Survivability Enhancement — Rapid Troubleshooting

An ideal network management system will advise the operator of one or more events,* reveal where in the network they occurred, and provide other handy troubleshooting data. In fact, the network management system should, in most cases, warn the operator in advance of an impending fault/faults. This would be a boon to survivability.

Oftentimes the "ideal" is unachievable or only partially achievable. Human intervention will be required. The troubleshooter should have available a number of units of test equipment, some more automated than others, to aid in the pinpointing the cause of an event.

Steve Dauber's article, "Finding Fault" [3], describes four steps to correct network faults. These are:

1. Observing symptoms
2. Developing an hypothesis
3. Testing the hypothesis
4. Forming conclusions

OBSERVING SYMPTOMS. It should be kept in mind that often many symptoms will appear at once, usually a chain reaction. We must be able to spot the real causal culprit or we may spend hours or even days chasing effects, not the root cause.

---

* We define an *event* as something out-of-the-ordinary that occurred.

The reference article suggests these four reminders:

1. Find the range and scope of the symptoms. Does the problem affect all stations (all users)? Does it affect random users or users in a given area? We're looking for an area pattern here.

2. Are there some temporal conditions to the problem? How often does it occur per day, per hour, and so on? Is the problem continuous or intermittent? What is its regularity? Can we set our watches by its occurrence or is it random in the temporal context?

3. Have there been recent rearrangements, additions to the network, reconfigurations?

4. Software and hardware have different vintages, which are called *release dates*. I write this on WordPerfect 7.0, but what was its release date? The question one should ask then would be, Are all items of a certain genre affected in the same way? Different release dates of a workstation may be affected differently, some release dates not at all. The problem may be peculiar to a certain release date. This is particularly true of software.

Before we can move forward in the troubleshooting analysis, we must have firmly at hand the troublesome network's *baseline* performance. What is meant here is that we must have a clear idea of "normal" operation of the network so that we can really qualify and quantify its anomalous operation. For example, what is the expected bit error rate (BER), and is the value related to a time distribution? Can we express error performance in EFS (error-free seconds), ES (errored seconds), and SES (severely errored seconds)? These are defined in ITU-T Rec. G.821 [4].

Steven Dauber [3] lists five network-specific characteristics that the troubleshooter should have familiarity with or data on.

A. *Network Utilization.* What is the average network utilization? How does it vary through the work day? Characteristics of congestion, if any, should be known, and where and under what circumstances might it be expected?

B. *Network Applications.* What are the dominant network applications on the network? What version numbers is it running?

C. *Network Protocol Software.* What protocols are running on the network? What are the performance characteristics of the software, and are these characteristics being achieved?

D. *Network Hardware.* Who manufactured the network interface controllers, media attachment units, servers, hubs, and other connection hardware? What versions are they? What are their performance characteristics? Expected? Met?

E. *Inter-networking Equipment.* Who manufactured the repeaters, bridges, routers, and gateways on the network? What versions of software and firmware are they running? What are the performance characteristics? What are the characteristics of the interfaced network that are of interest?

Developing a Hypothesis. In this second step, we make a statement as to the cause of the problem. We might say that T1 or E1 frame alignment is lost because of deep fades being experienced on the underlying microwave transport network. Or we might say that excessive frames being dropped on a frame relay network is due to congestion being experienced at node B.

Such statements cannot be made without some strong bases to support the opinion. Here is where the knowledge and experience of the troubleshooter really pays. Certainly, there could be other causes of E1 or T1 frame alignment, but if underlying microwave is involved, that would be a most obvious place to look. There could be other reasons for dropping frames in a frame relay system. Errored frames could be one strong reason.

Testing the Hypothesis. We made a statement, now we must back it up with tests. One test I like is correlation. Are the fades on the microwave correlated with the fade occurrences? That test can be done quite easily. If they are correlated, we have some very strong backup that the problem is with the microwave. The frame relay problem may be another matter. First, we could check the FECN and BECN bits to see if there was a change of state passing Node B. If there is no change, assuming that flow control is implemented, then congestion may not be the problem. Removing the frame relay from the system and carrying out a BERT (bit error rate test) over some period of time would prove or disprove the noise problem.

A network analyzer is certainly an excellent tool in assisting in the localization of faults. Some analyzers have preprogrammed tests which can save the troubleshooter time and effort. Many networks today have some sort of network monitoring equipment incorporated. This equipment may be used in lieu of network analyzers or in conjunction with network analyzers. Again we stress the importance of separating cause and effects. Many times network analyzers or network management monitors and testers will only show effects. The root cause may not show at all and must be inferred or separate tests must be carried out to pinpoint the cause.

We might digress here to talk about what is often called *tonterías* in Spanish. This refers to "silly things." Such tonterías are often brought about by careless installation or careless follow-up repair. Coaxial cable connectors are some of my favorites. Look for intermittents and cold solder joints. A good tool, but not necessarily foolproof, is a time-domain reflectometer (TDR). It can spot where a break in a conductor is down to a few feet or less. It can do the same for an intermittent, when in the fault state. In fact, intermittents can prove to be a nightmare to locate. An electrically noisy environment can also be very troublesome.

Conclusions. A conclusion or conclusions are drawn. As we say, "the proof of the pudding is in the eating." The best proof that we were right in our conclusion is to fix the purported fault. Does it disappear? If so, our job is done, and the network is returned to its "normal" (baseline) operation.

What "conclusions" can we draw from this exercise? There are two basic ingredients to elemental network troubleshooting: expertise built on experience of the troubleshooter and the availability of essential test equipment. Troubleshooting time can be reduced (degraded operation or out-of-service time reduction) by having online network management equipment. With ideal network management systems, this time can be cut to nearly zero.

5 SYSTEM DEPTH — A NETWORK MANAGEMENT PROBLEM

An isolated LAN is a fairly simple network management example. There is only a singular transmission medium and, under normal operating conditions, only one user is transmitting information to one or several recipients. It is limited to only two OSI layers. For troubleshooting, often a protocol analyzer will suffice, although much more elaborate network management schemes and equipment are available.

Now connect that LAN to the outside world by means of a bridge or router and network management becomes an interesting challenge. One example from experience was a VAX running DECNET which was a station on a CSMA/CD LAN. The LAN was bridged to a frame relay box which fed a 384-kbit/s channel with an E1 hierarchy (i.e., 6 E0 channels) via tandemed microwave links to a large facility at the distant end (550 km) with similar characteristics. This connectivity is shown diagrammatically in Figure 21.1. Such is typical of a fairly complex network requiring an overall network management system. To make it even more difficult, portions of the network were leased from the local common carrier, but soon to be cutover to own ownership.

5.1 Aids in Network Management Provisioning

Modern E1 and T1 digital systems* are provided a means for operational monitoring of performance. The monitoring is done in quasi-real time and while operational (i.e., in traffic).

In Chapter 8 we discussed PCM systems and the digital network including T1 (DS1) and E1 hierarchies. Let us quickly review their in-service performance monitoring capabilities.

DS1 or T1 has a frame rate of 8000 frames a second. Each frame is delineated with a framing bit or F-bit. With modern alignment and synchronization algorithms, to maintain framing repetition of the F-bit 8000 times a second is excessive and unnecessary. Advantage was taken of F-bit redundancy by the development of the extended superframe (ESF). The ESF consists of 24 consecutive DS1 frames. With 24 frames we expect to have 24 framing bits. Of these, only six bits need to be used for framing, six are used for a cyclic redundancy check (CRC-6) on the superframe, and the remaining 12 bits form a 4-kbit/s data link for network control and maintenance. It is this channel that can serve as transport for network management information. It can also serve as an ad hoc test link.

---

* This capability is not unique to T1- and E1-type systems. For example, SONET and SDH have even more sophisticated capabilities, often referred to as OA&M (operations, administration, and maintenance). ATM is (will be) rich in such network management aids.

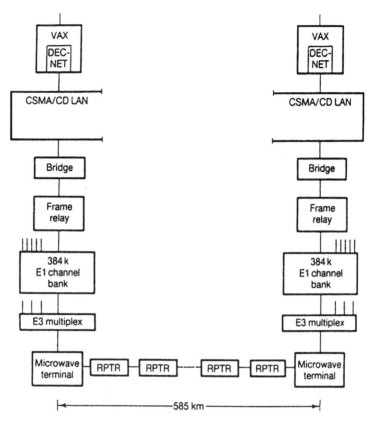

Figure 21.1. A typical multilevel network requiring a network management system. Note multiple convergences.

Our present concern is in-service monitoring. For instance, we can get real time error performance with the CRC-6, giving us a measure of errored seconds and severely errored seconds in accordance with ITU-T Rec. G.821 [4]. Such monitoring can be done with test equipment such as the HP* 37702A or HP 37741A.

Permanent monitoring can be carried out using Newbridge network monitoring equipment, which can monitor an entire T1 or E1 network for frame alignment loss, errored seconds, and severely errored seconds. One can "look" at each link and examine its performance in 15-minute windows for a 24-hour period. The Newbridge equipment can also monitor selected other data services such as frame relay, which, in our example above, rides on E1 aggregates. Such equipment can be a most important element in the network management suite.

The E1 digital network hierarchy also provides capability of in-service monitoring and test. We remember from Chapter 8 that E1 has 32 channels or time

---

* HP stands for Hewlett-Packard, a well-known manufacturer of electronics test equipment. HP spun off the test equipment line to Agilent Technologies.

slots: 30 are used for the payload and 2 channels or time slots serve as support channels. The first of these is channel (or time slot) 0, and the second is channel (time slot) 16. This latter is used for signaling. Time slot (channel) 0 is used from synchronization and framing. Figure 21.2 shows the E1 multiframe structure.

The sequence of bits in the frame alignment (TS0) signal of successive frames is illustrated in Figure 21.2. In frames not containing the frame alignment signal, the first bit is used to transmit the CRC multiframe signal (001011) which defines the start of the sub-multiframe (SMF). Alternate frames contain the frame alignment word (0011011) preceded by one of the CRC-4 bits. The CRC-4 remainder is calculated on all the 2048 bits of the previous sub-multiframe (SMF), and the 4-bit word is sent as C1, C2, C3, C4 of the current SMF. Note that the CRC-4 bits of the previous SMF are set to zero before the calculation is made.

At the receive end, the CRC remainder is recalculated for each SMF and the result is compared with the CRC-4 bits received in the next SMF. If they differ, then the checked SMF is in error. What this is telling us is that a block of 2048 bits had one or more errors. One thousand CRC-4 block error checks are made every second. It should be noted that this in-service error detection scheme does not indicate BER unless one assumes a certain error distribution (random or burst errors), to predict the average errors per block. Rather it provides a block error measurement.

This is very useful for estimating percentage of errored seconds (%ES), which is usually considered the best indication of quality for data transmission—itself a block or frame transmission process. CRC-4 error checking is fairly reliable with the ability of detecting 94% of errored blocks even under poor BER conditions (see ITU-T Rec. G.706 [5]).

	Sub-multiframe (SMF)	Frame number	Bits 1 to 8 of the frame in timeslot 0							
			1	2	3	4	5	6	7	8
Multiframe	I	0	C_1	0	0	1	1	0	1	1
		1	0	1	A	S_{a4}	S_{a5}	S_{a6}	S_{a7}	S_{a8}
		2	C_2	0	0	1	1	0	1	1
		3	0	1	A	S_{a4}	S_{a5}	S_{a6}	S_{a7}	S_{a8}
		4	C_3	0	0	1	1	0	1	1
		5	1	1	A	S_{a4}	S_{a5}	S_{a6}	S_{a7}	S_{a8}
		6	C_4	0	0	1	1	0	1	1
		7	0	1	A	S_{a4}	S_{a5}	S_{a6}	S_{a7}	S_{a8}
	II	8	C_1	0	0	1	1	0	1	1
		9	1	1	A	S_{a4}	S_{a5}	S_{a6}	S_{a7}	S_{a8}
		10	C_2	0	0	1	1	0	1	1
		11	1	1	A	S_{a4}	S_{a5}	S_{a6}	S_{a7}	S_{a8}
		12	C_3	0	0	1	1	0	1	1
		13	E	1	A	S_{a4}	S_{a5}	S_{a6}	S_{a7}	S_{a8}
		14	C_4	0	0	1	1	0	1	1
		15	E	1	A	S_{a4}	S_{a5}	S_{a6}	S_{a7}	S_{a8}

Figure 21.2. E1 multiframe, CRC-4 structure. *Notes*: E = CRC-4 error indication bits; S_{a4} to S_{a8} = spare bits. These bits may be used for a network management (maintenance) link. C_1 to C_4 = CRC-4 bits. A = remote alarm indication. From Table 4b/G.704, page 81, CCITT Rec. G.704, 1988 [10].

Another powerful feature of E1 channel 0 (when equipped) is the provision of local indication of alarms and errors detected at the far end. When an errored SMF is detected at the far end, one of the E-bits (see Figure 21.2) is changed from a 1 to a 0 in the return path multiframe (TS0). The local end, therefore, has exactly the same block error information as the far-end CRC-4 checker. Counting E-bit changes is equivalent to counting CRC-4 block errors. Thus the local end can monitor the performance of both the go and return paths. This can be carried out by the network equipment itself, or by a test set such as the HP 37722A monitoring the E1 2.048-Mbit/s data bit stream. In the same way, the A-bits return alarm signals for loss of frame or loss of signal from the remote end. Loopback testing is a fine old workhorse in our toolbox of digital data troubleshooting aids. There are two approaches for DS1 (T1) and E1 systems: intrusive and nonintrusive. Intrusive, of course, means that we interrupt traffic by taking one DS0 or E0 channel out of service, or the entire aggregate. We replace the channel with a PRBS (pseudo-random binary sequence) or other sequence specifically designed to stress the system. Commonly, we use conventional BERT (bit error rate test) techniques looping back. This idea of loopback is shown in Figures 21.3a and 21.3b.

ESF and channel 0 data channel testing is nonintrusive. It does not interfere with customer traffic. Trouble can also be isolated, whether in the "go" or "return" channel of the loopback. Both intrusive and nonintrusive testing could be automated in the network management suite.

Many frame relay equipments also have forms of in-service monitoring as well as a system of fault alarms. In fact, most complex telecommunication equipment has built-in monitoring and test features. The problem often is that these features are proprietary, whereas our discussion of DS1/E1 systems is they have been standardized by ITU-T and Telcordia recommendations and publications. This

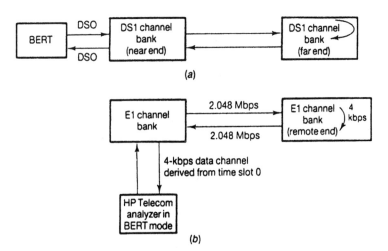

Figure 21.3. (a) Loopback of a DS0 channel with BERT test in place, intrusive, or on spare DS0. (b) Loopback of 4-kbit/s data channel derived from ESF on DS1 or channel 0 on E1.

is probably the stickiest problem facing network management systems—that is, handling, centralizing, and controlling network management features in a multi-vendor environment.

5.2 Communications Channels for the Network Management System

A network management facility is usually centrally located. It must monitor and control distant communications equipment. It must have some means of communicating with this equipment, which may be widely dispersed geographically. We have seen where DS1 and E1 systems provide a data channel for (OAM) operations and maintenance. Higher levels of the DS1 and E1 hierarchies have special communication channel(s) for OAM. So do SONET and SDH.

The solution for a LAN is comparatively straightforward. The network management facility/LAN protocol analyzer becomes just another active station on the LAN. Network management traffic remains as any other revenue-bearing traffic on the LAN. Of course, the network management traffic should not overpower the LAN with message unit quantity which we might call network management overhead.

WANs vary in their capacity to provide some form of communicating network management information. X.25 provides certain types of frames or messages dedicated for network control and management. However, these frames/messages are specific to X.25 and do not give data on, say, error rate at a particular point in the network. Frame relay provides none with the exception of flow control and the CLLM, which are specific to frame relay. For a true network management system, a separate network management communication channel may have to be provided. It would have to be sandwiched into the physical layer. However, SNMP (described below) was developed to typically use the transport services of UDP/IP. (See Chapter 11 for a discussion of TCP/IP and related protocols.) It is additional overhead, and care must be taken of the percentage of such overhead traffic compared to the percentage of "revenue bearing" traffic. Some use the term "in-band" when network management traffic is carried on separate frames on the same medium, and they use the term "out-of-band" when a separate channel or time slot is used such as with E1/T1.

6   NETWORK MANAGEMENT FROM A PSTN PERSPECTIVE

6.1   Objectives and Functions

The term used by Telcordia [12] for network management is "surveillance and control." The major objectives for network surveillance and control organizations are:

- Maintain a high level of network utilization.
- Minimize the effects of network overloads.
- Support the BOC's* National Security Emergency Preparedness commitment.

* BOC stands for Bell operating company, a local exchange carrier.

There are three important functions that contribute to attaining these objectives:

- Network traffic management (NTM)
- Network service
- Service evaluation

6.2 Network Traffic Management Center

An NTM center provides real-time surveillance and control of message traffic* in local access and transport area (LATA) telephone networks. The goal of an NTM center is to increase call completions and optimize the use of available trunks and switching equipment. Several dedicated OSSs (operational support systems) are employed by an NTM center to achieve this goal by accumulating information on both the flow of traffic and the manual and automatic/dynamic control capabilities provided by the network switching elements. Using this information and call control capability, a network traffic manager can optimize the call-carrying capacity of his/her network. The OSSs also enable the network traffic manager to interact with the network to minimize the adverse effects of traffic overloads and machine and/or facility failures.

6.3 Network Traffic Management Principles

NTM decisions are guided by four principles, which apply regardless of switching technology, network structure, signaling characteristics, or routing techniques. All NTM control actions are based on at least one of the following principles:

1. *Keep All Trunks Filled with Messages.*[†] Since the network is normally trunk-limited, it is important to optimize the ratio of messages to nonmessages on any trunk group. When unusual conditions occur in the network and cause increased short holding-time calls (nonmessages), the number of carried messages decreases because nonmessage traffic is occupying a larger percentage of network capacity.

   NTM controls are designed to reduce nonmessage traffic and allow more call completions. This results in higher customer satisfaction and increased revenues from the network.

2. *Give Priority to Single Link Connections.* In a network designed to automatically alternate route calls, the most efficient use of available trunks occurs when traffic loads are at or below normal engineered values. When the engineered traffic load is exceeded, more calls are alternate-routed, and therefore, must use more than one link to complete a call. During overload situations, the use of more than one link to complete a call occurs more often, and the possibility of a multilink call blocking other call attempts is greatly increased. Thus in some cases it becomes necessary to limit alternate routing to give first-routed traffic a reasonable chance to complete.

---

* Message traffic in this context means telephone traffic.
[†] Telcordia [12] defines a message as *a call that has a high probability of completion.*

Some NTM controls block a portion or all alternate-routed calls to give preference to first-routed traffic.

3. *Use of Available Trunking.* The network is normally engineered to accommodate average business day (ABD) busy-hour calling requirements. Focused overloads (e.g., due to storms, floods, civil disturbances) and holiday calling often result in greatly increased calling and drastic changes from the calling patterns for which the network is engineered. This aberration can also be caused by facility failures and switching system outages. In these cases, some trunk groups are greatly overloaded, while others may be virtually idle. NTM controls can be activated in many of these cases to use the temporary idle capacity in the network. These controls are known as *reroutes.*

4. *Inhibit Switching Congestion.* A switching system is engineered to handle the expected number of attempts generated over its trunk groups with little or no service degradation. However, large numbers of ineffective attempts that exceed the engineered capacity of the switching system can result in switching congestion. If this switching congestion is not relieved, it not only affects potential messages within the congested switching system, but also can cause connected switching systems to become congested. Therefore, NTM controls are available that can remove the ineffective attempts to a congested switching system. These controls will result in inhibiting the switching congestion and preventing its spread to adjacent switching systems.

6.4   Network Traffic Management Functions

The following are several types of overloads for which network traffic management controls can provide complete or partial relief.

1. *A general network overload* is caused by changes in traffic patterns and/or increased traffic load. These changes may be generated by a reduced business week (heavier calling before and after an extended weekend), holiday traffic, local or seasonal changes such as an increase in tourist traffic, unanticipated growth, and natural or man-made disasters. In cases of general network overload, a large amount of the network capacity may be used to switch calls that have a poor chance of completing. These calls are often regenerated many times by both the calling customers and switching systems before they are completed. This results in contention between the "poor completers" and those calls having a better chance of completing. Because of the volume of regeneration to the poor completers, much of the available trunking and switching capacity is used in switching these calls to an overflow condition.

As NTM personnel identify poor completers, appropriate measures are instituted, when necessary, to control congestion and remove some or all of these calls from the network. This is done by using code-blocking, call-gapping, and protective trunk group controls. The control of poor completers can greatly increase the number of messages handled by the network. Figure 21.4 illustrates the typical

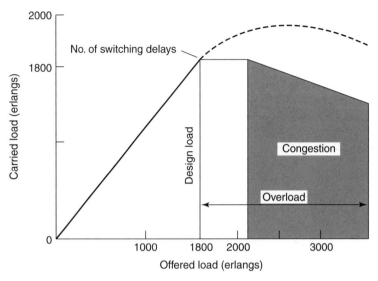

Figure 21.4. Network performance during overload conditions.

network performance under a general overload. This figure shows the decrease of completed traffic when the offered load exceeds the engineered capacity and congestion is present, with the increased number of calls encountering switching delay.

2. A *focused overload* is generally directed toward a particular location and may result from media stimulation (e.g., news programs, advertisements, call-in contests, telethons) or events that cause mass calling to government or public service agencies, weather bureaus or public utilities. Without the application of appropriate network controls, the effects of these types of overloads could spread throughout the network. Focused overloads are normally managed using code controls or, if anticipated, trunk-limited or choke networks.

3. A *switching system overload* occurs because each individual switch is engineered to handle a specific load which is known as *engineered capacity*. The engineered capacity is usually less than the total switching capacity. When the load is at or below engineered capacity, the switch handles calls in an efficient and reliable manner. However, when the load increases beyond the engineered capacity, delays can occur internal to the switch. This resulting internal congestion can spread, causing connected systems to wait for start-dial indications. This can also cause internal congestion in connected switching systems.

At the onset of overload, also known as circuit shortage, the dominant cause for customer blockage is the failure to find an idle circuit. Circuit blockage alone limits the number of extra calls that can be completed but does not cause a significant loss in call-carrying capacity of the network below its maximum. As the overload persists and the network enters a congested state, regeneration-calling pressure changes customer blockage from circuit shortage to switching delays.

Switching delays cause timeout conditions during call setup and occur when switches become severely overloaded. Timeouts are designed into switches to release common-control components after excessively long delay periods and provide the customer with a signal indicating call failure. Switching congestion timeouts with short holding-time attempts on circuit groups replace normal holding-time calls. Switching delays spread rapidly throughout the network.

A *trunk-group overload* usually occurs during general or focused overloads and/or atypical busy hours. Some of the overload causes not discussed above are facility outages, inadequate trunk provisioning, and routing errors. The results of trunk-group overload can be essentially the same as those previously discussed for general overloads. However, the adverse effects are usually confined to the particular trunk group or the apex area formed by the trunk group and those groups that alternate-route to the overloaded trunk-group. Trunk-group overload problems can often be minimized or handled completely by the use of temporary NTM reroute controls until a more permanent solution can be provided.

6.5 Network Traffic Management Controls

6.5.1 Circuit-Switched Network Controls. There are two broad categories of NTM controls:

- *Protective Controls.* These controls remove traffic from the network during overload conditions. This traffic is usually removed as close as possible to its origin, thus making more of the network available to other traffic with a higher probability of completion.
- *Expansive Controls.* These controls reroute traffic from routes experiencing overflows or failures to other parts of the network that are lightly loaded with traffic because of noncoincident trunk and switching system busy hours.

Implementation of either type of control can be accomplished on a manual or automatic basis. For example, manual controls are activated by network traffic managers, and automatic controls are activated by network components. In some switches, these controls are implemented on a planned control-response basis that is preprogrammed into the switch. In other systems, controls are available on a flexible basis, whereby any control can be assigned to any trunk group on a real-time basis.

The availability of any specific control, its allowable control percentages, and the method of operation can vary with the specific type of switch. In many instances, these network controls can be activated with variable percentages of traffic affected (e.g., 25%, 50%, 75%, and 100%) to fine-tune the control to match the magnitude of the problem. Some switches also allow further control selectivity by the use of hard-to-reach (HTR) code-determination algorithms or specification of alternate-routed traffic, direct-routed traffic, or combined direct- and alternate-routed traffic-control choices. The most common manual controls are described below:

- *Cancel controls* consist of two variations. *"Cancel from"* (CANF) potentially prevents overflow traffic from a selected trunk group from advancing to any alternate route. *"Cancel to"* (CANT) potentially prevents all sources of traffic from accessing a specific route. Some control arrangements permit CANF and CANT to be applied to alternate-routed or direct-routed traffic or both. All cancel controls are implemented on a percentage-of-traffic basis.

- *Skip route control* directs a percentage of traffic to bypass a specific circuit group and advance to the next route in its normal routing pattern. The control can be adjusted to affect alternate-routed or direct-routed traffic or both.

- *Code-block control* blocks a percentage of calls routed to a specific destination code. In most cases, a code-block control can also be specified to include the called-station address digits.

- *Call-gapping control*, like code-block control, limits routing to a specific code or station address. Call-gapping is more effective in controlling mass calling situations than the code-block control. Call-gapping consists of an adjustable timer that stops all calls to a specified code for a time interval selected from 16 different time intervals. After the expiration of the time interval, one call to the specified code or address is allowed access to the network, after which the call-gapping procedure is recycled for another time interval.

- *Circuit-directionalization control* changes 2-way circuits to 1-way operation.

- *Circuit-turndown control* removes 1- or 2-way circuits from service.

- *Reroute controls* serve in a variety of ways to redirect traffic from congested or failed routes to other circuit groups not normally included in the route advance chain but that have temporary idle capacity. Reroutes override the normal routing algorithms in a switch. Reroutes can be used on a planned basis, such as on a recurring peak-calling day, or in response to unexpected overloads or failures. "Regular reroute" affects traffic overflowing a trunk group. "Immediate reroute" (IRR) affects traffic before hunting the trunk group for an idle circuit. Reroute controls may redirect traffic to a single or to multiple routes. The multiple option is referred to as a "spray reroute."

6.5.2 Automatic Controls in Modern Digital Switches. Current, computer-based switches may include the following types of automatic controls:

- Selective dynamic overload control (SDOC)
- Selective trunk reservation (STR)
- DOC (dynamic overload control)
- Trunk reservation
- Selective incoming load control (SILC)

SDOC and STR are considered "selective" protective controls because they can selectively control traffic to HTR points* more severely than other traffic. If the

---

* HTR points are 3- or 6-digit destination codes to which calls have a very small chance of completing.

probability of completing through the network is very low and the outgoing trunk groups or connected switches are overloaded, selective protective controls can prevent wasted usage of these overloaded network resources for traffic to HTR points. SDOC responds to switching congestion by dynamically controlling the amount and type offered to an overloaded or failed switch. STR, conversely, responds to trunk congestion in the outgoing trunking field and is triggered on a particular trunk group when less than a certain number of circuits are idle in that group.

SDOC and STR are two-level control systems. The first level indicates less congestion than the second level. The first-level response is typically limited to control of traffic destined for HTR points, whereas the second level applies controls to both HTR points and other traffic, typically alternate-routed traffic. HTR traffic can also be manually enabled.

HTR traffic is automatically detected by the AT&T 4ESS switch based on an analysis of destination-code completion statistics. This analysis is performed on a 3- and 6-digit basis every five minutes. In Nortel DMS-100 and DMS-200/500 switches, HTR codes can also be manually selected and enabled.

Automatic controls, such as SDOC and STR, are intended to be activated by a switching system within a matter of seconds in response to a switch or trunk-group overload. These controls provide rapid protection for the network and, by their code-selective basis, attempt to restrict traffic that has a low probability of completion. When automatic controls trigger, network traffic managers monitor their operations and adjust system parameters to deal with the particular network condition, whether it is a general overload, a mass call-in, a natural disaster, or a major network-component failure. Among these parameters are call-completion determinations that designate a code "HTR" and control-response options that designate the amount of traffic to be controlled or trunks to be reserved at each triggering level. Since the optimum control response depends on the severity, geographical distribution, and type of overload, maximizing the calls carried by the network requires coordination and combination of automatic and manual control responses.

7 NETWORK MANAGEMENT SYSTEMS IN ENTERPRISE NETWORKS

7.1 What Are Network Management Systems?

Ostensibly a network management system provides an automated means of remotely monitoring a network for:

- Quantification of performance (e.g., BER, loss of synchronization, etc.)
- Equipment, module, subassembly, card failures; circuit outages
- Levels of traffic, network usage

The impetus of the systems described below has come from the enterprise network environment, typically from the developer of TCP/IP (US DoD) and later from the OSI world. There are now many proprietary network management systems available to the user. Among these are Hewlett-Packard's Openview, IBM's Netview, and Digital Equipment Corporation EMA (Enterprise Management Architecture). There has been a distinct trend toward distributed processing in the network management arena.

Such network management systems, especially in the distributed processing environment, require a means to communicate for monitoring and control of the enterprise network. Four network management protocols have evolved for this purpose, which we describe below.

7.2 Introduction to Network Management Protocols

7.2.1 There Are Four Management Protocols.
Two separate communities have been developing network management communication protocols:

- The TCP/IP (ARPANET) community: SNMP
- The ISO/OSI community: CMIP
- The ITU has fielded the Telecommunication Management Network (TMN)
- ILMI (Interim Local Management Interface) developed for ATM networks

The most mature and certainly the most implemented by far is SNMP (Simple Network Management Protocol). Certain weak points arose in the protocol, and a version two has been developed and fielded, called SNMPv.2. Even this version has been subsequently upgraded to SNMPv.3.

CMIP (common management information protocol) has been developed for the OSI environment. It is more versatile but requires about five times the memory of SNMP.

TMN provides a framework for network management and communication that is flexible, scalable, reliable, inexpensive to run, and easy to enhance. TMN defines standard ways of doing network management and communication across networks. The protocol allows processing to be distributed to appropriate levels for scalability, optimum performance, and communication efficiency.

7.2.2 An Overview of SNMP.
SNMP is probably the dominant method for devices on a network to relay network management information to centralized management consoles which are designed to provide a comprehensive operational view of the network. Having come on line in about 1990, literally thousands of SNMP systems have been deployed. The latest version of SNMP is v.3, which is described in RFC 3410, dated December 2002.

There are three components of the SNMP protocol:

- The management protocol itself
- The MIB (management information base)
- The SMI (structure management information)

Figure 21.5 shows the classic client–server model. The client runs the *managing* system. It makes requests and is typically called the *network management*

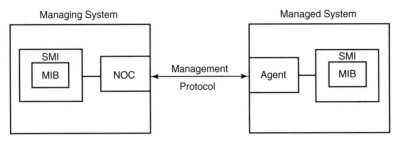

Figure 21.5. SNMP management architecture. SMI, structure of management information. NOC, network operations center. MIB, management information base.

system (NMS) or *network operation center* (NOC). The server is in the *managed* system. It executes requests and is called the *agent*.

SMI. Structure of management information (SMI) defines the general framework within which an MIB can be defined. In other words, SMI is the set of rules which define MIB objects, including generic types used to describe management information. The SNMP SMI uses a subset of Abstract Syntax Notation One (ASN.1) specification language that the ISO (International Standards Organization) developed for communications above the OSI presentation layer. Layer 7, for example, may use ASN.1 standards such as ITU-T Recs. X.400 and X.500. It was designed this way so that SNMP could be aligned with the OSI environment. The SMI organizes MIB objects into an upside-down tree for naming purposes.

MIB. Management information base (MIB) is the set of managed objects or variables that can be managed. Each data element, such as a node table, is modeled as an object and given a unique name and identifier for management purposes. The complete definition of a managed object includes its naming, syntax, definitions, and access method (such as read-only or read-write), which can be used to protect sensitive data and status. By allowing a status of "required" or "optional," the SNMP formulating committee allowed for the possibility that some vendors may not wish to support optional variables. Products are obligated to support required objects if they wish to be compliant with the SNMP standard.

Figure 21.5 also shows an *agent*. One can imagine an agent in every data equipment to be managed as shown in Figure 21.6. Here it shows that the (network) management console manages "agents."

SNMP utilizes an architecture that depends heavily upon communication between one or a small number of managers and a large number of remote agents scattered throughout the network. Agents use the MIB to provide a view of the local data that are available for manipulation by the network management station. In order for a variable, such as the CPU utilization of a remote Sun workstation, to be monitored by the network management station, it must be represented as an MIB object.

The management station sends *get* and *set* requests to remote agents. These agents initiate *traps* to the management station when an unexpected event occurs.

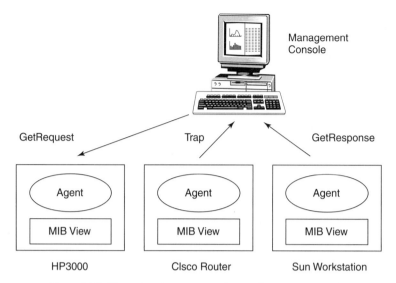

Management
Console

GetRequest

Trap

GetResponse

Agent

MIB View

HP3000

Agent

MIB View

CIsco Router

Agent

MIB View

Sun Workstation

Figure 21.6. The network management console manages agents.

In such a configuration, most of the burden for retrieving and analyzing data rests on the management application. Unless data are requested in a proactive way, little information will be shown at the management station. This poll-based approach increases network traffic, especially on the backbone where some users report a 5–10% increase in traffic due to SNMP network management packets (or messages).

SNMP is a connectionless protocol that initially was designed to run over a UDP/IP stack (UDP stands for user datagram protocol). Because of its design, it is traditionally high in overhead with the ratio of overhead to usable data running about ten bytes to one. A typical SNMP "message" (PDU) embedded in a local network frame is shown in Figure 21.7.

Inside the frame in Figure 21.7 we find an IP (internet protocol—see Chapter 11) datagram which has a header. The header has an IP destination address which directs the datagram to the intended recipient. Following the IP header, there is the UDP (user datagram protocol) which identifies the higher-layer protocol process. This is the SNMP message shown in Figure 21.7 as embedded in the UDP. The application here is typically for LANs. If the IP is too long for one frame, it is fragmented (segmented) into one or more additional frames.

Figure 21.8 shows a typical SNMP PDU structure which is valid for all messages but the *trap* format. This structure is embedded as the "message" field of Figure 21.7. The SNMP message itself is divided into two sections: a version identifier plus community name, and a PDU. The version identifier and community name are sometimes referred to as the SNMP *authentication header*.

The *Version* field assures that all parties in the management transaction are using the same version of SNMP protocol. We must remember the origins of

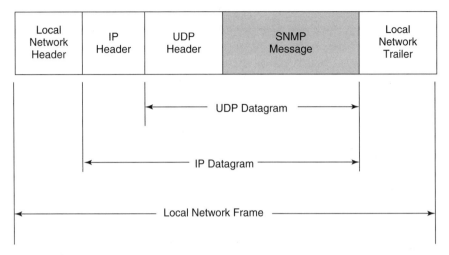

Figure 21.7. An SNMP "message" embedded in a local network frame.

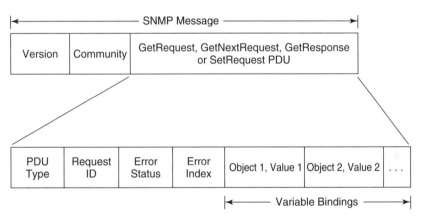

Figure 21.8. An SNMP PDU structure for GetRequest, GetNext-Request, GetResponse, and SetResponse. From Refs. 9, 13, and 14.

SNMP evolved from TCP/IP described in Chapter 11, where we have already seen the use of a "version" field.

Each SNMP message contains a *community name* which is one of the only security mechanisms in SNMP. The agent examines the community name to ensure it matches one of the authorized Community Strings loaded in its configuration files or nonvolatile memory. Each SNMP PDU is one of five types (sometimes called *verbs*): *GetRequest, GetNextRequest, SetRequest, GetResponse*, and *Trap*. The trap PDU is shown in Figure 21.9.

The PDU shown in Figure 21.8 has five initial fields. The first field is the *PDU type*. There are five types of PDU as we discussed previously. These are shown in Table 21.1.

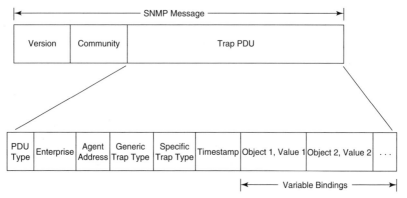

Figure 21.9. SNMP trap PDU format. From Refs. 9, 13, and 14.

TABLE 21.1   PDU-Type Field Values

GetRequest	0
GetNextRequest	1
GetResponse	2
SetRequest	3
Trap	4

The *Request ID* is the second field of the PDU field. It is an INTEGER-type field that correlates the manager's request with the agent's response. INTEGER type is a primitive type used in ASN.1.

The *Error Status* field is also an ASN.1 [11] primitive type. It indicates normal operation (noError) or one of five error conditions as shown in Table 21.2.

When an error occurs, the *Error Index* field identifies the entry within the variable bindings list that caused the error. If, for example, a readOnly error occurred, the error index returned would be 4.

A *variable binding* pairs a variable name with its value. A *VarBindList* is a list of such pairings. Note that within the Variable Binding field of the SNMP PDU, the word *Object* identifies the variable name [OID (object identifier) encoding of object type plus the instance] for which a value is being communicated.

A trap is an unsolicited packet sent from an agent to a manager after sensing a prespecified condition such as a cold start, link down, authentication failure,

TABLE 21.2   SNMP Error Codes

Error Type	Value	Description
noError	0	Success
tooBig	1	Response too large to fit in single datagram
noSuchName	2	Requested object unknown/unavailable
badValue	3	Object cannot be set to specified value
readOnly	4	Object cannot be set
genErr	5	Some other error occurred

TABLE 21.3   SNMP Trap Codes

Trap Type	Value
coldStart	0
warmStart	1
linkDown	2
linkup	3
authenticationFailure	4
egpNeighborLoss	5
enterpriseSpecific	6

or other such event. Agents always receive SNMP requests on UDP port 161, and network management consoles always receive traps on UDP port 162. This requirement means that multiple applications on the same management station that wish to receive traps must usually pass control of this port to an intermediate software process. This process receives traps and routes them to the appropriate application.

Figure 21.9 illustrates the SNMP trap PDU structure. We must appreciate that the Trap PDU structure differs from the structure of the other four PDUs shown in Figure 21.8. Like the other PDU structure, the first field, PDU type, will in this case be PDU Type = 4 (see Table 21.1).

The next field is called the *Enterprise field* and identifies the management enterprise under whose registration authority the trap was defined. As an example, the OID prefix {1.3.6.1.4.123} would identify Newbridge Networks Corporation as the enterprise sending the trap. Further identification is provided in the *Agent Address* field, which contains the IP address of the agent. In some circumstances a non-IP transport protocol is used. In this case the value 0.0.0.0 is returned.

After the *Agent Address* field in Figure 21.9, we find the *Generic Trap Type*, which provides more specific information on the event being reported. There are seven defined values (enumerated INTEGER types) for this field as shown in Table 21.3.

The next field is the *Time Stamp* field, which contains the value of the sysUpTime object. This represents the amount of time elapsed between the last re-initialization of the agent in question and the regeneration of the trap. As shown in Figure 21.8, the last field contains the Variable Bindings.

A trap is generated by an agent to alert the manager that a predefined event has occurred. To generate a trap the agent assigns PDU Type = 4 and has to fill in Enterprise, Agent Address, Generic Trap, Specific Trap Type, Time Stamp fields, and Variable Bindings list [8].

Figure 21.10 gives an excellent summary overview of SNMP.

7.3   Remote Monitoring (RMON)

The tendency toward more and more distributed networks has become apparent. The wide area distribution is both geographical and logical. One method of handling this situation is to place remote management devices on the remote segments. These remote management devices are sometimes called *probes*. These

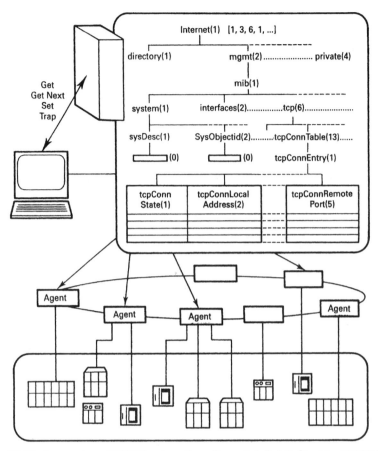

Figure 21.10. An overview of SNMP in place. From Figure 2-4, Ref. 6. Courtesy of IEEE Press.

probes are the remote sensors of the network management system, providing the centralized management station with the required monitoring data dealing with network operation. There is remote network monitoring (RMON) MIB, which standardizes the management information sent to/from these probes.

As an illustration, the Ethernet RMON MIB contains nine groups. One of these groups, for example, is titled "alarms." It compares statistical samples with preset thresholds, and it generates an alarm when a threshold is crossed. Another is "capture," which allows packets to be captured after they pass through a logical channel [17].

7.4   SNMP Version 2

SNMPv.2 design used the field experience gained by SNMP to sharpen and simplify the mappings to different transports. The management protocol has been separated from the transport environment, encouraging its use over practically any protocol stack [15].

One of the weaknesses of SNMP is in the unreliable manner of handling trap messages. Of course, management communications are most critical at times when the network is least reliable. A manager's communications with agents is vital. The use of UDP by SNMP means potentially unreliable transport. SNMP leaves the function of recovery from loss to the manager application. The GET-RESPONSE frame confirms, respectively, GET, GET-NEXT, and SET. A network manager can detect the loss of a request when a response does not return. For instance, it can repeat the request. Traps are another matter. Trap messages are generated by the agent and are not confirmed. If a trap is lost, the agent applications would not be aware that there is a problem, nor would the manager for that matter. Since trap messages signal information which is often of great significance, securing their reliable delivery is very important.

Accordingly, SNMPv.2 has pursued an improved mechanism to handle event notifications, namely, traps. For example, the trap primitive has been eliminated. It is replaced by an unsolicited GET-RESPONSE frame which is generated by an agent and directed to the *trap manager* (UDP port 162). Now event notifications can be unified as responses to virtual requests by event managers. In SNMPv.2, a special trap MIB has been added to unify event handling, subscription by managers to receive events, and repetitions to improve reliable delivery.

In the case of SNMP over TCP/IP, we are looking at what one might call in-band communications. In other words, we are using the same communications channel for network management as we use for "revenue-bearing" communications. This tends to defeat the purpose of network management. The problem can be somewhat alleviated if the access path to agents is somewhat protected from the entities that they manage.

Unlike SNMP, SNMPv.2 delivers an array of messaging options that enable agents to communicate more efficiently with management stations. Furthermore, SNMPv.2's bulk retrieval mechanism lets management stations obtain reports from agents about a range of variables without issuing repeated requests (i.e., GetBulkRequest). This feature should cut down the level of packet activity dedicated to network management and yet improve network management efficiency.

Another advantage of SNMPv.2 over its predecessor is manager-to-manager communications, which allows a station to act as either manager or agent. This allows SNMPv.2 systems to offer hierarchical management using mid-level managers to offload tasks from the central network management console.

SNMPv.2 does not have to rely on TCP/IP. It can run over a variety of protocol stacks including OSI and internet packet exchange (IPX). It also has notably enhanced security features over its predecessor.

Figure 21.11 shows a generic SNMPv.2-managed configuration.

7.5   SNMP Version 3

SNMP Version 3, as described in RFC 3410 [14], points up deficiencies in Version 2 of SNMP. These were unmet design goals of Version 2 and included

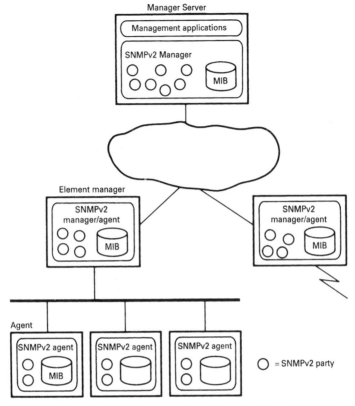

Figure 21.11. A generic SNMPv.2-managed configuration. From Ref. 2, Chapter 3.

provision of security and administration delivering so-called "commercial grade" security with:

- Authentication: origin identification, message integrity, and some aspects of replay protection
- Privacy: confidentiality
- Authorization and access control
- Suitable remote configuration and administration capabilities for these features

SNMPv.3 can be thought of as SNMPv.2 with additional security and administration capabilities. The documents which specify the SNMPv.3 management framework follow the same architecture as previous versions and for expository purposes are organized into four main categories as follows:

- The data definition language
- Management information base (MIB) modules
- Protocol operations
- Security and administration

The first three categories are covered in SNMPv.2. The last category is new, and at the publication date of RFC 3410 [14] it includes seven documents:

- RFC 3410, which is the referenced document of this section
- STD 62, RFC 3411, "An Architecture for Describing Simple Network Management Protocol (SNMP) Management Frameworks," describes the overall architecture for security and administration.
- STD 62, RFC 3412, "Message Processing and Dispatching for the Simple Network Management Protocol (SNMP)," describes the possibility of multiple message processing models and the dispatcher portion that can be part of the SNMP protocol engine.
- STD 62, RFC 3413, "Simple Network Management Protocol Application," describes the five initial types of applications that can be associated with SNMPv.3 engines and their elements of procedure.
- STD 62, RFC 3414, "User-Based Security Model (USM) for Version 3 of SNMP," describes the threats against which protection is provided, as well as the mechanisms, protocols, and supporting data used to provide SNMP message-level security with the user-based security model.
- STD 62, RFC 3415, "View-Based Access Control Model (VCAM) for the SNMP" describes how view-based access control can be applied with command responder and notification originator applications.
- RFC 2576, "SNMPv.3 Coexistence and Transition" describes coexistence between the SNMPv.3 Management Framework, and SNMPv.2 Management Framework, and the original SNMPv.1 Management Framework, and is in the process of being updated.

7.6   Common Management Information Protocol (CMIP)

SNMP was developed rapidly with an objective to serve as a network management communications standard until CMIP was issued. SNMP met such success that it now has a life of its own. CMIP remains with slow implementation. It has no where reached the popularity of SNMP.

CMIP is an ISO development and it is designed to operate in the OSI environment. It is considerably more complex than its SNMP counterpart. Figure 21.12 illustrates the OSI Management architecture, which uses CMIP to access managed information. In Figure 21.12 this managed information is provided by an agent in the LAN hub.

CMIP is part of the ITU-T X.700 OSI series of recommendations of the ITU. CMIP was developed and funded by government and corporations to replace and make up for the deficiencies of SNMP, thus improving the capabilities of network management systems.

CMIP uses different terminology than SNMP. An agent maintains a management information tree (MIT) as a database; it models platforms and devices using managed objects (MOs). These may represent LANs, ports, and interfaces. CMIP is used by a platform to change, create, retrieve, or delete MOs in the MIT. It can invoke actions or receive event notifications.

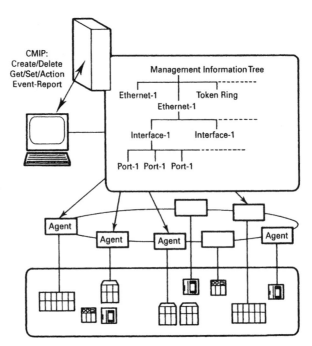

Figure 21.12. A typical overall architecture of an OSI network management system. From Figure 2-12, page 48, Ref. 6.

Object-oriented system concepts that are applied to the CMIP objects include *containment, inheritance*, and *allomorphism*. Containment refers to the characteristic of objects being a repository of other objects and/or attributes. A high-level object for a communication switch, for example, can contain several racks of equipment, each of which, in turn, can contain several slots for printed circuit boards. Here one might use the ITU-T M.3100 base class for a circuit pack to define the general features of modules within a communication switch. Object classes can then, in turn, be defined to represent the specific modules. Items including line interface cards, switching elements, and processors can be derived from the basic circuit pack definition. Each of these objects exhibits the behavior, actions, and attributes of both the derived classes and the base class. Allomorphism is a concept coined by the CMIP standards bodies to refer to the ability to interact with modules through a base set of interfaces, only to have the resulting behaviors coupled to the complete class definition. Disabling a power supply, for instance, may exhibit significantly different behavior than disabling a switching component.

Source: Ref. 16.

With CMIP and other OSI management schemes, there are three types of relationships between managed objects:

- *Inheritance Tree.* Defines the managed object class, super- and subclasses, as much as C++ base and derived classes are related. When a class is inherited from a superclass, it possesses all the characteristics of the superclass, with additional class-specific extensions (additional attributes, behaviors, and actions).
- *Containment Tree.* Defines which managed objects are contained in other managed objects. As an example, a subnetwork can contain several managed elements (ME).
- *Naming Tree.* Defines the way in which individual objects are referenced within the constraints of the management architecture.

Source: Ref. 16.

CMIP (i.e., OSI management communications) communications are very different from those found in SNMP. These communications are embedded in the OSI application environment, and they rely on conventional OSI peer layers for support. They use connection-oriented transport where SNMP uses the datagram (connectionless). In most cases these communications are acknowledged.

8 TELECOMMUNICATION MANAGEMENT NETWORK (TMN)

Figure 21.13 shows the general relationship between a TMN and a telecommunications network which it manages. A TMN is conceptually a separate

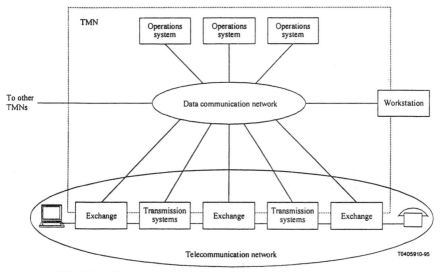

NOTE – The TMN boundary represented by the dotted line may extend to and manage customer/user services and equipment.

Figure 21.13. General relationship of a TMN to a telecommunication network. From Figure 1/M.3010, Ref. 20.

network that interfaces a telecommunications network at several different points to send/receive information to/from it and to control its operations. A TMN may use parts of the telecommunications network to provide its communications.

The objective of the TMN is to provide a framework for telecommunications management. By introducing the concept of generic network models for management, it is possible to perform general management of diverse equipment, network, and services using generic information models and standard interfaces.

A TMN is intended to support a wide variety of management areas which cover the planning, installation, operations, administration, maintenance, and provisioning of telecommunication networks and services. ITU-T Rec. M.3200 [18, 19] describes the scope of management through the following two main concepts: Telecommunications managed areas and TMN management services. The former relates to the grouping of telecommunications resources being managed, and the latter relates to the set of processes needed to achieve business objectives, namely TMN management goals.

The specification and development of the required range and functionality of applications to support the above management areas is a local matter and is not considered within the scope of ITU-T Rec. M.3200 series. Some guidance, however, has been provided by the ITU-T, which has categorized management into five broad management functional areas (Rec. X.700 [21]). These functional areas support the management scope described by ITU-T Rec. M.3020 [22]. They provide a framework through which the appropriate management services support the PTOs (Public Telecommunication Operators) business processes. Five management functional areas identified to date are as follows:

- Performance management
- Fault management
- Configuration management
- Accounting management
- Security management

The classification of the information exchange within the TMN is independent of the use that will be made of the information.

The TMN needs to be aware of telecommunications networks and services as collections of cooperating systems. The architecture is concerned with orchestrating the management of individual systems so as to have a coordinated effect upon the network. Introduction of TMNs gives PTOs the possibility to achieve a range of management objectives including the ability to:

- Minimize management reaction times to network events.
- Minimize load caused by management traffic where the telecommunication network is used to carry it.
- Allow for geographic dispersion of control over aspects of the network operation.
- Provide isolation mechanisms to minimize security risks.

- Provide isolation mechanisms to locate and contain network faults.
- Improve service assistance and interaction with customers.

TMN Functional Architecture. The TMN functional architecture is structured from the following functional elements:

- Function blocks
- Management application functions (MAFs)
- TMN management function sets and TMN management functions
- Reference points

TMN management functionality to be implemented can then be described in terms of these fundamental elements.

FUNCTION BLOCKS. Figure 21.14 illustrates the different types of TMN function blocks and indicates that only the functions that are directly involved in management are part of a TMN. Some of the function blocks are partly in and partly out of a TMN; these TMN function blocks also perform functions outside of the TMN functional boundaries as discussed below.

The TMN function block is the smallest *deployable* unit of TMN management functionality. If the TMN function block contains a management application function, it may only contain one management application function.

Operations Systems Function (OSF) Block. The OSF processes information related to the telecommunications management for the purpose of monitoring/coordinating and/or controlling telecommunication functions including management functions (i.e., the TMN itself).

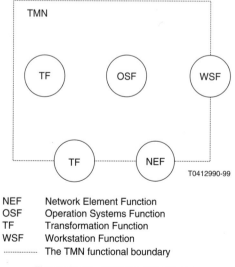

NEF	Network Element Function
OSF	Operation Systems Function
TF	Transformation Function
WSF	Workstation Function
....	The TMN functional boundary

Figure 21.14. TMN function blocks.

Network Element Function (NEF) Block. The NEF is a functional block which communicates with the TMN for the purpose of being monitored and/or controlled. The NEF provides the telecommunications and supports functions which are required by the telecommunications network being managed.

The NEF includes the telecommunications functions which are the subject of management. These functions are not part of he TMN but are represented to the TMN by the NEF. The part of the NEF that provides this representation in support of the TMN is part of the TMN itself, while the telecommunication functions themselves are outside.

Workstation Function (WSF) Block. The WSF provides the means to interpret TMN information for the human user, and vice versa. The responsibility of the WSF is to translate between a TMN reference point and a non-TMN reference point, and hence a portion of this function block is shown outside the TMN boundary.

Transformation Function (TF) Block. The transformation function block (TF) provides functionality to connect two functional entities with incompatible communication mechanisms. Such mechanisms may be protocols or information models, or both.

The TF may be used anywhere within a TMN or anywhere at the boundary of a TMN. When used within a TMN, the TF connects two function blocks, each of which supports a standardized, but different, communication mechanism. When used at a boundary of a TMN, the TF may be used as communication either between two TMNs or between a TMN and a non-TMN environment.

When used at a boundary at two TMNs, the TF connects two function blocks, one in each TMN, each of which supports a standardized, but different, communication mechanism.

When the TF is used between a TMN and a non-TMN environment, the TF connects a function block with a standardized communication mechanism in a TMN to a functional entity with a nonstandardized communication mechanism in the non-TMN environment.

Note: The TF consolidates and extends the functionality and scope associated with the mediation and Q adaption function blocks in ITU-T Rec. 3010 [20].

TMN FUNCTIONALITY

Management Application Functionality. The management application functionality (MAF) represents (part of) the functionality of one or more TMN management services. ITU-T Rec. M.32xx-series enumerates the MAFs with respect to the technologies and services supported by the TMN.

The management application functionality (MAF) may be identified with the type of TMN function block in which they are implemented. The following MAFs may be identified:

- Operations systems functionality (OSF)
- Management application function (OSF-MAF)

- Network element functionality—management application function (NEF-MAF)
- Transformation functionality (TF)
- Management application function (TF-MAF)
- Workstation functionality–management application function (WSF-MAF)

Support Functionality. Support functions may optionally be found in a TMN function block. The support functionality is potentially common to more than one TMN function block within an implemented TMN. Some support functionality assist the MAF within a TMN function block in its interactions with other function blocks.

TMN REFERENCE POINTS. A TMN reference point delineates one of several external views of functionality of a function block; it defines that function block's service boundary. This external view of functionality is captured in the set of TMN management functions that will have visibility from the function block.

Reference points have meaning in functional specification leading to an implementation. A reference point can represent the interactions between a particular pair of function blocks. Table 21.4 shows the relationships between the function blocks in terms of reference points between them. The reference point concept is important because it represents the aggregate of all of the abilities that a particular function block seeks from another particular function block, or the equivalent function blocks. It also represents the aggregate of all the operations and/or notifications (as defined in ITU-T Rec. X.703 [23]) that a function block can provide to a requesting function block.

TABLE 21.4   Relationships Between Logical Function Blocks Expressed as Reference Points

	NEF	OSF	TF	WSF	non-TMN
NEF		q	q		
OSF	q	q, x[a]	q	f	
TF	q	q	q	f	m[c]
WSF		f	f		g[b]
non-TMN			m[c]	g[b]	

[a] x reference point only applies when each OSF is in a different TMN.
[b] The g reference point lies between the WSF and the human user.
[c] The m reference point lies between the TF and the telecommunication functionality.
Note: Any function may communicate at a non-TMN reference point. These non-TMN reference points may be standardized by other groups/organizations for particular purposes.
Source: Table 1/M.3010, page 13, Ref. 20.

A TMN functionality specified reference point usually corresponds to a to-be-implemented physical interface, in the physical architecture, if and only if the function blocks are implemented in different physical blocks.

9   NETWORK MANAGEMENT IN ATM

The ITU-T (CCITT) organization and ANSI have left local network management procedures in the M-plane for "further study."

In the interim period until CCITT/ANSI have formulated such standards, SNMP and the ATM UNI Management Information Base (MIB) are required to provide any ATM user device with status and configuration information concerning virtual path and channel connections available at its UNI. (See Chapter 16 for a discussion of ATM and clarification of the acronyms used here.)

The ATM Forum has developed the ILMI (interim local management interface specification). The ILMI fits into the overall management model for an ATM device as shown in Figure 21.15 as clarified by the following principles and options.

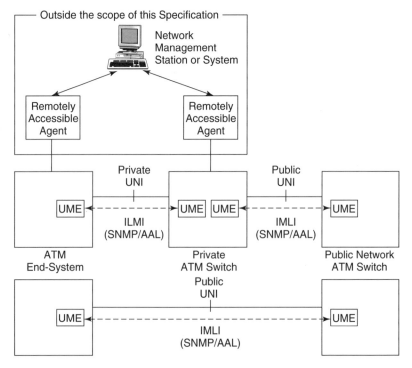

Figure 21.15. Definition and context of ILMI. UME = UNI management entity, Based on Ref. 6 and Figure 4-1, page 106, Ref. 7, ATM Forum.

1. Each ATM device supports one or more UNIs.
2. ILMI functions for a UNI provide status, configuration, and control information about link and physical layer parameters.
3. ILMI functions for a UNI also provide for address registration across the UNI.
4. There is a per-UNI set of managed objects, the UNI ILMI attributes, that is sufficient to support the ILMI functions for each UNI.
5. The UNI ILMI attributes are organized in a standard MIB structure; there is one UNI ILMI MIB structure instance for each UNI.
6. There is one MIB instance per ATM device, which contains one or more UNI ILMI MIB structures. This supports the need for general network management systems to have access to the information in the UNI ILMI MIB structures.
7. For any ATM device, there is a UNI management entity (UME) associated with each UNI that supports the ILMI functions for that UNI, including coordination between the physical and ATM layer management entities associated with that UNI.
8. When two ATM devices are connected across a (*point-to-point*) UNI, there are two UNI management entities (UMEs) associated with the UNI, one UME for each ATM device, and two such UMEs are defined as adjacent UMEs.
9. The ILMI communication takes place between adjacent ATM UMEs.
10. The ILMI communication protocol is an open management protocol (i.e., SNMP/AAL initially).
11. A UNI management entity (UME) can access, via the ILMI communication protocol, the UNI ILMI MIB information associated with its adjacent UME.
12. Separation of the MIB structure from the access methods allows for the use of multiple access methods for management information. For the ILMI function, the access method is an open management protocol (i.e., SNMP/AAL) over a well-known VPI/VCI value. For example, for general network management applications [e.g., from a network management station (NMS) performing generic customer network management (CNM) functions] the access method is also an open management protocol (e.g., SNMP/UDP/IP/AAL) over a specific VPI/VCI value (or a completely separate communications method) allocated to support the general management applications. The peer entity in an ATM device that communicates directly with an NMS is a management agent, not a UME; however, since the management agent can access the MIB instance for the ATM device, it can access all of the UNI ILMI MIB structure instances.

The simple network management protocol (SNMP) without UDP and IP addressing along with ATM UNI management information base (MIB) were chosen for the ILMI.

9.1 Interim Local Management Interface (ILMI) Functions

An ILMI supports bidirectional exchange of management information between UNI management entities (UMEs) related to UNI ATM layer and physical layer parameters. The communication across the ILMI is protocol symmetric. In addition, each of the adjacent UMEs supporting ILMI will contain an agent application

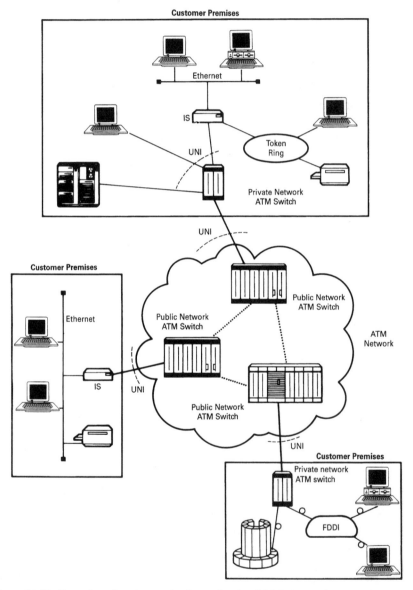

Figure 21.16. Examples of equipment implementing the ATM UNI ILMI. Based on Figure 4-2, page 109, Refs. 7 and 8, ATM Forum.

and may contain a management application. Unless stated otherwise for specific portions of the MIB, both of the adjacent UMEs contain the same management information base (MIB). However, semantics of some MIB objects may be interpreted differently. As shown in Figure 21.16, an example list of the equipment that will use the ATM UNI ILMI include:

- Higher-layer switches such as internet routers, frame relay switches, or LAN bridges that transfer their frames within ATM cells and forward the cells across an ATM UNI to an ATM switch.
- Workstations and computer with ATM interfaces which send their data in ATM cells across an ATM UNI to an ATM switch.
- ATM network switches which send ATM cells across an ATM UNI to other ATM devices.

9.2   ILMI Service Interface

The ILMI uses SNMP for monitoring and control operations of ATM management information across the UNI. The ATM UNI management information will be represented in a management information base (MIB). The types of management information that will be available in the ATM UNI MIB are as follows:

- Physical layer
- ATM layer

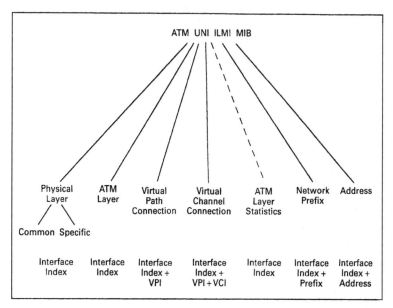

Figure 21.17. ILMI ATM UNI MIB tree structure. Based on Figure 4-3, page 111, Ref. 7, ATM Forum.

- ATM layer statistics
- Virtual path (VP) connections
- Virtual channel (VC) connections
- Address registration information

The tree structure of the ATM UNI ILMI MIB is shown in Figure 21.17.

REVIEW QUESTIONS

1. Give at least four major benefits to the network operator derived by implementing a network management system.

2. Name the five traditional tasks of network management.

3. Discuss fault management and describe some of the capability it should incorporate.

4. It has been said that a network management center is the front-line command post for network _____.

5. Give the four steps involved in finding a "fault" in a telecommunication network.

6. Describe how a well-engineered network management system can often cut the time almost to zero for isolating faults.

7. A network management system should be built around on-line monitoring systems that are nonintrusive and that are found in (list such systems).

8. Explain how BERT works.

9. Network management systems require communications, especially to/from the network management center and remote equipment. Discuss concerns you might have with such systems, particularly if they share revenue-traffic-bearing channels.

10. From a PSTN perspective, network management may be called "surveillance and control." What are the two major objectives of surveillance and control?

11. Distinguish "message traffic" from "non-message traffic."

12. Describe at least three NTM controls.

13. Give some of the causes of common causes of *general network overload*.

14. What is a *focused overload*? Name two measures to mitigate effects of focused overloads.

15. What are the two broad categories of NTM controls?

16. Describe the two variations of *cancel controls*.

17. List four of the automatic (traffic flow) controls one might encounter in a modern computer-controlled (SPC) switch.

18. There are really four distinguishable network management protocols. Name each along with its sponsoring agency.

19. What are the three components of SNMP? Use the acronyms and write out their meaning.

20. What does the MIB do for agents?

21. In SNMP, what does an agent do when an unexpected event occurs?

22. What is the efficiency of SNMP regarding overhead?

23. How does a *probe* fit into the network management operation?

24. The RAMON MIB contains nine groups. One of these groups is called "alarms." How do they work?

25. What is the function of a *trap*?

26. How does SNMPv.2 improve upon SNMPv.1?

27. What are the two general areas where SNMPv.3 improves upon SNMPv.2?

28. Give two major reasons why CMIP has not achieved acceptance in the network community?

29. In the data communications community there are really two separate worlds. One of these has been handed down by the U.S. Department of Defense. Name these two worlds. Name the network management protocols associated with each.

30. CMIP is based on what environment? (Think ISO.)

31. TMN is under the auspices of what international organization?

32. Identify the five fundamental functional areas of TMN. (Return to the basics of network management itself.)

33. Name at least two of the four TMN function blocks.

34. What is now providing network management in ATM?

35. Name at least five principles and options of ILMI.

36. What types of management information will be available in the ATM UNI MIB? List at least five items.

37. Give a list of example equipment that would use ATM UNI ILMI (name at least four items).

REFERENCES

1. G. Held, *Network Management: Techniques, Tools and Systems*, John Wiley & Sons, Chichester, UK, 1992.

2. W. Stallings, *Network Management*, IEEE Computer Society Press, Los Alamitos, CA, 1993.

3. S. M. Dauber, "Finding Fault," *BYTE Mag.* (March 1991).

4. *Error Performance of an International Digital Connection Forming Part of an Integrated Services Digital Network*, CCITT Rec. G.821, Fascicle III.5, IXth Plenary Assembly, Melbourne, 1988.

5. *Frame Alignment and Cyclic Redundancy Check (CRC) Procedures Relating to Basic Frame Structures Defined in Recommendation G.704*, CCITT Rec. G.706, CCITT, Geneva, 1991.

6. S. Aidarous and T. Plevyak, *Telecommunications Network Management into the 21st Century*, IEEE Press, Piscataway, NJ, 1993.

7. *ATM User–Network Interface Specification*, Version 3.0, The ATM Forum, PTR Prentice-Hall, Englewood Cliffs, NJ, 1993.

8. M. A. Miller, *Managing Internetworks with SNMP*, M&T Books, New York, 1993.

9. *A Simple Network Management Protocol*, RFC 1157, DDN Network Information Center, SRI International, Menlo Park, CA, May 1990.

10. *Synchronous Frame Structures Used at Primary and Secondary Hierarchical Levels*, CCITT Rec. G.704, Fascicle III.4, IXth Plenary Assembly, Melbourne, 1988.

11. *Information Processing Systems: Open Systems Interconnection—Abstract Syntax Notation One (ASN.1)*, ISO Std. 8824, Geneva, 1987.

12. *BOC Notes on the LEC Networks—1994*, Issue 2, Bellcore, Piscataway, NJ, April 1994.

13. G. Held, *Network Management*, John Wiley & Sons, Chichester, UK, 1992.

14. *Introduction and Applicability Statements for Internet Standard Management Framework*, (SNMPv.3) RFC 3410, IETF, ed., Ann Arbor, MI, December 2002.

15. "Version.2 of the Protocol Operations for the Simple Network Management Protocol (SNMP), STD 62, RFC 3416, IETF, ed., Ann Arbor, MI, December 2002.

16. *CMIP/CMIS—Object Oriented Network Management*, From the world wide web www.cellsoft..de/telecom/CMIP.htm, September 1, 2003.

17. *Structure of Management Information—Management Information Model*, CCITT Rec. X.720, ITU, Geneva, January 1992.

18. *Generic Network Information Model*, ITU-T Rec. M.3100, ITU, Geneva, July 1995 (with 5 amendments).

19. *Telecommunication Management Functions*, ITU-T Rec. N.3400, ITU, Geneva, February 2000.

20. *Principles of a Telecommunication Management Network*, ITU-T Rec. M.3010, ITU, Geneva, February 2000.

21. *Management Framework for Open System Interconnection (OSI) for CCITT Applications*, CCITT Rec. X.700, ITU, Geneva, 1992.

22. *TMN Interface Specifications Methodology*, ITU-T Rec. M.3020, ITU, Geneva, 2000.

23. *Information Technology—Open Distributed Management Architecture*, ITU-T Rec. X.703, ITU, Geneva, 1997.

APPENDIX I

ACRONYMS AND ABBREVIATIONS

A

AAL1, AAL2, AAL3/4	ATM adaptation layer
AAR	automatic alternative routing
ABD	average business day
ABM	asynchronous balanced mode
ABR	available bit rate (ATM)
AbS	analysis by synthesis
ABSBH	average busy season busy hour
ac	alternating current
AC	authentication center; access control
ACELP	algebraic code excited linear prediction
ACH	attempts per circuit per hour
ACK	acknowledge, acknowledgment
A/D	analog-to-digital (converter)
ADCCP	advanced data-link control communications procedure
ADM	add–drop multiplex
ADPCM	adaptive differential PCM
ADSL	asymmetric digital subscriber line
AES	iterated block cipher
AIOD	automatic identified outward dialing
AIS	alarm indication signal
ALBO	automatic line build-out
ALOHA	name originated at the University of Hawaii
AM	amplitude modulation
AMA	automatic message accounting
AM/CMP	administrative module/communications module processor
AMI	alternate mark inversion

Telecommunication System Engineering, by Roger L. Freeman
ISBN 0-471-45133-9   Copyright © 2004 Roger L. Freeman

AMPS	advanced mobile phone system
ANC	all number calling
ANI	automatic number identification
ANSI	American National Standards Institute
APD	avalanche photodiode
APIs	automatic programming interfaces, applications programming interface
AR	access rate
ARM	asynchronous response mode
ARP	address resolution protocol
ARQ	automatic repeat request
ARPA, ARPANET	Advanced Research Projects Agency, ARPA Network
ASCII	American Standard Code for Information Interchange
ASN.1	abstract syntax notation—1
ASON	automatically switched optical network
ATB	all trunks busy
ATIS	Alliance for Telecommunications Industry Solutions
AU	administrative unit
AUG	administrative unit group
AUI	attachment unit interface
AWG	American Wire Gauge
AWGN	additive white Gaussian noise

B

B3ZS, B6ZS, B8ZS	binary, 3-zero substitution; binary, 6-zero substitution; binary, 8-zero substitution
B&S	broadcast and select
BBE	background block error
BCC	block check count
BCH	Bose–Chaudhuri–Hocquenghem (names of developers of a family of block codes)
BECN	backward explicit congestion notification
BER	bit error rate (ratio)
BERT	bit error rate test (set)
BE-tag	beginning–end tag
BGP	border gateway protocol
BH	busy hour
BIB	backward indicator bit
BICSI	Building Industry Consulting Service(s) International
BIP	bit-interleaved parity
B-ISDN	broadband ISDN
BSN	backward sequence number
BIT	binary digit
BITS	building integrated timing supply
BIU	baseband interface unit

BNZS	binary N-zero(s) substitution
BOC	Bell operating company
BOM	beginning of message
BP	bandpass
bits/s	bits per second
BPSK	binary phase-shift keying
BRE	bridge relay encapsulation (frame relay)
BRI	basic rate interface (ISDN)
BSBH	busy season, busy hour
BSC	binary synchronous communications; base station controller
BSS	basic service set; base station subsystem
BSTJ	*Bell System Technical Journal*
BT	bridged tap
BTS	base transceiver station
BW	bandwidth

C	
CAC	connection admission control
CAD	computer-aided design
CAM	computer-aided manufacturing
CAMA	centralized automatic message accounting
CAMEL	customized application for mobile enhanced logic
CANF	cancel from
CANT	cancel to
CAP	CAMEL application part; carrierless amplitude/phase
CAT	category (often in reference to wire pair)
CATV	community antenna television (cable television)
CBDS	connectionless broadband data services
CBR	constant bit rate
CCH	connections per circuit per hour
CCIR	International Consultive Committee for Radio
CCITT	International Consultive Committee for Telephone and Telegraph
ccs, CCS	cent-call-second
CCS7	CCITT Signaling System No. 7
CD	collision detect(ion)
CDMA	code division multiple access
CDPD	cellular digital packet data
CDO	community dial office
CDV	cell delay variation
CELP	code-excited linear predictive; also code-book excitation linear predictive
CEPT	Conference European Post Telegraph
CEQ	customer equipment

CFR	code of federal regulations
CGSA	cellular geographic serving area
Ch, ch	call-hour
CHAP	challenge handshake authentication protocol
C/I	carrier (level) to interference (level) ratio
CIB	CRC indication bit
CIC	circuit identification code
CIR	committed information rate
CL	connectionless
CLEC	competitive local exchange carrier
CLLI	common language location identifier
CLLM	consolidated link layer management
CLNAP	connectionless network access protocol
CLP	cell loss priority
CLS	connectionless service
CLSF	connectionless service functions
cm	call minute; centimeter
CMIP	common management information protocol
CN	core network
C/N	carrier-to-noise ratio
CO	connection-oriented
CODEC, CoDec	coder, decoder (PCM)
comp	computer
CONUS	contiguous United States
COT	central office termination
CPCS	common part convergence sublayer
CPE	customer premise equipment
CPU, cpu	central processing unit
C/R	command/response (bit)
CRA	collision resolution algorithm
CRC	cyclic redundancy check (e.g., CRC4, CRC6, CRC16, CRC32, etc.)
CREG	concentration range extension with gain
CRF	central retransmission facility; connection related functions
CR-LDP	constraint routing—label distribution protocol
cs	call second
CS	circuit-switched
CS-ACELP	conjugate structure algebraic code excited linear prediction
CS_PDU	convergence sublayer protocol data unit
CSE	CAMEL service environment
CSI	convergence sublayer indicator
CSMA, CSMA/CD	carrier sense multiple access; carrier sense multiple access with collision detection
C/T	carrier level-to-thermal noise ratio

CUG	closed user group
CVSD	continuous variable slope delta modulation

D

DA	destination address
D/A	digital-to-analog (converter)
DAMA	demand assignment multiple access
DARPA	Defense Advanced Research Projects Agency
DAVIC	Digital Audio Video Council
dB	decibel
dBd	decibels referenced to a dipole (antenna)
dBi	decibels referenced to an isotropic (antenna)
dBm	decibel referenced to 1 milliwatt
dBmV	decibels referenced to 1 millivolt (75-Ω impedance assumed)
dBμ	decibels referenced to 1 microvolt (75-Ω impedance assumed)
dBr	dB "reference"
dBW	decibel referenced to 1 watt
dc	direct current
DCE	data communication equipment
DCS	digital cross-connect system
DDRP	domain-to-domain routing protocol
DDS	digital data system
DE	discard eligibility (bit) (frame relay)
DECT	digital European cordless telephone
demux	demultiplex
DEO	digital end-office
DF	don't fragment
DI	data interface
DL-CORE	refers to the "core" aspects of ISDN LAPD
DLC	digital link carrier; digital loop carrier
DLCI	data-link connection identifier
DLI	dual-link interface
DLL	data-link layer
DLTU	digital line and trunk unit
DM	delta modulation
DMT	discrete multitone
DN	directory number
DOCSIS	data over cable service interface specification
DoD	Department of Defense
DPC	destination point code
DQPSK	differential QPSK
D/R	distance/radius (frequency reuse in cellular)

DRNC	drift radio network controller
DSAP	destination service access point
DSB	double sideband
DSL	digital subscriber line
DSLAM	digital subscriber line access multiplexer
DSP	digital signal processing, processor
DSSS	direct sequence spread spectrum
DSU	digital service unit
DTC	digital trunk controller
DTE	data terminal equipment
DTMF	dual tone multiple frequency
DUP	data user part (SSN7)
DWDM	dense wavelength division multiplexing
DXC	digital cross-connect

E

EA	address field extension bit
EB	errored block
E and M	a type of supervisory signaling
EBCDIC	extended binary coded decimal interchange code
EBHC	equated busy hour call
E_b/N_0	energy per bit per noise density ratio (i.e., noise in 1 Hz of bandwidth)
EC	earth curvature
ED	end delimiter
EDD	envelope delay distortion
EDFA	erbium-doped fiber amplifier
EDGE	enhanced data rates for GSM evolution
EFR	enhanced full rate
EFS, efs	end-of-frame sequence; error-free seconds
EIA	Electronic Industries Association
EIR	equipment identity register
EIRP	effective (equivalent) isotropically radiated power
EMA	enterprise management architecture (proprietary to DEC)
Email	electronic mail
EMC	electromagnetic compatibility
EMI	electromagnetic interference
EN	equipment number (switching)
ENNI, E-NNI	exterior node-to-noise interface
EOM	end of message
EPIRB	emergency position indicating radio beacon
EPR, epr	equivalent peak load
erg	equivalent random group
ERP, erp	effective radiated power (see EIRP)
ES	end section; errored second

ESC	engineering service channel
ESF	extended superframe
ESR	errored second ratio
ETSI	European Telecommunications Standardization Institute

F

FAS	frame alignment signal
FC	frame control
FCC	Federal Communications Commission
FCS	frame check sequence
FDD	frequency division duplex
FDDI	fiber distributed data interface
FDM	frequency division multiplex
FDMA	frequency division multiple access
FEBE	far-end block error
FEC	forward error correction; forward equivalence class
FECN	forward explicit congestion notification
FEXT	far-end crosstalk
FFT	fast Fourier transform
FHSS	frequency-hop spread spectrum
FIB	forward information base; forward indicator bit
FIFO	first in, first out
FIN	"no more data"
FISU	fill-in signal unit
FM	frequency modulation
FRAD	frame relay access device
FRF	Frame Relay Forum
FS	frame status
FSC	fiber switch capable
FSK	frequency shift keying
FSL	free space loss
FSN	forward sequence number
FTN	FEC-to-NHLFE
FTP	file-transfer protocol

G

GbE	gigabit Ethernet
GBLC	Gaussian band-limited channel
GBSVC	general broadcast signaling virtual channel
GEO	geostationary orbit
GEOSAR	geostationary (satellite) search and rescue
GFC	generic flow control
GFSK	Gaussian frequency shift keying
GGSN	Gateway GPRS support node
GHz	gigahertz
GLOBALSTAR	name of a proprietary satellite system (LEO satellites)

GMDSS	Global Maritime Distress and Safety System
GMPLS	generalized MPLS
GMSK	Gaussian minimum shift keying
GOS	grade of service
GPRS	general packet radio service
GPS	Global Positioning System
GRE	generic routing encapsulation (protocol)
GSM	ground system mobile
G/T	gain-to-noise temperature ratio

H

HDB3	high-density binary 3
HDLC	high-level data-link control
HDSL	high-bit-rate digital subscriber line
HEC	header error control
HEL	header extension length
HFC	hybrid-fiber-coax
HFR	high-frequency regeneration
HL	high–low
HLR	home location register
HLPI	higher-layer protocol identifier
HP	Hewlett-Packard
HPA	high power amplifier
HSCSD	high-speed circuit switched data
HSP	high-speed printer
HSRC	hypothetical signaling reference connection
HTML	hypertext markup language
HTR	hard to reach
HTTP	hypertext transfer protocol

I

IAD	integrated access device
IAM	initial address message
IBM	International Business Machine(s)
IBS	International Business Systems
IC	integrated circuit
ICMP	internet control message protocol
ID	identification
IDLC	integrated digital loop carrier
IDR	intermediate data rate
IDU	interface data unit
IEC	International Engineering Consortium
IEEE	Institute of Electrical and Electronic Engineers
IETF	internet engineering task force
IF	intermediate frequency
IG	international gateway

IGRP	international gateway routing protocol
IKE	internet key exchange
ILM	incoming label map
ILMI	interim local management interface
IM	intermodulation (noise)
IMT	international mobile telecommunications
IN	intelligent network
INMARSAT	International Marine Satellite (consortium)
INNI, I-NNE	Interior—node-to-node interface
INTELSAT	International Telecommunication Satellite
INTRANET	interior network, usually interconnecting workstations of one enterprise
I/O	input/output (device)
IP	internet protocol
IPSec	IP security protocol
IPv.4, IPv.6	internet protocol version 4, version 6.
IPX	internet package exchange
IR	information rate; product of current and resistance
IRL	isotropic receive level
IRP	internal reference point
IRR	immediate reroute
ISC	international switching center
ISDN	integrated services digital network(s)
ISI	intersymbol interference
IS–IS–TE	intermediate system-to-intermediate system—traffic engineering
ISN	initial sequence number
ISO	International Standards Organization
ISP	internet service provider; international signaling point
ISPC	international signaling point code
ISUP	ISDN user part (SSN7)
ITA	international telegraph alphabet
ITM	irregular terrain model
ITU	International Telecommunications Union
IU	interface unit
IUA	IP user adaptation
IXC	interexchange carrier
K	
kbits/s	kilobits per second
L	
L2F	layer-2 forwarding
L2TP	layer-2 tunneling protocol
LAMA	local automatic message accounting
LAN	local area network

LAP	link access protocol
LAPB, LAP-B	link access protocol, B-channel
LAPD, LAP-D	link access protocol, D-channel
LAPF	link access protocol, frame relay
LATA	local access and transport area
LBO	line build-out
LCC	lost calls cleared
LCD	lost calls delayed
LCH	lost calls held
LCM	line concentrating module
LCN	logical channel number
LD	laser diode
LD-CELP	low delay−code excited linear predictive (coder)
LDP	label distribution protocol
LE	local exchange
LEC	local exchange carrier
LED	light emitting diode
LEN	length—a field in frame of TCP/IP family of protocols
LEO	low earth orbit
LEOSAR	LEO search and rescue (satellite)
LER	label edge router
LFA	loss of frame alignment
LGC	line group controller
LH	low−high
LI	length indicator
LIB	label information base
LIC	lowest incoming channel
LIFO	last in, first out
LIGA	lithographic plating and molding
LLC	logical link control
LMI	local management information (frame relay)
LMDS	local multipoint distribution service (system)
LMP	link management protocol
LNA	low noise amplifier
LOC	lowest outgoing channel
LOH	line overhead
LOS	line-of-sight
LP	linear predictive
LPC	linear predictive coder
LR	loudness rating
LRC	longitudinal redundancy check
LRD	long route design
LSC	lambda-switch capable
LSP	label switched path
LSR	label switch router

LSSU	link status signal unit
LTC	lowest two-way channel
LVDS	low-voltage differential signaling

M

MAC	medium access control
MAF(s)	managed (or management) application function(s)
MAN	metropolitan area network
MAU	medium attachment unit
Mbits/s	megabits per second
MDF	main distribution frame
MDS	multipoint distribution system
ME	managed element(s)
MeGaCo	a media gateway standard, also called ITU-T Rec. H.248
MEMS	micro-electromechanical systems
MEO	medium earth orbit
MF, mf	medium frequency (signaling); medium frequency (RF)
MG	media gateway
MGC	media gateway controller
MGCP	media gateway control protocol
mH	millihenries
MHz	megahertz
MIB	management information base
MID	message identification
MIT	management information tree
mlp	multilink procedures
MLRD	modified long route design
MMDS	microwave multipoint distribution system
MO	managed object
M2PA	MTP2 peer-to-peer adaptation
MPEG	Motion Picture Experts Group
MPLS	multiprotocol label switching
MPλS	multiprotocol wavelength switching
MPMLQ	multipulse maximum likelihood quantization
M-QAM	multi-value QAM, where M is any even number 4 or greater
MRTIE	maximum relative time interval error
ms	millisecond(s)
MSC	mobile switching center
MSPP	multiservice provisioning platform
MSS	mobile satellite service
MSU	message signal unit
MSVC	meta-signaling virtual channel
MTBF	mean time between failures
MTIE	maximum time interval error

MTP	message transfer part (SSN7)
MTSO	mobile telephone switching office
MTTR	mean time to repair
MTU	maximum transfer unit
M3UA	MTP3 user adaptation
MUX, mux	multiplex

N

NA	numerical aperture; not applicable
NACK	negative acknowledgment
NCT	network control and timing
NDF	new data flag
NE	network element
NEF	network element function
NEP	noise equivalent power
NEXT	near-end crosstalk
NF	noise figure
NGDLC	next generation DLC
NHLFE	next hop label forwarding entry
NIC	network interface card
NLPID	network level protocol ID
NMCC	network management control center
NMS	network management system; network management station
NMT	network management; Nordic mobile telephone (system)
NNI	network–node interface or network–network interface
NOC	network operations center
NOCC	network operations control center
NPA	numbering plan area
NPC	network parameter control
NPDU	network protocol data unit
NPE	network protection equipment
NRM	normal response mode; network resource management
NRZ	non-return to zero
NRZI	NRZ inverted ones
NSDU	network service data unit
NSS	network and switching subsystem
NSSDU	network service signaling data unit
NSP	national signaling point
NT1, NT2	network termination 1, network termination 2 (ISDN)
NTM	network traffic management
NTSC	National Television Systems Committee
NV	numeric value

O

OADM	optical add–drop multiplex
OA&M, OAM	operations, administration, and maintenance

OBLSR	optical bidirectional line-switched rings
OC	optical carrier
OCCS	optical cross-connect system(s)
OEO	optical–electrical–optical
OIF	Optical Internetworking Forum
OLR	overall loudness rating
O&M	operation and maintenance
OMC	operation and maintenance centers
ONU	optical network unit
OPC	originating point code
OPM	outside plant module
ORM	optical remote module
OSI	open systems interconnection
OSF	operation systems function
OSNR	optical signal-to-noise ratio
OSPF	open shortest path first
OSS	operations support subsystem
OUI	organizationally unique identifier
OW	orderwire
OXC	optical cross-connect

P

PA	preamble
PABX	private automatic branch exchange
PAD	padding, added bits to a frame to assure minimum number of bits, or a multiple of eight
PAM	pulse amplitude modulation
PAN	personal area network
PC	personal computer; protocol control
PCI	protocol control information
PCM	pulse code modulation
PCR	peak cell rate
PCS	personal communication service
PDC	personal digital communications
PDF	probability density function
PDN	public data network
PDU	protocol data unit
PDH	plesiochronous digital hierarchy
P/F	poll-final (bit)
PEP	path endpoint
PHY	in reference to the physical layer (OSI layer 1)
PIDB	peripheral interface data bus
PIN	$p-n$ junction with intrinsic material
PLC	packet loss concealment
PLCP	physical layer convergence protocol

PLMN	public land mobile network
PLS	physical layer signaling
PM	peripheral module
PMA	physical medium attachment
PN	pseudo-noise
PNNI	private network–node interface
POH	path overhead
POP	point-of-presence
POTS	"plain old telephone service"
PPM, ppm	pulse position modulation; parts per million
PPP	point-to-point protocol
PPTP	point-to-point tunneling protocol
PRBS	pseudo-random binary sequence
PRI	primary rate interface (ISDN)
PRS	primary reference source
PS	packet-switched
PSC	packet switch capable
PSK	phase-shift keying
PSN	public switched network
PSTN	public switched telecommunications network
PT	payload type
PTI	payload type identifier
PTO	public telecommunication operator(s)
PVAC	present value of annual charges
PVC	permanent virtual circuit
PWAC	present worth of annual charges
pW	picowatt(s)
pWp	picowatts psophometrically weighted

Q

QAM	quadrature amplitude modulation
QoS	quality of service
QPSK	quadrature phase-shift keying

R

RADIUS	remote authentication dial-in user service
RADSL	rate-adaptive DSL
RAM	read and write memory
RARP	reverse address resolution protocol
RBOC	regional bell operating company
RCC	remote cluster controller
RD	resistance design
RDT	remote digital terminal
Rej	reject
RELP	residual excited linear predictive (coder)
RF	radiofrequency

RFC	request for comments
RFI	radio-frequency interference
RI	routing information (field), route identifier
RLCM	remote line concentrating module
RLR	receive loudness rating
RMON	remote monitoring
rms	root mean square
RNC	radio network controller
RNR	receive not ready
RNS	radio network subsystem(s)
ROI	return on investment
ROM	read-only memory
RPC1	Radio Paging Code 1
rptr, RPTR	repeater
RQ	request (see ARQ)
RR	receive ready
RRD, rrd	revised resistance design
RSC	remote switching center
RSL	receive signal level
RSM	remote switching module
RSU	remote switching unit
RSVP-TE	resource reservation protocol—traffic engineering
RT	remote terminal
RTCP	real-time control protocol
RTP	real-time protocol
RZ	return-to-zero

S

SA	source address
SABM	set asynchronous balanced mode
SABME	set asynchronous balanced mode extended
SAD	speech activity detection
SANC	signaling area/network code
SAP	service access point
SAPI	service access point identifier
SAR	segmentation and reassembly; search and rescue
SARSAT	search and rescue satellite (service)
SBC	sub-band coding
SBS	selective broadcast signaling
SCN	switched circuit network
SCCP	signaling connection control part
SCPC	single channel per carrier
SCR	silicon-controlled rectifier; sustainable cell rate
SCTE	Society of Cable Telecommunications Engineers
SCTP	stream control transport protocol

SD	starting delimiter
S/D	sine wave power-to-quantizing distortion power ratio
SDH	synchronous digital hierarchy
SDLC	synchronous data-link control
SDOC	selective dynamic overload control
SDT	structured data transfer
SDU	service data unit
SEP	severely errored period
SEPI	severely errored period intensity
SES	severely errored second
SESR	severely errored second ratio
SF	single frequency (signaling)
SFD	start frame delimiter
SFS	start of frame sequence
SGCP	simple gateway control protocol
SIF	signaling information field
SigTran	signaling transport
SINAD	(signal + noise and distortion to interference + noise + distortion) ratio
SIO	service information octet
SIP	session initiation protocol
SLC	signaling link code
SLR	send loudness rating
SLS	signaling link selection
SM	switching module
SMF	sub-multiframe
SMI	structure of management information
SMPU	switching module processing unit
SMT	station management
SMTP	simple mail transfer protocol
SN	sequence number
S/N, SNR	signal-to-noise ratio
SNA	system network architecture
SNAP	subnetwork access protocol
SNMP	simple network management protocol
SNP	sequence number protection
SOA	semiconductor optical amplifier
SOH	section overhead
SONET	synchronous optical network
SP	signaling point
SPC	stored program control
SPE	synchronous payload envelope
SRB	source route bridge
SRNC	serving radio network controller
SRTS	synchronous residual time stamp

SSAP	source service access point
SSM	single segment message
ST	end of pulsing
STM	synchronous transport module
STP	shielded twisted pair; signal transfer point
STR	selective trunk reservation
STS	synchronous transport signal; space–time–space
SU	signal unit
SUA	SCCP user adaptation
SVC	switched virtual circuit
SYN	synchronize sequence numbers (TCP)
T	
TA	terminal adapter
TACS	total access communication system
TCP	transmission control protocol
TCXO	temperature compensated crystal oscillator
TDD	time division duplex
TDM	time division multiplex
TDMA	time division multiple access
TDR	time domain reflector
TD-SCDMA	time division—synchronous CDMA
TE	traffic engineering; terminal equipment
TEI	terminal endpoint identifier
TELR	talker echo loudness rating
TF	transformation function
THD	tandem high usage direct
THT	token holding time
THz	terahertz
TIA	Telecommunication Industry Association
TIE	time interval error
TKIP	temporal key integrity protocol
TLP	test level point
TMN	telecommunication management network
TMS	time multiplexed switch
TOS	type of service
TPC	transmission power control
TRAU	transcoder and rate adapter unit
TRT	token rotation timer
TSA	time-slot assignment
TSG	timing signal generator
TSGR	transport systems generic requirements
TSI	time-slot interchanger
TSPS	Traffic Service Position System
TST	time–space–time

TSTS	time–space–time–space
TS0	time slot zero
TTL	time to live
TU	traffic unit; trunk unit
TUG	tributary unit group
TUP	telephone user part
TVX	valid transmission timer
TWT, twt	traveling wave tube

U

UBR	unspecified bit rate (ATM)
UC	unit call
UDP	user datagram protocol
UE	user equipment
UG	unigauge (design)
UHF	ultrahigh frequency (the RF band from 300 MHz to 3000 MHz)
UI	unit interval; unnumbered information frame
UK	United Kingdom
ULP	upper-layer protocol
UME	UNI management entity
UMTS	Universal Mobile Telecommunications System(s)
UNI	user–network interface
UP	user part
UPC	usage parameter control
UPSR	unidirectional path-switched ring
URL	uniform resource locator
USM	user-based security model
UTC	universal time coordinated
UTP	unshielded twisted pair
UTRA	UMTS terrestrial radio access
UTRAN	UMTS radio access network
UWC	universal wireless communications

V

VAD	voice activity detector
VAX, Vax	trade name for a type of computer made by DEC
VBR	variable bit rate
VBR-NRT	variable bit rate—non-real time
VBR-RT	variable bit rate—real time
VC	virtual circuit, virtual connection
VCAM	view-based access control model
VCC	virtual channel connection
VCI	virtual channel identifier
VCSEL	vertical cavity surface emitting laser
VDC, V dc	voltage, DC (direct current)

VDSL	very (high bit rate) DSL
VERS	version (a field in frame of TCP/IP family of protocol)
VF	voice frequency
VHE	virtual home environment
VHF	very high frequency (the band from 30 to 300 MHz)
VHSIC	very high-speed integrated circuit
VLR	visitor location register
VLSI	very large scale integration
VMR	variance-to-mean ratio
VoIP	voice-over IP
VP	virtual path
VPC	virtual path connection
VPI	virtual path identifier
VPN	virtual private network
VRC	vertical redundancy check
VSAT	very small aperture terminal
VSB	vestigial sideband
VT	virtual tributary
VU	volume unit

W

WAN	wide area network
WAP	wireless access protocol
WDM	wave (length) division multiplex
WEP	wired equivalent privacy
W/G	waveguide
WLAN	wireless LAN
WLL	wireless local loop
WML	wireless markup language
WR	wavelength-routed
WS	workstation
WSF	workstation function

X

| XDSL | any digital subscriber line |
| XID | exchange identification (frame) |

Z

| Z_0 | characteristic impedance |
| 0 TLP | zero test level point |

INDEX

Boldface entries denotes in-depth coverage of a topic.

Italic entries denotes a definition of a term.

AAL (ATM adaptation layer), 648–654. *See also* ATM entries; SEAL (simple and efficient AAL layer)
  sublayering of, 648
AAL-0, 649
AAL-1, 649–650
AAL-2, 650
AAL-3/4, 650–653, 654
  SAR-PDU, 652f
AAL-5, 653–654, 654f
AAL categories or types, ATM, 649–654
AAL convergence sublayer, 651
AAL-PCI bit errors, 650
AAL-SDUs (service data units), variable length, 653
A bit, changed to, 1, 534
ABM (asynchronous balanced mode), 436
abort, 463
abort condition, 437
abort (flag) sequence, 589, 590
ABSBH (average busy season busy hour), 8
absence of transition, 541
absolute signal level, 145
absolute zero, 395
AC. *See access control (AC);* ac power line source; **AC signaling**
acceptable error rate, 399
access
  to lookup tables, 654
  SSN7 transport parameters for, 726
  VSAT techniques for, 234–235
access connection element model (ISDN), 574f
access control (AC), 475, 533

access manager, 331
access points, 556, 557
access protocol for packet transmission, 519
access rate (AR), 617
  at subscription, 618
accommodate clock offsets (SONET), 303
accounting management, 873
ACH (attempts per circuit hour), 9
acknowledged information, 588
acknowledgment
  explicit, 126
  single, 462
acknowledgment number, 467
acknowledgment (ACK) signal, 377, 378, 467
acoustic pressure, 3
ac power line source, 378
"acquired the medium," 521
acquisition time
  of the medium, 523
  versus bit error probability, 644f
AC signaling, 115–118
AC signaling systems, low-frequency, 115
actions
  ATM, 667
  at congested node, 616–617
  at destination exchange (ISUP SSN7), 725–726
  at incoming international exchange (SSN7), 725
  at intermediate international exchange (ISUP SSN7), 724–725
  at intermediate national exchange (ISUP SSN7), 722–723

Telecommunication System Engineering, by Roger L. Freeman
ISBN 0-471-45133-9   Copyright © 2004 Roger L. Freeman

WILEY SERIES IN TELECOMMUNICATIONS AND SIGNAL PROCESSING

John G. Proakis, Editor
Northeastern University

Introduction to Digital Mobile Communications
Yoshihiko Akaiwa

Digital Telephony, 3rd Edition
John Bellamy

ADSL, VDSL, and Multicarrier Modulation
John A. C. Bingham

Biomedical Signal Processing and Signal Modeling
Eugene N. Bruce

Elements of Information Theory
Thomas M. Cover and Joy A. Thomas

Erbium-Doped Fiber Amplifiers: Device and System Developments
Emmanuel Desurvire

Fiber-Optic Systems for Telecommunications
Roger L. Freeman

Practical Data Communications, 2nd Edition
Roger L. Freeman

Radio System Design for Telecommunications, 2nd Edition
Roger L. Freeman

Telecommunication System Engineering, 4th Edition
Roger L. Freeman

Telecommunications Transmission Handbook, 4th Edition
Roger L. Freeman

Introduction to Communications Engineering, 2nd Edition
Robert M. Gagliardi

Optical Communications, 2nd Edition
Robert M. Gagliardi and Sherman Karp

Efficient Algorithms for MPEG Video Compression
Dzung Tien Hoang and Jeffrey Scott Vitter

Active Noise Control Systems: Algorithms and DSP Implementations
Sen M. Kuo and Dennis R. Morgan

Mobile Communications Design Fundamentals, 2nd Edition
William C. Y. Lee

Expert System Applications for Telecommunications
Jay Liebowitz